Handbook of Sensory Physiology

Volume V/2

Auditory System

Physiology (CNS) · Behavioral Studies Psychoacoustics

By

M. Abeles G. Bredberg R. A. Butler J. H. Casseday
J. E. Desmedt I. T. Diamond S. D. Erulkar E. F. Evans
J. M. Goldberg M. H. Goldstein D. M. Green I. M. Hunter-Duvar
L. A. Jeffress W. D. Neff W. A. Yost E. Zwicker

Edited by

Wolf D. Keidel and William D. Neff

With 209 Figures

Springer-Verlag Berlin Heidelberg New York 1975

Wolf D. Keidel
I. Physiologisches Institut der Universität 852 Erlangen, Universitätsstraße 17 (Germany)

William D. Neff
Center for Neural Sciences and Department of Psychology, Indiana University, Bloomington, Indiana 47401 (USA)

ISBN-13: 978-3-642-65997-3 e-ISBN-13: 978-3-642-65995-9
DOI: 10.1007/978-3-642-65995-9

Library of Congress Cataloging in Publication Data (Revised)
Keidel, Wolf Dieter. Auditory system. (Handbook of sensory physiology; v. 5). Includes bibliographies and index. CONTENTS: 1. Anatomy, physiology (ear). 2. Physiology (CNS), behavioral studies, psychoacoustics. 1. Hearing. 2. Ear. I. Neff, William D., joint author. II. Ades, Harlow Whitting, 1911— . III. Title. IV. Series. [DNLM: 1. Neurophysiology. 2. Sensation. WL700 H236] QP351. H34 vol. 5 [QP461] 591.1'82'08s [596'.01'825] 74—415

Softcover reprint of the hardcover 1st edition 1975

Contents

Chapter 1 Cochlear Nerve and Cochlear Nucleus. By E. F. Evans. With 49 Figures 1

Chapter 2 Physiological Studies of Auditory Nuclei of the Pons. By J. M. Goldberg. With 20 Figures 109

Chapter 3 Physiological Studies of the Inferior Colliculus and Medial Geniculate Complex. By S. D. Erulkar. With 30 Figures 145

Chapter 4 Single Unit Activity of the Auditory Cortex. By M. H. Goldstein, Jr. and M. Abeles. With 10 Figures 199

Chapter 5 Physiological Studies of the Efferent Recurrent Auditory System. By J. E. Desmedt. With 11 Figures 219

Chapter 6 The Influence of the External and Middle Ear on Auditory Discriminations. By R. A. Butler. With 4 Figures 247

Chapter 7 Behavioral Tests of Hearing and Inner Ear Damage. By G. Bredberg and I. M. Hunter-Duvar. With 16 Figures 261

Chapter 8 Behavioral Studies of Auditory Discrimination: Central Nervous System. By W. D. Neff, I. T. Diamond and J. H. Casseday. With 25 Figures. . 307

Chapter 9 Scaling. By E. Zwicker. With 40 Figures 401

Chapter 10 Localization of Sound. By L. A. Jeffress. 449

Chapter 11 Binaural Analysis. By D. M. Green and W. A. Yost. With 4 Figures. . . 461

Author Index. 481

Subject Index. 503

List of Contributors

ABELES, MOSHE, Department of Physiology, Hadassah Medical School, Hebrew University, Jerusalem, Israel

BREDBERG, GÖRAN, Dept. of Otorhinolaryngology, University Hospital, S-75014 Uppsala 14, Sweden

BUTLER, ROBERT A., The University of Chicago, Department of Surgery (Otolaryngology), Chicago, Illinois 60637, USA

CASSEDAY, JOHN H., Departments of Surgery (Otolaryngology) and Psychology, Duke University, Durham, North Carolina 27710, USA

DESMEDT, JOHN E., Brain Research Unit, Université de Bruxelles, B-1000 Bruxelles, Belgium

DIAMOND, IRVING T., Department of Psychology, Duke University, Durham, North Carolina 27706, USA

ERULKAR, SOLOMON D., Department of Pharmacology, School of Medicine, University of Pennsylvania, Philadelphia, Pennsylvania 19174, USA

EVANS, E. F., Medical Research Council Group, Department of Communication, University of Keele, Keele, Staffordshire ST5 5BG, Great Britain

GOLDBERG, JAY M., Department of Pharmacological and Physiological Sciences, University of Chicago, Chicago, Illinois 60637, USA

GOLDSTEIN, MOISE H., Jr., The Johns Hopkins University, School of Medicine, Baltimore, Maryland 21205, USA

GREEN, DAVID M., Laboratory of Psychophysics and Social Relations, Harvard University, Cambridge, Massachusetts 02138, USA

HUNTER-DUVAR, IVAN M., Department of Otolaryngology, The Hospital for Sick Children, Toronto M5G 1X8, Canada

JEFFRESS, LLOYD, A., Department of Psychology and Applied Research Laboratories, University of Texas at Austin, Texas 78712, USA

NEFF, WILLIAM D., Center for Neural Sciences and Department of Psychology, Indiana University, Bloomington, Indiana 47401, USA

YOST, WILLIAM, A., Laboratories University of Florida, Communication Sciences, Grainesville, Florida 32601, USA

ZWICKER, E., Institut für Elektroakustik der Technischen Universität, D-8000 München, Germany

The editors and publishers dedicate
this volume to the memory of

Georg von Békésy
Russell R. Pfeiffer
S. Smith Stevens

Through their research, teaching, and writings
they have advanced significantly our understanding of the auditory system.
We are proud to have them as authors
of chapters in the Handbook of Sensory Physiology.

Chapter 1

Cochlear Nerve and Cochlear Nucleus

E. F. EVANS, Keele, Great Britain

With 49 Figures

Contents

I. Introduction . . . 2

II. Cochlear Nerve . . . 4

A. Tonotopic Organization . . . 4

B. Spontaneous Activity . . . 4

C. Response to Click Stimuli . . . 6

D. Response to Single Tonal Stimuli . . . 10

1. Threshold and Response as a Function of Frequency . . . 11
2. Response as a Function of Intensity . . . 21
3. Response as a Function of Time . . . 23
4. Response to Low Frequency Tones . . . 24

E. Response to Complex Sounds . . . 30

1. Two-tone Stimuli . . . 30
   a) Two-tone Suppression . . . 30
   b) Excitation by Tone Combinations . . . 33
   c) Interaction of Responses to Two Tones: Linear and Non-linear Aspects . 35
2. Noise Stimuli . . . 38
3. Stimuli with Multi-component Spectra: Comb-filtered Noise . . . 41
4. Effect of Noise on Responses to Click and Tone Stimuli . . . 43

III. Cochlear Nucleus . . . 43

A. Introduction . . . 43

B. Organization . . . 45

1. Tonotopic Organization . . . 45
2. Functional Organization . . . 46
   a) Ventral Cochlear Nucleus . . . 46
   b) Dorsal Cochlear Nucleus . . . 50
3. Nature of Sensory Input to Neurones of Cochlear Nucleus . . . 51

C. Spontaneous Activity . . . 52

D. Response to Click Stimuli . . . 52

E. Response to Single Tonal Stimuli . . . 53

1. Threshold and Response as a Function of Frequency . . . 53
2. Response as a Function of Intensity . . . 54
3. Response as a Function of Time . . . 56
4. Response to Low Frequency Tones . . . 57

F. Response to Complex Sounds . . . 58

1. Two-tone Stimuli . . . 58

2. Noise Stimuli . . . 58
3. Stimuli with Multi-component Spectra . . . 61
4. Frequency and Amplitude Modulated Tones . . . 63
5. Effect of Noise on Responses to Click and Tone Stimuli . . . 67

IV. Gross Nerve Action Potential Responses . . . 69

V. Effects of Activity of Efferent Pathways . . . 72

A. Cochlear Nerve . . . 72

B. Cochlear Nucleus . . . 76

VI. Discussion in Relation to . . . 77

A. Cochlear Mechanisms . . . 77
1. Cochlear Filtering . . . 77
2. Transducer Nonlinearity . . . 82
3. Generation of Cochlear Nerve Discharges . . . 83

B. Cochlear Innervation . . . 85

C. Psychophysical and Behavioural Phenomena . . . 87
1. Critical Band and Other Measures of Frequency Selectivity . . . 87
2. Masking . . . 89
3. Combination Tones . . . 91
4. Sensorineural Deafness . . . 91

D. Coding of Acoustic Stimulus Parameters . . . 92
1. Frequency . . . 92
2. Intensity . . . 94
3. Sound Complexes . . . 94

References . . . 96

Abbreviations

AP : Gross cochlear action potential.
CDT : Cubic difference tone ($2f_1 - f_2$).
CF : Characteristic ("best") frequency.
cn : Cochlear nerve (acoustic, auditory, nerve).
CN : Cochlear nucleus.
DCN, VCN, AVN, PVN : Dorsal, ventral, antero- and postero-ventral divisions of CN.
FRC : Frequency response curve (plot of firing rate versus frequency at constant SPL).
FTC : Frequency threshold curve ("tuning curve").
IHC : Inner hair cells of the organ of Corti.
OCB : Olivocochlear bundle.
OHC : Outer hair cells of the organ of Corti.
PST : Post or peri-stimulus time (applied to histogram).
$Q_{10\,dB}$: CF/bandwidth of FTC at 10 dB above threshold.
SON : Superior olivary nucleus
SPL : Sound pressure level relative to 2×10^{-4} dyne/cm^2 (2×10^{-5} N/m^2).

I. Introduction

The first recordings of single neurone activity in the auditory system were made from the cochlear nucleus of the cat, by Galambos and Davis (1943). In these experiments the authors were attempting to record from fibres in the cochlear nerve; subsequently, however, they concluded that the recordings had been from aberrant cells of the cochlear nucleus lying central to the glial margin of the VIII nerve (GALAMBOS and DAVIS, 1948). The first successful recordings from fibres of the cochlear nerve were made by TASAKI (1954) in the guinea pig. These classical but necessarily limited results were greatly extended by ROSE, GALAMBOS, and HUGHES (1959) in the cat cochlear nucleus and by KATSUKI and co-workers (KATSUKI *et al.*, 1958, 1961, 1962) in the cat and monkey cochlear nerve. Perhaps the most significant developments have been the introduction of techniques for precise control of the acoustic stimulus and the quantitative analysis of neuronal response patterns, notably by the laboratories of KIANG (*e.g.* GERSTEIN and KIANG, 1960; KIANG *et al.*, 1962b, 1965a, 1967) and ROSE (*e.g.* ROSE *et al.*, 1967; HIND *et al.*, 1967). These developments have made possible a large number of quantitative investigations of the behaviour of representative numbers of neurons at these levels of the peripheral auditory system under a wide variety of stimulus conditions.

Most of the findings discussed herein have been obtained on anaesthetized cats. Where comparative data are available, substantially similar results have been obtained in other mammalian species (*e.g.* guinea pig, monkey, rat). Certain significant differences have been noted in lizards, frogs and fish as would be expected from the different morphologies of their organs of hearing (*e.g.* see FLOCK, 1971; FRISHKOPF *et al.*, 1968; FURUKAWA and ISHII, 1967), and these will be discussed in the relevant sections. The direct effects of anaesthesia do not appear to be significant at the level of the cochlear nerve, as judged by a comparison with limited data obtained in the unanaesthetized cat (RUPERT *et al.*, 1963) and in one fibre studied for a long period under normal and anaesthetized conditions (SIMMONS and LINEHAN, 1968). This is not the case at the level of the cochlear nucleus, where barbiturate anaesthesia can have a profound effect (EVANS and NELSON, 1973a). Unfortunately, most cochlear nucleus studies have been carried out under pentobarbitone anaesthesia. Studies under anaesthesia do offer the advantage of being relatively free from the influences of the middle ear muscles and the efferent auditory system.

As a generalization, experimenters at this level of the auditory system now calibrate their acoustic system at the tympanic membrane, and except where noted, absolute stimulus levels will be given here in dB SPL at the tympanic membrane[1]. The experimental approach to the cochlear nerve and nucleus follows essentially that of KATSUKI *et al.* (1958), that is, removal or retraction of the cerebellum in order that a microelectrode may be inserted into the target under direct vision. With precautions to ensure freedom from movements of the brainstem, recordings can be routinely made from individual cochlear fibres for several tens of minutes, and from cochlear nucleus cells for several hours. For the criteria and techniques used to distinguish between fibres and cells, and for more detailed technical information on the methods involved, the reader is referred to ROSE *et al.* (1959), KIANG *et al.* (1965a), and EVANS (1972b).

[1] In some laboratories (*e.g.* KIANG *et al.*, 1967, 1970), it has been the practice to express data in terms of p—p stapes displacement, inferred from averaged measurements of stapes motion on other animals of the same species.

II. Cochlear Nerve

A. Tonotopic Organization

Individual fibres of the cochlear nerve innervate relatively restricted areas of the organ of Corti (*e.g.* Lorente de Nó, 1933a; Spoendlin, 1971, 1972). In spite of a certain degree of twisting of the cochlear nerve (Lorente de Nó, 1933a; Sando, 1965), the distribution of fibres along the cochlea is projected in a tonotopic fashion to the cochlear nucleus (see Section III.B.1). It is not surprising, therefore, that penetrations of the cochlear nerve encounter fibres responsive to a more or less restricted range of tone frequencies (see Section II.D.1), and arranged systematically according to their optimal, or characteristic, frequencies (CFs). In the cat, the fibres are arranged such that microelectrode penetrations in the posterodorsal to anteroventral direction generally encounter fibres with high CFs first, then low CFs, and subsequently with progressively higher CFs. Thus, fibres with higher CFs are located superficially in the nerve, and those with lower CFs more centrally (Kiang *et al.*, 1965a; Evans and Rosenberg, unpublished observations). In the guinea pig, on the other hand, the twisting of the nerve appears to be less complete, so that penetrations in the same direction as in the cat encounter fibres with CFs progressively decreasing from high to low values (Evans, 1972b). A progression of frequency sensitivities in the reverse direction to that in the guinea pig is found in the monkey (Katsuki *et al.*, 1962).

B. Spontaneous Activity

All studies of single fibres in the cochlear nerve have reported activity in the absence of intentional auditory stimulation. While for any given fibre the mean rate remains relatively steady for long periods of time, between fibres it varies from less than a few spikes/sec to 100–120 spikes/sec (*e.g.* Nomoto *et al.*, 1964 in the Macaque; Kiang *et al.*, 1965a, in the cat; Rose *et al.*, 1971, in the squirrel monkey; Evans, 1972b, in the guinea pig). This range does not appear to rrelate to any degree of injury to the fibres themselves, in view of (a) the stability of the rate over long periods, and (b) the finding of fibres with high and low rates apparently adjacent in an electrode penetration. On the other hand, there is some relation between the spontaneous discharge rate and the threshold sensitivity of a fibre. In normal animals, there is a tendency for the most sensitive units to have rates of spontaneous discharge in excess of 15/sec (Kiang *et al.*, 1965a, 1970; Rose *et al.*,

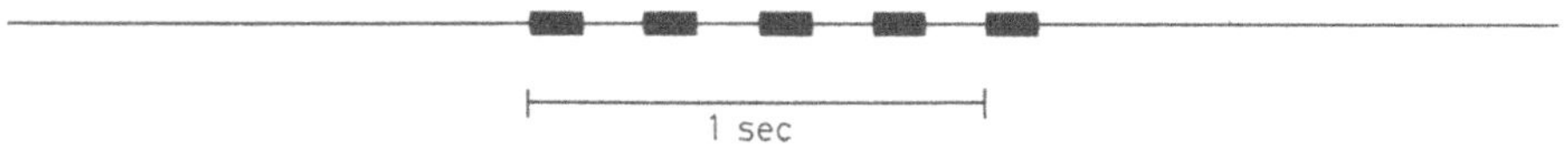

Fig. 1. Spontaneous and tone evoked activity of single cochlear nerve fibre. Cat. Tone at characteristic frequency, 9.5 kHz, 15 dB above threshold, indicated on lower trace. Positivity upward. Note: increase in spike discharge rate corresponding to duration of each tone burst; reduction in discharge rate at termination of stimuli

1971). In cats poisoned with kanamycin (KIANG *et al.*, 1970) and in guinea pigs in poor physiological condition (EVANS, 1972b), the affected fibres with abnormally high threshold have low or no spontaneous discharge. These abnormal fibres apart, the distribution of spontaneous rates tends to be bimodal: about a quarter of the fibres have rates below 10/sec; the majority of the remainder discharges at rates in excess of 30/sec. Forty percent of all fibres in cat and squirrel monkey discharge at rates above 50/sec (KIANG *et al.*, 1965a; ROSE *et al.*, 1971), and in the guinea pig, above 80/sec (EVANS, 1972b).

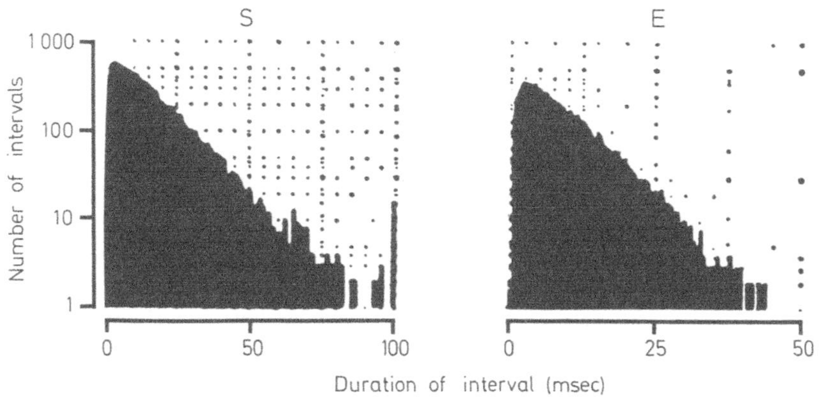

Fig. 2. Interspike interval histograms of spontaneous (S) and tone evoked (E) activity for a cat cochlear fibre. Tone at CF of fibre; 7.84 kHz. Histograms represent data collected from 5 min and 2 min of spontaneous and evoked activity, respectively. Note different time scales. (From KIANG, 1968)

The spontaneous discharge pattern of fibres in the cochlear division of the VIII nerve, in contrast to the vestibular division, is characteristically irregular (KIANG *et al.*, 1965a; WALSH *et al.*, 1972; see Fig. 1). Figure 2 shows the distribution of interspike intervals obtained from an analysis of several minutes of spontaneous discharge in a single cochlear fibre. The semi-logarithmic plot of the distribution has a nearly linear "decay", which may therefore be considered to be approximately exponential. This is characteristic of a Poisson process (KIANG *et al.*, 1965a). Furthermore, computations of joint interval statistics which indicate that the intervals between spikes are independent, are also compatible with this conclusion (KIANG *et al.*, 1965a). However, the observed interval distributions have fewer short intervals than would be expected on the basis of a Poisson process. This can be attributed to the absolute and relative refractory periods following each discharge, a conclusion supported by the observation that the modal values of the interspike interval histograms nearly all fall within the range: 4–7 msec irrespective of the rate of spontaneous activity (KIANG *et al.*, 1965a). A detailed analysis of the conditional probability of spontaneous discharge by GRAY (1967) suggests that a cochlear nerve fibre has not *completely* recovered from the effects of the last discharge until after an interval of about 20 msec.

The mechanism underlying the spontaneous discharge is unknown. Apart from the possible case of the most sensitive units (see Section V.A; WIEDERHOLD and KIANG, 1970), the influence of background acoustic noise can in the main be excluded.

WALSH *et al.* (1972) have suggested that the characteristic irregularity of the spontaneous discharge might relate to the stochastic excitation associated with chemical synaptic transmission. Most "blackbox" models of cochlear nerve activity incorporate a stochastic or probabilistic element in the neural spike generation process in order to account for the presence of spontaneous activity and the stochastic nature of the discharge pattern in response to steady tonal stimuli (see Section VI.A.3).

C. Response to Click Stimuli

In theory, a single click stimulus has uniform distribution of energy over the frequency spectrum, with energy minima at frequencies corresponding to multiples of the reciprocal of the pulse duration. Repetitive clicks generate concentrations of energy into spectral lines spaced at the repetition rate. At the rates of click presentation generally used in physiological experiments (10/sec or less), the spectrum can be treated as uniform.

In practice, however, clicks are generated by feeding an electrical pulse to an acoustical transducer. The spectral distribution is distorted by the frequency response of the system, and the time pattern of the stimulus is distorted likewise by delayed reflections and the "ringing" characteristics of the acoustic system. The introduction of carefully damped couplers and condenser microphones as transducers (*e.g.* KIANG *et al.*, 1965a; MOLNAR *et al.*, 1968) has minimised this problem.

A click stimulus evokes one or a few spikes discharges in a cochlear fibre (Fig. 3A). Although the time pattern of discharge differs from response to response even for stimuli well above threshold, on average the discharges occur at preferred times after the click, as is shown in the post-stimulus time (PST) histogram of Fig. 3B as a function of the time since the click stimulus. This average pattern of activity is characteristic of fibres with CFs below 3–4 kHz (Fig. 3B–D). For fibres with CF above 4 kHz, the periodic natuıe of the spike distribution is lost (Fig. 3E, F), apparently on account of the temporal "jitter" associated with the spike generation process (see VI.A.3). Figure 3 also indicates that the time interval between preferred periods is related to the CF of the fibre, and Fig. 4 shows that this relation is systematic: the time interval corresponds to the reciprocal of the CF. Furthermore, the latency of the first peak of the PST histogram is an inverse function of the CF (Fig. 5).

These findings are consistent with a travelling wave disturbance of the basilar membrane that takes time to propagate to the apical (lower frequency) regions of the cochlea and which exhibits a more or less damped oscillation to a click transient, the period of oscillation at a given location corresponding to the latter's optimum frequency (BÉKÉSY, 1960; FLANAGAN, 1962; ROBLES *et al.*, 1972; WILSON and JOHNSTONE, 1972).

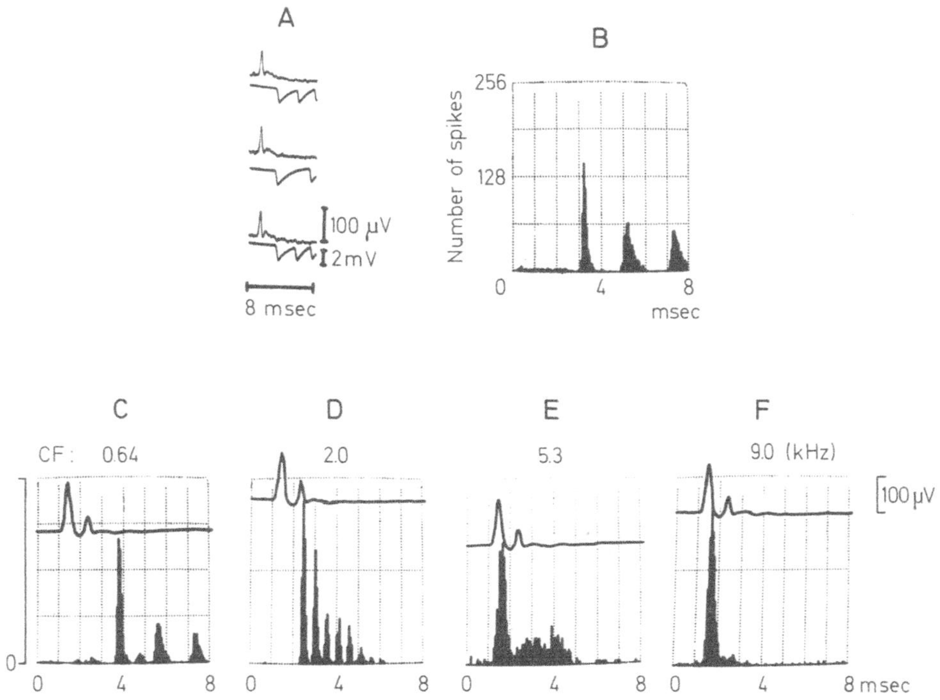

Fig. 3 A, B. Response of single cat cochlear fibre to click stimuli. A: Continuous film record of responses to 3 successive clicks. Upper traces, record from electrode at round window showing click evoked AP response. Lower traces, spike discharges of fibre. Negativity upwards. Note temporal patterning of spikes. B: Poststimulus time (PST) histogram showing averaged temporal pattern of activity in response to 600 clicks. CF of fibre 0.54 kHz. C-F: PST Histograms for 4 more cat cochlear fibres of differing CF, in comparison with the AP response recorded from the round window. CFs indicated above plots. Linear ordinate scale: number of spikes: 256, 128, 64 and 128 for C-F, respectively. Note periodic envelope of PST histograms for the fibres with lower CFs (B, C, and D). Data samples, 1 min; clicks presented at same level in each case, at 10/sec. (From Kiang *et al.*, 1965a)

The earliest observed peak in the PST histogram (vd. Fig. 6) is obtained with rarefaction acoustic transients, *i.e.*: from a movement of the cochlear partition toward the scala vestibuli. With condensation transients, the peaks of the PSTHs occur at times which interleave with those for clicks of the opposite polarity (lower halves of each histogram in Fig. 6 compared with upper halves). This is consistent with the conclusion (Kiang *et al.*, 1965a; Weiss, 1966; Brugge *et al.*, 1969; Goblick and Pfeiffer, 1969; Duifhuis, 1970) that excitation corresponds with movements of the basilar membrane (and therefore associated hair cell structures) in one direction only. In fibres with a sufficient rate of spontaneous activity, the probability of discharge is reduced during periods corresponding to deflections in the non-exciting direction. Refractory mechanisms are not entirely responsible for this: the reduction of discharge can appear as the earliest sign of oscillatory activity (see Fig. 6: histograms at +5 and 30 dB). In all fibres, the combination of

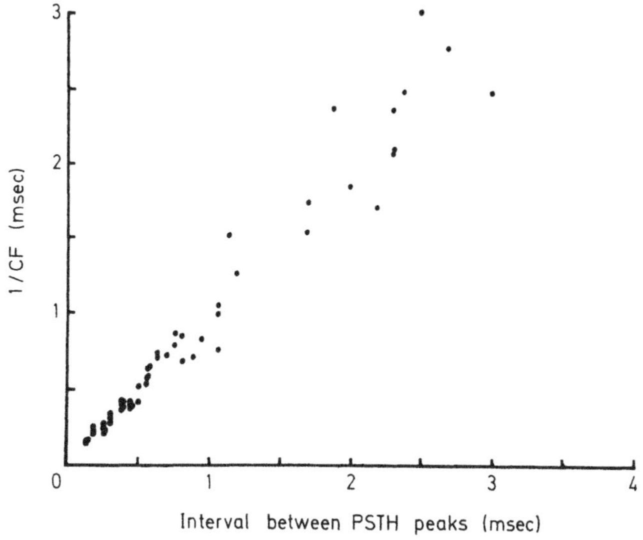

Fig. 4. Plot of reciprocal of CF against the interval between the PST histogram peaks in response to click stimuli, for 56 cat cochlear fibres (see text). (From Kiang *et al.*, 1965a)

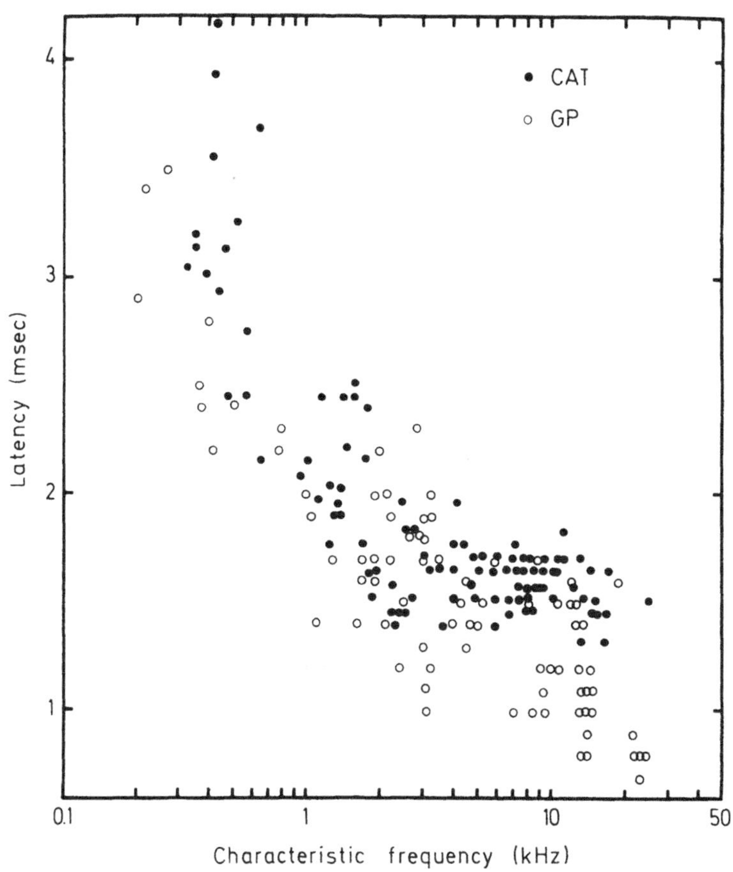

Fig. 5 (Legend see p. 9)

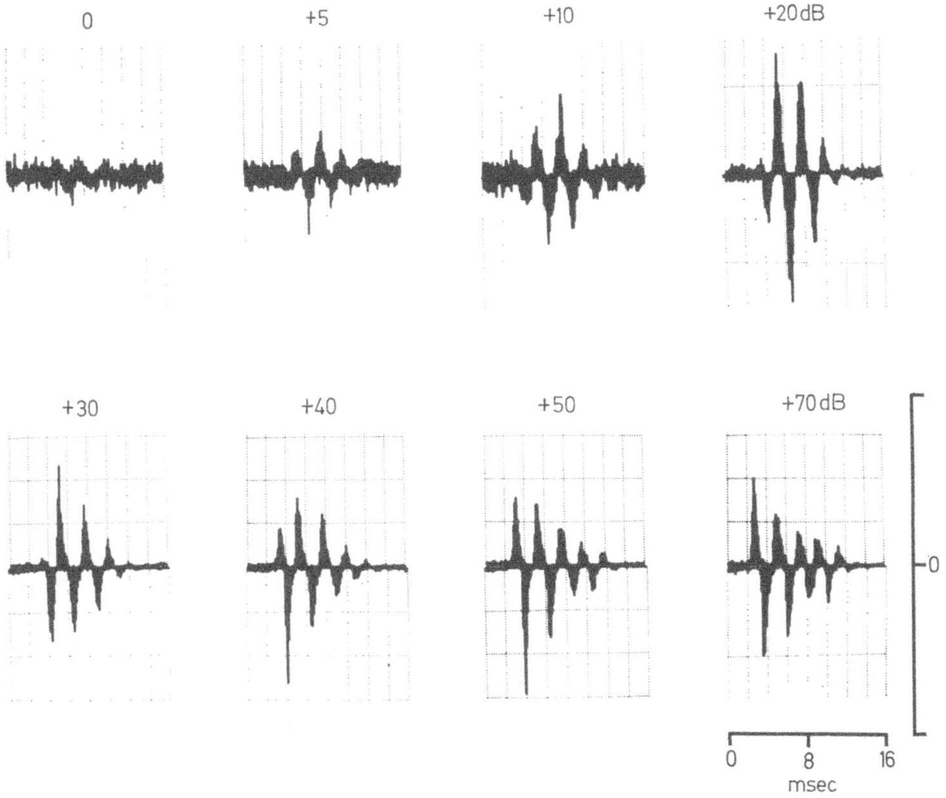

Fig. 6. Compound PST histograms of responses to click stimuli as a function of click level. Cat cochlear nerve. CF of fibre: 0.47 kHz. Click level expressed in dB relative to the "threshold" response level. Figure compiled from histograms of responses to rarefaction clicks and to condensation clicks (upper and lower halves of each compound histogram, respectively). Clicks: 100 μsec duration, 10/sec. Ordinate scales: linear number of spikes: 64 for 0—10 dB plots; 128 for 20 dB; 256 for 30—70 dB. (Data from Kiang *et al.*, 1965a)

PSTHs to transients of opposite polarity to form the compound PST histograms shown in Fig. 6, (Goblick and Pfeiffer, 1969) demonstrates most clearly the relationship between neural discharge pattern and presumed mechanical events.

Figure 6 also demonstrates how the pattern of discharge to a click is dependent upon the level of the stimulus. The positions of the peaks shift relatively little over an intensity range of 60 dB (Kiang *et al.*, 1965a; Goblick and Pfeiffer, 1969). On the other hand, at higher click levels, their relative heights change, an earlier peak with a latency of 2–4 msec appears, and the mode of the histogram occurs earlier.

Fig. 5. Latency of cochlear nerve response to click stimuli for fibres in the cat (filled circles, from Kiang *et al.*, 1965a) and the guinea pig (open circles, from Evans, 1972b) as a function of their CF. Rarefaction clicks of 100 μsec and 50 μsec duration, respectively, presented at 10/sec, at levels from 20—60 dB and from 20—80 dB, respectively, in electrical terms above minimum tone threshold (within 36 dB of click threshold for the cat data)

The number of spikes discharged in response to a click stimulus is a function of the stimulus level and the rate of click repetition (KIANG *et al.*, 1965a). With level, the average number of spikes discharged increases monotonically towards between one to a few spikes per click (fibres with low CF approaching the higher value) at low rates of repetition (10/sec). At higher rates, the number of spikes evoked per click decreases to an average value of about 0.1 at 1000 clicks/sec. The mean overall discharge rate, on the other hand, usually increases monotonically. The maximum *maintained* discharge rate obtained under optimum conditions rarely exceeds 100 spikes/sec for clicks (200/sec for tones) (KIANG *et al.*, 1965a; MOXON, 1968). Direct electrical pulse stimulation of the cochlea (MOXON, 1968) can evoke maintained discharge rates in excess of 500/sec. The limit on *maintained* discharge rate, therefore, appears to be set by the cochlear elements peripheral to the site of spike generation.

D. Response to Single Tonal Stimuli

In contrast to the responses of neurones at higher levels of the auditory system, the behaviour of single cochlear fibres is relatively simple as a function of frequency, stimulus level, and time, and the population of fibres is reasonably homogeneous in respect of these properties. The only response observed during continuous single tone stimulation is excitation (as in Fig. 1; cat: KIANG *et al.*, 1965a; squirrel monkey: ROSE *et al.*, 1969; guinea pig: EVANS, 1972b). The threshold stimulus level evoking this response is a relatively simple function of frequency; the firing rate is related monotonically to the stimulus level (except at very high levels, vd. later); the response adapts little with time; the interspike interval distribution is consistent with a Poisson process (Fig. 2E), superimposed on which, the temporal pattern of the discharges reflects the period of the sinusoidal stimulus at least for stimulus frequencies up to about 4 kHz. In short, the response properties of cochlear fibres to tone (and to click stimuli as observed above) can to a first approximation be predicted on the basis of the CF of the fibre (KIANG *et al.*, 1965a; KIANG, 1968).

Exceptions to these generalizations have been proposed, particularly from earlier studies. Thus, TASAKI, in the guinea pig (1954), and NOMOTO *et al.*, in the monkey (1964), obtained data which they considered to indicate that there were two populations of fibres. These fibre populations were distinguished on the basis of responsiveness, threshold, shape of frequency threshold curve (FTC) and the firing rate versus stimulus level function. Later experiments with more adequate control of stimulus parameters have failed to uncover such differences; rather, they have provided the explanations for the earlier discrepant data (*e.g.* KIANG, 1968; EVANS, 1972b). These will be discussed in Sections D.1. and 2. The only other contrary reports, as far as the present author is aware, is inhibition of the spontaneous activity of a few cochlear fibres described by RUPERT *et al.* (1963) in the unanaesthetized cat and by KATSUKI *et al.* (1962) in the monkey. Although RUPERT *et al.* concluded that, on latency grounds, their fibres were primary, the illustrated responses have latencies which substantially exceed (by more than 10 msec) those obtained in studies on fibres under anaesthesia, and the possibility exists that these fibres were part of, or were influenced by, the descending, efferent system. In the case of the latter study, it has been suggested by KIANG *et al.* (1965a) that the "spontaneous" activity was in fact evoked by ambient room noise and that the inhibition observed was an example of the suppression of stimulus-evoked activity discussed in Section II.E.1.a.

Against the above generalization of homogeneity of the mammalian cochlear nerve, recordings from the cochlear nerve of bullfrog and leopard frog (FRISHKOPF and GOLDSTEIN, 1963; LIFF and GOLDSTEIN, 1970), lizard (JOHNSTONE and JOHNSTONE, 1969) and fish (ENGER, 1963; FURUKAWA and ISHII, 1967) have demonstrated that at least two types of fibre can be clearly distinguished, which originate in anatomically and functionally separate transducer regions. Thus, in the frog and lizard, fibres can be separated into a "simple" population which cannot be inhibited by tonal stimuli and a "complex" population whose response to tonal or vibratory stimuli can be so inhibited. In fish, VIIIth nerve fibres have been sub-divided into two groups on the basis of their rate of adaptation to sound stimuli. Nevertheless, these different populations can also be distinguished on the basis of the CFs of the fibres, and fibres with similar CFs share similar properties.

1. Threshold and Response as a Function of Frequency

Rapid determinations of threshold of cochlear fibres as a function of frequency can be made by a variety of methods: by scanning across the relevant frequency range with continuous tones (Fig. 7), by the classical manner of successively approximating the frequency of a gated tone (with finite rise and fall time to avoid transients) towards the optimal frequency at each intensity until a change in rate of discharge is detected (Fig. 9), or by automatic or semi-automatic methods of "tracing" the threshold as the stimulating tone moves continuously across the responsive region (KIANG *et al.*, 1970; EVANS *et al.*, 1970). These methods yield consistent results (*e.g.* KIANG *et al.*, 1965a), although small inexplicable changes in pure tone and noise threshold from minute to minute have been observed in some fibres, in contrast to their "neighbours" (EVANS, ROSENBERG, and WILSON, unpublished results). The arbitrary choice of a "threshold" criterion (*e.g.* the boundary of the response area in Fig. 7) enables the frequency threshold ("tuning") curve (FTC) to be delineated. The frequency corresponding to the minimum threshold (tip) of the FTC defines the "characteristic frequency" (CF) of the fibre.

Early studies of the minimum thresholds of cochlear fibres led to the conclusions that they were widely distributed (over a range as great as 60 dB) at any frequency (*e.g.* KIANG *et al.*, 1965a) or that they fell into two groups of high and low threshold corresponding to fibres innervating inner and outer hair cells respectively (*e.g.* KATSUKI *et al.*, 1962). Subsequently, KIANG (1968) has shown that when sufficient data are collected from a single cat cochlear nerve (*i.e.* are not pooled across ears and animals) the distribution of thresholds becomes restricted to less than 20 dB at any frequency (Fig. 8). Within this limited range, there is some tendency for fibres with the lowest spontaneous discharge rates to have the highest thresholds (KIANG *et al.*, 1970). While there are considerable variations in distribution from animal to animal, the neural thresholds approach the average behavioral threshold for that species, with the exception of the higher frequencies (Fig. 8, interrupted line). These findings have been confirmed in the cat by EVANS *et al.* (1970, and to be published) and in the guinea pig (Evans, 1972b). Kiang and colleagues (Kiang *et al.*, 1967; Kiang, 1968) have pointed out that, unless the FTC is corrected for the frequency response of the sound system, it is possible to confound the low

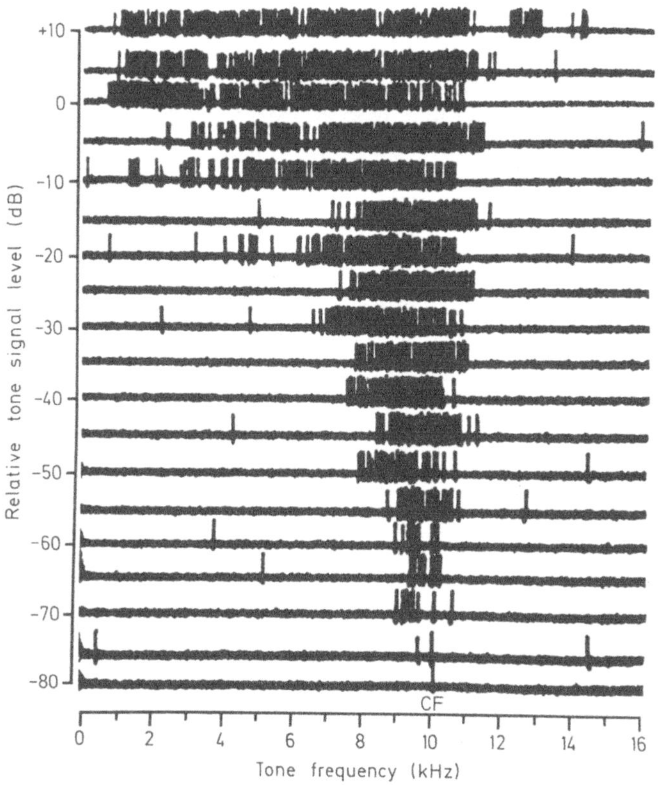

Fig. 7. Frequency sweep method of determination of frequency response area and Frequency Threshold Curve (FTC) of guinea pig cochlear fibre. A continuous tone is swept linearly in frequency, with 5 dB increments of signal level. Alternate sweeps are in the opposite direction. The pattern of spike discharges is built up on a storage oscilloscope. (Spikes are monophasically positive and 0.9 mV in amplitude.) Sweep rate: 14 kHz/sec. (From Evans, 1972b)

frequency high threshold segment of the FTC of a fibre with high CF (see Fig. 9) for a separate low frequency FTC of high threshold. This appears to be a satisfactory explanation for the high threshold population of fibres reported by Katsuki *et al.* (1962) in view of their being restricted to low CFs (below 6 kHz). In some guinea pigs, Evans (1972b) found a number of fibres which had abnormally high thresholds (above 70 dB SPL). Almost all of these, however, were derived from cochleas with evidence of pathological changes due to experimental interference or to circulatory insufficiency (from abnormally low systemic blood pressure or occlusion of the internal auditory artery) and were associated with a pathologically high AP threshold to clicks. The remainder had CFs above 12 kHz and were found in otherwise normal cochleas alongside low threshold fibres of similar

CFs. Like the other high threshold fibres, they had abnormally broad FTCs (see later, p. 18), and may relate (see Section VI.B) to the finding of a sporadic loss of outer hair cells in the basal turn of the cochleas of healthy guinea pigs (WERSÄLL, personal communication).

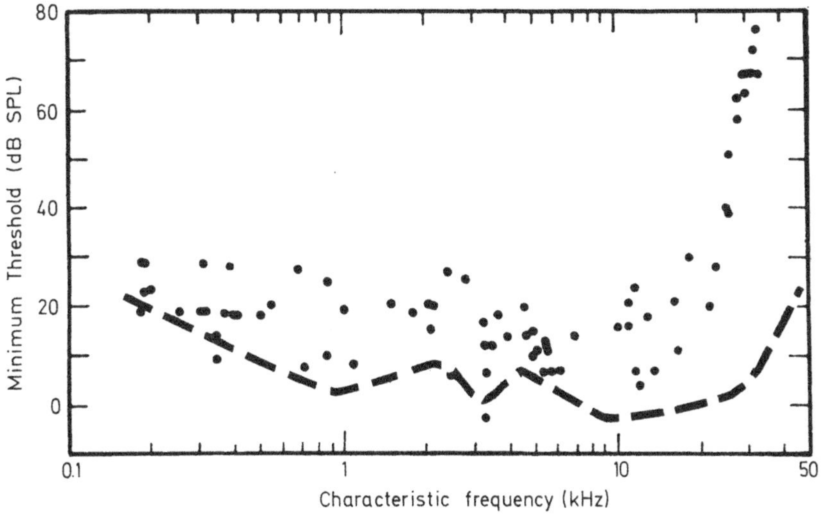

Fig. 8. Minimum pure tone thresholds of cochlear fibres from a single cat versus their CF (from KIANG, 1968), compared with a curve indicating the average behavioural threshold (interrupted line). Thresholds in dB SPL at the tympanic membrane. Open bulla. (Behavioural thresholds from the free-field measurements of NEFF and HIND, 1955, and corrected for the outer ear response and to open bulla conditions using the data of WIENER *et al.*, 1965, and GUINAN and PEAKE, 1967, respectively)

Figure 9 shows the similarities between the FTCs of single cochlear fibres of cat (KIANG *et al.*, 1967), guinea pig (EVANS, 1972b) and squirrel monkey (derived from the data of ROSE *et al.*, 1971). The thresholds in each case are given in SPL measured at the tympanic membrane. Plotted in this way on a logarithmic frequency scale, the shapes of the FTCs are systematically related to their CF. The curves become increasingly narrow and asymmetrical with higher CF, particularly for fibres with CFs above 2 kHz[2]. In these cases, the high frequency cut-off is steeper than the low-frequency cut-off and becomes steeper with increasing tone intensity, whereas the low frequency cut-off suddenly decreases to less than 10 dB/octave beyond about 30—60 dB above the minimum threshold. Fibres with CFs below 2 kHz have more nearly symmetrical FTCs and frequently exhibit an inflexion in the high frequency cut-off 20—30 dB above threshold. Discrete plotting of the FTCs as in Fig. 9 renders the tips artificially sharp; more detailed plots show that the tips are in fact rounded (KIANG *et al.*, 1970; EVANS *et al.*, 1970). The 3 dB bandwidths of the FTCs are approximately half the 10 dB bandwidths.

[2] These systematic differences in the shape with CF largely disappear if the data are plotted on a square-root frequency scale (ROSS, personal communication). Likewise, the 10 dB bandwidth values (Fig. 13C), when computed on a square root frequency basis, cluster about the same value (0.4) irrespective of CF (EVANS, 1972b).

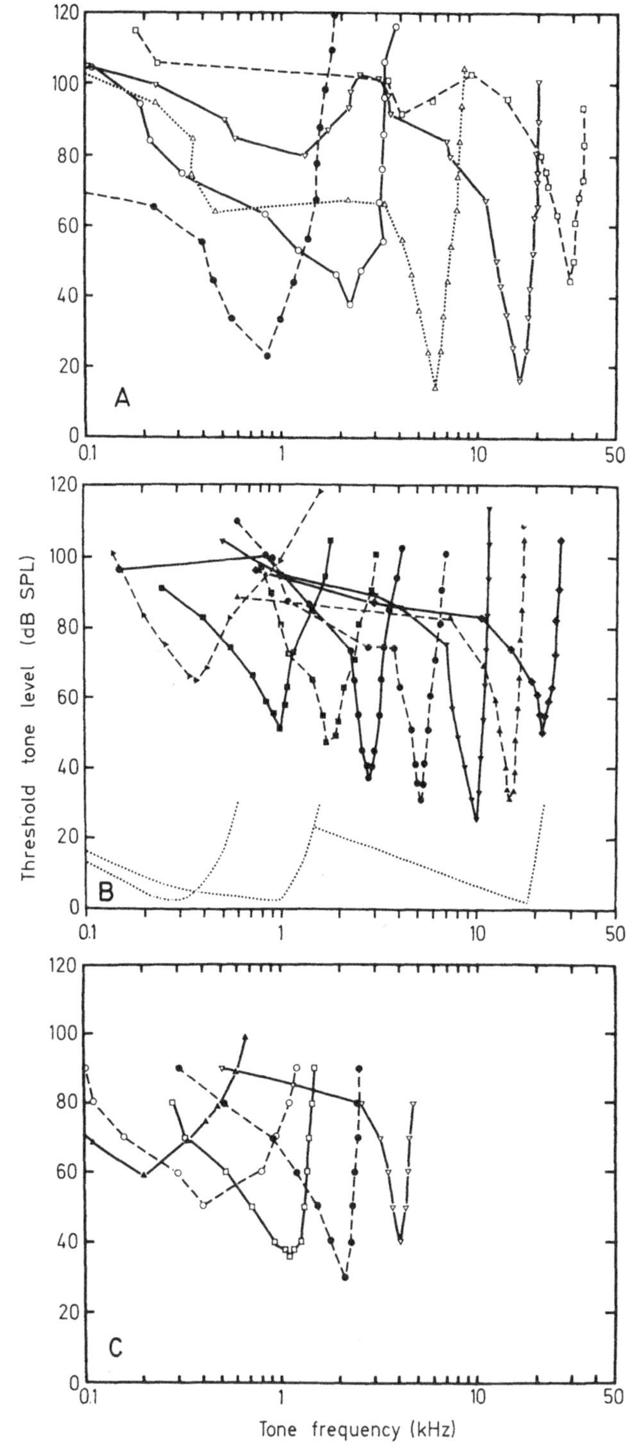

Fig. 9A—C (Legend see p. 15)

The values of the slopes of the low and high frequency cut-offs measured over the portion of the FTCs between 5 and 25 dB above threshold are shown in Fig. 10 A and B, respectively. In both cases, the slope values increase initially at least with CF and reach maximum values of between 100–250 dB/octave and between 100–600 dB/octave, respectively, at CFs of about 7–10 kHz. The slopes for the high frequency cut-offs represent the minimum slopes; as already mentioned, they become steeper with increase in signal level, approaching values of 1000 dB/octave, in some cases.

It is a common and convenient practice to express the relative sharpness of the FTCs in terms of the "$Q_{10\,dB}$" value. This is the CF divided by the bandwidth measured at 10 dB above minimum threshold. The values for cat, guinea pig and squirrel monkey are shown in Fig. 10 C. They reach a maximum of between 4–10 for fibres of intermediate CF, *i.e.*, of about 7–10 kHz.

The limited data available on fish (FURUKAWA and ISHII, 1967) and lizard (JOHNSTONE and JOHNSTONE, 1969) indicate that their FTCs are substantially broader than those of mammals. In the frog, however, fibres in both simple and complex populations can have FTCs as narrow as those of mammalian fibres of comparable CFs (0.1–1.5 kHz; FRISHKOPF and GOLDSTEIN, 1963: bullfrog). The $Q_{10\,dB}$ values of these fibres are as high as those of mammals: 1–3 (LIFF, 1969: leopard frog).

A discussion of the above various indices of frequency selectivity in comparison with psychophysical and behavioural data and basilar membrane data will be deferred to Section VI.A.1, C.1. However, two points regarding Fig. 10 deserve comment at this stage. The first concerns the finding of an intermediate range of CFs at which the slopes of the low and high frequency cut-offs, and the relative sharpness of the FTCs, are maximal. This range, approximately from 7–10 kHz, corresponds to the most sensitive region of the behavioural audiograms for the cat (*e.g.* MILLER *et al.*, 1963) and guinea pig (HEFFNER *et al.*, 1971); to the region where minimum tone thresholds are found for the cochlear fibres of the cat (Fig. 8; KIANG, 1968) and guinea pig (EVANS, 1972b); and to the region where the elevation of cochlear nerve threshold by stimulation of the olivocochlear bundle is greatest (WIEDERHOLD and KIANG, 1970; WIEDERHOLD, 1970; TEAS *et al.*, 1972; see Section V.A.). The second point is to emphasize that the data in Fig. 10 are pooled not only across species but animals. In a study of the slopes and bandwidths of populations of fibres in the cat, measured with a semi-automatic plotting technique, EVANS, WILSON, and ROSENBERG (1970) (also, EVANS and WILSON, 1971, and unpublished data) have observed small but systematic differences between animals. Of greater significance is the observation that even within a cochlear nerve of a single animal (cat and guinea pig), the range of slopes and bandwidths at a common CF is substantial and approaches a factor of 4 in some cases (EVANS, 1972b; EVANS and WILSON, 1973). This variation in sharpness from fibre to fibre at a

---

Fig. 9A—C. Representative Frequency Threshold Curves of cochlear fibres from A: cat (after KIANG *et al.*, 1967), B: guinea pig (from EVANS, 1972b), and C: squirrel monkey (derived from data of ROSE *et al.*, 1971). Dotted lines in B: basilar membrane frequency response curves (corrected to SPL at the tympanic membrane) from BÉKÉSY (1944) and JOHNSTONE *et al.* (1970), arbitrarily arranged on the ordinate scale common CF correlates to some extent (at least in the guinea pig) with the threshold of the fibre. Thus, the lower the threshold, the sharper the FTC tends to be (EVANS, 1972b; EVANS and WILSON, 1973).

On top of the systematic variation in shape of the FTC of cochlear fibres with their CF (as plotted on a logarithmic frequency scale), therefore, there are variations from fibre to fibre and from animal to animal. The latter, non-systematic, variations, (at least in the material of KIANG *et al.*, 1965a, 1967; EVANS *et al.*, 1970, in the cat; EVANS, 1972b, in the guinea pig) do not fall into two well-defined populations, as was reported by KATSUKI *et al.* (1959) for the cat. The latter authors described two kinds of response area: symmetrical and asymmetrical, which appeared to correlate to a certain extent with high and low threshold, respectively, and which were held to relate to the innervation of inner and outer hair cells. As has already been mentioned, it is clear that the high threshold, symmetrical, FTCs of low CF may be accounted for by the absence of correction of their FTCs for the response of the sound system at the tympanic membrane. Similar considerations may be responsible for the remaining differences in shape of FTCs of higher CF in their data, or they would appear to represent the *extremes* of the variation

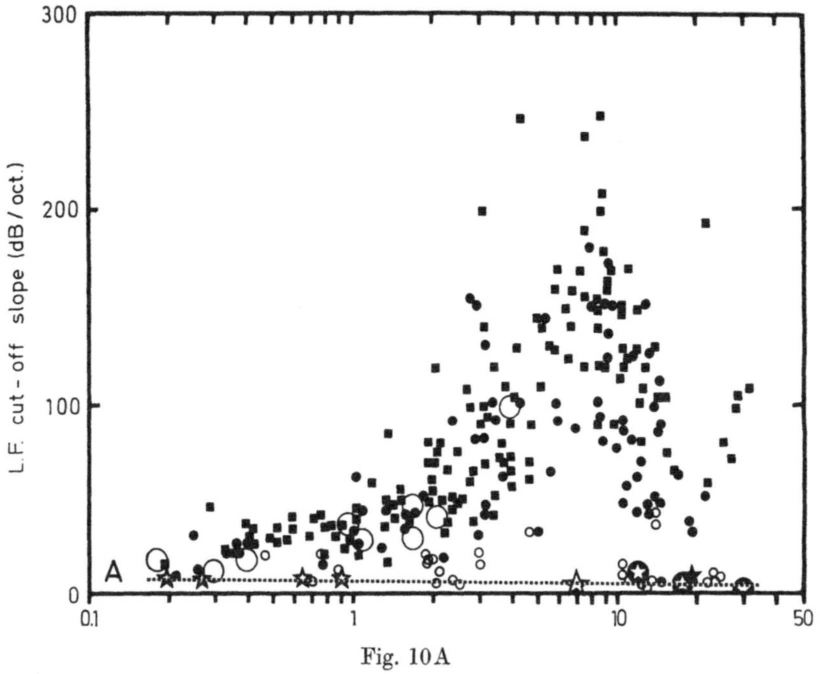

Fig. 10A

Fig. 10A—C. Measurements of slopes of low frequency cut-offs (A), high frequency cut-offs (B), and relative sharpness (C), of FTCs of cochlear fibres from cat, guinea pig, and squirrel monkey, versus their CF. Slopes measured over region 5—25 dB above minimum threshold. Relative sharpness measured as the $Q_{10\,dB}$ value, *i.e.*: CF/bandwidth of FTC at 10 dB above minimum threshold. Symbols indicate animal and source. Cn: cochlear nerve data for cat, from EVANS and WILSON, 1971; guinea pig (GP) from EVANS, 1972b, and squirrel monkey (SM) from Fig. 9C. Small open circles: values for fibres with pathologically high thresholds in guinea pig (see text). Star symbols: analogous measurements from basilar membrane frequency response data of VON BÉKÉSY (1944: B) in guinea pig, RHODE (1971, 1973: R) in squirrel monkey; WILSON and JOHNSTONE (1972: W) and JOHNSTONE *et al.* (1970): J) in guinea pig

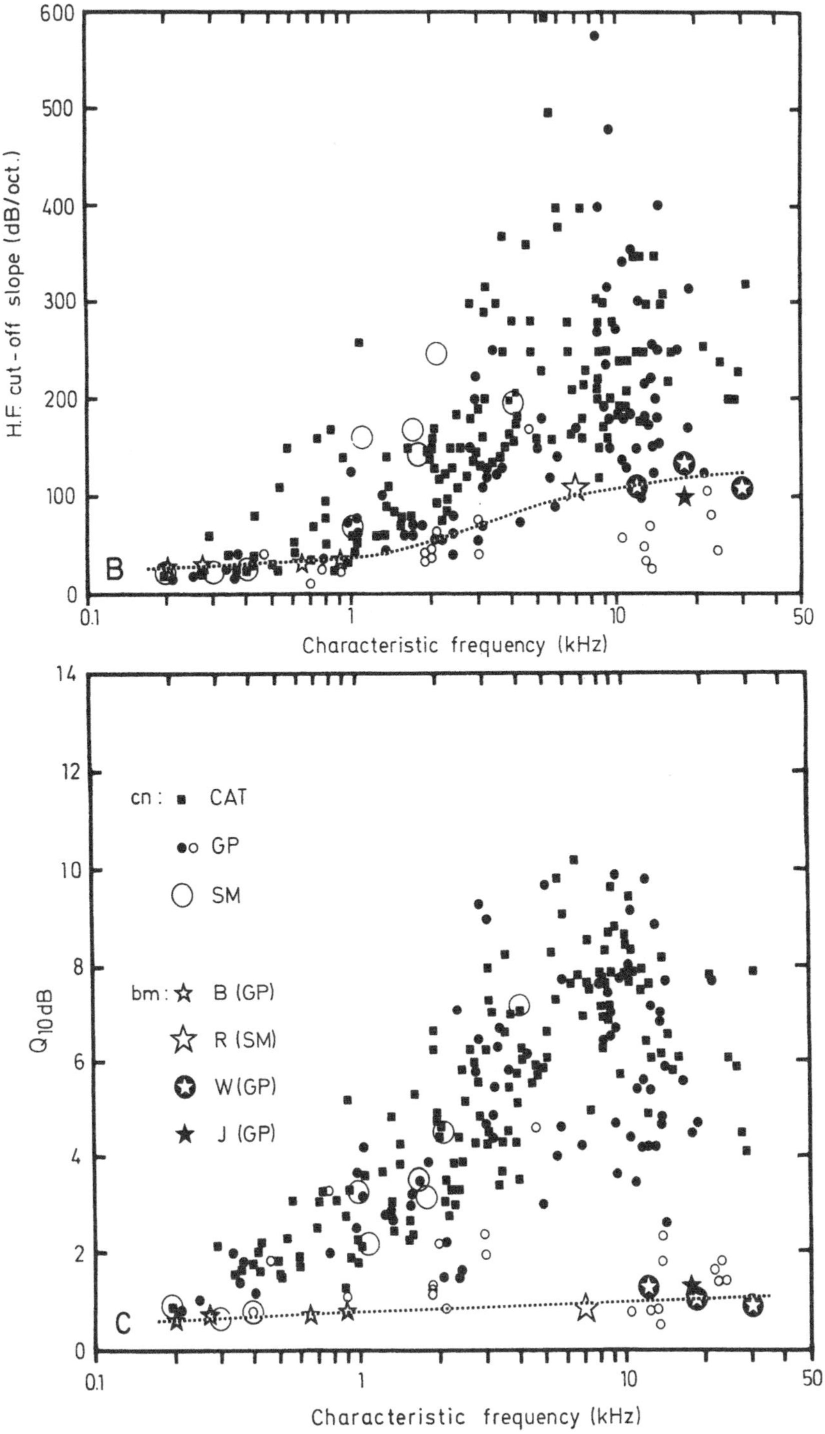

B
H.F. cut-off slope (dB/oct.)
Characteristic frequency (kHz)
C
Q10dB
Characteristic frequency (kHz)
cn: CAT
GP
SM
bm: B (GP)
R (SM)
W(GP)
J (GP)

Fig. 10B and C

noted above. In this connection, it is of interest that the same authors were unable to differentiate into two groups, on the basis of shape, cochlear fibres in the monkey, on account of the "variety of intermediate types" (KATSUKI *et al.*, 1962, p. 1402).

In all of the above considerations, the group of fibres mentioned earlier (p. 13) with pathologically high thresholds (*i.e.*: above 70 dB SPL) has been excluded. These were obtained in the guinea pig only, under conditions of apparent and presumed cochlear pathology (EVANS, 1972b). The FTCs of these fibres are clearly anomalous, being very broadly tuned, and resemble the normal FTCs shown in Fig. 9B but with the narrowly tuned lower threshold tip section (20—40 dB) missing. Measurements of the cut-off slopes and the relative sharpness of these fibres are given as the open circles in Fig. 10A—C, indicate their anomalous degree of tuning.

That anoxia can selectively decrease the sharpness of tuning of single cochlear fibres has been shown in recent experiments in the cat (EVANS, 1974a, c). *Reversible* loss of the low threshold, sharply tuned segment of the FTC, leaving behind the high threshold broadly tuned segment, can occur after only a few minutes of *hypoxia*. Local instillation of KCN (at concentrations of less than 10^{-3} M) into the scala tympani, or intra-arterial injection of Frusemide, a potent ototoxic diuretic, can produce identical reversible effects on the FTC without apparent action on the cochlear microphonic (EVANS and KLINKE, 1974; EVANS, 1974c, d). Other influences can also modify the threshold and shape of the FTC. Electrical stimulation of the olivocochlear bundle and prior exposure to long-term high level tonal stimuli at the CF cause a relatively greater elevation of the threshold at the CF than at other frequencies; that is to say, the FTC becomes less sharp (KIANG *et al.*, 1970). The addition of background noise at increasing levels produces a progressive elevation in the tip of the FTC (KIANG *et al.*, 1965), but without substantial loss of tip bandwidth until saturation of the discharge rate makes determination of response threshold difficult (EVANS, 1974d). It is interesting that under conditions of *continuous* noise stimulation, the saturation discharge rate becomes progressively reduced as the level of background noise is raised (EVANS, 1974d).

Complementary to the above descriptions of the frequency *sensitivity* of cochlear fibres, are representations of their frequency *response*, *i.e.* response rate versus frequency curves, (FRCs). These have been studied in most detail by ROSE and his coworkers in the squirrel monkey (ROSE *et al.*, 1967, 1971; HIND *et al.*, 1967; HIND, 1972), and examples are shown in Fig. 11. Each curve depicts the average *maintained* discharge rate (measured over several seconds continuous tone stimulation) as a function of frequency at a constant sound pressure level at the tympanic membrane. The FTCs of Fig. 9C were obtained from these iso-intensity rate curves by plotting the frequencies and stimulus levels evoking a constant rate of response near threshold, in other words by using as a criterion of "threshold", an isorate condition just above the spontaneous firing rate.

The considerably different appearance of these curves from the FTCs arises partly from the linear frequency scale, but mainly from the manifest non-linearity of the response rate as a function of stimulus level, with its limited dynamic range, which will be described in detail in the next section. Thus, comparing the FRCs of Figs. 11C and 12A, from two fibres of similar CF, the two sets of curves look very different. Fig. 12C, however, shows that the FTCs derivable from the two sets of data

(Curves *c* and *a* respectively) are in fact very similar. Figure 12B shows, from the same fibre as 12A, a plot based on the number of discharges synchronized to the stimulus waveform (see Section II.D.4). It has been claimed (ROSE *et al.*, 1971; MØLLER, 1972) that the differences between the plots of Fig. 12A and B indicate that the frequency selectivity is much more pronounced with respect to stimulus

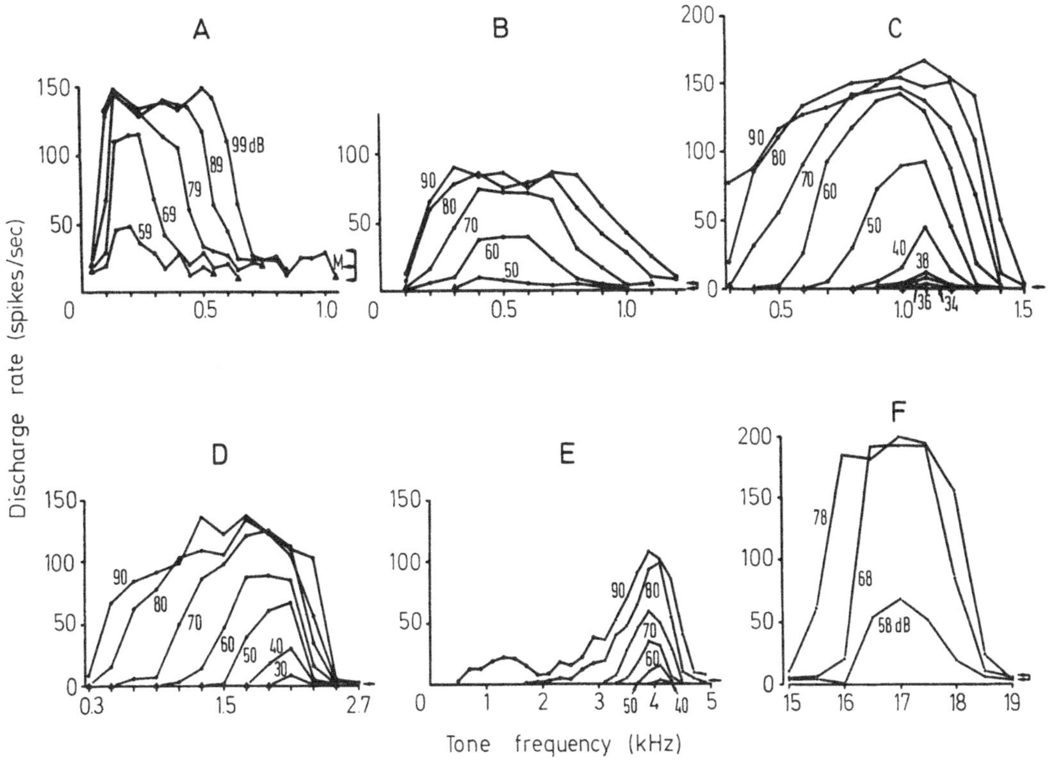

Fig. 11. Iso-intensity frequency response curves of 6 squirrel monkey cochlear fibres. Mean rate of discharge to several seconds of continuous stimulation at frequencies and intensities (in dB SPL) indicated. Note linear frequency scale and different expansions. Range of spontaneous discharge rates and means indicated by brackets and M (or arrow) respectively. Triangular points: no significant degree of synchronization of activity to cycles of stimulus. (From ROSE *et al.*, 1971)

locking than to mean discharge rate. Figure 12C, Curve *b*, however, shows a plot analogous to a FTC derived from the curves of Fig. 12B using a constant synchronized rate criterion (50 synchronized spikes/sec). (The ordinates of the two plots have been shifted for convenient comparison.) If anything, the "synchronized isorate contour" (Curve *b* in Fig. 12C) is less sharp than the FTC based on a mean rate criterion (Curve *a*).

Similarly, it has been claimed that the broadening of the FTCs with increase in stimulus level is an indication of degradation of cochlear filtering properties at suprathreshold levels. If, however, isorate contours with higher rate criteria are plotted as in Fig. 13, these do not exhibit systematic changes in shape and band-

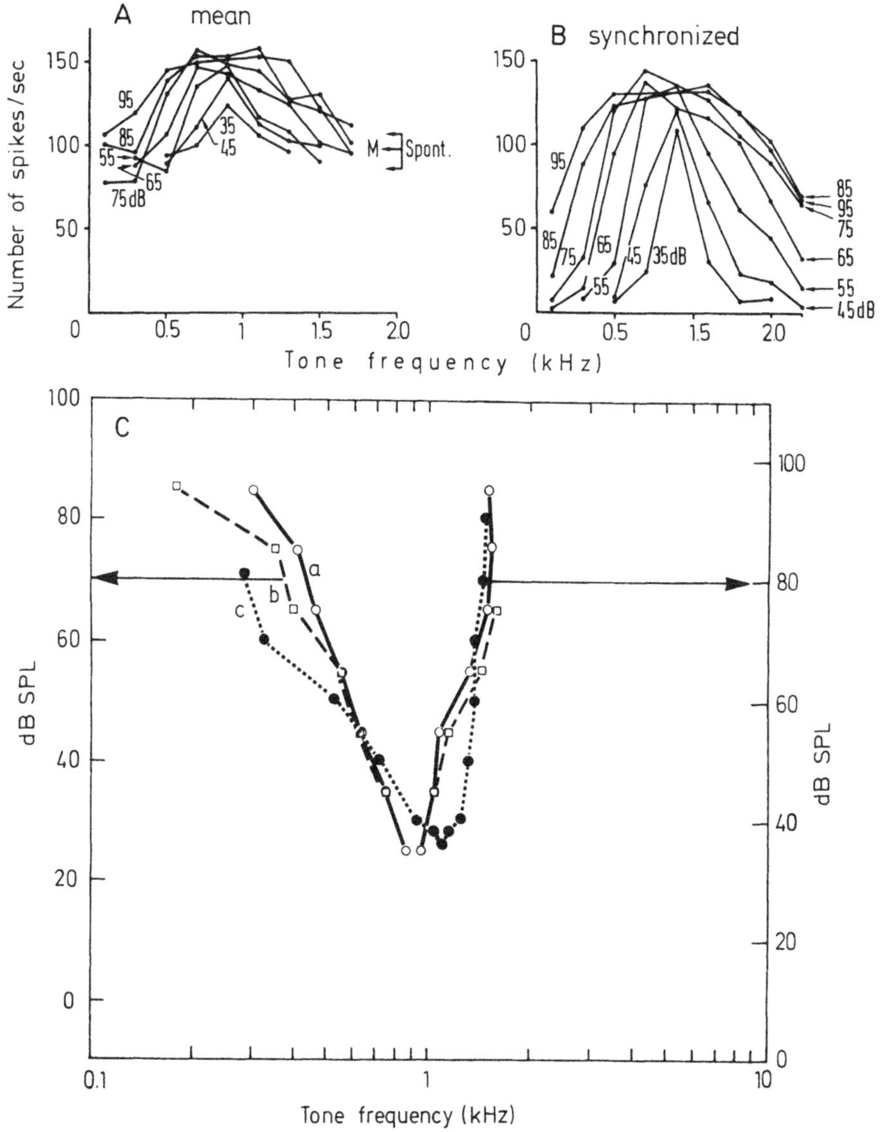

Fig. 12A—C. Relationship between FRCs (using mean and synchronized discharge rate measures), FTC and analogous plots. Squirrel monkey cochlear nerve. A: iso-intensity frequency response curves for a cochlear fibre with a high spontaneous discharge rate (range and mean indicated to right of plots). B: plots of the "synchronized" discharge rate for the same fibre, *i.e.*: the number of spikes discharged during the more effective half-cycle of the stimulus minus the number during the less effective half-cycle, expressed as a rate per sec. (From Rose *et al.*, 1971.) C: iso-rate intensity versus frequency curves (analogous to FTC but using supra-threshold response criteria) derived from FRCs of A and B compared with FTC of a fibre with low spontaneous discharge rate (derived from Fig. 11C). (a) (open circles): iso-rate contour for discharge rate of 120 spikes/sec (20% above spontaneous rate) for frequency response curves in A. (b) (open squares): iso-rate contour for "synchronized" discharge rate of 50 spikes/sec derived from frequency response curves in B. (c) (filled circles): FTC derived from iso-intensity frequency response curves of Fig. 11C (8 spikes/sec criterion of "threshold"). Curves a and c use the right hand ordinate. Note similarity between the 3 FTCs, compared with marked differences between the iso-intensity frequency response plots width, at least within the range of stimulus levels before saturation of the rate response precludes accurate measurement. The isorate data of SACHS and KIANG (1968) also support this conclusion. Above these relatively high levels, the filtering properties may indeed deteriorate; the evidence for this and the consequences will be discussed later (Section VI.A.1).

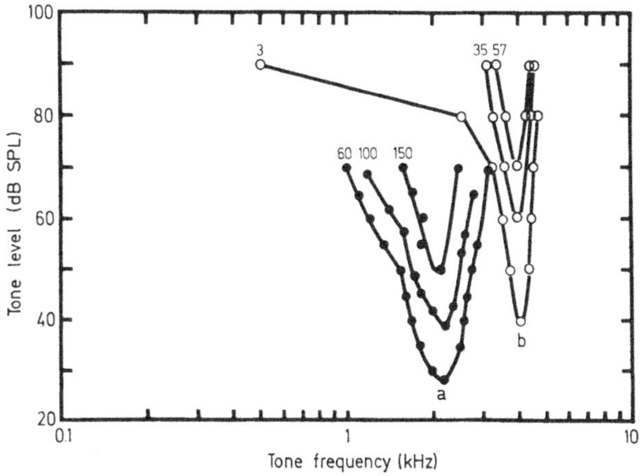

Fig. 13a, b. Iso-rate response contours for two cochlear fibres. Mean discharge rate, in spikes/sec, indicated above respective curve. a: Cat (EVANS, ROSENBERG, and WILSON, unpublished data). Mean spontaneous rate: 50 spikes/sec. b: Squirrel monkey; derived from Fig. 11E (data from ROSE *et al.*, 1971). Negligible spontaneous discharge rate. The difference in bandwidth between a and b probably reflects individual variation and that with CF, rather than any species difference (see Fig. 10)

In summary, the above considerations emphasise the care necessary to relate and interpret various measures of cochlear fibre activity as a function of frequency. Viewed thus, the data point more to systematic changes in shape of the cochlear filtering characteristic with the CF of a fibre than to differing individual characteristics suggestive of differences in cochlear innervation pattern, as suggested by HIND (1972) on the basis of FRC data alone. The FTC and suprathreshold isorate data indicate the characteristics of the cochlear filter function from which has been eliminated the non-linear response characteristics of the cochlear nerve excitation process. The FRC, on the other hand, indicates the limits which may be set upon the central representation of the cochlear filtering by the non-linear rate behaviour of the cochlear fibres. In particular, it indicates the surprisingly limited dynamic range of the system through which, in terms of discharge rate, the results of peripheral filtering are transmitted to the central regions of the auditory nervous system, at least in the anaesthetized animal (see Section VI.D.1).

2. Response as a Function of Intensity

To a first approximation, the discharge rate of a cochlear fibre is a monotonic function of stimulus intensity, as shown in Fig. 14 (KATSUKI *et al.*, 1962; NOMOTO *et al.*, 1964; KIANG *et al.*, 1965a; KIANG, 1968; EVANS, 1974d). The maximum discharge rate, and dynamic range, like the spontaneous discharge rate, differs from fibre to fibre, but, generally, the higher the spontaneous discharge rate, the higher the maximum discharge rate (Kiang *et al.*, 1965a). The dynamic range, in terms of stimulus level, ranges from 20—50 dB. Figure 14A shows that the rate-intensity function for a single fibre may also differ to a small extent depending on the stimulus frequency (Evans, Rosenberg, and Wilson, unpublished observations; Evans, 1974d). In general, for frequencies at and above the CF, or as in Fig. 14A only above the CF, the functions are less steep than those for frequencies below the CF. Not all fibres show these systematic dif-

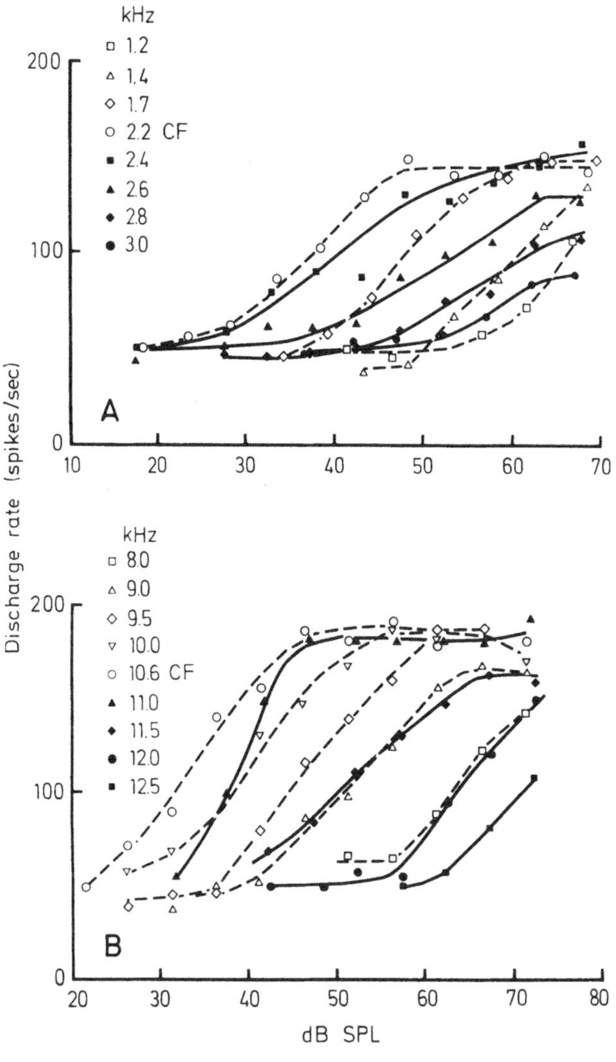

Fig. 14. Discharge rate versus stimulus intensity curves for two cat cochlear fibres, with CFs of 2.2 kHz (A, same fibre as in Fig. 13a) and 10.6 kHz (B). Open symbols and interrupted lines denote curves for frequencies at or below CF; filled symbols and continuous lines, for frequencies above CF. Note: in A, systematic differences in slopes of rate functions between frequencies above and below the CF. (Evans, Rosenberg, and Wilson, unpublished data)

ferencies with frequency (Fig. 14B). Similar rate functions are illustrated by WIEDERHOLD (1970; Fig. 6), and can be derived from the data of ROSE *et al.* (1971). This dependence of the rate function on frequency is likely to be responsible for the systematic shift of the peak discharge rate of some FRCs towards lower frequencies with increase in stimulus level above threshold (*e.g.* see Fig. 11C–E).

NOMOTO *et al.* (1964) described two groups of fibres possessing what was termed a "crossed ramp" and "parallel ramp" type of rate-intensity function respectively. These two types are consistent with those of Fig. 14A and B respectively, although the data of NOMOTO *et al.* are less complete. NOMOTO *et al.* regarded the existence of the two types as evidence for different kinds of cochlear fibres: specifically, the external spiral and internal radial fibres respectively. This speculation however, has not been well received (*e.g.* KIANG *et al.*, 1965a). Clearly, there is room for a more systematic study of these rate functions and their significance.

At very high stimulus levels (90 dB SPL and above) the maintained discharge rate can *decrease* with increase in stimulus level, *i.e.*: the rate response becomes non-monotonic. KIANG *et al.* (1969) and KIANG and MOXON (1972) have reported a curious phenomenon at these levels, where within a few dB, the discharge rate drops towards the spontaneous rate and rises again to a maximum. This notch in the rate-intensity function is found in fibres of widely differing CF at comparable sound pressure levels, although it is said to occur at somewhat lower levels for tones below the CF. Its characteristics could be accounted for by interference between two-out-of-phase excitation processes, one with low threshold accounting for the lower portion of the rate intensity function, and a higher threshold process responsible for the portion above approximately 90 dB SPL. In support of this hypothesis, for fibres of low CF and at low frequencies (where phase relations can be observed), substantial changes in the relative phase of the discharge pattern occur as the stimulus level is raised through the region of the notch. (See also discussion in Section VI.B).

3. Response as a Function of Time

Figure 15 shows the typical time-course of discharge of cochlear fibres to short duration tone (A) and noise stimuli (B). The rate of firing reaches a maximum within a few msec of stimulus onset and adapts at a increasingly slower rate with time. At the cessation of the stimulus, the firing rate drops below the spontaneous level transiently before recovery (as in Fig. 1). This pattern of response is characteristic of all cochlear fibres and, in qualitative terms, is relatively independent of the nature and parameters of the stimulus, *i.e.* tone, noise, frequency and intensity, and whether the efferent innervation of the cochlea is intact or not (Fig. 15B; KIANG *et al.*, 1965a; KIANG, 1968). However, the magnitudes of the transient excitation and suppression, following the onset and termination of the stimulus, depent on the level, and for the transient suppression, on the duration of the tonal stimulus. Thus, near threshold, the former becomes less marked (Fig. 15B, –70 dB), whereas the duration and degree of the latter increase with the level and duration of the signal.

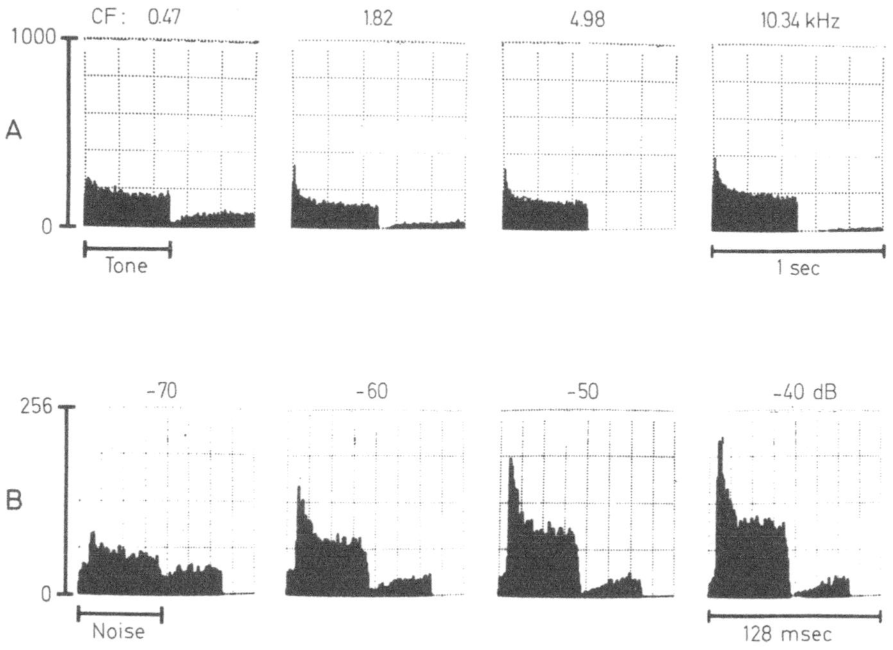

Fig. 15A, B. Time-course of response of cat cochlear fibres to tone and noise bursts. PST histograms. A: 4 different fibres of the CF indicated. Linear ordinate scale, number of spikes per bin; 2 min data; tone at CF, 0.5 sec duration, presented 1/sec. B: Effect of level of stimulus on time-course (similar for tones and noise bursts). 50 msec noise burst beginning approx. 2.5 msec after zero time of PST, repetition rate: 10/sec. Relative signal level indicated above each PST histogram. (From Kiang *et al.*, 1965a)

4. Response to Low Frequency Tones

Responses of cochlear fibres to tones of frequency lower than 4–5 kHz preferentially occur during a restricted segment of the cycle of the sinusoidal stimulus, *i.e.* the discharges are "phase-locked": Fig. 16, (Tasaki, 1954 in the guinea pig; Rupert *et al.*, 1963; Kiang *et al.*, 1965a in the cat; Katsuki *et al.*, 1962; Rose *et al.*, 1967 in the monkey; Frishkopff and Goldstein, 1963 in the bullfrog; Furukawa and Ishii, 1967 in the fish). This phenomenon has been studied most thoroughly by Rose and his colleagues in the squirrel monkey cochlear nerve (Rose *et al.*, 1967, 1968, 1971; Anderson *et al.*, 1971; Hind, 1972).

The probability of discharge of a fibre (irrespective of its CF) to a sufficiently intense low frequency tone appears to be a function of the displacement of the cochlear partition in one direction (Figs. 16B, 17, 18). Thus "folded" time histograms ("period histograms"), of the distribution of discharges relative to the period of the stimulus sinusoid, mirror the effective half-cycle of the stimulus (Figs. 17 and 18), and a compound histogram (Fig. 16B) can be constructed by repeating the analysis with inversion of the polarity of the stimulus (Arthur *et al.*, 1971). Thus, while the mean discharge rate is given by the FRC, the cadence of discharge is governed by the cadence of the stimulus cycles and the ability of the fibre to "follow" the stimulus (Fig. 17).

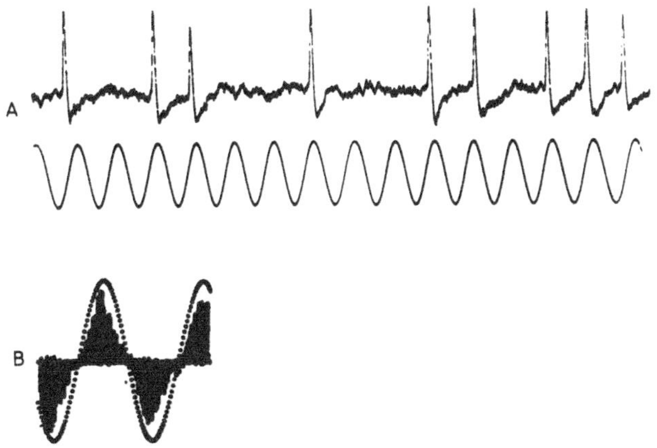

Fig. 16A, B. Synchronization of responses of cochlear fibres to low-frequency tones. A: continuous film record of response of guinea pig fibre to tone of 0.3 kHz, near threshold. B: compound period histogram (see text) of response of cat cochlear fibre to continuous tone at CF of 1.498 kHz, 63 dB SPL. (From ARTHUR *et al.*, 1971.) Waveform of stimulus tone superimposed. Continuous tone of 76 sec duration

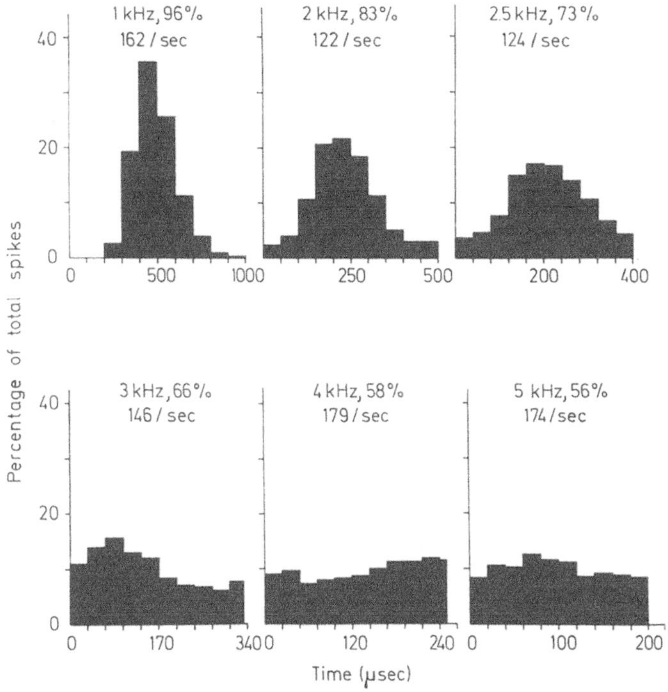

Fig. 17. Degree of synchronization of response of cochlear fibre to tones of differing frequency. Squirrel monkey. CF of fibre: 4 kHz. Tones presented at 90 dB SPL at frequencies indicated above period histograms. Scale of each histogram and bin width adjusted to span one period of stimulating tone. Ordinate scales: percentage of the total number of spikes falling in respective bin. Percentage figure above each histogram: "coefficient of synchronization" *i.e.* number of spikes in most effective stimulus half-cycle as percentage of total (100% = complete synchronization; 50% = no synchronization). Mean discharge rate in spikes/sec indicated over each histogram. Note maximum degree of synchronization to lowest frequency; progressive reduction in degree with increase in tone frequency. Spontaneous discharge rate: 64 spikes/sec. (From ROSE *et al.*, 1967)

Figure 17 shows the progressive loss of synchronization between discharge pattern and stimulus cycle as the stimulus frequency is increased. In the case illustrated, this loss occurs in spite of the greater effectiveness of the higher frequencies in terms of the mean discharge rate. In the experience of ROSE *et al.*, (1967), phase-locking fades out for most fibres for frequencies above 4–5 kHz though they do report significant synchronization in some fibres at frequencies up to 12 kHz.

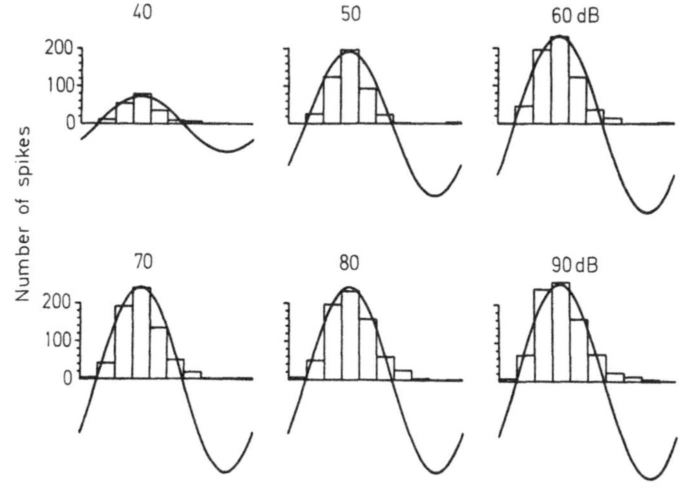

Fig. 18. Synchronization of response of cochlear fibre with level of tone at CF (1.1 kHz). Squirrel monkey. Stimulus level in dB SPL indicated above each period histogram. Period of histogram: 912 μsec. Histograms fitted by sinusoids of constant phase but differing amplitudes. Note preservation of stimulus waveform in period histogram even at stimulus levels above the level producing saturation of the rate response (70 dB SPL). The FRC of this fibre is shown in Fig. 11C. (From ROSE *et al.*, 1971)

Many spontaneously discharging fibres display a detectable degree of phase synchronization at stimulus levels 10–20 dB below the threshold for an increase in mean discharge rate (ROSE *et al.*, 1967; EVANS, 1972b). This synchronization, in addition, is maintained at stimulus levels above those producing a saturation of the mean discharge rate (Fig. 18). These considerations mean that the dynamic range over which phaselocking is obtained is far in excess of the 20–50 dB limit for the mean discharge rate.

It is clear from Fig. 16A and from interspike interval histogram analyses (Fig. 19) that a fibre does not discharge once every cycle of a low frequency stimulus waveform but predominantly to integral multiples of the waveform cycle (ROSE *et al.*, 1967). Exceptions to this generalization occur at very low stimulus frequencies where multiple discharges sometimes occur, and at higher frequencies, where the refractoriness of the fibre eliminates interspike intervals below about 0.7 msec (ROSE *et al.*, 1967). At lower stimulus levels (or at frequencies where the stimuli are less effective, as in Fig. 19A, F), there is an increase in the relative proportion of longer intervals, corresponding to the decrease in mean rate. It is clear that, for intervals longer than 0.5 msec, it is not refractoriness which determines whether or not a cycle is effective. ROSE *et al.* (1967) calculated the conditional probability of discharge and found it to be nearly constant at any one frequency *irrespective (after the refractory interval) of when the previous discharge occurred.* In their words (ROSE *et al.*, 1968) "... a sinusoidal stimulus acts in general as if it consisted of as many individual stimuli as there are cycles". They concluded, therefore, that the events determining the effectiveness of a cycle took place *peripheral* to the spike generation process which is assumed to be responsible for the properties of refractoriness.

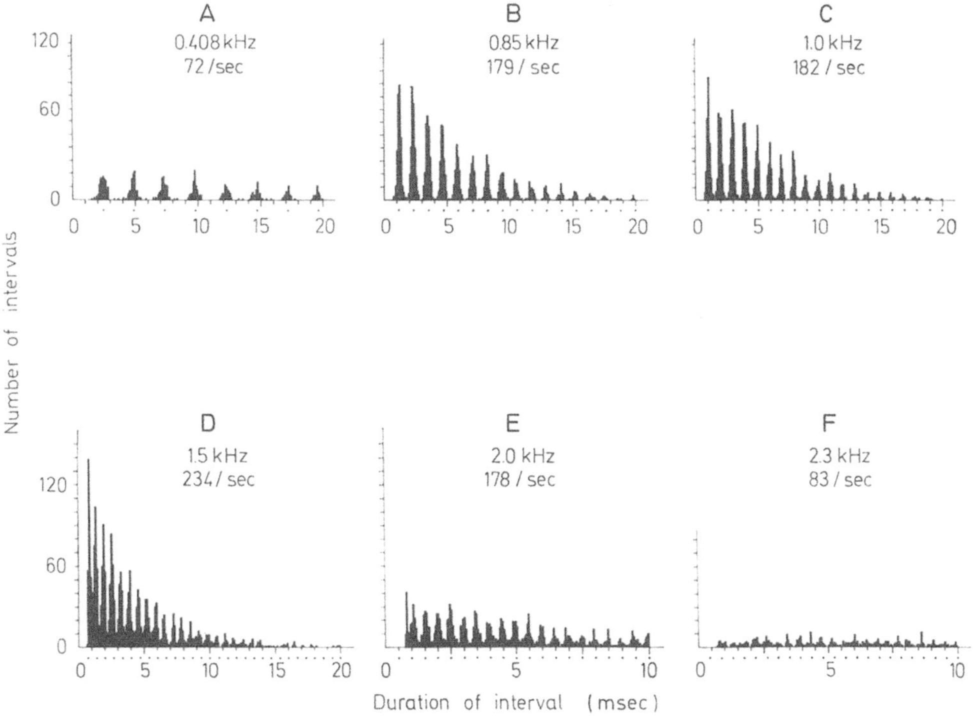

Fig. 19. Interspike interval distributions of response of single cochlear fibre to tones at different frequencies. Squirrel monkey. Tone frequency and mean response rate (in spikes/sec) indicated above each histogram. Intensity of all tones: 80 dB SPL; tone duration: 1 sec. Data comprise responses to 10 presentations of stimulus. Bin width: 100 μsec. Dots below abscissa indicate integral multiples of period of stimulating tone. CF: approx 1.6 kHz. Note different time scales of E and F. (From ROSE *et al.*, 1968)

The phase of the stimulus cycle at which the probability of discharge is maximum differs systematically from fibre to fibre according to the CF, and, for a single fibre, according to the stimulus frequency (PFEIFFER and MOLNAR, 1970; ANDERSON *et al.*, 1971). PFEIFFER and MOLNAR (1970) computed the phase lag (relative to the round window CM) of the fundamental component from Fourier analysis of period histograms obtained from cat cochlear fibres. For fibres of CF lower than 2 kHz, this was an approximately linear function of frequency, although a better fit to the data was obtained by two straight lines intersecting at a point close to the CF in some but not all cases (Fig. 20). For fibres with CF above about 1.1 kHz, the phase lag increased more rapidly with frequency for frequencies above the CF; for fibres with lower CF, the reverse tended to occur. Goldstein *et al.* (1971) have

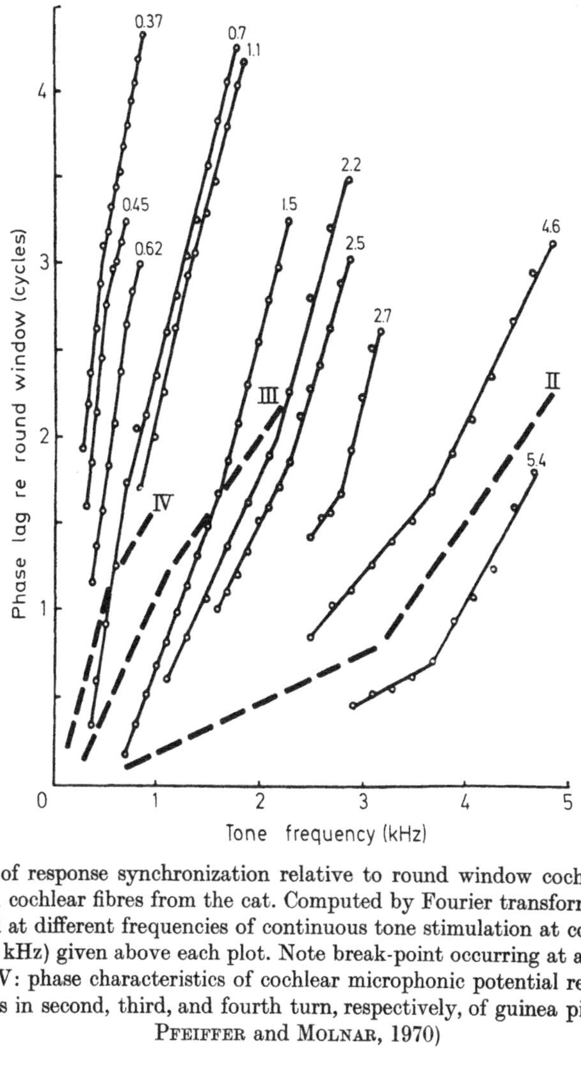

Fig. 20. Phase of response synchronization relative to round window cochlear microphonic potential for 11 cochlear fibres from the cat. Computed by Fourier transform of period histograms obtained at different frequencies of continuous tone stimulation at constant intensity. CF of fibres (in kHz) given above each plot. Note break-point occurring at about CF for each fibre. II, III, IV: phase characteristics of cochlear microphonic potential recorded by differential electrodes in second, third, and fourth turn, respectively, of guinea pig cochlea. (From Pfeiffer and Molnar, 1970)

reported similar findings. From more limited data in the squirrel monkey, Anderson *et al.* (1971) obtained linear phase versus frequency plots from which they derived a total time delay (acoustic plus cochlear plus neural) for each fibre (Fig. 21A). On the assumptions that the middle ear transmission delay could be neglected, and that the neural transmission amounted to 1 msec (the asymptote of Fig. 21A), they obtained the estimated "travel times" of the cochlear disturbance to the points of innervation of the cochlear partition shown in Fig. 21B. The interpretation of this "travel time" is, however, made difficult by the suggestion of GOLDSTEIN *et al.* (1971) and DUIFHUIS (1972) that the measured delay must include a factor corresponding to the "response time" of the cochlear filter (see Section VI.A.1).

In contrast to the behaviour with frequency, PFEIFFER and MOLNAR (1970) and ANDERSON *et al.*, (1971) showed that the phase lag between response and stimulus was relatively unaffected by stimulus level. For most fibres studied in the material of ANDERSON *et al.* (1971), the phase lag increased with stimulus level (less than

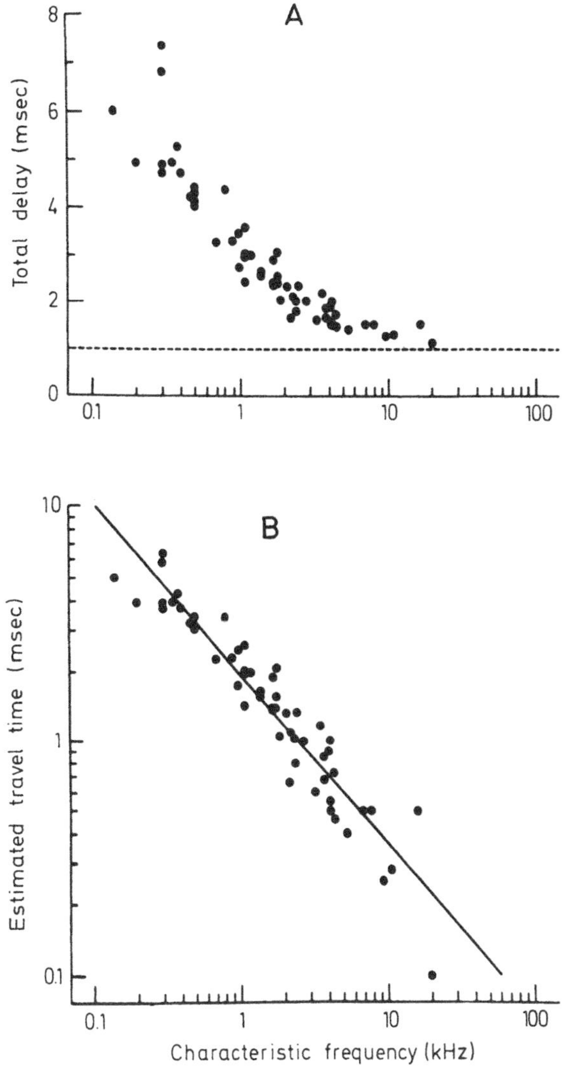

Fig. 21. A: Delay times of synchronized spike responses. 58 cochlear fibres in squirrel monkey. Delays computed from slope of phase versus frequency curves (as in Fig. 20) for each fibre, and plotted against CF of fibre. Values corrected for acoustic delay in sound system, and, therefore, represent travel time in cochlea plus synaptic and neural conduction delays. B: Cochlear travel times calculated from data of A, after subtracting 1 msec average neural delay estimated from asymptote of A (interrupted line), and plotted against each fibre's CF. (From ANDERSON *et al.*, 1971)

90°) for frequencies below the CF, and *vice-versa*. At frequencies at or near the CF, little or no change of relative phase with level occurred (*e.g.* Fig. 18). On the other hand, as has already been mentioned (Section II.D.2), at *very high* stimulus levels very large changes in response phase have been reported to accompany dramatic changes in the discharge rate (Kiang *et al.*, 1969; Pfeiffer and Molnar, 1970).

A consideration of these findings in relation to questions of cochlear mechanics and innervation will be deferred to Section VI.A. and B.

E. Response to Complex Sounds

1. Two-tone Stimuli

a) Two-tone Suppression. Perhaps the best known phenomenon arising from the interaction between two tones at this level is that known as two-tone inhibition or suppression, which is found in many species (*monkey:* Nomoto *et al.*, 1964; Hind *et al.*, 1967; *frog*: Frishkopf and Goldstein, 1963; Liff and Goldstein, 1970; *bat:* Frishkopf, 1964; *cat:* Kiang *et al.*, 1965a; Kiang, 1968; Sachs and Kiang, 1968; Sachs, 1969; Arthur *et al.*, 1971). As will be made clear subsequently, the term *suppression* is to be preferred, to distinguish the phenomenon at the cochlear nerve from that with different characteristics at the cochlear nucleus and higher levels of the auditory system, where lateral inhibitory influences can be clearly inferred (*e.g.* Galambos, 1944; Greenwood and Maruyama, 1965).

The phenomenon entails the suppression of activity evoked by one stimulus (tone or noise) by a second tone over a restricted range of frequencies and intensities (Fig. 22). Its properties have been described in detail by Sachs and Kiang (1968), and Arthur *et al.* (1971). The former authors were able to demonstrate suppression in every fibre examined (*cf.* Nomoto *et al.*, 1964). The situation most commonly examined is where the response to a continuous tone (CT) at the CF of a fibre is reduced by a second tone, generally of higher level, within a band of frequencies adjacent to or even slightly overlapping the excitatory response area of the fibre (Fig. 22B). This implies that the second, (suppressing) tone can, on its own, produce either no response or excitation (Fig. 22A, C). Figure 22B also indicates the asymmetry of the suppressive frequency bands, extending down towards the CT intensity only on the high frequency side. There is some evidence that with an exciting CT stimulus at a higher level, the suppressive frequency bands are shifted vertically upwards from the situation shown in Fig. 22B, so that the degree of overlap is greater (Nomoto *et al.*, 1964).

There is a non-monotonic relationship between the response to, and the intensity of, the "suppressing" tone (Fig. 22C; Arthur *et al.*, 1971). This would be expected as the latter progressed through the suppressive sideband into the excitatory response area (vertically upwards in Fig. 22B).

Hind *et al.* (1967), Brugge *et al.* (1969), and Arthur *et al.* (1971) have shown that the discharge pattern under conditions of two-tone suppression retains phase-locked information of *both* tones. Thus compound period histograms (*vd.* Fig. 16B) obtained under these conditions can be approximated by a waveform containing the two frequency components comprising the stimulus (as in Fig. 24).

The suppression has a latency of the same order as excitation, *i.e.* it occurs within a few msec of the onset of the suppressing tone (NOMOTO *et al.*, 1964; ARTHUR *et al.*, 1971). For the initial part of the suppressing tone, the suppression

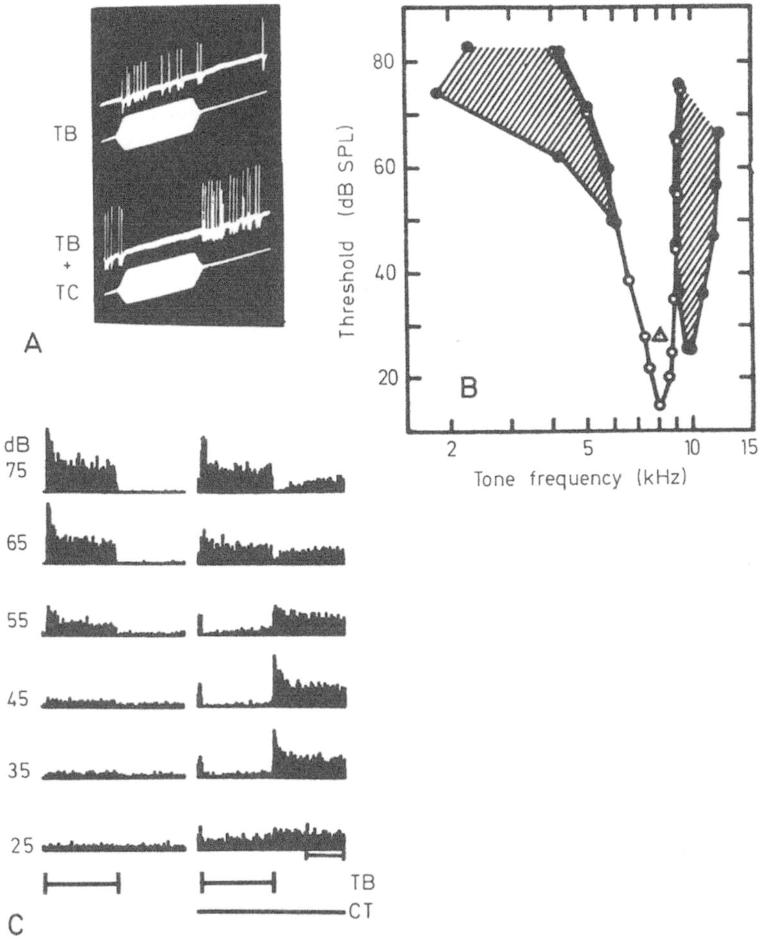

Fig. 22A—C. Two-tone suppression in cochlear fibres. A: continuous film records of response of fibre to tone burst (TB) at 0.8 kHz, 80 dB SPL (upper record) and to identical tone burst superimposed upon exciting continuous tone (CT) at 11.3 kHz, 77 dB SPL (lower record). Monkey. (From NOMOTO *et al.*, 1964.) B: Frequency response areas of single-tone excitation (open circles) and of tones which suppress response to continuous tone of the frequency and level indicated by the triangle (filled circles and hatched areas). Continuous outlines represent "threshold" response criteria of more than 20% above the spontaneous discharge rate and 20% below the response to the CT alone for excitatory and suppressive areas, respectively. Cat. (From ARTHUR *et al.*, 1971.) C: PST histograms of response to 100 msec tone bursts alone (left) and superimposed on continuous tone at CF (right) as a function of level of tone burst. Continuous tone: 8.08 kHz, 28 dB SPL; tone burst frequency: 8.893 kHz. Repetition rate: 5/sec. Average of 128 responses. Time bar: 50 msec. Cat. (From ARTHUR *et al.*, 1971)

may be complete, but a greater or lesser degree of adaptation ensues. The time course of this "adaptation" is generally slower than that of excitation of the fibre (Fig. 22; LIFF and GOLDSTEIN, 1970). On termination of the suppressing tone, a prominant rebound of discharge occurs (Fig. 22C) comparable to the excitatory transient at the beginning of the response to a steady tone.

A matter of some controversy (*e.g.* MOUSHEGIAN *et al.*, 1971) is whether two tone suppression can produce a *maintained* depression of the activity of a fibre *below the spontaneous discharge level.* SACHS and KIANG (1968) claimed that this did not occur for more than the initial several hundreds of msec, although certain of their records (*e.g.* Fig. 2, – 40 dB of SACHS and KIANG, 1968) suggest that it can occur for a few seconds, at least. All the above observations are, however, consistent with the hypothesis (ARTHUR *et al.*, 1971) that the initial suppression below the spontaneous discharge rate is characteristic of the cochlear fibre behaviour which follows the termination of excitation, *i.e.* postexcitatory depression (Section II.D.3). On this basis, the time-course of the suppression below the spontaneous level will depend upon the duration of the excitation preceding the onset of the two-tone suppression. The fact that, in the usual two-tone suppression paradigm, there has been at least several seconds of continous excitation before the application of the suppressing tone would ensure that in these circumstances the time course of the suppression below the spontaneous level would exceed that of excitation (as observed by LIFF and GOLDSTEIN, 1970). In the case of the data of SACHS and KIANG (1968), the unsuppressed excitation had been continuously present for over 20 sec before the onset of the suppressing tone. This explanation also enables an understanding of the observation of NOMOTO *et al.* (1964) that when the time order of the stimuli was interchanged, *i.e.* a tone burst at the CF was superimposed on a continuous tone which was suppressive in the usual two-tone paradigm, a transient suppression below the spontaneous level was not obtained. The suppression produced in the two-tone situation, therefore, occurs as if the exciting tone had merely been interrupted or abruptly attenuated. On this view, the transient total suppression and the transient excitatory "overshoot" corresponding to the onset and termination, respectively, of the suppressing tone are both effects identical to those occurring when a single exciting tone is turned off and on, respectively.

The mechanism by which the effects of an exciting tone are suppressed is still obscure. It is clear that it does not involve the descending efferent (inhibitory) pathway, because two-tone suppression can be recorded from fibres in the peripheral stump of the sectioned cochlear nerve (FRISHKOPF and GOLDSTEIN, 1963; KIANG *et al.*, 1965a) and after section of the crossed and uncrossed olivocochlear tracts and their subsequent degeneration (KIANG, 1968). The anatomical (*e.g.* SPOENDLIN, 1971) and physiological evidence is equally overwhelming against a lateral inhibitory mechanism involving synaptic connections between hair cells and/or afferent fibres. Thus, the latency, time-course, and quantitative characteristics differ from those associated with lateral inhibition in other systems, *e.g.* Limulus eye (FURMAN and FRISHKOPF, 1964; SACHS, 1969; LIFF and GOLDSTEIN, 1970; ARTHUR *et al.*, 1971). Furthermore, the application of strychnine is without effect (NOMOTO *et al.*, 1964). The possibility exists, however, of electrotonic interaction between the unmyelinated segments of outer spiral and inner radial fibres at the habenula perforata, where the two sets of fibres become intimately opposed (*e.g.* SPOENDLIN, 1972). ARTHUR *et al.* (1971) have suggested that any interaction must occur peripheral to the stage of rectification associated with generator or action potential initiation, because of the preservation of waveform interaction in the spike discharge patterns. However, SACHS (1969) claims that a shunting model of electrotonic interaction (*e.g.* FURMAN and FRISHKOPF, 1964) is not supported quantitatively by the data on two-tone suppression. An alternative possibility has been suggested by the similarities between two-tone suppression in the cochlear nerve and an analogous reduction of the cochlear microphonic, where the latter could be modelled qualitatively by a system comprising a non-linearity associated with one or two filtering processes (ENGERBRETSON and ELDREDGE, 1968; PFEIFFER, 1970). In PFEIFFER's model, designed to account for two-tone suppression at the cochlear nerve level, the nonlinearity is sandwiched between an input filter of low frequency selectivity (equivalent to that of the basilar membrane) and an output filter with a sharper frequency selectivity (equivalent to that of the FTC). The model generated asymmetric suppressive side bands overlapping the excitatory response area, and, in addition, a significant intermodulation distortion component at a frequency of $2f_1 - f_2$ (where f_1 and f_2 are the frequencies of the two primary tones). This product is in fact observed at the cochlear nerve level (GOLDSTEIN and KIANG, 1968; see below).

b) Excitation by Tone Combinations. NOMOTO *et al.* (1964) were the first to demonstrate that two tones of appropriately differing frequencies, neither of which alone excited a cochlear fibre, could, when presented together, nervertheless evoke a response. This phenomenon has since been investigated in detail by GOLDSTEIN and KIANG (1968) and by GOLDSTEIN (1970, 1972) who have shown its relation to the most conspicuous combination tone perceived under these conditions, namely one of frequency $2f_1 - f_2$, where f_1 and f_2 are the frequencies of the primary tones and where $1 < f_2/f_1 < 2$. Two examples are shown in Fig. 23, where the frequencies of two tones outside the response area of the fibre are so arranged that the frequency $2f_1 - f_2$ coincides with the CF of the fibre. A brisk response is obtained, which in the case of the fibre of low CF is time-locked to the cycle of a reference $2f_1 - f_2$ signal generated electrically from the two primary signals (Fig. 23B).

The response to the tone combination behaves as if a component of the $2f_1 - f_2$ frequency were actually present in the stimulus (GOLDSTEIN and KIANG, 1968). Thus, the responses are phase-locked to individual cycles of a reference $2f_1 - f_2$ signal (when the combination tone frequency was below 4 kHz) as in Fig. 23B; the synchronized responses can be cancelled by adding a tone of appropriate phase and frequency $2f_1 - f_2$ to the stimulus ensemble; the response discharge rate shows a similar dependence upon signal level as for single tone stimuli.

In other aspects also the cochlear fibre data are analogous to psychophysical data (ZWICKER, 1955; GOLDSTEIN, 1967, 1970). In particular, the degree of synchronization of the cochlear nerve activity with the $2f_1 - f_2$ frequency is not strongly dependent upon stimulus level, as if the stimulus contained a combination tone with nearly constant relative amplitude. Furthermore, the response to the combination tone decrements with separation of the frequency of the primaries at an estimated rate of the order of 100 dB per octave and above, in terms of equivalent stimulus change (GOLDSTEIN and KIANG, 1968) and the level of a cancelling $2f_1 - f_2$ tone (GOLDSTEIN, 1970). This frequency dependence indicates that the site

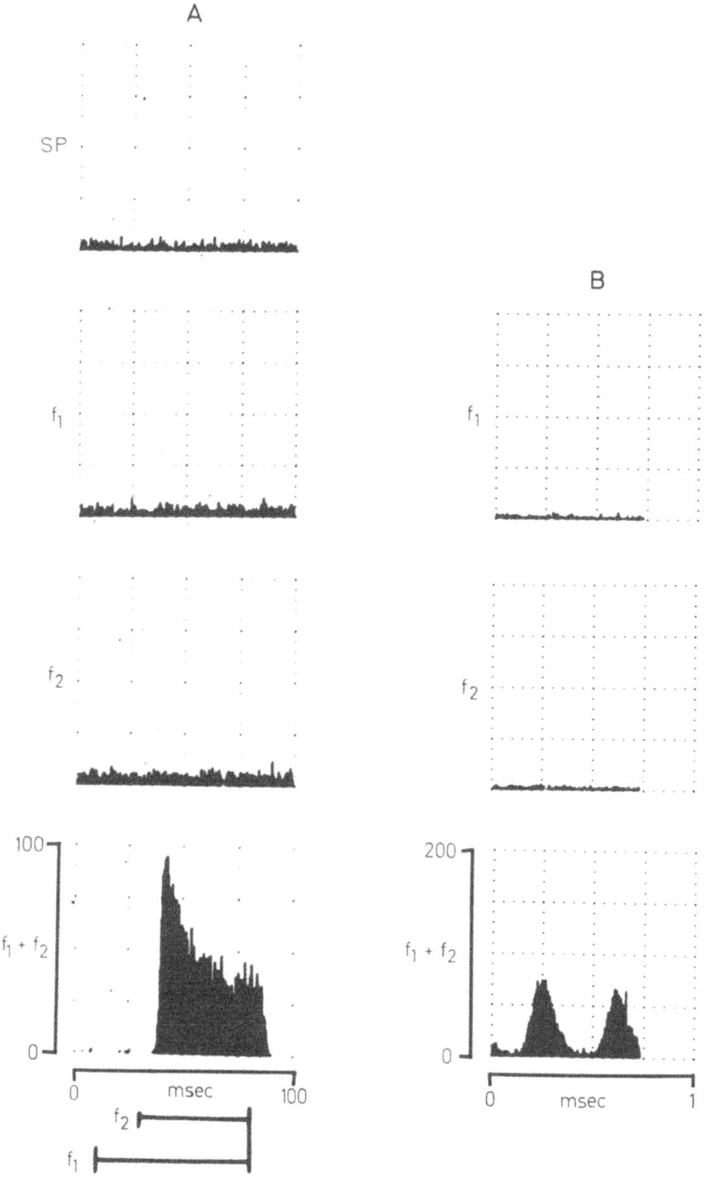

Fig. 23 A, B. Excitation of cat cochlear fibres by combination tone $2f_1 - f_2$ (see text). A: Time histograms of spike activity in presence of: no stimulation (SP); tone 1 alone, at 10.00 kHz, 59 dB SPL (f_1); tone 2 alone, at 12.15 kHz, 69 dB SPL (f_2); Tone 1 and tone 2 together ($f_1 + f_2$). Presentation time and duration of the two tone bursts indicated under lowest histogram. Repetition rate: 10/sec. Note vigorous response to combination of tones ($2f_1 - f_2$ fails on CF of fibre: 7.88 kHz) in comparison with negligible response to either tone alone. B: Period histograms of response to two tones alone (f_1 : 5.5 kHz and f_2 : 4.13 kHz) and to combined tones ($f_1 + f_2$), of fibre with low CF (2.69 kHz). Each histogram is synchronized to the fundamental frequency of the harmonically related (4 : 3) primary tones. The response to the tone combination exhibits a period of synchronization equal to about half the histogram period, *i.e.*: corresponding to the $2f_1 - f_2$ combination tone (2.76 kHz), close to the CF of the fibre. Spontaneous rate: 5 spikes/sec. (From Goldstein and Kiang, 1968)

of the generation of the distortion product must be central to the middle ear (because of the latter's relatively level frequency response) and that the process responsible for the distortion must be not simple overloading but an "essential" (*i.e.* level-normalized) odd-order nonlinearity (GOLDSTEIN and KIANG, 1968; GOLDSTEIN, 1970). In one curious respect, however, the neurophysiological and psychophysical data disagree: the measured phase of the psychophysical combination tone changes with the level of the primary stimuli, whereas that obtained in the cochlear nerve recordings does not (GOLDSTEIN, 1971: QPR No. 100, p. 202; GOLDSTEIN, 1972). There is as yet no explanation for this discrepancy.

There have been attempts to account for some, at least, of these phenomena on the basis of the non-linear threshold detector mechanism in the cochlear nerve spike generation process (DE BOER *et al.*, 1969). Such a process alone, could not, however, account for the excitation of a fibre by a combination of two tones, each outside the response area. In addition, the hypothesis of DE BOER *et al.* (1969) is not consistent with the other physiological findings, in particular: the absence of cubic intermodulation products higher than the primary frequencies predicted by the hypothesis (DE BOER, 1970, p. 246); the observed degree of synchrony of cochlear fibres to the combination tone component (GOLDSTEIN, 1972); and the interspike interval histograms obtained by GOLDSTEIN and KIANG (1968), (HALL 1971).

BOERGER and GRUBER (1970) have reported that some but not all cochlear fibres of low CF in the cat could be excited by multicomponent signals consisting of the 3rd to 6th harmonics of a fundamental which coincided with the CF, but which was absent from the signal (*i.e.* a "residue" or "missing fundamental" signal). None of the components, individually, evoked responses, whereas responses could be evoked by the composite signal at levels comparable to the threshold level for tones at the CF. These low relative thresholds were obtained from fibres with low CFs, *i.e.* about 0.2 kHz. The absolute levels were well below those expected to produce significant amounts of difference tone through distortion in the ear (see *e.g.* GOLDSTEIN, 1970). BOERGER and GRUBER have suggested that this phenomenon may be related to the perception of the "missing fundamental" with similar stimuli by man. On the face of it, this would seem unlikely, in view of the finding (LICKLIDER, 1954) that whereas the percept can be masked by signals at the frequencies of the components, it cannot be masked by signals at the frequency of the perceived fundamental.

c) Interaction of Responses to Two Tones: Linear and Non-linear Aspects. A detailed investigation of the effects of two tone stimulation on the discharge pattern of the cochlear fibres of the squirrel monkey, with analogous techniques to those described in Section II.D.4 above, has been made in ROSE's laboratory (HIND *et al.*, 1967; ROSE *et al.*, 1968; BRUGGE *et al.*, 1969; ROSE *et al.*, 1969; ROSE, 1970; HIND *et al.*, 1970). The findings appear to hold for tones which are related or unrelated harmonically. The above authors have shown that, within certain limits, the time-pattern of discharges reflects, to a first approximation, a rectified version of the waveform of the combined stimulus tones.

Figure 24 illustrates the nature of the interaction when the relative amplitude (A) and relative phase (B) of the two component tones is varied over appropriate

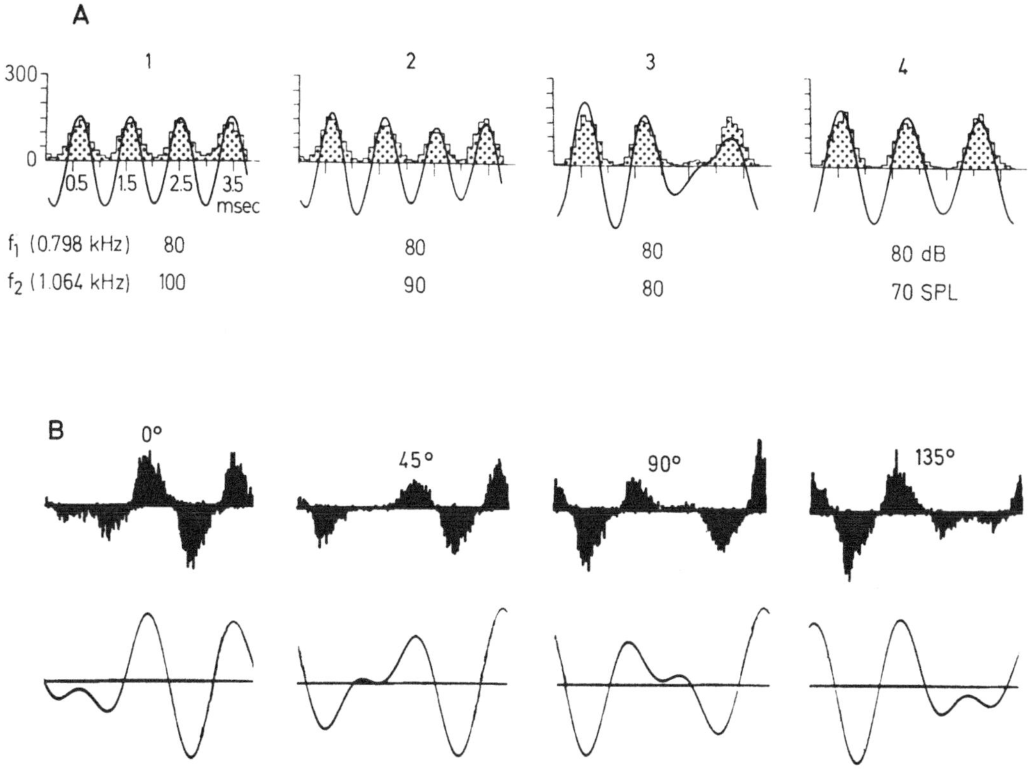

Fig. 24A, B. Interaction of two tones as a function of relative amplitude and phase. A: Period histograms of activity of a squirrel monkey cochlear fibre in response to a tone of 0.798 kHz at fixed intensity (80 dB SPL) combined with a 1.064 kHz tone at intensities: (1) 100 dB SPL, (2) 90 dB SPL, (3) 80 dB SPL, and (4) 70 dB SPL. Analysis of responses to two presentations of tone combination, each of 10 sec duration. Duration of histogram corresponds to the period of the fundamental frequency of the two harmonically related (3 : 4) tones (*i.e.*: 0.266 kHz). Ordinate: linear number of spikes in respective bin. Note almost complete domination of discharge patterns by more intense tone in 1 and 4. Each histogram is fitted by a waveform obtained by linear addition of the two primary waveforms (f_1 and f_2) in the same relative phase as that for the waveforms fitting the responses to each primary alone, but at arbitrary relative amplitudes (see text). (From Brugge *et al.*, 1969.) B: Compound period histograms of activity of a cat cochlear fibre in response to a combination of two exciting tones of frequencies 0.538 and 0.807 kHz. Relative phase between primary stimulus tones indicated above the histograms. Lower half of each histogram obtained by repeating analysis with polarity of stimulus inverted in a manner analogous to that of Figs. 6 and 16B. (Duration of the histograms corresponds to the period of the fundamental frequency of the two harmonically related (2 : 3) tones, *i.e.*: 0.269 kHz.) Waveforms beneath histograms obtained by linear addition of the two primary waveforms at fixed (arbitrary) amplitudes and of the same relative phase as in the stimulus, leading to the response indicated in the appropriate period histogram. (From Goblick and Pfeiffer, 1969)

values. Interaction occurs over a range of relative intensities restricted to 20 to 30 dB. At each end of the range, the discharge pattern (as indicated by the period histogram) is dominated by the more effective tone. Thus, in the case illustrated by Fig. 24, this occurs when the f_2 tone is 20 dB higher in level than the f_1 tone (Fig. 24, A.1) and when the level of f_1 is 10 dB higher than that of f_2 (Fig. 24, A.4).

At intermediate relative levels, the discharge pattern reflects the interaction between the two component tones. These discharge patterns can be fitted, to a first approximation, by a linear addition of two waveforms of the same frequencies of the stimulus tones but of arbitrarily chosen relative amplitude and phase, as shown by the superimposed waveforms of Fig. 24A. Within this limited range of relative stimulus levels, ROSE *et al.* (1971) showed that a reasonable fit could be obtained by adjusting the amplitudes of the matching sinusoidal components in the same ratio as in the stimulus. From this, it would appear that the amplitude ratio of the effective stimulating waveform of the cochlear transduction process changes very nearly as the ratio in the delivered sound.

Likewise, changes in the relative phase of the two stimulus components (Fig. 24B) introduces changes in the discharge pattern which can be fitted by linear addition of two sinusoids with the same changes in relative phase (BRUGGE *et al.*, 1969; GOBLICK and PFEIFFER, 1969). In many cases, good fits could be obtained to the empirical combined tone period histograms by using as a starting point the phase of fits to the responses to the individual tones.

These patterns of interaction described above are maintained over a wide range of absolute levels of the two component stimuli, even at high levels where the discharge rate of the fibre is saturated. Under the latter conditions, the period histograms indicate that the complex waveform is preserved at the excitation process (as in the single tone situation illustrated in Fig. 18) in spite of saturation of the spike generation process (BRUGGE *et al.*, 1969).

Interaction between two tones can be demonstrated at frequencies inside and outside the frequency response area for the fibre. The absolute stimulus levels required for interaction to occur, however, appear to be related to the frequency response area, but not in a simple fashion (HIND *et al.*, 1967). It appears from the data of HIND *et al.* (1967) that, for frequencies well inside the response area, the levels of a tone required for it to achieve a certain degree of synchronization of the response in the presence of another effective tone are to a large extent governed by its effectiveness in driving the fibre on its own. This is not the case for peripheral frequencies, however, which can be more effective in terms of their dominating of the discharge pattern that their rate response would predict.

Indeed, tones at peripheral frequencies which have no effect on their own on the discharge rate can nevertheless dominate the discharge *pattern*. In this connection, HIND *et al.* (1970) made the interesting observation that under these conditions, the discharge rate evoked by the combined stimuli tends towards the rate for the tone alone which is dominating the discharge pattern. Thus, the discharge rate to combined stimuli can be substantially less than that for the most effective tone alone, if the latter is paired with a tone of peripheral frequency which alone is relatively ineffective. The possible relationship of this to the two-tone suppression phenomenon discussed above has been suggested by HIND *et al.* (1970). It is clear from their data, however, that domination of the discharge *rate* by a second tone of peripheral frequency can occur even when the discharge *pattern* has not been dominated by the second tone.

Thus, stimulation of a fibre (66-107-2 in HIND *et al.*, 1967, Table I) with a second tone of 3 kHz at 90 dB SPL reduced the discharge rate to 54%, whereas the coefficient of synchronization of the discharge pattern to the second tone was only 57.9%, compared with a value of

87% for the other tone. (50% = no synchronization; "100%" = complete synchronization.) For comparison, a tone of 3 kHz at 80 dB SPL had no significant effect on discharge rate and gave a synchronization coefficient of 55%.

A clear understanding of these interaction phenomena obviously awaits further data.

Another curious feature of the interactions is that the relative effectiveness of the constituents of a two-tone combination depends on the absolute level of the two. If a given relative level difference is maintained while the absolute levels are changed, both the relative phase and amplitudes of matching constituent waveforms have to be adjusted for the combination to fit the period histograms. In particular, Hind *et al.* (1967), Brugge *et al.* (1969), and Rose *et al.* (1971) observed that the component of a tone pair which was less effective at lower stimulus levels could dominate the discharge pattern of the response at higher levels. This non-linearity in relative effectiveness of phase-locking amounted to 5–12 dB when the levels of both components tones were changed by 50–70 dB (Rose *et al.*, 1971). Over a smaller dynamic range, *e.g.* 20 dB, these amplitude and phase non-linearities become negligible (as already noted). It is interesting to note that, in the illustrated examples of this phenomenon in the above reports, the initially "less effective" tone was the *lower* frequency tone in each case.

From a study of the interspike interval distributions obtained under two tone stimulation, Rose *et al.* (1969) observed that the predominant intervals corresponded, as would be expected, to that of the period of the stimulating waveform excursions. It was concluded that the probability of occurrence of a spike discharge was a function of the amplitude of the excursion (in one direction only). Thus, when one or other primary dominated the response to the combined tones, the predominant intervals were grouped around the integral multiples of the period of that primary. On the other hand, when the two tones contributed about equally to the combined response, the predominant intervals corresponded to the overall period of the complex sound (*i.e.* that of the fundamental frequency of two harmonically related tones), or integral multiples thereof. In addition, smaller numbers of intervals were observed corresponding to frequencies higher than the lower primary. The values of these latter intervals were critically dependent upon the relative phase and amplitude of the primaries, in contrast to those corresponding to the fundamental period.

The possible significance of the above findings, on the patterns of response evoked by stimulation with more than one tone, in relation to the coding of information pertaining to perceived combination tones will be discussed later (Section VI.C.3., D.1).

It must be emphasised that the above data were obtained under conditions of maintained (ca. 10 sec) stimulation (including the response to the onset of the tone). While it does seem unlikely that the conclusions derived will not hold for stimuli of shorter duration, some of the more puzzling features (*e.g.* in relation to the rate response) could conceivably be determined by adaptation phenomena.

2. Noise Stimuli

The responses of cochlear fibres to broadband noise stimuli have been analysed in two different ways, both of which have given rise to similar conclusions on the nature of the cochlear filter.

The first approach has examined the time-structure of the suprathreshold discharge pattern of cat cochlear fibres and its correlation with the waveform of the noise stimulus (DE BOER, 1968, 1969, 1970; DE BOER and JONGKEES, 1968). By an ingenious procedure of "reverse" or "triggered" correlation, where the cross-correlation function between the waveform of the noise stimulus and the ensuing spike discharge pattern is computed, these authors have been able to derive the impulse response of an hypothetical linear cochlear filtering system "seen through" the non-linear excitation process of the cochlear nerve (DE BOER and KUYPER, 1968). From the impulse response, the frequency response of an equivalent linear filter was computed and found to approximate to the observed pure tone FTC within the 10–20 dB range of the method. In other words, under broadband conditions, the discharge pattern is determined by a filtered version of the noise signal, the characteristics of the filter approximating to that of the FTC of the fibre. This result is of course limited to fibres with CFs below 4 kHz, as beyond this frequency, the phase-locking of the discharge pattern to the stimulus drops out. This correspondence between the computed filter function and the FTC obtained for a relatively wide range of noise levels, in one case, over a range to 60 dB above threshold (DE BOER 1969, 1970). DE BOER and DE JONGH (1971) and DE JONGH (1972) have simulated the hypothetical filter (computed from the cross-correlation function between noise stimulus and spike discharge pattern), used this simulated filter to filter a pseudorandom (*i.e.* repetitive) noise waveform, and then compared the actual moments of discharge with this filtered waveform. A result of such a

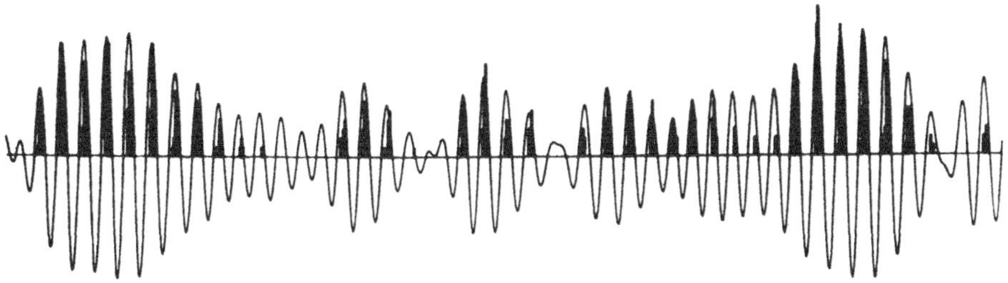

Fig. 25. Time-pattern of discharges of cat cochlear fibre to pseudo-random broadband noise stimulus. Bars: time histogram of average number of spike discharges related of the period of repetition of the pseudo-random noise stimulus. Waveform: filtered version to the stimulus waveform. Characteristics of the filtering operation simulate those of the particular cochlear fibre whose responses are indicated in the superimposed histogram (see text). (From DE JONGH, 1972)

comparison averaged over many repetitions of the waveform is shown in Fig. 25. While the authors interpret their data in terms of the efficacy of "zero crossings" of the waveform, the similarity between this situation and that described in Section II.D.4 and E.1 above is such to suggest the more obvious explanation that, to a first approximation, the probability of discharge is proportional to the amplitude of unidirectional excursions of the filtered waveform presumed to be acting at the site of excitation (see Section VI.A.3).

In this connection, it is interesting to note that a better fit between the filtered waveform and the spike discharge would have been obtained if the waveform had been subjected to a downward dc shift. Rose *et al.* (1971) comment that this was sometimes necessary to obtain an adequate fit between the spike discharge patterns and simulating waveforms in their two-tone interaction data. The significance of this dc shift is not clear.

The second approach has been to examine the relation between the threshold of response to broad band noise and to pure tone stimuli (Evans *et al.*, 1970; Evans and Wilson, 1971, 1973). Using the same criterion of "threshold" for tone and noise stimuli (an audio-visual indication of increase in discharge rate above the spontaneous rate, evoked by gated stimuli of 100 msec duration), the data of Fig. 26M were obtained for cat cochlear fibres of CF ranging from 0.2 to over

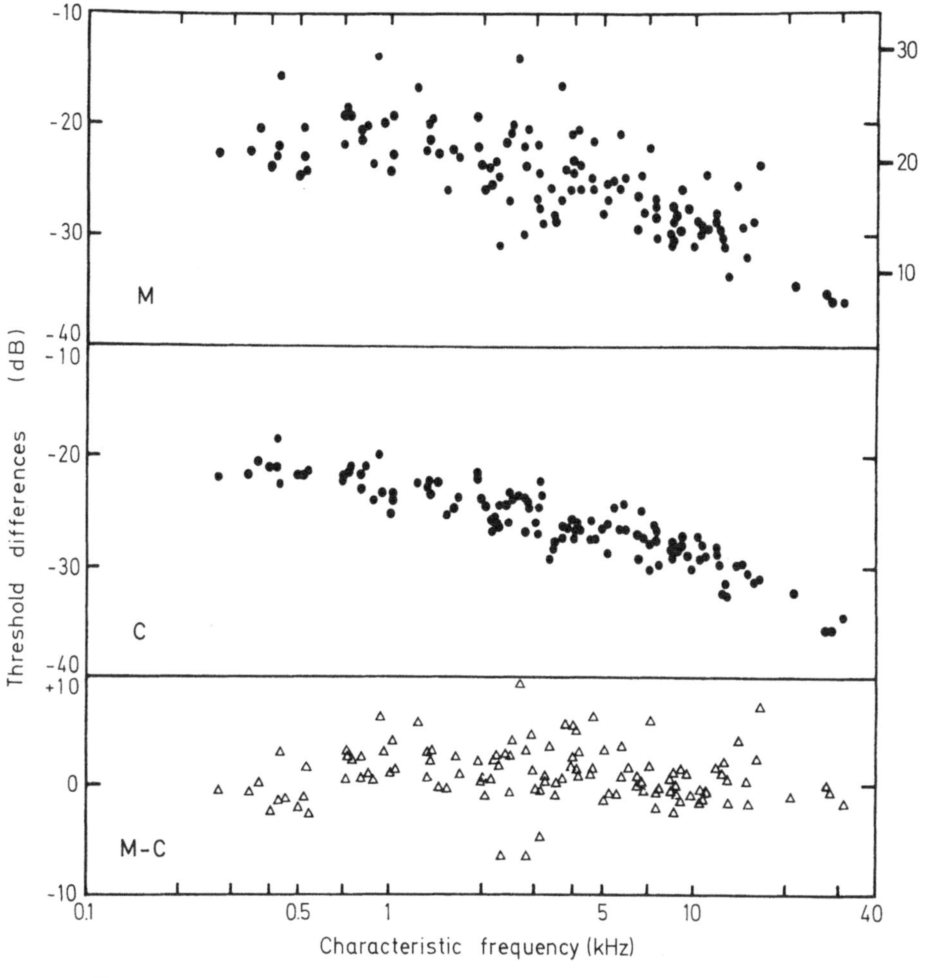

Fig. 26. Difference between noise and tone thresholds for 118 cochlear fibres from 5 cats, plotted against the CF of each fibre. M: measured values of noise threshold intensity relative to threshold of tone at CF, noise intensity being expressed in terms of r.m.s. power in 1 Hz bandwidth. (Right ordinate gives the values measured using a full noise bandwidth of 21.5 kHz.) C: corresponding values computed from FTC of each fibre using expression: 10 $\log_{10}$ ("effective bandwidth" of FTC). The "effective bandwidth" is the bandwidth of the equivalent rectangular filter (equal in area to the FTC plotted on linear power and frequency coordinates as in Fig. 27). M—C: differences between corresponding measured and computed values. (After Evans and Wilson, 1971, 1973)

30 kHz. These show that, with noise level expressed as spectral level (*i.e.*, power in 1 Hz bandwidth), the noise threshold was 13–36 dB below the threshold of a tone at the CF. (Measured with a noise bandwidth of 21.5 kHz, the r.m.s. noise threshold was 7–30 dB higher than that of the tone). On the assumption that the pure tone FTC represented the frequency response function of a *linear* filter, the predicted threshold difference between noise and pure tone stimuli could be computed from the FTC of each fibre (see Fig. 26C and legend). The similarity between the observed and computed threshold differences (Fig. 26M–C) suggests that, within the limits of experimental error, and within the limited range of the method (10–20 dB), the overall frequency filtering characteristics of the cochlea act as a linear filter of bandwidth consistent with that of the pure tone FTC. The latter conclusion thus extends that of DE BOER to fibres of all CFs.

These two studies therefore indicate that, within limits, the relative threshold and the discharge pattern of the response to noise can be predicted in principle from the filtering characteristics of a fibre, as defined by the pure tone FTC.

3. Stimuli with Multiple-Component Spectra: "Comb-filtered" Noise

When a stimulus with a wide-band spectrum (noise or click) is mixed with a delayed version of itself, a signal results which has as its spectrum a sinusoidal distribution of energy with regard to frequency, *i.e.* a series of peaks and valleys of energy evenly spaced on a linear frequency scale, a so-called "comb-filtered" noise spectrum (see Fig. 27). The spacings of the peaks and valleys of energy depend upon the delay (*e.g.* see WILSON, 1967, 1970). Multicomponent stimuli of this kind give rise to a sensation of pitch (WILSON, 1967, 1970), are generated whenever broadband signals reach the ear via direct and reflected routes, and are utilized by the blind for providing ranging cues (BASSETT and EASTMOND, 1964; WILSON, 1967). Experiments utilizing comb-filtered noise to stimulate single cochlear fibres (EVANS *et al.*, 1971; WILSON and EVANS, 1971; EVANS and WILSON, 1973) have demonstrated that fibres of the cochlear nerve of the cat respond to multicomponent stimuli in the manner of linear band-pass filters of the same width and shape as the FTC; that lateral inhibition is not involved in the filtering; and that the frequency selectivity of these filters corresponds to that observed in psychophysical measurements with the same stimuli.

Figure 27 illustrates the response of a single cochlear fibre to comb-filtered noise spectra adjusted and switched so that a peak and valley of the spectrum alternately coincides with the CF of the frequency threshold curve, for two different spacings of the spectrum (A and B, respectively). The mean level of the noise spectrum is about 10 dB above the threshold level for noise. At the widest spacing of the spectrum (Fig. 27A), there is a marked difference in the discharge rate between the periods of time (100 msec in each case) when an energy peak (thin continuous line) and an energy "valley" (interrupted line) overlie the inverted FTC (thick continuous line). At progressively closer spacings of the spectral peaks and valleys, the ratio of discharge corresponding to peak and trough responses decreases until a spacing is reached (Fig. 27B) where this ratio becomes unity; *i.e.* the discharge is unaffected by the position of the spectral peaks and troughs relative to the FTC. This point indicates that the fibre is no longer capable of resolving the separate peaks of the spectrum. Inasmuch as these stimuli are analogous to the luminosity gratings employed in studies of the visual system (*e.g.* CAMPBELL *et al.*, 1969), we can say that the fibre has reached the limit of its "grating acuity". These responses can be quantitatively matched by the responses of a linear band-pass filter of identical frequency response characteristics to that of the fibre's FTC. The

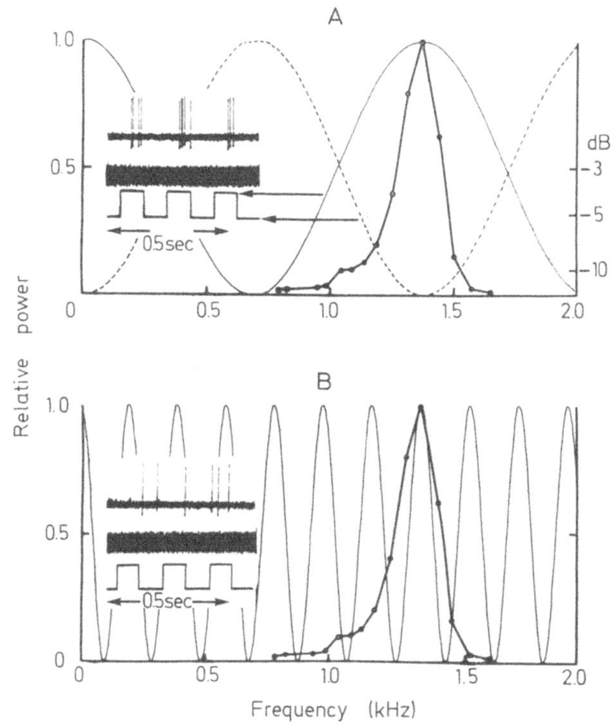

Fig. 27A, B. Response of cat cochlear fibre to comb-filtered noise stimuli, widely (A) and narrowly (B) spaced. Filled circles and thick continuous line: FTC of fibre plotted as a bandpass filter function in linear relative power and frequency coordinates. A: Widest spacing. Thin continuous sinusoid: envelope of "normal" comb-filtered noise spectrum, with the first energy peak of the spectrum coincident with CF of fibre (1.37 kHz). Dashed sinusoid: envelope of "inverted" spectrum, *i.e.*: with energy minimum at CF of fibre. Inset: continuous film record of spike discharges in response to comb-filtered noise, alternated every 100 msec between normal and inverted conditions according to the switching waveform shown on the lowest trace. Middle trace shows oscillogram of comb-filtered noise stimulus. B: Narrowly spaced spectrum, with peak spacing of 0.2 kHz. Envelope of normal spectrum only shown (thin sinusoid). Otherwise as in A. Note discharge unrelated to alternation of normal and inverted spectra (see text). (From EVANS and WILSON, 1973)

fact that this match obtains for all spacings is evidence against the involvement of lateral inhibition in the production of the fibre's narrowly tuned FTC. Lateral inhibition would be expected to produce systematic deviations from the predicted linear filter values such that the response to widely spaced spectra was reduced and that to intermediate spacings was enhanced (vd. RATLIFF, 1965 for an analogous case in vision).

A comparison of the limit of "grating acuity" (established as above) for cat cochlear fibres covering a wide range of CFs, with analogous psychophysical measurements (Wilson and Evans, 1971; Evans and Wilson, 1973) suggests that the limits upon the frequency resolving power of the auditory system for multicomponent signals are already largely determined at the level of the cochlear nerve (see Section VI.C.1).

4. Effect of Noise on Responses to Click and Tone Stimuli

Kiang *et al.* (1965a) have demonstrated that the responses to click and tone burst stimuli can be attenuated and finally abolished as the level of a background broadband noise stimulus is progressively raised. In both cases, the initial component of the click and tone responses is attenuated relatively more than the later components.

From the data of Kiang *et al.*, the total "masking" of the click and tone responses appeared to be obtained when the level of the continuous noise stimulus was high enough to cause saturation of the fibres' *maintained* response. Interestingly, this rate was substantially lower than the discharge rate evoked by the *onset* of the tone alone (see p. 18 and Evans, 1974d). Whether this represents an "exhaustion" of some metabolic process, or what the nature of the interaction is, waits upon more data.

III. Cochlear Nucleus

A. Introduction

While all fibres of the cochlear nerve terminate in the cochlear nucleus complex (Stotler, 1953; Powell and Cowan, 1962), the cells of the nucleus receive greatly differing numbers of terminals and modes of termination (*e.g.* Cajal, 1909; Poljak, 1926; Lorente de Nó, 1933a, b; Stotler, 1949; Rasmussen, 1957; Harrison and Warr, 1962; Powell and Erulkar, 1962; Harrison and Irving, 1965, 1966; Osen, 1969; Cohen *et al.*, 1972). For the purpose of this review the simplest and morphologically most conspicuous anatomical subdivision (Fig. 28) will be utilized, namely into dorsal (DCN) and ventral (VCN) divisions, with the latter further subdivided into anteroventral (AVN) and posteroventral (PVN) nuclei. It must nevertheless be remembered that there are at least 9 distinct cell types within the complex (Osen, 1969, in the cat), and that their distribution is not limited by the boundaries of these divisions.

In contrast to the relatively homogeneous properties of cochlear fibres, the behaviours of different neurones of the cochlear nucleus exhibit great variety. To a first approximation, the properties of cells in the VCN differ little from those of the primary fibres, whereas cells in the DCN show a great variety of differences. While such differences were noted in the earlier studies of the cochlear nucleus (*e.g.* Galambos and Davis, 1943, 1944; Rose *et al.*, 1959; Moushegian *et al.*, 1962) it is only recently that they have been related systematically to location within the nucleus. Pfeiffer and Kiang and their colleagues (Pfeiffer and Kiang, 1965; Kiang *et al.*, 1965b; Kiang, 1965; Pfeiffer, 1966b; Kiang *et al.*, 1973) have

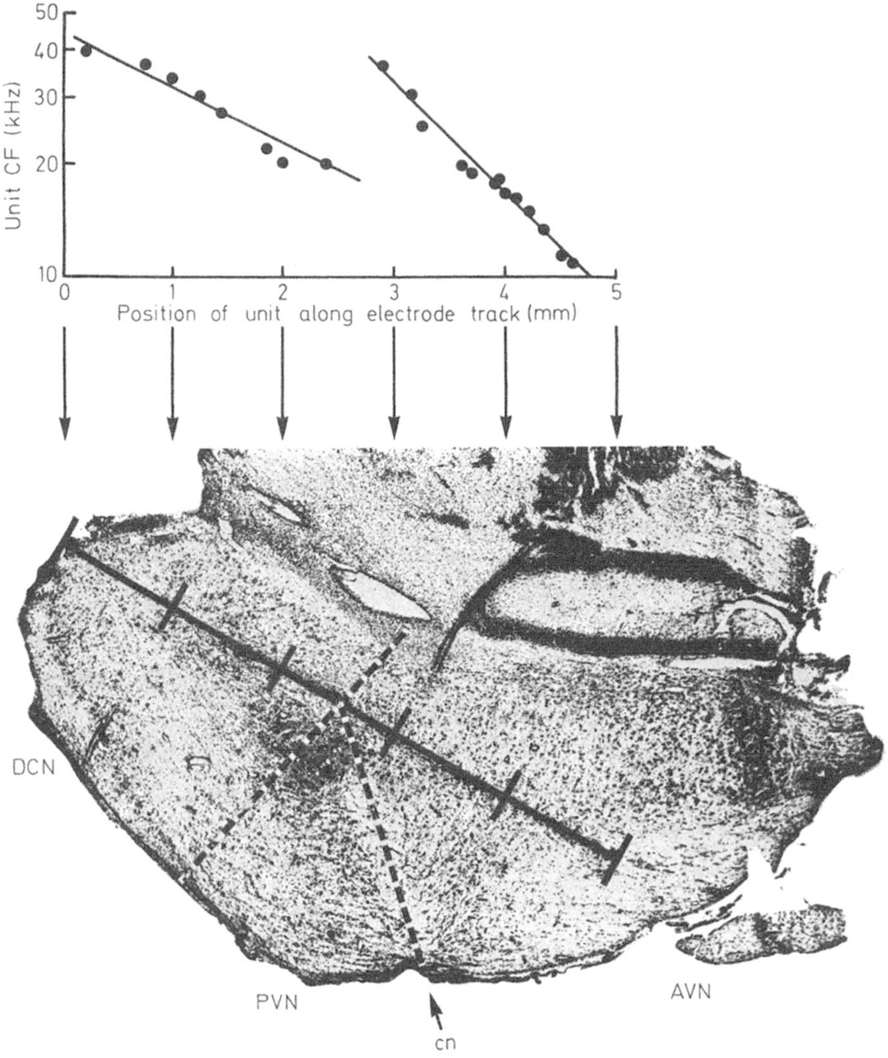

Fig. 28. Gross and tonotopic organization of the cochlear nucleus (cat). Sagittal section through medial region of CN (and adherent cerebellum dorsally) stained with Thionine. Caudal and rostral to left and right respectively. Major subdivisions into dorsal (DCN) and antero- (AVN) and posteroventral (PVN) nuclei indicated approximately by interrupted lines. Incoming and branching cochlear fibres indicated by cn and arrow. Solid line and cross-bars indicate electrode track and mm depth from dorsal surface. Above is plotted the CFs of 21 units found at the indicated depths

undertaken a classification based on the interspike interval statistics of the spontaneous activity and on the time course of the response to tones at the CF about 20 dB above threshold. EVANS and NELSON (1966a, b; 1973a, b), in a complementary study of unit responsiveness as a function of stimulus frequency (*e.g.* Figs. 30—32), have shown that there are systematic differences in the distribution of inhibitory and excitatory influences. Their survey was carried out with tones within 20–40 dB of threshold.

All attempts at classification of response properties face a number of difficulties. In the first place, many anaesthetic agents (including barbiturate and urethane) have a profound effect on the responses of neurones in the DCN (EVANS and NELSON, 1973a, b). Such anaesthetics tend to obliterate the differences in properties between the DCN and VCN, and it is unfortunate that all studies of single units in the DCN, with the exception of those by EVANS and NELSON (1973a, b) have been carried out under barbiturate or urethane anaesthesia. Secondly, the response properties of many cells in the cochlear nucleus, unlike those of the cochlear nerve, are modified *qualitatively* by changes in stimulus parameters such as frequency, intensity, and repetition rate. Thirdly, different microelectrodes may sample different populations of cells, and, in any case, less accessible and less extensive regions (such as the nuclei lateralis and interstitialis of LORENTE DE NÓ, 1933b) and small-celled areas (*e.g.* the molecular and granular layers of the DCN) are likely to be under-represented. On the other hand, the use of large tipped electrodes by most workers who have explored the CN minimised the chance of recording from primary afferent fibres (KIANG, 1965).

Such technical variables as the above inevitably make impossible the *complete* reconciliation of the findings by different researchers; nevertheless, there are sufficient parallels that a synthesis is possible and useful. While response types cannot be correlated *exclusively* with regions, they can be related at least to subdivision and may turn out to be related to cell type (see, *e.g.* KIANG *et al.*, 1973). A general survey of single unit behaviour in relation to the organization of the cochlear nucleus will therefore be attempted, before proceeding with a more detailed account which will seek to emphasise the major differences in unit properties between the cochlear nucleus and cochlear nerve.

B. Organization

The fibres of the cochlear nerve divide within the VCN into an anterior "ascending" branch which innervates the AVN, and a posterior or "descending" branch which innervates the PVN. The innervation of the DCN is less clear, but appears to arise from a continuation of the posterior branch (SANDO, 1965; OSEN, 1970; COHEN *et al.*, 1972). In addition, the DCN receives a large projection of "association fibres" from the AVN (LORENTE DE NÓ, 1933b; HARRISON and WARR, 1962; OSEN, 1970), and the DCN and VCN terminals from the descending, efferent system (RASMUSSEN, 1960, 1964, 1967).

1. Tonotopic Organization

ROSE *et al.* (1959, 1960) were the first to show that there was a triple representation of the cochlea in the cat CN, namely in each of the three major subdivisions. Within each subdivision the CFs of the cells are arranged in strict sequence from high to low frequency in the dorsal to ventral, posterior to anterior, and medial to lateral directions (Fig. 28). This means that as a microelectrode traverses the boundary between two divisions, a discontinuity in CFs will be observed (*e.g.*

between DCN and AVN in Fig. 28). In the granular cell region between the DCN and AVN, the frequency representation is haphazard or reversed.

This strict, thrice-repeated tonotopic organization results from the orderly branching and distribution of the cochlear fibres (LORENTE DE NÓ, 1933a; SANDO, 1965). It has been confirmed by other workers and in other species: cat (PFEIFFER and KIANG, 1965; EVANS and NELSON, 1973a), chinchilla (MAST, 1970a), gerbil (SMITH and ZWISLOCKI, 1971), bird (KONISHI, 1970), crocodile (CAIMAN: MANLEY, 1970a).

2. Functional Organization

a) VCN. To a first approximation, the responses to clicks and tones of neurones in VCN resemble those of the cochlear nerve, and are relatively unaffected by anaesthesia (EVANS and NELSON, 1973a). The latencies of the responses to supra-threshold click and tone stimuli are about 1 msec longer than those of cochlear fibres of comparable CF, *i.e.* range from 2–5 msec (RADIONOVA and POPOV, 1965; EVANS and NELSON, 1973b). Their FTCs closely resemble those of primary fibres, as in Fig. 9 (KIANG *et al.*, 1965b). The predominant response to steady tones is excitation, and the discharge rate is a monotonic function of intensity. In the majority of units, the time course of the response resembles that of primary neurones (Fig. 29F, G, K, L), but, in a small number, the rate of adaptation is very rapid (Fig. 29I, N). In less than 10% of the neurones in the survey of EVANS and NELSON (1973a) were inhibitory responses to pure tones observed (as in Fig. 30).

By the criteria of PFEIFFER and KIANG and their colleagues (KIANG *et al.*, 1965b; PFEIFFER, 1966b), neurones located in the anterodorsal region of the AVN can be distinguished from those in the remainder of the AVN and in the PVN. The cells of the *anterodorsal region of AVN* are the large spherical cells of OSEN (1970) which receive only a few (three or so) very large terminals, the so-called calyces (bulbs) of HELD. Characteristic prepotentials are observed in extracellular recordings from these neurones (Fig. 29P; PFEIFFER, 1966a). These appear to relate to discharges in the cochlear fibre terminals, and their constant relation to the cell discharges suggests that each input spike in any one afferent terminal generates an output spike in the CN cell providing that the input spike does not fall within the refractory period of spikes generated by the other terminals (MOLNAR and PFEIFFER, 1968; see (3) below). The spontaneous discharges are irregular, the interspike

Fig. 29A—R. Categorization of CN units on basis of time course of response. In each column, the first three rows represent the same unit whose responses are typical of the location indicated below plots. A—E: interspike interval histograms of spontaneous activity. F—J: High resolution PST histograms of response to 25 msec tone burst at CF of unit. K—O: PST histograms of response to long duration tone (0.9 sec) at CF. P: waveform of extracellularly recorded potential from unit in anterodorsal cochlear nucleus (A—D AVN). Note positive pre-synaptic potential at beginning of time bar preceding the negative post-synaptic unit spike by about 0.5 msec. Q: "dot display" of individual responses of a "chopper type" unit. Each dot represents one spike, and adjacent horizontal rows response to successive tone bursts commencing at time zero. R: PST histogram to 0.9 sec tone of a different unit in DCN from O, with "build-up" time course. Ordinates: numbers of intervals (A—E) and spikes (F—R) per bin. Abscissae indicate linear time in msec (A—J, Q) and sec (K—O, R); tone bursts indicated by bars. (After KIANG *et al.*, 1965b; PFEIFFER, 1966)

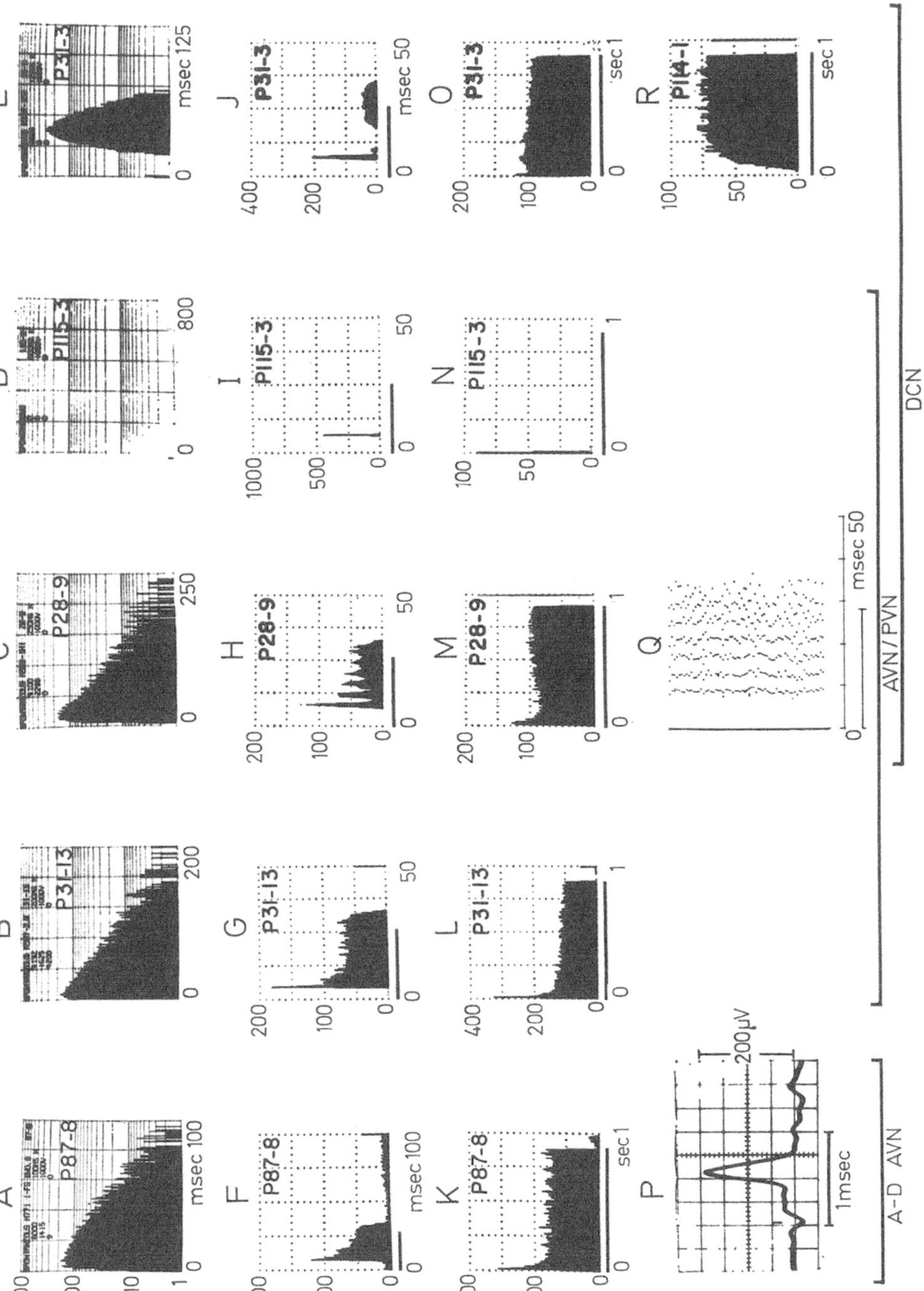

Fig. 29 A—R (Legend see p. 46)

interval statistics resembling those of cochlear fibres, with the exception that the rate of decay of the histogram function is somewhat greater than exponential (Fig. 29A; see (3) below). In all other respects the responses are "primary-like" (Fig. 29F, K).

The cells of the *remainder of the AVN*, *PVN* and the interstitial nucleus (in the root of the cochlear nerve) have similar latencies to the anterodorsal AVN units but can be divided into 3 categories on the basis of the interval statistics of the spontaneous activity and high resolution PST histograms of the responses to brief tone bursts (Fig. 29B–D, G–I). The first category, termed "primary-like", is characterised by an exponential decay of the interval histogram analysis of spontaneous activity (Fig. 29B) and a tone response pattern resembling that of primary

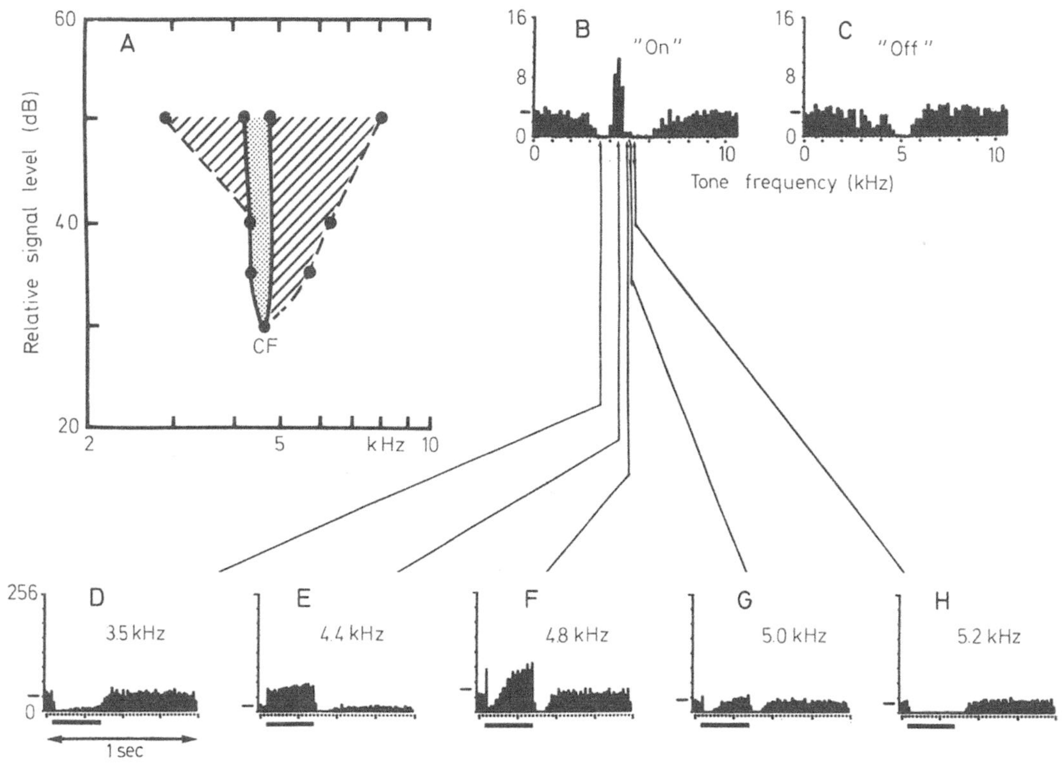

Fig. 30A—H. Frequency response area, frequency responses, and time courses of unit in DCN. A: Excitatory (stippled) and inhibitory (cross-hatched) response areas. Threshold expressed in terms of relative signal level to transducer. The values given represent dB SPL at the CF. The SPLs at other frequencies are within 5 dB of these values. B and C: frequency response histograms of mean response in a 100 msec period during (B) and immediately following (C) 100 msec tone bursts of randomly presented frequency at 20 dB above threshold. Spontaneous discharge rate indicated by bars against ordinate. Note upper and lower frequency inhibitory side-bands, and off-inhibition occurring over similar band as the on-inhibition. D—H: PST histograms of response to 0.3 sec tone burst (bar) at the frequencies indicated (in kHz), at same stimulus level as B and C. Ordinates: total number of spikes per bin, as a result of 53, 22, 79, 41, and 36 tone presentations respectively. Note "primary-like", "pauser", and "build-up" time courses in E, F, and G respectively, and inhibition of spontaneous activity during and following the tone in D and H. (From EVANS and NELSON, 1973a)

fibres (Fig. 29G, L). Units of the second category have a variety of interval statistics (including modes longer than 12 msec and slower than exponential histogram decays, *e.g.* Fig. 29C). They are, however, characterised by a regularity of discharge to tones (Fig. 29 Q) which produces a periodic "ripple" in the PST histogram pattern to brief tones (Fig. 29H) – hence the descriptive term "chopper-type"

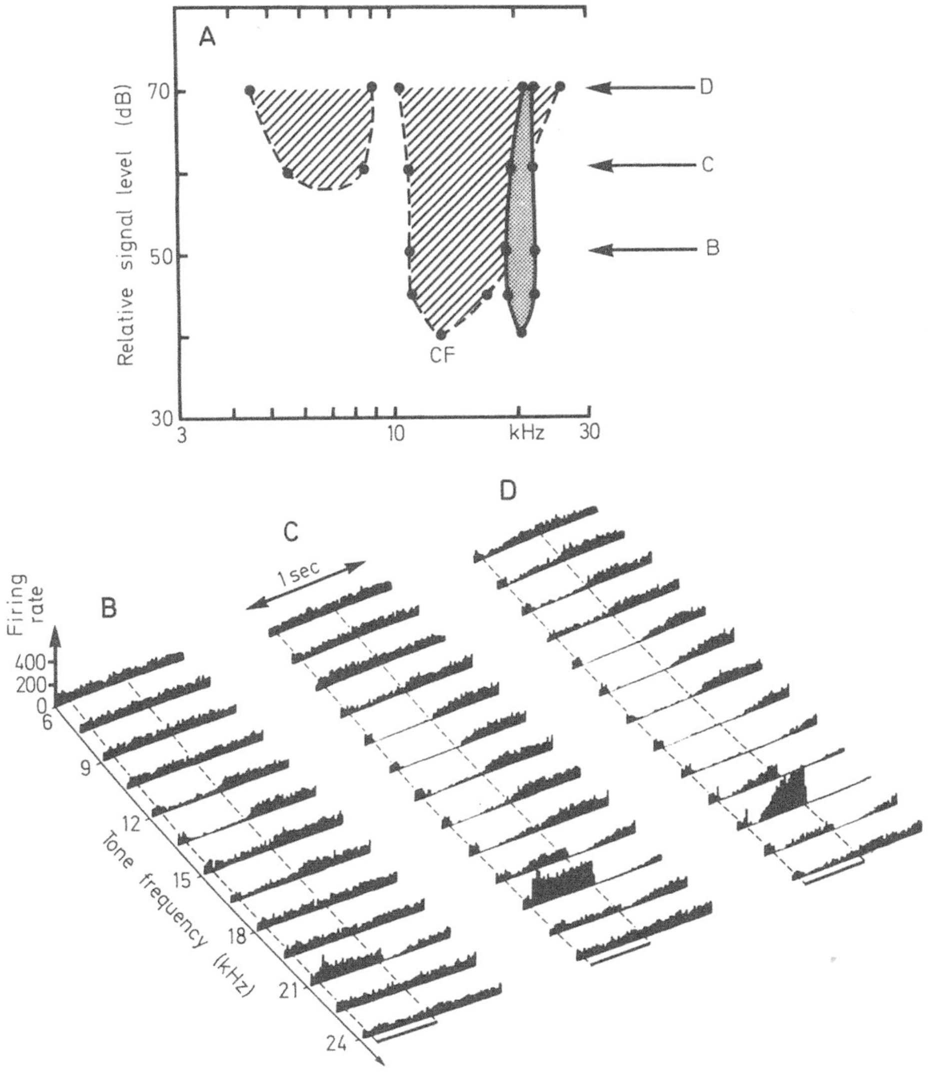

Fig. 31A—D. Frequency response and time courses of predominantly inhibitory DCN unit (cat under chloralose). A: Excitatory (stippled) and inhibitory (cross-hatched) response areas. Relative signal level equals dB SPL at CF (see Fig. 30). B—D: three-dimensional arrays of PST histograms ordered in frequency (6—24 kHz) and intensity (10, 20, and 30 dB above threshold respectively as indicated in A). Tone burst (0.4 sec) indicated by bar and interrupted lines. Ordinate: firing rate in spikes/sec. Note extensive "sea" of inhibition in D with narrow band "island" of delayed ("build-up") excitation at 21 kHz. (From EVANS and NELSON, 1973a)

neurones. Units of the third category have little or no spontaneous activity and exhibit a transient response to tones (Fig. 29D, I, N). These are termed "on" units. They are particularly found in the "octopus cell" region of the PVN (Kiang *et al.*, 1973).

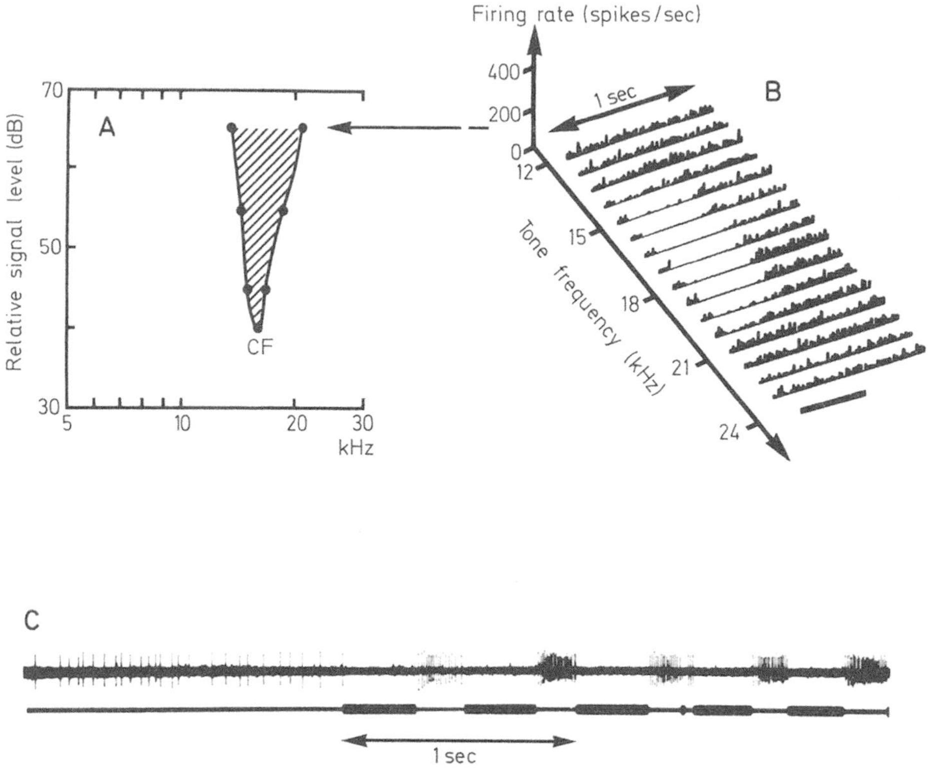

Fig. 32A—C. Frequency response and time course of wholly inhibitory DCN unit (unanaesthetized decerebrate cat). A: Inhibitory response area. Relative signal level equals dB SPL at CF (see Fig. 30). B: three-dimensional array of PST histograms of response to 0.4 sec tone at 25 dB above threshold. Note inhibition of spontaneous activity during, and following, tone. C: Different unit from A, B, but also found in unanaesthetized DCN. Inhibition of spontaneous activity during tone with off or rebound excitation. Tones at CF, 15 dB above threshold. (From Evans and Nelson, 1973a)

b) DCN. In the unanaesthetized cat, less than 10% of DCN units exhibit behaviour resembling that of primary fibres (Evans and Nelson, 1973a). The responses of the remainder arise from greater or lesser admixtures of inhibitory and excitatory influences, as indicated by inhibition of spontaneous activity and non-monotonic response relationships with stimulus level. The latency of the responses to click and tone stimuli are generally longer than those in the VCN, and range from 4–12 msec (Radionova and Popov, 1965; Evans and Nelson, 1973b). The frequency response areas range from central excitation with inhibition over one or more adjacent frequency side-bands (Fig. 30), through predominant inhibitory response bands associated with a narrow excitatory response band (Fig. 31), to inhibition only (Fig. 32). The complex time-courses of the responses associated with apparently contiguous inhibitory and excitatory bands indicate that the response regions in fact overlap (Figs. 30F, G and 31C, D). The proportions of DCN units exhibiting side-band, predominant, or total inhibition were about 20, 20, and 50% respectively in the unanaesthetized cat, compared with 40, 0, and 10% in pentobarbitone or halothane anaesthetized cats (EVANS and NELSON, 1973a). These anaesthetic agents thus exert a profound effect on the DCN in reducing the predominance of inhibition.

On the basis of their PST histogram analyses of tone evoked activity, KIANG *et al.* (1965b) and PFEIFFER (1966b) observed in the anaesthetized cat a category of unit not found in the VCN, which they termed a "pauser type" (Fig. 29J). This type of PST histogram was also common in the DCN data of EVANS and NELSON (1973a), but, as Figs. 30 and 31 indicate, it is not invariant of stimulus frequency or level. Figure 29R and Fig. 30G illustrate another type of time-course that is characteristic of some units in the DCN, namely, where the discharge rate increases (instead of adapting) with time. This pattern of behaviour was termed "build-up" by ROSE *et al.* (1959); it is dependent upon the repetition rate as shown in Fig. 34 (also see PFEIFFER, 1966b). Units of the "chopper" and "on" categories were also found in the DCN. Analyses of the interval statistics of the spontaneous activity revealed almost symmetrical distributions in some DCN units (Fig. 29E) and that the majority of the DCN interspike interval histograms had modal values over 10–20 msec, in contrast to the situation in the VCN (PFEIFFER and KIANG, 1965).

3. Nature of Sensory Input to Neurones of CN

With its diversity of response, related to the morphology, the cochlear nucleus is ideally suited to studies of the functional significance of different patterns of innervation.

PFEIFFER and MOLNAR (1968) have been able to fit to the observed interspike interval statistics of primary and VCN neurones, input-output models with assumptions which correlate well with the anatomy. In the anterodorsal AVN, the statistics are consistent with a superposition of 3–4 input renewal processes, where (apart from its refractory period) the cell discharges each time a discharge appears on one of the 3–4 input fibres. These cells are innervated via the large calyces of Held by 3 or so cochlear fibres (CAJAL, 1909). Certain cells in the PVN have discharge patterns consistent with a model where each output discharge requires the accumulation of a number of discharges from a large number of input fibres, a situation consistent with the morphology of the region.

In a remarkable series of experiments, KOERBER *et al.* (1966) have demonstrated that the spontaneous discharge of units in the VCN is dependent upon the activity of the cochlear nerve input. Acute (and chronic) interruption of the nerve eliminated the spontaneous activity of all units in the AVN and of almost all in the PVN. In contrast, the spontaneous rates and interval statistics of units in the DCN were unaffected. This finding correlates with the conclusion, on anatomical grounds (STOTLER, 1949; RASMUSSEN, 1957; POWELL and COWAN, 1962; HARRISON and WARR, 1962; OSEN, 1970; COHEN *et al.*, 1972), that the cells of the DCN receive at most only a small proportion of the total number of their afferents from the cochlear nerve. In a series of experiments, EVANS and NELSON (1968, 1973b) have implicated the intranuclear fibre pathway running between the AVN and the DCN (LORENTE DE NÓ, 1933b; see Fig. 49) as the mediator of the predominant inhibitory sensory input to the DCN, seen particularly in the unanaesthetized preparation. This inhibition has too short a latency (4–6 msec) for it to be mediated by the efferent auditory system, and it can be evoked by tones after interruption of descending pathways. Furthermore, it can be evoked by direct electrical stimulation of the cells of the AVN or the fibres of the intranuclear pathway itself. These data give further grounds for the conclusion, from the greater complexity of response, the longer latency, and susceptability to anaesthetic agents of cells in the DCN compared with those of the VCN, that the DCN is a higher than second-order nucleus.

C. Spontaneous Activity

The range of spontaneous rates from cell to cell in the CN is at least as great as that in the cochlear nerve, from 0–150 spikes/sec (PFEIFFER and MOLNAR, 1968; EVANS and NELSON, 1973a). The latter authors found a slightly higher mean rate in cells of the VCN (29/sec) compared with those of the DCN (20/sec) in the anaesthetized cat, consistent with the former authors' report of a relative dearth of units in the AVN with rates below 10/sec.

Mention has already been made (Section III.B.2,3) of the finding and possible significance of the variety of interspike interval distributions in the CN. In addition to distributions identical or similar to those obtained in the cochlear nerve, *i.e.* quasi-Poisson (Fig. 29A–C; VIERNSTEIN and GROSSMAN, 1961; RODIECK *et al.*, 1962; PFEIFFER and KIANG, 1965), more symmetrical and restricted distributions are encountered, particularly in the DCN (Fig. 29E; PFEIFFER and KIANG, 1965). This means that some units in the DCN have a comparatively regular spontaneous activity, and adds further weight to the argument that the activity of DCN cells is less dependent than that of the VCN on input from the cochlear nerve (Section III.B.3 above). This argument is based partly on the finding of KOERBER *et al.* (1966) that the spontaneous activity of cells in the VCN is abolished by section of the cochlear nerve, while that of the DCN is unaffected.

D. Response to Click Stimuli

Not all units in the DCN are driven by click stimuli (ROSE *et al.*, 1959); others are very sensitive. In the sample of ROSE *et al.* (1959), thresholds of response to click stimuli ranged from 10–75 dB above those to CF tones. (The click stimuli had approximately the same threshold for normal human hearing as 2 kHz tones). Some of their units in the DCN were exquisitely sensitive to the click level, and required only a 4 dB increase to raise the probability of a response from zero to 100%.

The ability of CN units to "follow" repetitive clicks has received a good deal of attention. At least two types of behaviour have been found. In the first, resembling that of the cochlear nerve, the discharge rate increases monotonically with click rate, while the probability of discharge per click decreases with click rates in excess of 10–100/sec. The second group of units, which exhibit transient responses to tones, generate one spike for every click at click rates of up to 100–800/sec (KIANG *et al.*, 1962 a in the cat; MØLLER, 1969 b in the rat). Thereafter, the probability of discharge and the mean discharge rate, drop off abruptly towards zero. KONISHI (1969 a) has pointed out that this 1 : 1 response can be maintained in some units in the bird up to click rates in excess of 1000/sec.

The PST histograms of responses to clicks of "primary-like" cells with CFs below 4 kHz in the AVN exhibit the multiple peaks characteristic of cochlear fibres. On the other hand, cells of the "chopper", "on" and "pauser" types had histograms with single peaks only, irrespective of their CF (KIANG *et al.*, 1965 b).

The latencies of response to click stimuli have already been mentioned (III.B.2).

E. Response to Single Tonal Stimuli

The varieties of response and their relative preponderance in the different divisions of the CN have already been outlined (III.B.2; Figs. 29–32). It will be necessary, however, to add a qualification in respect of the group of units exhibiting transient ("on") responses to tones. The proportion of these units found ranged from 1–20% as follows: 1% EVANS and NELSON (1973 a); 8% PFEIFFER (1966 b); 20% RADIONOVA (1965, 1971), MØLLER (1969 b). In the latter studies, where these units accounted for one fifth of the total encountered, the possibility exists that some at least of these responses represent those of high CF units stimulated by the energy dispersion associated with the very short tone onset times used in the studies. This could account for the finding by MØLLER (1969 b) and RADIONOVA (1971) that the units had very high threshold, broadly tuned FTCs, and for the otherwise surprising independence found by RADIONOVA (1971) of the threshold upon tone duration (for durations as short as 1 msec) compared with that of units where the response to tones was sustained. Furthermore, the thresholds of RADIONOVA's transient units were sensitive to the rise time of the tonal stimuli, and transient responses were often present at the abrupt termination of the tones. The transiently responding units in the surveys by PFEIFFER (1966 b) and EVANS and NELSON (1973 a), on the other hand, had relatively normal thresholds.

1. Threshold and Response as a Function of Frequency

With the possible exception of the "transient units" mentioned above, there is no evidence to suggest that the range of minimum thresholds of CN cells is any less restricted than in the cochlear nerve (MAST, 1970 a). Similarly, the minimum unit thresholds approach the threshold of behavioural response ("audiogram") of the animal as in Fig. 8 (cat: GALAMBOS and DAVIS, 1943; rat: MØLLER, 1969 a; chinchilla: MAST, 1970 a; bird: KONISHI, 1969 b, 1970). KONISHI (1970) has shown that a striking correlation exists between the range of unit CFs in different birds with the frequency components of their song. MANLEY (*e.g.* 1970 b) has demonstrated a relationship between the range of CFs and basilar membrane measurements in different species of reptile.

Under anaesthesia, the excitatory FTCs of CN cells are not substantially different from those of cochlear fibres (KIANG, 1965) nor are marked differences observed between divisions of the nucleus (KIANG *et al.*, 1965b). Unfortunately, there are insufficient data available obtained under identical and calibrated conditions to allow a quantitative comparison of excitatory bandwidths between cells of the cochlear nerve and nucleus, but if anything, the CN bandwidths at 10 dB above threshold are on average larger than those of the cochlear nerve (*e.g.* data of KIANG, 1965). The variation in bandwidth on the other hand may be much greater in the CN than in the cochlear nerve. ROSE *et al.* (1959) and MØLLER (1969a) reported the finding of wide and narrow bandwidths in their data from the cat CN. Taking all the available data, it is clear that, at one extreme, one can find some FTCs which are substantially broader (especially on the low-frequency side) than those of cochlear fibres of corresponding CF, and, at the other extreme, some excitatory FTCs (particularly in the DCN) which do not increase in bandwidth with increase in stimulus level or which actually decrease (*e.g.* Fig 31A; KATSUKI *et al.*, 1958; KIANG *et al.*, 1965b). In other words, the latter FTCs are narrower (for higher stimulus levels) than those of corresponding cochlear fibres.

The FTCs of a few CN cells are multiple, *i.e.*, they have two or more definite peaks (ROSE *et al.*, 1959; MOUSHEGIAN *et al.*, 1962; MANLEY, 1970; EVANS and NELSON, 1973a).

The major innovation in frequency response characteristics is the presence, in certain cells of the CN, of response regions where the spontaneous activity is inhibited by single tones, as has been outlined above (III.B.2). Cells with entirely inhibitory response areas (as in Fig. 32) have been found in 20% of cells of the N. magnocellularis of the anaesthetized pigeon (STOPP and WHITFIELD, 1961), in occasional cells of the unanaesthetized cat VCN (MOUSHEGIAN *et al.*, 1962; EVANS and NELSON, 1973a) and in cells of the DCN of cat (GERSTEIN *et al.*, 1968; EVANS and NELSON, 1973a) and chinchilla (MAST, 1970a). In other cells, extensive (Fig. 31) or relatively restricted (Fig. 30) "side-bands" of inhibition are associated with and are usually contiguous with excitatory response regions (cat: GALAMBOS, 1944; KATSUKI *et al.*, 1958; MOUSHEGIAN *et al.*, 1962; GREENWOOD and MARUYAMA, 1965; GERSTEIN *et al.*, 1968; EVANS and NELSON, 1973a; chinchilla: MAST, 1970a). In some cells, frequencies, which near threshold are excitatory, become inhibitory at higher levels (*e.g.* 20 kHz in Fig. 31; ROSE *et al.*, 1959; GREENWOOD and MARUYAMA, 1965). Reports on these inhibitory regions are almost entirely confined to findings in the DCN (see also TASAKI, 1957). Where they have been systematically looked for, inhibitory response regions were found in less than 10% of cells in the VCN but in over 90% of DCN cells in the *unanaesthetized* cat (EVANS and NELSON, 1973a). About half of the DCN cells in the latter study had entirely inhibitory response regions.

2. Response as a Function of Intensity

From the fore-going, it will be apparent that the spike rate versus intensity functions differ from cell to cell and at different frequencies. While VCN units generally have monotonic rate functions resembling those of primary fibres (Fig. 14; GALAMBOS and DAVIS, 1943; ROSE *et al.*, 1959; GREENWOOD and MARUYAMA, 1965;

MØLLER, 1969a), non-monotonic functions are found in the DCN (Fig. 33; ROSE *et al.*, 1959; GREENWOOD and MARUYAMA, 1965; GOLDBERG and GREENWOOD, 1966). In the latter case, the ultimate reduction in firing rate with increase in intensity can occur at quite low stimulus levels (*e.g.* 30 dB SPL as in Fig. 33) and can be associated with inhibition of spontaneous discharges at higher stimulus levels still (as in Fig. 31). The converse occurs in other cases, *i.e.* inhibition at low stimulus levels is superseded by excitation at higher levels (GALAMBOS, 1944).

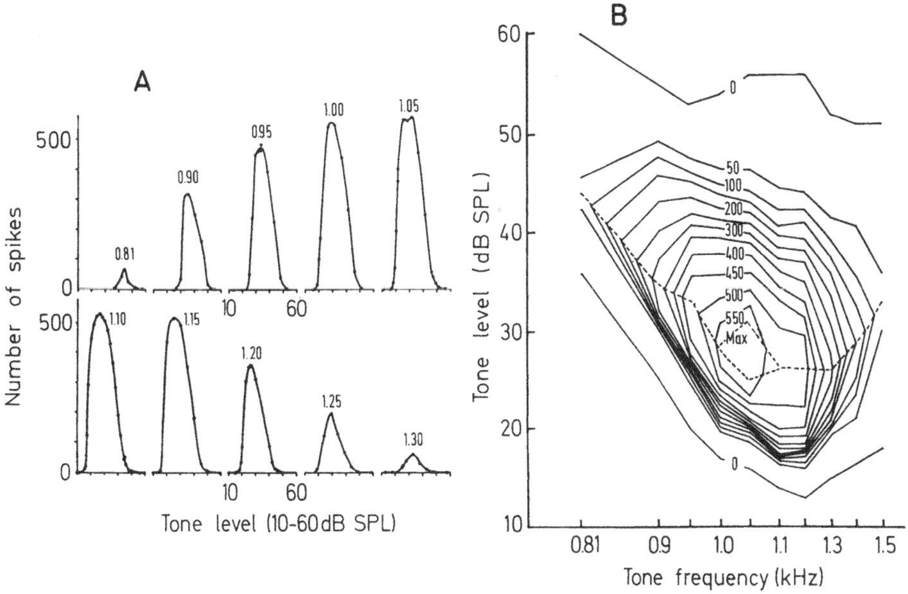

Fig. 33A, B. Non-monotonic intensity functions of unit in anaesthetized DCN. A: spike discharge versus tone intensity functions at each of the frequencies indicated (in kHz). Number of spikes counted over 20 trials during initial 0.2 sec of tone presentation. Each plot relates to its own abscissa marked off in 10 dB steps, from 10—60 dB SPL. B: Iso-rate contours derived from A. Numbers on contours correspond to counts in A. The lowest curve (0 spikes) represents the FTC for excitation. (From GREENWOOD and MARUYAMA, 1965)

SMITH and ZWISLOCKI (1971) have addressed themselves to the interesting question whether the rate functions for the initial onset burst of primary-like monotonic cell discharges differ from those for the later, adapted portion of the response to tones. Comparing the firing rates at the tone onset and after 200 msec of adaptation, they concluded that the adapted response saturated at a lower discharge rate. Unfortunately, their experiments were conducted only over a range of intensities restricted to 10 dB, an insufficient range to demonstrate that the unit illustrated had actually reached saturation. On the contrary, in additional experiments, by superimposing short increments in intensity upon the adapting tone "pedestal", they demonstrated that the rate limited at essentially the same saturation rate as for the onset response, irrespective of the length of adaptation.

3. Response as a Function of Time

In Section III.B.2, and Figs. 29–31, the different time courses of responses to steady tones ("primary-like", "chopper", "on", and "pauser" types) and their relationship to location within the CN have been outlined according to the classification of Kiang *et al.* (1965); Pfeiffer (1966b). With the exception of the chopper-type time-courses, the different types have been noted by other workers (*e.g.* Rose *et al.*, 1959; Moushegian *et al.*, 1962; Greenwood and Maruyama, 1965; Evans and Nelson, 1973a). In addition, as noted, purely inhibitory responses with time courses the inverse of those of primary-like excitation (Fig. 32) and with (Fig. 32C) and without (Fig. 32B) an "off" response have been found in the DCN (Galambos, 1944; Moushegian *et al.*, 1962; Greenwood and Maruyama, 1965; Mast, 1970; Evans and Nelson, 1973a).

The "chopper" type of PST histogram is produced by a discharge pattern which is relatively regular in time (*i.e.* the intervals between successive spike discharges are approximately constant) and which is synchronized to the onset of the tone (Fig. 29Q). Since the preferred interspike interval is of the order of 3 msec, high resolution histograms are required to demonstrate the "chopper" appearance (compare Fig. 29H with M). This preferred interval is unrelated to the tone frequency, phase of tone inset, and the CF of the cell, and is relatively invariant of tone level (Pfeiffer, 1966b).

The "primary-like", "chopper" and "on" time-courses are maintained over a wide range of effective stimulus frequencies and levels (Pfeiffer, 1966b; Evans and Nelson, 1973a). This is not, however, the case for the DCN units where excitatory-inhibitory admixtures are evident in their responses (Figs. 30, 31; Rose

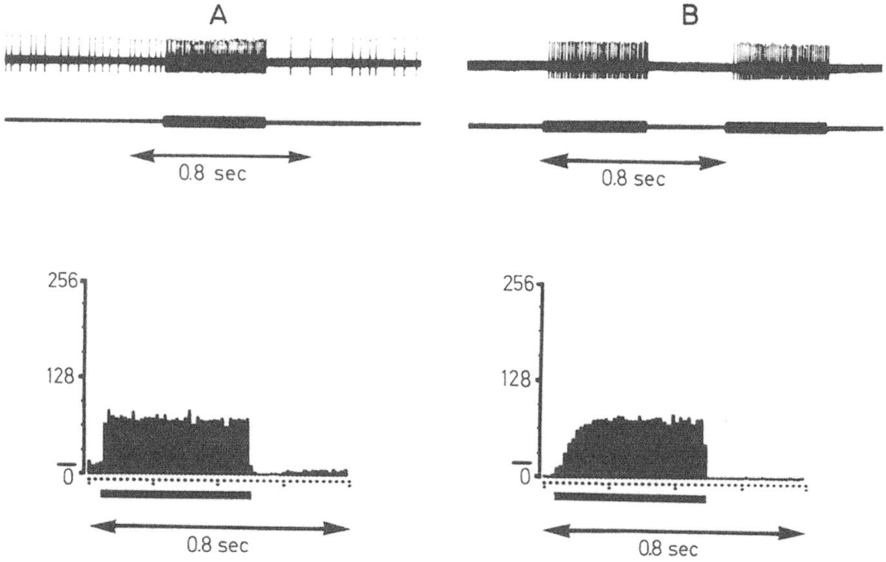

Fig. 34A, B. Effect of repetition rate on time course of response. Unit in DCN (anaesthetized cat). 0.5 sec tone at CF, 20 dB above threshold. Repetition rate: 1 per 3 sec in A; 1 per 0.8 sec in B. Above are shown continuous film records taken during PST histograms (below) of 32 and 43 tone presentations respectively *et al.*, 1959; GREENWOOD and MARUYAMA, 1965; PFEIFFER, 1966b; EVANS and NELSON, 1973a). Pre-exposure to stimuli can modify the rate of adaptation of response (ROSE *et al.*, 1959) or convert a "primary-like" time course into one characteristic of "build-up" units (Fig. 29R). The latter effect appears to result from a more powerful and long-lasting off-suppression seen in the responses of many CN cells (particularly in those in the DCN) compared with those of cochlear fibres (Fig. 34; EVANS, 1971; EVANS and NELSON, 1973a). From the relation observed between the duration of off-suppression and the duration and level of the stimulus tone, STARR (1965) concluded that the off suppression was a form of post-excitatory depression representing some metabolic effect of activity. However, this cannot be so for many units, particularly those in the DCN. Here, the frequency range over which the off-suppression is observed is not identical with that which evokes the preceding excitation (*e.g.*, Fig. 30C) and, in many cases, the off-suppression follows a response which is inhibitory (*e.g.*, Fig. 30C, G, H; Fig. 31D: responses between 16.5 and 22.5 kHz). Furthermore, in some units the duration of off-suppression is relatively invariant of tone duration. These observations strongly suggest that the powerful off-suppression seen in many CN units results from active inhibitory mechanisms, and STARR and BRITT (1970) have succeeded in demonstrating, by intracellular recording, hyperpolarizing potentials during the post-stimulus period and inhibition, during these periods, of discharges evoked by intracellular current injection.

4. Response to Low-frequency Tones

"Phase-locked" responses (*cf.* II.D.4) in the cat CN were first reported by GALAMBOS and DAVIS (1943) and were consistent with an upper limit of neural following of about 4 kHz. Unlike the situation in the cochlear nerve, however, not all CN units of low CF exhibit phase locking to the cycles of low frequency tones within their response areas (RUPERT and MOUSHEGIAN, 1970: kangaroo rat VCN; LAVINE, 1971: cat). RUPERT and MOUSHEGIAN (1970) described a group of units in the VCN which appeared to give phase-locked responses only over a narrow range of low frequencies (*e.g.* less than 0.5 kHz). (It is not clear, however, whether in some of their cases phase-locked responses were being confused with or were receiving interference from chopper types of response.) On the basis of these studies, RUPERT and MOUSHEGIAN (1970) have suggested that this response diversity may correlate with the heterogeneous morphology of the kangaroo rat VCN.

In the cat, on the other hand, LAVINE (1971) found the heterogeneity of phase-locking properties to occur only in the DCN. Those units which showed phase-locking exhibited a variation in the preferred phase of discharge with stimulus frequency, which was relatively unaffected by changes in stimulus levels, as found in the cochlear nerve (II.D.4).

Large electrodes have been used to record gross potentials which correspond to the cycles of stimuli up to frequencies of 2 to 3 kHz – the so-called "frequency following response" of MARSH and WORDEN (1968; see also MARSH *et al.*, 1970). STARR and HELLERSTEIN (1971) have mapped out these responses in the cat CN and found them to have larger amplitudes in the VCN than in the DCN. This may relate to the narrower distribution of response latencies in the VCN.

F. Response to Complex Sounds

1. Two-tone Stimuli

Suppression of the excitation due to one tone, by tones at other frequencies has been observed in many studies of the CN in the cat (Galambos, 1944; Katsuki *et al.*, 1959; Moushegian *et al.*, 1962) and in the pigeon (Stopp and Whitfield, 1961). The characteristics of this two-tone suppression are identical in many respects (particularly for cells in the VCN) with those already described for the cochlear nerve (Section II.E.1.a). Thus, the frequencies of suppressing tones adjoin and overlap the upper and lower frequency borders of the excitatory response area. Such properties are presumably handed on to the CN neurones by the primary fibres.

In addition, there is considerable evidence to indicate that neural inhibition is involved in the two-tone suppression of some cells at least. As originally described by Galambos (1944), the inhibition in some cases extended well below the spontaneous discharge rate, even for tones of long (20 sec) duration; the two-tone inhibitory frequency bands were often associated with the frequency bands in which inhibition of the spontaneous discharge could be evoked by single tones.

Greenwood and Maruyama (1965) have demonstrated that a narrow band of noise can be as effective as a second tone in suppressing tone evoked activity. Again, the results obtained under these conditions can be interpreted in terms of two stimulus interaction effects in the primary afferents to the CN cells, with the addition of local inhibitory influences (Greenwood and Goldberg, 1970).

2. Noise Stimuli

As might be expected, the responses of neurones in the CN to noise stimuli depend upon the type of neurone and upon the frequencies covered by the noise spectrum.

Møller (1970a) has carried out a comparison of noise and tone thresholds in the rat CN, analogous to that described for the cochlear nerve in Section II.E.2 (Fig. 35M, solid points). In addition, he has made a similar comparison between *suprathreshold* noise and tone intensities, *i.e.*, those required to evoke the same firing rate as a tone at 10 dB above threshold (Fig. 35M, circles). The data have been replotted in Fig. 35 for convenient comparison with the analogous cochlear nerve data in Fig. 26. It will be seen that the cochlear nerve and nucleus measurements *at threshold* are very comparable (in spite of the species difference). This is also the case for most of the suprathreshold data. Some of the latter, however, appear to form a separate population (circles above main body of data) with behaviour substantially different from the remainder. This is what would be expected if the separate population was drawn from DCN cells, where lateral inhibitory processes could be expected to produce the effect of raising the noise threshold relative to that of a tone at the CF. In the lower panel (C) of Fig. 35 are shown the corresponding values derived, as indicated in Fig. 26, from the FTCs of the units represented in M. The effective bandwidths of the FTCs were not computed directly by Møller but were estimated by taking half the value of the bandwidths measured at 10 dB above minimum threshold. A comparison of

Fig. 35M with C indicates that, apart from the "separate population" referred to above, there is good agreement between the noise thresholds actually measured and those derived on the basis that the FTC represents a linear filter function. Pooling and averaging his data, MØLLER himself concluded that the suprathreshold spectral resolution of CN units was higher than would be indicated by the FTC

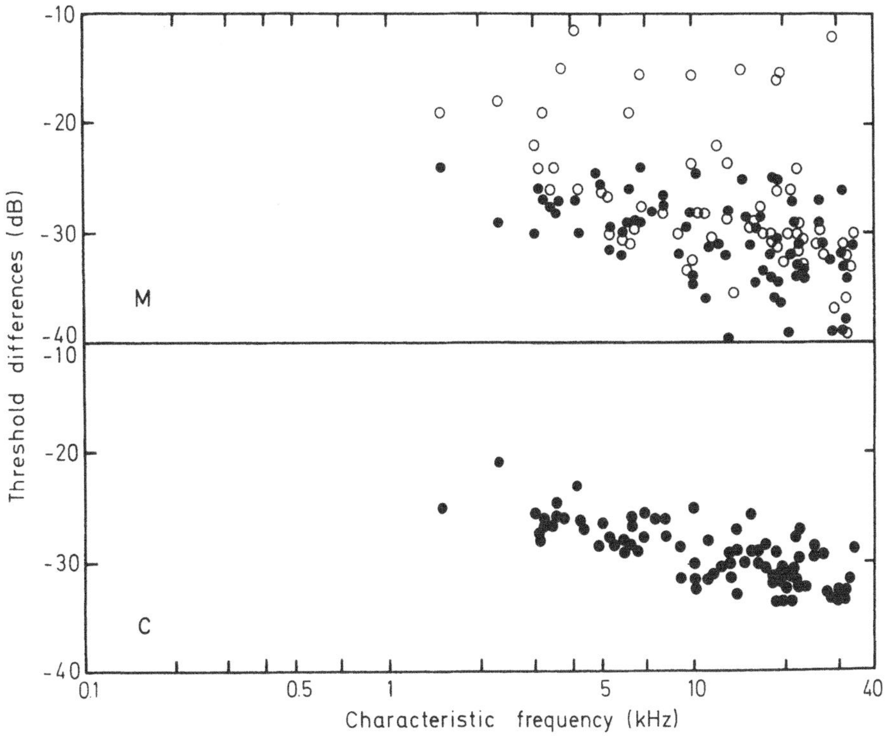

Fig. 35. Difference between noise and tone thresholds for CN units in anaesthetized rat, plotted against CF for each unit. See legend of Fig. 26 and text. Filled circles: determinations at threshold; open circles: determinations of signal levels for equal response at 10 dB above threshold. (Data from MØLLER, 1970a)

and by the near threshold noise: tone comparisons. The point being made here is that MØLLER's conclusion only holds for *part* of his data, namely, from a population of cells which might be drawn from the DCN. In contrast, the majority of his units behaved as did the cochlear fibres, *i.e.* their spectral resolution was consistent with that predicted from the pure tone FTC. The CN cannot be treated as if it were a homogeneous nucleus.

GOLDBERG and GREENWOOD (1966, 1970) found that the responses of CN neurones to narrow bands (100–200 Hz wide) of noise centred at the CF were generally similar to the response to CF tones, but differed slightly in some details (Fig. 36A). The thresholds for the noise stimuli (open circles) were slightly lower than for the CF tones (open squares); the rate of increase of response with increase in intensity was not as rapid with the noise as with the tones; and the maximum, saturated, discharge rates in some cases was lower with noise compared with tones. These effects are presumably related to the considerable fluctuations in amplitude, characteristic of such narrow noise bands.

In many units, widening the bandwidth of the noise stimuli about the CF produced summation effects as would be expected from the considerations of Section II.E.2. That is, doubling the bandwidth (at constant spectrum level) had

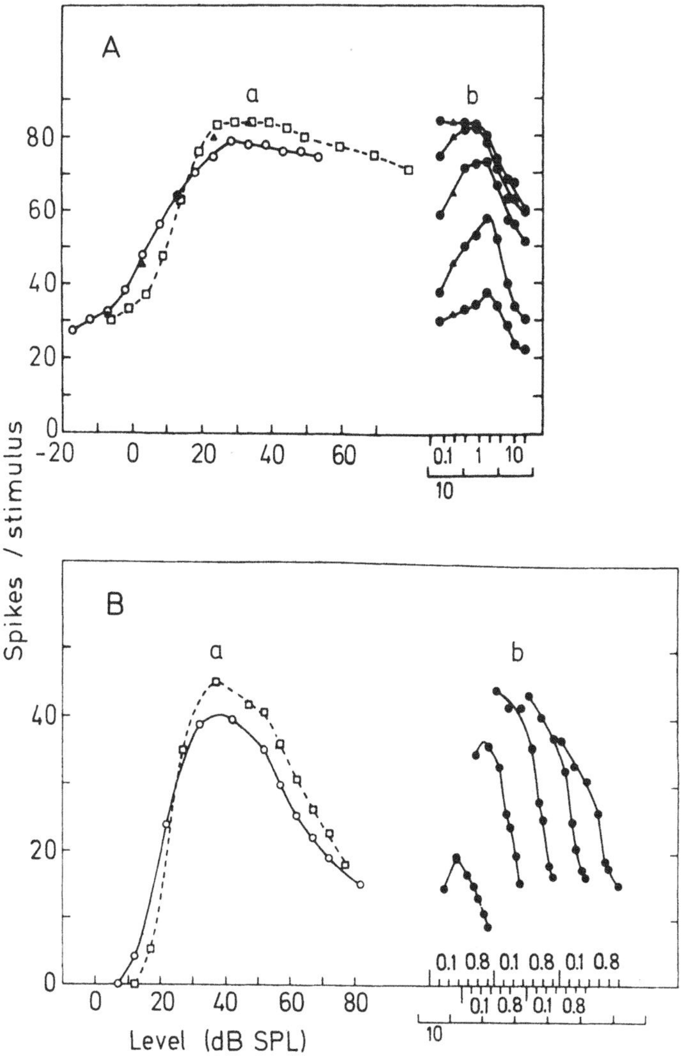

Fig. 36 A, B. Effect of widening noise bandwidth on two units in the cat CN with A, monotonic and B, nonmonotonic intensity functions. A in PVN; B in DCN. Left-hand plots (a): Number of spikes per 200 msec stimulus, square symbols: tones at CF, open circles: narrow band (A: 200 Hz; B: 60 Hz) noise centred on CF (13.4 kHz in A; 0.8 kHz in B). Righthand plots (b: solid circles): number of spikes per 200 msec stimulus of band-limited noise (centred on the CF) of increasing bandwidth at constant spectrum level (left to right: expressed in kHz in upper abscissae and as relative intensity in 10 dB steps in the lower abscissae) and increasing spectrum level (lower to upper plots). (From GREENWOOD and GOLDBERG, 1970)

the same effect on the response as increasing the spectrum level by 3 dB, until the noise bandwidth approached the effective bandwidth of the FTC. Thereafter, widening the noise band had progressively less effect on the response (as the increments in bandwidth and, therefore, power lay outside the FTC bandwidth). This was not the case however in all units: in some, summation occurred only up to a certain bandwidth beyond which the neurone's discharge rate actually *decreased* towards, and in some cases below, the spontaneous level. For the latter units which had monotonic rate-intensity functions (drawn from DCN and PVN), the bandwidths at which suppression began to appear ranged from about 1–4 kHz, tending towards the lower values at higher spectrum levels (Fig. 36A). For non-monotonic units (found in the DCN), on the other hand, these bandwidths were below 0.4 kHz (Fig. 36B). In these cases, presumably, the widening noise bands are encroaching on lateral inhibitory side-bands (Section III.B.2.b). It is doubtful whether a substantial reduction in the input from primary fibres is involved by a mechanism analogous to that of two-tone suppression, in view of the fact many CN neurones do not show this suppression with increase in bandwidth, and in view of the findings discussed in Section II.E.2.

The narrower bandwidths of summation found in nonmonotonic DCN cells presumably reflect the greater incursions of the inhibitory sidebands in these units.

3. Stimuli with Multicomponent Spectra

MØLLER (1970b) has carried out an ingenious experiment in which the responses to paired click stimuli of units in the rat CN were measured over a wide range of separation of the clicks. He interpreted his results in temporal terms (*i.e.* interference between two damped oscillatory responses of the peripheral auditory analyser), but they could equally be looked at in spectral terms, for paired click stimuli generate short term comb-filtered spectra where the spacing of the spectral peaks and valleys is related to the click pair separation (vd. Section II.E.3). Figure 37A shows the responses of a unit to paired click stimulation, where the interval is varied from about 50–1000 μsec. After converting the response (number of discharges per click pair) into terms of the equivalent stimulus level, an autocorrelation function was computed of the system's impulse response (Fig. 37B, interrupted line). The thin continuous line in Fig. 37B represents the autocorrelation function computed, by inverse Fourier transform, of a near threshold isorate function for the unit (analogous to the FTC). MØLLER found that a better fit could be obtained between observed and computed autocorrelation functions if the response area used for the computed continuous line in Fig. 37B was halved in bandwidth. From this he suggested that the peripheral analyser had a higher spectral resolution with regard to double click stimulation than it had for tones. Such a conclusion conflicts with findings in the cochlear nerve (discussed in II.E.3), where the results of comb-filtered noise stimulation were consistent with the cochlea acting as a simple linear filter of *equal* bandwidth to tones and multicomponent spectral stimuli. However, it is clear from Fig. 7 of MØLLER (1969a) that the isorate contour chosen from the data on that unit by MØLLER (1970b) was somewhat anomalous and was wider in bandwidth than the suprathreshold isorate contours obtained for the same unit and than the average for the rat CN (MØLLER, 1972b: Fig. 20). If he had used one of the suprathreshold isorate contours (which in fact would seem to be a more appropriate choice in view of the click results being obtained at suprathreshold levels), it appears likely that a more satisfactory fit still would have been obtained with the observed data in Fig. 37 B, in terms of both amplitude and phase. This, then would be consistent with cochlear nerve findings. Møller's demonstration that the frequency selectivity represented by his results was available immediately and persisted unchanged over a click intensity range of 30 dB is also consistent with the cochlear nerve data and affords

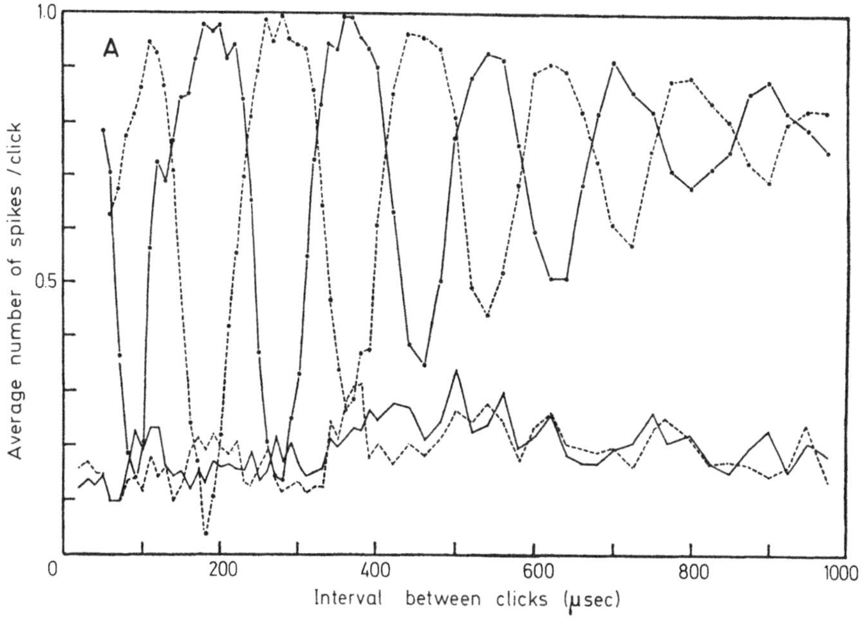

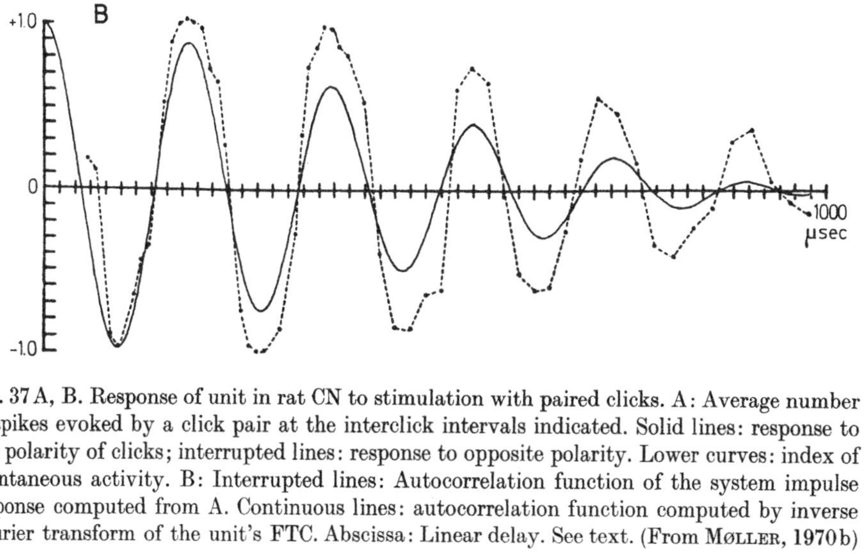

Fig. 37 A, B. Response of unit in rat CN to stimulation with paired clicks. A: Average number of spikes evoked by a click pair at the interclick intervals indicated. Solid lines: response to one polarity of clicks; interrupted lines: response to opposite polarity. Lower curves: index of spontaneous activity. B: Interrupted lines: Autocorrelation function of the system impulse response computed from A. Continuous lines: autocorrelation function computed by inverse Fourier transform of the unit's FTC. Abscissa: Linear delay. See text. (From Møller, 1970b)

further evidence that the frequency selectivity is not accomplished by sharpening mechanisms (such as lateral inhibition) which take a significant amount of time to operate (see Section VI.A.1).

4. Frequency and Amplitude Modulated Tones

Since the finding, in the upper levels of the auditory system, of units selectively sensitive to frequency-modulated sounds (BOGDANSKI and GALAMBOS, 1960; EVANS and WHITFIELD, 1964; SUGA, 1964, 1965b; WATENABE and OHGUSHI, 1968), there have been many studies of the responses to modulated tones in the cochlear nucleus. In contrast to the former data, there is little evidence for *specific* sensitivities for modulation in the CN. In fact, little difference exists in the thresholds of CN cells for both steady and modulated stimuli (SUGA, 1965a: bat; EVANS and NELSON, 1966a, b: cat; MØLLER, 1971: rat).

To a first approximation, the responses of most CN cells to modulated tones can be predicted from the response, as a function of frequency and time, to steady tones (Fig. 38). Thus, sinusoidal modulation of frequency from the high frequency border into the excitatory response area (bar C under frequency response histogram A and modulation histograms C, G, I) produces a roughly sinusoidal distribution of unit discharge density, the maximum of which corresponds with minimum tone frequency. In fact, the phase of maximum firing density "leads" the modulation waveform slightly, as would be expected on the basis of the adaptation with time shown to a steady tone (B). Modulation at the low frequency border (bar D and histogram D) evokes a similar pattern but with inverted relationship to the modulation. Modulation of frequency across the bandwidth of response (bar E and histogram E) evokes a symmetrical bimodal distribution, as expected. Amplitude modulation (at the CF marked by the arrow, and histograms F, H, J) likewise evokes a nearly sinusoidal distribution. These correspondences between firing density and modulation are maintained over a wide range of modulation rates (C, G, I; F, H, J). The main divergences occur at higher modulation rates (G, I; H, J) where the flanks of the histograms (corresponding to minimum excitation) drop *below* the spontaneous discharge level (horizontal bars). This is a common feature and appears to result from the cumulative off-inhibition characteristic of many cells in the CN (see Fig. 34 for another example of this effect). In a few cells (particularly in the DCN), however, the sensitivities to upward and downward frequency sweeps are not symmetrical (see Fig. 39 for an extreme example: EVANS and NELSON, unpublished data). Lesser degrees of asymmetry have been reported by ERULKAR *et al.* (1968) and MØLLER (1971) using linearly frequency-modulated tones. This asymmetry was often related to asymmetrical disposition of inhibitory side-bands about the central excitatory response area in the data of EVANS and NELSON (NELSON and EVANS, 1971). Thus, explanatory models based on asymmetrical inhibition (analogous to that of BARLOW and LEVICK (1965) for the sensitivity of units in the retina to direction of movement) may be more appropriate than ones based on a tonotopic organization of synapses on the dendrites of single cells (FERNALD, 1971).

MØLLER (1969c, 1971, 1972a) has made a detailed quantitative study of the characteristics exhibited by CN cells in the rat to linear and sinusoidal FM and AM at higher rates of frequency change than so far discussed. Under these conditions, two main departures occur from a discharge pattern which simply corresponds to the modulation. Firstly, the bandwidth of the units' response area appears to become narrower at intermediate modulation rates (Fig. 40A: 12.8 and 25 sweeps/

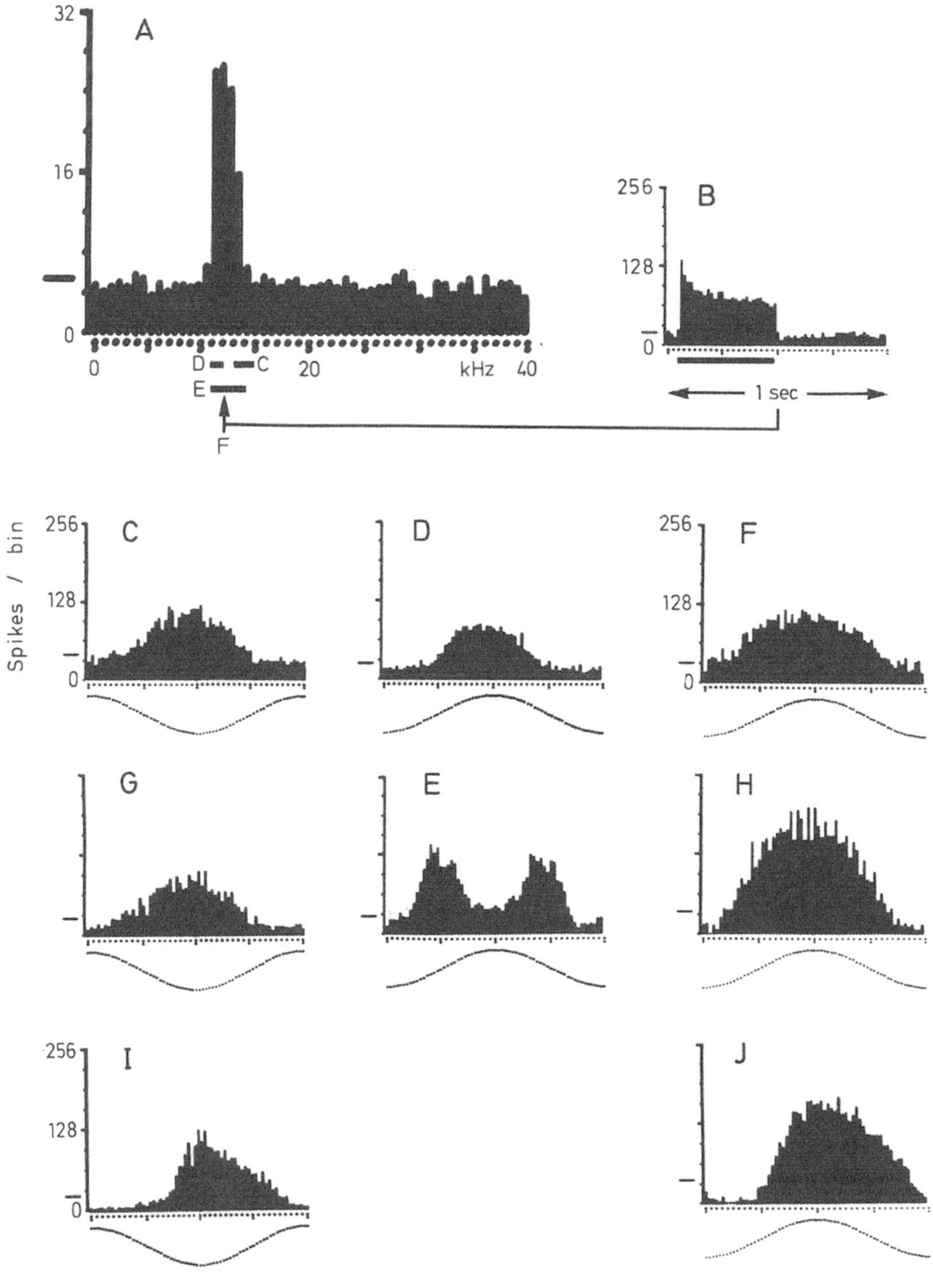

Fig. 38A—J. (Legend see p. 65)

sec). This is probably a result of cumulative off-inhibition preferentially diminishing the responses to the less effective (flanking) frequencies. Secondly, at intermediate modulation rates (50–300 c/s for sinusoidal modulation) the modulation of the response discharge density reaches a maximum, for both FM and AM

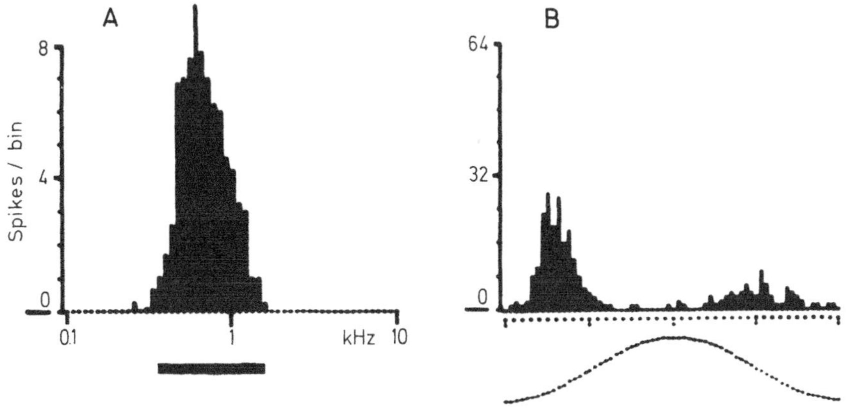

Fig. 39 A, B. Uncommon type of response of CN unit to frequency modulation. Cat AVN. A: Frequency response histogram of response to 100 msec tones at 30 dB above threshold. B: Asymmetrical response distribution to frequency modulation positioned across response area (bar below A). Modulation rate: 1 c/s. Optimum response to upward frequency changes. Data collected for 30 sec

(Fig. 40B), and, for AM, exceeds the stimulus modulation depth by factors of 2–15 dB, as in Fig. 40B (Møller, 1972a). Sinusoidally amplitude-modulated noise had similar effects. Fig. 40B, C indicates that, for many units, amplitude and phase characteristic of the response to FM and AM are very similar, at least when the FM is not such as to produce an inverted or bimodal relationship between response firing density and the modulation waveform (as in Fig. 38C and E respectively).

Glattke (1969) has looked at the responses of CN units to square wave modulated AM tones of high frequency for evidence of "de-modulation" of the signal, *i.e.* responses to the modulation in units of CF corresponding to the modu-

---

Fig. 38 A—J. Common types of response of CN unit to frequency and amplitude modulated tones. Cat PVN. All analyses at 20 dB above threshold. A: Frequency response histogram (see Fig. 30) of response to 100 msec tones as a function of frequency. B: PST histogram at CF. C, G, I: Modulation histograms of averaged firing density in response to sinusoidal frequency modulation over band of frequencies indicated by bar C under A. Waveform of modulation indicated under modulation histogram. Modulation rate: 1, 10, 40 c/s in C, G, I respectively. D: Frequency modulation over band of frequencies D; Modulation rate: 1 c/s. E: Frequency modulation symmetrically across excitatory response area (E). Note symmetrical double peaked distribution of discharge, corresponding to modulated frequency crossing CF twice per modulation cycle, F, H, J: Amplitude modulation (depth 80%) at CF (F) at modulation rates of 1, 10, and 40 c/s respectively. At the higher rates of modulation (G, I, H, J) note depression of "tails" of distributions below spontaneous level (indicated by bars against ordinates). Data collected for about 30 sec in each analysis

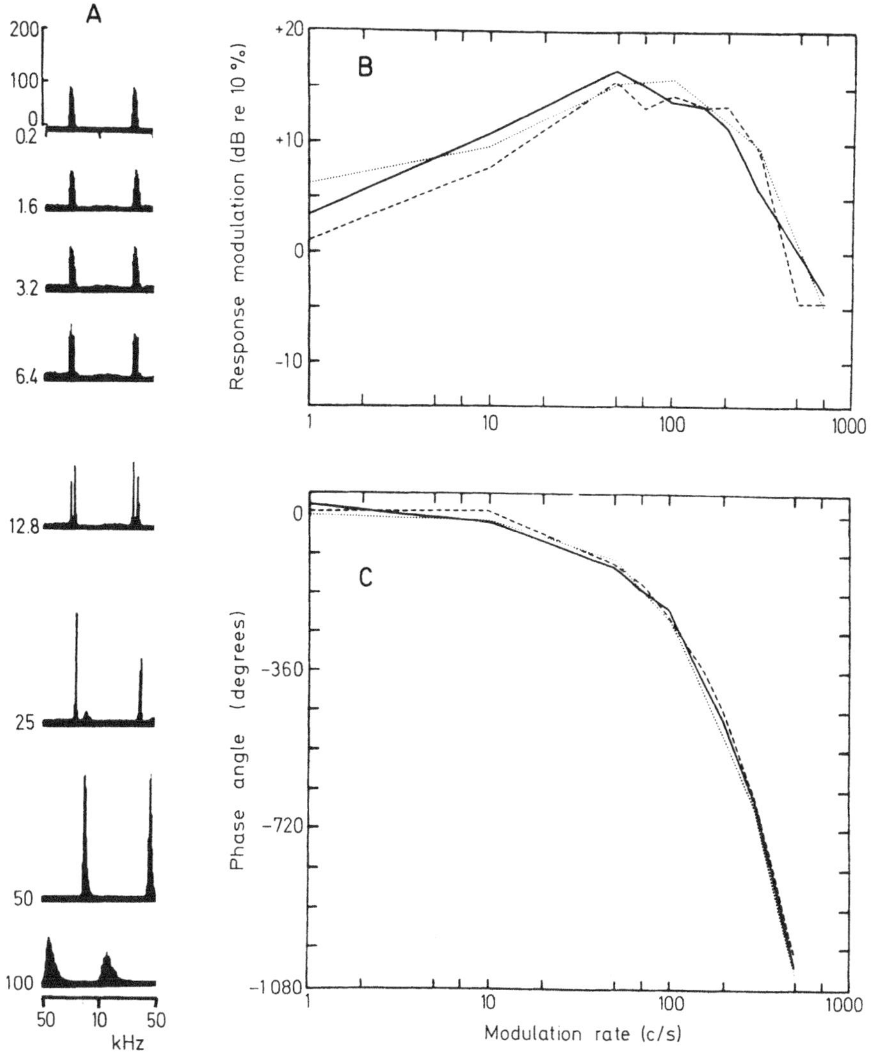

Fig. 40A—C. Effect of rapid rates of frequency and amplitude modulation on response of CN units. Rat. A: linear (triangular) frequency modulation between 10 and 50 kHz across CF of unit (29.3 kHz), at 20 dB above threshold. Note narrowing of firing distributions at sweep rates of 12.8 and 25 c/s (rates in c/s indicated to left of modulation histograms). The progressive shift of histogram peaks to the right with increasing modulation rate (which causes the right-hand peak at 50 c/s to appear as the left one at 100 c/s results from the latency of response. B: Modulation of discharge rate in response to sinusoidal frequency-modulation (dotted line) and amplitude modulation of tone at CF (solid line) and of broad-band noise (dashed line) as a function of the rate of modulation. Response modulation expressed in terms of ratio of observed modulation of response relative to 10%, which in the case of amplitude modulation equals the depth of stimulus modulation. (100% response modulation is represented by + 20 dB.) Measurements obtained from the sinusoids which best fit the modulation histograms. Frequency modulation: ± 45 Hz at CF of 1.5 Hz. C: Phase of discharge modulation relative to that of stimulus modulation. (From Møller, 1969c, 1972a)

lation frequency. No such responses were found[3]. GLATTKE claimed to have found two classes of CN units: one class where the modulation of activity became negligible with modulation rates in excess of about 200 Hz; another where the response and stimulus modulations corresponded to rates in excess of 500–800 Hz. It appears, however, from the published data, that the latter class of units are of the "chopper" type where the regular interspike repetition period of 3–4 msec can be confused for "following" of the modulation by the unit. With the latter qualification, however, GLATTKE's results are consistent with those of MØLLER (1972a).

5. Effect of Noise on Responses to Click and Tone Stimuli

The presentation of wide band noise reduces the responses of CN units to click and tone stimuli (GALAMBOS, 1944; MØLLER, 1969b; MARSH *et al.*, 1972). To what extent this effect differs from that found at the level of the cochlear nerve (Section II.E.4) is not known, in the absence of quantitative comparisons. Certainly, the desynchronization by noise stimuli of unit activity in response to low frequency tones (MARSH *et al.*, 1972) is not unexpected. On the other hand, the results of GREENWOOD and GOLDBERG (1970) where progressive inhibition was found in certain units with widening bands of noise, suggest that the inhibitory response regions characteristic of many CN units must contribute to the reduction in response found under conditions of broadband noise stimulation.

GOLDBERG and GREENWOOD (1966) and GREENWOOD and GOLDBERG (1970) have made a careful study of the responses of CN units to combinations of tones and narrow (ca. 0.1 kHz) bands of noise centred (a) at the CF and (b) off the CF of the units. In the former case, where the two stimuli were excitatory the *intensities of the stimuli* (rather than the responses) appeared to sum. Thus, the discharge rate and certain statistics of the discharge to the stimulus combination were determined by the more intense of the two stimuli. The addition or subtraction of the less intense stimulus had little effect (see Section VI. C.2). In the second case, where the narrow noise band was displaced from the CF and centred over an inhibitory sideband, the response to the tone at the CF was *reduced* (Fig. 41). The degree of reduction increased with increase in intensity of the noise band, although the relationship was not always monotonic (Fig. 41B). In addition, the response pattern (*i.e.* statistics) could shift from one corresponding to response to the tone alone, to that corresponding to the inhibiting noise signal alone.

It appears that in many respects, the behaviour of the CN neurones described by GREENWOOD and GOLDBERG (1970) to combinations of tone and noise bands, resembles that of the cochlear nerve to two-tone combinations (as discussed in II.E.1c). In particular, the response discharge rate, and the pattern of response were generally dominated by the more effective component of the two stimuli, whether the latter produced a lesser (or in the case of the CN an inhibitory) response than of tone alone (as in Fig. 41A, a–d; 41B, a–c), or whether it produced a greater response (Fig. 41Ac; 41Ba). In some cases, however, the situation was less straightforward, in that a narrow band of noise in an inhibitory side band could augment or suppress the response to a tone at the CF depending upon the level of

[3] Incidentally, this result appears to conflict with that of BOERGER and GRUBER (1970) obtained in the cochlear nerve and discussed in Section II.E.1.b.

the latter. (Response to tone in combination with noise at 60 dB SPL in Fig. 41 B; compare with response to tone alone at each tone level.)

In summary, therefore, the addition of a band of noise to a tone stimulus at the CN level may render the tone without effect upon the unit either by inhibiting or interfering with the response to the tone or by activating the unit itself. "In either case, the effect of the noise is to mask the tone as far as this unit is concerned" (Greenwood and Maruyama, 1965).

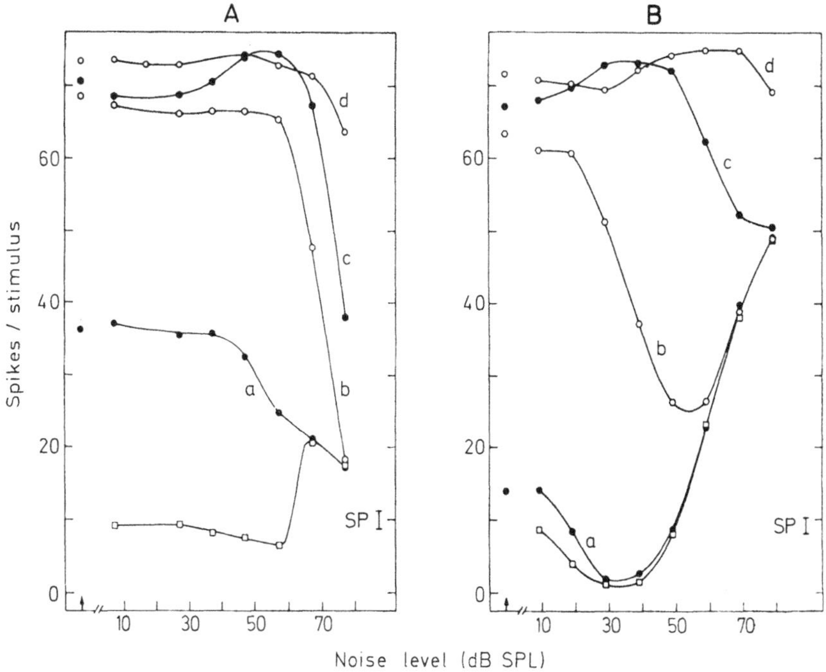

Fig. 41 A, B. Response of CN unit to noise bands placed at frequencies (A) below and (B) above CF and to combinations of noise bands and CF tone. Cat PVN. CF of unit: 11.6 kHz. Noise bandwidth: 200 Hz, centred at 6 kHz and 13.5 kHz in A and B respectively. Square symbols: noise alone; a—d: noise, plus tone at CF of intensity 4, 24, 44, and 64 dB respectively. Ordinates: spikes per 200 msec stimulus. Responses to tone alone indicated by arrow. SP indicates level of spontaneous activity. (From Greenwood and Goldberg, 1970)

In a study directed at the question of the neural mechanisms underlying non-simultaneous psychophysical masking, Watenabe and Simada (1971) have demonstrated some correlation between the time course of psychophysical forward masking and that of off-inhibition in CN cells. Forward masking occurs when a preceding noise ("masking") stimulus interferes with the response to a succeeding ("test") tone stimulus. Figure 42 indicates the elevation of threshold for a 10 msec tone burst at the CF, produced by the off-inhibition (Section III.E.3.), and the time course of its recovery. This magnitude and time course of the threshold elevation resembles that which can be obtained psychophysically in forward masking paradigms (*e.g.* Plomp, 1964a). Watenabe and Simada did not find any correlates of backward masking in the CN.

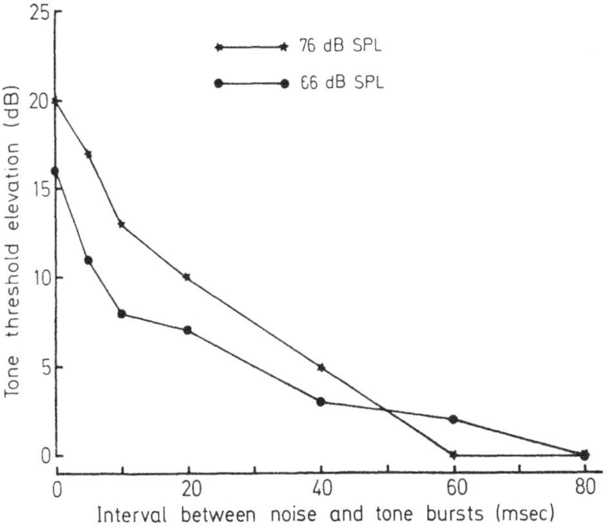

Fig. 42. "Forward masking" of unit in CN of cat. Elevation in threshold of a 10 msec test tone at CF by 40 msec broadband noise bursts at 66 or 76 dB SPL preceding the test tone by the interval indicated on the abscissa. (From WATENABE and SIMADA, 1971)

IV. Gross Nerve Action Potential Responses

It has long been known (DERBYSHIRE and DAVIS, 1935) that an electrode in the region of the cochlea can record a succession of potential waves in response to acoustic clicks or tone bursts. The close relation of the waves to gross potentials recorded directly from the cochlear nerve and their lack of similarity with the cochlear microphonic potentials (*e.g.* DERBYSHIRE and DAVIS, 1935; PEAKE and KIANG, 1962) have led to their being termed the gross nerve action potential or AP. Most, but not all, of the data pertaining to the AP response have been superseded and made clear by studies at the single fibre level in the cochlear nerve. However, it has recently been shown that these APs can be reliably recorded in normal and deaf human subjects from an electrode in the outer or middle ear and be related to normal and pathological subjective thresholds (YOSHIE *et al.*, 1967; YOSHIE, 1968, 1971; PORTMANN *et al.*, 1967; SOHMER and FEINMESSER, 1967; SPRENG and KEIDEL, 1967; COATS and DICKEY, 1970; ARAN, 1971; CULLEN *et al.*, 1972). Hence, their study and interpretation has taken on a new lease of life. The present account will be limited to the question of the origin of the components of the AP and the effects of stimulus intensity.

At suprathreshold stimulus levels, the AP response consists of two or occasionally three negative waves, the N_1, N_2, and N_3 waves (Fig. 3C–F shows the N_1 and N_2 waves). The N_1 response appears to originate predominantly from the basal turn of the cochlea (TASAKI, 1954; DEATHERAGE *et al.*, 1959; PEAKE and KIANG, 1962; KIANG *et al.*, 1965a). Its latency corresponds, with allowance of approximately 0.2 msec (presumably representing conduction from a more peripheral location to the internal auditory meatus) with that of single cochlear fibres of high CF (Fig. 3E, F). It does, however, receive some contribution from the second turn of the cochlea in the guinea pig at least (TEAS *et al.*, 1962). There is less unanimity for the origin of the later AP waves. The N_2 wave has been considered to arise in part from the cochlear nucleus (KIANG *et al.*, 1962a) or from repetitive activity in the cochlear nerve (TASAKI, 1954; TEAS *et al.*, 1962). Against the former suggestion is the demonstration that the N_2 component can still be recorded after section of the cochlear nerve (DAIGNEAULT *et al.*, 1968). Against the second suggestion (repetitive activity in cochlear nerve), PST histograms of the activity of cochlear fibres of high CF do not show an appropriate second peak (Fig. 3E, F; KIANG *et al.*, 1965a). In fact, there is positive evidence that the later waves originate in the upper turns of the cochlea (PUGH *et al.*, 1973).

To broad-band signals such as clicks, the cochlea acts as a delay line. Hence, the AP recorded by a gross electrode must represent the spatially weighted sum over time of the progressive activation of endocochlear or modiolar all-or-none synaptic or nerve potentials which may be diphasic in nature (TEAS *et al.*, 1962). Inasmuch as the thresholds of these individual component potentials are a function of frequency (*e.g.*, Fig. 8) and therefore delay along the cochlea (Fig. 21), and the time of initiation is dependent to a greater or lesser extent upon the level and phase (Fig. 6) of the stimulus, the form and latency of the AP will depend in a complex manner upon the level and phase of a broadband stimulus and also to a certain extent on the frequency response of the transducer used.

Some of these features are illustrated in Fig. 43. Almost all investigations of the relation between the amplitude of the N_1 response (measured as peak-to-peak amplitude of first negative and positive waves or as baseline to N_1 peak) and the intensity of click stimuli in cat (DERBYSHIRE and DAVIS, 1935; ROSENBLITH, 1954; KIANG *et al.*, 1962a; PEAKE and KIANG, 1962; RADIONOVA, 1963; WIEDERHOLD and KIANG, 1970), guinea pig (TEAS *et al.*, 1962; TAUB and RAAB, 1969; NIEDER and NIEDER, 1970; SPOOR and EGGERMONT, 1971) and normal man (YOSHIE *et al.*, 1967, 1971; EBERLING and SOLOMON, 1971) have found the function to have an inflexion or even a dip at about 50–60 dB above the threshold for visual detection of the AP, in conjunction with a discontinuity in the latency function, as in Fig. 43. This finding has almost invariably been interpreted in terms of the participation of two populations of receptor elements or neural units, a small low threshold set responsible for the low threshold segment of Fig. 43 (continuous curve) and a larger high threshold set responsible for the remainder (ROSENBLITH, 1954; DAVIS *et al.*, 1958; KEIDEL, 1965, 1970; RADIONOVA, 1963, YOSHIE, 1968; ARAN, 1971). With the demonstration that in the cat the number of cochlear fibres innervating the inner hair cells exceeds those for the outer hair cells by a factor as much as 20:1 (SPOENDLIN, 1966, 1970, 1971, 1972), it has appeared more reasonable to relate the low and high threshold populations with the two populations of cochlear fibres innervating the outer and inner hair cells respectively than to the receptors themselves, where the numerical relations are reversed (SPOENDLIN, 1971). Attempts to produce differential damage to the outer hair cells by exposure to noise or ototoxic agents in animals (*e.g.* DAVIS *et al.*, 1958; KEIDEL, 1965) also support this notion. Against this interpretation, it must be emphasised (a) that two populations of cochlear fibres distinguishable by their thresholds have not been found in the normal cochlear nerve (see Section II.D.1); (b) that the AP is a compound potential and its amplitude a complex sum of potentials differing in time, space, and possibly polarity. That a drastic change is introduced in the intensity and latency function by merely a change in phase of the click (Fig. 43 interrupted line) illustrates the latter point, and consequently many workers are reluctant to press an anatomical interpretation (*e.g.* ARAN and DELAUNAY, 1969; YOSHIE, 1971). It is also possible that the inflexion present in almost all AP intensity functions is

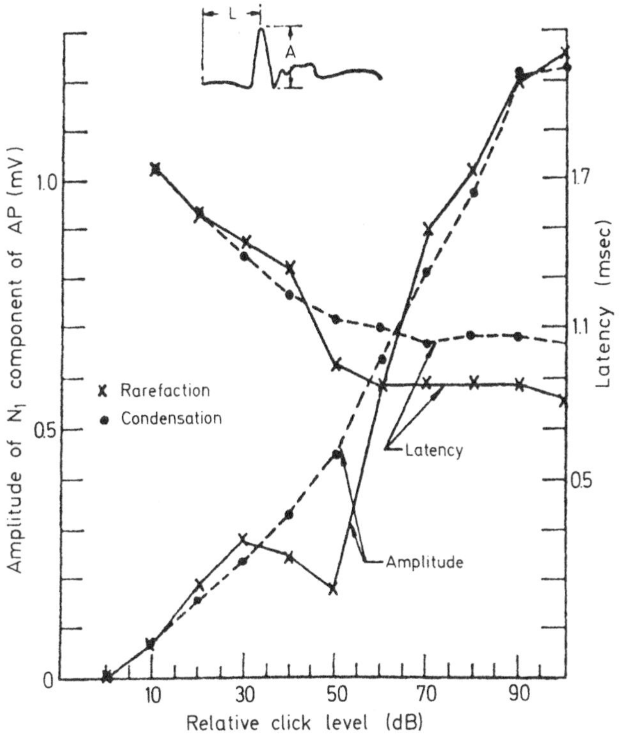

Fig. 43. Amplitude and latency of gross AP response (recorded by concentric electrodes in the cochlear nerve) as a function of relative click intensity. Cat. Latency corrected for acoustic delay. Solid curves: rarefaction clicks; interrupted lines: condensation clicks. Measurements of amplitude and latency made as indicated in inset. (From PEAKE and KIANG, 1962)

related to the sudden change in bandwidth of cochlear fibre FTCs about 50 dB above their minimum threshold (Fig. 9), and that the loss of the low threshold segment in cochlear pathology is related to the uniformly high threshold, broadly tuned, FTCs obtained in such preparations (KIANG *et al.*, 1970; EVANS, 1972b). Clearly more correlative studies are required between single fibre and gross AP recordings.

Most recently it has been claimed that potentials reflecting activity in the cochlear nucleus and higher auditory centres as well as the cochlear nerve can be recorded from the region of the pinna in man and animals (LEV and SOHMER, 1972).

V. Effects of Activity in Efferent Pathways

As the anatomy and functional properties of the recurrent, efferent, auditory system are fully discussed in Chapters 8 and 11 of Vol. V/1 of the Handbook, only a brief account of the direct effects on the cochlear nerve and nucleus will be given here.

A. Cochlear Nerve

The efferent fibres which project to the cochlea originate in the superior olivary nucleus (SON) of the same and contralateral sides, forming the uncrossed and crossed olivocochlear bundles (OCB) (RASMUSSEN, 1960, 1964). The effects of electrical stimulation of the OCB on the cochlear nerve were first noted by GALAMBOS (1956) with gross nerve AP recordings. Subsequent studies of the AP response (DESMEDT and MONACO, 1962; WIEDERHOLD and PEAKE, 1966; DEWSON, 1967; NIEDER and NIEDER, 1970; GALLEY *et al.*, 1971) and of single cochlear fibres (FEX, 1962; WIEDERHOLD and KIANG, 1970; WIEDERHOLD, 1970; TEAS *et al.*, 1972) have established a consensus of opinion in regard to the effect of artificial stimulation of the efferent system at this level, while paradoxically its action in the behaving animal remains obscure (GALAMBOS, 1960; IRVINE *et al.*, 1972). Because of its accessibility, the crossed OCB has been the component studied most thoroughly.

Electrical stimulation of the crossed OCB results in an inhibition (never an increase) in the activity of most, but not all cochlear fibres. Individual fibres show considerable differences in sensitivity. It takes 40–110 msec for the maximum effect to occur, while fibres take 100–400 msec to recover after the termination of electrical stimulation, fibres with higher CF having a longer latency of effect yet taking a shorter time to recover. The inhibition involves activity evoked by acoustic stimuli, but there is some uncertainty whether the spontaneous activity can be so affected. In a careful study on this point, WIEDERHOLD and KIANG (1970) found definite *maintained* reductions in the spontaneous activity of a small proportion of their fibres (*e.g.*, Fig. 44A), but these were restricted to the lowest threshold fibres, and the question remains whether, in spite of precautions, these fibres were being subjected to low level background acoustic stimulation such as may arise from the respiration and circulation. The time course of the inhibition and recovery is illustrated in Fig. 44A.

The magnitude of the suppression by stimulation of OCB depends on many factors, including the parameters of acoustic and electrical stimulation (WIEDERHOLD and KIANG, 1970). In a systematic study, where the latter parameters were optimized, WIEDERHOLD (1970) examined the effects of prior electrical stimulation of the crossed OCB on the responses of cochlear fibres to tones. In nearly all cases, the effects could be summarised as a greater or lesser degree of shift of a fibre's discharge rate versus intensity function to higher sound levels (Fig. 44B). In other words, the suppression was maximal (and nearly constant) over an intermediate range of tone intensities, and minimal at subthreshold and supra-saturation intensities. The magnitude of the shift is maximal when the fibre is being stimulated at its CF. Thus, OCB stimulation reduces the sharpness of the FTC (Fig. 45A), by a factor of 7–31% expressed in terms of $Q_{10\,dB}$. In addition, the magnitude of the shift differs from fibre to fibre over a range of 1–25 dB, maximum values being obtained for fibres in the intermediate range of CFs (Fig. 45B). Similar results were obtained in the guinea pig by TEAS *et al.* (1972). These findings are consistent with those from studies of the gross AP response to acoustic transients, where the degree of suppression observed differs from animal to

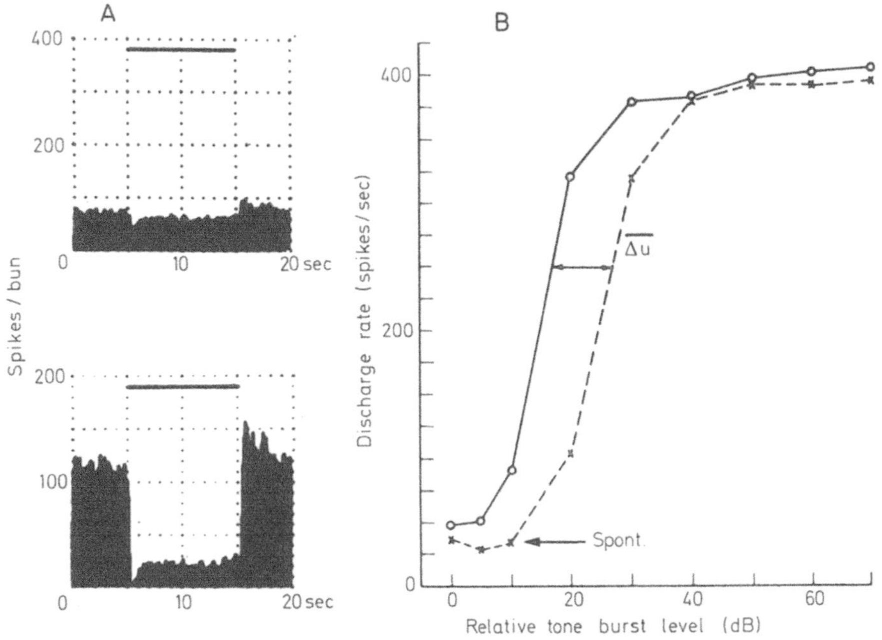

Fig. 44 A, B. Effect of crossed OCB stimulation on spontaneous and tone evoked activity of cochlear fibres. Cat. A: PST histograms of effect of electrical stimulation of the crossed OCB (during bar) on spontaneous activity (above) and on activity evoked by a continuous tone at the fibre's CF (11.58 kHz), 10 dB above threshold (below). Ordinate scale (spikes per bin) and number of presentations represented by the two histograms arranged so that equal bar heights represent equal discharge rates. (From WIEDERHOLD and KIANG, 1970.) B: Mean discharge rate (spikes/sec) evoked by 40 msec tone bursts at the CF of a cochlear fibre at the relative stimulus levels indicated, without (continuous line) and immediately preceded by electrical stimulation of the crossed OCB (interrupted line). Spont. indicates spontaneous discharge rate. Upper arrow indicates one determination for estimate of mean shift of the rate-intensity function, $\overline{\Delta u}$ used in Fig. 45B. (From WIEDERHOLD, 1970)

animal over a range of 13–23 dB, is minimal at high click levels, and is maximal for high frequency acoustic transients (WIEDERHOLD and PEAKE, 1966). The dependence of the magnitude of the inhibition upon the CF of the fibres may correlate with the finding that the density of the efferent innervation of the outer hair cells is maximal in the basal turn of the cochlea and decreases towards the apex and in the extreme basal end (ISHII and BALOGH, 1968). This point will be further discussed in relation to the question of the cochlear innervation in general in Section VI.B.

Whereas histochemical findings suggest strongly that the efferent endings are cholinergic, pharmacological evidence is incomplete (for reviews see Whitfield, 1967; Klinke and Galley, 1974).

The functional role of the cochlear efferent system is surprisingly unclear, in spite of the above data. No effects of section of the crossed OCB have been obser-

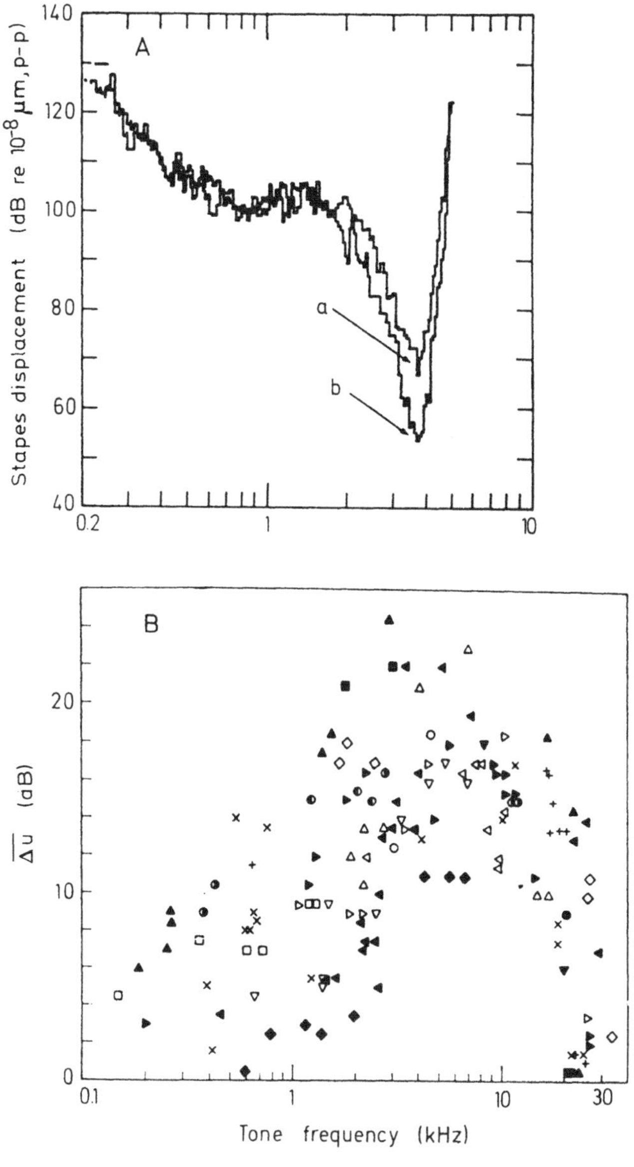

Fig. 45 A, B. Effect of crossed OCB stimulation on cochlear fibres as a function of stimulus frequency (A) and fibre CF (B). Cat. A: FTCs (where tone threshold level is expressed in terms of stapes displacement) of single cochlear fibre with (a) and without (b) electrical stimulation of crossed OCB. (From Kiang *et al.*, 1970.) B: Plot of mean shift of rate intensity function, $\overline{\Delta u}$ (see Fig. 44 B) for 141 cochlear fibres against their CF. Different symbols represent different animals. (From Wiederhold, 1970)

ved on absolute behavioural threshold (DEWSON, 1968; IGARASHI *et al.*, 1972). Of considerable interest has been the question whether efferent activity can increase the detection of acoustic signals in the presence of background noise. Such an effect was inferred by DEWSON (1967) and apparently demonstrated by NIEDER and NIEDER (1970) from the effect of electrical stimulation of the OCB on the gross AP

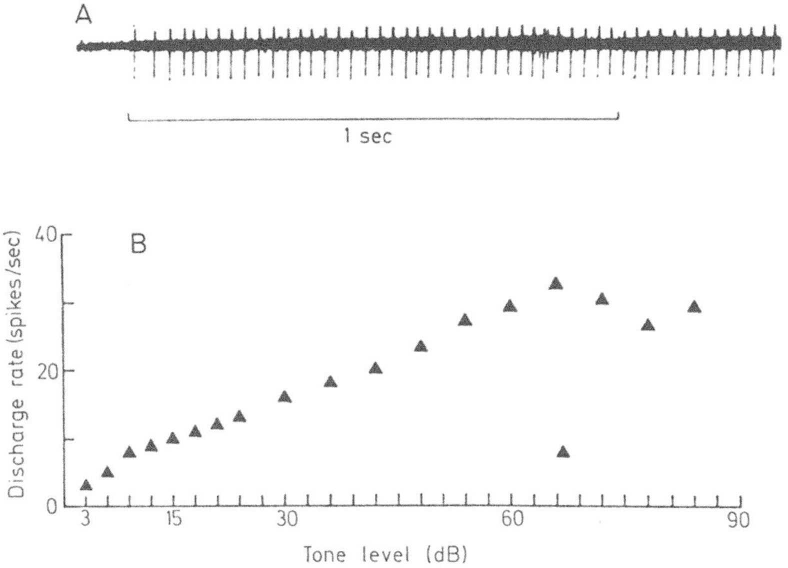

Fig. 46 A, B. Response of cochlear efferent fibres to tones. Cat. A: Response to tone at 66 dB above threshold. Note regular discharge and absence of adaptation. Tone commences at time marker. B: Rate: intensity function for the fibre shown in A. (From FEX, 1962)

responses to clicks in background noise. Behavioural evidence, however, is conflicting (compare DEWSON, 1968, and COMIS and PICKLES, 1972; with TRAHIOTIS and ELLIOTT, 1970, and IGARASHI *et al.*, 1972). In fact, apart from the inexplicable finding of CAPPS and ADES (1968) that the frequency discrimination of squirrel monkeys deteriorated after crossed OCB section (contrary to what would be expected from the effect of the OCB on cochlear tuning, Fig. 45 A), investigations looking for systematic effects of the OCB at the cochlear level related to attention, conditioning etc. in behaving animals (with their middle ear muscles cut) have proved negative (GALAMBOS, 1960; IRVINE *et al.*, 1972).

FEX (1962, 1965) has recorded the activity of the crossed and uncrossed efferent fibres themselves in the vestibulo-cochlear anastomosis in unanaesthetized decerebrate cats with the ipsilateral ear destroyed. Most of the fibres were silent in the absence of acoustic stimulation to the contralateral ear, but when activated by tones gave a low, regular, discharge with little adaptation, after a latency of 5–40 msec (Fig. 46 A). The FTCs of the crossed fibres resembled those of VCN cells, whereas those of most of the uncrossed fibres had both excitatory and inhibitory response bands (FEX, 1965). Interestingly, the discharge rate versus tone intensity functions, for some of these units at least, have dynamic ranges in excess of 60 dB (Fig. 46B). In other experiments, FEX (1963) found that the crossed efferents were best activated from the ipsilateral ear.

B. Cochlear Nucleus

At least four efferent pathways to the CN have been described (*e.g.* RASMUSSEN, 1960). The first takes origin from the OCB. The second enters the DCN via the dorsal acoustic stria and takes part origin from the ventral nucleus of the contralateral lateral lemniscus. The third enters via the intermediate stria, originates in the *S* segment of the ipsilateral SON, and terminates chiefly in the VCN. The fourth takes origin in the ipsilateral inferior colliculus and dorsal nucleus of the lateral lemniscus and enters the DCN via the trapezoid body. Investigations of the actions of these pathways, with the exception of the last, have been made by electrical stimulation of the nuclei of origin.

DESMEDT (1960) was the first to describe an inhibitory efferent effect on gross potentials recorded in the CN of the unanaesthetized decerebrate cat after electrical stimulation near the ventral nucleus of the lateral lemniscus. Recording from single units in the anaesthetized cat, COMIS and WHITFIELD (1968) found that electrical stimulation of the ventral nucleus of the contralateral lateral lemniscus evoked inhibition of spontaneous and tone evoked activity in the DCN. Stimulation of the *dorsal* nucleus of the contralateral lemniscus, on the other hand, generally evoked inhibition in the DCN and excitation in the VCN.

COMIS and WHITFIELD (1968) and COMIS (1970) have studied the effects of electrical stimulation of the SON on the responses of cells in the ipsilateral CN. They used direct-current stimulation to minimise the effects on afferent and efferent fibres of passage. Stimulation of the medial portion of the *S* segment produced an increase in spontaneous and tone evoked activity in AVN cells with, in some cases, a reduction in threshold of up to 15 dB. The minimum latency of the effects was 25–50 msec. There is a considerable body of evidence suggesting that this pathway is cholinergic (RASMUSSEN, 1964; COMIS and WHITFIELD, 1966, 1968; COMIS and DAVIES, 1969). Stimulation of the *lateral* portion of the *S*-segment, in contrast, evoked inhibition in cells of the DCN and in some cells of the VCN (COMIS, 1970).

STARR and WERNICK (1968) also found excitatory and inhibitory effects on single units of the CN, but resulting from stimulation of the crossed OCB in decerebrate cats. Both the spontaneous and tone evoked activity were affected. Inhibitory effects were the most commonly encountered, and the effects on the spontaneous activity persisted after destruction of the ipsilateral cochlea. Nevertheless, the question remains how far the effects on tone evoked activity were mediated via the cochlea.

Inhibition of cells in the CN can be obtained by acoustic stimulation of the contralateral ear (PFALZ, 1962; DUNKER, GRUBEL, and PFALZ, 1964; KLINKE *et al.*, 1969; MAST, 1970b). The effect is not mediated by the crossed OCB, but disappears when lesions are made in the trapezoid body (GRUBEL *et al.*, 1964). PFALZ and his co-workers demonstrated the inhibition on the spontaneous activity after destruction of the ipsilateral cochlea, a finding that suggests that it was the DCN that was being recorded from (see Section III.B.3), in keeping with the findings of MAST (1970b). The effect can be obtained with contralateral tones within 40 dB of the animals' behavioural threshold, where interaural cross-talk can be excluded (BOERGER *et al.*, 1968), thus enabling a study of contralateral inhibition with both cochleas intact (KLINKE *et al.*, 1969). While the effect on the responses to ipsilateral acoustic stimulation was small at these levels, it was frequency dependent: frequencies immediately on either side of the unit's CF were most effective in depressing the response to ipsilateral tones at the CF. Again, the latency of inhibition was relatively long: from 30–50 msec (KLINKE *et al.*, 1969).

It is not at all clear what the functional relevance of the efferent pathways actually is. In spite of encouraging early reports, later studies of the grossly recorded CN activity in behaving animals under conditions of directed attention and conditioning have not found consistent effects attributable to a descending control system (see WHITFIELD, 1967 for review), with the exception of the series of experiments by PIDDINGTON (*e.g.* 1971) in the fish.

VI. Discussion

The large body of data which has been collected at this peripheral level of the auditory system, and particularly those quantitative studies conducted under controlled conditions, have given rise to considerable efforts to relate them to cochlear mechanisms on the one hand, and psychophysics, including disorders of hearing, on the other. This section will be limited to these topics, and to a brief account of their relevance to the problem of coding of frequency and intensity information.

A. In Relation to Cochlear Mechanisms

In the wake of the quantitative studies, summarised in earlier sections, particularly those of the cochlear nerve, a number of "blackbox" models of the inner ear mechanisms have been developed (*e.g.* SIEBERT, 1965, 1968, 1970; WEISS, 1966; GEISLER, 1968; DE BOER, 1969; DUIFHUIS, 1970, 1972; models restricted to cochlear mechanics are dealt with in Chapter Mathematical Models in Vol. V/3 of this Handbook). While it is evident that many of the components of these models cannot be identified with presently available physiological and anatomical data nor presented with the same degree of confidence, nevertheless the models continue to serve useful explanatory and heuristic purposes. One of the earliest and probably the most comprehensive of these models, that of WEISS (1966) will serve as a useful point of departure (Fig. 47).

1. Cochlear Filtering

The first compartment of the model accounts for the mechanical properties of the outer, middle and inner ears. In Weiss's model, these were considered to be linear over a large intensity range (80 dB), and the frequency response characteristics of the outer and middle ears were neglected. For the purpose of accounting for the dependence of threshold on frequency represented by the audiogram, it is clear now that the effect of the outer and middle ear response can be quite significant

(PRICE, 1971; DALLOS, 1971; JOHNSTONE and SELLICK, 1972). There is, however, still uncertainty on whether the audiogram can be entirely accounted for in this manner, or whether inner ear mechanisms are also to be implicated, as WEISS assumed (MANLEY, 1970b, 1971; EVANS, 1972b; EVANS and WILSON, 1973).

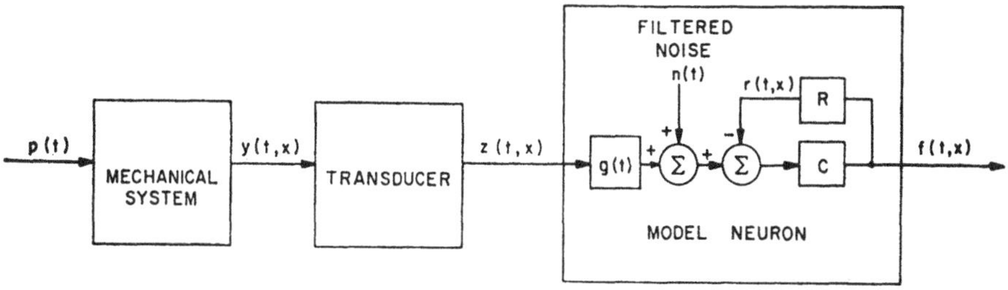

Fig. 47. Compartmental outline of model of WEISS (1966) relating the discharge of cochlear nerve fibres to acoustic stimuli. $p(t)$: pressure at ear; $y(t, x)$: displacement of cochlear partition; $z(t, x)$: output of transducer; $f(t, x)$: sequence of discharges; $r(t, x)$ threshold potential; $g(t)$: linear filter; t: time; x: distance along cochlear partition

The major component of the first compartment is the band-pass filtering characteristic existing at any point along the cochlear partition. The nature of this filter function, and its relation to the form of the neural FTCs continues to be the subject of much experiment and discussion. The similarity between the first successfully measured cochlear fibre FTC (TASAKI, 1954 in the guinea pig) and the so-called resonance curves of the cochlear partition (*e.g.* BÉKÉSY, 1944) has prompted the conclusion that the relatively broadly tuned characteristics of the basilar membrane were handed on to the auditory nervous system relatively unchanged (*e.g.* BÉKÉSY, 1969, 1970; WHITFIELD, 1967, 1968). However, the more recent FTC measurements in cat (KIANG *et al.*, 1965a, 1967) indicated that its cochlear nerve frequency responses were substantially sharper than BÉKÉSY's curves. It has now been shown (EVANS, 1970a, 1972b) that this is in fact true for the guinea pig, so that comparisons can be made (Fig. 9B) between the neural and mechanical tuning in the same species, but more importantly at the same frequencies. For the latter purpose we now have measurements of the frequency response at the basal end of the guinea pig basilar membrane by JOHNSTONE *et al.* (1970) using the Mössbauer technique and by WILSON and JOHNSTONE (1972) using a capacitive probe. Both mechanical and neural functions show an increase in sharpness with frequency, but at all frequencies the two sets of data are substantially different. These differences can be examined quantitatively in Fig. 10 (filled circles versus star symbols). The most conspicuous difference is in the slopes of the low frequency cut-off of the cochlear nerve FTCs, which are steeper than the analogous mechanical values by factors of between 2 and 12 for fibres with CF below 2 kHz, and between 6 and 36 for those above 2 kHz. Above 2 kHz the neural high-frequency cut-offs are 1–7 times as steep as the mechanical. (Slopes here measured are of the cut-off segment between 5 and 25 dB above the tip of the curves). In terms of 10 dB bandwidth (Fig. 10C), the neural FTCs were narrower than the mechanical functions by a factor approaching an order of magnitude at intermediate CFs (3–10 kHz).

Unfortunately, no actual measurements of the guinea pig basilar membrane motion exist at these intermediate regions.

The question arises whether one is justified in making these comparisons between the neural frequency response functions determined at threshold and the mechanical functions determined at much higher sound levels (above, and in most cases well above, 70 dB SPL). This question has received added force from the recent Mössbauer measurements of RHODE (1971, 1973) in the squirrel monkey. While his published slope values for the basilar membrane frequency response of a region subserving frequencies of 6–8 kHz are in substantial agreement with the guinea pig data (large open star in Fig. 10A–C), RHODE found substantial non-linearities in the response at frequencies near the peak of vibration. These were such that, if a constant low amplitude criterion (*e.g.* 0.003 μm, Fig. 7; RHODE, 1971) is used for constructing a frequency response function, then the frequency response function has cut-offs and bandwidth approaching those of cochlear fibres of comparable CF. However, the fact that this derivation depends upon a level dependent non-linearity which was specifically *not* found by BÉKÉSY, JOHNSTONE *et al.*, and WILSON and JOHNSTONE (the latter over SPLs from 40–120 dB at the peak) makes it difficult to interpret. More mechanical data are evidently required in this intermediate frequency region to resolve this discrepancy between the data of RHODE and the other investigations. In this connection, it is appropriate to note that the frequency response inferred from the differential cochlear microphonic study of the guinea pig by DALLOS (1973), is in close agreement with that of WILSON and JOHNSTONE in both relative amplitude and phase. The correspondence between the neural data of Figs. 9 and 10 make a substantial species difference seem unlikely. As far as the question of using threshold criteria for estimating neural tuning is concerned, there is no evidence that isorate contours constructed for cochlear fibres become less sharp as higher rate criteria are used (Fig. 13) at least over the 30–40 dB range available before saturation of the discharge rate is reached (above these levels, there is evidence of a decrease in sharpness of the cochlear filter: see later in this section). Furthermore, the slopes of the high frequency cut-offs of the neural FTCs become steeper at higher levels, whereas the measurements of RHODE, WILSON and JOHNSTONE, and JOHNSTONE (personal communication) have in common an as yet unexplained plateau of the vibration amplitude at about 30–40 dB from the tip on the high frequency side.

HUXLEY (1969) raised the question whether the discrepancy between neural and mechanical measurements could arise as a result of the surgical opening of the cochlea necessary to make the mechanical measurements thus disturbing a mechanical resonant system. However, EVANS (1970b) has shown that normally sharp FTCs can be recorded from cochlear fibres emanating from the region of the guinea pig basilar membrane exposed as in the experiments of JOHNSTONE *et al.* and WILSON and JOHNSTONE.

To reconcile the sharp neural tuning with the relatively broadly tuned properties of the basilar membrane, WEISS inserted a second linear filter into his model, but in the neurone compartment. He used a filter which was essentially low-pass, but with a constant slope of decrease in sensitivity with increasing frequency which effected some improvement in the slope of the low frequency cut-offs. It is clear, however, from the data of Fig. 10, that such a second filter needs to be *band-pass* in its characteristics. This notion of a second, more sharply tuned, filter succeeding a relatively broad basilar membrane filter has received support from a number of sources. Firstly, a feature of the neural FTCs is the considerable variation which exists in the cut-off slope, and bandwidth values, from fibre to fibre even at similar CFs and in the same animal (EVANS, 1972a, b; EVANS and WILSON, 1973). This variation may amount to differences in bandwidth up to a ratio of 4 : 1. Secondly, in the guinea pig (EVANS, 1972a, b) fibres recorded under conditions of cochlear circulatory insufficiency have high threshold, broadly tuned characteristics resembling those of the basilar membrane (small open circles in Fig. 10). In addition, such fibres are occasionally found apparently adjacent to and sharing a common CF (above 12 kHz) with fibres of normal threshold and tuning. These findings led us to propose the existence of a second filter "private" to each cochlear fibre, whose normally sharply tuned properties weie thereby physiologically vulnerable and the differences between which could account for the lack of homogeneity observed in the tuning properties of cochlear fibres. It is difficult to reconcile such properties with any filtering properties of the basilar membrane, which would be expected to the *evenly distributed.* Direct evidence for such a physiologically vulnerable second filter is in fact provided by the demonstration that the sharply tuned segment of the FTC can be eliminated and restored within a few minutes by the action of hypoxia and certain agents (cyanide, Frusemide) on the cochlea *without substantial effects on the cochlear microphonic* (EVANS, 1974b, c, d; EVANS and KLINKE, 1974; vd. p. 18). The time scale of these effects is much shorter, and their magnitude much larger than those that have been observed in Mössbauer measurements of the squirrel monkey basilar membrane vibration *after death* by RHODE (1973), where all normal cochlear potentials would have been eliminated (*e.g.* BUTLER *et al.*, 1962). The third line of support for the notion of a second sharply tuned filter comes from studies of two phenomena which are related to cochlear non-linearities, *i.e.* the combination tone $2f_1 - f_2$ and two-tone suppression. Psychophysical (GOLDSTEIN, 1967, 1970; SMOORENBURG, 1972b) and neurophysiological (GOLDSTEIN and KIANG, 1968) investigations of the $2f_1 - f_2$ cubic difference tone (CDT) indicate that the "essential" non-linearity must be preceded by a frequency selective process (which accounts for the dependence of the level of the CDT on the frequency separation of the primary tones) and also followed by a frequency selective process (to account for the frequency analysis of the CDT to separate it from the primaries). Two possible ways by which this double filtering process may be accomplished have been proposed by the above authors. In the first, the coupling hypothesis, the nonlinear process is tightly coupled to a sharply tuned distributed mechanical frequency analyser (the basilar membrane), so that the CDT generated at the region of overlap of the primaries is fed back to the basilar membrane and analysed at the place appropriate to its frequency. In the second (two-filter) hypothesis, the non-linearity follows a broadly tuned basilar membrane filter, and the CDT it generates is not coupled back to the basilar membrane but is analysed by a separate, more sharply tuned filter. The ingenious study by SMOORENBURG (1972b) of a subject with a sharply defined hearing defect (likely to be of hair cell origin) excludes a third possibility, perhaps implicated in the results of RHODE (1971), in which the non-linearity resides in a sharply tuned basilar membrane. The two filter hypothesis is in fact favoured by the demonstration that no $2f_1 - f_2$ component of the magnitude predicted by the coupling hypothesis is present in the basilar membrane vibration pattern, according to the differential cochlear microphonic data of DALLOS (1969) and the direct measurements of basilar membrane vibration by WILSON and JOHNSTONE (1972, 1973), using a capacitance probe technique. Lastly, PFEIFFER (1970) has demonstrated that such a two filter model can also account, qualitatively at least, for the characteristics of the two tone suppression encountered in the cochlear nerve (Section II.E.1.a).

In spite of the presence of these cochlear non-linearities, there is now a good deal of evidence that, for a wide variety of stimuli, the sharply tuned cochlear filtering process(es) act as if they were linear, at least over a 30–60 dB range above threshold. This holds for wideband noise assessed by two different methods (Section II.E.2), for click stimuli (Section III.F.3), and for comb-filtered noise (multicomponent spectra: Section II.E.3), and in each case the effective bandwidth of the filtering approximates to that of the pure tone FTC. As will be noted later (Fig. 48; Section VI.C.1) the magnitude of the effective bandwidth of the cochlear filter corresponds to the analogous psychophysical measure, the critical band. GOLDSTEIN *et al.* (1971) have made an interesting comparison of cochlear nerve response times estimated from click responses and pure tone data with those computed on the basis that the cochlear filter responsible for the FTC was a minimum phase linear filter. The tolerable degree of agreement led them to conclude that the cochlear nerve FTC, phase and click data were interrelated in a manner similar to that which would be predicted for a linear filter system. The important point that they make is that the response time (which is approximately the delay to maximum response of the system) of any *linear* filter system *cannot be less* than that for the equivalent minimum phase filter. In the case of their own (cat) data it appears that the agreement was obtained only by neglecting the finite synaptic delay and neural conduction time. The former could be about 0.5 msec, and the latter, from the latency of direct electrical stimulation of the cochlea (KIANG and MOXON, 1972) 0.5 msec also. Taken at face value, this then suggests that the response of the cochlear filter is faster than that of a minimum phase network by approximately 1 msec. If this is the case, it is serious evidence against the sharply tuned second filter being linear. On the other hand, the comparison of GOLDSTEIN *et al.* (1971) may be rendered inaccurate if, as appears to have been the case, the tone phase and click data from which the estimates of response time were derived, were obtained at high stimulus levels. As Fig. 6 demonstrates, the response time indicated by the click PST histogram becomes substantially shorter at higher levels. In other words, comparisons of response times computed from the frequency threshold curves should be made with estimates derived from click and tone phase data within 20–30 dB of threshold.

The sharp filtering characteristics of cochlear fibres appear to remain relatively unchanged over at least an intensity range of 30–60 dB above threshold (see Section II.D.1; II.E.2; III.F.3; see also EVANS and WILSON, 1973). At higher levels, however, there is some indication that the cochlear filter becomes nonlinear, or less sharply tuned, or both. The increasingly nonlinear behaviour with stimulus level of cochlear fibres in respect of two tone interactions has already been noted (Section II.E.1.c). In addition, the investigation by GOBLICK and PFEIFFER (1969) of the extent of cochlear linearity (by an ingenious method involving the nulling of individual peaks of the cochlear nerve PST histogram in response to a click by a second click stimulus) has demonstrated the existence of an amplitude non-linearity increasing from zero to about 6 dB over the upper 20 dB of the 40 dB dynamic range tested (probably representing levels of 40–60 dB above threshold). Furthermore, the fact that fewer than expected later peaks appear in the click PST histogram with increase in stimulus levels (*e.g.* 30–70 dB in Fig. 6, where the scale is constant) has been interpreted as an indication of a decrease in the system's "Q" at higher levels (Pfeiffer and Kim, 1972; compare response of Weiss' (1966) model). The progressive shift of the mode of the PST histogram to the left with increase in stimulus level (from ca. 8 msec at +10 dB to ca. 3 msec at +70 dB above "threshold" in Fig. 6), implying a decrease in the "response time" of the filter, would also appear to support this conclusion. An increase in damping of the cochlear filter with stimulus level has also been proposed by Anderson *et al.* (1971) as a possible explanation of the changes in phase of cochlear fibre discharge with level of low frequency tone stimulation (Section II.D.4). All of these findings may, of course, be consistent with the basilar membrane vibration patterns observed by Rhode (1971; see discussion by Pfeiffer and Kim, 1973), but, as discussed above, the interpretation of his data remains a difficult question.

Many mechanisms have been proposed to account for the sharpening of the basilar membrane frequency response discussed above. These range from mechanical processes (*e.g.* "beam" hypothesis of Huggins and Licklider, 1951; longitudinal shear model of Tonndorf, 1962, Khanna *et al.*. 1968; fluid-hair cell coupling, Steele, 1973), through innervation pattern models (*e.g.* Nieder, 1971), positive feedback models (Gold, 1948), to lateral inhibition (*e.g.* Békésy, 1960, 1970; Furman and Frishkopf, 1964). Unfortunately, none of these models appears to account quantitatively for the extent of sharpening required and at the same time is consistent with the available physiological and anatomical data. Evidence against lateral inhibition as a mechanism in the cochlea has been presented above (Section II.D., II.E.1.a, II.E.3, III.F.3). Gold's model, involving a positive feedback mechanism, has some attractions in relation to the physiologically vulnerable second filter hypothesis and the mechanisms underlying tone-like tinnitus, in spite of it being based on estimates of cochlear "Q" which were incorrect by over an order of magnitude and which were derived from incorrectly interpreted data (vd. Hiesey and Schubert, 1972). Another class of second filter type mechanism, as yet not explicitly treated, involves some form of active "resonance" of electrochemical nature within the hair cells themselves. Such a mechanism, however, would appear to have too high a noise limitation set by Brownian motion for a system loosely coupled to the basilar membrane or other structures, if the arguments of Harris (1968) can be applied to this situation.

Clearly, further experimental data on the properties of the basilar membrane and of the hypothetical frequency sharpening process are required before further assessment of these and other models is appropriate.

2. Transducer Nonlinearity

The second compartment of Weiss' model incorporated a non-linear no-memory transducer. The non-linearity was invoked to account for the greater number of peaks encountered in the cochlear nerve PST histogram in response to a click, compared with that computed from a model based on BÉKÉSY's basilar membrane data. (In fact, data from recent measurements of basilar membrane motion (ROBLES *et al.*, 1973; WILSON and JOHNSTONE, 1972) indicate that the (more basal) cochlear partition is less damped than was thought by BÉKÉSY.) DE BOER (1969) first suggested that such a nonlinearity could be dispensed with if the linear cochlear filter was made as sharp as that required to account for the FTC. The finite "response time" (see GOLDSTEIN *et al.*, 1971; above) of such a filter could also account for the appearance of earlier peaks in the PST histogram on raising click levels above threshold, which were inexplicable on WEISS' model.

Two types of cochlear non-linearity, both of which have received some attention above, may well, however, reside in the transducer stage. The first is the so-called "essential" nonlinearity (ZWICKER, 1955; GOLDSTEIN, 1967, 1970) responsible for generation of the $2f_1 - f_2$ combination tone (CDT). Psychophysical measurements of this CDT indicate that it is generated at stimulus levels close to threshold, and that its level is proportional to the level of the primaries (GOLDSTEIN, 1967; SMOORENBURG, 1972a; see Section II.E.1b). This nonlinearity is therefore not a level-dependent saturating type of non-linearity (such as that responsible for the difference tone, $f_2 - f_1$, heard only at much higher stimulus levels, GOLDSTEIN, 1967). It could also be responsible for the generation of the two-tone suppression phenomenon observed in the cochlear nerve, the effects of which are also observable close to threshold (Fig. 22; Section II.E.1.a).

The second type of non-linearity is that discussed in the section above, responsible for the deviation of certain cochlear nerve properties from linear behaviour at stimulus levels higher than 40–60 dB above threshold (*e.g.* Section II.E.1.c).

3. Generation of Cochlear Nerve Discharges

In the third compartment of WEISS' model, the transducer output was added to internally generated low-pass filtered noise to produce a noisy membrane potential which, whenever it exceeded a threshold, triggered a neuronal spike. Immediately after each spike, the threshold was reset to some higher value (by the component R in Fig. 47) from which it decayed to a resting value. These features were incorporated to account respectively for the spontaneous discharge and probabilistic suprathreshold behaviour, and the refractory period exhibited by cochlear fibres.

A threshold trigger mechanism was also incorporated in the models of GEISLER (1968) and DE BOER (1969) which modified the overall model of WEISS to account for cochlear nerve behaviour in response to low frequency tones and broadband noise. A problem with all these threshold trigger type models is that they fail to describe adequately the responses of cochlear fibres to high level click and low frequency tone stimulation (DE BOER, 1969; GEISLER, 1968; DUIFHUIS, 1972). In particular, at high stimulus amplitudes, the effects of the internal noise would be expected to become less significant, and the temporal "jitter" observed in the PST histograms of activity evoked by clicks (*e.g.*, Fig. 6) and tones (*e.g.*, Fig. 18) should show a tendency to disappear, which is manifestly not the case. Similarly the phase of discharge should correspond increasingly with the hypothetical zero crossing at higher levels, which again does not occur with noise (Fig. 25) or with tones at the CF of a cochlear fibre (Fig. 18).

For these reasons, Siebert (*e.g.*, 1965, 1970) and Duifhuis (*e.g.*, 1972) have instead utilised in their models a stochastic spike generator where the probability of discharge at any instant is a *continuous* function of the unidirectional amplitude of the transducer output (*i.e.*: there is no threshold trigger level). The properties of adaptation and saturation of discharge can either be incorporated in the mathematical relation connecting the discharge probability to the transducer output waveform (Siebert), or, as in the case of refractoriness, as a multiplicative feedback from the spike generator output to the input to the Poisson probabilistic generator (Duifhuis, 1972).

An outcome of both classes of model is the effective rectification of the stimulating waveform, which is demonstrated most clearly in response to click, tone, and noise stimulation in Figs. 6, 16B, 17, 18, 24, 25, all of which show a discharge pattern more clearly correlated with the stimulus waveform (stochastic model) than a threshold crossing (threshold trigger) model.

Duifhuis' (1972) model incorporates a linear filter with a finite "response time" (consistent with the cochlear nerve FTC properties) to precede the Poisson process, and accounts satisfactorily for the statistics of cochlear fibre spontaneous and tone evoked discharges (*e.g.* Fig. 2), for interspike interval and period histograms of the response to low frequency tones (*e.g.* Figs. 19 and 18 respectively), and for PST histograms to click stimuli (including the shift of the centre of gravity to the left with level introduced in the model by a reduction of filter "Q" at higher stimulus levels).

The incorporation of a linear filter with finite "response time" in Duifhuis' model (as in Goldstein *et al.*, 1971) eliminates the need for some form of feedback control, as in the model of Lynn and Sayers (1970), to account for the apparent "suppression" of the early peaks of the click evoked PST histogram.

All of the models referred to above may have to be revised in the light of very recent findings which run counter to the generally accepted conclusion that the spike generator discharged only in response to "upward" movements of the basilar membrane, *i.e.* towards the scala vestibuli, (*e.g.* in response to the predominant component of a rarefaction click, Kiang *et al.*, 1965a). Thus, under low frequency tone (Kiang and Moxon, 1972) or noise (de Boer, 1969) stimulation, different cochlear fibres exhibit differing phases of discharge inspite of their having a similar CF. More strikingly still, Konishi and Nielsen (1973) claim to have found a majority of guinea pig cochlear fibres which were excited by static "*downwards*" displacement of the basilar membrane (*i.e.*: toward the *scala tympani*), a minority whose spontaneous discharge was inhibited by the same displacement, and a substantial proportion of fibres which were excited during dynamic changes of membrane position. It is unlikely that these differences can be accounted for by different hair cells having different "polarities" to movement of their hairs, in view of their basal bodies being oriented in the same direction (see Flock, 1971). Perhaps the differences may arise from different modes of bending of the hair cells, as suggested by the recent findings of Dallos *et al.* (1972), who infer that the outer hair cells (mainly responsible for generating the cochlear microphonic) are sensitive to displacement, while the inner hair cells are sensitive to the velocity of endolymphatic flow (consistent with the anatomical evidence which suggests that, the inner hair cells, unlike the outer hair cells, are *not* attached to the tectorial membrane, *e.g.* LIM, 1972). At least, all of these findings underline our almost complete ignorance of the processes lying between the displacement of the basilar membrane and the generation of nerve impulses in the cochlear nerve (see also FEX, volume V/1).

B. In Relation to Cochlear Innervation

The question remains open as to what part the complex innervation patterns of the cochlea might take in the mechanisms discussed in the last section. There is good evidence that the probabilistic generator responsible for the "jitter" of discharge to low frequency tones (vd. Section II.D.4) and which is sensitive to non-excitatory outputs of the transducer, thus mediating static (vd. above) and dynamic (vd. Section II.C) depression of spontaneous activity, is located peripheral to the generation of action potentials, and, therefore, presumably in the hair cell-synapse area. Similarly, the process underlying the saturation of discharge rate is not likely to be in the nerve fibres, in view of KIANG and MOXON'S (1972) observation that direct electrical stimulation of the cochlea (and therefore presumably of the cochlear fibre initial segment) could evoke higher rates of discharge than could be evoked by acoustic stimulation. In keeping with this, the passage of direct current through the cochlear partition augments the "spontaneous" and "evoked" activity without affecting the maximal (saturation) discharge rate (KONISHI *et al.*, 1970). Perhaps the most economical hypotheses are to locate the transducer process in the hair cell (as in the generally accepted hypothesis of DAVIS, 1965, for CM), and the probabilistic generator in a chemical transmission process in the synaptic junction. Refractory processes could be located in the nerve itself. Unfortunately, chemical transmission has been confirmed only in the fish sacculus, in an important series of experiments by ISHII *et al.* (1971) and by FURUKAWA *et al.* (1972).

One problem which has attracted considerable attention is the rôle played in cochlear nerve responses by the outer spiral fibres innervating the outer hair cells (OHC). The OHC have been held responsible for the low thresholds of cochlear fibres in contrast to the presumed higher thresholds of the inner hair cells (IHC), inferred from the assumed relative mechanical disadvantage of the latter. However, the detailed electron microscopic studies of SPOENDLIN (1966, 1968, 1970, 1971, 1972) in the cat indicate that no more than 5% of the the fibres leaving the cochlear partition originate in the OHC. In other words, the great majority of fibres recorded by microelectrodes in the cochlear nerve must innervate the IHC. Yet it is these fibres which have uniformly low thresholds approximating to behavioural threshold (Fig. 8). Apart from the very recent description by PFEIFFER and KIM (1972) of "anomalous" PST histograms to click stimuli in 7% of their cat cochlear fibre population (where the number of peaks increased dramatically up to 34 with increase in level), there is no real evidence for the separation of the cochlear nerve into two fibre populations, particularly on the basis of threshold and tuning properties (see discussion of cochlear nerve homogeneity in Section II.D). This is in spite of assertions to the contrary from observations on threshold tuning and rate-intensity properties of cochlear fibres (see detailed discussion in Section II.D.1,2) and from studies of the gross cochlear AP (see detailed discussion in Section IV).

On the other hand, there is some evidence suggesting that the cochlear fibres innervating the IHC may receive interaction from outer spiral fibres or actually be dependent upon the latter for their normal function. Firstly, KIANG *et al.* (1970), recording from the cochlear nerve of cats poisoned with an ototoxic antibiotic, were unable to obtain normal threshold responses in regions of extensive OHC loss from fibres innervating IHC which were (to light microscopy at least) normal in appearance. Secondly, as has been pointed out in Section V.A, maximal crossed OCB inhibition of cochlear fibres occurs in fibres whose CFs correspond with the region of maximal density of efferent innervation of the *OHC*. In contrast, the density of efferent innervation of the IHC afferents appears to be more evenly distributed (ISHII and BALOGH, 1968). Consequently, it has been concluded (WIEDERHOLD, 1970; TEAS *et al.*, 1972) that the effectiveness of crossed OCB stimulation corresponds more closely with the pattern of efferent innervation of the OHC than the IHC, (although it must be admitted that the data on the efferent distribution to the IHC afferents are far from complete). In fact, whereas the activity of nearly all cat cochlear fibres can be suppressed by crossed OCB stimulation (FEX, 1962; WIEDERHOLD and KIANG, 1970), it has been reported that in the rat the whole crossed OCB innervation goes to the OHC (IURATO, 1964). Recent anatomical studies of the cat cochlea (SPOENDLIN, 1970), however, suggest that at least part of the IHC efferent innervation comes from the crossed OCB.

If the fibres innervating the IHC do depend upon the integrity of the OHC for their normal function, this could relate the finding of occasional high threshold high frequency cochlear fibres in otherwise normal cochleas (Section II.D.1; EVANS, 1972b), to the presence of sporadic OHC loss in normal guinea pigs (WERSÄLL, personal communication).

In the absence of signs of synaptic contact or branching (*contra* NIEDER and NIEDER, 1970) between outer spiral and inner radial fibres, the above considerations lead to the speculation that interaction between one of the former and 20 of the latter takes place in each habenular opening, where they become intimately apposed (SMITH and DEMPSEY, 1957; SPOENDLIN, 1970; ARTHUR *et al.*, 1971; LYNN and SAYERS, 1970). Considerations of the length and diameter of the outer spiral fibres appear to rule out electrotonic conduction (*contra* LYNN and SAYERS, 1970), and lead to the suggestion that action potentials propagated along them could produce sufficient local current to generate action potentials at the inner segment region of the inner radial fibres. On this basis, and considering for a moment that the OHC are sharply tuned, whereas the IHC mirror the basilar membrane motion, the low threshold sharply tuned properties of cochlear fibres would be accounted for by OHC activity, whereas the higher threshold, more broadly tuned, and non-linear neural properties would depend on IHC activity. Under conditions where OHC function was impaired, broadly tuned, high threshold FTCs would result (as in Section II.D.1 and from exposure to intense noise and ototoxic agents). Furthermore, the fact that each outer spiral fibre innervates OHC in a higher frequency region than that of the inner radial fibres with which it comes into contact could account for the asymmetry of the upper section of the normal FTC and for the relatively greater effectiveness of low frequency compared with high frequency tones at high levels (see Section II.E.1.c). Similarly, it could account for the shift of the CF towards lower frequencies which accompanies the loss of the sharply tuned low threshold segment of the FTC during the action on the cochlea of hypoxia (EVANS, 1974b, c, d) and certain ototoxic agents (EVANS and KLINKE, 1974; EVANS, 1974c, d)

Unfortunately, this speculation is not supported by the close correspondence between CM and basilar membrane filtering properties (Section VI.A) if as DALLOS (1973) suggests, the OHC are almost entirely responsible for the CM generation; and by the fact that the period of the click evoked PST histogram does not show signs of increasing with stimulus level (*e.g.* Fig. 6).

An alternative possibility is interaction, at the IHC level, of local currents generated by the OHC. These are not likely to be in phase with the IHC currents (DALLOS *et al.*, 1972). If sufficient phase differences could exist between these IHC and OHC "currents", the otherwise curious abrupt reduction in cochlear nerve spike discharge (see Section II.D.2) and in cochlear microphonic (KARLAN *et al.*, 1972) at high stimulus levels could be accounted for in terms of cancellation in the region of overlap of the lower threshold OHC and high threshold IHC "currents".

The participation, in cochlear nerve excitation, of two different excitatory processes (KIANG and PEAKE, 1960) or of the local efferent innervation (NIEDER and NIEDER, 1970) has also been suggested. Clearly the need is for more data, before the roles of the IHC, OHC, their innervations, and that of other structures such as the giant fibres linking 10 or so IHC (SPOENDLIN, 1971, 1972), can be established.

C. In Relation to Psychophysical and Behavioural Phenomena

1. Critical Band and Other Measures of Frequency Selectivity

Figure 48 plots a number of psychophysical and behavioral measures of frequency selectivity in comparison with the effective bandwidths of cat cochlear fibres (EVANS and WILSON, 1973). "Effective bandwidth" is approximately the half-power bandwidth of the cochlear fibre filter function expressed by the FTC (and defined and measured exactly as in the legend of Fig. 26). It should be emphasised that by frequency selectivity we do not mean frequency discrimination – the ability to distinguish a change or difference in frequency. What is meant is the ability of the ear to select or separate information arriving at one frequency from that at another. This is expressed in the "critical band", which at any frequency is about 30 times wider than the difference limen for frequency discrimination (ZWICKER, 1970). The critical band expresses the frequency band: within which simultaneously sounded tones cannot be separated (PLOMP, 1964b; MCCLELLAND and BRANDT, 1969); within which power summation occurs for threshold measurements in the presence (GREENWOOD, 1961) and absence (GÄSSLER, 1954) of masking noise; within which loudness is determined by the summed power of the components (ZWICKER *et al.*, 1957); within which tones are effective in masking a narrow band of noise centred between them (ZWICKER, 1954); within which the effects of the phase relations between components of modulated tones can be detected (ZWICKER, 1952); and within which "roughness" arising from a complex of tones can be detected (TERHARDT, 1968, 1970). The critical band can thus be understood by analogy with the bandwidth of a linear filter (GREENWOOD, 1961; ZWICKER, 1971). That this measure (the critical band) determined in man corresponds tolerably well over a wide frequency range with the analogous measure for individual cat cochlear fibres (Fig. 48) leads to the conclusion that the ear's critical band property derives directly from the filtering properties of cochlear

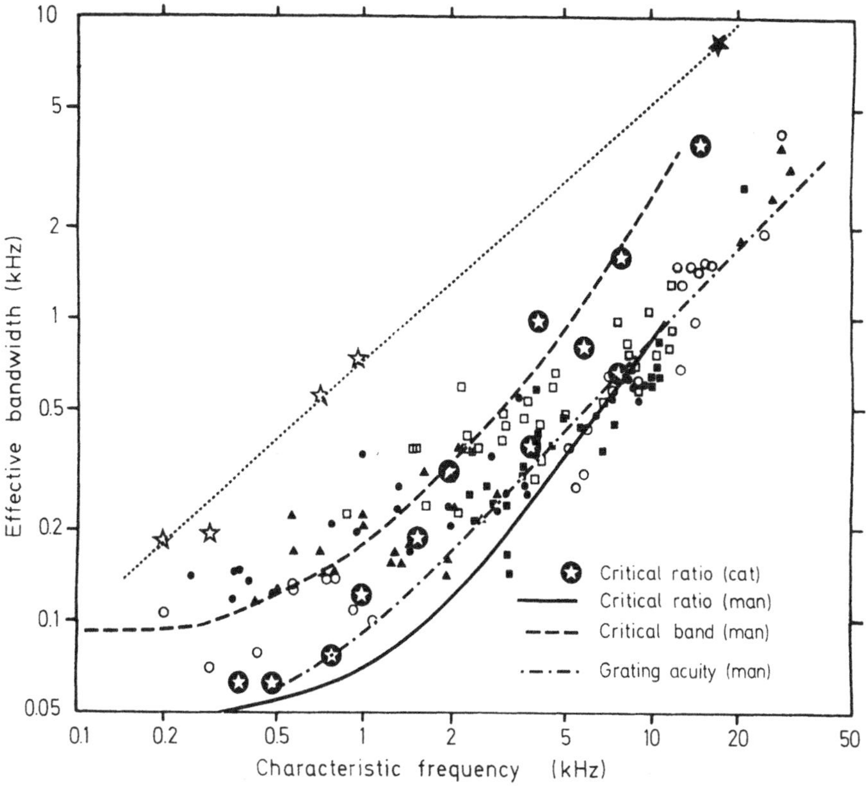

Fig. 48. Comparison of effective bandwidths of cat cochlear fibres with human psychophysical and cat behavioural measures of frequency selectivity. Circle, triangle, and square symbols: effective bandwidths computed for 140 cochlear fibres from 5 cats, plotted against their CFs. Dotted line: effective bandwidths computed for basilar membrane response curves of BÉKÉSY (1944; open stars) and JOHNSTONE *et al.* (1970; solid star). Dashed line: critical band values for man (ZWICKER *et al.*, 1957). Solid line: critical ratio values for man (HAWKINS and STEVENS, 1950). Dotted and dashed line: resolving power expressed as effective bandwidths for acoustic grating measurements (WILSON and EVANS, 1971). Enclosed stars: behavioural measurements of critical ratio in cat (WATSON, 1963). (From EVANS and WILSON, 1973)

fibres and is therefore already determined at the cochlear level (EVANS and WILSON, 1973). This conclusion is supported by the correspondence between direct measurements of the frequency resolving power of cat cochlear fibres (vd. Section II.E.3) and of human subjects for acoustic "gratings" *i.e.*: comb-filtered noise (WILSON and EVANS, 1971; EVANS and WILSON, 1973). These measurements determine the limit of the ability of single cat cochlear fibres and of human subjects (in psychophysical experiments) to resolve between peaks and valleys of the comb-filtered noise spectrum at successively closer spacings (vd. Fig. 27 and text). The effective bandwidths derived from the psychophysical measurements are shown as the dot-dash curve in Fig. 48, although there are indications that the values below 2 kHz may be too low (WILSON and SEELMAN, data to be published), and therefore may agree better with the cat neural data.

The cochlear nerve effective bandwidths represent a filter "Q" (*i.e.*: CF/effective bandwidth) of about 10 from 1–10 kHz (Fig. 48). This figure agrees well with that derived by DUIFHUIS (1971) from measurements of the peripheral filtering characteristics of the human ear.

A possible objection to comparison between cat neurophysiological and human critical band data arises from the one existing series of behavioural measurements of an analogous function, the critical ratio (WATSON, 1963; enclosed stars, Fig. 48). In man, this function is approximately 2.5 times smaller than the critical band (continuous versus dashed lines, Fig. 48; ZWICKER *et al.*, 1957), hence the values of critical band for the cat might be expected to be well above the points representing the measured cochlear nerve effective bandwidth. However, SEATON and TRAHIOTIS (1973) have recently suggested that such behavioural indications of poorer frequency resolving ability in animals may in fact not be the case, but may be related to the high criterion adopted in the training paradigms.

The correspondence proposed between the critical band measures and the effective bandwidth of cochlear fibre FTCs would appear to account for all of the characteristics of the critical band so far mentioned. In addition, it would account: for the findings which indicate that the critical band does not require any significant amount of time to be established apart from the "response time" of the cochlear filter (ZWICKER and FASTL, 1972) because the cochlear filter characteristics hold for click stimuli also (Section III.F.3); for the increase in the critical band with stimulus level (corresponding to the inferred decrease in "Q" of the FTCs at levels 40 dB and above minimum threshold, vd. Section VI.A.1); and for the frequency asymmetry of masking (*e.g.* SMALL, 1959; SCHARF, 1971).

2. Masking

While the arguments of the previous section enable an understanding to be made of the energy and frequency relations which exist between masker and maskee signals, the mechanisms responsible for the actual elevation of maskee threshold are less clear. At least four mechanisms have been proposed to explain how peripheral auditory neurones are prevented from transmitting or processing information that would be adequately dealt with in the unmasked state. (That masking is not primarily a cochlear mechanical effect is suggested by its absence in the cochlear microphonic, *e.g.* DERBYSHIRE and DAVIS, 1935.) On the other hand, it must be emphasised that explanations based on peripheral mechanisms may be inappropriate or at best incomplete for high levels of masking, in view of the phenomenon of binaural unmasking (first extensively described by HIRSH, 1948), where a monaural tone signal, masked completely by a noise signal in the same ear, can be rendered *audible* by the presentation of a correlated noise signal to the other ear. This reduction in masked threshold can amount to 10–20 dB under certain conditions of interaural phase of tone and masking noise signals.

The earliest explanation of masking was a "*line-busy*" effect (DERBYSHIRE and DAVIS, 1935). Here, a neurone is prevented from responding to the maskee stimulus because its discharge rate has already saturated due the masker, or because of refractory effects from the latter stimulus. While there is no evidence for the latter, the "masking" of tone and click responses of the cochlear nerve by noise stimuli mentioned in Section II.E.4 does appear to represent the inability of the fibre to discharge at any greater rate.

The second, and most compelling explanation of masking, that of stimulus *dominance*, is able to deal with stimulus levels below those producing saturated discharge rates (GREENWOOD, 1961; GOLDBERG and GREENWOOD, 1966). In this case, the response of a neurone is dominated by the more effective stimulus, in the sense that the response to the two stimuli is virtually identical to that to the more effective (masking) stimulus alone. For stimuli of similar frequency, this effect will arise because the response of cochlear nerve and nucleus units (of appropriate CF) is related to the *logarithm* of the combined stimulus power (Section II.D.2, III.E.2), hence the addition of a second stimulus of equal power does not double the response but only increases it by the equivalent of an extra 3 dB of stimulus level; similarly, the addition of a second stimulus 10 dB below the first, however effective on its own, will have a negligible effect in combination, given a noisy system which has to detect increments in overall signal intensity. For masking stimuli displaced in frequency from the CF of a unit, the situation can be complicated by the two tone suppression phenomenon, but the net effect is the same as in the case above. Thus, the pattern of response (mean rate and temporal statistics) is dominated by the more effective component of the two stimuli whether the addition of the latter produces an increase, decrease, or no change in the discharge rate (vd. Section II.E.1.a, c; III.F.1, 2, 5).

Thirdly, at the cochlear nucleus level, a masking signal can inhibit the response to a test signal if it is situated in the inhibitory side bands (Section III.E.1, III.F.2, 5). The fact that the inhibitory side bands often extend over a wider frequency range compared to the excitatory band suggests that a wide range of masking frequencies could prevent any response to an exciting tone arising in the population of cells with such side-bands (GALAMBOS, 1944; GREENWOOD and MARUYAMA, 1965).

Fourthly, evidence obtained from studies of the cochlear AP response, and the "frequency following" gross response of the CN, suggests that masking represents a disruption of the synchronization of individual neurone activity to the masked stimulus (BONE *et al.*, 1972; MARSH *et al.*, 1972). While it is the case that the statistics of the activity of individual neurones in the CN are dominated by the more effective stimulus of a stimulus pair, this cannot serve as an explanation for masking of responses with masker and maskee at frequencies higher than 4 kHz (observed even at the AP level FINCK, 1966).

The fact that broadband noise masking is restricted by critical band considerations suggests that the second of the above mechanisms is the most likely one to be involved. There is evidence that lateral inhibition is not involved when broadband signals are used (vd. RAINBOLT and SMALL, 1972). In the case of tone on tone masking however (*e.g.* SMALL, 1959), lateral inhibitory effects could well be involved, at the level of the CN and above.

The above considerations apply to simultaneous masking. A good deal of interest has been aroused by the non-simultaneous psychophysical masking procedures of HOUTGAST (1972) which appear to produce definite evidence for lateral inhibitory effects. These effects, however, are unlikely to be related to cochlear nerve activity but may represent temporal facilitation and inhibition due to the different latencies of responses at different frequencies at higher levels of the auditory system. WATENABE and SIMADA (1971) were able to find evidence of single unit behaviour corresponding to forward masking in the CN; whereas evidence of backward masking effects were only found in the inferior colliculus (vd. Section III.F.5).

ZWISLOCKI (*e.g.* 1971) has shown that a number of aspects of central masking (*i.e.*: with binaural stimuli) can be related to the behaviour of CN neurones.

3. Combination Tones

The most prominent combination tone, particularly at low stimulus levels, is that corresponding in frequency to $2f_1 - f_2$. This has been discussed above in terms of cochlear nerve activity (Section II.E.1.b) and cochlear mechanisms (Section VI.A.1, 2). It is only necessary here to emphasise that, except in one respect, the psychophysical properties of this combination tone correspond to the behaviour of single cochlear fibres to the same stimulus combination. In terms of mean discharge rate and synchronization in relation to the level and frequency separation of the primaries, cochlear fibres of CF corresponding to the combination tone frequency respond as if there was a component of frequency $2f_1 - f_2$ actually present in the stimulus. The possible site of generation of the CT is discussed in Section VI.A.1, 2. Arguments against the suggestion that the effects are produced simply as a result of the rectifier action of the cochlear nerve spike generator are given in Section II.E.1.b. Further work is necessary to account for the significant contrast between the phase behaviour with level of the combination tone observed psychophysically and the absence of such effects at the cochlear nerve (Section II.E.1.b).

4. Sensorineural Deafness

The more recently obtained cochlear nerve fibre data have allowed a revision of hypotheses on the mechanisms underlying the perceptual deficits produced by pathological conditions of the inner ear (vd. EVANS, 1972b; 1974c).

Formerly, it was held (and still is, from misinterpretations of AP data, *e.g.* PORTMANN *et al.*, 1973, *cf.* Section IV) that the phenomenon of *loudness recruitment* (which is associated with threshold elevation of cochlear origin, DIX *et al.*, 1948), arose because of a selective loss of activity in lower threshold cochlear fibres (assumed to originate in the outer hair cells) out of a population of fibres with widely differing thresholds (*e.g.* KIANG *et al.*, 1965a). The loss was related in turn to outer hair cell damage. On the contrary, it has now been established that most cochlear fibres innervate the inner hair cells (Section VI.B.), and, in the normal animal, all have uniformly low thresholds consistent with behavioural threshold (Fig. 8). Furthermore, the data from guinea pig cochleas with evidence of circulatory insufficiency (Section II.D.1) and from cats poisoned with ototoxic antibiotics (KIANG *et al.*, 1970) indicate that hair cell damage causes an elevation in the threshold of the *same* fibres, and in addition, a substantial broadening of their FTCs. It is as if there was a selective loss of the more sharply tuned, lower threshold segment of the FTC to leave the high threshold wide-band segment (*cf.* Fig. 9). Under conditions where the FTCs of a fibre population are abnormally broad, a correspondingly greater proportion of fibres will be recruited for a given increase in suprathreshold tone level than when the fibres are normally sharply tuned. If the sensation of loudness is related (among other factors) to the proportion of fibres active, then loudness should grow at a greater rate under conditions where the fibres have abnormally broad frequency response properties than under normal conditions (EVANS, 1972b). There is in fact evidence of abnormally broad critical bands in patients with cochlear deafness (*e.g.* DE BOER, 1959; SCHARF and HELLMAN, 1966). If, as appears to be the case from Fig. 10 of KIANG *et al.* (1970), the properties of the cochlear fibres innervating apparently normal inner hair cells are affected by loss of outer hair cells, the additional assumption of some form of interaction between inner and outer hair cells and/or their innervations is required (see Section VI.B.).

It is possible to speculate further that the impaired speech perception which patients with cochlear deafness possess even after correction for the elevated thresholds may reflect the inability of their abnormally broadly tuned cochlear fibres to resolve (separate) the formant frequencies of the speech signal (EVANS and WILSON, 1973).

D. In Relation to the Coding of Acoustic Stimulus Parameters

1. Frequency

Abundant quantitative data are now available on the peripheral auditory nervous system to indicate that it carries information which could be analysed by either "place" or "periodicity" mechanisms for the coding of stimulus frequency. In order to discover the relative role of such putative codes, it is of course necessary to determine whether the information they carry can be utilized at higher levels of the auditory system (*e.g.* see WHITFIELD, 1970). On the other hand, it can be instructive to examine some of the possible obstacles to the acceptance of presently formulated "place" and "periodicity" coding theories set by the properties of the auditory periphery.

Naively, it may be throught implicit in the discussion above that if cochlear fibre FTCs can account quantitatively for frequency selectivity (critical band) they cannot, as filters, account for our acuity of frequency discrimination, which is some 30 times finer. The problem appears to be analogous to that between grating and vernier acuity in vision. However, slopes of the FTCs, and particularly the high frequency slopes are sufficiently steep that at least the low frequency border of the activity in a tonotopic array of cochlear fibres would be very sharp. Furthermore, this border should become sharper at higher tone levels (as the high frequency slopes become steeper: see Fig. 9) and could therefore account for the observed improvement in the difference limen. In fact, SIEBERT (1968, 1970) has estimated that an "ideal observer", having access to this sort of information from an array of fibres with the threshold, FTC, and rate-intensity properties measured for cat cochlear nerve fibres, would have a difference limen for frequency approximating to that observed psychophysically. ZWICKER (1970) has similarly obtained a correspondence between a difference limen estimated from the shape of the cochlear "excitation function" (inferred from psychophysical masking measurements) and that observed psychophysically.

Perhaps the greatest problem facing an exclusive place theory is the limited dynamic range implicit in the cochlear nerve and nucleus data. Not only is the increase in mean discharge rate of a single neurone limited to an intensity range of less than 50 dB (Section II.D.2, III.E.2), but the distribution of thresholds of different neurones at any one CF in an animal is less than 20 dB (Section II.D.1, III.E.1). While this is not necessarily a problem for pure tone discrimination (where the low frequency border of an active fibre array will remain sharp, and the high frequency border could still convey intensity information, as in the scheme of ALLANSON and WHITFIELD, 1956), it is not clear how the frequency components of complex waveforms such as speech could be separated. At average speech levels, the energy in a critical bandwidth is 50–60 dB above threshold for frequencies between 0.5–2 kHz. This means that most fibres will be saturated, and differences of energy between formants could not be signalled, nor presumably the formants themselves coded in separable groups of fibres (KIANG and MOXON, 1972). This problem of course rests on the contrast between human perceptual performance and that of anaesthetized cat cochlear fibres, where the activity of middle ear muscles and efferent nervous system are inoperative. However, the results described in Section V suggest that any effects of the latter would be to broaden rather than to sharpen the frequency selectivity of the cochlear fibres. The studies of SIMMONS and LINEHAN (1968) and of RUPERT *et al.* (1963) of cochlear fibres in awake cats do not provide any further clues.

It is clear from the data described in Section II.D.4 that the phase-locking properties of cochlear fibres are not limited in this way. Phase-locking to tonal stimuli can be observed for 10 dB or so below the mean rate "threshold" and this is retained at levels in excess of 90 dB SPL (Fig. 18). On the other hand, SIEBERT (1968, 1970) has shown than "an ideal listener" who utilizes all the time clues transmitted in his cochlear nerve (inferred from the cat's nerve) would have a difference limen two orders of magnitude smaller than that actually observed. But as the relationship of the limen to many stimulus parameters would also be substantially different a complicated type of inefficiency would have to be involved. There are however, a number of more serious problems for an exclusive periodicity theory. Firstly, the ability of cochlear nerve and nucleus discharges to retain periodic information concerning tones appears to be lost above 4–5 kHz. Furthermore, while the degree of neural phase locking improves with *decrease* in frequency (see Fig. 19), there is no such improvement in the difference limen; on the contrary dF/F *increases* with frequency reduction below about 0.5 kHz (but this could be expected on considerations of the high frequency slopes of FTCs). Secondly, from the data of ROSE and colleagues (vd. Section II.E.1.c), it does not look as if periodicity information in the discharge patterns will disappear for combinations of tones separated in frequency by less than the critical band; whereas the ability of the ear to "hear out" (PLOMP, 1964b) or match in pitch (McCLELLAND and BRANDT, 1969) the individual frequency components is lost. Thirdly, as ROSE *et al.*

(1969) point out, the time structure of cochlear fibre discharges does not correlate well with the perception of combination tones. In particular, apart from the fundamental of a harmonically related pair of tones, their data predicted the presence of combination tones of frequencies above the primaries, but none below. This conflicts with the psychophysical finding that the $2f_1 - f_2$ tone (see Section VI.C.3) is the most prominent combination tone heard (GOLDSTEIN, 1967). Furthermore, GOLDSTEIN and KIANG (1968) demonstrated that responses to the $2f_1 - f_2$ component were present in the appropriate place in the cochlear fibre spectrum at frequencies well above that at which phase locking becomes insignificant (Section II.E.1.b). Fourthly, BRUGGE *et al.* (1969), GOLDSTEIN (1970, p. 191), and SMOORENBURG (1972b) have pointed out that there is a discrepancy between (a) the marked sensitivity of cochlear fibre temporal discharge patterns to the phase of interaction of the tone components of a combination, and (b) the lack of corresponding phase sensitivity of the ear. Fifthly, the transmission of periodic information in the discharge patterns of fibres, irrespective of their CF, makes it difficult to understand why a patient with a sharply defined tonal gap and focal cochlear lesion should have evidence of a diplacusis related to stimulus level (SMOORENBURG, 1972b). Lastly, one classical argument for periodicity coding theory – residue pitch – has been called into question by the psychophysical investigations of HOUTSMA and GOLDSTEIN (1972), GOLDSTEIN (1972), and also by the data of GRUBER and BOERGER (1970; but see the reservations on these data in Section II.E.1.b). In contrast, an attractive theory of pitch perception is emerging which accounts for these and other data by proposing a peripheral spectral frequency analysis and a central pitch extraction mechanism (*e.g.* DE BOER, 1956; WILSON, 1970, 1974; WHITFIELD, 1970; WIGHTMAN, 1972).

2. Intensity

The problem of the limited dynamic range demonstrated in the present data from cochlear nerve and nucleus neurones has been detailed in (1) above. For stimulus levels above those which saturate the neurones with corresponding CFs, it appears that we must conclude that the intensity information is carried by neurones of higher and lower frequencies, *i.e.*: on the outskirts of the active array as well as by the width of the array (ALLANSON and WHITFIELD, 1956), although the perception of multicomponent stimuli such as speech cannot be explained in this way (see above section).

3. Sound Complexes

The great majority of data on cochlear nerve and nucleus have been obtained with pure tones, clicks, and noise as stimuli. Experiments with more complex stimuli suggest that in contrast to the situation at higher levels of the auditory system (vd. EVANS, 1968), the responses of neurones in the periphery to more complex and behaviourally meaningful sounds are relatively simply predicted in the spectral and temporal terms described above. Thus, no evidence of specific or even substantial sensitivity to the rate of change of frequency and amplitude of sounds has been obtained in cochlear nerve (data of WATENABE, 1972) or cochlear nucleus (see Section III.F.4). What evidence there is appears to be predominantly in the dorsal cochlear nucleus, where much more complex excitatory-inhibitory interactions abound compared with the ventral division (Section III.B.2, 3, III.F.4). This latter dichotomy has led to the proposal that in the diversity of the CN we may have the earliest evidence of a "division of labour" within the auditory system into neural mechanisms responsible for complementary information processing tasks. Thus, in Fig. 49, the division of the cochlear nucleus and its ascending projections is crudely represented into functionally separable dorsal and ventral systems. The conjecture has been made elsewhere (EVANS, 1971; EVANS and NELSON, 1973b) that this subdivision may be analogous to that apparently present in the visual system: namely into (at least) two systems responsible for place and form information processing respectively (HELD *et al.*, 1967, 1968). In the case of Fig. 49, the ventral pathway would be ideally suited, by its preservation of temporal information, for the transmission of information pertaining to sound localization, while the dorsal pathway with its more complex responses may represent the first stages in the neural processing of acoustic patterns (EVANS, 1971, 1974a).

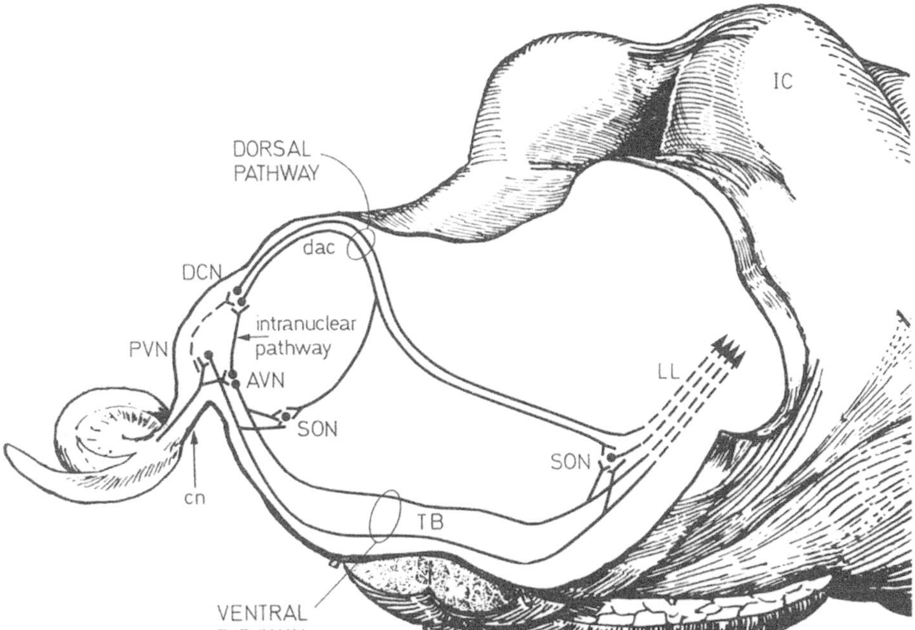

Fig. 49. Dorsal and ventral ascending acoustic pathways. Transverse section of brainstem at level of cochlear nuclei (after PAPEZ, 1929). cn: cochlear nerve; DCN, AVN, PVN: dorsal, anteroventral and postero-ventral divisions of cochlear nucleus. dac: dorsal acoustic stria. SON: superior olivary complex (includes nuclei of trapezoid body). TB: Trapezoid body. LL: lateral lemniscus. IC: inferior colliculus. This diagram represents a gross simplification of the output fibre pathways of the cochlear nuclei, after POLJAK (1926); LORENTE DE NÓ (1933b), WARR (1966, 1967) and JUNGERT (1968). The dorsal pathway represents the classical dorsal acoustic stria of Monakow. The ventral pathway represents the intermediate acoustic stria of HELD and the trapezoid body. The SON is not subdivided into the trapezoid nuclei and medial, lateral, peri- and pre-olivary nuclei of the superior olive proper. (From EVANS and NELSON, 1973b)

Acknowledgements

I am indebted to Dr. J. P. WILSON for extensive discussion of many of the topics dealt with herein and for his helpful comments and those of Dr. R. KLINKE on the manuscript. I also wish to thank Mr. J. S. CORBETT, for photographic assistance, and Miss H. O. HENRY and Mrs. J. E. CLARKE for their patient typing of the manuscript.

References

ALLANSON, J. T., WHITFIELD, I. C.: The cochlear nucleus and its relation to theories of hearing. 3rd London Symposium on Information Theory, pp. 269—284. London: Butterworth 1956.

ANDERSON, D. J., ROSE, J. E., HIND, J. E., BRUGGE, J. F.: Temporal position of discharges in single auditory nerve fibres within the cycle of a sine-wave stimulus: frequency and intensity effects. J. acoust. Soc. Amer. **49**, 1131—1139 (1971).

ARAN, J. M.: The electrocochleogram: recent results in children and in some pathological cases Arch. Ohr-, Nas.- u. Kehlk.-Heilk. **198**, 128—141 (1971).

ARAN, J. M., DELAUNAY, J.: Neural responses of end organ of the guinea pig. VIIIth nerve action potentials evoked by click and tone pips (filtered clicks) of different frequencies. Rev. Laryng. (Bordeaux) **90**, 598—614 (1969).

ARTHUR, R. M., PFEIFFER, R. R., SUGA, N.: Properties of "two-tone inhibition" in primary auditory neurones. J. Physiol. (Lond.) **212**, 593—609 (1971).

BARLOW, H. B., LEVICK, W. R.: The mechanism of directionally sensitive units in rabbits' retina. J. Physiol. (Lond.) **178**, 477—504 (1965).

BASSETT, I. G., EASTMOND, E. J.: Echolocation: measurement of pitch versus distance for sounds reflected from a flat surface. J. acoust. Soc. Amer. **36**, 911—916 (1964).

BÉKÉSY, G. VON: Über die mechanische Frequenzanalyse in der Schnecke verschiedener Tiere. Akust. Z. **9**, 3—11 (1944).

BÉKÉSY, G. VON: Experiments in hearing. New York: McGraw-Hill 1960.

BÉKÉSY, G. VON: Resonance in the cochlea? Sound **3**, 86—91 (1969).

BÉKÉSY, G. VON: Travelling waves as frequency analysers in the cochlea. Nature (Lond.) **225**, 1207—1209 (1970).

BOERGER, G., GRUBER, J.: In: Frequency Analysis and Periodicity Detection in Hearing, pp. 147—149. Leiden: Sijthoff 1970.

BOERGER, G., GRUBER, J., KLINKE, R.: Interaurales Übersprechen bei der Katze. Naturwissenschaften **55**, 234 (1968).

BOGDANSKI, D. F., GALAMBOS: In: Neural Mechanisms of Auditory and Vestibular systems, pp. 143—148. Springfield, Ill.: Thomas 1960.

BONE, R. C., CROWLEY, D., RAUCHBACH, E.: VIIIth nerve round window action potentials masked by high frequency noise in rats and guinea pigs: a comparative study. Laryngoscope (St. Louis) **82**, 1499—1513 (1972).

BRUGGE, J. F., ANDERSON, D. J., HIND, J. E., ROSE, J. E.: Time structure of discharges in single auditory nerve fibres of the squirrel monkey in response to complex periodic sounds. J. Neurophysiol. **32**, 386—407 (1969).

BUTLER, R. A., HONRUBIA, V., JOHNSTONE, B. M., FERNANDEZ, C.: Cochlear function under metabolic impairment. Ann. Otol. (St. Louis) **71**, 648—656 (1962).

CAJAL, S. R. y.: Histologie du système nerveux, Vol. 1. Madrid: Instituto Ramon Y Cajal 1909.

CAMPBELL, F. W., COOPER, G. F., ENROTH-CUGELL, C.: The spatial selectivity of the visual cells of the cat. J. Physiol. (Lond.) **203**, 223—235 (1969).

CAPPS, M. J., ADES, H. W.: Auditory frequency discrimination after transection of olivocochlear bundle in squirrel monkeys. Exp. Neurol **21**, 147—158 (1968).

COATS, A. C., DICKEY, J. R.: Non-surgical recording of human auditory nerve action potentials and cochlear microphonics. Ann. Otol. (St. Louis) **79**, 844—852 (1970).

COHEN, E. S., BRAWER, J. R., MOREST, D. K.: Projections of the cochlea to the dorsal cochlear nucleus in the cat. Exp. Neurol. **35**, 470—479 (1972).

COMIS, S. D.: Centrifugal inhibitory processes affecting neurones in the cat cochlear nucleus. J. Physiol. (Lond.) **210**, 751—760 (1970).

COMIS, S. D., DAVIES, W. E.: Acetylcholine as a transmitter in the cat auditory system. J. Neurochem. **16**, 423—429 (1969).

COMIS, S. D., PICKLES, V. O.: Centrifugal pathways and the detection of signals in noise. J. Physiol. (Lond.) **226**, 58P (1972).

COMIS, S. D., WHITFIELD I. C.: The effect of acetylcholine on neurones of the cochlear nucleus. J. Physiol. (Lond.) **183**, 22—23P (1966).

COMIS, S. D., WHITFIELD I. C.: Influence of centrifugal pathways on unit activity in cochlear nucleus. J. Neurophysiol. **31**, 62—68 (1968).

CULLEN, J. K., JR., ELLIS, M. S., BERLIN, C. I., LOUSTEAU, R. J.: Human acoustic nerve action potential recordings from the tympanic membrane without anaesthesia. Acta Oto-laryng. (Stockh.) **74**, 15—22 (1972).

DAIGNEAULT, E. A., BROWN, R. D., PRUETT, J.: Cochlear round window recorded responses to acetylcholine and click stimulation following decentralization. Acta Oto-laryng. (Stockh.) **66**. 10—16 (1968).

DALLOS, P.: Combination tone $2f_1 - f_h$ in microphonic potentials. J. acoust. Soc. Amer. **46**, 1437—1444 (1969).

DALLOS, P.: Comments on "correspondence between cochlear microphonic sensitivity and behavioural threshold in the cat". J. acoust. Soc. Amer. **50**, 1554 (1971).

DALLOS, P.: Cochlear potentials and cochlear mechanics. In: Basic Mechanisms in Hearing, pp. 335—372. New York: Academic Press 1973.

DALLOS, P., BILLONE, M. C., DURRANT, J. D., WANG, C-Y., RAYNOR, S.: Cochlear inner and outer hair cells: functional differences. Science **177**, 356—358 (1972)

DAVIS, H.: A model for transducer action in the cochlea. Cold Spr. Harb. Symp. quant. Biol. **30**, 181—189 (1965).

DAVIS, H., DEATHERAGE, B. H., ROSENBLUT, B., FERNANDEZ, C., KIMURA, R., SMITH, C. A.: Modifications of cochlear potentials produced by streptomycin poisoning and by venous obstruction. Laryngoscope (St. Louis) **68**, 596—627 (1958).

DEATHERAGE, B. H., ELDREDGE, D. H., DAVIS, H.: Latency of action potential in the cochlea of the guinea pig. J. acoust. Soc. Amer. **31**, 479—486 (1959).

DE BOER, E.: On the "residue" in hearing (Thesis). Netherlands: Excelsior 1956.

DE BOER, E.: Measurement of critical band-width in cases of perception deafness. Proc. 3rd Internat. Congr. on Acoustics Stuttgart **1**, pp. 100—102. Amsterdam: Elsevier 1959.

DE BOER, E.: Reverse correlation. I. A heuristic introduction to the technique of triggered correlation with applications to the analysis of compound systems. Proc. kon. nederl. Akad. Wet. **71**, 472—486 (1968).

DE BOER, E.: Reverse correlation. II. Initiation of nerve impulses in the inner ear. Proc. kon. Nederl. Akad. Wet. **72**, 129—151 (1969).

DE BOER, E.: Synchrony between acoustic stimuli and nerve fibre discharges. In: Frequency Analysis and Periodicity Detection in Hearing, pp. 204—212. Leiden: Sijthoff 1970.

DE BOER, E., DE JONGH, H. R.: Computer simulation of cochlear filtering. Proc. 7th Internat. Congr. on Acoustics, pp. 393—396. Budapest: Akademiai Kiado 1971.

DE BOER, E., JONGKEES, L. B. W.: On cochlear sharpening and cross-correlation methods. Acta Oto-laryng. (Stockh.) **65**, 97—104 (1968).

DE BOER, E., KUYPER, P.: Triggered correlation. IEEE Trans. Biomed. Eng. **15**, 169—179 (1968).

DE BOER, E., SMOORENBURG, G. F., KUYPER, P.: Proposed explanation of synchrony of auditory-nerve impulses to combination tones. J. acoust. Soc. Amer. **46**, 1579—1581 (1969).

DE JONGH, H. R.: About coding in the VIIIth nerve. In: Hearing Theory 1972, pp. 70—77. Eindhoven: IPO 1972.

DERBYSHIRE, A. J., DAVIS, H.: The action potentials of the auditory nerve. Amer. J. Physiol. **113**, 476—504 (1935).

DESMEDT, J. E.: Neurophysiological mechanisms controlling acoustic input. In: Neural Mechanisms of Auditory and Vestibular Systems, pp. 152—164. Springfield, Ill.: Thomas 1960.

DESMEDT, J. E., MONACO, P.: The pharmacology of a centrifugal inhibitory pathway in the cat's acoustic system. In: Pharmacological Analysis of Central nervous action. Proc. 1st Internat. Pharmacol. Meeting. London: Pergamon 1962.

DEWSON, J. H. III.: Efferent olivocochlear bundle: some relationships to noise masking and to stimulus attenuation. J. Neurophysiol. **30**, 817—832 (1967).

DEWSON, J. H. III: Efferent olivocochlear bundle: some relationships to stimulus discrimination in noise. J. Neurophysiol. **31**, 122—130 (1968).

DIX, M. R., HALLPIKE, C. S., HOOD, J. D.: Observations upon the loudness recruitment phenomenon with especial reference to the differential diagnosis of disorders of the internal ear and VIII nerve. Proc. roy. Soc. Med. **41**, 516—526 (1948).

DUIFHUIS, H.: A tentative firing model for the auditory receptor. I.P.O.-A.P.R. **5**, 18—24 (1970).

DUIFHUIS, H.: Audibility of high harmonics in a periodic pulse. II. Time effect. J. acoust. Soc. Amer. **49**, 1155—1162 (1971).

DUIFHUIS, H.: Perceptual Analysis of Sound (Thesis). Eindhoven: 1972.

DUNKER, E. G., GRUBEL, G., PFALZ, R.: Beeinflussung von spontanaktiven, deafferentierten Einzelneuronen des Nucleus cochlearis der Katze durch Tonreizung der Gegenseite. Pflügers Arch. ges. Physiol. **278**, 610—623 (1964).

ELBERLING, C., SALOMON, G.: Electrical potentials from the inner ear in man, in response to transient sounds generated in a closed acoustic system. Rev. Laryng. Suppl. pp. 691—706 (1971).

ENGEBRETSON, A. M., ELDREDGE, D. H.: Model for the nonlinear characteristics of cochlear potentials. J. acoust. Soc. Amer. **44**, 548—554 (1968).

ENGER, P. S.: Single unit activity in the peripheral auditory system of a teleost fish. Acta physiol. scand. **59**, Suppl. 210 (1963).

ERULKAR, S. D., BUTLER, R. A., GERSTEIN, G. L.: Excitation and inhibition in cochlear nucleus. II. Frequency modulated tones. J. Neurophysiol. **31**, 537—548 (1968).

EVANS, E. F.: Upper and lower levels of the auditory system: a contrast of structure and function. In: Neural Networks, pp. 24—33. Berlin-Heidelberg-New York: Springer 1968.

EVANS, E. F.: Narrow "tuning" of cochlear nerve fibre responses in the guinea pig. J. Physiol. (Lond.) **206**, 14—15P (1970a).

EVANS, E. F.: Narrow tuning of the responses of cochlear nerve fibres emanating from the exposed basilar membrane. J. Physiol. (Lond.) **208**, 75P—76P (1970b).

EVANS, E. F.: Central mechanisms relevant to the neural analysis of simple and complex sounds. In: Pattern Recognition in Biological and Technical Systems, pp. 328—343. Berlin-Heidelberg-New York: Springer 1971.

EVANS, E. F.: Does frequency sharpening occur in the cochlea? In: Hearing Theory 1972, pp. 27—34. Eindhoven: IPO 1972a.

EVANS, E. F.: The frequency response and other properties of single fibres in the guinea-pig cochlear nerve. J. Physiol. (Lond.) **226**, 263—287 (1972b).

EVANS, E. F.: Neural processes for the detection of acoustic patterns and for sound localization. In: The Neurosciences: Third Study Program, pp. 131—145. Boston: M.I.T. Press 1974a.

EVANS, E. F.: The effects of hypoxia on the tuning of single cochlear nerve fibres. J. Physiol. (Lond.) **238**, 65—67P (1974b).

EVANS, E. F.: Normal and abnormal functioning of the cochlear nerve. In: Sound Reception in Mammals. Symp. Zoo. Soc. Lond. In Press. London: Academic Press. 1974c.

EVANS, E. F.: Auditory Frequency selectivity and the cochlear nerve. In: Facts and Models in Hearing, pp. 118—129. Berlin-Heidelberg-New York: Springer. 1974d.

EVANS, E. F., KLINKE, R.: Reversible effects of cyanide and Furosemide on the tuning of single cochlear fibres. J. Physiol. (Lond.) **242**, 129—131 P (1974).

EVANS, E. F., NELSON, P. G.: Behaviour of neurones in cochlear nucleus under steady and modulated tonal stimulation. Fed. Proc. **25**, 463 (1966a).

EVANS, E. F., NELSON, P. G.: Responses of neurones in cat cochlear nucleus to modulated tonal stimuli. J. acoust. Soc. Amer. **40**, 1275—1276 (1966b).

EVANS, E. F., NELSON, P. G.: An intranuclear pathway to the dorsal division of the cochlear nucleus of the cat. J. Physiol. (Lond.) **196**, 76P—78P (1968).

EVANS, E. F., NELSON, P. G.: The responses of single neurones in the cochlear nucleus of the cat as a function of their location and the anaesthetic state. Exp. Brain Res. **17**, 402—427 (1973a).

EVANS, E. F., NELSON, P. G.: On the relationship between the dorsal and ventral cochlear nucleus. Exp. Brain Res. **17**, 428—442 (1973b).

EVANS, E. F., WHITFIELD, I. C.: Classification of unit responses in the auditory cortex of the unanaesthetised and unrestrained cat. J. Physiol. (Lond.) **171**, 476—493 (1964).

EVANS, E. F., WILSON, J. P.: Frequency sharpening of the cochlea: the effective bandwidth of cochlear nerve fibres. Proc. 7th Internat. Congr. on Acoustics, Vol. 3, pp. 453—456. Budapest: Akademiai Kiado 1971.

EVANS, E. F., WILSON, J. P.: Frequency selectivity of the cochlea. In: Basic Mechanisms in Hearing, pp. 519—551. New York: Academic Press 1973.

EVANS, E. F., ROSENBERG, J., WILSON, J. P.: The effective bandwidth of cochlear nerve fibres. J. Physiol. (Lond.) **207**, 62P—63P (1970).

EVANS, E. F., ROSENBERG, J., WILSON, J. P.: The frequency resolving power of the cochlea. J. Physiol. (Lond.) **216**, 58P—59P (1971).

FERNALD, R.: A neuron model with spatially distributed synaptic input. Biophys. J. **11**, 323—340 (1971).

FEX, J.: Auditory activity in centrifugal and centripetal cochlear fibres in cat. Acta physiol. scand. Suppl. **189**, 1—68 (1962).

FEX, J.: Crossed cochlear efferents activated by sound through both ears. Acta physiol. scand. Suppl. **213**, 41 (1963).

FEX, J.: Auditory activity in uncrossed centrifugal cochlear fibres in cat. Acta physiol. scand. **64**, 43—57 (1965).

FINCK, A.: Physiological correlate of tonal masking. J. acoust. Soc. Amer. **39**, 1056—1062 (1966).

FLANAGAN, J. L.: Computational models for basilar-membrane displacement. J. acoust. Soc. Amer. **34**, 1370—1376 (1962).

FLOCK, Å.: Sensory transduction in hair cells. In: Handbook of Sensory Physiology, Vol. 1. Principles of Receptor Physiology, Chap. 14. Berlin-Heidelberg-New York: Springer 1971.

FRISHKOPF, L. S.: Excitation and inhibition of primary auditory neurones in the little brown bat. J. acoust. Soc. Amer. **36**, 1016A (1964).

FRISHKOPF, L. S., GOLDSTEIN, M. H.: Responses to acoustic stimuli from single units in the eighth nerve of the bullfrog. J. acoust. Soc. Amer. **35**, 1219—1228 (1963).

FRISHKOPF, L. S., CAPRANICA, R. R., GOLDSTEIN, M. H., JR.: Neural coding in the bullfrog's auditory system: a teleological approach. Proc. IEEE **56**, 969—980 (1968).

FURMAN, G. G., FRISHKOPF, L. S.: Model of neural inhibition in the mammalian cochlea. J. acoust. Soc. Amer. **36**, 2194—2201 (1964).

FURUKAWA, T., ISHII, Y.: Neurophysiological studies on hearing in goldfish. J. Neurophysiol. **30**, 1377—1403 (1967).

FURUKAWA, T., ISHII, Y., MATSUURA, S.: Synaptic delay and time course of synaptic potentials at the junction between hair cells and eighth nerve fibres in the goldfish. Jap. J. Physiol. **22**, 617—636 (1972).

GALAMBOS, R.: Inhibition of activity in single auditory nerve fibres by acoustic stimulation. J. Neurophysiol. **7**, 287—303 (1944).

GALAMBOS, R.: Suppression of auditory nerve activity by stimulation of efferent fibers to cochlea. J. Neurophysiol. **19**, 424—437 (1956).

GALAMBOS, R.: Studies of the auditory system with implanted electrodes. In: Neural Mechanisms of the Auditory and Vestibular Systems, pp. 137—151. Springfield, Ill.: Thomas 1960.

GALAMBOS, R., DAVIS, H.: Responses of single auditory nerve fibres to acoustic stimulation. J. Neurophysiol. **6**, 39—57 (1943).

GALMBOS, R., DAVIS, H.: Action potentials from single auditory nerve fibres? Science **108**, 513 (1948).

GALLEY, N., KLINKE, R., PAUSE, M., STORCH, W.-H.: The effect of flaxedil (gallamine triethiodide) on the efferent endings in the cochlea. Pflügers Arch. **330**, 1—4 (1971).

Gässler, G.: Über die Hörschwelle für Schallereignisse mit verschieden breitem Frequenzspektrum. Acustica **4**, 408 (1954).

Geisler, C. D.: A model of the peripheral auditory system responding to low frequency tones. Biophys. J. **8**, 1—15 (1968).

Gerstein, G. L., Kiang, N. Y. S.: An approach to the quantitative analysis of electrophysiological data from single neurons. Biophys. J. **1**, 15—28 (1960).

Gerstein, G. L., Butler, R. A., Erulkar, S. D.: Excitation and inhibition in cochlear nucleus. I. Tone burst stimulation. J. Neurophysiol. **31**, 526—536 (1958).

Glattke, T. J.: Unit responses of the cat cochlear nucleus to amplitude-modulated stimuli. J. acoust. Soc. Amer. **45**, 419—425 (1969).

Goblick, T. J., Jr., Pfeiffer, R. R.: Time-domain measurements of cochlear non-linearities using combination click stimuli. J. acoust. Soc. Amer. **46**, 924—938 (1969).

Gold, T.: Hearing. II. The physical basis of the action of the cochlea. Proc. roy. Soc. B **135**, 492—498 (1948).

Goldberg, J. M., Greenwood, D. D.: Response of the neurons of the dorsal and postoventral cochlear nuclei of the cat to acoustic stimuli of long duration. J. Neurophysiol. **29**, 72—93 (1966).

Goldstein, J. L.: Auditory non-linearity. J. acoust. Soc. Amer. **41**, 676—689 (1967).

Goldstein, J. L.: In: Frequency Analysis and Periodicity Detection in Hearing, pp. 230—245. Leiden: Sijthoff 1970.

Goldstein, J. L.: Evidence from aural combination tones and musical notes against classical temporal periodicity theory. In: Hearing Theory, 1972, pp. 186—208. Eindhoven: IPO 1972.

Goldstein, J. L., Kiang, N. Y. S.: Neural correlates of the aural combination tone $2f_1 - f_2$. Proc. IEEE **56**, 981—992 (1968).

Goldstein, J. L., Baer, T., Kiang, N. Y. S.: A theoretical treatment of latency, group-delay and tuning characteristics for auditory-nerve responses to clicks and tones. In: Physiology of the Auditory System, pp. 133—141. Baltimore, Md.: National Educational Consultants 1971.

Gray, P. R.: Conditional probability analyses of the spike activity of single neurons. Biophys. J. **7**, 759—777 (1967).

Greenwood, D. D.: Auditory masking and the critical band. J. acoust. Soc. Amer. **33**, 484—502 (1961).

Greenwood, D. D., Goldberg, J. M.: Response of neurons in the cochlear nuclei to variations in noise bandwidth and to tone-noise combinations. J. acoust. Soc. Amer. **47**, 1022—1040 (1970).

Greenwood, D. D., Maruyama, N.: Excitatory and inhibitory response areas of auditory neurons in the cochlear nucleus. J. Neurophysiol. **28**, 863—892 (1965).

Grubel, G., Dunker, D., Rehren, D. V.: Zur Funktionsweise des efferenten auditorischen Systems. I. Mitteilung. Pflügers Arch. ges. Physiol. **281**, 109 (1964).

Guinan, J. J. jr., Peake, W. T.: Middle ear characteristics of anaesthetized cats. J. acoust. Soc. Amer. **41**, 1237—1261 (1967).

Hall, J. L.: Comment on "Proposed explanation of synchrony of auditory-nerve impulses to combination tones". J. acoust. Soc. Amer. **50**, 1555—1556 (1971).

Harris, G. G.: Brownian motion and the threshold of hearing. Int. Audiol. **7**, 111—120 (1968).

Harrison, J. M., Irving, R.: The anterior ventral cochlear nucleus. J. comp. Neurol **124**, 15—42 (1965).

Harrison, J. M., Irving, R.: The organization of the posterior ventral cochlear nucleus in the rat. J. comp. Neurol. **126**, 391—402 (1966).

Harrison, J. M., Warr, B.: A study of the cochlear nuclei and ascending pathways of the medulla. J. comp. Neurol. **119**, 341—380 (1962).

Hawkins, J. E. jr., Stevens, S. S.: The masking of pure tones and of speech by white noise. J. acoust. Soc. Amer. **22**, 6—13 (1950).

Heffner, R., Heffner, H., Masterton, B.: Behavioural measurements of absolute and frequency difference thresholds in guinea pig. J. acoust. Soc. Amer. **49**, 1888—1895 (1971).

Held, R., Ingle, D., Schneider, G. E., Trevarthen, C. B.: Locating and identifying: two modes of visual processing. Psychol. Forsch. **31**, 44—62; 299—348 (1967/68).

HIESEY, R. W., SCHUBERT, E. D.: Cochlear resonance and phase-reversal signals. J. acoust. Soc. Amer. **51**, 518—519 (1972).

HIND, J. E.: Physiological correlates of auditory stimulus periodicity. Audiol. **11**, 42—57 (1972).

HIND, J. E., ANDERSON, D. J., BRUGGE, J. F., ROSE, J. E.: Coding of information pertaining to paired low-frequency tones in single auditory nerve fibres of the squirrel monkey. J. Neurophysiol. **30**, 794—816 (1967).

HIND, J. E., ROSE, J. E., BRUGGE, J. F., ANDERSON, D. J.: Two-tone masking effects in squirrel monkey auditory nerve fibres. In: Frequency Analysis and Periodicity Detection in Hearing, pp. 193—200. Leiden: Sijthoff 1970.

HIRSH, I. J.: The influence of interaural phase on interaural summation and inhibition. J. acoust. Soc. Amer. **20**, 536—544 (1948).

HOUTGAST, T.: Psychophysical evidence for lateral inhibition in hearing. J. acoust. Soc. Amer. **51**, 1885—1894 (1972).

HOUTSMA, A. J. M., GOLDSTEIN, J. L.: The central origin of the pitch of complex tones: evidence from musical interval recognition. J. acoust. Soc. Amer. **51**, 520—529 (1972).

HUGGINS, W. H., LICKLIDER, J. C. R.: Place mechanisms of auditory frequency analysis. J. acoust. Soc. Amer. **23**, 290—299 (1951).

HUXLEY, A. F.: Is resonance possible in the cochlea after all? Nature (Lond.) **221**, 935—940 (1969).

IGARASHI, M., ALFORD, B. R., NAKAI, Y., GORDON, W. P.: Behavioural auditory function after transection of crossed olivo-cochlear bundle in the cat. I. Pure-tone threshold and perceptual signal-to-noise ratio. Acta Oto-laryng. (Stockh.) **73**, 455—466 (1972).

IRVINE, D. R. F., WEBSTER, W. R.: Studies of peripheral gating in the auditory system of cats. Electroenceph. clin. Neurophysiol. **32**, 545—556 (1972).

ISHII, D., BALOGH, K., JR.: Distribution of efferent nerve endings in the organ of Corti. Acta Oto-laryng. (Stockh.) **66**, 282—288 (1968).

ISHII, Y., MATSUURA, S., FURUKAWA, T.: An input-output relation at the synapse between hair cells and eighth nerve fibres in goldfish. Jap. J. Physiol. **21**, 91—98 (1971).

IURATO, S.: Fibre efferente dirette e crociate alle cellule acustiche dell'organo del Corti. Monit. Zool. ital. Suppl. **72**, 62—63 (1964).

JOHNSTONE, J. R., JOHNSTONE, B. M.: Unit responses from the lizard auditory nerve. Exp. Neurol. **24**, 528—537 (1969).

JOHNSTONE, B. M., SELLICK, P. M.: The peripheral auditory apparatus. Quart Rev. Biophys. **5**, 1—58 (1972).

JOHNSTONE, B. M., TAYLOR, K. J., BOYLE, A. J.: Mechanics of the guinea pig cochlea. J. acoust. Soc. Amer. **47**, 504—509 (1970).

JUNGERT, S.: Auditory pathways in the brainstem. A neurophysiological study. Acta oto-laryng. (Stockh.) Suppl. **138**, 5—67 (1958).

KARLAN, M. S., TONNDORF, J., KHANNA, S. M.: Dual origin of the cochlear microphonics: inner and outer hair cells. Ann. Otol. (St. Louis) **81**, 696—704 (1972).

KATSUKI, Y., SUMI, T., UCHIYAMA, H., WATENABE, T.: Electric responses of auditory neurones in cat to sound stimulation. J. Neurophysiol. **21**, 569—588 (1958).

KATSUKI, Y., WATENABE, T., SUGA, N.: Interaction of auditory neurons in response to two sound stimuli in cat. J. Neurophysiol. **22**, 603—623 (1959).

KATSUKI, Y., KANNO, Y., SUGA, N.: Primary auditory neurons of the monkey. Jap. J. Physiol. **11**, 678—683 (1961).

KATSUKI, Y., SUGA, N., KANNO, Y.: Neural mechanisms of the peripheral and central auditory system in monkeys. J. acoust. Soc. Amer. **34**, 1396—1410 (1962).

KEIDEL, W. D.: Physiologie des Innenrohres. In: Hals-Nasen-Ohren-Heilkunde, Vol. III/1, pp. 235—310 (1965).

KEIDEL, W. D.: Biophysics, mechanics and electrophysiology of the cochlea. In: Frequency Analysis and Periodicity Detection in Hearing, pp. 60—79. Leiden: Sijthoff 1970.

KHANNA, S. M., SEARS, R. E., TONNDORF, J.: Some properties of longitudinal shear waves: a study of computer stimulation. J. acoust. Soc. Amer. **43**, 1077—1084 (1968).

KIANG, N. Y.-S: Stimulus coding in the auditory nerve and cochlear nucleus. Acta Oto-laryng. (Stockh.) **59**, 186—200 (1965).

Kiang, N. Y.-s: A survey of recent developments in the study of auditory physiology. Ann. Otol. (St. Louis) **77**, 656—676 (1968).

Kiang, N. Y.-s, Moxon, E. C.: Physiological considerations in artificial stimulation of the inner ear. Ann. Otol. (St. Louis) **81**, 714—730 (1972).

Kiang, N. Y.-s., Peake, W. T.: Components of electrical responses recorded from the cochlea. Ann. Otol. (St. Louis) **69**, 448—458 (1960).

Kiang, N. Y.-s., Goldstein, M. H., Jr., Peake, W. T.: Temporal coding of neural responses to acoustic stimuli. IRE Trans. Inf. Theory **8**, 113—119 (1962a).

Kiang, N. Y.-s., Watenabe, T., Thomas, E. C., Clark, L. F.: Stimulus coding in the cat's auditory nerve. Ann. Otol. (St. Louis) **71**, 1009—1026 (1962b).

Kiang, N. Y.-s., Watenabe, T., Thomas, E. C., Clark, L. F.: Discharge patterns of single fibres in the cat's auditory nerve. Cambridge, Mass.: M.I.T. Press 1965a.

Kiang, N. Y.-s., Pfeiffer, R. R., Warr, W. B., Backus, A. S. N.: Stimulus coding in the cochlear nucleus. Ann. Otol. (St. Louis) **74**, 463—485 (1965b).

Kiang, N. Y.-s., Sachs, M. B., Peake, W. T.: Shapes of tuning curves for single auditory nerve fibres. J. acoust. Soc. Amer. **42**, 1341—1342 (1967).

Kiang, N. Y.-s., Baer, T., Marr, E. M., Demont, D.: Discharge rates of single auditory nerve fibres as functions of tone level. J. acoust. Soc. Amer. **46**, 106 (1969).

Kiang, N. Y.-s., Moxon, E. C., Levine, R. A.: Auditory-nerve activity in cats with normal and abnormal cochleas. In: Sensorineural Hearing Loss, pp. 241—268. London: Churchill 1970.

Kiang, N. Y.-s., Morest, D. K., Godfrey, D. A., Guinan, J. J., Jr., Kane, E. C.: Stimulus coding at caudal levels of the cat's auditory system. I. Response characteristics of single units. In: Basic Mechanisms in Hearing, pp. 455—475. New York: Academic Press 1973.

Klinke, R., Galley, N.: The efferent innervation of the vestibular and auditory receptors. Physiol. Rev. **54**, 316—357 (1974).

Klinke, R., Boerger, G., Gruber, J.: Studies of the functional significance of efferent innervation in the auditory system: efferent neuronal activity as influenced by contralaterally applied sound. Pflügers Arch. **306**, 165—175 (1969).

Koerber, K. L., Pfeiffer, R. R., Warr, W. B., Kiang, N. Y.-s.: Spontaneous spike discharges from single units in the cochlear nucleus after destruction of the cochlea. Exp. Neurol. **16**, 119—130 (1966).

Konishi, M.: Time resolution by single auditory neurones in birds. Nature (Lond.) **222**, 566—567 (1969a).

Konishi, M.: Comparative neurophysiological studies of hearing and vocalizations in songbirds. Z. vergl. Physiol. **66**, 257—272 (1970).

Konishi, M.: Hearing, single-unit analysis and vocalizations in songbirds. Science **166**, 1178—1181 (1969b).

Konishi, T., Nielsen, D. W.: The temporal relationship between motion of the basilar membrane and initiation of nerve impulses in the auditory nerve fibres. J. acoust. Soc. Amer. **53**, 325 (1973).

Konishi, T., Teas, D. C., Wernick, J. S.: Effects of electrical current applied to cochlear partition on discharges in individual auditory nerve fibres. I. Prolonged direct current polarization. J. acoust. Soc. Amer. **47**, 1519—1526 (1970).

Levine, R. A.: Phase locking in response of single neuron in cat cochlear nucleus to low-frequency tonal stimuli. J. Neurophysiol. **34**, 467—483 (1971).

Lev, A., Sohmer, H.: Sources of averaged neural responses recorded in animal and human subjects during cochlear audiometry (electrocochleogram). Arch. Ohr.-, Nas.-, Kehlk.-Heilk. **201**, 79—90 (1972).

Licklider, J. C. R.: "Periodicity" pitch and "phase" pitch. J. acoust. Soc. Amer. **26**, 945 (1954).

Liff, H.: Responses from single auditory units in the eighth nerve of the leopard frog. J. acoust. Soc. Amer. **45**, 512—513 (1969).

Liff, H. J., Goldstein, M. H., Jr.: Peripheral inhibition in auditory fibres in the frog. J. acoust. Soc. Amer. **47**, 1538—1547 (1970).

Lim, D. J.: Fine morphology of the tectorial membrane. Arch. Otolaryng. **96**, 199—215 (1972).

Lorente de Nó, R.: Anatomy of eighth nerve. The central projection of the nerve ending of the internal ear. Laryngoscope **43**, 1—38 (1933a).

LORENTE DE NÓ, R.: Anatomy of the eighth nerve: III. General plan of structure of primary cochlear nucleus. Laryngoscope **43**, 327—350 (1933b).

LYNN, P. A., SAYERS, B. McA.: Cochlear innervation, signal processing, and their relation to auditory time-intensity effects. J. acoust. Soc. Amer. **47**, 525—533 (1970).

MANLEY, G. A.: Frequency sensitivity of auditory neurons in the caiman cochlear nucleus. Z. vergl. Physiol. **66**, 251—256 (1970a).

MANLEY, G. A.: Comparative studies of auditory physiology in reptiles. Z. vergl. Physiol. **67**, 363—387 (1970b).

MANLEY, G. A.: Some aspects of the evolution of hearing in vertebrates. Nature **230**, 506—509 (1971).

MARSH, J. T., WORDEN, F. G.: Sound evoked frequency-following responses in the central auditory pathway. Laryngoscope **78**, 1149—1163 (1968).

MARSH, J. T., WORDEN, F. G., SMITH, J. C.: Auditory frequency-following response: neural or artifact? Science **169**, 1222—1223 (1970).

MARSH, J. T., SMITH, J. C., WORDEN, F. G.: Receptor and neural responses in auditory masking of low frequency tones. EEG and clin. Neurophysiol. **32**, 63—74 (1972).

MAST, T. E.: Study of single units of the cochlear nucleus of the chinchilla. J. acoust. Soc. Amer. **48**, 505—512 (1970a).

MAST, T. E.: Binaural interaction and contralateral inhibition in dorsal cochlear nucleus of chinchilla. J. Neurophysiol. **33**, 108—115 (1970b).

McCLELLAND, K. D., BRANDT, J. F.: Pitch of frequency modulated sinusoids. J. acoust. Soc. Amer. **45**, 1489—1498 (1969).

MILLER, J. P., WATSON, C. S., COVELL, W. P.: Deafening effects of noise on the cat. Acta Otolaryng. (Stockh.) Suppl. **176**, 1—91 (1963).

MØLLER, A. R.: Unit responses in the cochlear nucleus of the rat to pure tones. Acta physiol. scand. **75**, 530—541 (1969a).

MØLLER, A. R.: Unit responses in the cat cochlear nucleus to repetitive transient sounds. Acta physiol. scand. **75**, 542—551 (1969b).

MØLLER, A. R.: Unit responses in the cochlear nucleus of the rat to sweep tones. Acta physiol. scand. **76**, 503—512 (1969c).

MØLLER, A. R.: Unit responses in the cochlear nucleus of the rat to noise and tones. Acta physiol. scand. **78**, 289—298 (1970a).

MØLLER, A. R.: Studies of the damped oscillatory response of the auditory frequency analyser. Acta physiol. scand. **78**, 299—314 (1970b).

MØLLER, A. R.: Unit responses in the rat cochlear nucleus to tones of rapidly varying frequency and amplitude. Acta physiol. scand. **81**, 540—556 (1971).

MØLLER, A. R.: Coding of amplitude and frequency modulated sounds in the cochlear nucleus of the rat. Acta physiol. scand. **186**, 223—238 (1972a).

MØLLER, A. R.: Coding of sounds in lower levels of the auditory system. Quart. Rev. Biophys. **53**, 59—155 (1972b).

MOLNAR, C. E., PFEIFFER, R. R.: Interpretation of spontaneous spike discharge patterns in the cochlear nucleus. Proc. IEEE **56**, 993—1004 (1968).

MOLNAR, C. E., LOEFFEL, R. G., PFEIFFER, R. R.: Distortion compensating condenser earphone driver for physiological studies. J. acoust. Soc. Amer. **43**, 1177—1178 (1968).

MOUSHEGIAN, G., RUPERT, A., GALAMBOS, R.: Microelectrode study of ventral cochlear nucleus of the cat. J. Neurophysiol. **25**, 515—529 (1962).

MOUSHEGIAN, G., RUPERT, R. L., STILLMAN, R. D., WEISS, I. P.: Inhibition in the auditory nerve. J. acoust. Soc. Amer. **50**, 1558—1560 (L), (1971).

MOXON, E. C.: Auditory nerve responses to electric stimuli. M.I.T. Q.P.R. **90**, 270—275 (1968).

NELSON, P. G., EVANS, E. F.: Relationship between dorsal and ventral cochlear nuclei. In: Physiology of the Auditory System, pp. 169—174. Baltimore: National Educational Consultants Inc. 1971.

NIEDER, P.: Addressed exponential delay line theory of cochlear organization. Nature **230**, 255—257 (1971).

NIEDER, P., NIEDER, I.: Antimasking effect of crossed olivocochlear bundle stimulation with loud clicks in guinea pig. Exp. Neurol. **28**, 179—188 (1970).

NOMOTO, M., SUGA, N., KATSUKI, Y.: Discharge pattern and inhibition of primary auditory nerve fibres in the monkey. J. Neurophysiol. **27**, 768—787 (1964).
OSEN, K. K.: Cytoarchitecture of the cochlear nuclei in the cat. J. comp. Neurol. **136**, 453—483 (1969).
OSEN, K. K.: Course and termination of the primary afferents in the cochlear nuclei of the cat. Arch. ital. Biol. **108**, 21—51 (1970).
PAPEZ, J. W.: Comparative Neurology (1929). Reprinted. New York: Hafner Publ. Co. 1967.
PEAKE, W. T., KIANG, N. Y.-S.: Cochlear responses to condensation and rarefaction clicks. Biophys. J. **2**, 23—34 (1962).
PFALZ, R. K. J.: Centrifugal inhibition of afferent secondary neurons in the cochlear nucleus by sound. J. acoust. Soc. Amer. **34**, 1472—1477 (1962).
PFEIFFER, R. R.: Anteroventral cochlear nucleus: waveforms of extracellularly recorded spike potentials. Science **154**, 667—668 (1966a).
PFEIFFER, R. R.: Classification of response patterns of spike discharges for units in the cochlear nucleus: tone-burst stimulation. Exp. Brain Res. **1**, 220—235 (1966b).
PFEIFFER, R. R.: A model for two-tone inhibition of single cochlear nerve fibres. J. acoust. Soc. Amer. **48**, 1373—1378 (1970).
PFEIFFER, R. R., KIANG, N. Y.-S.: Spike discharge patterns of spontaneously and continuously stimulated activity in the cochlear nucleus of anaesthetized cats. Biophys. J. **5**, 301—316 (1965).
PFEIFFER, R., KIM, D. O.: Response patterns of single cochlear nerve fibres to click stimuli: descriptions for cat. J. acoust. Soc. Amer. **52**, 1669—1677 (1972).
PFEIFFER, R. R., KIM, D. O.: Considerations of nonlinear response properties of single cochlear nerve fibers. In: Basic Mechanisms in Hearing, pp. 555—587, New York: Academic Press. 1973.
PFEIFFER, R. R., MOLNAR, C. E.: Cochlear nerve fibre discharge patterns: relationship to the cochlear microphonic. Science **167**, 1614—1616 (1970).
PIDDINGTON, R. W.: Central control of auditory input in the goldfish. II. Evidence of action in the free-swimming animal. J. exp. Biol. **55**, 585—610 (1971).
PLOMP, R.: Rate of decay of auditory sensation. J. acoust. Soc. Amer. **36**, 277—282 (1964a).
PLOMP, R.: The ear as a frequency analyser. J. acoust. Soc. Amer. **36**, 1628—1636 (1964b).
POLJAK, S.: The connections of the acoustic nerve. J. Anat. **60**, 465—469 (1926).
PORTMANN, M., LE BERT, G., ARAN, J.-M.: Potentials cochléaires obtenus chez l'homme en dehors de toute intervention chirurgicale. Rev. Laryng. Bordeaux **88**, 157—164 (1967).
PORTMANN, M., ARAN, J.-M., LAGOURGUE, P.: Testing for "recruitment" by electrocochleography: preliminary results. Ann. Otol St. Louis. **82**, 36—43 (1973).
POWELL, T. P. S., COWAN, W. M.: An experimental study of the projection of the cochlea. J. Anat. (Lond.) **96**, 269—284 (1962).
POWELL, T. P. S., ERULKAR, S. D.: Transneuronal cell degeneration in the auditory relay nuclei of the cat. J. Anat. (Lond.) **96**, 249—268 (1962).
PRICE, G. R.: Correspondence between cochlear microphonic sensitivity and behavioural threshold in the cat. J. acoust. Soc. Amer. **49**, 1899—1901 (1971).
PUGH, J. E., JR., ANDERSON, D. J., BURGIO, P. A., HORWITZ, M. R.: The origin of N_2 of the cochlear whole-nerve action potential. J. acoust. Soc. Amer. **53**, 325 (1973).
RADIONOVA, E. A.: Measuring the intensity of a brief sound signal at the first neuron level of the auditory system. Soviet Physics-Acoustics 8, 350—355 (1963).
RADIONOVA, E. A.: Reactions of neurons in cochlear nucleus to acoustic signals of varying duration. J. Higher Nervous Activity USSR **15**, 739 (1965).
RADIONOVA, E. A.: Two types of neurons in the cat's cochlear nuclei and their role in audition. In: Sensory Processes at the neuronal and behavioural levels, pp. 135—155. New York: Academic Press 1971.
RADIONOVA, E. A., POKOV, A. V.: Electrophysiological examination of neurons in cochlear nucleus of the cat. FASEB Trans. Suppl. **25**, T231—235 (1965).
RAINBOLT, H., SMALL, A. M.: Mach bands in auditory masking: an attempted replication. J. acoust. Soc. Amer. **51**, 567—574 (1972).
RASMUSSEN, G. L.: Selective silver impregnation of synaptic endings. In: New Research Techniques of Neuroanatomy, pp. 27—39. Springfield, Ill.: Thomas 1957.

RASMUSSEN, G. L.: Efferent fibres of the cochlear nerve and cochlear nucleus. In: Neural Mechanisms of the Auditory and Vestibular Systems, pp. 105—115. Springfield, Ill.: Thomas 1960.

RASMUSSEN, G. L.: Anatomic relationships of the ascending and descending auditory systems. In: Neurological Aspects of Auditory and Vestibular Disorders, pp. 5—19. Springfield, Ill.: Thomas 1964.

RASMUSSEN, G. L.: Efferent connections of the cochlear nucleus. In: Sensorineural Hearing Processes and Disorders, pp. 61—75. Boston, Mass: Little, Brown & Co. 1967.

RATLIFF, F.: Mach Bands, p. 41. San Francisco: Holden-Day 1965.

RHODE, W. S.: Observations of the vibration of the basilar membrane in squirrel monkeys using the Mössbauer technique. J. acoust. Soc. Amer. **49**, 1218—1231 (1971).

RHODE, W.: An investigation of cochlear mechanics using the Mössbauer effect. In: Basic Mechanisms in Hearing, pp. 49—63. New York: Academic Press 1973.

ROBLES, L., RHODE, W. S., GEISLER, C. D.: Measurements of the transient response of the basilar membrane using the Mössbauer technique. J. acoust. Soc. Amer. **53**, 292 (1973).

RODIECK, R. W., KIANG, N. Y.-S., GERSTEIN, G. L.: Some quantitative methods for the study of spontaneous activity of single neurons. Biophys. J. **2**, 351—368 (1962).

ROSE, J. E.: Organization of frequency sensitive neurons in the cochlear nucleus complex of the cat. In: Neural Mechanisms of the Auditory and Vestibular Systems, pp. 116—136. Springfield, Ill.: Thomas 1960.

ROSE, J. E.: Discharges of single fibres in the mammalion auditory nerve. In: Frequency Analysis and Periodicity Detection in Hearing, pp. 176—188. Leiden: Sijthoff 1970.

ROSE, J. E., GALAMBOS, R., HUGHES, J. R.: Microelectrode studies of the cochlear nuclei of the cat. Johns Hopkins Hosp. Bull. **104**, 211—251 (1959).

ROSE, J. E., BRUGGE, J. F., ANDERSON, D. J., HIND, J. E.: Phase-locked response to low frequency tones in single auditory nerve fibres of the squirrel monkey. J. Neurophysiol. **30**, 769—793 (1967).

ROSE, J. E., BRUGGE, J. F., ANDERSON, D. J., HIND, J. E.: Patterns of activity in single auditory nerve fibres of the squirrel monkey. In: Hearing Mechanisms in Vertebrates, pp. 144—157. London: Churchill 1968.

ROSE, J. E., BRUGGE, J. F., ANDERSON, D. J., HIND, J. F.: Some possible neural correlates of combination tones. J. Neurophysiol. **32**, 402—423 (1969).

ROSE, J. E., HIND, J. E., ANDERSON, D. J., BRUGGE, J. F.: Some effects of stimulus intensity on response of auditory nerve fibres in the squirrel monkey. J. Neurophysiol. **34**, 685—699 (1971).

ROSENBLITH, W. A.: Electrical responses from the auditory nervous system. Ann. Otolaryng. (St. Louis) **63**, 839—860 (1954).

RUPERT, A. L., MOUSHEGIAN, G.: Neuronal responses of kangaroo rat ventral cochlear nucleus to low frequency tones. Exp. Neurol. **26**, 84—102 (1970).

RUPERT, A., MOUSHEGIAN, G., GALAMBOS, R.: Unit responses to sound from auditory nerve of the cat. J. Neurophysiol. **26**, 449—465 (1963).

SACHS, M. B.: Stimulus-response relation for auditory nerve fibres: two-tone stimuli. J. acoust. Soc. Amer. **45**, 1025—1036 (1969).

SACHS, M. B., KIANG, N. Y.-S.: Two-tone inhibition in auditory nerve fibres. J. acoust. Soc. Amer. **43**, 1120—1128 (1968).

SANDO, I.: The anatomical interrelationships of the cochlear nerve fibres. Acta Oto-laryng. (Stockh.) **59**, 417—436 (1965).

SCHARF, B.: Fundamentals of auditory masking. Audiol. **10**, 30—40 (1971).

SCHARF, B., HELLMAN, R. P.: Model of loudness summation applied to impaired ears. J. acoust. Soc. Amer. **40**, 71—78 (1966).

SEATON, W. H., TRAHIOTIS, C.: Comparison of direct and indirect measures of critical bands of monaural chinchilla. J. acoust. Soc. Amer. **53**, 376 (1973).

SIEBERT, W. M.: Some implications of the stochastic behaviour of primary auditory neurones. Kybernetik **2**, 206—215 (1965).

SIEBERT, W. M.: Stimulus transformations in the peripheral auditory system. In: Recognizing Patterns, pp. 104—133. Cambridge: M.I.T. Press 1968.

Siebert, W. M.: Frequency discrimination in the auditory system: place or periodicity mechanisms? Proc. IEEE **58**, 723—730 (1970).
Simmons, F. B., Linehan, J. A.: Observations on a single auditory nerve fibre over a six week period. J. Neurophysiol. **31**, 799—805 (1968).
Small, A. M., Jr.: Pure tone masking. J. acoust. Soc. Amer. **31**, 1619—1625 (1959).
Smith, C. A., Dempsey, E. W.: Electron microscopy of the organ of Corti. Amer. J. Anat. **100**, 337 (1957).
Smith, R. L., Zwislocki, J. J.: Responses of some neurones of the cochlear nucleus to tone intensity increments. J. acoust. Soc. Amer. **50**, 1520—1525 (1971).
Smoorenburg, G. F.: Audibility region of combination tones. J. acoust. Soc. Amer. **52**, 603—614 (1972a).
Smoorenburg, G. F.: Combination tones and their origin. J. acoust. Soc. Amer. **52**, 615—632 (1972b).
Sohmer, H., Feinmesser, M.: Cochlear potentials recorded from the external ear in man. Ann. Otolaryng. (St. Louis) **76**, 427 (1967).
Spoendlin, H.: The organization of the cochlear receptor. In: Fortsch. Hals-, Nas-., Ohrenheilk. **13**, 1—227 (1966).
Spoendlin, H.: Ultrastructure and peripheral innervation pattern of the receptor in relation to the first coding of the acoustic message. In: Hearing Mechanisms in Vertebrates, pp. 89—119. London: Churchill 1968.
Spoendlin, H.: Structural basis of peripheral frequency analysis. In: Frequency Analysis and Periodicity Detection in Hearing, pp. 2—36. Leiden: Sijthoff 1970.
Spoendlin, H.: Degeneration behaviour of cochlear nerve. Arch. Ohr.-, Nas.-, Kehlk.-Heilk. **200**, 275—291 (1971).
Spoendlin, H.: Innervation densities of the cochlea. Acta Otolaryng. (Stockh.) **73**, 235—248 (1972).
Spoor, A., Eggermont, J. J.: Action potentials of the cochlea. Masking, adaptation and recruitment. Audiol. **10**, 340—352 (1971).
Spreng, M., Keidel, W. D.: Separierung von Cerebroaudiogramm (CAG), Neuroaudiogramm (NAG) und Otoaudiogramm (OAG) in der objektiven Audiometrie. Arch. Ohr.-, Nas.-, Kehl.-Heilk. **189**, 225 (1967).
Starr, A.: Suppression of single unit activity in cochlear nucleus of the cat following sound stimulation. J. Neurophysiol. **28**, 850—862 (1965).
Starr, A., Britt, R.: Intracellular recordings from cat cochlear nucleus during tone stimulation. J. Neurophysiol. **33**, 137—147 (1970).
Starr, A., Hellerstein, D.: Distribution of frequency following responses in cat cochlear nucleus to sinusoidal acoustic signals. Brain Res. **33**, 367—378 (1971).
Starr, A., Wernick, J. S.: Olivocochlear bundle stimulation: effects on spontaneous and tone evoked activities of single units in cat cochlear nucleus. J. Neurophysiol. **31**, 549—564 (1968).
Steele, C. R.: A possibility for sub-tectorial membrane fluid motion. In: Basic Mechanisms in Hearing, pp. 69—90. New York: Academic Press 1973.
Stopp, P. E., Whitfield, I. C.: Unit responses from the brainstem nuclei in the pigeon. J. Physiol. (Lond.) **158**, 165—177 (1961).
Stotler, W. A.: The projection of the cochlear nerve on the acoustic relay nuclei of the medulla. Anat. Rec. **103**, 561 (1949).
Stotler, W. A.: An experimental study of the cells and connections of the superior olivary complex of the cat. J. comp. Neurol. **98**, 401—432 (1953).
Suga, N.: Single unit activity in cochlear nucleus and inferior colliculus of echo-locating bats. J. Physiol. (Lond.) **172**, 449—474 (1963).
Suga, N.: Analysis of frequency-modulated sounds by auditory neurones of echo-locating bats. J. Physiol. (Lond.) **179**, 26—53 (1965a).
Suga, N.: Functional properties of auditory neurones in the cortex of echolocating bats. J. Physiol. **181**, 671—700 (1965b).
Tasaki, I.: Nerve impulses in individual auditory nerve fibres of guinea pig. J. Neurophysiol. **17**, 97—122 (1954).
Tasaki, I.: Hearing. Ann. Rev. Physiol. **19**, 417—438 (1957).

TAUB, H. B., RAAB, D. H.: Fluctuations of N_1 amplitude in relation to click-intensity discrimination. J. acoust. Soc. Amer. **46**, 969—978 (1969).

TEAS, D. C., ELDREDGE, D. H., DAVIS, H.: Cochlear responses to acoustic transients: an interpretation of whole-nerve action potentials. J. acoust. Soc. Amer. **34**, 1438—1459 (1962).

TEAS, D. C., KONISHI, T., NIELSEN, D. W.: Electrophysiological studies on the spatial distribution of the crossed olivocochlear bundle along the guinea pig cochlea. J. acoust. Soc. Amer. **51**, 1256—1264 (1972).

TERHARDT, E.: Über akustische Rauhigkeit und Schwankungsstärke. Acustica **20**, 215—224 (1968).

TERHARDT, E.: Frequency analysis and periodicity detection in the sensations of roughness and periodicity pitch. In: Frequency Analysis and Periodicity Detection in Hearing. Leiden: Sijthoff 1970.

TONNDORF, J.: Time/frequency analysis along the partition of cochlear models: a modified place concept. J. acoust. Soc. Amer. **34**, 1337—1350 (1962).

TRAHIOTIS, C., ELLIOTT, D. N.: Behavioural investigation of some possible effects of sectioning the crossed olivo-cochlear bundle. J. acoust. Soc. Amer. **47**, 592—596 (1970).

VIERNSTEIN, L. J., GROSSMAN, R. G.: Neural discharge patterns in the transmission of sensory information. Symp. on Inform. Th. **4**, 252—269 (1961).

WALSH, B. T., MILLER, J. B., GACEK, R. R., KIANG, N. Y.-S.: Spontaneous activity in the eight cranial nerve of the cat. Int. J. Neurosci. **3**, 221—236 (1972).

WARR, W. B.: Fiber degeneration following lesions in the anterior ventral cochlear nucleus of the cat. Exptl. Neurol. **14**, 453—474 (1966).

WARR, W. B.: Fiber degeneration following lesions in the postero-ventral cochlear nucleus in the cat. Anat. Rec. **157**, 338 (1962).

WATENABE, T.: Fundamental study of the neural mechanism in cats subserving the feature extraction process of complex sounds. Jap. J. Physiol. **22**, 569—583 (1972).

WATENABE, T., OHGUSHI, K.: FM sensitive auditory neuron. Proc. Jap. Acad. **44**, 968—973 (1968).

WATENABE, T., SIMADA, Z.: Auditory temporal masking: an electrophysiological study of single neurons in the cat cochlear nucleus and inferior colliculus. Jap. J. Physiol. **21**, 537—550 (1971).

WATSON, C. S.: Masking of tones by noise for the cat. J. acoust. Soc. Amer. **35**, 167—172 (1963).

WEISS, T. F.: A model of the peripheral auditory system. Kybernetik **3**, 153—175 (1966).

WIGHTMAN, F. L.: Pitch as auditory pattern recognition. In: Hearing Theory 1972, pp. 161—171. Eindhoven: IPO 1972.

WHITFIELD, I. C.: The Auditory Pathway. London: Arnold 1967.

WHITFIELD, I. C.: The functional organization of the auditory pathways. J. Sound Vib. **8**, 108—117 (1968).

WHITFIELD, I. C.: Central nervous processing in relation to spatio-temporal discrimination of auditory patterns. In: Frequency Analysis and Periodicity Detection in Hearing, pp. 136—147. Leiden: Sijthoff 1970.

WIEDERHOLD, M. L.: Variations in the effects of electric stimulation of the crossed olivocochlear bundle on cat single auditory-nerve fibre responses to tone bursts. J. acoust. Soc. Amer. **48**, 966—977 (1970).

WIEDERHOLD, M. L., KIANG, N. Y.-S.: Effects of electric stimulation of the crossed olivocochlear bundle on single auditory-nerve fibres in the cat. J. acoust. Soc. Amer. **48**, 950—965 (1970).

WIEDERHOLD, M. L., PEAKE, W. T.: Efferent inhibition of auditory-nerve responses: dependence on acoustic-stimulus parameters. J. acoust. Soc. Amer. **40**, 1427—1430 (1966).

WIENER, F. M., PFEIFFER, R. R., BACKUS, A. S. N.: On the sound pressure transformation by the head and auditory meatus of the cat. Acta oto-laryng. (Stockh.) **61**, 255—269 (1965).

WILSON, J. P.: Psychoacoustics of obstacle detection using ambient or self-generated noise. In: Les Systèmes Sonars Animaux, pp. 89—114, 1967.

WILSON, J. P.: An auditory after-image. In: Frequency Analysis and Periodicity Detection in Hearing, pp. 303—315. Leiden: Sijthoff 1970.

WILSON, J. P.: Psychoacoustical and neurophysiological aspects of auditory pattern recognition. In: The Neurosciences Third Study Program, pp. 147—153. Boston: M.I.T. Press 1974.

WILSON, J. P., EVANS, E. F.: Grating acuity of the ear: Psychophysical and neurophysiological measures of frequency resolving power. Proc. 7th Internat. Congr. on Acoustics, Vol. 3, pp. 397—400. Budapest: Akademiai Kiado 1971.

WILSON, J. P., JOHNSTONE, J. R.: Capacitive probe measures of basilar membrane vibration. In: Hearing Theory, pp. 172—181. Eindhoven: IPO 1972.

WILSON, J. P., JOHNSTONE, J. R.: Basilar membrane correlates of the combination tone $2f_1—f_2$. Nature **241**, 206—207 (1973).

YOSHIE, N.: Auditory nerve action potential responses to clicks in man. Laryngoscope **78**, 198—215 (1968).

YOSHIE, N.: Clinical cochlear response audiometry by means of an average response computer: non-surgical technique and clinical use. Rev. Laryng. Suppl., 646—670 (1971).

YOSHIE, N., OHASHI, T., SUZUKI, T.: Non-surgical recording of auditory nerve action potentials in man. Laryngoscope **77**, 76 (1967).

ZWICKER, E.: Die Grenzen der Hörbarkeit der Amplitudenmodulation und der Frequenzmodulation eines Tones. Acustica **2**, 125 (1952).

ZWICKER, E.: Die Verdeckung von Schmalbandgeräuschen durch Sinustöne. Acustica **4**, 415 (1954).

ZWICKER, E.: Der ungewöhnliche Amplitudengang der nichtlinearen Verzerrungen des Ohres. Acustica **5**, 67—74 (1955).

ZWICKER, E.: Masking and physiological excitation as consequences of the ear's frequency analysis. In: Frequency Analysis and Periodicity Detection in Hearing, pp. 376—394. Leiden: Sijthoff 1970.

ZWICKER, E.: Introduction to round-table-discussion on "critical bands". Proc. 7th Internat. Cong. on Acoustics, Vol. 1, pp. 189—192. Budapest: Akademiai Kiado 1971.

ZWICKER, E., FASTL, H.: On the development of the critical band. J. acoust. Soc. Amer. **52**, 699—702 (1972).

ZWICKER, E., FLOTTORP, G., STEVENS, S. S.: Critical bandwidth in loudness summation. J. acoust. Soc. Amer. **29**, 548—557 (1957).

ZWISLOCKI, J. J.: Central masking and neural activity in the cochlear nucleus. Audiol. **10**, 48—59 (1971).

Chapter 2

Physiological Studies of Auditory Nuclei of the Pons*

J. M. GOLDBERG, Chicago, Illinois (USA)

With 20 Figures

Contents

I. Introduction 109
II. Functional Organization 110
A. Morphology of the Pontine Auditory Nuclei 110
B. Evoked-Potential Studies 111
1. Anatomical Pathways 111
2. Medial Superior Olive 112
3. Evidence of Binaural Interactions 113
C. Single-Unit Studies 113
1. Medial Superior Olive 113
2. Lateral Superior Olive 116
3. Nucleus of the Trapezoid Body 117
4. Other Nuclei of the Superior Olivary Complex 120
5. Nuclei of the Lateral Lemniscus 120
III. Discharge Properties of the Neurons 122
A. Monaural Response Areas 122
B. Binaural Response Areas 125
C. Monaural Intensity Functions 127
D. Binaural Intensity Functions 130
E. Response to Clicks 131
IV. Mechanisms of Sound Localization 133
A. Response to High-Frequency Tones 133
B. Response to Low-Frequency Tones 134
C. Response to Transients 139
References 141

I. Introduction

The superior olivary complex and the nuclei of the lateral lemniscus are a group of more or less distinct nuclei, located in the pons and lying in close association with the major tracts interconnecting the cochlear nuclei with the inferior colliculus. The tracts, the acoustic striae and the lateral lemniscus, provide afferent

* Preparation of this chapter supported by Grant NS-05237, National Institutes of Health, and by a Sloan Foundation Grant.

inputs to the pontine auditory nuclei and the nuclei, in turn, make major contributions to the ascending pathways. In considering the physiology of the pontine auditory nuclei, three facts should be kept in mind. First, each nucleus has a distinctive set of anatomical connections. One would then expect, and this has been amply confirmed, that the neurons in the several nuclei would differ in their physiological characteristics. Second, the olivary and lemniscal nuclei constitute the lowest level of the auditory system receiving ascending inputs from the two ears. This has suggested that the nuclei might play an important, if not unique, role in the mechanisms of binaural hearing. Accordingly, several studies have been devoted to elucidating the mechanisms of binaural interaction at the single-unit level. Not surprisingly, many neurons are affected by stimulation of either ear and are sensitive to the binaural cues involved in the localization of sound sources. And third, the nuclei are important links not only in the ascending auditory pathways, but also in the descending pathways which ultimately lead back to the cochlea. This last topic will be considered by DESMEDT (this volume).

II. Functional Organization

A. Morphology of the Pontine Auditory Nuclei

The structure of the superior olivary complex (SOC) and the nuclei of the lateral lemniscus (LLN) is illustrated in Fig. 1. The terminology employed here is similar to that used by CAJAL (1909) and by STOTLER (1953). The three most distinctive nuclei of the SOC (Fig. 1A) are the medial superior olive (MSO),

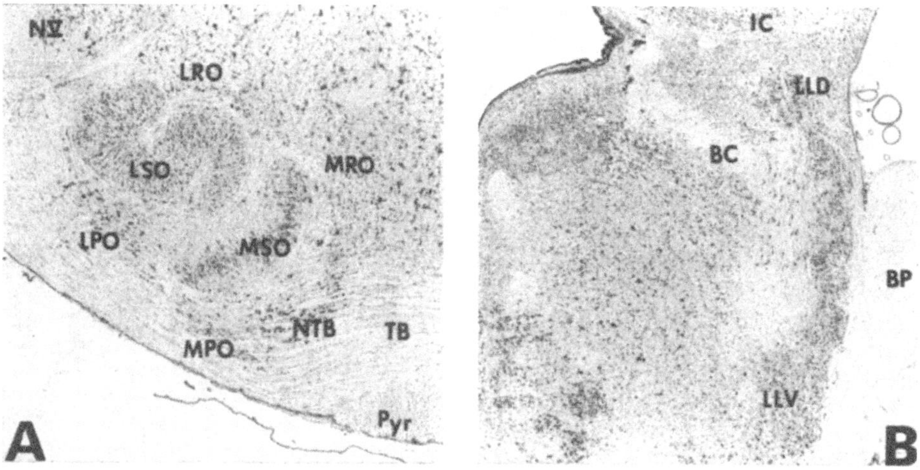

Fig. 1A, B. Frontal sections through the superior olivary complex (A) and the lateral lemniscus (B) in the cat. Abbreviations: *BC* brachium conjunctivum; *BP* brachium pontis; *IC* inferior colliculus; *LLD* dorsal nucleus of lateral lemniscus; *LLV* ventral nucleus of lateral lemniscus; *LPO* lateral preolivary nucleus; *LRO* lateral retroolivary region; *LSO* lateral superior olive; *MPO* medial preolivary nucleus; *MRO* medial retroolivary region; *MSO* medial superior olive; *NTB* nucleus of trapezoid body; *NV* spinal trigeminal nucleus; *Pyr* medullary pyramid; *TB* trapezoid body the lateral superior olive (LSO), and the nucleus of the trapezoid body (NTB). The MSO is a thin ribbon of spindle-shaped cells. Even in Nissl-stained sections, it is apparent that the neurons possess two primary dendrites, one pointing ventromedially, the other dorsolaterally. The cells of the S-shaped LSO are similar in appearance to those of the MSO and have their dendrites directed towards the curved border of the nucleus. The NTB is an ovoid mass of cells lying among the fibers of the trapezoid body (TB); the predominant cell type has a round or oval shape and is characterized by an eccentric nucleus. The remaining regions of the SOC contain more heterogeneous populations of cells. The medial preolivary nucleus (MPO) is situated ventral to the MSO and NTB, the lateral preolivary nucleus (LPO) ventral to the LSO. Capping the complex on its dorsal surface is the retroolivary region, which may be divided into medial (MRO) and lateral zones (LRO).

The lemniscal nuclei (Fig. 1B) consist of two distinct groups, a dorsal nucleus (LLD) largely composed of medium-sized, deeply staining neurons and a ventral nucleus (LLV) containing somewhat smaller and more palely staining cells. Joining the two nuclei, and situated among the dorsoventrally oriented fibers of the lemniscus, is a more loosely arranged aggregate of cells, similar in appearance to the neurons of the LLV.

The morphology, particularly that of the SOC, displays considerable variation among species (Irving and Harrison, 1967; Moore and Moore, 1971). Unfortunately, physiological studies have been confined to carnivores and we can only guess as to the physiological implications of these morphological diversities.

B. Evoked-Potential Studies

Early studies of the physiology of brainstem auditory pathways employed relatively large electrodes, which recorded from populations of cells and fibers. These studies, though necessarily crude when compared to modern single-unit studies, did provide information about the pathways taken by auditory impulses in travelling from the cochlea to the midbrain and about possible sites of binaural convergence. In the case of one nucleus, the MSO, evoked-potential studies led to relatively precise statements as to the kinds of synaptic inputs.

1. Anatomical Pathways

The first electrophysiological studies of the brainstem auditory pathways were those of Saul and Davis (1932). Subsequently, more systematic studies were undertaken by Kemp *et al.* (1937a, 1937b), Ades and Brookhart (1950) and Jungert (1958). Figure 2 shows typical evoked potentials recorded from different locations within the auditory pathway. At successively higher levels, there is an increase in both the latent period and the duration of the response. The increase in latent period can be attributed to nuclear and conduction delays. The prolongation of the response may, in part, reflect the diverse origins of the fibers providing inputs to a region. To take one example, an electrode placed in the lateral lemniscus should record from second-order fibers coming from the contralateral cochlear nuclei, from third-order fibers originating in the olivary complex, and from third- and fourth-order elements arising within the lemniscal nuclei. The lengthening of the responses may also reflect the nature of the synaptic potentials occurring at higher levels of the system.

Anatomical studies (BARNES *et al.*, 1943; WARR, 1966, 1969; FERNANDEZ and KARAPAS, 1967; VAN NOORT, 1969) demonstrate that the direct pathway from the cochlear nucleus to the lateral lemniscus is almost entirely crossed. The fibers cross in the three acoustic striae, of which the most prominent is the ventral stria or trapezoid body. The major indirect pathway involves an additional synapse in the superior olive — more particularly, in the MSO and LSO — and provides to the lemniscus both its main ipsilateral input and an additional contralateral input (STOTLER, 1953; VAN NOORT, 1969). Responses can be recorded from the lemniscus or from the inferior colliculus on separate stimulation of the contralateral and ipsilateral ears. The relative magnitude of the ipsilateral response should provide some measure of the importance of the indirect pathway. Such a comparison indicates that the ipsilateral and contralateral responses are of similar magnitude, though usually the contralateral response is somewhat larger and has a slightly shorter latent period and an appreciably faster rise time (ADES and BROOKHART, 1950; ROSENZWEIG and WYERS, 1955; ROSENZWEIG and SUTTON, 1958). The results confirm that the indirect pathway, and by implication the superior olive, forms a major link in the transmission of ascending auditory impulses.

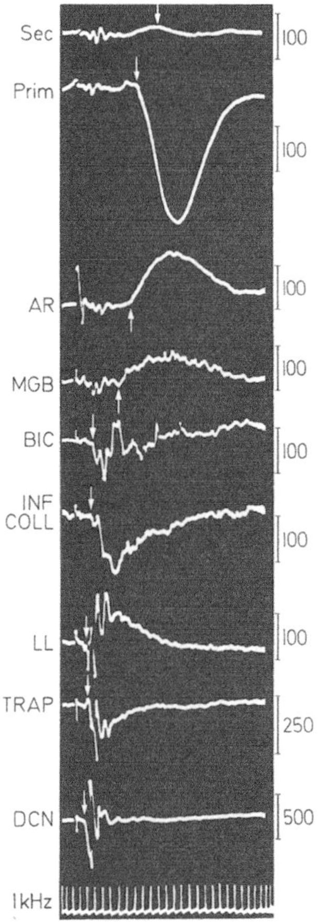

Fig. 2. Potentials evoked by binaural click stimulation from dorsal cochlear nucleus (*DCN*), trapezoid body (*TRAP*), lateral lemniscus (*LL*), inferior colliculus (*INF COLL*), inferior brachium (*BIC*), medial geniculate (*MGB*), auditory radiation (*AR*), primary auditory cortex (*PRIM*), and a second auditory area (*SEC*). Voltage calibrations in μV. (From ADES and BROOKHART, 1950)

2. Medial Superior Olive

As first described by STOTLER (1953), the MSO receives almost symmetrical projections from the ipsilateral and contralateral cochlear nuclei. The ipsilateral afferents mainly innervate laterally directed dendrites, the contralateral afferents medially directed dendrites. A striking confirmation of this anatomical arrangement was provided by GALAMBOS *et al.* (1959), who studied the pattern of slow waves recorded as a microelectrode traversed the MSO. When the electrode was medial to the MSO, the potential evoked by ipsilateral stimulation was predominantly positive, the contralateral potential predominantly negative. Lateral to the MSO, the polarities reversed. The ipsilateral potential turned negative, the contralateral potential positive. The explanation offered by GALAMBOS *et al.* (1959), and agreed to by subsequent workers

(BIEDENBACH and FREEMAN, 1964; TSUCHITANI and BOUDREAU, 1964), is that ipsilateral stimulation results in a depolarization of the lateral dendrites, contralateral stimulation in a depolarization of the medial dendrites. The depolarizations should have an excitatory influence on the postsynaptic neuron. The results thus suggest that the majority of MSO neurons receive a predominantly excitatory input from each ear. Single-unit studies (GOLDBERG and BROWN, 1968; GUINAN, 1968) confirm this conclusion.

3. Evidence of Binaural Interactions

KEMP and ROBINSON (1937b) investigated the possibility of a binaural convergence onto single neurons. Evoked potentials were recorded from the lateral lemniscus and it was shown that the response to binaural clicks had a smaller amplitude than would be expected from the algebraic sum of the two monaural responses. Other experiments, also indicating the existence of binaural convergence, were subsequently performed in the superior olive (ROSENZWEIG and AMON, 1955; WERNICK and STARR, 1968), the lateral lemniscus (ROSENZWEIG and SUTTON, 1958), the inferior colliculus (ROSENZWEIG and WYERS, 1955), and the auditory cortex (ROSENZWEIG, 1954; HIRSCH, 1968).

Such experiments suffer from several limitations. First, the slow waves recorded from most auditory regions represent activity derived from functionally inhomogeneous neuronal populations. Second, it is usually difficult to determine the kinds of binaural interactions that are occurring. And third, the amount of occlusion may only poorly reflect the degree of binaural convergence. This last point may be illustrated by considering the MSO. The slow waves recorded from the MSO most probably represent excitatory postsynaptic potentials (EPSPs). Simultaneous stimulation of the two ears would result in a spatial summation of the EPSPs, a summation which is in most cases approximately linear (ECCLES, 1964). Because of these drawbacks, slow-wave studies have been largely supplanted by single-unit studies. This is somewhat unfortunate since slow-wave potentials, when properly interpreted, can provide important clues as to what is happening in large populations of neurons.

C. Single-Unit Studies

Microelectrode experiments have been of particular use in determining the tonotopic organization of the various brainstem auditory nuclei and in defining the nature of their afferent inputs. In this section, we shall summarize pertinent physiological observations and, where possible, relate them to the results of anatomical studies. Most of the nuclei to be considered possess a precise tonotopic organization and many of them have been implicated in the mechanisms of binaural hearing.

1. Medial Superior Olive

The main afferent projections to the MSO come from the rostral (or spherical-cell) region of the anteroventral cochlear nucleus. The projections are bilateral and are topographically organized in both the cat (WARR, 1966; OSEN, 1969; VAN NOORT, 1969) and dog (GOLDBERG and BROWN, 1968). The tonotopic organization of the nucleus (Goldberg and Brown, 1968; Guinan, 1968) is consistent with the aforementioned anatomical findings. Low-frequency neurons are found in the dorsal part of the MSO, high-frequency neurons in the ventral part. There is some suggestion that very high frequencies (> 10 kHz) may have a disproportionately small representation.

Most MSO neurons can be affected by stimulation of either ear (Goldberg and Brown, 1968). In considering such binaural neurons, it will be useful to classify them into two categories. *Excitatory-excitatory* or *EE* neurons receive a predominantly excitatory input from each ear. *Excitatory-inhibitory* or *EI* neurons receive a

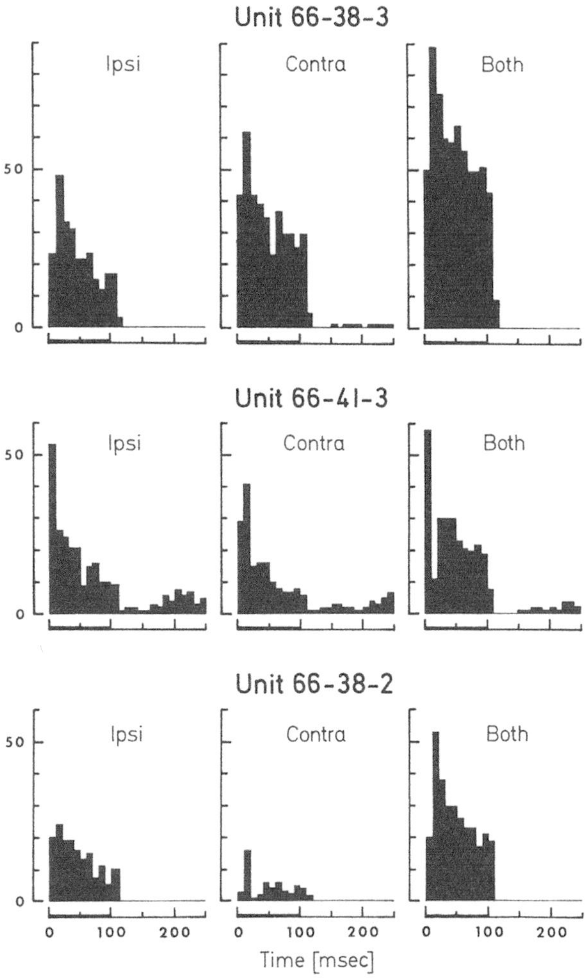

Fig. 3. Average-response histograms for three EE cells of the MSO. In Figs. 3, 4 and 11, ordinate is number of spikes cumulated for 20 or 50 stimulus presentations. Abscissa is time after start of stimulus. Binwidth is 10 msec. Solid bar denotes stimulus duration. All stimuli, 100-msec duration, 500-msec repetition period. Best-frequency tone bursts unless otherwise stated. Intensities in dB SPL. *Unit 66-38-3:* 1.7 kHz, 70 dB, 50 trials. *Unit 66-41-3:* 1.75 kHz, 60 dB, 20 trials. *Unit 66-38-2:* 4.7 kHz, 60 dB, 20 trials. (From Goldberg and Brown, 1968)

predominantly excitatory input from one ear and a predominantly inhibitory input from the other ear. The classification is justified by the observation that, for many binaural neurons, the response to stimulation of a particular ear, whether it be excitatory or inhibitory, is qualitatively similar over a wide range of frequencies to either side of best frequency. It must be recognized, nevertheless, that the classification represents an oversimplification. Later in the chapter, we shall review evidence which indicates that monaural stimulation may, under appropriate circumstances, activate both excitatory and inhibitory inputs to individual binaural neurons.

In their study of the dog MSO, GOLDBERG and BROWN (1968) found that roughly three-fourths of the binaural neurons were EE cells. The responses of three such neurons are illustrated in Fig. 3. Several features may be noted. First, the pattern of response to either monaural or binaural stimulation is relatively simple,

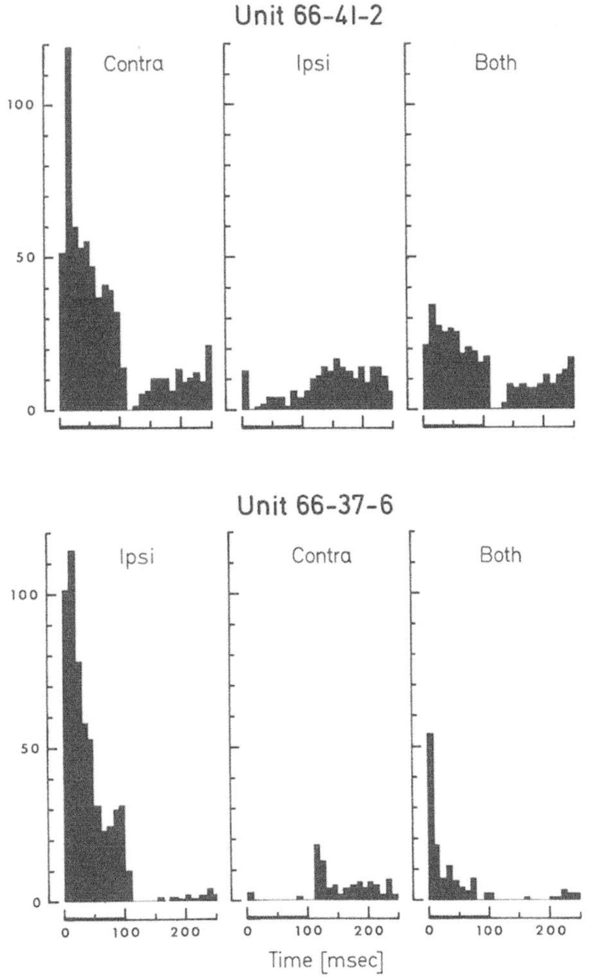

Fig. 4. Average-response histograms for two EI cells of the MSO. *Unit 66-41-2:* white-noise bursts, 50 trials. *Unit 66-37-6:* 7.6 kHz, 90 dB, 50 trials. (From GOLDBERG and BROWN, 1968)

consisting of a high-frequency discharge followed by a gradual decline to a maintained firing level. Second, though stimulation of either ear results in a discharge, one of the ears is usually more effective. And third, the most common forms of binaural interaction involve either a summation of discharge (*unit 66–38–3*) or a facilitation (*unit 66–38–2*). Occlusion was consistently found (*unit 66–41–3*) only at the onset of discharge, and then only when the firing rate was quite high.

The remaining binaural cells were of the EI variety. For most of these, contralateral stimulation was excitatory and ipsilateral stimulation inhibitory (Fig. 4, *unit 66–41–2*). But the reverse situation was also encountered (Fig. 4, *unit 66–37–6*). The response of EI cells to stimulation of the excitatory ear was similar to the excitatory responses observed in EE cells. Stimulation of the inhibitory ear resulted in a suppression of spontaneous discharge and a marked reduction in the response to simultaneous stimulation of the excitatory ear.

Both EE and EI neurons have been encountered in the cat MSO (GALAMBOS *et al.*, 1959; MOUSHEGIAN *et al.*, 1964a, 1964b, 1967; HALL, 1965; RUPERT *et al.*, 1966; GUINAN, 1968), but there have been no systematic studies of the relative proportions of the two kinds of units. The slow-wave studies, reviewed above, would suggest that EE neurons predominate.

2. Lateral Superior Olive

The ipsilateral afferents to the LSO come from the rostral (spherical-cell) region of the anteroventral cochlear nucleus and are topographically organized (WARR, 1966; OSEN, 1969; VAN NOORT, 1969). TSUCHITANI and BOUDREAU (1966) have demonstrated that the LSO has a precise tonotopic organization. High-frequency neurons are encountered in the ventromedial limb of the *S*-shaped nucleus, low-frequency neurons in the dorsolateral limb. The whole range of frequencies is represented in each frontal plane and the frequency gradient follows the contours of the nucleus. Here, in contrast to the MSO, low frequencies may have a disproportionately small representation.

Neurons of the LSO are activated by ipsilateral stimulation (GALAMBOS *et al.*, 1959; GOLDBERG *et al.*, 1963; TSUCHITANI and BOUDREAU, 1966), the great majority also being inhibited by contralateral stimulation (BOUDREAU and TSUCHITANI, 1968, 1970; TSUCHITANI and BOUDREAU, 1969). The behavior of one such EI cell is illustrated in Fig. 5. The neuron, like most of those studied by BOUDREAU and TSUCHITANI, had no spontaneous activity. Stimulation of the ipsilateral ear led to a maintained response. Simultaneous contralateral stimulation resulted in a response suppression, which grew more marked the greater the intensity of the contralateral tone.

The contralateral inhibition involves an inhibitory interneuron, most probably located in the ipsilateral NTB. The cells of the latter nucleus are activated by contralateral stimulation (see below) and they send their axons to the ipsilateral LSO (RASMUSSEN, 1967; VAN NOORT, 1969). The contralateral inhibitory pathway must be precisely organized. For any given LSO neuron, the best frequencies (BFs) for ipsilateral excitation and contralateral inhibition are virtually identical (BOUDREAU and TSUCHITANI, 1968 and 1970; TSUCHITANI and BOUDREAU, 1969). The only LSO neurons apparently not inhibited by contralateral stimulation are those with BFs less than 1.0 kHz (Boudreau and Tsuchitani, 1970). This may be related to the observation that NTB lesions result in anterograde degeneration which is most marked in the ventromedial (high-frequency) limb of the LSO and least pronounced in the dorsolateral (low-frequency) limb (Fernandez and Goldberg, unpublished findings).

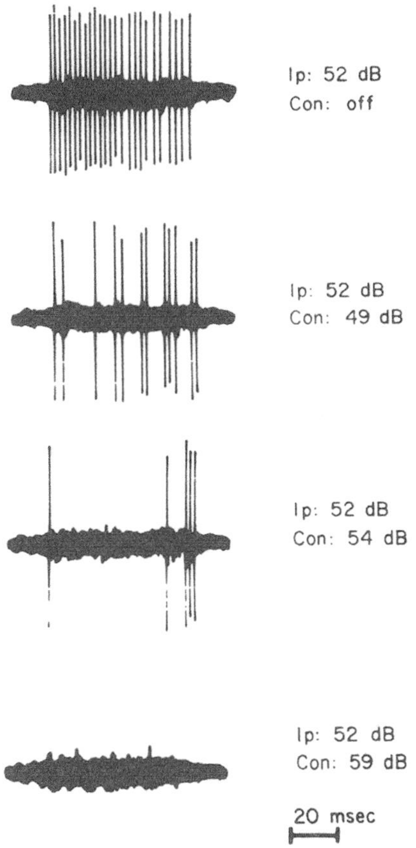

Fig. 5. Discharge of LSO cell to monaural and binaural best-frequency tones, 35 kHz, 60-msec duration. Intensity in dB SPL. (From Boudreau and Tsuchitani, 1970)

3. Nucleus of the Trapezoid Body

The NTB receives a predominantly contralateral projection from the cochlear complex (Stotler, 1953; Harrison and Irving, 1964; van Noort, 1969). The great majority of NTB neurons respond only to stimulation of the contralateral ear (Galambos *et al.*, 1959; Goldberg *et al.*, 1963; Goldberg and Brown, 1968; Guinan, 1968). A tonotopic organization has been demonstrated by Guinan (1968). High frequencies are represented ventromedially, low frequencies dorsolaterally.

A distinctive feature of the NTB is that its afferents include the largest fibers in the trapezoid body and these end as calyces of Held (CAJAL, 1909; STOTLER, 1953; MOREST, 1968). Each calyx makes a series of synaptic contacts of more or less typical morphology with the soma of the postsynaptic cell (LENN and REESE, 1966). The neurons innervated by the calyces – the principal cells of MOREST (1968) – are the most numerous cells of the NTB. The action potentials recorded from the principal cells may have triphasic waveforms (PFEIFFER, 1966a; GUINAN, 1968), similar to those seen in regions of the anteroventral cochlear nucleus also characterized by calyceal endings (PFEIFFER, 1966a). The potential consists of a positive-negative component, followed some 0.4–0.6 msec later by a second, usually larger and predominantly negative, component. PFEIFFER (1966a) presented circumstantial evidence that the first component represented activity in the presynaptic ending, the second component an impulse in the postsynaptic neuron. LI and GUINAN (1971) have offered conclusive proof that this is so.

The results indicate that the triphasic waveforms observed in the NTB are recorded from principal cells. Other units encountered in the NTB have more conventional diphasic waveforms and many of these have discharge characteristics similar to those of triphasic units (GOLDBERG *et al.*, 1964; GUINAN, 1968). The behavior of both kinds of units may be illustrated by Fig. 6 (left). The unit was diphasic and responded only to contralateral stimulation. Its response to tone bursts took a simple form. At low intensities, there was an irregular, but maintained discharge. As intensity was increased, the discharge became more vigorous and the latent period became shorter and more precisely timed. The latent period was some 1 msec less than that encountered in other nuclei of the SOC. GUINAN (1968) has shown that, in their response to clicks, the principal cells may have latent periods 0.5–1.0 msec shorter than those observed elsewhere in the complex.

The short latent period of NTB cells may reflect their calycoid innervation. The large myelinated fibers should have a rapid conduction velocity. More importantly, the small interval between the presynaptic and postsynaptic components of the triphasic action potential indicates a minimal nuclear delay, presumably the result of the nearly simultaneous activation of the multiple contacts of the calycoid ending. The rapid transmission through the calyceal synapse has an obvious functional consequence. NTB neurons, most likely principal cells, function as inhibitory interneurons in the contralateral pathway reaching the LSO. Given the short latent periods of the principal cells, contralateral and ipsilateral activity could still arrive in the LSO at virtually the same time, despite the interposition of an additional synapse in the contralateral pathway.

Certain anatomical complications, summarized by MOREST (1968), are of importance in considering the physiology of the NTB. First, the principal cell, in addition to its receiving a calycoid ending, is innervated by a peridendritic plexus partly derived from collaterals of axons passing to the MSO. How this dendritic input modifies the activity of the principal cell is not known. Second, while the main axon of the principal cell most probably projects to the LSO, collaterals are distributed to other regions. MOREST (1968) describes such collaterals as arborizing within the NTB or projecting to the medial retroolivary region. Many NTB axons pass through the MSO on their way to the LSO (VAN NOORT,

1969) and it is possible that synaptic contacts are made with MSO neurons. Collaterals may also pass to the ipsilateral MPO and LPO (CAJAL, 1909; FERNANDEZ and GOLDBERG, unpublished observations). In short, the principal cells may function as inhibitory interneurons, not only for the LSO, but also for the remaining nuclei of the complex. Third, MOREST (1968) describes two other kinds of neurons in the NTB, the stellate cell and the elongate cell. Perhaps related to this,

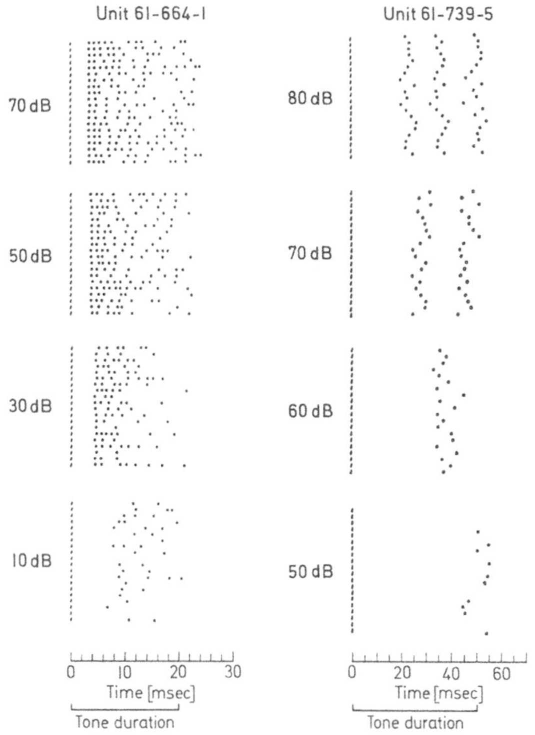

Fig. 6. Firing (dot) patterns for two NTB cells. Left-hand hash marks denote start of stimulus, dots to right of mark indicate discharge times during single stimulus. Each block, response to 20 repetitions of same stimulus. For both units, stimuli were 9.0 kHz (best-frequency) tone bursts of indicated duration. Intensities in dB SPL. (From GOLDBERG *et al.*, 1964)

single units have been encountered in the NTB whose discharge properties are distinguishable from those of the presumed principal cells. One of these is illustrated in Fig. 6 (right) (GOLDBERG *et al.*, 1964). It differs from the unit of Fig. 6 (left) in having a longer latent period, a lower range of discharge frequencies, and a more regular discharge pattern. Such neurons are similar in many respects to olivocochlear fibers (FEX, 1963 and 1965) and have been seen in both the NTB and the MPO (GOLDBERG *et al.*, 1964; GOLDBERG and BROWN, 1968). GUINAN (1968) has also described NTB units which discharge, throughout some or all of their effective frequency range, only at stimulus termination. Both kinds of neurons may be affected by ipsilateral, as well as contralateral, stimulation and are less frequently encountered than the presumed principal cells.

4. Other Nuclei of the Superior Olivary Complex

Relatively little information exists concerning the remaining nuclei of the SOC, including the medial and lateral preolivary nuclei (MPO and LPO) and the medial and lateral retroolivary regions (MRO and LRO). Their afferent connections from the cochlear nuclei have been traced in anatomical studies (STOTLER, 1953; WARR, 1966, 1969; FERNANDEZ and KARAPAS, 1967; VAN NOORT, 1969). The preolivary and retroolivary nuclei may be important links in the descending auditory pathways (RASMUSSEN, 1946, 1964, and 1967; ROSSI and CORTESINA, 1963; VAN NOORT, 1969), but their role in the ascending pathways is unclear. In particular, there is no anatomical evidence that they send fibers into the lateral lemniscus (STOTLER, 1953; VAN NOORT, 1969). Physiological observations are also meager, being mainly related to the laterality of afferent projections (GOLDBERG *et al.*, 1963; GOLDBERG and BROWN, 1968; GUINAN, 1968; BOUDREAU and TSUCHITANI, 1970). Most of the neurons of the MPO and MRO respond to contralateral stimulation, fewer to ipsilateral stimulation. For the LPO and LRO the reverse is true. These findings are consistent with the aforementioned anatomical studies. Binaural neurons have been encountered in all four nuclei, but the data are too fragmentary to permit generalizations.

5. Nuclei of the Lateral Lemniscus

The LLN possess a tonotopic organization (AITKIN *et al.*, 1970). Within both the LLD and the LLV, there is an orderly dorsoventral sequence of best frequencies from low to high. The two nuclei differ, however, in the laterality of their inputs.

The great majority of LLV neurons are affected only by stimulation of the contralateral ear (AITKIN *et al.*, 1970). None are solely influenced by ipsilateral stimulation. The few binaural neurons encoundtered are located at the margins of the nucleus. The relative paucity of binaural neurons within the LLV would seem to preclude any substantial innervation from the SOC, since the main lemniscal inputs from the olive come from the binaural neurons of the MSO and LSO (STOTLER, 1953; VAN NOORT, 1969). Anatomical evidence indicates that the LLV receives a major projection from the contralateral cochlear complex (BARNES *et al.*, 1943; WARR, 1966 and 1969; FERNANDEZ and KARAPAS, 1967; VAN NOORT, 1969) and a smaller, restricted projection from the ipsilateral cochlear complex (WARR, 1969).

A large fraction of the LLD neurons are affected by stimulation of both ears, fewer solely by stimulation of the contralateral ear, and still fewer solely by stimulation of the ipsilateral ear (AITKIN *et al.*, 1970; BRUGGE *et al.*, 1970). Low-frequency binaural neurons can be activated by stimulation of either ear, though contralateral stimulation is usually more effective; the units are sensitive to variations in interaural phase. High-frequency binaural neurons are usually excited by contralateral stimulation and inhibited by ipsilateral stimulation. The LLD (BARNES *et al.*, 1943; WARR, 1966 and 1969; FERNANDEZ and KARAPAS, 1967; VAN NOORT, 1969) receives an exclusively contralateral input from the cochlear complex. Binaural inputs may be provided by the ipsilateral MSO and by the LSO of both sides (STOTLER, 1953; VAN NOORT, 1969). Another source of binaural convergence is the commissure of Probst, which links the LLD of the two sides (WOOLLARD and HARPMAN, 1940; GOLDBERG and MOORE, 1967).

A wide variety of discharge patterns are seen in the lemniscal nuclei (Aitkin *et al.*, 1970; Brugge *et al.*, 1970). In some cells, the pattern is simple, consisting of a maintained discharge. In others, complicated patterns are observed and the particular pattern obtained may depend on the frequency and intensity of the stimulating tone. Figure 7 provides one of several possible examples. The best

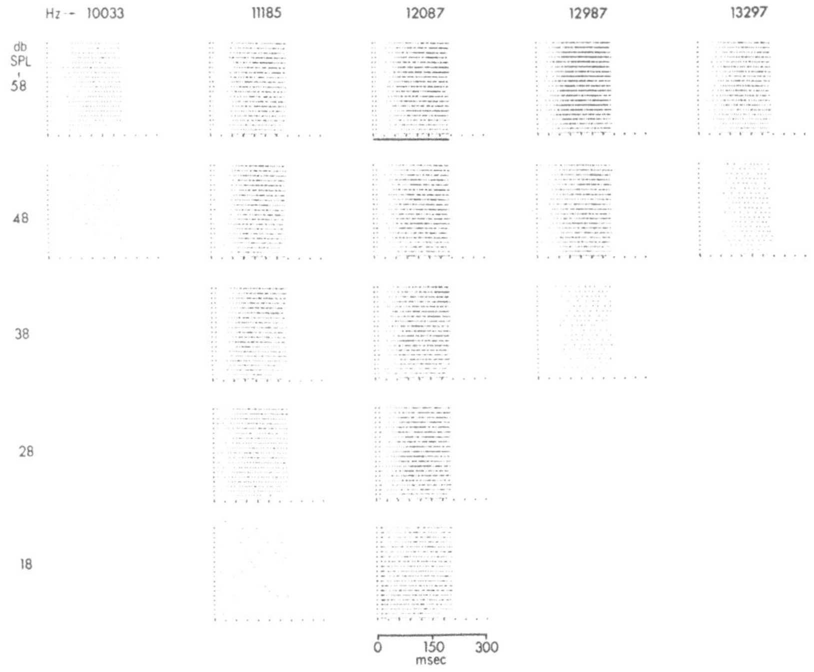

Fig. 7. Firing (dot) patterns for an LLD neuron. Tone bursts, 200 msec, 1/sec. Best-frequency: near 12087 Hz. (From Aitkin *et al.*, 1970)

frequency of the neuron was near 12087 Hz. Best-frequency tones of 18 dB evoke a sustained discharge. At 28 dB, a gap appears, the discharge now consisting of an onset spike, a silent pause, and then a resumption of firing. With further intensity increases, the latent period of the onset spike decreases, while the duration of the gap increases. Tones below best frequency evoke a similar sequence of patterns. At higher frequencies, the onset spike is only sporadically present and the predominant pattern involves a gradual buildup of activity.

Complicated discharge patterns have been observed in the cochlear complex (Kiang *et al.*, 1965a; Greenwood and Maruyama, 1965; Pfeiffer, 1966b; Gerstein *et al.*, 1968; Erulkar *et al.*, 1968), in the lemniscal nuclei (Aitkin *et al.*, 1970; Brugge *et al.*, 1970), and in the inferior colliculus (Rose *et al.*, 1963; Hind *et al.*, 1963; Nelson and Erulkar, 1963). The patterns have been interpreted as indicating that the firing of individual neurons is governed by a complex interplay of excitatory and inhibitory processes. Much simpler patterns are obtained in the MSO, the LSO, and from the presumed principal cells of the NTB. The apparent difference between the main nuclei of the SOC, on the one hand, and the lemniscal nuclei and inferior colliculus, on the other, may be related to their afferent inputs. In particular the main input from the cochlear complex to the LSO and MSO comes from the spherical-cell portion of the anteroventral nucleus (WARR, 1966; OSEN, 1969; VAN NOORT, 1969). The discharge characteristics of spherical-cell neurons are relatively simple and in most respects resemble those of auditory-nerve fibers (KIANG *et al.*, 1965a; PFEIFFER and KIANG, 1965; PFEIFFER, 1966b). Inputs to the lemniscal nuclei and inferior colliculus, in contrast, come from all divisions of the cochlear complex (WARR, 1966 and 1969; FERNANDEZ and KARAPAS, 1967; OSEN, 1969; VAN NOORT, 1969). Many cochlear-nucleus neurons, particularly those from its dorsal nucleus, have response patterns as complicated as those seen in the lemniscus and inferior colliculus (KIANG *et al.*, 1965a; GREENWOOD and MARUYAMA, 1965; PFEIFFER, 1966b; GERSTEIN *et al.*, 1968; ERULKAR *et al.*, 1968).

Stated more generally, the discharge of lemniscal and colliculus neurons is influenced by activity originating from many distinct regions of the lower auditory system and the neurons of these regions themselves exhibit a wide variety of response characteristics. Lemniscal and colliculus neurons are also under the control of descending pathways (RASMUSSEN, 1964; VAN NOORT, 1969). Finally, although there is little concrete information, it may be supposed that intrinsic connections of the lemniscal nuclei and of the inferior colliculus – including connections between closely adjacent neurons as well as connections provided by the lemniscal and intercollicular commissures – serve to modify activity triggered by the ascending and descending pathways. How these various influences cooperate to determine the complicated firing patterns of single units remains a mystery, as does the functional significance of the firing patterns themselves.

III. Discharge Properties of the Neurons

A. Monaural Response Areas

Neurons of the SOC and LLN, like those at other levels of the auditory system, are activated by tonal stimuli only within a limited domain of frequencies and intensities. Threshold-tuning curves, which define the boundaries of this domain, have been described for neurons of the MSO (GALAMBOS *et al.*, 1959; MOUSHEGIAN *et al.*, 1964a; GOLDBERG and BROWN, 1968), the LSO (TSUCHITANI and BOUDREAU, 1966; BOUDREAU and TSUCHITANI, 1970), and the NTB (GALAMBOS *et al.*, 1959; GUINAN, 1968).

Figure 8A (GUINAN, 1968) presents tuning curves for units with triphasic action potentials. Though these units are confined almost exclusively to the NTB, results are similar to those obtained throughout the SOC. Each curve has a best (or characteristic) frequency (BF). The minimum or BF-thresholds are low and, indeed, are similar to those obtained in the auditory nerve (KIANG *et al.*, 1965b). We may conclude, from this and other studies, that most brainstem auditory neurons are quite sensitive, their thresholds essentially being limited by the sensitivity of the peripheral auditory mechanism.

The shapes of the tuning curves depend on BF (Fig. 8A). Figure 8B (GUINAN, 1968) plots the Q_{10}, a measure of the sharpness of tuning, against BF. Again, the results resemble those for auditory-nerve fibers (KIANG *et al.*, 1965b). The increase in Q_{10} with increase in BF reflects the broader tuning characteristics of low-frequency units. Whether there is a frequency-dependent increase in Q_{10} for units with BFs above 4 kHz is not clear. The results of TSUCHITANI and BOUDREAU

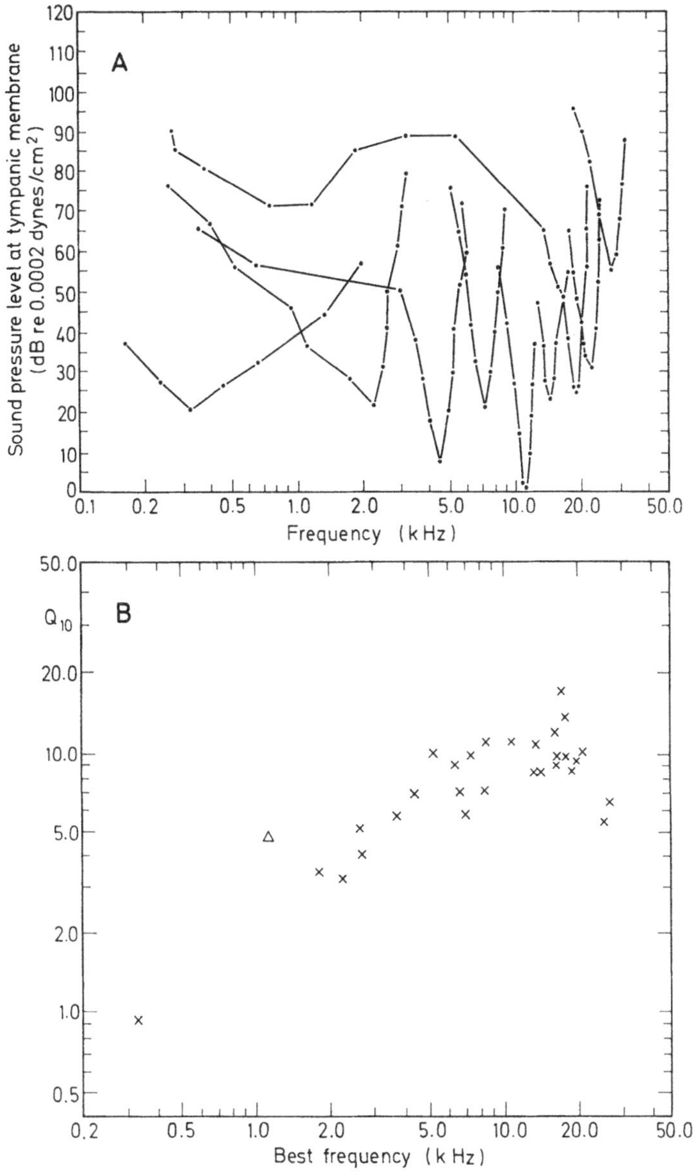

Fig. 8A, B. Tuning characteristics for NTB units with triphasic action potentials. (A) Typical threshold-tuning curves. (B) Q_{10} (ratio obtained on dividing BF by width of tuning curve 10 dB above BF-threshold) for several units, plotted against best-frequency. (From GUINAN, 1968)

(1970) indicate that the tuning curves of LSO neurons with BFs above 4 kHz are virtually superimposable, provided that the curves are suitably normalized. The normalization involves expressing the frequency scale in octaves re BF and the intensity scale in dB above BF-threshold.

There are no great differences in the widths of the excitatory response areas obtained from auditory-nerve fibers (KIANG *et al.*, 1965b) and from many central auditory neurons (see, for example, KIANG, 1965; GUINAN, 1968; BOUDREAU and TSUCHITANI, 1970). This is, at first glance, difficult to understand. Most central neurons are innervated by several ascending fibers. In some neurons — *e.g.*, the principal cells of the NTB — the convergence is minimal. But even the principal cell receives, in addition to its axosomatic calycoid innervation, a convergent axodendritic input (CAJAL, 1909; MOREST, 1968). Unless the converging afferents have almost identical BFs, one might suppose that the response areas of central neurons would be broader than those of typical auditory-nerve fibers. Actually such is the case. Many central neurons have receptive fields consisting of an excitatory region flanked on one or both sides by inhibitory areas. The inhibitory sidebands may subtend a large fraction of the frequency scale. The relative sharpness of the excitatory tuning curve should reflect the balance between excitatory and inhibitory processes. An example is shown in Fig. 9 (AITKIN *et al.*, 1970). The neuron, located in the LLD, had a narrow excitatory region, extending at moderate intensities from 8 to 10 kHz. Above and below this range, stimulation resulted in a suppression of spontaneous discharge. The total range of effective frequencies is seen to be considerably larger than are the limits of the excitatory tuning curve.

Unfortunately, there have been few studies of inhibitory sidebands in the SOC. Inhibitory areas have been described for MSO neurons (MOUSHEGIAN *et al.*, 1964a; GOLDBERG and BROWN, 1969). Our own, somewhat casual, observations indicate

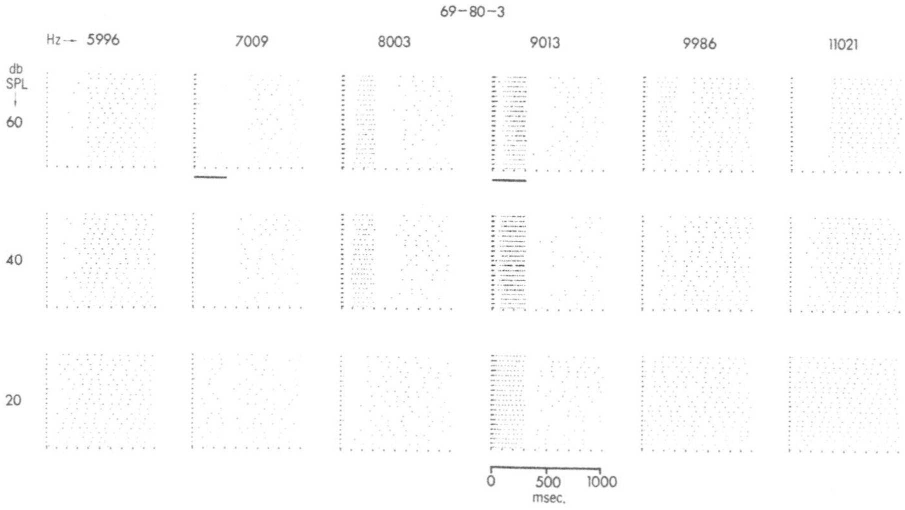

Fig. 9. Firing (dot) patterns for an LLD neuron. Tone-bursts, 300 msec, 1/sec. Best-frequency: near 9013 Hz. (From AITKIN *et al.*, 1970)

that similar sidebands can also be demonstrated in NTB and LSO neurons. The problem deserves further study.

B. Binaural Response Areas

In binaural neurons, it is possible to delimit two response areas, one for each ear. This has been done in the MSO (MOUSHEGIAN *et al.*, 1964a; GOLDBERG and BROWN, 1969). Within this nucleus both EE and EI neurons are encountered. For the majority of EE neurons investigated, the BFs and the shapes of the two response areas are similar (see also WATANABE *et al.*, 1968). In EI neurons, the BF for the excitatory ear is usually near the most effective frequency for the inhibitory ear. Moreover, the range of frequencies that lead to suppression of discharge when presented to the inhibitory ear is more or less coextensive with the frequency range that evokes discharge when presented to the excitatory ear.

Our most detailed knowledge of binaural response areas comes from the work of BOUDREAU and TSUCHITANI (1970) in the LSO. Figure 10 illustrates the close correspondence usually found between ipsilateral (excitatory) and contralateral (inhibitory) response areas. Comparisons made in a large number of LSO neurons

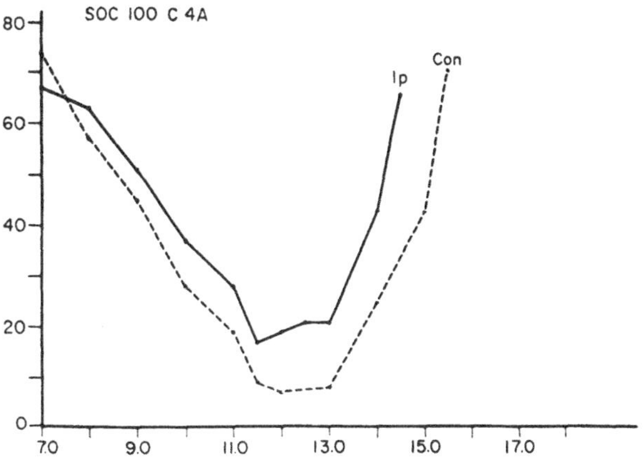

Fig. 10. Ipsilateral (excitatory) and contralateral (inhibitory) tuning curves for an LSO neuron. Inhibitory tuning curves measured by determining intensity levels at contralateral ear which suppress a one-spike discharge elicited by an ipsilateral best-frequency tone. *Abscissa:* frequency (kHz). Ordinate: dB SPL. (From BOUDREAU and TSUCHITANI, 1970)

indicate that the BFs and the BF-thresholds are almost always similar for the two ears. The only systematic difference between the two response areas was a tendency for the inhibitory area to be slightly broader than its excitatory counterpart.

The relation between ipsilateral and contralateral response areas need not be as simple as that described above. One complication may arise in neurons where each of the two response areas consists of a central excitatory region flanked by inhibitory sidebands. Most of these cells can be considered EE neurons, since the

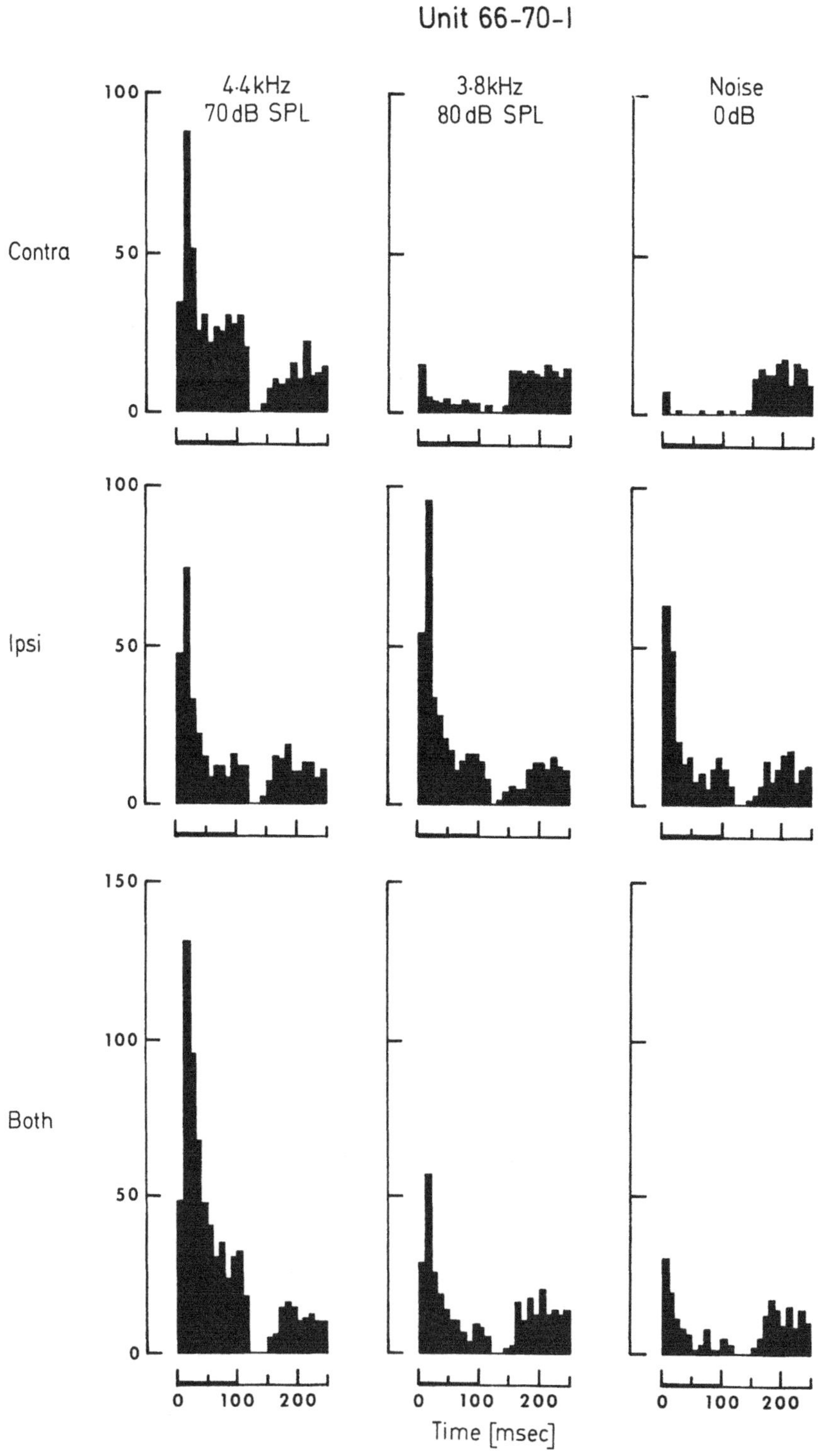

Fig. 11. Average-response histograms for tones at two frequencies and for white-noise. 50 trials. Best-frequency: 4.4 kHz. (From Goldberg and Brown, 1968)

excitatory areas for the two ears are similar and extend for reasonable distances to either side of the BF. In an occasional neuron, however, the excitatory region for one ear is considerably broader than that for the other ear and actually overlaps the inhibitory sideband for the second ear. The result is that the kind of binaural interaction obtained depends on tonal frequency. One such SOC neuron is illustrated in Fig. 11 (GOLDBERG and BROWN, 1968). Both ears were excitatory for tones extending from 4.0–4.8 kHz, including the BF of 4.4 kHz (Fig. 11, left column). Tones below 4.0 kHz were still within the ipsilateral excitatory area but fell within the contralateral inhibitory sideband. At 3.8 kHz (Fig. 11, middle column), for example, contralateral stimulation suppressed spontaneous activity and reduced the response to an ipsilateral stimulus. The effectiveness of the contralateral sideband is further indicated by the fact that a white noise presented to the contralateral ear had a net inhibitory effect (Fig. 11, right column).

C. Monaural Intensity Functions

For most SOC neurons, discharge rate increases with increase in the intensity of tones delivered to the excitatory ear(s). The quantitative relation between discharge rate and the intensity of BF tones has been studied in the NTB (GOLDBERG *et al.*, 1964), the MSO (GOLDBERG and BROWN, 1969), and the LSO (BOUDREAU and TSUCHITANI, 1970). In addition, the latter workers have extended the analysis to non-BF tones.

Figure 12 (BOUDREAU and TSUCHITANI, 1970) illustrates salient observations made in LSO neurons. The curves relating firing rate to intensity have sigmoid shapes. As intensity is increased from its threshold value, discharge rate increases rapidly, then more slowly, and eventually reaches a plateau. The functions are similar in shape for BF and non-BF tones. There are two effects of displacing the frequency from BF. There is, as expected, an increase in the threshold intensity. Further, there is a decrease in the maximum rate of discharge. Most interestingly, the latter decrease is seen for frequencies above the BF, but not for those below it.

The relation between discharge rate and intensity depends not only on tonal frequency, but also on stimulus duration. Most SOC neurons exhibit adaptation in their response to maintained stimulation (GOLDBERG *et al.*, 1964; MOUSHEGIAN *et al.*, 1964a; TSUCHITANI and BOUDREAU, 1966). As the duration of a tone is increased, the discharge rate decreases (Fig. 13, GOLDBERG *et al.*, 1964). The decrease is most rapid near the beginning of the response and is more conspicuous, the higher the intensity. The response to near-threshold stimulation shows only slight adaptation. Increasing the stimulus duration thus has little effect on response thresholds. But such a procedure reduces both the slopes of the rate-intensity function and the value of the maximum response. A further aspect of adaptation, illustrated in Fig. 13, is the decrease in background discharge following stimulus termination, a decrease which is more pronounced after high-intensity stimulation.

In neurons with relatively simple discharge characteristics, spike-count functions provide a straightforward measure of response. If, however, the discharge pattern is complicated, such as is commonly observed in LLN neurons, the functions may be deceptive (AITKIN *et al.*, 1970; BRUGGE *et al.*, 1970). Figure 14A (AITKIN *et al.*, 1970) plots the spike counts for the LLN neuron whose discharge patterns were presented in Fig. 7. At 12087 Hz, the spike counts vary by less than 20% over the intensity range from 18–58 dB. An inspection of Fig. 7 reveals, nevertheless, a striking change in response. As intensity increases, a gap

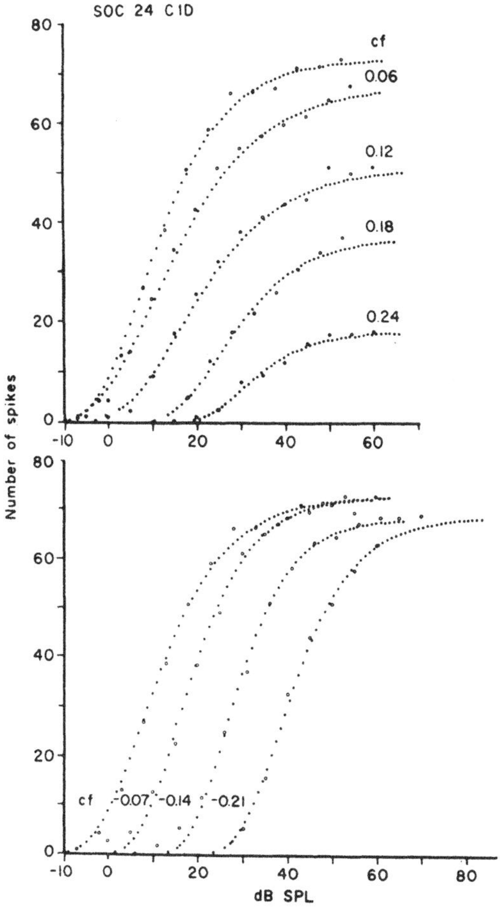

Fig. 12. Intensity functions for an LSO neuron. *Above:* Functions for frequencies above best (characteristic) frequency (*cf*). *Below:* Functions for frequencies below *cf*. Frequency in octave units re *cf*. Stimulus duration, 200 msec. (From BOUDREAU and TSUCHITANI, 1970)

in discharge appears and becomes more conspicuous; at the same time, the discharge rate during the maintained portion of the response increases. The relative constancy of spike counts is the result of these two opposing tendencies. Another effect of increasing intensity, and one that is seen in most central auditory neurons, is a decrease in the latent period to the first spike. Figure 14B indicates that the latent period is a function of frequency as well as intensity. At a given intensity, the latent period is least at or near BF.

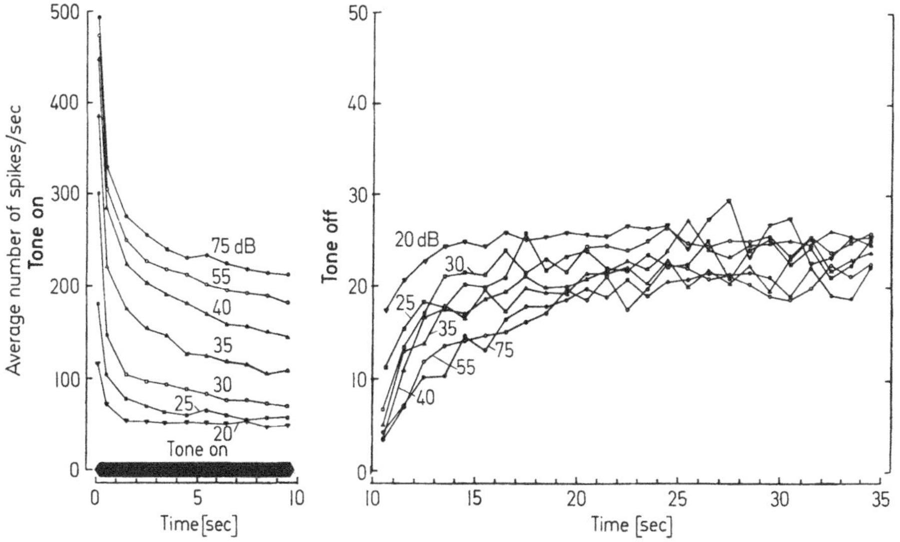

Fig. 13. Discharge rate for NTB neuron during (left) and after (right) stimulation with best-frequency tones (15.6 kHz), 10-sec duration. Intensities, dB SPL. (From Goldberg *et al.*, 1964)

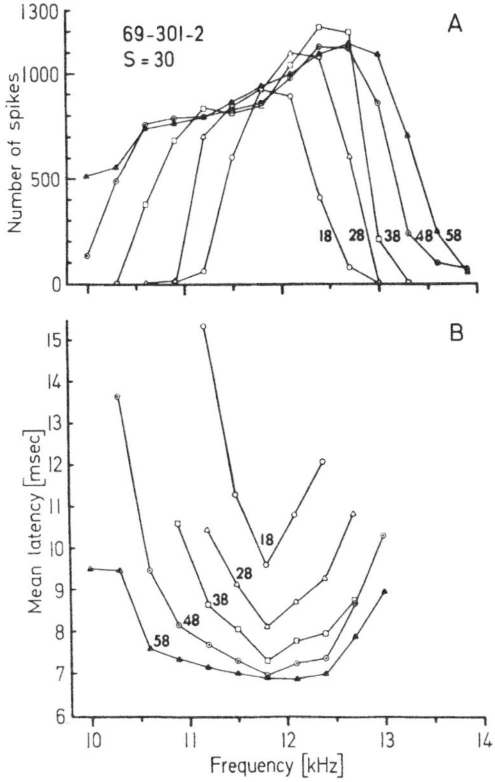

Fig. 14 A, B. Total number of spikes, 30 stimulus presentations (A) and mean latent period (B) plotted against frequency for 5 sound-pressure levels. Tone bursts, 200 msec. Same LLD neuron as in Fig. 7. (From Aitkin *et al.*, 1970)

D. Binaural Intensity Functions

In binaural neurons, the discharge rate and the latent period are both determined by the intensities of the tones delivered to the two ears. Figure 15 A presents a series of rate-intensity functions for an EE neuron located in the MSO (Goldberg and Brown, 1969). Each curve was obtained by presenting a tone of fixed (parametric) intensity to one ear, while varying the intensity of the tone delivered

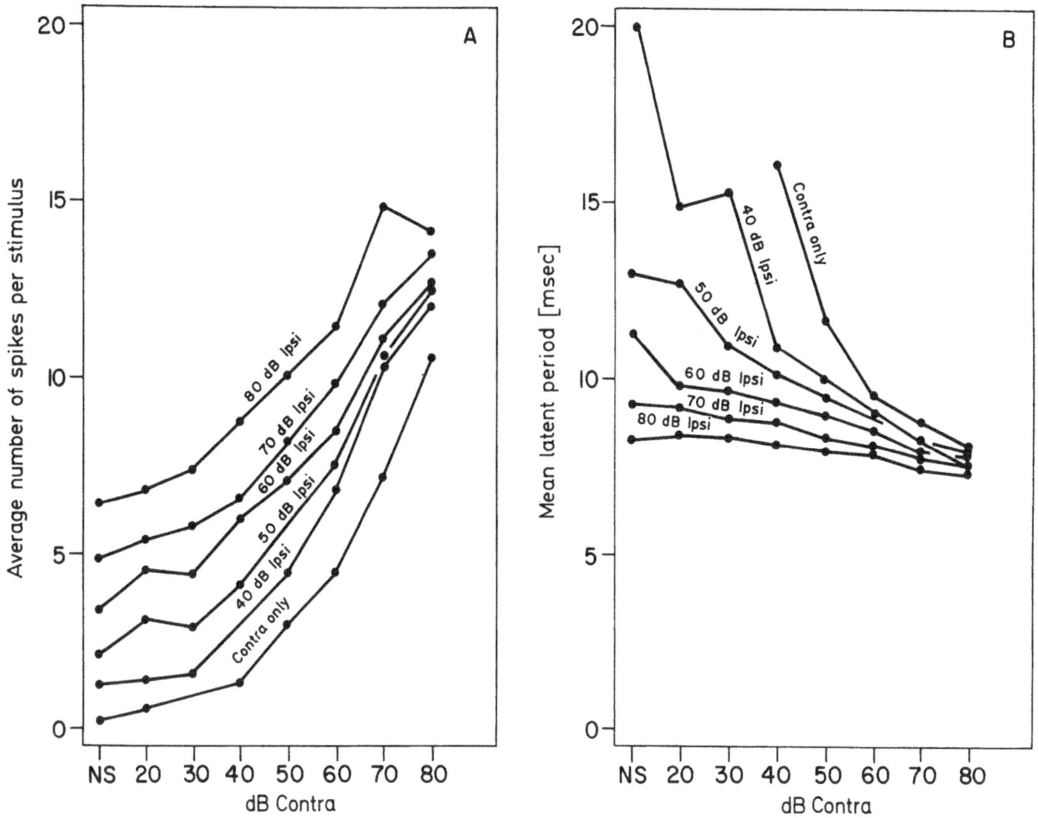

Fig. 15A, B. Binaural intensity functions for an EE cell in MSO. (A) spike counts; (B) mean latent period. Best-frequency (1.7 kHz) tone bursts, 100- msec duration, 2/sec, 50 repetitions. *Abscissa* is intensity in contralateral ear. Each curve for a fixed intensity in ipsilateral ear. Intensity in dB SPL. NS is no stimulus. (Modified from Goldberg and Brown, 1969)

to the other ear. The result is a family of more or less typical functions, which are displaced upward as the parametrie tone is itself raised in intensity. The latent period (Fig. 15 B) is, as expected, reduced as the intensity of the stimulus delivered to either ear is increased. Further, the binaural latent period may be significantly shorter than either of the monaural latent periods. The decrease in binaural latent period is most obvious at weak or moderate intensities and is more pronounced when the two monaural stimuli are set so that the corresponding latent periods are of comparable value.

Quantitative studies of the discharge of EE neurons have been largely confined to the MSO. Similar studies of EI neurons have been performed in the LSO (TSUCHITANI and BOUDREAU, 1969; BOUDREAU and TSUCHITANI, 1970), the MSO (GOLDBERG and BROWN, 1969), and the LLD (BRUGGE *et al.*, 1970). Results for a typical LSO neuron are shown in Fig. 16 (BOUDREAU and TSUCHITANI, 1970).

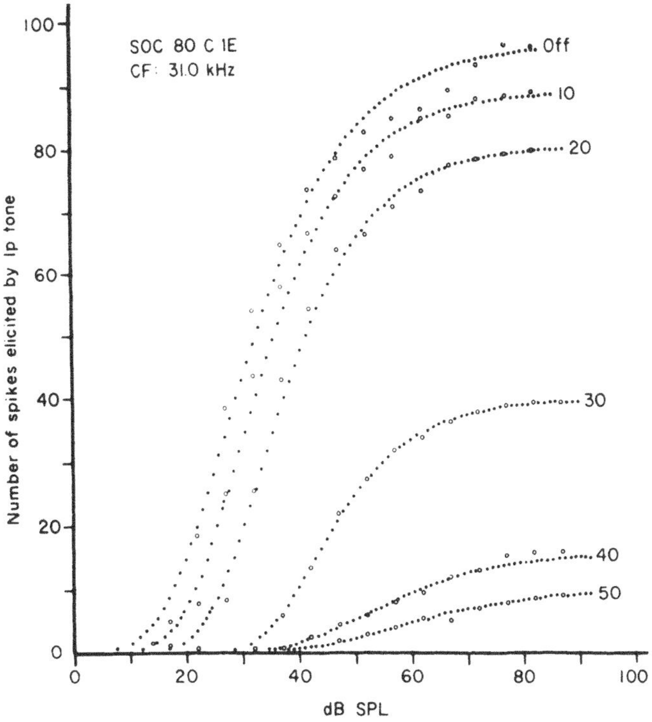

Fig. 16. Binaural intensity functions for an EI cell in LSO. Best-frequency tone bursts, 200 msec. *Abscissa* is intensity in ipsilateral (excitatory) ear. Each curve for a different intensity (dB SPL) in contralateral (inhibitory) ear. (From BOUDREAU and TSUCHITANI, 1970)

Each curve depicts the effects of increasing the intensity of an ipsilateral (excitatory) stimulus when presented in conjunction with a contralateral (inhibitory) stimulus of fixed intensity. The family of parametric rate-intensity functions are displaced downward as the intensity of the inhibitory stimulus is raised. The more effective the inhibitory stimulus, the higher the threshold of activation and the lower the maximum discharge elicited by an ipsilateral or excitatory tone.

E. Response to Clicks

There have been several studies of the response of MSO neurons to monaural and binaural clicks (GALAMBOS *et al.*, 1959; MOUSHEGIAN *et al.*, 1964b and 1967; HALL, 1965; RUPERT *et al.*, 1966). GUINAN (1968) has described the click response of triphasic units of the NTB, as well as those of units located elsewhere in the SOC.

Most SOC neurons respond to clicks with one or two spikes (RUPERT *et al.*, 1966), although repetitive trains of up to six spikes have also been seen (GALAMBOS *et al.*, 1959). In many cells, the effects of increasing click intensity are relatively simple. The discharge becomes more secure, the latent period decreases, and the timing of the first discharge becomes more precise. All of these changes occur as a continuous function of intensity. In other units, the discharge takes place only at a series of preferred times, and the shift in latent period may be discontinuous

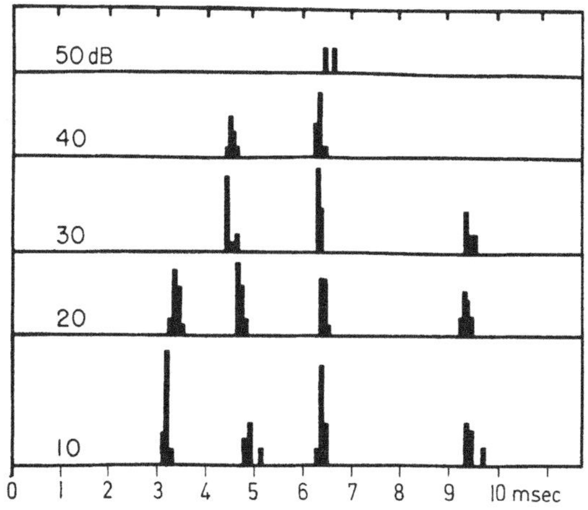

Fig. 17. Average-response histograms for an MSO neuron responding to contralateral clicks. Intensity in dB attenuation. (From GALAMBOS *et al.*, 1959)

(GALAMBOS *et al.*, 1959; MOUSHEGIAN *et al.*, 1967; GUINAN, 1968). The average-response histograms of Fig. 17 (GALAMBOS *et al.*, 1959), for example, exhibit several, approximately equally spaced, peaks. Variations in intensity may influence the relative prominence of a peak, without affecting its time of occurrence. Only low-frequency neurons exhibit such responses. Almost certainly, the periodic nature of the response reflects events taking place within the cochlea. A commonly held view is that the discharge is related to the mechanical response of the cochlear partition to impulsive stimuli, the latter response itself consisting of a series of damped oscillations. In any case, KIANG *et al.* (1965b) observed similar response patterns in auditory-nerve fibers with BFs less than 4–5 kHz. Consistent with the above explanation, the spacing between peaks in the response histogram is reciprocally related to the BF of the unit; this is true both for auditory-nerve fibers (KIANG *et al.*, 1965b) and for SOC neurons (MOUSHEGIAN *et al.*, 1967; GUINAN, 1968). GUINAN (1968) could demonstrate such a reciprocal relation for NTB neurons with BFs of up to 3 kHz. This last observation demonstrates the remarkably secure transmission of temporal information, at least across some of the synapses of the lower auditory system.

Two kinds of binaural response patterns have been observed in the MSO. Stimulation of either ear may result in a net excitatory effect. When this is so, binaural stimulation may evoke a discharge in situations where monaural stimulation does not (GALAMBOS *et al.*, 1959). Other units discharge in response to stimulation of either ear; binaural stimulation may then result in complex interactions which depend on interaural-time differences (MOUSHEGIAN *et al.*, 1964b; HALL, 1965; RUPERT *et al.*, 1966). The most thoroughly studied units are those receiving an excitatory input from one ear and an inhibitory input from the other ear (GALAMBOS *et al.*, 1959; MOUSHEGIAN *et al.*, 1964b, 1967; HALL, 1965; RUPERT *et al.*, 1966; WATANABE *et al.*, 1968). In such EI neurons, the discharge evoked by stimulation of the excitatory ear is suppressed by stimulation of the inhibitory ear. The degree of suppression may be a function of interaural-time and intensity differences. Because of their obvious bearing on the mechanisms of sound localization, we shall consider these latter units in the next section of the chapter.

IV. Mechanisms of Sound Localization

Psychophysical studies (JEFFRESS, this volume) demonstrate that the localization of sound sources in the horizontal plane involves several binaural cues. The localization of high-frequency sinusoids depends on interaural-intensity differences, that of low-frequency sinusoids on interaural-phase differences. Impulsive stimuli are localized on the basis of interaural-time and/or intensity cues. We now consider how neurons of the SOC and the LLN respond to such binaural stimuli.

A. Response to High-Frequency Tones

As we have seen, two kinds of binaural neurons are commonly found within the central auditory pathways. Excitatory-excitatory (EE) neurons have been encountered in the MSO (GALAMBOS *et al.*, 1959; MOUSHEGIAN *et al.*, 1964a, 1964b, 1967; HALL, 1965; RUPERT *et al.*, 1966; GOLDBERG and BROWN, 1968, 1969; GUINAN, 1968), in other nuclei of the SOC (GOLDBERG and BROWN, 1968, 1969; GUINAN, 1968), in the inferior colliculus (HIND *et al.*, 1963), in the medial geniculate (ADRIAN *et al.*, 1966) and in auditory cortex (HALL and GOLDSTEIN, 1968). Excitatory-inhibitory (EI) neurons predominate in the LSO (BOUDREAU and TSUCHITANI, 1970), and in the LLD (AITKIN *et al.*, 1970; BRUGGE *et al.*, 1970); they have also been found in the MSO (GALAMBOS *et al.*, 1959; MOUSHEGIAN *et al.*, 1964a, 1964b, 1967; HALL, 1965; RUPERT *et al.*, 1966; GOLDBERG and BROWN, 1968, 1969; GUINAN, 1968), in the preolivary and retroolivary nuclei (GOLDBERG and BROWN, 1968; GUINAN, 1968; BOUDREAU and TSUCHITANI, 1970) and in higher auditory centers (HIND *et al.*, 1963; ADRIAN *et al.*, 1966; HALL and GOLDSTEIN, 1968). Quite simple reasoning leads to the suggestion that these two kinds of neurons respond differently to changes in average binaural intensity and to variations in interaural-intensity differences.

Consider first a variation in the overall intensity of a binaural stimulus. In the simplest imaginable situation, there will be parallel changes in the activity of the two monaural pathways leading to a binaural neuron. For EE cells, the changes will summate; in EI cells, they will neutralize one another. Introduction of an interaural-intensity difference should lead to opposite changes in the two pathways. In EI cells, the oppositely directed changes will reinforce one another; in EE cells, they will cancel. Hence EE cells should be more sensitive to variations in binaural intensity than to the introduction of interaural-intensity differences. For EI cells, the reverse should be true. Evidence supporting these ideas comes from several studies (Goldberg and Brown, 1969; Brugge *et al.*, 1969, 1970; Rose *et al.*, 1966; Geisler *et al.*, 1969).

The differential sensitivity exhibited by any particular neuron depends on the balance of influences arriving from the two ears. An EE neuron should be virtually unresponsive to interaural-intensity differences provided that both ears are almost equally effective in activating the neuron. If one ear is dominant, introduction of an interaural-intensity difference will result in a change in firing rate. This is so because an intensity variation in the dominant ear will not be perfectly compensated for by the oppositely directed variation in the other ear (Goldberg and Brown, 1969).

Similarly the response of EI cells to binaural intensity variations depends on the relative potencies of the excitatory input from one ear and the inhibitory input from the other ear. If the inhibition is relatively weak, the neuron may be only slightly less sensitive to variations in average binaural intensity than it is to the introduction of an interaural-intensity difference. Such an imbalance has been seen in the MSO and MPO (Goldberg and Brown, 1969) and in the inferior colliculus (Geisler *et al.*, 1969). The LSO neurons studied by Boudreau and Tsuchitani (1970) provide an example of a second kind of imbalance, namely one in which the inhibition predominates. The discharge of these neurons is almost completely suppressed when equal-intensity stimuli are delivered to the two ears. When the intensity presented to the ipsilateral (excitatory) ear is made 10 dB more intense than the contralateral (inhibitory) stimulus and a rate-intensity function plotted, it is found that the discharge rate reaches a maximum some 20–30 dB above threshold and then declines back to zero at higher intensities. The maximum rate attained is usually less than one-third that elicited by a high-intensity monaural stimulus. Further, interaural-intensity disparities of 30 dB or more are needed before the neuron begins to discharge vigorously to high-intensity binaural stimuli. It may be concluded, then, that such neurons are insensitive both to binaural intensity variations and, as is particularly evident at high intensities, to the introduction of small interaural-intensity variations. Neurons with similar properties have also been observed elsewhere in the SOC (Goldberg and Brown, 1969) and in the inferior colliculus (Geisler *et al.*, 1969).

At first glance, it may seem puzzling that a binaural neuron would be insensitive to the two kinds of intensity variations important in binaural hearing. It may be argued, however, that those EI neurons in which inhibition predominates would be ideally suited in the handling of the small interaural-time differences involved in the localization of transients. This is so because the potent inhibitory input could, when correctly timed, obliterate the response to stimulation of the excitatory ear.

B. Response to Low-Frequency Tones

Binaural neurons may, when low-frequency tones are simultaneously delivered to the two ears, be quite sensitive to variations of interaural phase. Phase-sensitive neurons, first described in the inferior colliculus (HIND *et al.*, 1963; ROSE *et al.*, 1966; GEISLER *et al.*, 1969), have subsequently been encountered in the SOC (MOUSHEGIAN *et al.*, 1964a and 1967; GOLDBERG and BROWN, 1969), in the LLD (BRUGGE *et al.*, 1970), and in the auditory cortex (BRUGGE *et al.*, 1969). Phase sensitivity has been observed for tonal frequencies as high as 2.5–3.0 kHz (ROSE *et al.*, 1966; GOLDBERG and BROWN, 1969; BRUGGE *et al.*, 1969) and also with noise stimulation (GEISLER *et al.*, 1969).

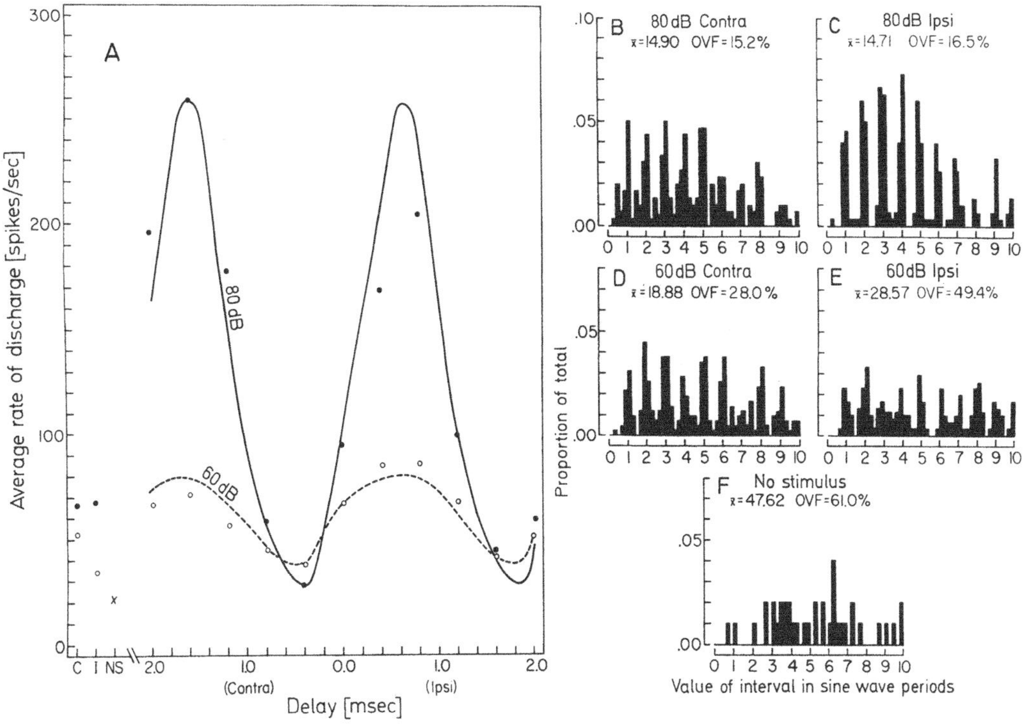

Fig. 18A, B. MSO neuron stimulated with best frequency (444.5 Hz) tones, 1-sec duration. (A) Effects on discharge rate of varying interaural delay. Points on left, monaural stimuli: contralateral (*C*), ipsilateral (*I*), and no stimulus (NS). Curves on right, binaural stimulation at two intensities (dB SPL); *abscissa*, interaural delay. (B) Interspike-interval histograms, monaural stimuli. (From GOLDBERG and BROWN, 1969)

The behavior of a phase-sensitive MSO neuron is illustrated in Fig. 18A (GOLDBERG and BROWN, 1969). Separate stimulation of either ear resulted in a sustained discharge. During binaural stimulation, the discharge rate was greatly affected, particularly at the higher intensity, by variations in the interaural delay. The functions relating discharge rate to interaural delay had a period equal to that of the stimulating sinusoid, indicating that the response was being influenced by interaural phase rather than by the delay *per se*. The discharge at the most favorable delay was considerably greater, and that at the least favorable delay considerably less, than that resulting from monaural stimulation. The values of the most and least favorable delays vary from neuron to neuron, but for any one unit, the values are relatively unaffected by variations of stimulus intensity or by the introduction of an interaural-intensity difference (ROSE *et al.*, 1966; MOUSHEGIAN *et al.*, 1967; GOLDBERG and BROWN, 1969; GEISLER *et al.*, 1969; BRUGGE *et al.*, 1969). Further, in any particular neuron, the value of the most favorable delay (or, in some cases, that of the least favorable delay) may remain constant when tonal frequency is changed (ROSE *et al.*, 1966; GEISLER *et al.*, 1969; BRUGGE *et al.*, 1969) or when noise is substituted for tones (GEISLER *et al.*, 1969). This suggests that each phase-sensitive neuron functions in some sense as a detector of a constant (or characteristic) delay (ROSE *et al.*, 1966).

A reasonable explanation of these observations was offered by ROSE *et al.* (1966). Stimulation of either ear may be assumed to evoke a periodic train of interleaved excitatory and inhibitory processes. When binaural tones are presented at favorable delays, the excitatory events emanating from the two ears arrive at the binaural neuron in phase and will reinforce one another. If the interaural delay is unfavorable, the inhibitory events from each ear will arrive simultaneously with the excitatory events from the other ear and will neutralize them. Stated another way, phase sensitivity is envisioned as involving an in-phase facilitation and an out-of-phase inhibition. Note that the terms "in-phase" and "out-of-phase" refer to the time of arrival of the monaural excitatory volleys at the binaural neuron and not to the phase relations existing at the two ears. The characteristic delay of any unit would be determined by the delays involved in transmitting the activity from each ear to the binaural neuron. Conceivably the transmission delays might be fixed by anatomical arrangement.

Confirmation of some of these ideas comes from studies in the MSO (GOLDBERG and BROWN, 1969). First, the discharge of phase-sensitive MSO neurons is phase-locked when low-frequency tones are presented to either ear. This is indicated (Fig. 18B–E) by the fact that the relevant interspike-interval histograms are polymodal, with peaks falling at integral multiples of the sine-wave period. Thus, the two monaural inputs are periodic. Second, the timing of the monaural phase-locked discharges is consistent with theory. Figure 19 presents monaural phase histograms for the unit shown in Fig. 18. The mean phase of the ipsilateral discharge led that of the contralateral discharge by 650 μsec. In other words, it would be necessary to delay the ipsilateral stimulus by 650 μsec in order that the excitatory inputs from the two ears arrive in phase at the binaural neuron. The value of 650 μsec (ipsilateral delay) corresponds closely to the most favorable delay determined by binaural stimulation (*cf.* Fig. 18A). A similar correspondence was observed in neurons which differed widely from one another in their characteristic binaural delays. And finally, it was possible to identify two modes of activation during binaural discharge, each mode corresponding to the main excitatory input coming from one of the ears. An analysis of the probability of discharge occurring during the two modes confirmed the existence of both an in-phase facilitation and an out-of-phase inhibition.

Most phase-sensitive neurons described in the literature (ROSE *et al.*, 1966; GOLDBERG and BROWN, 1969; BRUGGE *et al.*, 1969, 1970) resemble the unit illustrated in Fig. 18 in at least one respect. Separate stimulation of the more effective ear results in a discharge rate which is less than the binaural discharge obtained at the most favorable delay and greater than that seen at the least favorable delay. A simpler situation has been observed in some phase-sensitive MSO neurons (MOUSHEGIAN *et al.*, 1964a, 1967). In these, one ear has a predominantly excitatory effect, the other ear a predominantly inhibitory effect. The

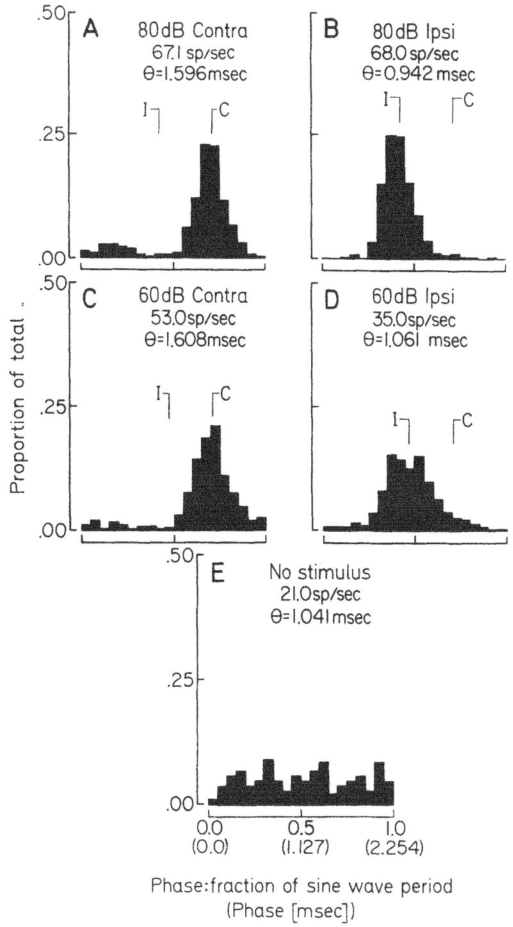

Fig. 19A—E. Same unit as in Fig. 18. Histograms depicting the phase relations between stimulating sinusoid and neuron's discharge on separate stimulation of contralateral (A, C) and ipsilateral (B, D) ears. (E) no stimulus condition. (From GOLDBERG and BROWN, 1969)

discharge is quite sensitive to variations in interaural phase, but at all delays the binaural response is less than that evoked by monaural stimulation of the excitatory ear. Such neurons have also been found in the MPO (GOLDBERG and BROWN, 1969) and in the LLD (BRUGGE *et al.*, 1970). The mechanisms underlying the phase-sensitivity of these neurons are probably similar to those described above. In particular, it may be supposed that the discharge rate is determined by the relative times of arrival of a phase-locked excitatory input from one ear and a phase-locked inhibitory input from the other ear. The suggestion, though reasonable, lacks experimental verification.

Another problem requiring further study relates to the existence of a characteristic delay. It must be emphasized that the relevant evidence is based on the investigation of a relatively small number of neurons. One of the more thorough studies was conducted by GEISLER *et al.* (1969) in the inferior colliculus. Critical observations were made on only 10 neurons. Nine of these could be said to possess a characteristic delay in the sense that each displayed a maximum (or minimum) in its discharge at an interaural delay whose value did not change with variations in the frequency content and/or intensity of the stimulus. The one seemingly aberrant neuron was taken by GEISLER *et al.* (1969) to mean that not all phase-sensitive neurons possess a characteristic delay. Similarly one of the four neurons studied by ROSE *et al.* (1966) could not be said to possess a characteristic delay.

Despite these uncertainties, the notion of a characteristic delay remains fruitful. It is instructive, therefore, to consider how such a characteristic delay might arise. In Fig. 19, the mean values of the monaural phase histograms were, at most, slightly altered as stimulus intensity was raised. A similar phase constancy in monaural response was observed in other phase-sensitive SOC neurons (GOLDBERG and BROWN, 1969) and in cochlear-nucleus neurons (LAVINE, 1971). The most complete data on this point come from auditory-nerve studies (PFEIFFER and MOLNAR, 1970; ANDERSON *et al.*, 1971). For BF-tones, the mean phase remains more or less fixed throughout most of the intensity range. For non-BF tones, phase shifts may occur, but these seldom amount to more than 90°. The relative constancy of phase, together with the timing mechanisms described above, provide a basis for the fact that the most and least favorable delays are relatively unaffected by variations in binaural intensity or by the introduction of an interaural-intensity difference. For the most favorable delay not to be affected by variations in tonal frequency, it would further be necessary that the temporal disparity ΔT between the ipsilateral and contralateral inputs remain constant. This can be accomplished if the mean phases of discharge, Θ_C and Θ_I, obtained on stimulation, respectively, of the contralateral and ipsilateral ears were linearly related to tonal frequency, f, *i.e.*,

$$\left.\begin{aligned} \Theta_C &= 2\,\pi f T_C + K \\ \Theta_I &= 2\,\pi f T_I + K \\ \Delta T &= T_C - T_I \end{aligned}\right\} .$$

Approximately linear relations have been seen in auditory-nerve fibers (PFEIFFER and MOLNAR, 1970; ANDERSON *et al.*, 1971) and in cochlear-nucleus neurons (LAVINE, 1971). To the extent that the equations might hold for the monaural inputs to phase-sensitive binaural neurons, and the evidence is at best indirect, the afferent pathways would behave as if they were characterized by fixed transmission delays, T_C and T_I.

A final problem relates to the primary site of binaural convergence. As mentioned above, phase-sensitive MSO neurons exhibit a phase-locked discharge when either ear is stimulated alone (GOLDBERG and BROWN, 1969). The situation in the LLD (BRUGGE *et al.*, 1970) and in the inferior colliculus (ROSE *et al.*, 1966; GEISLER *et al.*, 1969) is more complicated. Some of the phase-sensitive neurons of these regions are characterized by a phase-locked discharge; others are not. In the auditory cortex, BRUGGE *et al.* (1969) were unable to find any aspect of the monaural discharge of phase-sensitive neurons which was synchronized to individual sine-wave cycles. The failure of a phase-sensitive neuron to exhibit phase-locking may be taken as evidence that its discharge reflects a primary convergence occurring at another (presumably lower) level of the auditory system.

The central mechanisms of phase-sensitivity involve a transformation from a time code to a place code. In the periphery and in the central monaural pathways, interaural phase relations are represented by small time differences in the discharge of the corresponding neurons of the two sides. Within the central pathways, these temporal disparities are converted into a place representation, *i.e.*, a representation involving the relative discharge rates of a population of binaural neurons. The MSO is one site where such a transformation takes place. Whether or not the inferior colliculus and the lemniscal nuclei represent similar sites cannot be decided with certainty. The evidence indicates that the original temporal information is discarded, if not at the level of the midbrain, then at the level of the auditory cortex. It is not clear why the time code is not retained. Perhaps its retention would require more precise synaptic arrangements than exist at higher levels of the auditory system.

C. Response to Transients

Several studies have demonstrated that EI neurons are sensitive to the interaural-time differences involved in the localization of transients. The process may be viewed as again involving a conversion from a time code in the monaural pathways to a place code in populations of binaural neurons. Stimulation of the excitatory ear with, say, a click results in a discharge which may be suppressed by similar stimulation of the inhibitory ear. Maximum suppression should occur when the excitatory and inhibitory inputs reach the binaural neuron in close temporal proximity. The timing of the inputs, and hence the binaural discharge, should thus be dependent on the interaural-time separation.

Figure 20 (MOUSHEGIAN *et al.*, 1967), provides a typical example. The unit, located in the MSO, responded to clicks delivered to the ipsilateral (left) ear. The discharge could be suppressed by a suitably timed contralateral click. Maximum suppression was obtained when the inhibitory stimulus preceded the excitatory stimulus. This would appear to be the most common situation, both in the SOC (HALL, 1965) and in the inferior colliculus (BENEVENTO and COLEMAN, 1970). Nevertheless, in some units, maximum suppression occurs when the excitatory click leads the inhibitory click (GALAMBOS *et al.*, 1959; MOUSHEGIAN *et al.*, 1964b) or when the two stimuli are presented simultaneously (MOUSHEGIAN *et al.*, 1964b, 1967).

Time-sensitive EI neurons can also be affected by small interaural-intensity differences (HALL, 1965; RUPERT *et al.*, 1966; BENEVENTO and COLEMAN, 1970). The work of HALL (1965) is most interesting in this regard. He studied a selected population of SOC neurons that were activated by contralateral stimuli and inhibited by ipsilateral stimuli. The cells responded to clicks in such a way that their probability of discharge was high when the excitatory stimulus led the inhibitory stimulus and low when these time relations were reversed. Further, the intro duction of an interaural-intensity difference favoring, for example, the inhibitory ear decreased the likelihood of firing and thus had the same effect as did a time lead at that ear. In this sense, the neurons exhibited a time-intensity trading bearing a formal resemblance to that observed in psychophysical studies (JEFFRESS, this volume).

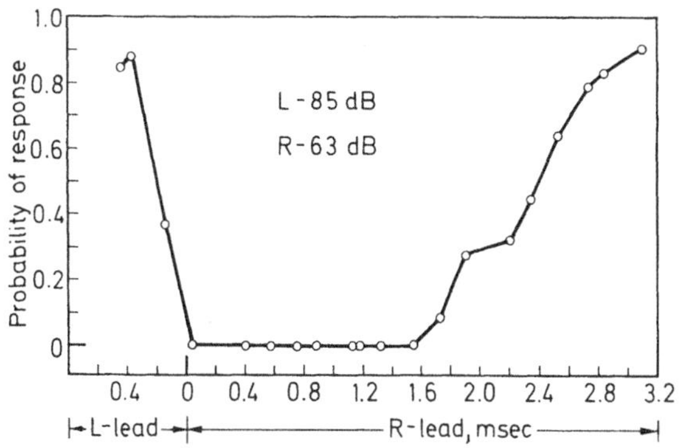

Fig. 20. Response of EI neuron of MSO to binaural clicks. Left (ipsilateral) ear excitatory; right (contralateral) ear inhibitory. Intensities in dB attenuation. Differences on abscissa, time differences between the peaks of the NI potentials recorded near the two round windows. (From MOUSHEGIAN *et al.*, 1967)

Two factors may be operative in the time-intensity trading observed by HALL (1965). First, an intensity increase will result in a decreased latent period. We may assume, though experimental proof is lacking, that this would be true for inhibitory, as well as excitatory, inputs. One can introduce an intensity increase or a time lead at a given ear. Both procedures will have similar effects on the relative timing of the monaural inputs to the binaural neuron. Functions relating latent period to stimulus intensity are steep near threshold, but flatten out as intensity is increased (GALAMBOS *et al.*, 1959; MOUSHEGIAN *et al.*, 1964b; RUPERT *et al.*, 1966). Consonant with these observations, HALL (1965) found that the time-intensity trading ratio was large (approximately 100 μsec/dB) for weak clicks and had declined to less than 20 μsec/dB for clicks 50 dB above threshold, a trend also observed in psychophysical experiments (DAVID *et al.*, 1959).

An intensity increase will result not only in a decreased latent period, but also in an increase in the potency of the corresponding input. This second factor will result in an appropriate time-intensity trading in neurons, such as those studied by HALL (1965), where maximum suppression is obtained when the inhibitory ear leads. For such neurons, then, the two factors would tend to reinforce one another. This would not be the case, however, for neurons which exhibit a maximum suppression when the excitatory ear leads. For the latter, the first factor would lead to an appropriate time-intensity trading, but the second factor would result in a reversed or paradoxical trading. That is, raising the intensity at the inhibitory ear should, by virute of its increasing the potency of the corresponding input, cause a decreased firing of the binaural neuron and hence be equivalent to a time lead at the excitatory ear. Presumably the behavior of the neuron would depend on the relative importance of the two factors.

A final complication arises when we consider the balance between the excitatory and the inhibitory inputs. As was discussed previously (see p. 134), EI neurons in which inhibition predominates should be insensitive to interaural-intensity differences, but could still be exquisitely sensitive to interaural-time differences. Such neurons, assuming that they exist, would not exhibit time-intensity trading. The existence of two populations of time-sensitive neurons, those that are sensitive to interaural-intensity variations and those that are not, would provide a striking parallel to the two-image lateralization phenomenon demonstrated psychophysically by HAFTER and JEFFRESS (1968).

References

ADES, H.W., BROOKHART, J.M.: The central auditory pathway. J. Neurophysiol. **13**, 189—205 (1950).

ADRIAN, H.O., LIFSCHITZ, W.M., TAVITAS, R.J., GALLI, F.P.: Activity of neural units in medial geniculate body of cat and rabbit. J. Neurophysiol. **29**, 1046—1060 (1966).

AITKIN, L.M., ANDERSON, D.J., BRUGGE, J.F.: Tonotopic organization and discharge characteristics of single neurons in nuclei of the lateral lemniscus of the cat. J. Neurophysiol. **33**, 421—440 (1970).

ANDERSON, D.J., ROSE, J.E., HIND, J.E., BRUGGE, J.F.: Temporal position of discharges in single auditory nerve fibers within the cycle of a sine-wave stimulus: frequency and intensity effects. J. acoust. Soc. Amer. **49**, 1131—1139 (1971).

BARNES, W.T., MAGOUN, H.W., RANSON, S.W.: The ascending auditory pathway in the brain stem of the monkey. J. comp. Neurol. **79**, 129—152 (1943).

BENEVENTO, L.A., COLEMAN, P.D.: Responses of single cells in cat inferior colliculus to binaural click stimuli: combinations of intensity levels, time differences and intensity differences. Brain Res. **17**, 387—405 (1970).

BIEDENBACH, M.A., FREEMAN, W.J.: Click-evoked potential map from the superior olivary nucleus. Amer. J. Physiol. **206**, 1408—1414 (1964).

BOUDREAU, J.C., TSUCHITANI, C.: Binaural interaction in the cat superior olive S-segment. J. Neurophysiol. **31**, 442—454 (1968).

BOUDREAU, J.C., TSUCHITANI, C.: Cat superior olive S-segment cell discharge to tonal stimulation. Contrib. Sensory Physiol. **4**, 143—213 (1970).

BRUGGE, J.F., ANDERSON, D.J., AITKIN, L.M.: Response of neurons in the dorsal nucleus of the lateral lemniscus of the cat to binaural tonal stimulation. J. Neurophysiol. **33**, 441—458 (1970).

BRUGGE, J.F., DUBROVSKY, N.A., AITKIN, L.M., ANDERSON, D.J.: Sensitivity of single neurons in auditory cortex of cat to binaural tonal stimulation; effects of varying interaural time and intensity. J. Neurophysiol. **32**, 1005—1024 (1969).

CAJAL, S.R.: Histologie du système nerveux de l'homme et des vertébrés, Vol. 1. Paris: Maloine 1909.

DAVID, E.E., Jr., GUTTMAN, N., VAN BERGEIJK, W.A.: Binaural interaction of high-frequency complex stimuli. J. acoust. Soc. Amer. **31**, 774—782 (1959).

DESMEDT, J.E.: Recurrent efferent system and relations of auditory to other systems. This Volume.

ECCLES, J.C.: The physiology of synapses. Berlin-Göttingen-Heidelberg-New York: Springer 1964.

ERULKAR, S.D., BUTLER, R.A., GERSTEIN, G.L.: Excitation and inhibition in cochlear nucleus. II. Frequency-modulated tones. J. Neurophysiol. **31**, 537—548 (1968).

FERNANDEZ, C., KARAPAS, F.: The course and termination of the striae of Monakow and Held in the cat. J. comp. Neurol. **131**, 371—386 (1967).

FEX, J.: Auditory activity in centrifugal and centripetal cochlear fibres in cat. Acta physiol. scand. **55**, Suppl. 189, 1—68 (1962).

FEX, J.: Auditory activity in uncrossed centrifugal cochlear fibres in cat. Acta physiol. scand. **64**, 43—57 (1965).

GALAMBOS, R., SCHWARTZKOPFF, J., RUPERT, A.: Microelectrode study of superior olivary nuclei. Amer. J. Physiol. **197**, 527—536 (1959).

GEISLER, C. D., RHODE, W. S., HAZELTON, D. W.: Responses of inferior colliculus neurons in the cat to binaural acoustic stimuli having wide-band spectra. J. Neurophysiol. **32**, 960—974 (1969).

GERSTEIN, G. L., BUTLER, R. A., ERULKAR, S. D.: Excitation and inhibition in cochlear nucleus. I. Tone-burst stimulation. J. Neurophysiol. **31**, 526—536 (1968).

GOLDBERG, J. M., ADRIAN, H. O., SMITH, F. D.: Response of neurons of the superior olivary complex of the cat to acoustic stimuli of long duration. J. Neurophysiol. **27**, 706—749 (1964).

GOLDBERG, J. M., BROWN, P. B.: Functional organization of the dog superior olivary complex: an anatomical and electrophysiological study. J. Neurophysiol. **31**, 639—656 (1968).

GOLDBERG, J. M., BROWN, P. B.: Response of binaural neurons of dog superior olivary complex to dichotic tonal stimuli: some physiological mechanisms of sound localization. J. Neurophysiol. **32**, 613—636 (1969).

GOLDBERG, J. M., MOORE, R. Y.: Ascending projections of the lateral lemniscus in the cat and monkey. J. comp. Neurol. **129**, 143—156 (1967).

GOLDBERG, J. M., SMITH, F. D., ADRIAN, H. O.: Response of single units of the superior olivary complex of the cat to acoustic stimuli: laterality of afferent projections. Anat. Rec. **145**, 232 (1963).

GREENWOOD, D. D., MARUYAMA, N.: Excitatory and inhibitory response areas of auditory neurons in the cochlear nucleus. J. Neurophysiol. **28**, 863—892 (1965).

GUINAN J. J. Jr.: Firing patterns and locations of single auditory neurons in the brainstem (superior olivary complex) of anesthetized cats. Sc. D. Thesis, Mass. Inst. Technol. 1968.

HAFTER, E. R., JEFFRESS, L. A.: Two-image lateralization of tones and clicks. J. acoust. Soc. Amer. **44**, 563—569 (1968).

HALL J. L. II: Binaural interaction in the accessory superior-olivary nucleus of the cat. J. acoust. Soc. Amer. **37** 814—823 (1965).

HALL, J. L. II., GOLDSTEIN, M. H., Jr.: Representation of binaural stimuli by single units in primary auditory cortex of unanesthetized cats. J. acoust. Soc. Amer. **43**, 456—461 (1968).

HARRISON, J. M., IRVING, R.: Nucleus of the trapezoid body: dual afferent innervation. Science **143** 473—474 (1964).

HIND, J. E., GOLDBERG, J. M., GREENWOOD, D. D., ROSE, J. E.: Some discharge characteristics of single neurons in the inferior colliculus of the cat. II. Timing of the discharges and observations on binaural stimulation. J. Neurophysiol. **26** 321—341 (1963).

HIRSCH, J. E.: Effect of interaural time delay on amplitude of cortical responses evoked by tones. J. Neurophysiol. **31**, 916—927 (1968).

IRVING, R., HARRISON, J. M.: The superior olivary complex and audition: a comparative study. J. comp. Neurol. **130** 77—86 (1967).

JEFFRESS, L. A.: Binaural hearing. 1. Localization and lateralization. This Volume.

JUNGERT, S.: Auditory pathways in the brain stem. Acta oto-laryng. (Stockh.) Suppl. 138 (1958).

KEMP, E. H., COPPÉE, G. E., ROBINSON, E. H.: Electric responses of the brain stem to unilateral auditory stimulation. Amer. J. Physiol. **120**, 304—315 (1937a).

KEMP, E. H., ROBINSON, E. H.: Electric responses of the brain stem to bilateral auditory stimulation. Amer. J. Physiol. **120**, 316—322 (1937b).

KIANG, N. Y.-S.: Stimulus coding in the auditory nerve and cochlear nucleus. Acta oto-laryng. (Stockh.) **59**, 186—200 (1965).

KIANG, N. Y.-S., PFEIFFER, R. R., WARR, W. B., BACKUS, A. S. N.: Stimulus coding in the cochlear nucleus. Ann. Otol. **74**, 463—485 (1965a).

KIANG, N. Y.-S., WATANABE, T., THOMAS, E. C., CLARK, L. F.: Discharge patterns of single fibers in the cat's auditory nerve. Cambridge (Mass.): Mass. Inst. Technol. Press 1965b.

LAVINE, R. A.: Phase-locking in response of single neurons in cochlear nuclear complex of the cat to low-frequency tonal stimuli. J. Neurophysiol. **34**, 467—483 (1971).

LENN, N. J., REESE, T. S.: The fine structure of nerve endings in the nucleus of the trapezoid body and the ventral cochlear nucleus. Amer. J. Anat. **118**, 375—390 (1966).

LI, R. Y.-S., GUINAN, J. J., Jr.: Antidromic and orthodromic stimulation of neurons receiving calyces of Held. Mass. Inst. Technol., Quart. Progr. Report **100**, 227—234 (1971).

MOORE, J. K., MOORE, R. Y.: A comparative study of the superior olivary complex in the primate brain. Folia primat. **16**, 35—51 (1971).

MOREST, D. K.: The collateral system of the medial nucleus of the trapezoid body of the cat, its neuronal architecture and relation to the olivo-cochlear bundle. Brain Res. **9**, 288—311 (1968).

MOUSHEGIAN, G., RUPERT, A. L., LANGFORD, T. L.: Stimulus coding by medial superior olivary neurons. J. Neurophysiol. **30**, 1239—1261 (1967).

MOUSHEGIAN, G., RUPERT, A., WHITCOMB, M. A.: Brain-stem neuronal response patterns to monaural and binaural tones. J. Neurophysiol. **27**, 1174—1191 (1964a).

MOUSHEGIAN, G., RUPERT, A., WHITCOMB, M. A.: Medial superior-olivary-unit response patterns to monaural and binaural clicks. J. acoust. Soc. Amer. **36**, 196—202 (1964b).

NELSON, P. G., ERULKAR, S. D.: Synaptic mechanisms of excitation and inhibition in the central auditory pathway. J. Neurophysiol. **26**, 908—923 (1963).

OSEN, K. K.: Cytoarchitecture of the cochlear nuclei in the cat. J. comp. Neurol. **136**, 453—484 (1969).

PFEIFFER, R. R.: Anteroventral cochlear nucleus: wave forms of extracellularly recorded spike potentials. Science **154**, 667—668 (1966a).

PFEIFFER, R. R.: Classification of response patterns of spike discharges for units in the cochlear nucleus; tone-burst stimulation. Exp. Brain Res. **1**, 220—235 (1966b).

PFEIFFER, R. R., KIANG, N. Y.-S.: Spike discharge patterns of spontaneous and continuously stimulated activity in the cochlear nucleus of anesthetized cats. Biophys. J. **5**, 301—316 (1965).

PFEIFFER, R. R., MOLNAR, C. E.: Cochlear nerve fiber discharge patterns: relationship to the cochlear microphonic. Science **167**, 1614—1616 (1970).

RASMUSSEN, G. L.: The olivary peduncle and other fiber projections of the superior olivary complex. J. comp. Neurol. **84**, 141—219 (1946).

RASMUSSEN, G. L.: Anatomic relationships of ascending and descending auditory systems. In: Neurological aspects of auditory and vestibular disorders. Springfield (Ill.): Ch. C Thomas 1964.

RASMUSSEN, G. L.: Efferent connections of the cochlear nucleus. In: Sensorineural hearing processes and disorders. Boston: Little, Brown 1967.

ROSE, J. E., GREENWOOD, D. D., GOLDBERG, J. M., HIND, J. E.: Some discharge characteristics of single neurons in the inferior colliculus of the cat. I. Tonotopical organization, relation of spike-counts to tone intensity, and firing patterns of single elements. J. Neurophysiol. **26**, 294—320 (1963).

ROSE, J. E., GROSS, N. B., GEISLER, C. D., HIND, J. E.: Some neural mechanisms in the inferior colliculus of the cat which may be relevant to localization of a sound source. J. Neurophysiol. **29**, 288—314 (1966).

ROSENZWEIG, M. R.: Cortical correlates of auditory localization and of related perceptual phenomena. J. comp. physiol. Psychol. **47**, 269—276 (1954).

ROSENZWEIG, M. R., AMON, A. H.: Binaural interaction in the medulla of the cat. Experientia (Basel) **11**, 498—499 (1955).

ROSENZWEIG, M. R., SUTTON, D.: Binaural interaction in lateral lemniscus of cat. J. Neurophysiol. **21**, 17—23 (1958).

ROSENZWEIG, M. R., WYERS, E. J.: Binaural interaction at the inferior colliculi. J. comp. physiol. Psychol. **48**, 426—431 (1955).

ROSSI, G., CORTESINA, G.: Research on the efferent innervation of the inner ear. J. Laryng. **77**, 202—233 (1963).

RUPERT, A., MOUSHEGIAN, G., WHITCOMB, M. A.: Superior-olivary response patterns to monaural and binaural clicks. J. acoust. Soc. Amer. **39**, 1069—1076 (1966).

Saul, L. J., Davis, H.: Action currents in the central nervous system: I. Action currents of the auditory tracts. Arch. Neurol. Psychiat. (Chicago) **28**, 1104—1116 (1932).

Stotler, W. A.: An experimental study of the cells and connections of the superior olivary complex of the cat. J. comp. Neurol. **98**, 401—432 (1953).

Tsuchitani, C., Boudreau, J. C.: Wave activity in the superior olivary complex of the cat. J. Neurophysiol. **27**, 814—827 (1964).

Tsuchitani, C., Boudreau, J. C.: Single unit analysis of cat superior olive S segment with tonal stimuli. J. Neurophysiol. **29**, 684—697 (1966).

Tsuchitani, C., Boudreau, J. C.: Stimulus level of dichotically presented tones and cat superior olive S-segment cell discharge. J. acoust. Soc. Amer. **46**, 979—988 (1969).

van Noort, J.: The structure and connections of the inferior colliculus. An investigation of the lower auditory system. Leiden: Van Gorcum 1969.

Warr, W. B.: Fiber degeneration following lesions in the anterior ventral cochlear nucleus of the cat. Exp. Neurol. **14**, 453—474 (1966).

Warr, W. B.: Fiber degeneration following lesions in the posteroventral cochlear nucleus of of the cat. Exp. Neurol. **23**, 140—155 (1969).

Watanabe, T., Liao, T.-T., Katsuki, Y.: Neuronal response patterns in the superior olivary complex of the cat to sound stimulation. Jap. J. Physiol. **18**, 267—287 (1968).

Wernick, J. S., Starr, A.: Binaural interaction in the superior olivary complex of the cat: an analysis of field potentials evoked by binaural-beat stimuli. J. Neurophysiol. **31**, 428—441 (1968).

Woollard, H. H., Harpman, J. A.: The connexions of the inferior colliculus and of the dorsal nucleus of the lateral lemniscus. J. Anat. (Lond.) **74**, 441—458 (1940).

Chapter 3

Physiological Studies of the Inferior Colliculus and Medial Geniculate Complex

SOLOMON D. ERULKAR, Philadelphia, Pennsylvania (USA)

With 30 Figures

Contents

I. Morphology 147
A. Neuronal Structure of the Inferior Colliculus 147
B. Neuronal Structure of the Medial Geniculate Body 153
C. Other Thalamic Regions of the Auditory Pathway 158
II. Electrophysiology 161
A. The Inferior Colliculus 162
1. Tonotopical Organization and Tuning Curves 162
2. The Nature of Unitary Response Patterns and their Relationship to Excitatory-Inhibitory Influences 164
3. Binaural Phenomena 172
B. Medial Geniculate Body 178
1. Tonotopical Organization and Tuning Curves 178
2. Response Patterns and Excitatory-Inhibitory Interactions 180
3. Binaural Phenomena 186
C. Other Thalamic Groups 188
III. Conclusions 189
References 191

Within the last decade there have emerged several new concepts which have re-defined the roles of the inferior colliculus and medial geniculate body in the central auditory pathway. These concepts have been derived from both anatomical and physiological studies and results from these two disciplines must be considered together if a functionally meaningful interpretation is to be obtained.

The importance of the inferior colliculus and medial geniculate within the auditory system must depend on their connections to neurons at both higher and lower levels of the central nervous system as well as on interconnections within the nuclei of these regions. At a more discrete level, we cannot ignore the geometrical distribution of synaptic inputs onto individual neurons or indeed any ordered pattern of distribution that may be related to a particular auditory parameter.

We have assumed that order exists within this system, and this assumption allows us to separate different components and define their functional roles under different stimulus conditions. For instance, the question may be asked whether the patterns of input seen morphologically reflect specific functional characteristics, such as frequency discrimination. Do certain neurons with typical structural features receive only one type of input? What are the patterns of discharge of neurons at these levels to different types of acoustic stimuli and how do these patterns differ from those recorded at other levels? What are the pharmacological properties of the synapses? The answers to these questions lead to the ultimate question of the role of the inferior colliculus and medial geniculate in "hearing". In this chapter, I shall attempt to set the stage for this final question.

Although the anatomy of the central auditory pathway and descending pathways have been described in previous chapters (Chapters 9 and 11, Vol. V/1), several aspects of the morphology which are pertinent to the physiology will be given here.

As introduction, it must be remembered that by the time acoustic signals – in the form of nerve impulses – reach the level of the inferior colliculus, they have undergone several modifications in pattern. This is not only due to synaptic transformation occurring between the cochlea and the inferior colliculus, but also due to the convergence of fibers originating from the two sides onto the same neurons at lower levels. Obviously patterns of excitation and inhibition become increasingly complex at higher levels of the central auditory pathway and this increasing complexity suggests specific information processing at these levels. Our task is to decide how the signal is abstracted at these levels before being relayed to the cortex.

Although the inferior colliculus and its brachium leading to the medial geniculate nucleus provide a pathway for auditory information to the cortex, this may not be the only pathway. GALAMBOS, MYERS, and SHEATZ in 1961 found that in unanesthetized cats, or in those anesthetized with chloralose, potentials at the auditory cortex elicited by acoustic stimuli remained unchanged following section of the brachium of the inferior colliculus (BIC). In the same year, GOLDBERG and NEFF (1961 b) found that small foci of activity remained after complete section of BIC in cats anesthetized with sodium pentobarbital. On the other hand, MAJKOWSKI (1964) and NIEDER and STROMINGER (1965) found that complete section of the BIC abolished evoked potentials in auditory cortex. Later ADRIAN, GOLDBERG, and BRUGGE (1966) concluded from their studies that the normal pattern of elicited activity in the classical auditory cortex "depends upon the integrity of the inferior brachium". On the other hand, the authors point out that there exist other structures within the lateral part of the midbrain which can also transmit impulses to the auditory cortex, a conclusion also reached by NIEDER and STROMINGER. These results then would not disagree with those from later studies of MAJKOWSKI and MORGADES (1969) who state that section of BIC eliminated responses in the auditory cortex to acoustic stimulation and also in the auditory cortex and medial geniculate body to electrical stimulation of the inferior colliculus. There is general agreement that the BIC is not necessarily involved in transmission of impulses to association areas and sensorimotor cortex and that, in these cases, the pathways involved may lie medially to the main auditory pathway.

While there may, therefore, be more than one pathway for acoustic information to reach the cortex, it has also become clear that, at thalamic levels, considerable integration of input from polysensory systems occurs. We cannot think simply in terms of auditory information relayed to the inferior colliculus and thence to the medial geniculate body and finally to the AI region of the cortex. Rather, we should consider the total picture that the cortex receives, and this involves relays from areas other than the classical auditory pathway relay regions.

The question whether fibers of the primary auditory pathway reach thalamic levels without synapsing in the inferior colliculus has caused considerable controversy. Originally CAJAL (1911) had stated that fibers of the lateral lemniscus entered the "voie acoustique centrale" and projected to the medial geniculate body. His conclusions were confirmed by PAPEZ (1929) whose data suggested that fibers of "the central acoustic tract terminated in the dorsal and medial portions of the medial geniculate body". These results could not be confirmed by later investigators (WOOLLARD and HARPMAN, 1939; BARNES *et al.*, 1943; RASMUSSEN, 1946; MOORE and GOLDBERG, 1961; RASMUSSEN, 1964; GOLDBERG and MOORE, 1967) who reported that few, if any, fibers from the lateral lemniscus reached the medial geniculate body by way of the inferior brachium. GOLDBERG and MOORE (1967) found no lemniscal fibers ascending in the inferior brachium, but did identify fibers which may have been those of the central acoustic tract. These fibers terminated "... among other places, in the medial part of the magnocellular medial geniculate of the cat and the corresponding portion of the medial geniculate of the squirrel monkey" (GOLDBERG and MOORE, 1967). Whether these fibers are a part of the central auditory pathway is uncertain and GOLDBERG and MOORE contend that the central acoustic tract may well be "... a component of the spinothalamic pathways". They point out that the fact that lemniscal fibers end in the medial geniculate does not necessarily mean that these fibers belong to the auditory system. Certainly, no lemniscal fibers terminate in the principal part (pars parvocellularis) which is known to be the primary relay nucleus to AI of the auditory cortex. Those fibers that do reach this level terminate at a point medial to the area which receives a dense projection from the inferior colliculus (MOORE and GOLDBERG, 1963, 1966; TARLOV and MOORE, 1966; GOLDBERG and MOORE, 1967). This particular area in the medial geniculate does not undergo retrograde degeneration after removal of the auditory fields of the temporal cortex (DIAMOND and NEFF, 1957; ROSE and WOOLSEY, 1958; GOLDBERG and NEFF, 1961a; LOCKE, 1961) and as will be discussed later, may represent a site for convergence of several different sensory systems.

These observations suggest that the inferior colliculus is an obligatory synaptic region interposed in the central auditory pathway.

I. Morphology

A. Neuronal Structure of the Inferior Colliculus

Studies on the neuronal architecture of the inferior colliculus were first done by CAJAL (1911) on the mouse, cat and dog. He clearly distinguished three regions – the central nucleus, the internuclear cortex and the lateral cortex. Since that time, the terminology has been changed and has increased as different investigators discover different nuclear subdivisions. I shall attempt to give an integrated description, biased towards physiological interpretation.

All investigators agree that there is a main central nucleus of the inferior colliculus (MOREST, 1964b, 1966b; GOLDBERG and MOORE, 1967; BERMAN, 1968; GENIEC and MOREST, 1971; ROCKEL, 1971; ROCKEL and JONES, 1973a, b) (Fig. 1A, Fig. 1B). MOREST and GENIEC and MOREST have stressed that, in the cat and in

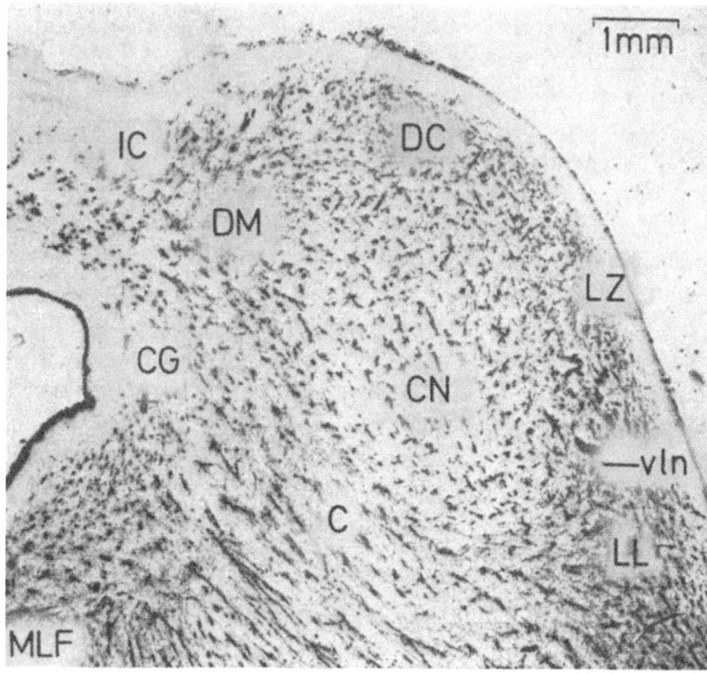

Fig. 1A

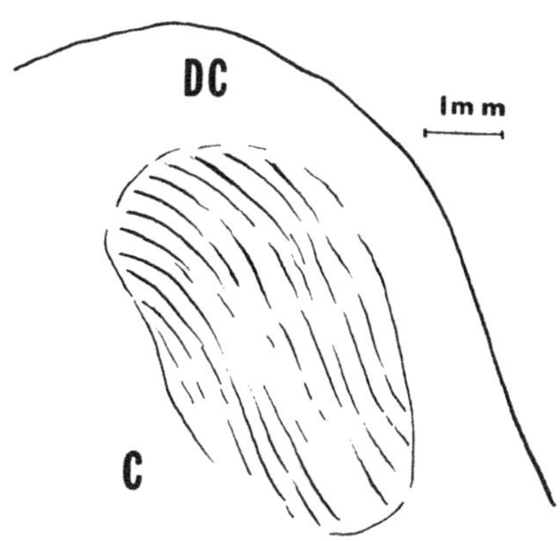

Fig. 1B

Fig. 1. (A) Frontal section at junction of the caudal and middle thirds of the posterior colliculus of a 55 year old man. Golgi-Cox method (B) Orientation of the dendritic laminae from the central nucleus as seen in frontal section in (A). *C* cuneiform area, *CG* central gray, *CN* central nucleus, *DC* dorsal cortex, *DM* dorsomedial nucleus, *IC* intercollicular commissural zone, *LL* lateral lemniscus and dorsal nucleus of lateral lemniscus, *LZ* lateral zone, MLF medial longitudinal fasciculus, *V* ln ventrolateral nucleus. (Taken from GENIEC and MOREST, 1971) man, neurons in this nucleus may be distinguished on the basis of their dendritic morphology. The stereotyped branching patterns of the dendrites from different neuronal types are consistent in different mammalian species and may be considered to subserve similar physiological functions. Thus, one type of neuron in the central nucleus has disc-shaped dendritic fields. These neurons line up in a laminated manner (see Fig. 2) and are paralleled by incoming fibers of the lateral lemniscus. The authors maintain that this laminar arrangement of the "disc-shaped neurons" in parallel with the afferent input may be a structural means of

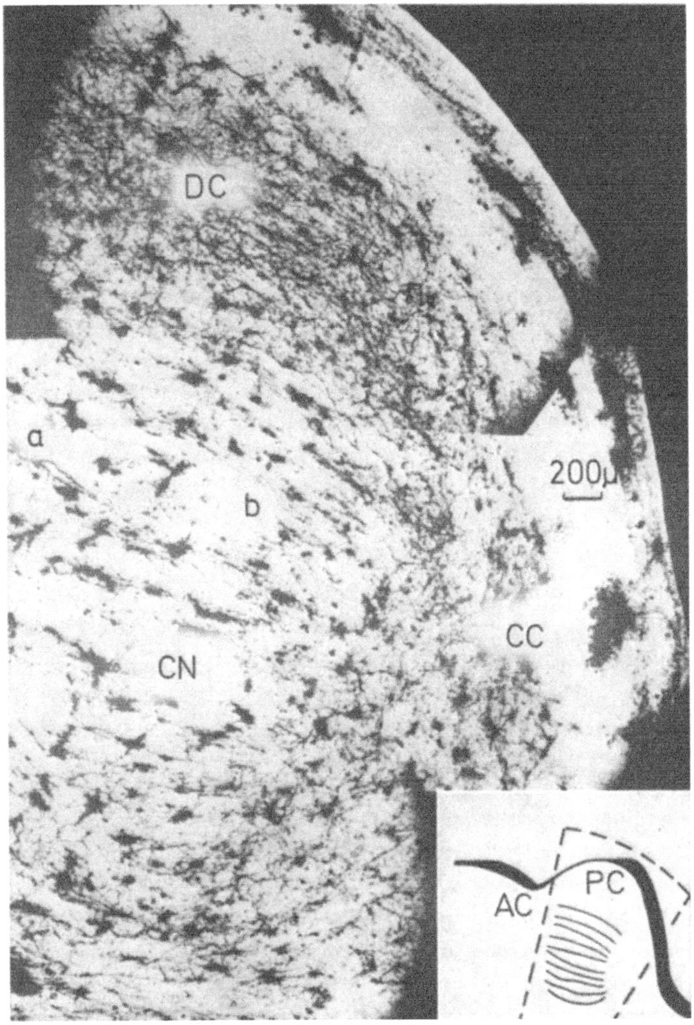

Fig. 2a and b. Parasagittal section at junction of medial and middle thirds of the posterior colliculus. Note impregnation of dendritic laminae of central nucleus and cellular layers of the cortex. *a* Large disc-shaped neuron. *b* Small disc-shaped neuron. Other abbreviations as in Fig. 1. Insert: Orientation of dendritic laminae in central nucleus. The dashed line gives the limits of the sector shown in the photomicrograph. *AC* anterior colliculus, *PC* posterior colliculus. (Taken from GENIEC and MOREST, 1971)

preserving functional tonotopic localization within the central nucleus (MOREST, 1964b).

In 1972, OSEN described the topographic projection of the cochlear nuclei upon the central nucleus of the inferior colliculus. OSEN found that "strict point to point projections" from the cochlear nuclei occurred onto the central nucleus of the colliculus and, therefore, inferred the presence of tonotopic localization with low tonal frequencies represented dorsolaterally and high tonal frequencies represented rostroventrally. OSEN also suggested that "some relationship" existed between the tonotopic arrangement and the laminar structure of the central nucleus. As will be described later, tonotopic localization has been shown to exist at this level by physiological techniques (ROSE *et al.*, 1963).

Stellate cells are also present in the central nucleus (MOREST, 1964c; GENIEC and MOREST, 1971) but occur less frequently than disc-shaped neurons. These cells have dendrites which may parallel or cross the dendritic laminae sometimes perpendicularly. In the cat, these dendrites receive small punctate endings from the ascending afferents. The axons of these cells have many branched collaterals that can reach many different laminae. GENIEC and MOREST (1971) speculate that these fibers would "intercept a wide sector of the afferent fiber spectrum", and, therefore, the cells would have a broad frequency-response curve. There are indeed units in the inferior colliculus which have extremely wide response curves (ERULKAR, 1959; ROSE *et al.*, 1963). Such cells might provide an integrative function in relation to their inputs. They could integrate afferent activity and thus modulate by excitatory and inhibitory influence the overall activity of the central nucleus. Complex sequences of excitation and inhibition have been recorded from cells of the inferior colliculus (NELSON and ERULKAR, 1963). Obviously one can speculate on a wide variety of interactions, such as binaural interaction, also known to occur at inferior colliculus cells (ERULKAR, 1959) or summation of stimulus intensities, but, as GENIEC and MOREST point out, there is no direct proof that a particular cell in the central nucleus subserves a particular function.

The recent extensive studies of ROCKEL (1971) and ROCKEL and JONES (1973a, b) have added greatly to our knowledge of the structure of the inferior colliculus. They divided the central nucleus into 1) a dorsomedial division which receives fibers from the auditory cortex and 2) a ventrolateral division receiving fibers exclusively from the lateral lemniscus. They agree with MOREST that a laminar arrangement of cells, dendrites, and axons exists within the nucleus – especially within the ventro-lateral two-thirds. There is, however, some irregularity in arrangement. The laminae are not flat but curved and are arranged about a center of curvature in the dorsolateral part of the nucleus (Fig. 3). ROCKEL and JONES maintain that entering lateral lemniscal fibers run along the full length of each lamina, have "multiple synaptic knobs" throughout their extent, and can branch into several parallel fibers which may run along different laminae. These authors agree with other investigators that the laminar organization in the central nucleus may represent some form of tonotopic arrangement.

CAJAL (1911) also described the presence of an external nucleus or lateral cortex as well as an internuclear or roof cortex. GOLDBERG and MOORE (1967) state that the external nucleus can be divided into lateral, dorsal and caudal sectors which can be differentiated morphologically and which receive discrete projections from different pathways. MOREST (1966a) and GENIEC and MOREST (1971) define a cortex and pericollicular tegmentum surrounding the central nucleus (Fig. 2). The cortex, which separates the central nucleus from the collicular surface dorsally and caudally can be shown in both cat and man to contain morphologically distinct layers of neuropil. The dorsal and caudal cortices are thus equivalent to the dorsal and caudal sectors of the external nucleus. GENIEC and MOREST, however, stress

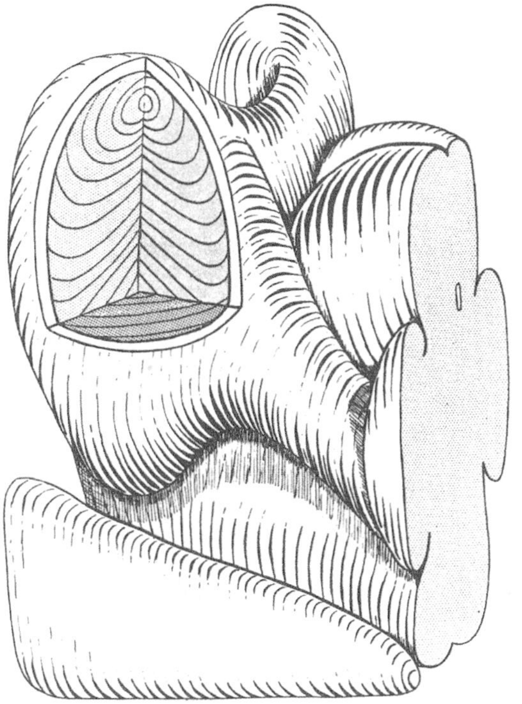

Fig. 3. Artist's impression of the arrangement of the laminar organization of the central nucleus of the inferior colliculus. Laminar contours through the right colliculus are shown as they appear in horizontal, sagittal and coronal planes. Anterior is to the right and dorsal towards the top. (Taken from ROCKEL and JONES, 1973a)

that this structure is indeed a cortex, and that the different layers may receive different inputs. Furthermore, the layers are crossed by dendrites whose cell bodies are situated in other layers thus allowing for integration of the different inputs.

The internuclear cortex of CAJAL (1911) and the cortex described by GENIEC and MOREST (1971) are called the pericentral nucleus by BERMAN (1968) and ROCKEL and JONES (1973c), who distinguish this region from the external and central nuclei. ROCKEL and JONES (1973c) in their detailed studies, differ from MOREST (1966a) and GENIEC and MOREST (1971) in defining a laminated structure in the pericentral region. They report that the appearance of any discrete layering is "obscured" and that this region receives a heavy descending projection of fibers from the auditory cortex. This pericentral nucleus thus differs from the external nucleus in its connections. Indirect evidence suggests that the pericentral nucleus and central nucleus receive ascending fibers from the dorsal nucleus of the lateral lemniscus but that other lemniscal afferent fibers project to the central nucleus alone. As there are separate systems of ascending afferents to the pericentral and central nuclei, ROCKEL and JONES (1973c) suggest the possibility of a separate frequency representation in these two nuclei and believe that in the pericentral nucleus this would be in the form of "parallel but overlapping bands of best frequency, running from ventro-lateral to dorso-medial over the posterior surface of the inferior colliculus and onto its dorsal surface, ...".

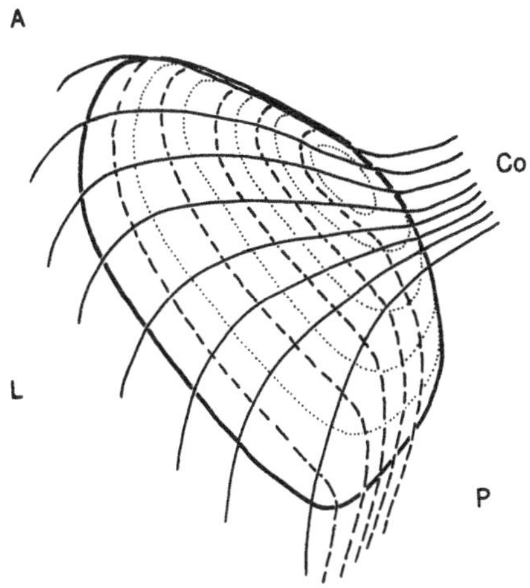

Fig. 4. Schematic representation of postero-dorsal aspect of the left inferior colliculus to show how afferent corticofugal fibers (continuous lines) entering the pericentral nucleus from the brachium of the inferior colliculus and running toward the intercollicular commissure (Co) intersect with those (interrupted lines) entering from below and passing up over the posterior face of the colliculus. Note also the relationships of these fibers to the underlying laminar pattern of organization (dotted lines) of the central nucleus. (Taken from ROCKEL and JONES, 1973c)

The same authors also discuss the functional implications of the "rectangular grid" found in the pericentral nucleus by intersections of ascending afferent fibers and descending cortical afferents (Fig. 4). In view of the fact that fibers from both systems give rise to short collaterals along their course, ROCKEL and JONES speculate that there would be in these regions "... a small, localized region of combined ascending and descending influence, ...". Furthermore, "the position of this focus would be uniquely defined". It is, however, difficult to speculate any further than this on the functional implications of an anatomical arrangement.

The pericollicular tegmentum as defined by GENIEC and MOREST separates the central nucleus from the superior colliculus, from the lateral surface of the tectum, and from the underlying regions of the midbrain tegmentum. The so-called lateral zone is equivalent to the lateral sector of the external nucleus of GOLDBERG and MOORE (see Fig. 1A). These regions may be concerned with the integration of somato-sensory and auditory inputs. The authors also describe dorsomedial and ventrolateral nuclei (Fig. 1A) which, in the cat, receive descending fibers from the auditory cortex (RASMUSSEN, 1964).

GOLDBERG and MOORE (1967) showed that fibers of the lateral lemniscus mainly terminate in the ventral and dorsolateral part of the central nucleus. Very few fibers terminate in the external nucleus. The dorsomedial portion of the central nucleus which is equivalent to the dorsal cortex and commissural zone of MOREST, is the major site of termination of fibers of the commissure of the inferior colliculus which originate in the dorsal part of the contralateral central nucleus. In the cat, there is relatively little overlap between projections of the lateral lemniscus to the central nucleus and that of the commissure, but, in the squirrel monkey, there is a widespread distribution of both throughout the central nucleus.

Both GOLDBERG and MOORE (1967), GENIEC and MOREST (1971) and ROCKEL and JONES (1973a) agree that the majority of fibers from the lateral lemniscus end in the central nucleus. GOLDBERG and MOORE contend that few lemniscal fibers enter and terminate in the external nucleus but that the nucleus receives projections from the major fiber tracts emerging from the central nucleus (MOORE and GOLDBERG, 1963, 1966; TARLOV and MOORE, 1966). MOREST (1966a) and GENIEC and MOREST (1971) describe lemniscal afferents which send collaterals primarily to layer IV of the cortex, along with collaterals of efferent fibers of the central nucleus. This arrangement allows the integration of input and output activity of the inferior colliculus in layer IV. No lemniscal fibers project to layers I to III. The superficial layers receive projections from more anterior parts of the brain, and the authors believe that both in the cat and in the human, layer III is "best situated to integrate elements of both the ascending and descending pathways, especially by way of intrinsic connections of the neurons of layers II and IV, which receive the heaviest exogenous projections" (GENIEC and MOREST, 1971). In fact, they state that layer III by way of its connections can monitor the activity of all other layers. Thus the cortex (or pericentral nucleus) and the dorsomedial and ventrolateral nuclei, which receive projections from the descending pathway from the auditory cortex (RASMUSSEN, 1964), provide an area for integration of activity, perhaps especially for modulation of ascending patterns of activity by descending pathways.

As stated earlier, the neurons of the pericollicular tegmentum may serve to integrate activity of different sensory modalities. Projections from the spinal cord and from the dorsal column nuclei are known to terminate in these regions especially at the lateral zone (MOREST, 1966a; GENIEC and MOREST, 1971).

B. Neuronal Structure of the Medial Geniculate Body

It is appropriate at this time to describe the neuronal architecture of the medial geniculate body. The connections to the various divisions of this region from the inferior colliculus will be discussed later. It is insufficient, however, to consider the medial geniculate nucleus purely in terms of classical morphology and physiology. Rather one must consider those thalamic structures which not only provide relays to the cortical auditory areas but which may subserve auditory or polysensory integrative functions at this level.

Although (or perhaps because) the morphology of the medial geniculate body has been extensively studied by many investigators, the terminology to describe it has become confused. An excellent review of the historical aspects of the structure of the medial geniculate has been given by Morest in 1964a. The confusion arises as a result of the different ways the medial geniculate body has been subdivided

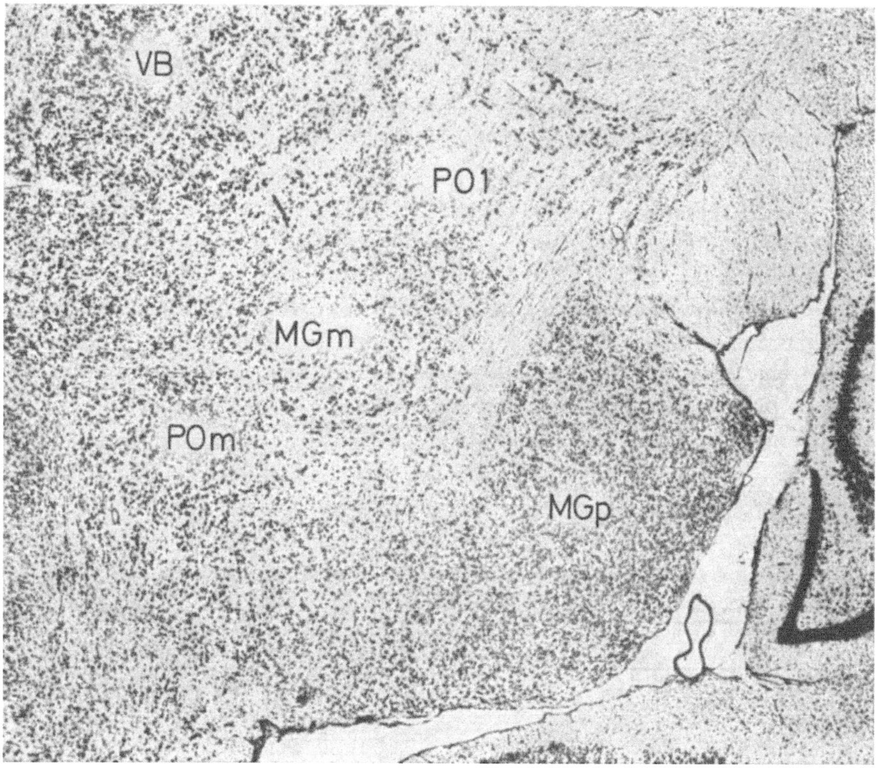

Fig. 5. Horizontal section through medial geniculate body and posterior thalamic group. *MGp* principal division of medial geniculate body, *MGm* magnocellular division of medial geniculate body, *POl* lateral part of posterior thalamic group, *POm* medial part of posterior thalamic group, *VB* ventrobasal complex. (Taken from Moore and Goldberg, 1963)

according to the different techniques used for the analysis. Early morphological descriptions of the medial geniculate body recognized three basic divisions – ventral, dorsal and medial divisions (Gudden, 1881; Munzer and Wiener, 1902; Cajal, 1911; Rose, 1942). In 1929, Rioch described, in carnivores, a small cellular division that he called the lateral pars principalis and pars magnocellularis (Fig. 5) with large loosely-packed cells placed medioventrally. This nomenclature became widely accepted especially after Rose and Woolsey (1949, 1958) showed that the primary auditory cortical area as delimited by acoustically evoked electrical activity receives essential projections from the pars principalis of the medial geniculate body (also see Woolsey, 1961, 1964; Woolsey *et al.*, 1962). Rose and Woolsey (1949) showed that following an extensive lesion of the whole auditory region in the cortex there was severe degeneration in the pars principalis of the medial geniculate body with the exception of the posterodorsal sector. Later, however, ROSE and WOOLSEY (1958) showed that the pars principalis projects in a sustaining fashion upon a much larger cortical region than AI and that the caudal parts of this thalamic division project upon temporal and possibly insular fields probably also in a sustaining manner. Electrophysiological evidence obtained from recordings in the medial geniculate also suggested that the pars principalis was the main projection area for auditory input (ROSE and GALAMBOS, 1952).

Recently, however, the early classification by CAJAL and others has come back into prominence. This is not because investigators believe that the above classification is wrong; rather, the emphasis has now been placed upon functional relationships. MOREST (1964a) divided the medial geniculate on the basis of dendritic patterns, as described earlier for the inferior colliculus, and maintained that

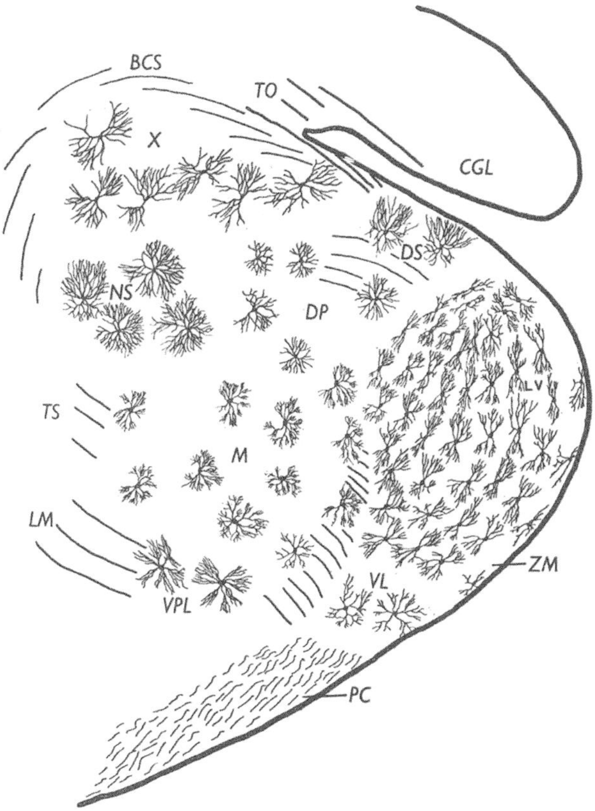

Fig. 6. Distribution of principal neuron types at the junction of the anterior and middle thirds of the medial geniculate body of a 15 day old cat. Transverse section. Golgi-Cox. *BCS* brachium of the superior colliculus, *CGL* lateral geniculate body, *DP* deep dorsal nucleus of the medial geniculate, *DS* superficial dorsal nucleus of the medial geniculate, *LM* medial lemniscus, *M* medial devision of the medial geniculate, *NS* suprageniculate nucleus, *PC* cerebral peduncle, *TO* optic tract, *TS* spinothalamic tract, *VL* ventro-lateral nucleus of the medial geniculate, *VPL* postero-lateral ventral nucleus of the thalamus, *X* lateral posterior nucleus of the thalamus, *ZM* marginal zone. (Taken from MOREST, 1964a)

the ventral, dorsal and medial divisions could be distinguished from each other on this basis. The ventral division is typified by neurons with tufted dendrites, while the dorsal division is typified by neurons with radiate dendrites (Figs. 6 and 7), and MOREST (1964a) believes that the pars principalis embodies both dorsal and ventral divisions, and not exclusively the ventral divisions as other authors have claimed. Indeed, MOREST comments on the dorsal division "In the cat it is an integral part of the medial geniculate that includes the posterior tip of this body."

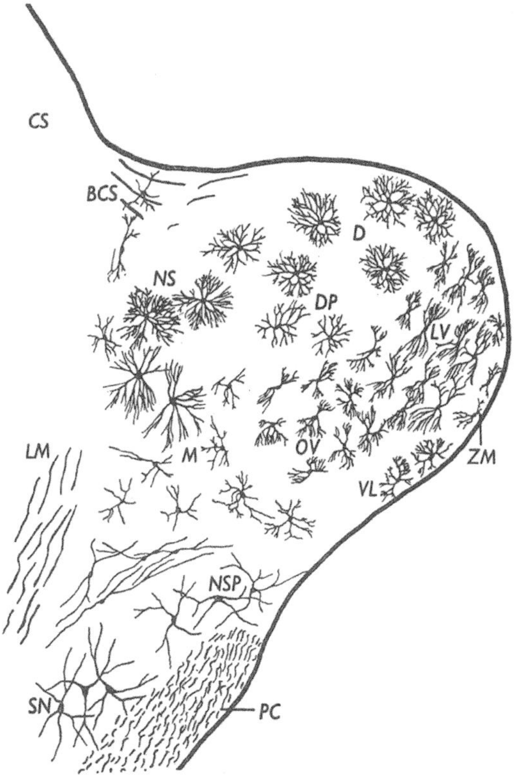

Fig. 7. Principal neuron types at the junction of the posterior and middle thirds of the medial geniculate body of a 15 day old cat. Transverse section. Golgi-Cox. *CS* superior colliculus, *D* dorsal nucleus, *NSP* suprapeduncular nucleus, *OV* pars ovoidea, *SN* substantia nigra. Other abbreviations as in Fig. 4. (Taken from MOREST, 1964a)

The ventral nucleus may be divided into an antero-lateral pars lateralis and a ventro-medial pars ovoidea (Figs. 7 and 8), and, as in the central nucleus of the inferior colliculus, the dendrites of the principal neurons show a laminated pattern (MOREST, 1964a, c, 1965a; JONES and ROCKEL, 1971). Importantly, ascending fibers of the BIC parallel the dendritic layers and make numerous contacts with the principal neurons (Fig. 8). The implication here is that there exists some topographically organized function and, indeed, as will be shown later, a tonotopic organization is present (Aitkin and Webster, 1971, 1972). Morest (1965a) expands his thesis by pointing out that the tufted dendrite pattern is seen in many sensory nuclei and may be associated with large ascending fibers. Ramon-Moliner (1962) believed that these tufted dendrites are associated with axons from neurons related to specific sense organs. Jones and Rockel (1971) also point out that the synaptic organization of the medial geniculate nucleus is similar to that of other thalamic relay nuclei, especially the ventroposterior and pulvinar nuclei. They suggest that there may be "a common plan" for organization of the main thalamic

Fig. 8. Transverse section of medial geniculate body. Ventral nucleus of 15-day old cat. Heavy lines drawn parallel to predominant orientation of the dendrites. Golgi-Cox. (Taken from Morest, 1965a)

nuclei; thus, "ascending afferents to each of them give a small number of large terminals to glomerular aggregations in which they terminate . . . on the dendrites of thalamo-cortical relay cells . . .".

Morest (1971) also reported the presence of a large population of Golgi type II cells in the ventral nucleus of the cat. Not only do the dendrites of these cells receive afferent auditory axons, but they also form connections with dendrites of thalamo-cortical neurons. Morest speculates that dendrodendritic synapses may involve the Golgi type II neurons in an inhibitory role in the medial geniculate, modulating activity being relayed through the geniculate to the cortex.

The dorsal division has radiate dendrites; these are typical, according to Ramon-Moliner (1962), of those regions with many dissimilar inputs. Morest (1964a) feels that this region may be involved in the integration of sensory inputs "that are heterogeneous with respect to the modality, the parameters, or the temporal or spatial sequence of stimuli".

The medial division has considerable morphological variation and the dendritic patterns and projections of the neurons to adjacent nuclei suggest to Morest that "this zone distributes diverse systems to functionally appropriate subcortical nuclei".

C. Other Thalamic Regions of the Auditory Pathway

Several years ago, the above stated considerations would have been adequate to describe the auditory pathway at thalamic level. It has become clear however that this description is inadequate. In fact, both morphological and electrophysiological techniques have clearly defined other thalamic areas of importance in processing auditory information.

The first point concerns the role of the medial division of the geniculate which includes the pars magnocellularis (Fig. 9). ROSE and WOOLSEY (1949), as mentioned before, showed that the pars principalis provides an essential projection to the primary auditory area, but they also felt that only a small fraction of the auditory thalamocortical connections are essential projections. They showed that the pars

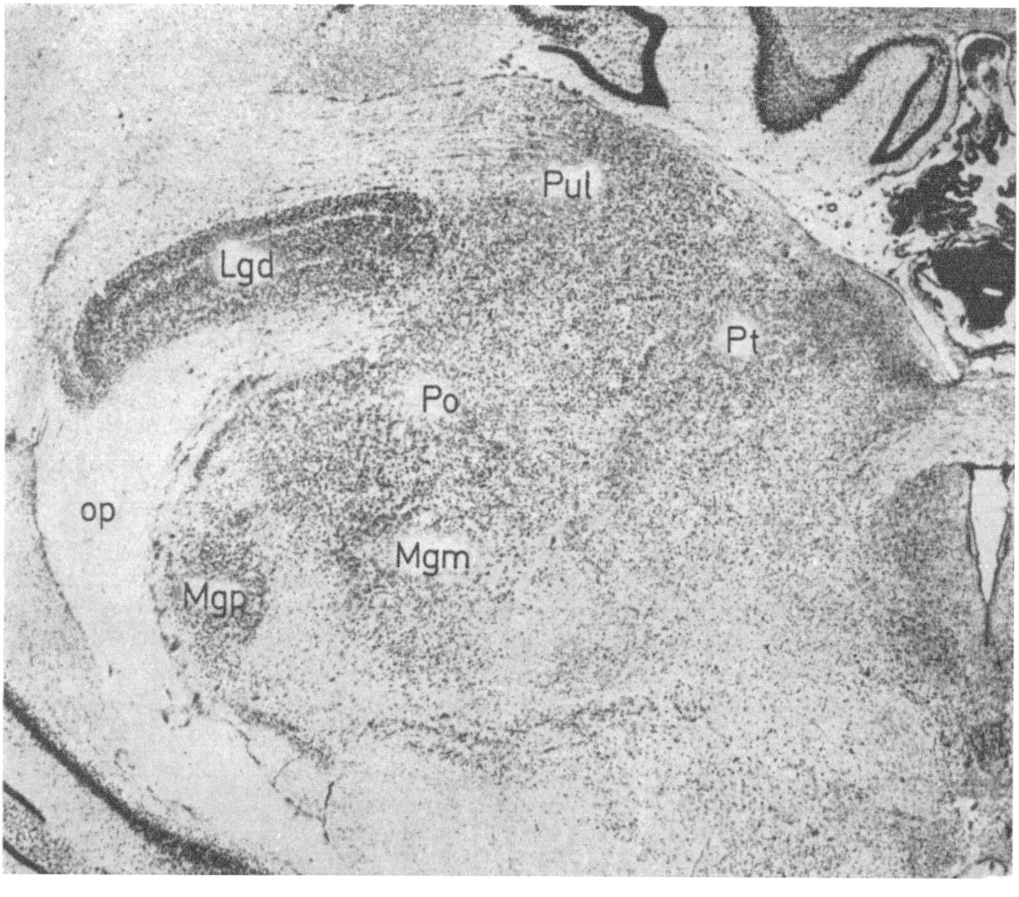

Fig. 9. Frontal section at oral end of medial geniculate body. The principal division is just in the section and the magnocellular division is large and fusing dorsally with the region of the posterior group. This latter group adjoins the pulvinar system and lateral group of thalamic nuclei. *Mgp* principal division, *Mgm* magnocellular division, *Lgd* dorsal lateral geniculate body, *Po* posterior group, *Pt* pretectal region, *Pul* pulvinar. (Taken from ROSE and WOOLSEY, 1958)

magnocellularis emitted a sustaining projection to AI, probably to AII and Ep as well, but definitely to temporal and insular fields (ROSE and WOOLSEY, 1958). Others had also shown by retrograde degeneration studies that temporal cortex, including AI, AII, Ep and insular-temporal cortex receive sustaining projections from this division of the thalamus (WALLER, 1939; NEFF *et al.*, 1956; BUTLER, DIAMOND, and NEFF, 1957; DIAMOND, CHOW, and NEFF, 1958; GOLDBERG and NEFF, 1961a; LOCKE, 1961). Furthermore WOOLSEY and his colleagues (1962) found that electrical stimulation of the pars magnocellularis elicits potentials from AII. Finally, POGGIO and MOUNTCASTLE (1960) isolated single units in the pars magnocellularis which were activated by acoustic stimulation. These data suggest that the medial division, specifically the pars magnocellularis, does form part of the central auditory pathway.

One further aspect should be mentioned. ROSE and WOOLSEY (1958) found that SII in the cortex receives sustaining projections from the anterior part of the posterior group of the thalamus. This part of the group is interpolated between the ventrobasal complex and the medial geniculate (Fig. 9). They pointed out that this locus would be ideal for "tactile and auditory interconnections". GOLDBERG and NEFF (1961a) confirmed these anatomical findings. HEATH and JONES (1971) included the pars magnocellularis in the posterior group complex and concluded that, in this complex, the suprageniculate nucleus and the pars magnocellularis and particularly "the region of transition between the two" send fibers to a band of cortex virtually surrounding and partly overlapping SII but not to a large portion of SII and not at all to SI (Fig. 10). By examining Fig. 10, taken from HEATH and JONES' paper, it can be seen that several polysensory areas appear to be involved; these include the insular cortex which may be a region of interaction between auditory and visual stimuli (DESMEDT and MECHELSE, 1959; LOC and BENEVENTO, 1969), the suprasylvian fringe (which has complete representations of the cochlea (WOOLSEY, 1961)) and parts of the ectosylvian gyrus overlapping SII. Note that AIII, receiving auditory input and possibly somatic input, is involved. Furthermore, zones A and B of CARRERAS and ANDERSSON (1963) in which sensory

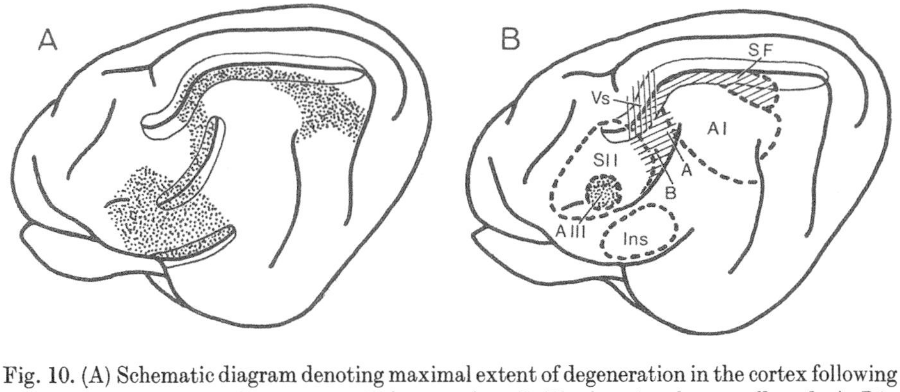

Fig. 10. (A) Schematic diagram denoting maximal extent of degeneration in the cortex following lesions of the suprageniculate-magnocellular complex. (B) The functional areas affected. *A*, *B* in the figure denote zones *A* and *B* of CARRERAS and ANDERSSON, 1963. *AI* first auditory area, *AIII* third auditory area, Ins insular area, *SII* second somatic sensory area, *SF* suprasylvian fringe, *Vs* vestibular area. (Taken from HEATH and JONES, 1971)

and auditory responses can be recorded are also involved. Finally, an area involving vestibular and cochlear projections in the anterior suprasylvian sulcus also shows projections from this suprageniculate-magnocellular complex (WALZL and MOUNTCASTLE, 1949; MICKLE and ADES, 1952; ANDERSSON and GERNANDT, 1954; LANDGREN, SILFVENIUS and WOLSK, 1967).

To summarize: On the basis of morphological techniques, it is clear that the auditory thalamus contains ventral, dorsal and medial divisions of the medial geniculate body which include a pars principalis in the ventral division (and in part of the dorsal division) and a pars magnocellularis in the medial division. The auditory thalamus also includes the posterior group complex. The dendritic patterns and the projections suggest different functional roles for these different regions.

What interactions take place in these different areas of the thalamus? The inputs to these regions may provide some indication. MOORE and GOLDBERG (1963) described ascending fibers in the brachium of the inferior colliculus which originated in the inferior colliculus and projected to the rostral part of the pars principalis of the medial geniculate body – indeed to the ventral lobe. The pars principalis is the area which provides an essential projection to AI. The pars principalis, however, also receives descending fibers from the auditory cortex (DIAMOND, JONES, and POWELL, 1969). Similar findings were obtained by WOOLLARD and HARPMAN (1939), RASMUSSEN (1961), POWELL and HATTON (1969), JONES and POWELL (1971), MOORE and GOLDBERG (1963). POWELL and HATTON (1969) agree that the caudal part of the principal division receives only a small projection from the inferior colliculus. This region also does not provide an essential projection to AI. It is the region called the dorsal division by CAJAL and others. WOOLSEY and his colleagues (1962) found on the basis of electrical stimulation that this region projects to the suprasylvian gyrus. The pars principalis of the ventral lobe thus appears to be organized as a relay for acoustic information.

The pars magnocellularis also receives strong projections from the inferior colliculus (MOORE and GOLDBERG, 1963; POWELL and HATTON, 1969; JONES and POWELL, 1971). NIIMI, MIKI, and KAWAMURA (1970) found that, after lesions in the superior colliculus, degenerated fibers could be traced to the pars magnocellularis as well as to dorsal and dorsomedial parts of the pars principalis. Furthermore although magnocellularis does not send essential projections to AI, AII, or Ep it may receive corticofugal fibers from these areas (DIAMOND, JONES, and POWELL, 1969).

The so-called posterior group also receives fibers from both cortex and lower levels. HEATH and JONES (1971) showed that the insular cortex projects almost exclusively to the suprageniculate-magnocellular complex. Furthermore, zone A of CARRERAS and ANDERSSON (1963) also projects to this area (RINVIK, 1968; DIAMOND, JONES, and POWELL, 1969).

It should now be clear that the suprageniculate-magnocellular complex receives multiple inputs, the polysensory nature of which has been shown clearly to exist. The rostral portion of the pars magnocellularis and the adjoining suprageniculate nucleus, especially in the junctional region, receives ascending somatic fibers from the anterolateral and lemniscal pathways (WHITLOCK and PERL, 1959; MEHLER, FEFERMAN, and NAUTA, 1960; MEHLER, 1966) and auditory fibers from the lateral tegmental pathway (MOREST, 1965b) and from the inferior colliculus (MOORE and GOLDBERG, 1963). Descending fibers project here not only from the auditory regions of the cortex but also from the somatic sensory cortex (MEHLER, 1966; DIAMOND, JONES, and POWELL, 1969; JONES and POWELL, 1971).HEATH and JONES (1971) maintain that, although no part of the posterior group has entirely neurons of one type, somato-sensory neurons are concentrated in the rostral parts, auditory in the caudal parts and polysensory neurons in between. MOORE and GOLDBERG subdivided the posterior group into medial and lateral components and stated that the lateral component alone receives fibers from the auditory system, while other investigators concluded (MEHLER, 1966; RINVIK, 1968; JONES and POWELL, 1968) that the somato-sensory system projects mainly to the medial component.

JONES and POWELL (1971) suggest on the basis of studies after lesions in the cortex and ascending sensory pathways, that "the posterior group forms the caudal continuation of the intralaminar system of nuclei; also that there is a close interrelationship between: (i) the site of termination of afferent fibers to the nuclei, (ii) the principal nuclei in which these same afferent fibers end, and (iii) the region of cortex to which the principal nuclei project and which in turn sends fibers back to them and to the intralaminar-posterior group of nuclei".

II. Electrophysiology

Recent electrophysiological studies have emphasized results obtained either extracellularly or intracellularly from single neuronal elements. The goal of these experiments is to correlate stimulus parameters and the characteristics of the neuronal response. A problem arises in choosing a meaningful stimulus and a measure of neural activity. Clicks, pure tones and even modulated tones are poor substitutes for natural stimuli, yet they do provide well-defined physical properties, whose attributes can be varied over a measurable range of values. Neural activity may also be defined in a variety of ways: latency, spike number, probability of response, interspike intervals, and density of firing as well as the temporal coefficient of unit discharge, have all been recorded and correlated with stimulus parameters at different levels of the auditory pathway. These measures of neural activity are especially amenable to extracellular single unit recording techniques, where recording can be done over considerable periods of time without apparent damage to the neuron. On the other hand, intracellularly-placed electrodes provide information on both neuronal membrane properties and on synaptic relationships at the cell itself. The time during which stable recordings can be obtained, however, is limited.

Both intracellular and extracellular recording techniques may be used to determine the sensitivity of a neuron to a particular frequency and the "response curve" to different frequencies of that neuron, although the time required to obtain these data makes the problem more amenable to extracellular recording. The geographic location of a particular neuron may be obtained and the presence or absence of tonotopic localization within a particular nucleus can be determined.

A. The Inferior Colliculus

1. Tonotopical Organization and Tuning Curves

The first suggestion that tonotopic organization was present at the inferior colliculus was provided by Katsuki, Sumi, Uchiyama, and Watanabe (1958) who stated that in the inferior colliculus all elements "at the rostral part responded to high tones while those at the caudal part responded to tones below 2000 c/sec".

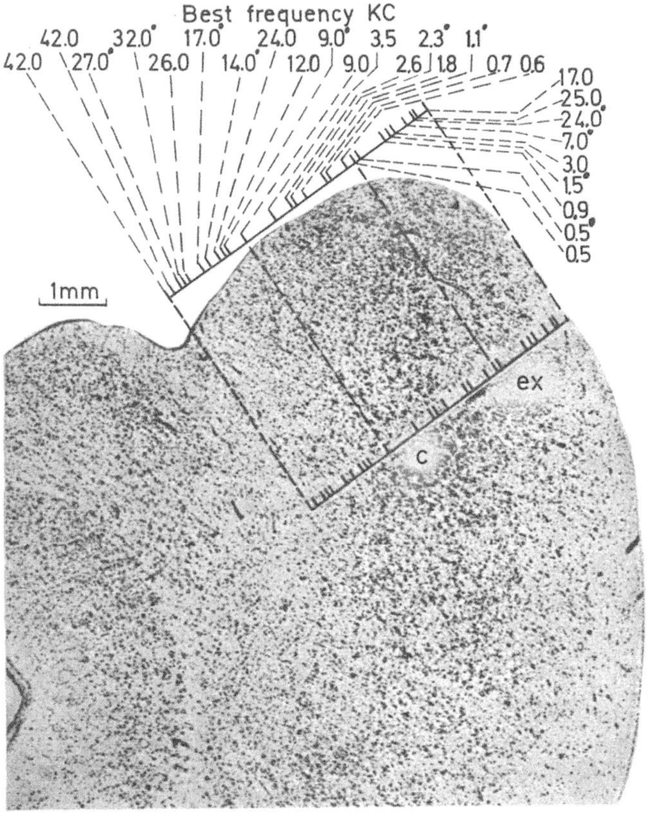

Fig. 11. Frontal section at the level of the inferior colliculus of the cat. Nissl stain. The solid black line is a projection of the electrode track on the photograph. Bars indicate depth of the successive points at which a single unit or cluster of units was isolated and the best frequency determined. Values marked by a dot indicate best frequencies of single neurons. Other values show the upper limit of the effective frequency range, which for small clusters of units, the authors took as an approximation of the best frequency. (Taken from Rose *et al.*, 1963)

Definitive evidence of tonotopic localization was later provided by Rose, Greenwood, Goldberg, and Hind (1963). They showed that the neurons are arranged in an orderly manner with respect to frequency in both central and external nuclei of the inferior colliculus (Fig. 11). They stated that the orientation was such that when the electrodes were inserted "at a point which was usually dorsal, caudal, and lateral to the point of deepest penetration" the sequence of best frequencies was from high to low when the external nucleus was penetrated and from low to high as the central nucleus was transversed (Fig. 9).

Recently DIXON (1973) used bipolar electrodes to record responses to pure tones from the inferior colliculus; he found that the polarity of the response at a particular recording site depended upon the frequency of the tone-burst stimulus. This parameter allowed DIXON to assign a characteristic or best frequency for units at different locations in the inferior colliculus. His results showed mediolateral and anteroposterior tonotopic arrangements in addition to the vertical arrangement described by ROSE *et al.* (1963). Furthermore, when points of similar best frequency were connected, a laminar pattern similar to that described by ROCKEL and JONES (1973a) resulted. In fact, points located medially, posteriorly, or ventrally to a reference point in the central nucleus had higher best frequencies than the unit at the reference point itself. These data would thus fit the concepts of MOREST (1964b), GENIEC and MOREST (1971), ROCKEL (1971), and ROCKEL and JONES (1973a) for the central nucleus that laminar arrangement of neurons in parallel with ascending auditory afferents may be the morphological substrate for tonotopic organization. Although the cortex region of the inferior colliculus may contain distinct layers, it is difficult to equate tonotopic organization with these layers; therefore, the morphological basis for the localization of best frequencies in this structure is still unknown.

As ROSE and his colleagues (1963) point out, the existence of a tonotopical organization implies that many individual neurons may be "tuned" to some frequency. The range of tuning may be deduced by determining the intensities of different tonal frequencies which evoke a discernible change in the probability of unitary firing – a threshold response. As stated earlier, however, there are many methods of characterizing unitary responses in the central auditory pathway. Thus, while a tuning curve may have a particular range and a certain "best frequency" when determined by the threshold response method, the same unit may show both a different range of responsiveness and a different best frequency when determined by another criterion for neuronal responsiveness. For instance, if the best frequency is determined by measuring the latency of a unit's response as a function of frequency, so that the minimal latency occurs at the "best frequency", the value may differ from that obtained by measuring the threshold response. Usually, the values obtained by these different methods tend to be nearly the same (HIND *et al.*, 1963). When the responsiveness of a unit in terms of density of firing (number of spikes per unit time) in response to different frequencies is measured in contrast to the sensitivity of the unit, striking differences are seen (Fig. 12). For instance, Figs. 12A and 12B show that for one and the same unit, the best frequency expressed as intensity of firing, can change from 436 Hz to 566 Hz when the intensity of the stimuli is increased by 15 db. Figure 12C shows results from the paper by ROSE *et al.* (1963) where an intensity increase of 70 db caused the "best frequency" to decrease from 20 KHz to 4 KHz. ERULKAR, NELSON, and BRYAN (1968) point out that both sensitivity and responsiveness are valid but different indicators of unitary activity; on the other hand, they state "The problems arising in terms of tonotopic localization from such differences are obvious, and at the present time no answers to these problems are at hand."

There has been considerable work devoted to the analysis and interpretation of threshold response curves at different levels of the central auditory pathway. Although KATSUKI (1966a) has claimed that there occurs a funnelling mechanism at successive auditory relay stations, so that the frequency range activating a unit becomes progressively narrower up to thalamic levels, there is relatively little

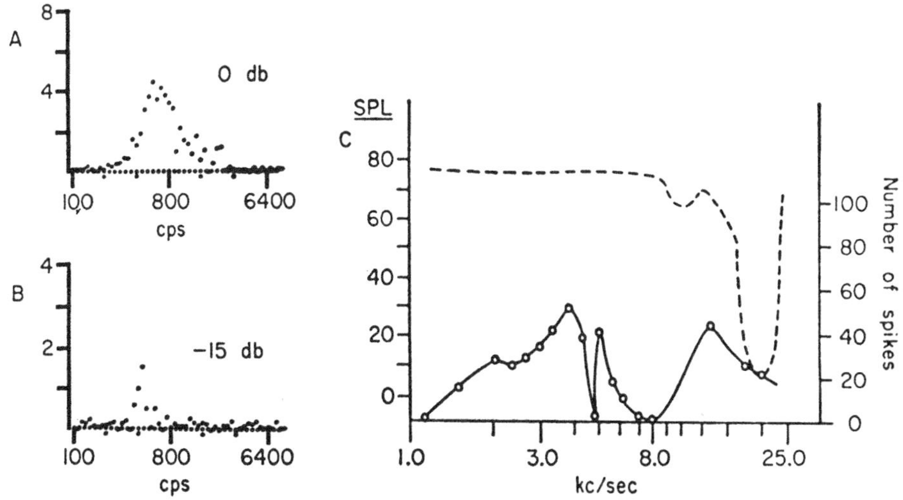

Fig. 12. A and B Frequency-response plots for a unit taken at two different stimulus intensities. Ordinate: Number of spikes elicited per presentation of the tonal stimulus. Abscissa: Logarithmic scale of tonal frequencies presented. When stimulus intensity attenuated by 15 db in (B), peak response occurs at a lower frequency. (C) Illustration taken from ROSE *et al.* (1963). Ordinate at the left of the graph: Stimulus intensity in db. This relates to the dashed line. Abscissa: Tonal frequency. Dashed line is a threshold-intensity curve. Best frequency 20 KHz. Ordinate at the right of the graph: Number of spikes elicited by tonal frequencies at 80 db SPL. Replotted from figures given by ROSE and his colleagues. Now the best frequency is 4 KHz. (Taken from ERULKAR *et al.*, 1968)

evidence to confirm this. Thus, while KATSUKI *et al.* (1958) and ERULKAR (1959) reported that at the inferior colliculus there were some threshold response curves that were extremely narrow in range, both ERULKAR (1959) and ROSE *et al.* (1963) showed that there were some extremely broad threshold response curves at the same level.

It is in fact this author's view that the analysis of the static properties of auditory unitary discharge is of limited functional significance – except mechanistically. Thus while excitatory and inhibitory interactions in relation to different frequencies are of great value to interpretation of neural mechanisms of audition, the role of these frequency response curves in frequency discrimination is still unclear.

2. The Nature of Unitary Response Patterns and their Relationship to Excitatory-Inhibitory Influences

ERULKAR (1959) described two patterns of responses of single units in the inferior colliculus to tonal stimuli at the unit's best frequencies. The two patterns consisted of an onset response with no further activity to the tonal stimulus and a sustained discharge throughout the stimulus. Response patterns were described in greater detail by Rose, Greenwood, Goldberg, and Hind (1963), and Fig. 13,

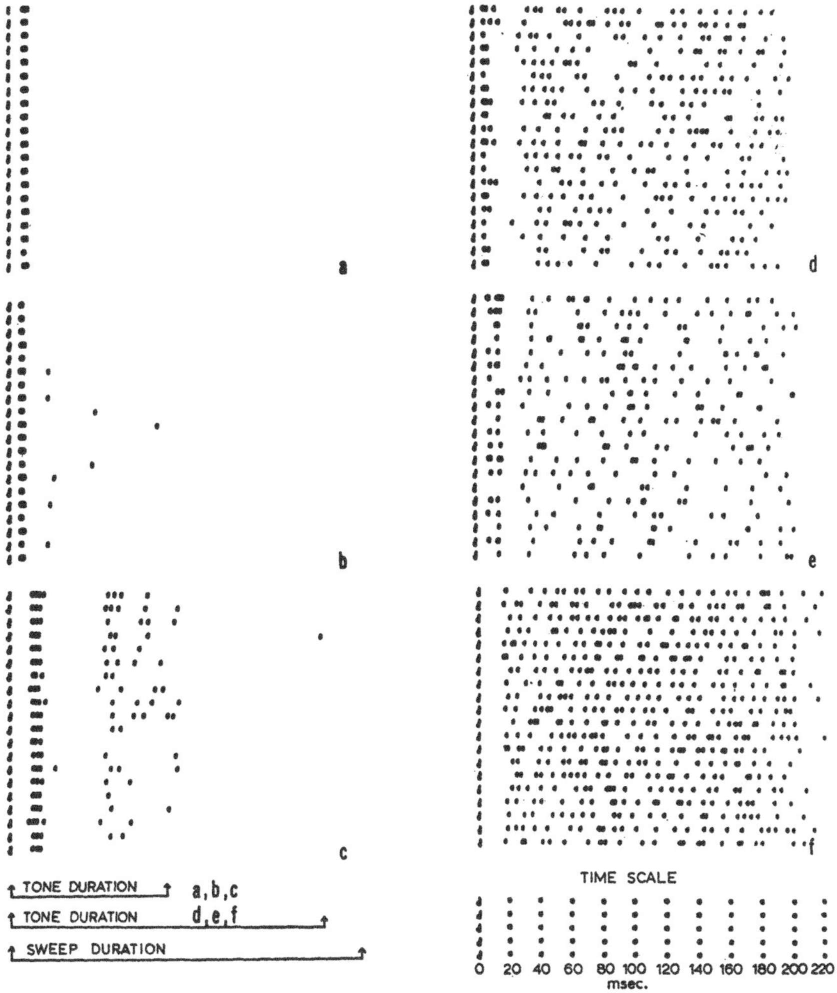

Fig. 13a—f. Firing patterns of six neurons from the inferior colliculus of the cat showing different types of response patterns. Each sample consists of twenty repetitions of the same tone delivered to the contralateral ear. (a) 14 KHz, 50 db SPL, (b) 19 KHz, 40 db SPL, (c) 6.1 KHz, 70 db SPL, (d) 2.6 KHz, 60 db SPL, (e) 1.34 KHz, 80 db SPL, (f) 3.8 KHz, 80 db SPL. (Taken from Rose *et al.*, 1963)

taken from their paper shows the varieties of discharge patterns encountered in response to tonal stimuli. The response sequences vary from pure onset patterns (Fig. 13a) to sustained discharges throughout the duration of the tone (Fig. 13b). Rose *et al.* (1963) found that a given neuron could display a firing pattern that is characteristic of that particular neuron. Nearly 50% of the units recorded by Rose and his colleagues showed pure onset patterns, and they speculate that these may be the result of "interactions within the synaptic chain since such patterns seem to occur with increasing frequency in higher synaptic regions of the auditory system".

Indeed it is now well documented that in contrast to responses at cochlear nucleus and superior olivary nucleus, onset patterns are more common at the inferior colliculus and at higher levels of the central auditory pathway (ROSE *et al.*, 1963; ERULKAR, NELSON, and BRYAN, 1968). This finding would suggest that inhibitory processes become more predominant at these higher levels. There is some evidence for this. ROSE *et al.* (1963) found that a sustained pattern of response to tonal stimuli from some units may consist of an onset burst, a silent period which follows it and a period of sustained firing. The silent period lengthens as the stimulus intensity is increased, and this would suggest that the inhibitory input occurring during the excitatory response varies with stimulus strength. Furthermore, ROSE *et al.* (1963) reported that in response to a tone burst the spike count could vary as a function of either frequency or intensity of the stimulus. Over 50% of the total number of units investigated responded to intensity increase with a non-monotonic spike function. The authors suggested that the "simplest hypothesis" would be that a strong stimulus could activate both excitatory and inhibitory circuits and that the interaction of these at a single neuron would result in the relationship seen. Obviously, if inhibitory circuits were activated more strongly with stronger stimuli, then the non-monotonic relationship would occur. There is pharmacological evidence for both presynaptic and postsynaptic inhibitory mechanisms being present in the inferior colliculus. WATANABE and SIMADA (1971) showed that a phasic onset response of a collicular neuron to a sustained tonal stimulus could be converted into a tonic sustained discharge to the same stimulus by microinjection of picrotoxin at the site. When picrotoxin was administered at sites of units which normally responded with sustained discharges to tonal stimuli, only a slight increase in the number of spikes was recorded. This suggests that presynaptic inhibition may be present in collicular neurons as well as postsynaptic inhibition as shown by NELSON and ERULKAR (1963) (see later). The abolition of synaptic inhibition due to the action of picrotoxin appeared to be limited only to the discharges at the beginning of the stimulation period which lasted from 40–80 msec. ERULKAR (1959) has reported "unresponsive periods" following initial excitation and lasting up to 120 msec. Perhaps, both pre- and postsynaptic inhibition may be involved. In fact, GABA has been found in collicular neurons (OTSUKA, M. quoted by WATANBE and SIMADA, 1971) and both GABA and glycine were shown to depress tone-induced responses in inferior collicular units.

Excitation may be mediated in some neurons by acetylcholine. WATANABE and SIMADA (1970) applied dihydro-β-erythroidine iontophoretically in the vicinity of collicular neurons that responded to click stimuli with multiple spike burst responses and also to those neurons with sustained slowly adapting responses to tonal stimuli. In both cases the activity was depressed, although in the latter case, the initial discharge remained. This suggests the presence of nicotinic cholinoceptive receptors in collicular neurons.

Direct evidence for these different synaptic relationships was obtained by NELSON and ERULKAR (1963). They reported that a wide variety of synaptic sequences could be recorded from different cells of the inferior colliculus by means of intracellularly placed electrodes. In many cells, an initial depolarizing excitatory potential, which would cause the cell to fire, might be followed by a powerful inhibitory input which terminated the cell discharge (Fig. 14). This pattern could be responsible, as suggested, for onset responses recorded by extracellularly-placed electrodes. NELSON and ERULKAR (1963) also showed by intracellular current

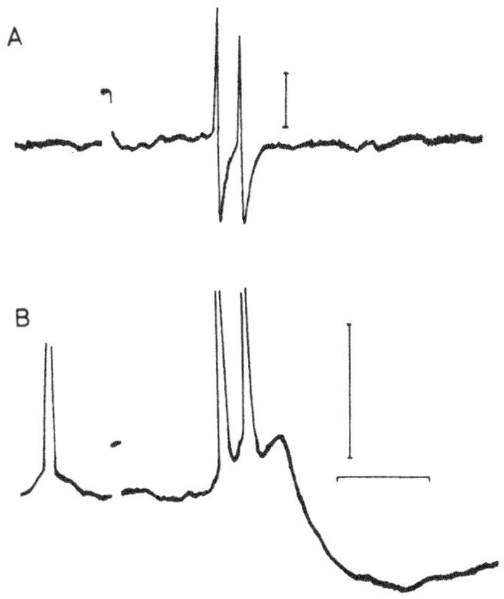

Fig. 14. (A) Response to click stimulation from unit in the inferior colliculus of the cat. Calibration 1 mV. (B) Transmembrane recording after development of 42 mV resting potential. Calibration 10 mV. Spikes truncated. Time line for both traces 5 msec. (Taken from NELSON and ERULKAR, 1963)

injection that these inhibitory mechanisms may be a result of active postsynaptic inhibition for some cells and not the result of mere withdrawal of excitation. The complex interactions between excitation and inhibition which changed with change in tonal frequency or stimulus intensity are shown in Fig. 15, and provide direct evidence for the concepts which were initially derived on the basis of extracellular recordings. The presence of hyperpolarization in so many recordings at this level is in contrast to the situation at the anteroventral cochlear nucleus where hyperpolarizing responses could rarely be recorded to acoustic stimulation (GERSTEIN, BUTLER, and ERULKAR, 1968). On the other hand, pure depolarizing responses to tonal stimuli were also recorded from some cells of the inferior colliculus and the authors concluded that these cells conveyed information on total auditory input to other regions in the central auditory pathway.

These complex patterns of synaptic interaction must also find functional significance in the processing of more intricate acoustic signals. As a step in this direction, NELSON, ERULKAR, and BRYAN (1966) recorded responses of single units in the inferior colliculus to time-varying stimuli, specifically, to frequency (fm) and amplitude (am) modulated tones. These authors used density of firing as the criterion for neuronal activity. Some units were directionally sensitive to fm and am tones so that responses were elicited as the stimulus changed in one direction but

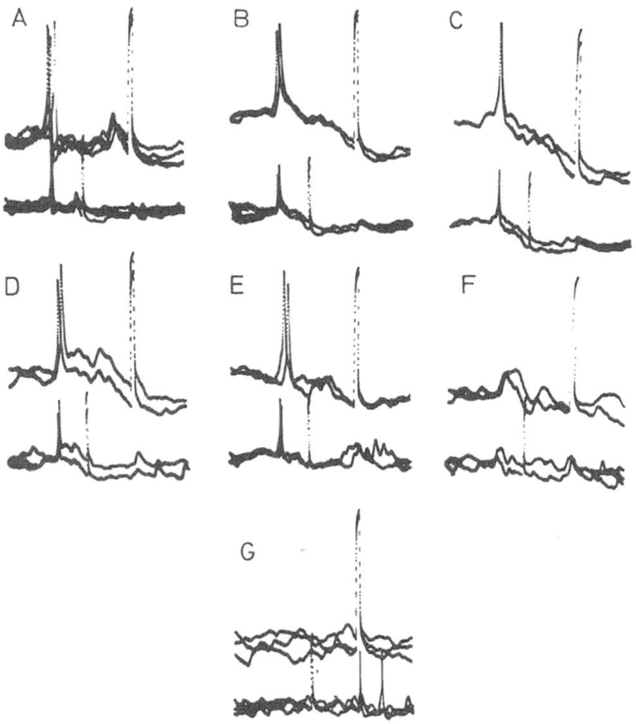

Fig. 15. Inferior colliculus of the cat. (A). Multiple sweeps showing initial spike response and late depolarizing potential in response to 1 KHz pulse. Calibration 10 mV, 1 msec in all records. Upper record shows expanded section of trace of lower record. (B) through (G). Responses of same unit as in A to 400 Hz, with decreasing intensity in 10 db steps. Note that in (E) hyperpolarization has almost disappeared, while the initial spike is still elicited. (Taken from NELSON and ERULKAR, 1963)

not the other (Fig. 16). Other units showed no directional selectivity and showed a linear relationship between the modulated stimulus and the density of firing. Although the majority of units responded to both fm and am stimulation, some units showed greater activation by one stimulus than by the other.

In some cells, the response of the cell to fm stimulation could be predicted from the frequency response curve to pure tones, but in other cells no such predictions could be made. Indeed, in some units responses to fm stimuli occurred at frequencies lying completely outside the pure tone frequency response plot (Fig. 17).

Other stimulus parameters to which a unit could be sensitive included the rate of change and depth of modulation. While some units responded selectively to the rate of change of the stimulus, modulation rate was also an important parameter. Indeed, in some units, no activation occurred at lower rates of modulation (e.g. 1/sec) but as the modulation rate was increased, marked activation took place. Modulation depth could also be critical, and ERULKAR, NELSON, and BRYAN (1968) state that within limits for a given peak amplitude of response, a reciprocal relationship may exist between modulation depth and modulation rate. They consider that "the use of these stimulus parameters serves to characterize the responsiveness of the unit in a functionally significant way and may be compared to parameters of intensity and frequency as used in conventional steady-state frequency-threshold plots" (ERULKAR, NELSON, and BRYAN, 1968).

The data reported by ERULKAR *et al.* must also reflect different excitatory-inhibitory interactions occurring in units of the inferior colliculus. Thus it was

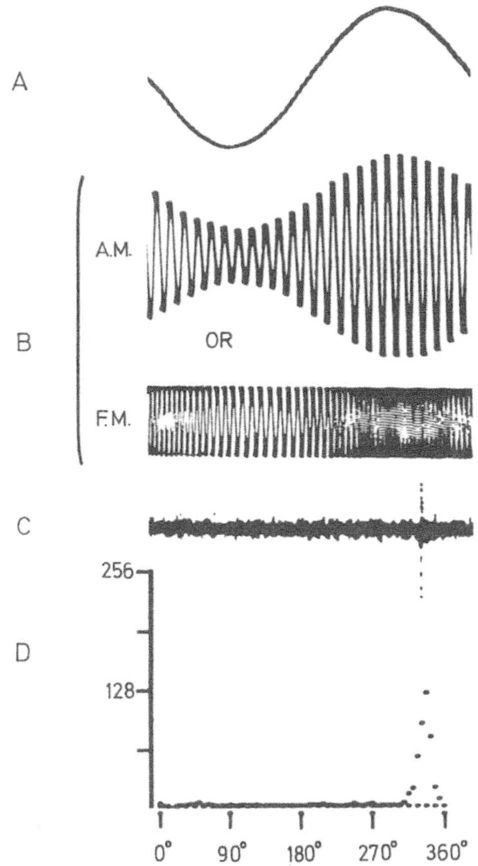

Fig. 16. An illustration to show the relationship between the sinusoidal control voltage (A) generated by the LINC computer which modulates either the amplitude or the frequency of the audio signal (B). The relationship of the unitary discharge to this signal during one cycle of the frequency-modulated stimulus is shown in (C). The LINC display of the number of spikes occurring during many modulation cycles at various phases of the frequency-modulated stimulus cycle shown in (B) is given in (D). The scale on the ordinate corresponds to the number of spikes occurring during the period of stimulus presentation at the phases of the modulation cycle shown on the abscissa. Each point on the cycle plot display corresponds to the points on the modulation cycles aligned above it. The 0°, 180°, and 360° points on the cycle plot correspond to the center frequency or center amplitude. 90° corresponds to minimum frequency or amplitude and 270° corresponds to maximum frequency or amplitude. (Taken from NELSON *et al.*, 1966)

shown that the patterns of activity recorded extracellularly and intracellularly, and results obtained by pharmacological techniques, are due to interaction of converging synaptic inputs to the collicular neurons. Another way of showing this is by the use of models; BRYAN developed a model which incorporated different features of the synaptic organization and cellular mechanisms in the central

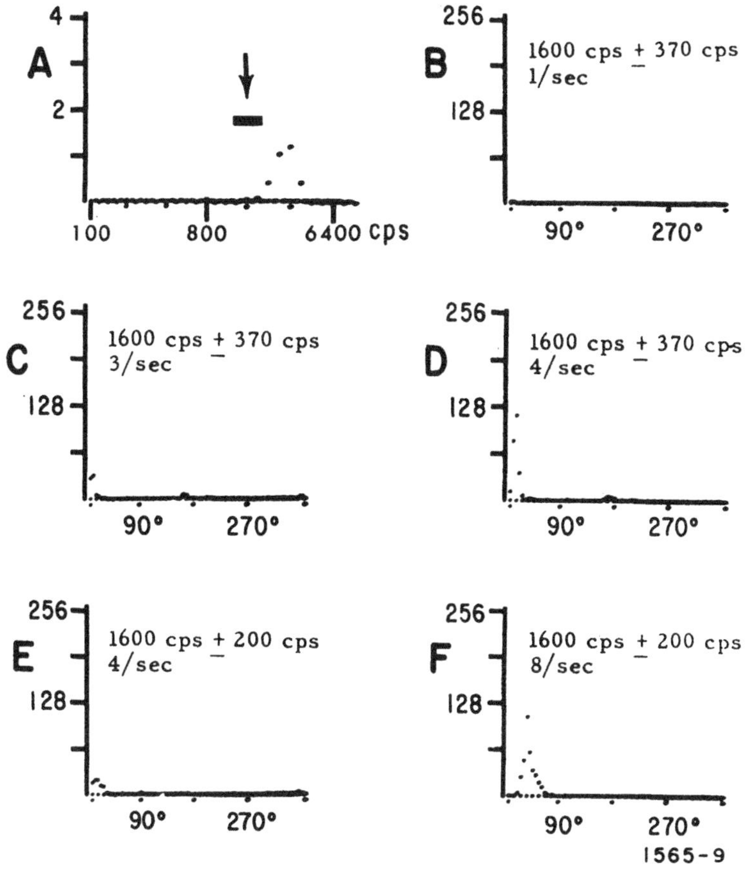

Fig. 17. (A) Frequency-response plot. (B—F) Frequency-modulated cycle plots at stimulus parameters indicated. Note lack of response in B, and development of response in (C) and (D). Arrow and bar above (A) indicate center frequency of 1.6 KHz and modulation depth of ± 370 Hz respectively. (Taken from NELSON *et al.*, 1966)

auditory pathway. The purpose of the model was to provide a quantitative test of different qualitative models of converging synaptic inputs to a given cell and also to study temporal aspects of this synaptic convergence. Although no model can say how a system functions, it may be able to predict quantitatively the behavior of a particular neuron.

The model used is described by ERULKAR, NELSON, and BRYAN (1968); it consists of three "neurons" essentially the same as those described by HILTZ (1962, 1965). For simplification, two driving units converging upon a terminal unit were chosen to represent different multineuronal pathways to the collicular (terminal) unit. By the use of this relatively simple electronic model system, it was possible to obtain a virtually close simulation of the behavior of cells in the inferior colliculus in response to complex tonal stimuli. It was found necessary in order to obtain a fit to the experimental physiological data to include noise in the system; this noise may represent many synaptic inputs from different units with different average firing rates. Adding noise appeared to permit a more linear transfer characteristic above threshold.

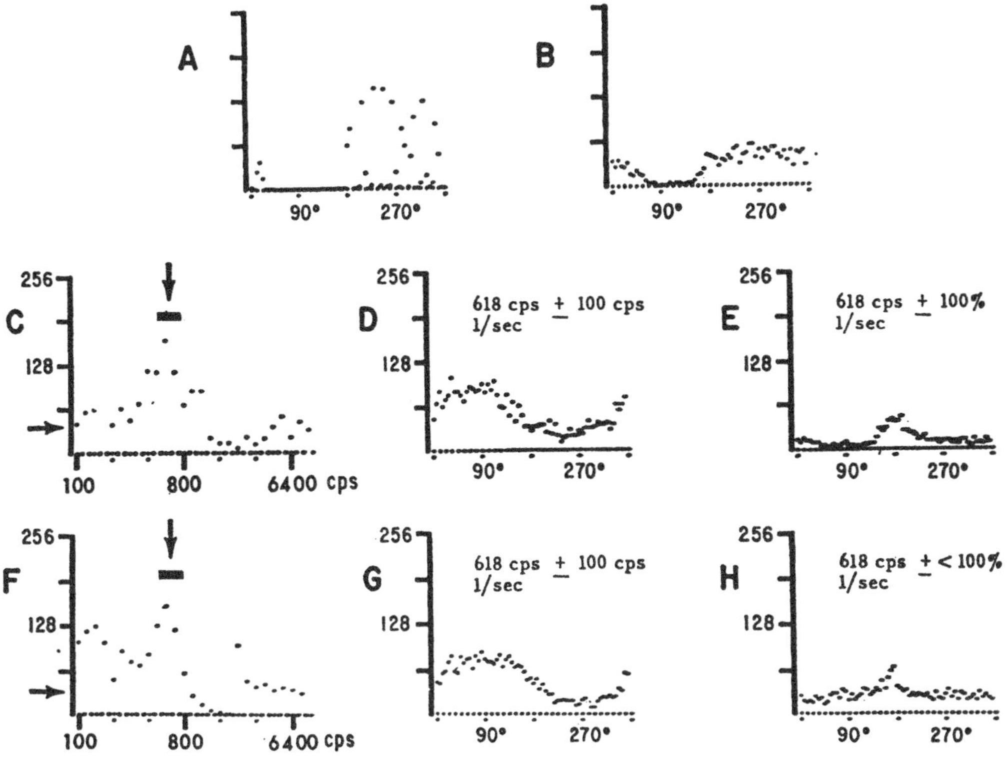

Fig. 18. (A) Modulation cycle plot of the response of a single simulated cell to frequency modulation. Note the tendency of spikes to lock at certain times of the modulation cycle. (B) Same system as in (A) but with noise added to the system, which results in the elimination of the locked pattern and a more sinusoidal response (C, D, E) Response patterns for a simulated three-neuron system. (F, G, H) Responses obtained experimentally from a unit in the inferior colliculus. (C, F) Frequency-response plots. Horizontal arrows at the left of the graph show levels of spontaneous firing. (D, G) Modulation cycle plots of responses to frequency-modulated stimuli with parameters as indicated. (E, H) Modulation cycle plots of responses to amplitude-modulated stimuli with parameters as indicated. (Taken from Erulkar *et al.*, 1968)

The closeness of the fit between the physiological and simulated data is shown in Fig. 18. This example (taken from Erulkar, Nelson, and Bryan, 1968) serves to show that highly complex mechanisms are not required to produce complex response patterns. It was shown that excitation and inhibition act synergistically during frequency modulation, but antagonistically during amplitude modulation and also with a time difference between their onset of action. In order to obtain the characteristic single peak curve shown for amplitude modulation, it was found necessary to use different delays, different accommodation time constants and different thresholds in the excitatory and inhibitory pathways. It was not possible to match the experimental data with a model system involving excitatory and inhibitory pathways differing in only one of these parameters.

This is just one example, however, and other combinations of input were shown to fit other patterns of response recorded experimentally. The reader is referred to the original paper (ERULKAR, NELSON, and BRYAN, 1968).

3. Binaural Phenomena

The studies described above give some insight into the mechanisms involved in the response of a neuron at collicular level which receives inputs in a multineuronal pathway. The pathway may originate from the same or from the opposite side. ADES and BROOKHART in 1950 suggested that the inferior colliculus may be involved in the analysis of the spatial localization of sound because slow wave activity could be elicited in this region by click stimulation presented separately to the two sides. This was followed in 1954 by the report of ROSENZWEIG and WYERS who showed that interaction of the elicited responses from the two sides occurred at this level.

In 1959, ERULKAR using extracellularly-placed microelectrodes, showed that acoustic stimulation presented separately to the two ears could activate the same

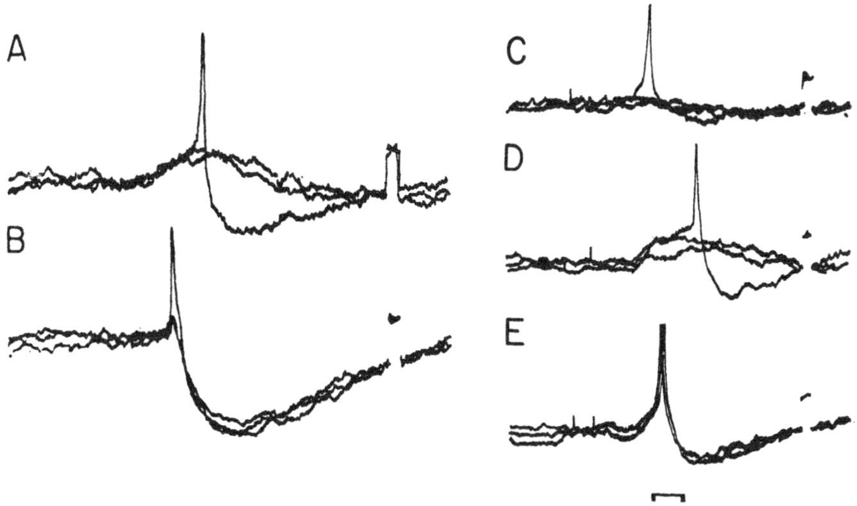

Fig. 19A—E. Responses from single unit in the inferior colliculus of the cat to click stimulation. Total sweep duration for (A and B) — 50 msec. Time line for (C, D and E) — 5 msec. Calibration for all traces — 1 mV. (A) supramaximal ipsilateral stimulation, (B) supramaximal contralateral stimulation; (C) contralateral stimulation at near threshold intensities. Probability of firing = 2/12; (D) ipsilateral stimulation at near threshold intensities. Probability of firing = 3/9; (E) stimulation of both ears, contralateral preceding ipsilateral by 3.53 msec. Probability of firing = 8/9. (Taken from NELSON and ERULKAR, 1963)

neural unit in the inferior colliculus. If the click stimuli were presented to both ears at different time intervals, the response of the unit to the first stimulus could influence the response to the second stimulus. The characteristics of the second response such as the number of spikes per discharge, latency of the first spike and probability of response varied dependent upon the time between the two stimuli. ERULKAR described the presence of an "absolutely unresponsive period" during which no response could be elicited by the second stimulus. The duration of this unresponsive period depended upon which ear was stimulated first as well as on the relative intensities of the stimuli at the two ears. In the majority of experiments, the unresponsive period was longer when the contralateral ear was stimulated first. If the stimuli were separated by short intervals, summation of the two responses could take place. These responses must reflect the interaction of excitatory and inhibitory inputs from the two sides. Binaural interaction at the inferior colliculus was studied by NELSON and ERULKAR (1963) by means of intracellularly-placed electrodes. They showed that identical click stimuli presented separately to the two ears could elicit different patterns of response from the same cell. For instance, Fig. 19 shows that a click stimulus delivered to one ear produced a depolarizing potential while the same stimulus to the other ear elicited a predominantly hyperpolarizing potential. In some cells, clicks could elicit postsynaptic potentials (psps) which were similar in configuration; in others, the

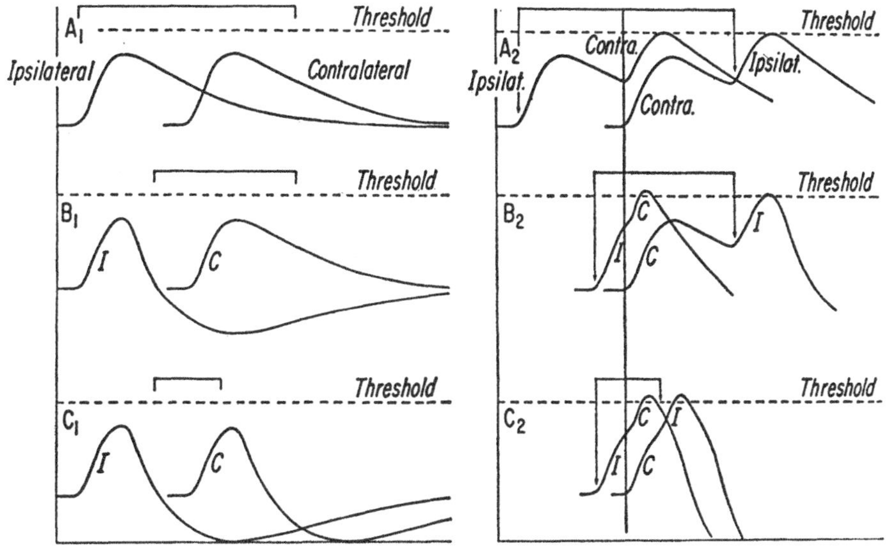

Fig. 20. Diagrams of three hypothetical patterns of synaptic activity elicited by transient binaural stimulation (clicks or tone bursts). A_1, B_1, and C_1: The complete time course of each synaptic event is given. A_2, B_2, and C_2: Interactions of the synaptic potentials leading to spike generation are shown. The response to contralateral stimulation is fixed in time, and the time of occurrence of the response to ipsilateral stimulation varied about it. The bars above each diagram denote the intervals during which spike generation to ipsilateral stimulation may be initiated by summation of the synaptic potentials. Note: The time for spike generation in C_1 and C_2 is much shorter than for the other two patterns: if the cell fires, therefore, binaural stimulation within an accurately specified interval has occurred. This could possibly indicate the spatial localization of a sound source. (Taken from ERULKAR *et al.*, 1968)

sequence of excitatory to inhibitory psps would differ. Figure 19 also shows that in order to bring about optimal facilitory interactions, precise timing was required between the two stimuli presented to the two ears. In fact, ERULKAR, NELSON, and BRYAN (1968) described three hypothetical patterns of synaptic activity elicited by transient binaural stimulation, and showed that a configuration of depolarization followed by hyperpolarization elicited from both sides would provide the optimum basis for a neuronal mechanism of sound localization (Fig. 20). Summation of the depolarizing components could only occur within a limited time interval, so that if the cell fired, it would mean that binaural stimulation within a specified time had occurred. This specification of interaural time difference would then serve to localize the source of a given sound. A complete discussion of possible mechanisms of sound localization has been given by ERULKAR in 1972.

These excitatory-inhibitory interactions were studied in great detail by ROSE, HIND and their colleagues (ROSE *et al.*, 1963; HIND *et al.*, 1963; ROSE *et al.*, 1966). They found (HIND *et al.*, 1963) that the spike count could be increased in response to simultaneous binaural stimulation over that which occurred when each ear was stimulated separately. They also found, however, that the spike count could be *decreased*, by simultaneous binaural stimulation, for some units which were activat-

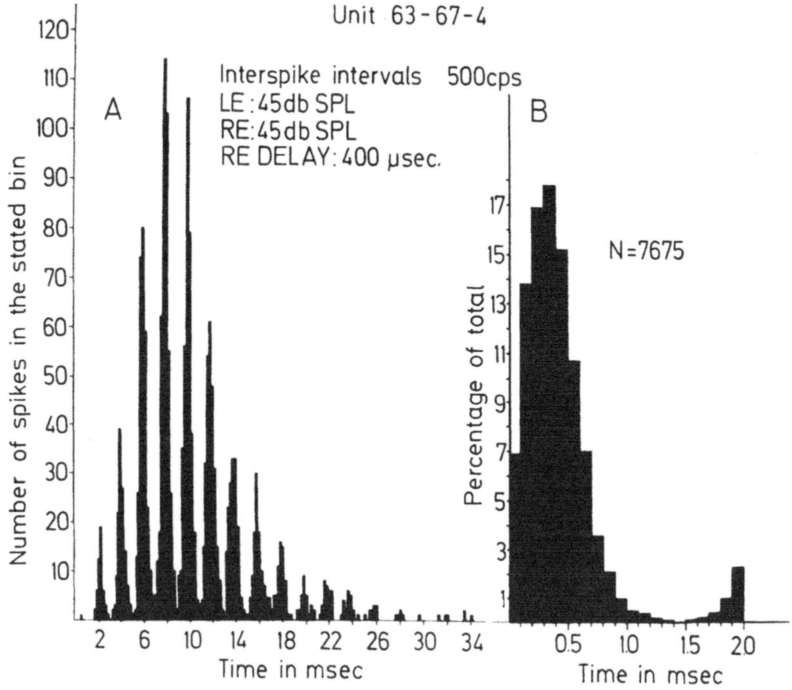

Fig. 21 A and B

Fig. 21. (A) Distribution of interspike intervals for a unit in the cat inferior colliculus when the phase angle between binaural stimuli was most favorable for discharge of spikes. Abscissa: Time in msec. Each time bin = 0.2 msec. Ordinate: Number of spikes per bin. The stimulus to the right ear was delayed by 400 μsec. Binaural stimulation — 10 sec; 500 Hz; 45 db SPL. N = Number of interspike intervals in the sample. (B) Folded histogram showing the distribution of discharge times throughout a 2-msec period of the sinusoidal stimulus. Interaural phase relation is nominally the same as in (A) but was achieved by delaying the left ear stimulus by ed by stimulation of one ear alone. This would suggest that stimulation of the second ear alone suppressed or inhibited the activity of that unit. Changes in latency and in firing patterns of the neural response were reported for simultaneous binaural stimulation. Certain cells, which responded to binaural tonal stimuli of low frequency, gave markedly different spike counts for a given intensity of the tone depending upon whether the tonal stimuli were in phase or 180° out of phase at the two ears. Rose and his colleagues (1966) confirmed these findings and also found that when the stimulus to one or the other ear was delayed (even though frequency and intensity remained constant), the spike-count function was periodic, and the period was equal to the period of the applied tone. Figure 21 shows the periodic discharge curve which illustrates these features. Of interest is the fact that, with a favorable delay, facilitation occurred rather than mere summation; with an unfavorable delay, suppression of activity took place. Thus, interaural time relationships once again appear to bring about these excitatory-inhibitory interactions. For the unit whose results are given in Fig. 21, separate stimulation of the ears caused a net excitation. Yet, interaction brings out the fact that stimulation to at least one ear must have elicited both excitation and inhibition. When

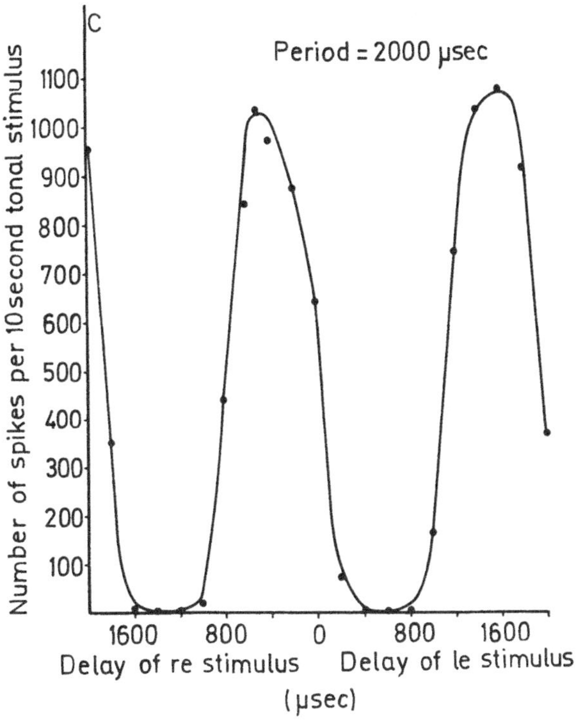

Fig. 21 C

1600 μsec. Abscissa: Time in msec. Each time bin = 0.2 msec. Ordinate: Percentage of spikes per bin. (C) Same unit as above. Number of spikes discharged to binaural stimuli when tone to the right (RE) or to the left (LE) ear is successively delayed in time with respect to the stimulus presented to the other ear. All stimuli: 10 sec duration; 500 Hz; 45 db SPL. Abscissa: Delay of the stimulus to the appropriate ear in microseconds. Ordinate: Number of spikes per trial. N = Total number of trials. (Taken from Rose *et al.*, 1966. The reader is referred to the original paper for details) a neuron was activated by different frequencies, the periodic discharge curve reached the same relative amplitude at the same absolute interaural delay of the stimuli (Fig. 22). This delay has been designated "the characteristic delay of the unit". Thus a unit is capable of detecting an absolute time difference between impinging stimuli regardless of their frequency. ROSE *et al.* (1966) suggest that the results of binaural interaction depend on at least two variables: the interaural time

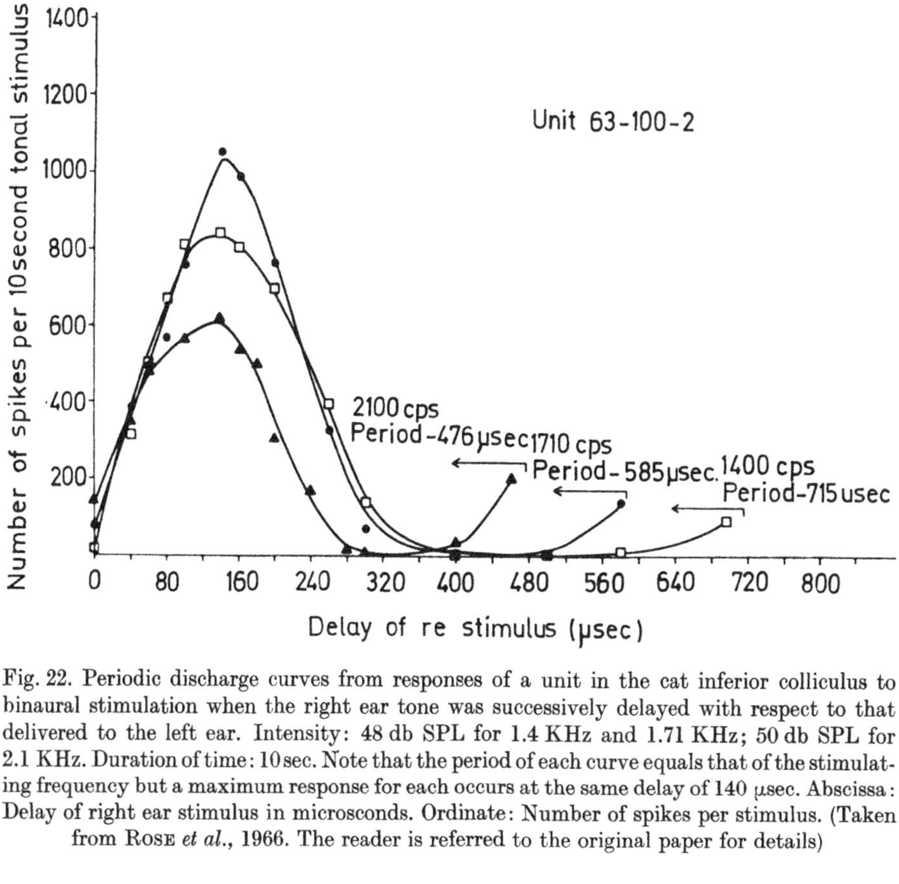

Fig. 22. Periodic discharge curves from responses of a unit in the cat inferior colliculus to binaural stimulation when the right ear tone was successively delayed with respect to that delivered to the left ear. Intensity: 48 db SPL for 1.4 KHz and 1.71 KHz; 50 db SPL for 2.1 KHz. Duration of time: 10 sec. Note that the period of each curve equals that of the stimulating frequency but a maximum response for each occurs at the same delay of 140 μsec. Abscissa: Delay of right ear stimulus in microsconds. Ordinate: Number of spikes per stimulus. (Taken from ROSE *et al.*, 1966. The reader is referred to the original paper for details)

relationship of the binaural stimuli, which will determine the temporal relationship of inhibitory and excitatory events; and, the stimulus strength which will determine the amplitude as well as the duration and rise time of the excitatory and inhibitory events. Thus, differences in excitatory-inhibitory balance were responsible for the reported effects. In addition to the neural units which were sensitive to interaural time relationships, other units were found which were sensitive to small interaural intensity differences. For the latter units, contralateral ear stimulation usually was effective in causing the neuron to fire, while ipsilateral stimulation appeared to be ineffective. Stimulation of both ears, however, caused "a net inhibitory effect" (ROSE *et al.*, 1966). BENEVENTO and COLEMAN (1970) described four different populations of units in the inferior colliculus: those which were sensitive to 1) interaural intensity differences, 2) interaural time differences,

3) neither of these, and 4) both of these. Furthermore, they found considerable overlap between the four populations so that, if the intensity level was changed, a unit could change its properties and "shift from one of these populations to another". For instance, units which were sensitive to both interaural time and interaural intensity differences, could alter their characteristics dependent upon the intensity level of the interaural intensity difference. They could respond in a way typical of population 1 (interaural intensity sensitive) or population 2 (interaural time sensitive). On the other hand, while the function of an interaural time sensitive unit could be changed by change in intensity level, that of an interaural intensity sensitive unit could not. The authors failed to find any systematic arrangement of colliculus neurons sensitive to time or intensity differences and suggested that this failure may result from the "complex coding characteristic of these units". Those units sensitive to time appeared to have more complex relationships than those sensitive to intensity, and it was proposed that differential mechanisms may be used by these two populations during the process of localization. Furthermore "the complex types of unit interaural time codings suggest a convergence of lower order units, with simple codings, onto higher order complex units". BENEVENTO and COLEMAN used clicks as stimuli, and suggested that ROSE's findings of two populations of neurons stemmed from the fact that he used pure tones – a kind of signal not readily localized.

Time-intensity trading, which had previously been described for units of the medial superior olive (HALL, 1964; MOUSHEGIAN, RUPERT, and WHITCOMB, 1964), was shown to exist also for units of the inferior colliculus.

GEISLER, RHODE, and HAZELTON (1969) found that interaural time delay could be signaled in terms of spike counts, not only for pure tones as described by ROSE *et al.* (1966) but also for non-sinusoidal stimuli, specifically for band-pass noise. Although it was not possible, they contended, to predict from the data of ROSE *et al.* the behavior of neurons exposed to wide-band stimuli, some of the conclusions reached in the previous study were applicable. For instance, they confirmed the finding that some inferior colliculus neurons exhibited "characteristic delays" regardless of spectral content and/or intensity levels. They showed that even with noise stimuli, when the interaural delay was progressively delayed, the time-delay curves were periodic; this had been shown previously by ROSE *et al.* (1966) using tonal stimuli. In most cases, the periodicity corresponded to that of the neuron's most sensitive frequency. GEISLER and his colleagues proposed a model which incorporated the characteristic delay and a cross-correlogram calculation. They also presented data from several neurons which could be activated by stimuli to one ear alone and inhibited by stimuli to the other ear. These neurons were usually sensitive to interaural intensity differences even though the discharge rates of most of these units were influenced by absolute intensity level regardless of the spectral content.

ALTMAN has studied binaural interactions in neurons of the inferior colliculus of the cat (ALTMAN, 1966, 1967, 1968, 1969). He presented dichotic click stimuli and varied either interaural delay or interaural intensity. He characterized the units according to the changes in activity brought about by change in one or another of these parameters. In response to interaural intensity change, he found that collicular neurons that were activated by stimulation of one ear but suppressed by stimulation of the other were indeed sensitive to intensity change but that, in contrast to those of the superior olive, such units responded only to restricted ranges of intensity. Results with interaural time changes were similar, in that the neurons responded nonlinearly to restricted ranges of time intervals. Not all neurons of the colliculus responded in this manner, as some units responded to the interaural change in a similar manner to units of the superior olive.

B. Medial Geniculate Body

It is surprising that although extensive electrophysiological studies have been made of most of the relays of the central auditory pathway, the medial geniculate nucleus stands out as one region which until recently has only received limited attention.

1. Tonotopical Organization and Tuning Curves

ADES, METTLER, and CULLER (1939) were the first to show experimentally that there was an orderly arrangement of frequency representation in the medial geniculate body. They reported that following discrete lesions in this structure functional deficits in frequency discrimination could be seen. They stated that high frequencies were localized dorsally in the medial geniculate whereas lower frequencies were localized more ventrally in a circular fashion until the lowest frequencies were located in the most ventral section. COAKLEY and CULLER (1940) followed up this investigation with an electrophysiological study and reported that neural fibers carrying different frequencies were organized in a spiral manner in the medial geniculate body with the lowest frequencies represented in the center of the structure.

ROSE and WOOLSEY (1958) have described experiments which suggested that stimulation of cochlear fibers from base to apex respectively activated the principal division of the medial geniculate body in "mediolateral succession".

Recently, AITKIN and WEBSTER (1971) clearly demonstrated a tonotopic organization in the medial geniculate body of the cat with low frequency elements located laterally and high frequency elements placed medially in the principal division. They extended these findings (AITKIN and WEBSTER, 1972) using MOREST'S (1964a) classification of the different subdivisions of the medial geniculate. They found that in the ventral lobe the anterolateral pars lateralis was organized with lateromedial orientation as described above (Fig. 23). They confirmed the lateromedial orientation by showing that a dorsal-ventral penetration by the electrode recorded from units with similar best frequencies. The authors were unable to demonstrate a tonotopic organization in pars ovoidea but they felt that the coiled nature of the fiber laminae in this structure would preclude a simple arrangement and analysis would therefore be difficult.

The results of the experiments of AITKIN and WEBSTER support MOREST'S ideas that the laminar structure of the pars lateralis is the morphological basis for tonotopic organization and that each dendrite lamina has a common best frequency.

AITKIN and WEBSTER (1972) raise an interesting point concerning thalamocortical relations and I should like to expand on their ideas. The available evidence indicates that a point to point projection exists from the medial geniculate body to AI in the cortex. ROSE and WOOLSEY (1949) and more recently WOOLSEY (1964) have provided strong evidence for discrete localization. If the pars principalis is tonotopically organized then one must expect a similar organization in AI. Microelectrode studies at auditory cortex have shown such organization to be equivocal (ERULKAR, ROSE, and DAVIES, 1956; HIND, 1960; EVANS, ROSS, and WHITFIELD,

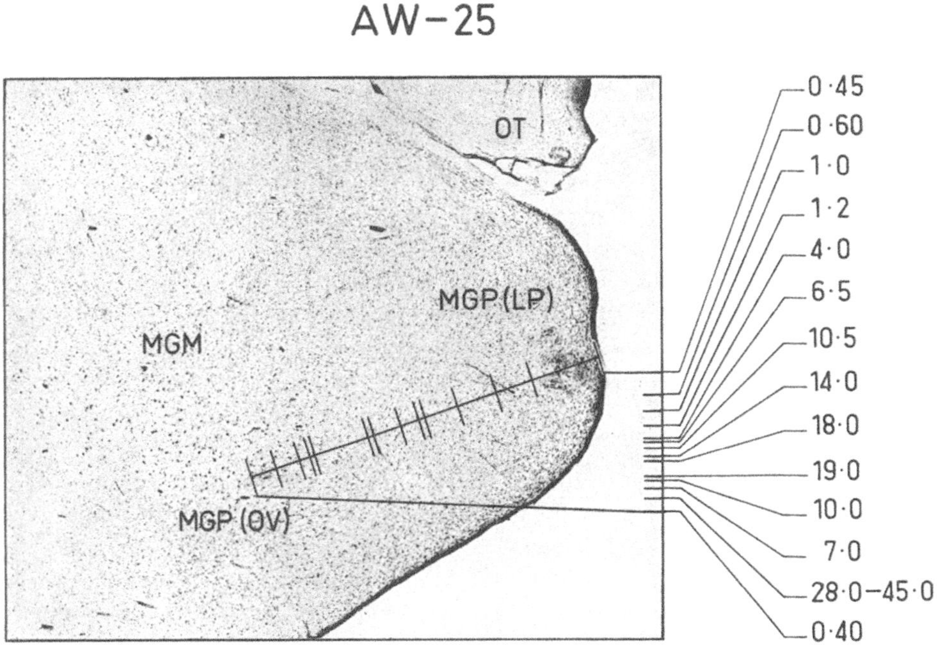

Fig. 23. Near frontal section through posterior part of anterior third of the medial geniculate body of the cat. Numbers to the right of the section denote best frequencies in KHz. OT optic tract, *MGP* (*LP*) pars principalis (pars lateralis), *MGP* (*OV*) pars principalis (pars ovoidea), *MGM* pars magnocellularis. (Taken from AITKIN and WEBSTER, 1972)

1965; ABELES and GOLDSTEIN, 1970). AITKIN and WEBSTER point out that in view of the variety of patterns of termination of ascending auditory fibers within different layers of the cortex, perhaps after numerous relays, that it may be erroneous to expect similar best frequencies within a given vertical penetration of the cortex.

We feel that there is an underlying framework of tonotopic organization in AI which relies upon the projection of the pars principalis to AI. This organization may be obscured by other inputs but it exists nevertheless. Order in terms of tonotopic projection exists therefore at all levels of the primary auditory pathway, and at the inferior colliculus and medial geniculate body, it is expressed structurally by the laminar arrangement of the neuronal dendrites.

KATSUKI and his colleagues, in their studies of tuning curves, (KATSUKI *et al.*, 1958; KATSUKI, WATANABE, and MARUYAMA, 1959; KATSUKI, 1966a) reported that neurons of the medial geniculate were "responsive to sounds in a restricted frequency range irrespective of intensity". They felt therefore that thalamic neurons show "frequency specificity" and that "frequency analyses . . . are completed at the medial geniculate body, . . ." As mentioned earlier, they maintained that a progressive narrowing of the tuning curves occurred at progressively higher levels and that frequency discrimination was finally carried out at the medial geniculate body.

AITKIN and WEBSTER (1972) could not confirm KATSUKI'S observations and showed that at the ventral medial geniculate body, tuning curves from different units could vary from "very broad to narrow". One interesting point was that those units whose best frequencies were relatively low were more broadly tuned than those with high best frequencies. The functional significance of this is not clear.

2. Response Patterns and Excitatory-Inhibitory Interactions

GROSS and THURLOW (1951) were the first to record responses to clicks and tones from single units in the medial geniculate body. They described responses consisting of sustained discharges with rapid adaptation during the tonal stimulus. In response to click stimuli, they recorded onset discharges with latencies of 8–13 msec and in some cases reverberatory responses.

GALAMBOS (1952) recorded from single units of the medial geniculate of the cat and described two components of the responses to sustained tonal stimulation. The two components consisted of an initial onset response followed by an interval of inactivity which in turn was followed by a discharge that was maintained as long as the tone was delivered. In response to click stimuli GALAMBOS and his colleagues (1952) noted that single spike or multiple spike discharges could be elicited with either short (8–24 msec) or long (50–200 msec) latencies, but for some units the responses consisted of rhythmic type discharges with alternating periods of spike activity and silence.

It was not until 1966 that a detailed study of unitary activity in the medial geniculate was started by the Australian workers, DUNLOP, AITKIN, and WEBSTER, and it is almost entirely their work which provides the material in the following account.

AITKIN, DUNLOP, and WEBSTER (1966) studied the temporal firing patterns of units in the medial geniculate body of the cat in response to click stimuli and described three broad categories of unitary response: there were units that were excited by the stimulus, units that were suppressed, and units that responded with reverberatory responses; examples are shown in Figs. 24, 25, and 26. Those units that were excited responded with either single or multiple spikes with latencies of 8–16 msec. Furthermore, following the response, unitary activity was suppressed for periods from 20–250 msec. Units that were categorized as being suppressed were those in which the spontaneous activity was abolished by click stimulation for periods from 40–200 msec. Following the suppression, there often occurred a "rebound"-like phenomenon (Fig. 25). The most common type of response obtained was the reverberatory response (Fig. 26); this occurred especially in anesthetized animals. The authors suggest that the reverberatory responses may result from inhibition mediated by recurrent collaterals from thalamocortical axons which project to interneurons and thence back to the geniculate cell. Later work by

AITKIN and DUNLOP (1969) suggested that the origin of these responses lay in the medial geniculate itself. Reverberatory responses resulted from stimulation of the BIC and, therefore, did not originate in the inferior colliculus. Cortical ablation did not abolish these responses, and cortical stimulation elicited these patterns of

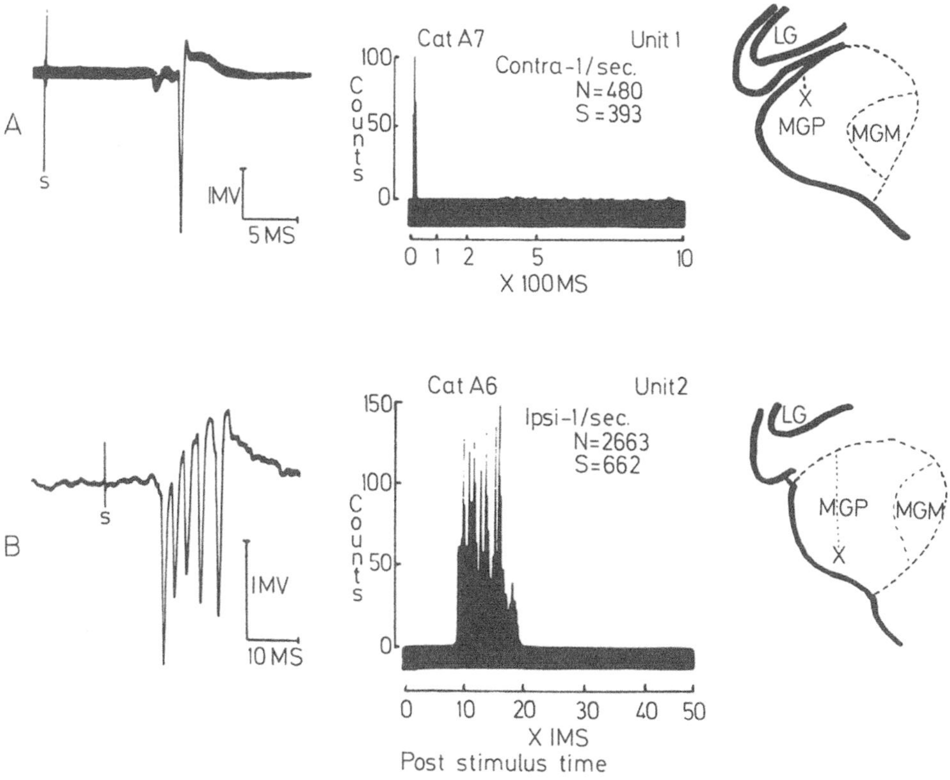

Fig. 24. Responses of units of the medial geniculate body of the cat to click stimuli. Responses are shown on the left of the figure, averaged response histograms in the center, and histological location on the right. Abscissa: Period of time after presentation of stimulus. Ordinate: Number of spike discharges. N total spike count, S total number of stimuli, LG lateral geniculate body, *MGP* pars principalis of medial geniculate body. (Taken from AITKIN *et al.*, 1966)

responses in a large number of units. Thus recurrent inhibition may be responsible for the reverberatory discharges; AITKIN and DUNLOP compare this property of the medial geniculate with similar responses recorded from the cat ventrobasal complex (ANDERSEN, ECCLES, and SEARS, 1964) and the rat lateral geniculate body (SEFTON and BURKE, 1965). In the medial geniculate, AITKIN and DUNLOP maintain that the anatomical substrate for these responses exists in the Golgi type II cells described by MOREST (1964a, 1971). It will be remembered that dendrites of the Golgi cells not only make synaptic contacts, within synaptic nests, with dendrites of thalamo-cortical cells, but they also receive synaptic contacts from afferent axons. The Golgi type II cells could thus modulate the activity of the thalamo-cortical cells. AITKIN and WEBSTER (1972) point out that it is in this synaptic nest structure and, in particular, the dendrodendritic endings which occur in the nests, that distinguishes functionally neurons in the ventral division of MGB from those of lower auditory centers. Furthermore, similar synaptic aggregates or nests have

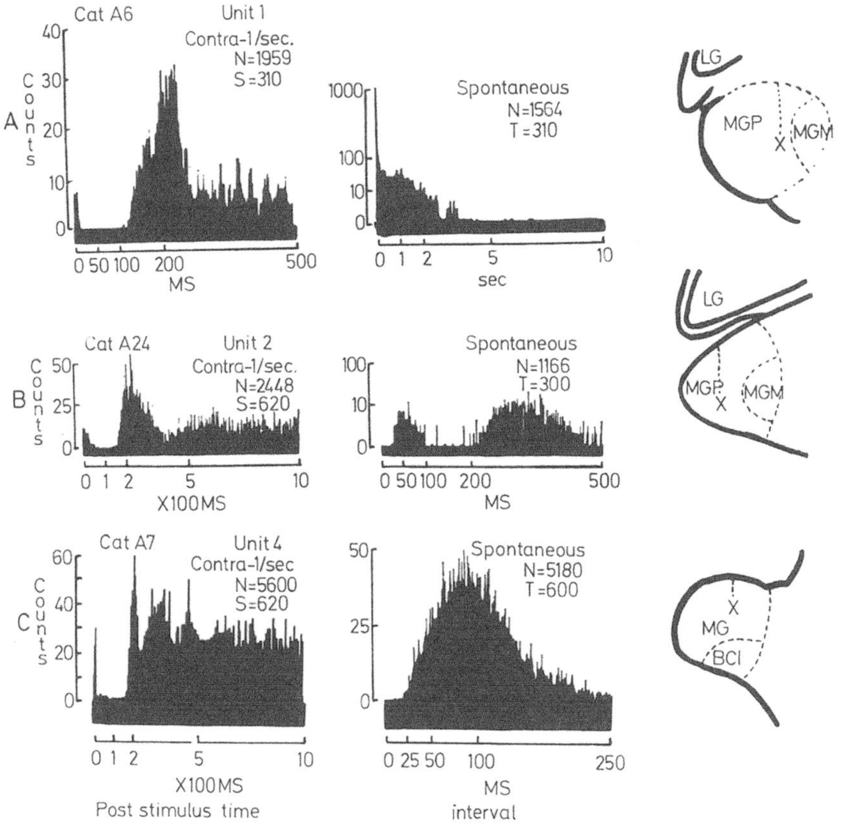

Fig. 25. Spontaneously active units suppressed by the click stimulus in the medial geniculate body of the cat. Averaged response histograms are on the left of the figure, interval histograms in the center, and histological location on the right. Note the different scales used for the abscissae and ordinates of the histograms. (A and C) data obtained from unanesthetized cats, (B) data obtained from cat anesthetized with sodium pentobarbital, *MG* medial geniculate body, BCL brachium of the inferior colliculus. (Taken from AITKIN *et al.*, 1966)

been described in other specific sensory relay nuclei of the thalamus (SZENTAGOTHAI, 1963; JONES and POWELL, 1969; JONES and ROCKEL, 1971).

In response to binaural tonal stimuli, DUNLOP, ITZKOWIC, and AITKIN (1969) found a variety of patterns of single unit activity. Usually the response was similar to that recorded to clicks, that is an onset response followed by a "silent period" (Fig. 27). The silent periods were rarely shorter than 50 msec and could last for 200–490 msec, which is longer than silent periods reported for responses of units in the inferior colliculus. In some units, there occurred an initial period of sup-

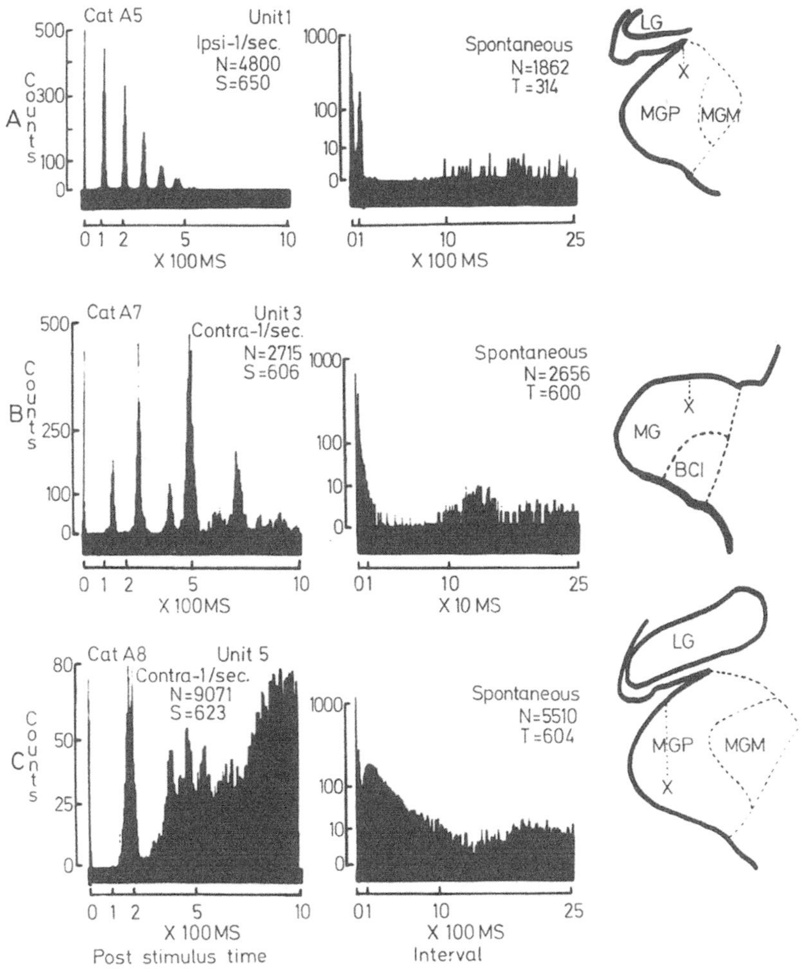

Fig. 26. Units giving reverberatory responses. Averaged response histograms are on the left of the figure, interval histograms are in the center, and histological location on the right. (A) and (B) are from cats anesthetized with sodium pentobarbital, while (C) is obtained from an unanesthetized paralyzed cat. *LG* lateral geniculate body, *MG* medial geniculate body, *MGM* medial geniculate, pars magnocellularis, *MGP* medial geniculate, pars principalis. (Taken from Aitkin *et al.*, 1966)

pression of activity (Fig. 28). It was of interest that a period of suppression could be followed by a "rebound" discharge, which is rarely seen in recordings from collicular units. One final point should be mentioned, namely that sustained discharge to tones could rarely be recorded from units at the geniculate level.

The patterns of discharge recorded from medial geniculate units must represent strong excitatory inhibitory interactions at this level. Nelson and Erulkar (1963) showed by intracellular recordings that complex synaptic interactions occurred in the geniculate as well as in the inferior colliculus. They point out, however, that pure excitation with no detectable inhibition can also occur (Fig. 29), and it would seem logical that neurons with this type of input play no integrative role but rather relay to the cortex information on the total auditory input to that cell or, in fact, information about the intensity of the acoustic signal. Aitkin and Dunlop (1968) varied click intensity while recording from units of the medial geniculate.

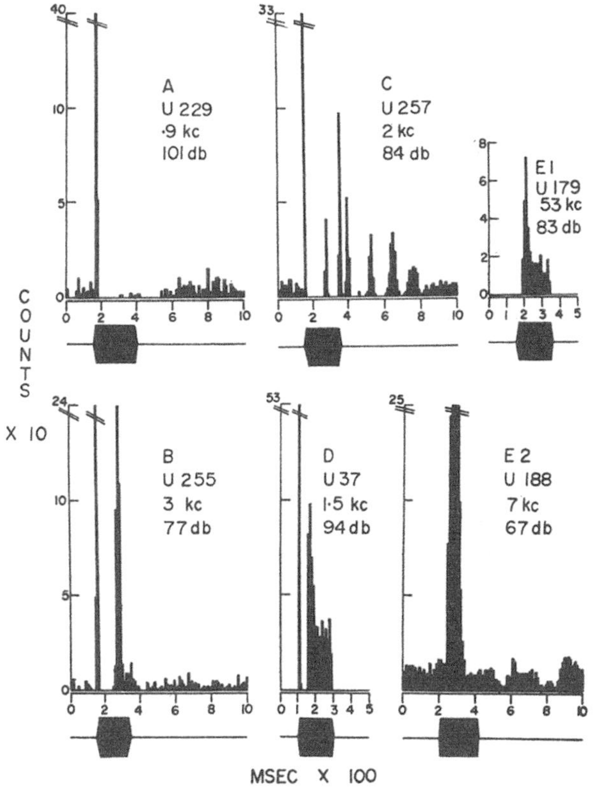

Fig. 27. Response histograms of onset excitation units. Abscissa: Period of time prior to, during and after stimulus presentation. Ordinate: Number of spike discharges. Histograms derived from 200 tone-burst presentations at 1/sec with duration of each tone burst being 200 msec. Stimulus intensity and frequency for each unit at top right hand corner of each histogram. (Taken from Dunlop *et al.*, 1969)

They agreed with Hind *et al.* (1963) that as stimulus intensity increases, the single unit responses become more precisely timed, the population response becomes more synchronous, and, thus, the response increases in amplitude. The reader is referred to Fig. 29 to see how precise and invariant the purely excitatory response of a neuron in the medial geniculate can be to a high intensity click stimulus.

There obviously is considerable inhibition at the geniculate level as discussed earlier in relation to the reverberatory responses. Aitkin and Dunlop (1969) found that the unresponsive period following a click-evoked response was much more profound and longer in the MGB than in the inferior colliculus. As with reverberatory responses, the origin of this depression is within the MGB itself and once again Golgi type II cells may be the elements responsible.

TEBECIS (1970a) found that iontophoretically administered ACh had excitatory or depressive effects on a large proportion of neurons in the MGB of the cat. ACh

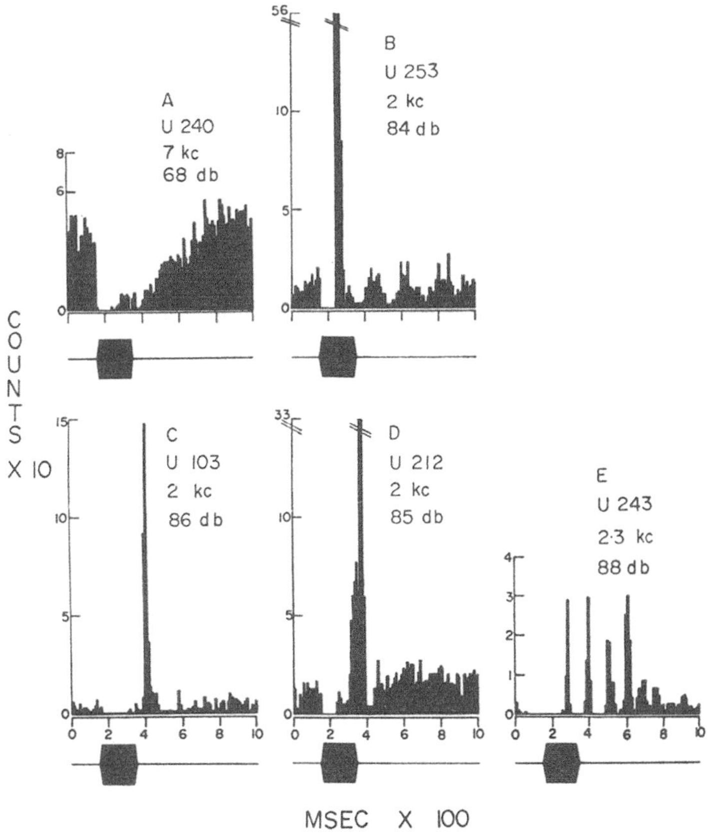

Fig. 28. Types of onset inhibition response patterns. Same arrangement as in Fig. 27. (Taken from DUNLOP *et al.*, 1969)

excited most of the neurons that projected to the auditory cortex, and the majority of the neurons appeared to be of the intermediate nicotinic-muscarinic type. In further studies (TEBECIS, 1970b), he suggested that the medial geniculate body received excitatory cholinergic, as well as non-cholinergic pathways from the auditory cortex, inferior colliculus and lower brain-stem. In an extension of this work, TEBECIS (1972) confirmed that ACh is an excitatory transmitter at neurons of the MGB; he also showed that the cholinergic fibers originate in or traverse the mesencephalon. Auditory afferents and cortico-geniculate fibers are not cholinergic; unitary discharges in the MGB, which were blocked by antagonists of ACh, in response to stimulation of the auditory cortex, could be explained (TEBECIS believed) by assuming an indirect activation of MGB neurons via fibers of the brain-stem.

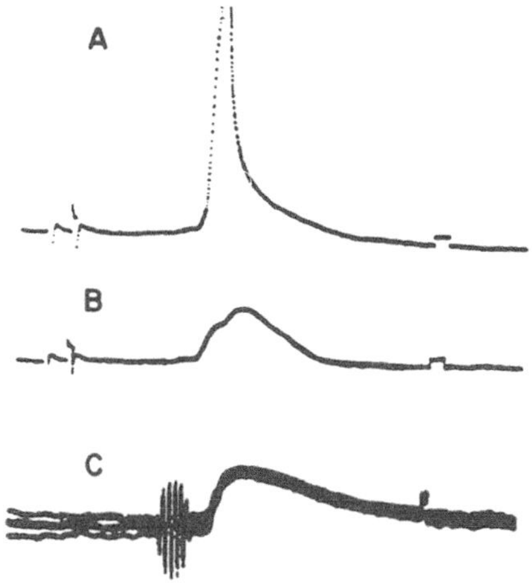

Fig. 29A—C. Responses to click and tone stimulation from a unit in the medial geniculate body of a cat. (A) Transmembrane recording soon after development of 70 mV resting potential. Click stimulation. (B) EPSP recorded several minutes after trace (A). Resting potential had declined to 20 mV. No spikes could be elicited. Click stimulation. (C) Multiple sweeps of EPSPs in response to 1 KHz tone pip. Note the relative invariance of the responses. Resting potential 20 mV. Calibration at end of each sweep — 1 mV, 1 msec. (Taken from NELSON and ERULKAR, 1966)

3. Binaural Phenomena

GALAMBOS *et al.* (1952) were among the first to recognize and report on binaural interaction at the medial geniculate body. They stated that "... it is evident that the same element can be aroused by stimulation of either ear". On the other hand, they also reported that the contralateral ear appeared to project more strongly to the medial geniculate body. ADRIAN *et al.* (1966) agreed that the majority of geniculate units were activated by contralateral ear stimulation; other units were activated by separate stimulation of both ears; the smallest number of units were activated by stimuli delivered to the ipsilateral ear alone. The interaction of excitation and inhibition was perhaps best shown for those units in which stimulation of one ear was capable of inhibiting activity evoked by stimulation to the other ear.

Similar activity in response to binaural stimulation was reported by AITKIN and DUNLOP (1968), and by AITKIN and WEBSTER (1972). AITKIN and DUNLOP reported similar effects to those shown by ADRIAN *et al.* (1966). They also showed that simultaneous stimulation of both ears could elicit three types of response from different units: one which was greater than the largest monaural response (facilitation), the same as the monaural response, or less than the least effective excitatory monaural response (occlusion). Change in interaural intensity difference could change the magnitude of the binaural response. For instance, the responses of one cell which was activated by both ipsilateral and contralateral stimulation separately, were reduced when the two ears were stimulated simultaneously. If the intensity to the contralateral ear was increased, the spike count was increased as was the scatter of latencies about the modal value (see Fig. 30). The implications here are important. Although there is excitatory input from both ears there appears

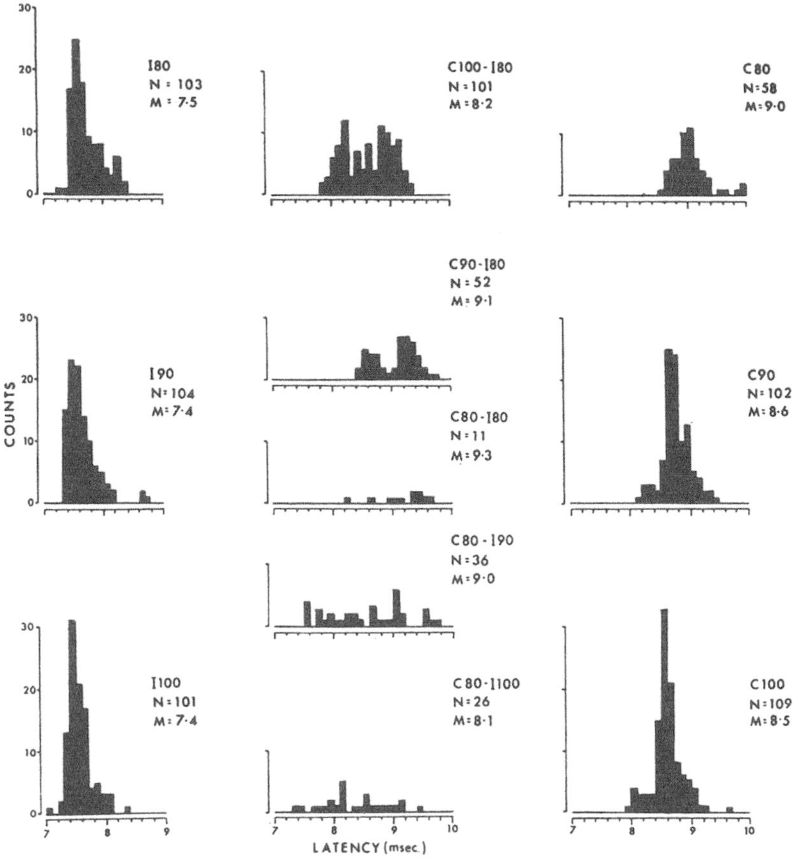

Fig. 30. Binaural occlusion at five interaural intensity differences. This unit was excited by stimulation of either ear. The scales of the abscissae and ordinates of all histograms are the same as those below and to the left of the figure respectively. I ipsilateral, C contralateral, C80 — I80 simultaneous binaural stimulation of 80 db at each ear. Location: ventral-medial geniculate body. (Taken from Aitkin and Dunlop, 1968)

also to be a "subliminal fringe of inhibition" from both sides which only becomes evident (with these extracellular recording techniques) when both ears are excited. Each ear, therefore, provides excitatory and inhibitory inputs to the cell. This was not invariably the case, for sometimes a cell, which was excited by stimulation of one ear but not by stimulation of the other ear, showed facilitation at some interaural intensity differences but suppression of activity at others (see Fig. 30). In this case, there must have been a subliminal fringe of excitation from one side.

Aitkin and Webster (1972) extended these studies. They found that a high proportion of recorded units in both pars lateralis (LP) and pars ovoidea (OV) of MGB were influenced by binaural stimulation. In agreement with previous findings at the dorsal nucleus of the lateral lemniscus (Brugge, Anderson, and Aitkin, 1970), Aitkin and Webster reported that those neurons that are sensitive to interaural time delay "tend to have low best frequencies while those influenced by interaural intensity differences have high best frequencies". Furthermore, those units sensitive to delay showed a periodic relationship between spike content and interaural time delay. Characteristic delays were also encountered in some units of the medial geniculate as in units of the inferior colliculus. However, the strong inhibition at unfavorable delays which was described for the inferior colliculus (Rose *et al.*, 1966) was not commonly seen in units of the medial geniculate body.

Altman *et al.* (1970a, b) recorded response characteristics of units in the MGB during monaural and binaural stimulation. Three types of responses were obtained: 1) onset response with short latency, 2) onset response followed by decreased activity followed by a late discharge, and 3) a late discharge only. Both initial and late responses changed characteristically to interaural time and intensity changes, but not necessarily in the same way. When the sound signals simulated a moving source, some units responded specifically to the direction of the movement. The majority of units, in contrast to those of the inferior colliculus, responded over only a restricted range of the movement (Altman, 1968).

C. Other Thalamic Groups

The anatomical evidence described earlier suggests that the posterior thalamic group may be a region in which different afferent systems converge. Functional confirmation of this suggestion was provided by Poggio and Mountcastle (1960) who showed that single units located within pars magnocellularis and the posterior thalamic group could be activated by both auditory and somatic stimuli. Perl and Whitlock (1961) later reported that in the posterior nuclear group a few units that responded to mechanical stimulation of the skin could also be activated by sound stimuli. Hotta and Kameda (1963) recorded interactions between responses to somatic and auditory stimulation from units in the suprageniculate nucleus, the posterolateral nucleus and from parts of the medial geniculate nuclei. They reported that of twenty-six units that showed somatic-auditory interaction, six units showed that the somatic input had an inhibitory effect on the response to auditory stimulation while the remainder showed occlusion effects.

More recently Raffaele, Santangelo, Sapienza, and Urbano (1969) have shown the presence of a strong somatic input to the pars magnocellularis throughout its caudo-ventral extent. They felt that the organization reflects convergence of multiple projection systems with fast conduction.

The posterior thalamic group is not the only region of the thalamus at which integration of responses to stimulation of different sensory modalities takes place. There is, for instance, evidence of convergence of different afferent inputs on cells of the centrum medianum (Albe-Fessard and Fessard, 1963; Albe-Fessard and GILLETTE, 1961; BUSER *et al.*, 1963), while ARDEN and SODEBERG (1961) showed that, in the lateral geniculate nucleus, the characteristics of responses of single units to visual stimulation could be modified by the sound made by clapping the hands.

HOTTA and TERASHIMA (1965) recorded single unit activity from the antero-lateral region of the thalamus of the cat, especially the n. reticularis and reported facilitatory or reciprocal suppression effects of interaction of auditory and visual inputs. The same units responded to stimulation of occipital cortex, suprasylvian gyrus and ectosylvian gyrus. The authors believe that the facilitatory effect of sound stimulation on the visual response may be due to "synaptically induced depolarization from the auditory afferent pathway".

III. Conclusions

In this chapter, I have tried to show from both morphological and physiological evidence that the patterns of input to the inferior colliculus and medial geniculate body are more diverse than have been hitherto suggested. This diversity is ordered and not chaotic and different components of the system can be dissected out by the techniques that have been described earlier. One point of importance is that recent experiments suggest that the system is a dynamic one which can vary its pattern according to different stimulus conditions. We have already discussed response characteristics and their relationship to different parameters of stimulation; another aspect of this dynamic response to stimulation is revealed by experiments using repetitive stimulation. Studies (HERNANDEZ-PEON and SCHERRER, 1955; SHEATZ, VERNIER, and GALAMBOS, 1955; GALAMBOS, SHEATZ, and VERNIER, 1956; GALAMBOS, 1959) have shown that the responses elicited at different neuronal levels by an acoustic stimulus changed as the signal lost its significance through monotonous repetition; or the response changed if the animal was conditioned to attend to another sound. If the animal's attention was made to shift from the auditory stimulus, the response was again modified. It appeared as if a "gating" process occurred as a result of such changes as shift of attention, but the levels at which the neuronal effects first occurred were unknown. The studies cited above led to the hypothesis that the nervous system could block irrelevant sensory input at some level of the auditory pathway – perhaps at the receptor or soon thereafter (HERNANDEZ-PEON, 1961). One theory, the central hypothesis, states that the process occurs at higher levels of the auditory pathway (THOMPSON and SPENCER, 1966). Evidence in favor of this idea was provided by WICKELGREN (1968) who found that the click-evoked response in the cat, during behavioral acoustic habituation, was significantly attenuated at the medial geniculate body and auditory cortex, but not at the cochlear nucleus, superior olive or inferior colliculus. Similar evidence was provided by HALL (1968), although KITZES and BUCHWALD (1969) did find significant changes at more peripheral relay areas. HOLSTEIN *et al.* (1969) also found that progressive changes do occur to a constant repeated stimulus at peripheral levels although such changes are more prominent in more central nuclei. WEBSTER (1971) found that the decrements in the response did not have the accepted characteristics of habituation (THOMPSON and SPENCER,

1966) and suggested that the decrements were due to an accumulating local synaptic inhibitory process. WEBSTER and AITKIN (1971) also felt that local synaptic inhibitory activity was involved in the decrement of the response to repetitive acoustic stimulation and felt that either pre- or postsynaptic inhibition may be responsible.

Although many of the response patterns described may be due to local synaptic activity, it is known that modulation of activity by descending pathways can influence activity in both the inferior colliculus and medial geniculate body. AMATO, LA GRUTTA, and ENIA (1970) have shown that stimulation at well-defined points of AI can inhibit potentials elicited by click stimulation in both of these nuclei. KATSUKI (1966b) demonstrated facilitation and inhibition of single unit firing in the medial geniculate body after stimulation of the auditory cortex. It is not only the cortex that is involved in this modulation. POWELL, FURLONG, and HATTON (1970) reported that different units in the medial geniculate body could be excited or depressed by stimulation of different sites in the septum. Similar effects were reported as a result of stimulation of the inferior colliculus, and interaction between the effects of stimulation of the two sites was noticed. They concluded that interaction between sensory and limbic influences may occur at the thalamic level, and some interaction may take place at the inferior colliculus.

Finally, LA GRUTTA, GIAMMANCO, and AMATO (1968) demonstrated that stimulation of the caudate nucleus inhibited click-evoked activity of neurons in the medial geniculate body.

The inferior colliculus and medial geniculate body therefore act not only as relays to transfer to the cortex information about the acoustic signal, but they also serve as regions of integration that abstract and extract different components of that signal. It would seem that much of the analysis of acoustic signals occurs at these levels but that the mechanisms responsible for the analysis may be partly under higher control. Up to the present time, the different components of the analysis (such as responses to pure tones, responses to frequency-modulated tones, etc.) are being studied, but they will only be understood when each component can be placed in appropriate context in relation to other components. This may occur at cortical levels but that is another story.

Since this manuscript was submitted a number of important contributions have come to the author's attention. These studies relate to the patterns of neuronal activity within the inferior colliculus and medial geniculate body to linguistics and especially to data processing and possible detection of phoneme-related responses (for reviews see KEIDEL, 1974).

KALLERT (1972) in continuation of work by KEIDEL (1966, 1970) and by DAVID, FINKENZELLER, KALLERT, and KEIDEL (1968, 1969), showed that a wide variety of response patterns to tone bursts, noise bursts and clicks could be recorded from different units of the medial geniculate body and inferior colliculus. These patterns included onset responses, on-off responses, suppression of activity and periodicity. These investigators felt that units which respond with periodic responses, which are manifest as multiple peak responses in peristimulus histograms, may act as vowel detectors. These responses are related, they postulate, to the fundamental and its formant in vowels. The combination of different frequencies in the vowel could be separated in the multiple peak responses. The mechanism of formation of these periodic responses may be in the spontaneous activity of some units. These units could serve as a clock whose outputs converge onto units which also receive outputs from stimulus-driven units. If the final unit acts as a coincidence detector between the spontaneous (click) and stimulus-driven signals, multiple periodicities could be formed (KALLERT *et al.*, 1970).

Some units in the inferior colliculus and medial geniculate bodies respond only to dynamic aspects of the stimulus — such as frequency-modulated (FM) tones, with asymmetric response patterns (see NELSON, ERULKAR, and BRYAN, 1966). These units the authors labelled as "consonant detectors" and suggested that they could distinguish and process phonemes such as "ja" and "ai". Transient detectors responded to FM tonal stimuli with symmetric response patterns and the authors felt that these units responded to transients between vowels.

In order to show the relationship of these neuronal patterns to linguistic communication, KALLERT and KEIDEL (1973), used special telemetric methods in recording from single units in the medial geniculate body of unrestrained awake cats. They found that different units responded differently to words with different consonants or different words. For instance, one unit responded to the word „Fein" with a burst of impulses, but failed to respond to the words „Dein" and „Mein". Another unit responded to the vowel "a" but not to other vowels. DAVID and BUNDZEN (1974) have now applied these techniques to human subjects recording from regions other than the medial geniculate body. They concluded that the "semantic content" of known and unknown words in human beings when communicating with each other can be recorded neurophysiologically.

These studies provide new concepts in interpretation of neurophysiological activity and it should be interesting indeed to read of future developments in this field.

References

ABELES, M., GOLDSTEIN, M. H., JR.: Functional architecture in cat primary auditory cortex: columnar organization and organization according to depth. J. Neurophysiol. **33**, 172—187 (1970).

ADES, H. W., METTLER, F. A., CULLER, E. A.: Effect of lesions in the medial geniculate bodies upon hearing in the cat. Amer. J. Physiol. **125**, 15—23 (1939).

ADES, H. W., BROOKHART, J. M.: The central auditory pathway. J. Neurophysiol. **13**, 189—205 (1950).

ADRIAN, H. O., GOLDBERG, J. M., BRUGGE, J. F.: Auditory evoked potentials after lesions of brachium of inferior colliculus. J. Neurophysiol. **29**, 456—466 (1966).

ADRIAN, H. O., LIFSCHITZ, W. M., TAVITAS, R. J., GALLI, F. P.: Activity of neural units in medial geniculate body of cat and rabbit. J. Neurophysiol. **29**, 1046—1060 (1966).

AITKIN, L. M., DUNLOP, C. W., WEBSTER, W. R.: Click-evoked response patterns of single units in the medial geniculate body of the cat. J. Neurophysiol. **29**, 109—123 (1966).

AITKIN, L. M., DUNLOP, C. W.: Interplay of excitation and inhibition in the cat medial geniculate body. J. Neurophysiol. **31**, 44—61 (1968).

AITKIN, L. M., DUNLOP, C. W.: Inhibition in the medial geniculate body of the cat. Exp. Brain Res. **7**, 68—83 (1969).

AITKIN, L. M., WEBSTER, W. R.: Tonotopic organization in the medial geniculate body of the cat. Brain Res. **26**, 402—405 (1971).

Aitkin, L. M., Webster, W. R.: Medial geniculate body of the cat: organization and responses to tonal stimuli of neurons in ventral division. J. Neurophysiol. **35**, 365—380 (1972).
Albe-Fessard, D., Gillette, E.: Convergences d'afférentes d'origine corticale et périphérique vers le centre median du chat anesthesié ou éveillé. Electroenceph. clin. Neurophysiol. **13**, 257—269 (1961).
Albe-Fessard, D., Fessard, A.: Thalamic integrations and their consequences at the telencephalic level. In: Brain Mechanisms. Progress in Brain Research, Vol. 1, pp. 115—154. Amsterdam: Elsevier 1963.
Altman, Ya. A.: Responses of the inferior colliculus neurones of the cat by changing time intervals between binaural presented stimuli. Biophysics **11**, 554—565 (1966).
Altman, Ya. A.: Response in units of inferior colliculus in cats to unilateral change of intensity of a binaural stimulus. Neurosci. Transl. **3**, 239—247 (1967).
Altman, Ya. A.: Are there neurons detecting direction of sound source motion? Exp. Neurol. **22**, 13—25 (1968).
Altman, Ya. A.: Responses of the inferior colliculus neurones to lateralization of rhythmical sound stimulation. J. Higher Nervous Activity **19**, 59—70 (1969).
Altman, Ya. A., Syka, J., Schmigidina, G. N.: Neuronal activity in the medial geniculate body of the cat during monaural and binaural stimulation. Exp. Brain Res. **10**, 81—93 (1970a).
Altman, Ya. A., Syka, J., Schmigidina, G. N.: Characteristics of the neuronal activity in the medial geniculate body of the cat during monaural and binaural stimulation. Physiol. bohemoslov. **19**, 177—184 (1970b).
Amato, G., Lagrutta, V., Enia, F.: The control of acoustic input in the medial geniculate body and inferior colliculus by auditory cortex. Experientia (Basel) **26**, 55—56 (1970).
Andersen, P., Eccles, J. C., Sears, T. A.: The ventrobasal complex of the thalamus: types of cells, their responses and their functional organization. J. Physiol. (Lond.) **174**, 370—399 (1964).
Andersson, S., Gernandt, B.: Cortical projection of vestibular nerve in cat. Acta oto-laryng. (Stockh.) Suppl. **116**, 10—16 (1954).
Arden, G. B., Sodeberg, V.: The transfer of optic information through the lateral geniculate body of the rabbit. In: Sensory Communication, pp. 521—544. Cambridge, Mass.: M. I. T. Press 1961.
Barnes, W. T., Magoun, H. W., Ranson, S. W.: The ascending auditory pathway in the brainstem of the monkey. J. comp. Neurol. **79**, 129—152 (1943).
Benevento, L. A., Coleman, P. D.: Responses of single cells in cat inferior colliculus to binaural click stimuli: combinations of intensity levels, time differences and intensity differences. Brain Res. **17**, 387—405 (1970).
Berman, A. L.: The brain stem of the cat. A cytoarchitectonic atlas with stereotaxic coordinates. Madison: Univ. of Wisconsin Press 1968.
Brugge, J. F., Anderson, D. J., Aitkin, L. M.: Responses of neurons in the dorsal nucleus of the lateral lemniscus of cat to binaural tonal stimulation. J. Neurophysiol. **33**, 441—458 (1970).
Buser, P., Ascher, P., Bruner, J., Jassik-Gerschenfeld, D., Sindberg, R.: Aspects of sensorimotor reverberation to acoustic and visual stimuli. The role of primary specific cortical areas. In: Brain Mechanisms. Progress in Brain Research, Vol. I, pp. 294—324. Amsterdam: Elsevier 1963.
Butler, R. A., Diamond, I. T., Neff, W. D.: Role of auditory cortex in discrimination of changes in frequency. J. Neurophysiol. **20**, 108—120 (1957).
Cajal, S. R.: Histologie du système nerveux de l'Homme et des Vertébrés, Tome II. Paris: Maloine 1911.
Carreras, M., Andersson, S. A.: Functional properties of neurons of anterior ectosylvian gyrus of the cat. J. Neurophysiol. **26**, 100—126 (1963).
Coakley, J. D., Culler, E.: Distribution of auditory action-potentials in the medial geniculate body of the cat. Amer. J. Physiol. **129**, 335—336 (1940).
David, E., Finkenzeller, P., Kallert, S., Keidel, W. D.: Die mit Mikroelektroden ableitbare Reaktion einzelner Elemente des colliculus inferior und des corpus geniculatum mediale auf akustische Reize verschiedener Form und verschiedener Intensität. Pflügers Arch. ges. Physiol. **299**, 83—93 (1968).

DAVID, E., FINKENZELLER, P., KALLERT, S., KEIDEL, W. D.: Reizfrequenzkorrelierte „untersetze" neuronale Entladungsperiodizität im colliculus inferior und im corpus geniculatum mediale. Pflügers Arch. **309**, 11—20 (1969).

DAVID, E., BUNDZEN, P. V.: Neuronale Kodierungsmechanismen des verbalen Kurzzeitgedächtnisses. Pflügers Arch. (in press) 1974.

DESMEDT, J. E., MECHELSE, K.: Mise en évidence d'une quatrième aire de projection dans l'écorce cérébrale du chat. J. Physiol. (Paris) **51**, 448—449 (1959).

DIAMOND, I. T., NEFF, W. D.: Ablation of temporal cortex and discrimination of auditory patterns. J. Neurophysiol. **20**, 300—315 (1957).

DIAMOND, I. T., CHOW, K. L., NEFF, W. D.: Degeneration of caudal medial geniculate body following cortical lesion ventral to auditory area AII in the cat. J. comp. Neurol. **109**, 349—362 (1958).

DIAMOND, I. T., JONES, E. G., POWELL, T. P. S.: The projection of the auditory cortex upon the diencephalon and brain stem in the cat. Brain Res. **15**, 305—340 (1969).

DIXON, N. T.: Tonotopic organization of the inferior colliculus and the effects of lateral lemniscus lesions on hearing sensitivity in the cat. Ph. D. Thesis. Univ. of Indiana 1973.

DUNLOP, C. W., ITZKOWIC, D. J., AITKIN, L. M.: Tone-burst response patterns of single units in the cat medial geniculate body. Brain Res. **16**, 149—164 (1969).

ERULKAR, S. D.: The responses of single units of the inferior colliculus of the cat to acoustic stimulation. Proc. roy. Soc. B **150**, 336—355 (1959).

ERULKAR, S. D.: Comparative aspects of spatial localization of sound. Physiol. Rev. **52**, 237—360 (1972).

ERULKAR, S. D., RÔSE, J. E., DAVIES, P. W.: Single unit activity in the auditory cortex of the cat. Bull. Johns Hopk. Hosp. **99**, 55—86 (1956).

ERULKAR, S. D., NELSON, P. G., BRYAN, J. S.: Experimental and theoretical approaches to neural processing in the central auditory pathway. Contrib. Sensory Physiol. **3**, 149—189 (1968).

EVANS, E. F., ROSS, H. F., WHITFIELD, I. C.: The spatial distribution of unit characteristic frequency in the primary auditory cortex of the cat. J. Physiol. (Lond.) **179**, 238—247 (1965).

GALAMBOS, R.: Microelectrode studies on medial geniculate body of cat. III. Response to pure tones. J. Neurophysiol. **15**, 381—400 (1952).

GALAMBOS, R.: Electrical correlates of conditioned learning. In: The Central Nervous System and Behavior, Vol. 1, pp. 375—415. New York: Macy Foundation 1959.

GALAMBOS, R., ROSE, J. E., BROMILEY, R. B., HUGHES, J. R.: Microelectrode studies on medial geniculate body of cat. II. Response to clicks. J. Neurophysiol. **15**, 359—380 (1952).

GALAMBOS, R., SHEATZ, G., VERNIER, V. G.: Electrophysiological correlates of a conditioned response in cats. Science **123**, 376—377 (1956).

GALAMBOS, R., MYERS, R. E., SHEATZ, G. C.: Extralemniscal activation of auditory cortex in cats. Amer. J. Physiol. **200**, 23—28 (1961).

GEISLER, C. D., RHODE, W. S., HAZELTON, D. W.: Responses of inferior colliculus neurons in the cat to binaural acoustic stimuli having wide-band spectra. J. Neurophysiol. **32**, 960—974 (1969).

GENIEC, P., MOREST, D. K.: The neuronal architecture of the human posterior colliculus: a study with the Golgi method. Acta. oto-laryngol. (Stockh.) Suppl. **295**, 1—33 (1971).

GERSTEIN, G. L., BUTLER, R. A., ERULKAR, S. D.: Excitation and inhibition in cochlear nucleus. I. Tone-burst stimulation. J. Neurophysiol. **31**, 526—536 (1968).

GOLDBERG, J. M., MOORE, R. Y.: Ascending projections of the lateral lemniscus in the cat and monkey. J. comp. Neurol. **129**, 143—156 (1967).

GOLDBERG, J. M., NEFF, W. D.: Frequency discrimination after bilateral ablation of cortical auditory areas. J. Neurophysiol. **24**, 119—128 (1961a).

GOLDBERG, J. M., NEFF, W. D.: Frequency discrimination after bilateral section of the brachium of the inferior colliculus. J. comp. Neurol. **116**, 256—290 (1961b).

Gross, N. B., Thurlow, W. R.: Microelectrode study of neuronal auditory activity of cat. II. Medial geniculate body. J. Neurophysiol. **14**, 409—422 (1951).

Gudden, B.: Beitrag zur Kenntnis des Corpus mammillare und der sogenannten Schenkel des Fornix. Arch. Psychiat. Nervenkr. **11**, 428—452 (1881).

Hall, J. L., II: Binaural interaction in the accessory superior olivary nucleus of the cat — an electrophysiological study of single neurons. M. I. T. Res. Lab. Electron Techn. Rep. 416. Cambridge, Mass.: 1964.

Hall, R. D.: Habituation of evoked potentials in the rat under conditions of behavioral control. Electroenceph. clin. Neurophysiol. **24**, 155—165 (1968).

Heath, C. J., Jones, E. G.: An experimental study of ascending connections from the posterior group of thalamic nuclei in the cat. J. comp. Neurol. **141**, 397—426 (1971).

Hernandez-Péon, R.: Reticular mechanisms of sensory control. In: Sensory Communication, pp. 467—520. New York: Wiley 1961.

Hernandez-Péon, R., Scherrer, H.: Habituation to acoustic stimuli in cochlear nucleus. Fed. Proc. **14**, 71 (1955).

Hiltz, F. F.: Analog computer simulation of a neural element. IRE Trans. Bio-Med. Electron BME **9**, 12—20 (1962).

Hiltz, F. F.: Simulated membrane junctions and additional feedback characteristics for an artificial neuron. IEEE Trans. Bio-Med. Eng. BME **12**, 94—104 (1965).

Hind, J. E.: Unit activity in the auditory cortex. In: Neural Mechanisms of the Auditory and Vestibular Systems, pp. 201—210. Springfield: Thomas 1960.

Hind, J. E., Goldberg, J. M., Greenwood, D. D., Rose, J. E.: Some discharge characteristics of single neurons in the inferior colliculus of the cat. II. Timing of the discharges and observations on binaural stimulation. J. Neurophysiol. **26**, 321—341 (1963).

Holstein, S. B., Buchwald, J. S., Schwafel, J. A.: Progressive changes in auditory response patterns to repeated tone during normal wakefulness and paralysis. Brain Res. **16**, 133—148 (1969).

Hotta, T., Kameda, K.: Interactions between somatic and visual or auditory responses in the thalamus of the cat. Exp. Neurol. **8**, 1—13 (1963).

Hotta, T., Terashima, S.: Audio-visual interaction and its correlation with cortical stimulation in the lateral thalamus. Exp. Neurol. **12**, 146—158 (1965).

Jones, E. G., Powell, T. P. S.: The projection of the somatic sensory cortex upon the thalamus in the cat. Brain Res. **10**, 369—391 (1968).

Jones, E. G., Powell, T. P. S.: Electron microscopy of synaptic glomeruli in the thalamic relay nuclei of the cat. Proc. roy. Soc. B **172**, 153—171 (1969).

Jones, E. G., Powell, T. P. S.: An analysis of the posterior group of thalamic nuclei on the basis of its afferent connections. J. comp. Neurol. **143**, 185—216 (1971).

Jones, E. G., Rockel, A. J.: The synaptic organization in the medial geniculate body of afferent fibres ascending from the inferior colliculus. Z. Zellforsch. **113**, 44—66 (1971).

Kallert, S.: Über die Reizantwort einzelner Zellen im Corpus geniculatum mediale der Katze bei Untersuchung mit Mikroelektroden. Dissertation, Erlangen, 1972.

Kallert, S., David, E., Finkenzeller, P., Keidel, W. D.: Two different neuronal discharge periodicities in the acoustical channel. In: Plomp, R., Smoorenburg, G. F. (Eds.): Frequency analysis and periodicity detection in hearing, pp. 153—160. Leiden: A. W. Sijthoff 1970.

Kallert, S., Keidel, W. D.: Telemetrical microelectrode study of the upper parts of the auditory pathway in free moving cats. Pflügers Arch. **343**, R 79, 1973.

Katsuki, Y.: Neural mechanisms of hearing in cats and monkeys. In: Progress in Brain Research, Vol. 21 A, pp. 71—97. Amsterdam: Elsevier 1966a.

Katsuki, Y.: Integrative organization in the thalamic and cortical auditory centers. In: The Thalamus, pp. 349—364. New York: Columbia Univ. Press 1966b.

KATSUKI, Y., SUMI, T., UCHIYAMA, H., WATANABE, T.: Electric responses of auditory neurons in cat to sound stimulation. J. Neurophysiol. **21**, 569—588 (1958).

KATSUKI, Y., WATANABE, T., MARUYAMA, N.: Activity of auditory neurons in upper levels of the brain of cat. J. Neurophysiol. **22**, 343—359 (1959).

KEIDEL, W. D.: Anatomie und Elektrophysiologie der zentralen akustischen Bahnen. In: BERENDES, J., LINK, R., ZOLLNER, F. (Hrsg.): Hals-Nasen-Ohren-Heilkunde. Ein kurzgefaßtes Handbuch in 3 Bänden; Bd. III/3. Stuttgart: G. Thieme 1966.

KEIDEL, W. D.: Neuere Ergebnisse der akustischen Informationsverarbeitung. In: Ergebnisse der experimentellen Medizin Band 3; Hrsg.: Präsidium der Deutschen Gesellschaft für experimentelle Medizin. Berlin: VEB Verlag Volk und Gesundheit 1970.

KEIDEL, W. D.: Information processing in the higher parts of the auditory pathway. In: ZWICKER, E., TERHARDT, E. (Eds.): Psychophysical models and Physiological facts in hearing, pp. 1—12. Berlin-Heidelberg-New York: Springer 1974.

KITZES, M., BUCHWALD, J.: Progressive alterations in cochlear nucleus, inferior colliculus and medial geniculate responses during acoustic habituation. Exp. Neurol. **25**, 85—105 (1969).

LA GRUTTA, V., GIAMMANCO, S., AMATO, G.: Azione inibitoria del nucleo caudato. II. Alcune caratteristiche dell'azione inibitoria del nucleo caudato sull'area auditiva primaria e sul corpo genicolato mediale. Boll. Soc. ital. Biol. sper. **44**, 1810—1812 (1968).

LANDGREN, S., SILFVENIUS, H., WOLSK, D.: Vestibular, cochlear and trigeminal projections to the cortex in the anterior suprasylvian sulcus of the cat. J. Physiol. (Lond.) **191**, 561—573 (1967).

LOC, P. R., BENEVENTO, L. A.: Auditory-visual interaction of single units in the orbito-insular cortex of the cat. Electroenceph. clin. Neurophysiol. **26**, 395—398 (1969).

LOCKE, S.: The projection of the magnocellular medial geniculate body. J. comp. Neurol. **116**, 179—194 (1961).

MAJKOWSKI, J.: Wplyw niszczenia dróg sluchowych kota na proces uczenia sie odpowiedzi na bodźce dzwiękowe. Rozpr. Wydz. Nauk. med. pol. Akad. Nauk **9**, 59—108 (1964).

MAJKOWSKI, J., MORGADES, P. P.: Primary auditory evoked potentials after section of brachium of the inferior colliculus. Electroenceph. Clin. Neurophysiol. **27**, 587—593 (1969).

MEHLER, W. R.: Some observations on secondary ascending afferent systems in the central nervous system. In: Pain, pp. 11—32. Boston: Little, Brown & Co. 1966.

MEHLER, W. R., FEFERMAN, M. E., NAUTA, W. J. H.: Ascending axon degeneration following anterolateral cordotomy. An experimental study in the monkey. Brain **83**, 718—750 (1960).

MICKLE, W. A., ADES, H. W.: A composite projection area in the cerebral cortex of the cat. Amer. J. Physiol. **170**, 682—689 (1952).

MOORE, R. Y., GOLDBERG, J. M.: Midbrain auditory connections in the cat. Anat. Rec. **139**, 256 (1961).

MOORE, R. Y., GOLDBERG, J. M.: Ascending projections of the inferior colliculus in the cat. J. comp. Neurol. **121**, 109—136 (1963).

MOORE, R. Y., GOLDBERG, J. M.: Projections of the inferior colliculus in the monkey. Exp. Neurol. **14**, 429—438 (1966).

MOREST, D. K.: The neuronal architecture of the medial geniculate body of the cat. J. Anat. (Lond.) **98**, 611—630 (1964a).

MOREST, D. K.: The laminar structure of the inferior colliculus of the cat. Anat. Rec. **148**, 314 (1964b).

MOREST, D. K.: The probable significance of synaptic and dendritic patterns of the thalamic and midbrain auditory system. Anat. Rec. **148**, 390—391 (1964c).

MOREST, D. K.: The laminar structure of the medial geniculate body of the cat. J. Anat. (Lond.) **99**, 143—160 (1965a).

MOREST, D. K.: The lateral tegmental system of the midbrain and the medial geniculate body: study with Golgi and Nauta methods in cat. J. Anat. (Lond.) **99**, 611—634 (1965b).

MOREST, D. K.: The cortical structure of the inferior quadrigeminal lamina of the cat. Anat. Rec. **154**, 389—390 (1966a).

Morest, D. K.: The non-cortical neuronal architecture of the inferior colliculus of the cat. Anat. Rec. **154**, 477 (1966b).

Morest, D. K.: Dendrodendritic synapses of cells that have axons: the fine structure of the Golgi Type II cell in the medial geniculate body of the cat. Z. Anat. Entwickl.-Gesch. **133**, 216—246 (1971).

Moushegian, G., Rupert, A. L., Whitcomb, M. A.: Medial superior-olivary unit response patterns to monaural and binaural clicks. J. acoust. Soc. Amer. **36**, 196—202 (1964).

Münzer, E., Wiener, H.: Das Zwischen- und Mittelhirn des Kaninchens und die Beziehungen dieser Teile zum übrigen Centralnervensystem mit besonderer Berücksichtigung der Pyramidenbahn und Schleife. Mschr. Psychiat. Neurol. **12**, 241—279 (1902).

Neff, W. D., Fisher, J. E., Diamond, I. T., Yela, M.: Role of auditory cortex in discrimination requiring localization of sound in space. J. Neurophysiol. **19**, 500—512 (1956).

Nelson, P. G., Erulkar, S. D.: Synaptic mechanisms of excitation and inhibition in the central auditory pathway. J. Neurophysiol. **26**, 908—923 (1963).

Nelson, P. G., Erulkar, S. D., Bryan, J. S.: Responses of units of the inferior colliculus to time-varying acoustic stimuli. J. Neurophysiol. **29**, 834—860 (1966).

Nieder, P. C., Strominger, N. L.: Evoked potentials in auditory cortex after bilateral transection of the brachium of the inferior colliculus in the cat. J. Neurophysiol. **28**, 1185—1194 (1965).

Niimi, K., Miki, M., Kawamura, S.: Ascending projections of the superior colliculus in the cat. Okajimas Fol. anat. jap. **47**, 269—287 (1970).

Osen, K. K.: Projection of the cochlear nuclei on the inferior colliculus in the cat. J. Comp. Neurol. **144**, 355—372 (1972).

Papez, J. W.: Central acoustic tract in cat and man. Anat. Rec. **42**, 60 (1929).

Perl, E. R., Whitlock, D. G.: Somatic stimuli exciting spinothalamic projection to thalamic neurons in cat and monkey. Exp. Neurol. **3**, 256—296 (1961).

Poggio, G. F., Mountcastle, V. B.: A study of the functional contributions of the lemniscal and spinothalamic system to somatic sensibility. Central nervous systems in pain. Bull. Johns Hopk. Hosp. **106**, 266—316 (1960).

Powell, E. W., Hatton, J. B.: Projections of the inferior colliculus in cat. J. comp. Neurol. **136**, 183—192 (1969).

Powell, E. W., Furlong, L. D., Hatton, J. B.: Influence of the septum and inferior colliculus on medial geniculate body units. Electroenceph. Clin. Neurophysiol. **29**, 74—82 (1970).

Raffaele, R., Santangelo, F., Sapienza, S., Urbano, A.: Analisi elettrofisiologica della afferenze somatiche alla porzione magnocellulare del corpo genicolato mediale nel gatto. Boll. Soc. ital. Biol. sper. **45**, 1456—1458 (1969).

Ramon-Moliner, E.: An attempt at classifying nerve cells on the basis of their dendritic patterns. J. comp. Neurol. **119**, 211—227 (1962).

Rasmussen, G. L.: The olivary peduncle and other fiber projections of the superior olivary complex. J. comp. Neurol. **84**, 141—219 (1946).

Rasmussen, G. L.: Distribution of fibers originating from the inferior colliculus. Anat. Rec. **139**, 266 (1961).

Rasmussen, G. L.: Anatomic relationships of the ascending and descending auditory systems. In: Neurological Aspects of Auditory and Vestibular Disorders, pp. 5—19. Springfield: Thomas 1964.

Rinvik, E.: The corticothalamic projection from the second somatosensory cortical area in the cat. An experimental study with silver impregnation methods. Exp. Brain Res. **5**, 153—172 (1968).

Rioch, D. McK.: Studies on the diencephalon of carnivora. Part I. The nuclear configuration of the thalamus, epithalamus, and hypothalamus of the dog and cat. J. comp. Neurol. **49**, 1—119 (1929).

Rockel, A. J.: Observations on the inferior colliculus of the adult cat stained by the Golgi technique. Brain Research **30**, 407—410 (1971).

ROCKEL, A. J., JONES, E. G.: The neuronal organization of the inferior colliculus of the adult cat. J. comp. Neurol. **147**, 11—60 (1973a).

ROCKEL, A. J., JONES, E. G.: Observations on the fine structure of the central nucleus of the inferior colliculus of the cat. J. comp. Neurol. **147**, 61—92 (1973b).

ROCKEL, A. J., JONES, E. G.: The neuronal organization of the inferior colliculus of the adult cat. II. The pericentral nucleus. J. comp. Neurol. **149**, 301—333 (1973c).

ROSE, J. E.: The thalamus of the sheep: cellular and fibrous structure and comparison with pig, rabbit and cat. J. comp. Neurol. **77**, 469—523 (1942).

ROSE, J. E., WOOLSEY, C. N.: The relations of thalamic connections, cellular structure and evocable electrical activity in the auditory region of the cat. J. comp. Neurol. **91**, 441—466 (1949).

ROSE, J. E., GALAMBOS, R.: Microelectrode studies on medial geniculate body of cat. I. Thalamic region activated by click stimuli. J. Neurophysiol. **15**, 343—357 (1952).

ROSE, J. E., WOOLSEY, C. N.: Cortical connections and functional organization of the thalamic auditory system of the cat. In: Biological and Biochemical Bases of Behavior, pp. 127—150. Madison: Univ. of Wisconsin Press 1958.

ROSE, J. E., GREENWOOD, D. D., GOLDBERG, J. M., HIND, J. E.: Some discharge characteristics of single neurons in the inferior colliculus of the cat. I. Tonotopical organization, relation of spike counts to tone intensity, and firing patterns of single elements. J. Neurophysiol. **26**, 294—320 (1963).

ROSE, J. E., GROSS, N. B., GEISLER, C. D., HIND, J. E.: Some neural mechanisms in the inferior colliculus of the cat which may be relevant to localization of a sound source. J. Neurophysiol. **29**, 288—314 (1966).

ROSENZWEIG, M. R., WYERS, E. J.: Binaural interaction in the midbrain of the cat. Amer. Psychol. **9**, 460—461 (1954).

SEFTON, A. J., BURKE, W.: Reverberatory inhibitory circuits in the lateral geniculate nucleus of the rat. Nature (Lond.) **205**, 1325—1326 (1965).

SHEATZ, G. C., VERNIER, V. G., GALAMBOS, R.: E. E. G. study of conditioning. Amer. J. Physiol. **183**, 660—661 (1955).

SZENTÁGOTHAI, J.: The structure of the synapse in the lateral geniculate body. Acta Anat. **55**, 166—185 (1963).

TARLOV, E. C., MOORE, R. Y.: The tectothalamic connections in the brain of the rabbit. J. comp. Neurol. **126**, 403—421 (1966).

TEBECIS, A. K.: Properties of cholinoceptive neurones in the medial geniculate nucleus. Brit. J. Pharmacol. **38**, 117—137 (1970a).

TEBECIS, A. K.: Studies on cholinergic transmission in the medial geniculate nucleus. Brit. J. Pharmacol. **38**, 138—147 (1970b).

TEBECIS, A. K.: Cholinergic and non-cholinergic transmission in the medial geniculate nucleus of the cat. J. Physiol. **226**, 153—172 (1972).

THOMPSON, R. F., SPENCER, W. A.: Habituation: a model phenomenon for the study of neuronal substrates of behavior. Psychol. Rev. **73**, 16—43 (1966).

WALLER, W. H.: Thalamic degeneration induced by temporal lesions in the cat. J. Anat. (Lond.) **74**, 528—536 (1939).

WALZL, E. M., MOUNTCASTLE, V. B.: Projection of vestibular nerve to cerebral cortex of cat. Amer. J. Physiol. **159**, 595 (1949).

WATANABE, T., SIMADA, Z.: Collicular auditory interneuron and its blocking by dihydro-β-erythroidene hydrobromide. Proc. Jap. Acad. **46**, 983—988 (1970).

WATANABE, T., SIMADA, Z.: Picrotoxin: effect on collicular auditory neurons. Brain Research **28**, 582—585 (1971).

WEBSTER, W. R.: The effects of repetitive stimulation on auditory evoked potentials. Electroenceph. Clin. Neurophysiol. **30**, 318—330 (1971).

WEBSTER, W. R., AITKIN, L. M.: Evoked potential and single unit studies of neural mechanisms underlying the effects of repetitive stimulation in the auditory pathway. Electroenceph. Clin. Neurophysiol. **31**, 581—592 (1971).

WHITLOCK, D. G., PERL, E. R.: Afferent projections through ventrolateral funiculi to thalamus of cat. J. Neurophysiol. **22**, 133—148 (1959).

WICKELGREN, W. O.: Effect of acoustic habituation on click-evoked responses in cats. J. Neurophysiol. **31**, 777—784 (1968).

WOOLLARD, H. H., HARPMAN, J. A.: The connexions of the inferior colliculus and of the dorsal nucleus of the lateral lemniscus. J. Anat. (Lond.) **74**, 441—458 (1939).

WOOLSEY, C. N.: Organization of cortical auditory system. In: Sensory Communication, pp. 235—257. Cambridge, Mass.: M. I. T. Press 1961.

WOOLSEY, C. N.: Electrophysiological studies on thalamocortical relations in the auditory system. In: Unfinished Tasks in the Behavioral Sciences, pp. 45—57. Baltimore: Williams and Wilkins 1964.

WOOLSEY, C. N., ADRIAN, H. O., GROSS, N. B., HIRSCH, J. E., LANE, J. F., LIFSCHITZ, W. S., SMITH, F. D.: Thalamic projections to subdivisions of auditory cortex. XXIInd International Congress of Physiology, Leiden 1962.

Chapter 4

Single Unit Activity of the Auditory Cortex

Moïse H. Goldstein, Jr., Baltimore, Maryland (USA) and Moshe Abeles, Jerusalem (Israel)

With 10 Figures

Contents

I. Introduction . . . 199
II. Neural Coding . . . 200
A. General Observations . . . 200
B. Coding of Tonal Stimuli . . . 201
1. Patterns of Response Activity . . . 201
2. Tuning Curves . . . 205
3. Responses to Binaural Stimulation . . . 206
C. Coding of Clicks . . . 208
D. Coding of other Stimuli . . . 210
III. Functional Architecture . . . 212
A. Organization According to Depth . . . 213
B. Tonotopic Organization . . . 213
C. Columnar Organization . . . 214
IV. Conclusion . . . 216
References . . . 217

I. Introduction

The single unit technique is a relatively recent development which involves recording the temporal pattern of nerve impulse discharges from individual neurons. Its use in investigating the auditory cortex has provided a more detailed picture of the activity of the population of cortical neurons than that obtainable by previous techniques. The advantage of a more detailed look at cortical neural activity is tempered by the need for the collection and processing of great volumes of data. The researcher is usually aided by a computer, but even so, as we shall see, the diversity of the properties of the individual elements is such that the task of obtaining a reasonably systematic picture of the functional aspects of the cortical population appears Herculean. Developing theories or finding models for the functional properties of the auditory cortical neurons is equally difficult.

We have chosen to organize the material of this chapter by topics rather than historically. Recent work will be stressed and the reader is referred to the handbook chapter of Harlow Ades (1959) for a review of earlier work. Most of the studies of the auditory cortex used cats, and a great majority of studies have involved the primary auditory cortex, AI. Exceptions will be noted. This chapter focuses on two aspects of single unit studies of the auditory cortex: neural coding and functional architecture.

II. Neural Coding

A. General Observations

The way in which acoustic input is coded in the activity of the nerve cells is called neural coding. For practical reasons, neural coding is not studied by presenting a given sound and observing responses of units of the neural population. Rather, one unit is studied with a set of stimuli, then another unit is studied with the same sounds, etc., until the population has been sampled for responses to the set of stimuli.

The response properties of cortical neurons are so diverse that it is inappropriate to consider them as a population in the sense that physicists and mathematicians treat a group of elements with homogeneous properties. We are more in the position of the social scientist who is investigating a population of individuals. If the questions asked are well defined and sufficiently restricted, statistical measures can be given (*e.g.*, for percent responding and how they respond) without denying the diversity of the individuals in the population. Collection of data is time consuming and expensive. An extensive study comprises responses from several hundred units from a population of many millions.

To date, studies of cortical units have made extensive use of tonal stimuli. Signals of broad spectral content, specifically clicks and noise, have also been used. Some studies (WHITFIELD and EVANS, 1965) have featured frequency modulated tones (sometimes called swept tones), and one group has employed the vocalizations of the squirrel monkey in studying the cortex of that animal (FUNKENSTEIN *et al.*, 1971; FUNKENSTEIN and WINTER, 1973; WOLLBERG and NEWMAN, 1972).

Unit studies of the visual cortex (HUBEL and WIESEL, 1962) have revealed a class of stimuli to be especially appropriate. Linear patterns (slits, edges, or corners) lighter or darker than the surround evoke stronger responses than other stimuli. An analogous situation, *i.e.*, a class of stimuli more appropriate than others for affecting activity of cortical neurons, has not been found for the auditory system (ABELES and GOLDSTEIN, 1972). Although it is not unusual to find single cells in the auditory cortex that will respond strongly only to a particular acoustic stimulus, it does not appear to us that there is any one special class of acoustic stimuli that will evoke especially strong responses or have especially low thresholds for the majority of the units in the auditory cortex. This finding is consistent with a picture of diversity of coding properties for neurons of the auditory cortex.

Early unit studies reported a sizeable fraction of units in the anesthetized cat's primary auditory cortex (AI) to be unresponsive to acoustic stimulation (ERULKAR *et al.*, 1956). However, in the muscle-relaxed, unanesthetized cat the proportion of units unresponsive to sound was below 5%; a larger fraction, about 20%, exhibited pronounced lability of response properties (GOLDSTEIN *et al.*, 1968). Among these units with labile response properties are those showing properties such as habituation, response to novel stimuli, and "attention" (HUBEL *et al.*, 1959). Needless to say, systematic study of such units is difficult and has not been accomplished in cats. Quite recent studies with behaviorally trained monkeys (MILLER, 1971) represent an effort to explore these interesting aspects of neural coding. We will be content to discuss in this chapter "the other 80%" which give reasonably repeatable responses to acoustic stimuli during the period of study.

Ongoing or spontaneous activity is that recorded when there is no acoustic stimulus presented to the animal. A sizeable portion of cortical units in AI have little or no spontaneous activity. In one study of unanesthetized, muscle-relaxed cats (GOLDSTEIN *et al.*, 1968), about 40% of the units had ongoing rates less than 1 spike/sec. For the rest, ongoing rates were low and rates above 35/sec were quite unusual. A special class of cortical units having narrower or thinner spikes than most units, however, do exhibit high rates of ongoing activity (DE RIBAUPIERRE *et al.*, 1972a).

For units exhibiting spontaneous spike activity, an acoustic stimulus can evoke either an increase or a decrease in activity, called "excitation" or "suppression", respectively. Units with no spontaneous activity can normally exhibit only excitatory responses; however, with two-tone stimulation suppressive effects can be demonstrated in these units (ABELES and GOLDSTEIN, 1972).

How does the experimenter define an evoked response? This is not a trivial question. Through the use of computer techniques, very weak and subtle changes in temporal patterns of nerve spike activity can be detected. ABELES and GOLDSTEIN (1970) defined a response as secure for excitatory responses if in sequential presentations of similar stimuli, time-locked spike activity occurs in over half of the presentations; a similar definition does not hold for suppressive responses.

Although a small minority of cortical units respond vigorously in a sustained manner to steady-state stimulation, this is the exception rather than the rule, and it is usual to use transient stimuli. Thus, tonal stimuli are usually presented as tone bursts and noise stimuli as noise bursts.

B. Coding of Tonal Stimuli

1. Patterns of Response Activity

In a series of studies, a standard set of stimuli was used to investigate the response properties of units in AI of muscle-relaxed, unanesthetized cats. The standard set included tone bursts, noise bursts, and clicks of moderate intensity. About four-fifths of the units which could be studied with our standard set of stimuli responded to tone bursts, with a smaller proportion responding to noise bursts, and just over half the units responding to clicks (GOLDSTEIN *et al.*, 1968; ABELES and GOLDSTEIN, 1970). In a preparation such as that used in the experiments described above a variety of temporal patterns of nerve spike activity can be observed. Similar patterns are obtained from AI of unanesthetized and unrestrained cats (EVANS and WHITFIELD, 1964). Samples of patterns of response to tone burst stimulation are given in Fig. 1. Usually terms such as *on*, *off*, *through*, and combinations of such terms are used to describe patterns of response. It is clear that in responses to tone bursts, units in AI exhibit characteristic diversity. A given unit often shows diverse response patterns as the parameters of tone burst stimulation are changed.

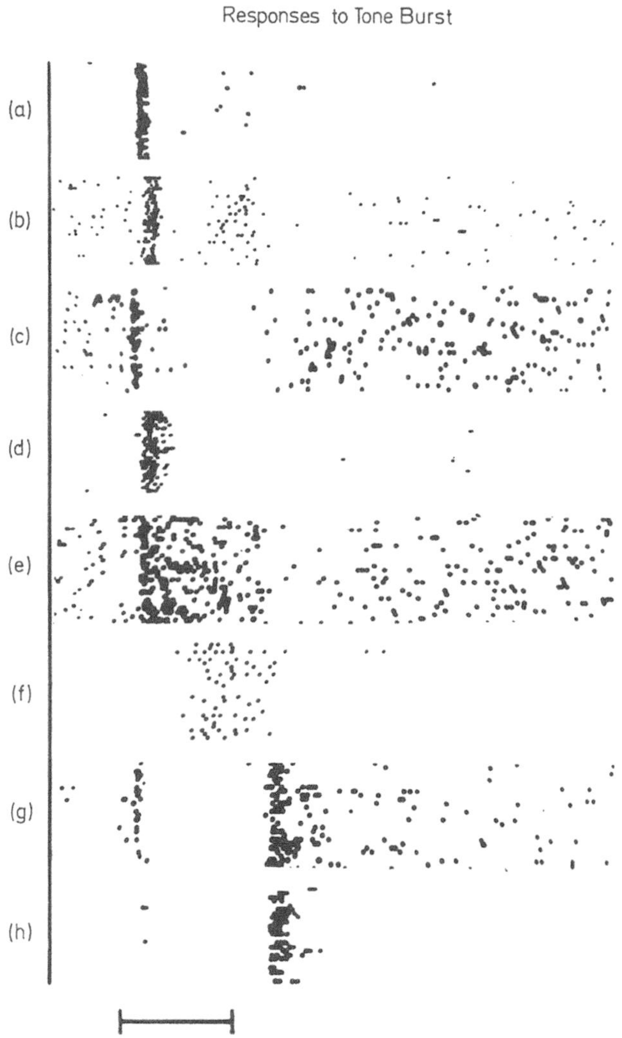

Fig. 1a—h. Patterns of response activity evoked by tone bursts. Each subset is a dot display of spike activity in response to 25 consecutive presentations of 100 msec duration tone bursts. Rise and fall time was 10 msec, and bursts were presented 1/sec. Bar at bottom of the figure indicates timing of 100 msec tone bursts. Frequency, ears stimulated, and intensity (dB SPL) were: (a) 50 kHz, both, 30 dB; (b) 33 kHz, contralateral, 50 dB; (c) 26.7 kHz, contralateral, 50 dB; (d) 9.7 kHz, both, 20 dB; (e) 26.7 kHz, contralateral, 20 dB; 9.5 kHz, both, 20 dB; (f) 9.5 kHz, both, 20 dB; (g) 26.7 kHz, both, 50 dB; (h) 22.7 kHz, both, 50 dB. Units from different cats except for (c), (e), and (g); (c) and (e) are the same unit for different intensities, and (g) is a nearby unit in the same track

Examples of a changing pattern of spike discharge with changing frequency of tone burst stimulation are shown in Fig. 2. In both records *on*, *off*, and *through suppression* are found in different frequency ranges.

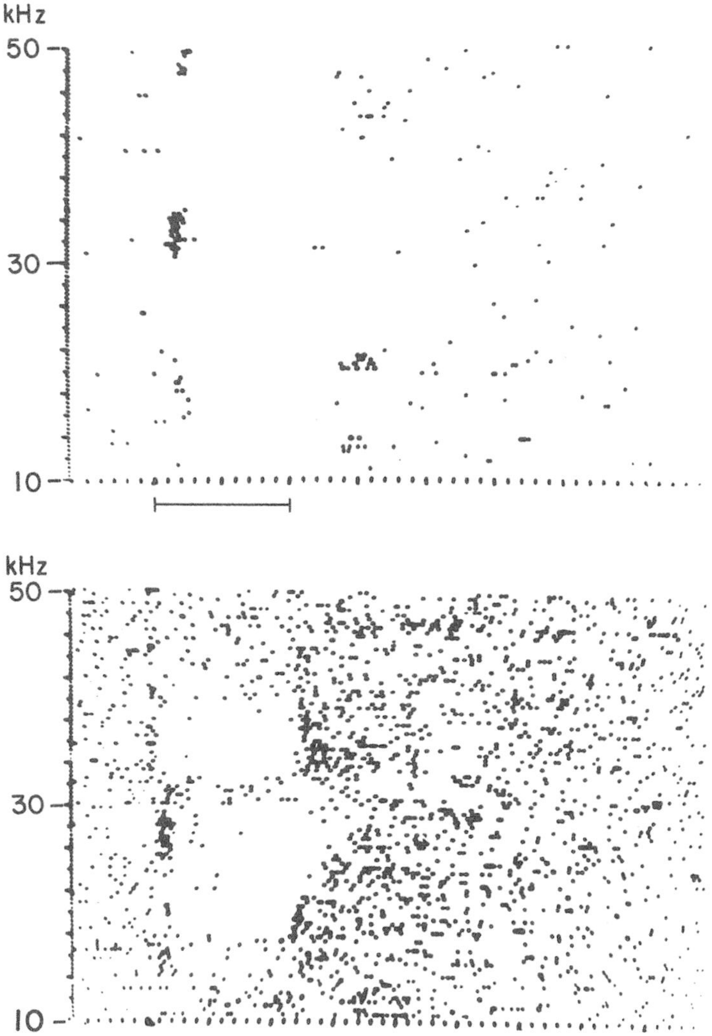

Fig. 2. Changing pattern of response with changing frequency. Each record displays 100 responses to tone bursts with frequencies ranging linearly from 10—50 kHz. Different units from two animals. Tone bursts as in Fig. 1; bar shows timing of the 100 msec bursts. Stimulus intensity 50 dB SPL. Presentations of different frequencies random in upper record, sequential in lower

Changes in pattern of response with tone-burst intensity are illustrated in Fig. 3. The pattern of Fig. 3 was typical of the majority of units studied (ABELES and GOLDSTEIN, 1972). There was a *through* response near threshold. At higher intensities, an *on* component was added, the peak size of which increased rapidly with intensity and reached a plateau at 10 to

30 dB above threshold. The *through* component decreased rapidly, and usually disappeared completely at about 20 dB above threshold. Fig. 1c and e also illustrate change of response pattern with intensity. The records give responses from a unit to 26.7 kHz tone bursts. The intensity level in (e) was 20 dB SPL resulting in a *through* response; the intensity level in (c) was 50 dB SPL resulting in an *on* response followed by suppression.

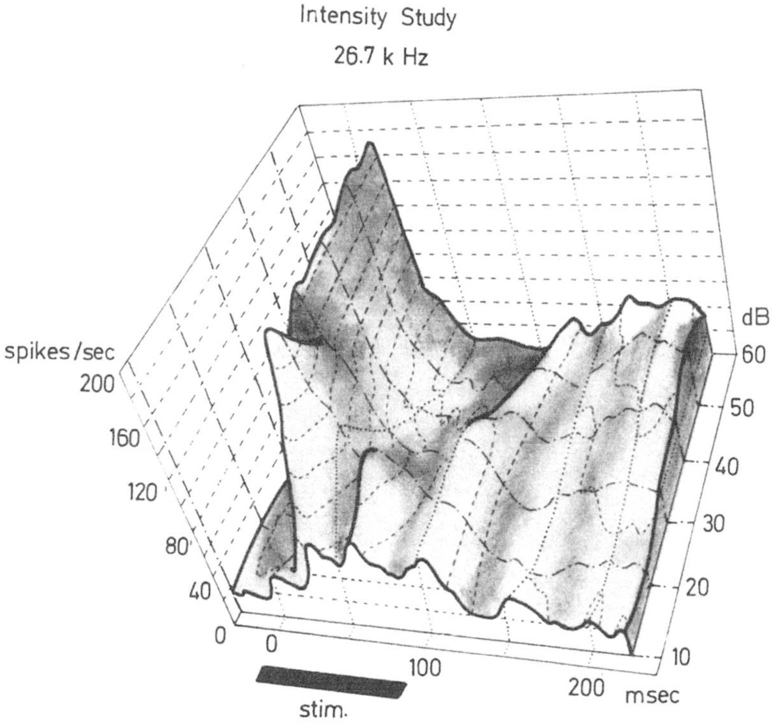

Fig. 3. Graph shows instantaneous rate of firing as a function of time and intensity of the stimulus. Tone bursts, as in Fig. 1, were used. Responses to tone bursts of fixed frequency (26.7 kHz) were obtained at intensities from 10 dB to 60 dB SPL. The intensity range was covered in 5 dB steps. Twenty-five responses at each level were recorded and a PST histogram constructed. The three dimensional graph was constructed by linear interpolation between these smoothed histograms. Axis calibrations are indicated on the figure

We have been most concerned in the above discussion with gross changes in the pattern of spike activity related to different parameters of tonal stimuli. It is worth asking if the fine structure of spike activity of cortical units is related to the fine structure of the tone, that is, to the individual cycles in the sinusoidal stimuli. For frequencies up to 4 or 5 kHz, tonal stimuli evoke spike discharges from single auditory nerve fibers that exhibit "phase locking"; that is, there is a greater probability of firing for a certain range of phases of each cycle of the sinusoidal stimulus than for other phases (ROSE *et al.*, 1967; also see the present Handbook chapter on the auditory nerve).

Phase locking is also observed for more central cells in the auditory pathway, although the limiting frequency of phase locking and the proportion of cells exhibiting it decrease for the more central levels. For example, in studies at the level of the inferior colliculus (ROSE *et al.*, 1966), only a few of the neurons studied showed phase locking. It has not been described for any cortical units, even those responding to low frequency sinusoidal stimuli (BRUGGE *et al.*, 1969). One should bear in mind that these experiments were done on anesthetized animals and that the anesthetic may affect firing patterns of neurons in the more central locations of the pathway more than in neurons in the peripheral part of the auditory pathway.

2. Tuning Curves

The tuning curve indicates the frequencies and intensities for which some feature of the response pattern to single tone stimuli is obtained. We have seen that, for all eighth nerve units and for many of those in the subcortical auditory pathway,

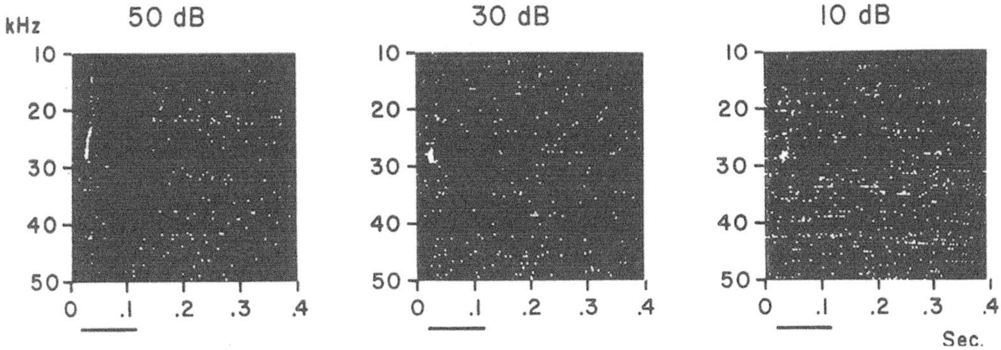

Fig. 4. Responses to tone bursts at three intensity levels. Each dot display shows the timing of spike activity in response to 100 sequential presentations of tone bursts. The tone frequency was increased by 400 Hz between each presentation. Tone burst frequency is indicated on the ordinate. Lines below the displays show timing of the 100 msec bursts. (From ABELES and GOLDSTEIN, 1970)

tuning curves are characteristically "V" shaped. (See pp. 14 and 123.) At the cortex especially, it is important to specify which feature of the response pattern is represented by the tuning curve. Figure 4 shows the spike pattern in response to tone stimuli for 3 intensities of tone bursts. The excitatory aspects of the response pattern of this unit yielded a narrow tuning curve. There is a through suppression in the response pattern which would have a much broader tuning curve.

Three main types of tuning curves have been obtained for cortical units (see Fig. 5). They are referred to as "narrow", "multi-peaked", and "broad" tuning curves and were first described for the barbiturate anesthetized cat in which the units have little ongoing activity (OONISHI and KATSUKI, 1965). If attention is on the excitatory features of the response, the classification is also appropriate for units in the unanesthetized muscle-relaxed cat (ABELES and GOLDSTEIN, 1970).

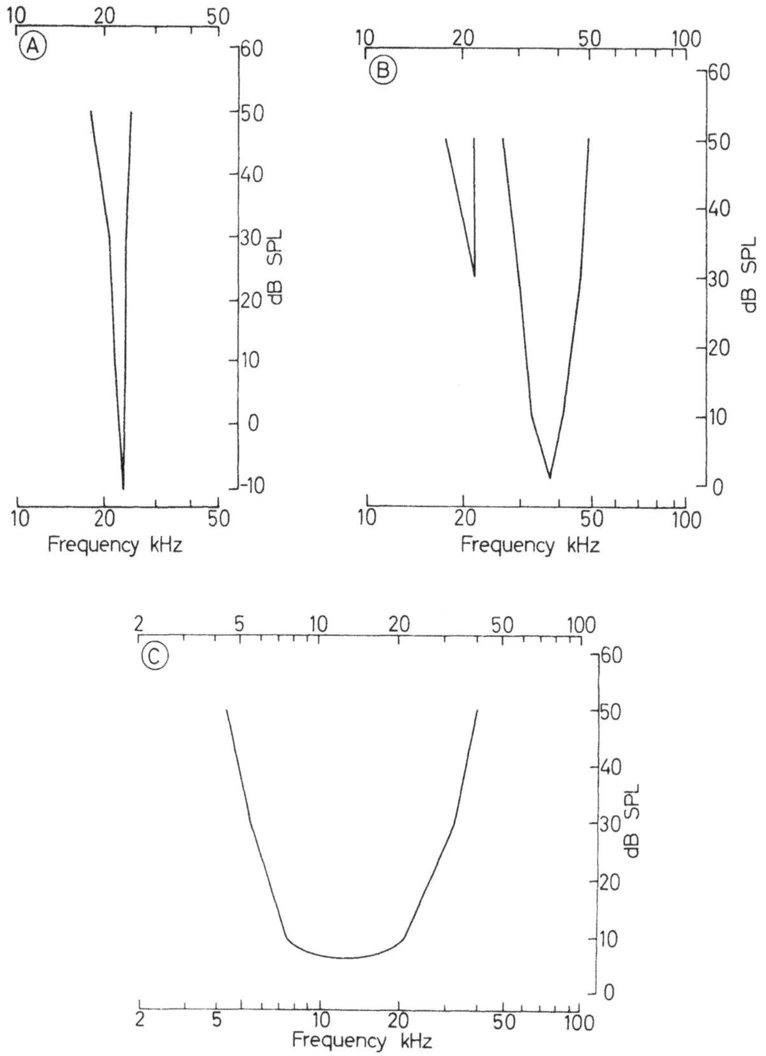

Fig. 5A—C. Examples of three types of tuning curves: (A) narrow; (B) multi-peaked; and (C) broad. The curves were obtained from analysis of dot displays obtained using tone bursts of different frequencies. For all tuning curves, the units were studied for at least three intensity levels. From displays such as those of Fig. 4, ranges of excitatory response at each intensity level were obtained. Tuning curves were plotted by connecting by straight lines the corresponding ends of the response ranges at different intensity levels

3. Responses to Binaural Stimulation

Representation of the two ears in cortical units was investigated by using three modes of stimulation: contralateral ear alone, ipsilateral ear alone, and stimulation of both ears with identical signals. Although nearly all units in AI are influenced by both ears, the population of units in the cortex represents the contralateral ear more strongly in that more cells respond to stimulation of the contralateral ear alone than to stimulation of the ipsilateral ear. Various interactions were demonstrated, the most usual being summation; but inhibition, occlusion, and the more complex interactions were also observed (HALL and GOLDSTEIN, 1968).

Interaural interactions are of interest because they may be used for localization of a source of sound in space. The locus of sound source in respect to the head determines the relative intensity and timing of the sound at the two ears. Relative timing is usually expressed as phase difference for sinusoidal stimuli and time difference for sharp transients. It was shown that, for units showing interaural inhibitory interaction, it is usual for the contralateral ear to excite the unit and the ipsilateral ear to inhibit it. Such units have been shown to be particularly sensitive to interaural intensity differences over a range of 20—30 dB but quite insensitive to changes in the level of the binaural stimulus (BRUGGE *et al.*, 1969).

In Section IIB 1, we indicated that there is no evidence for phase locking by cortical units. However, it has been demonstrated that some of the units in AI and AII, which respond to tones of frequency below 2500 Hz, were sensitive to the interaural phase relation between stimuli delivered to the two ears (BRUGGE *et al.*, 1969). This dependence of response on interaural phase is illustrated in Fig. 6.

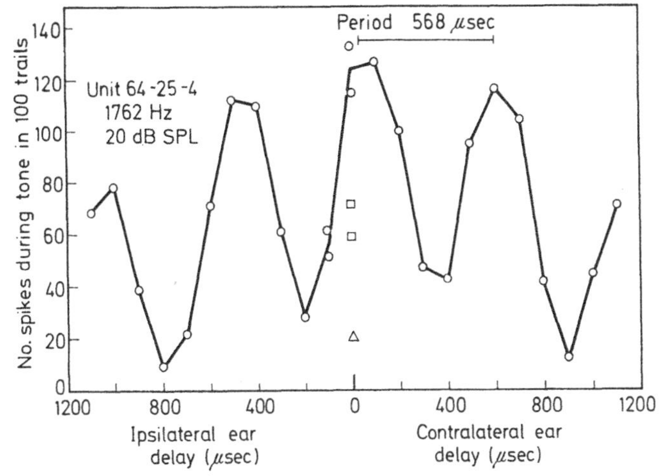

Fig. 6. Dependence of response on interaural phase. Curves show effects of interaural time delay on number of spikes in response to tone bursts. Left half of each graph corresponds to a delay of ipsilateral ear stimulus, the right half to delay of contralateral ear stimulus. Closed circles correspond to responses to binaural stimulation; open squares and triangle correspond, respectively, to response to contralateral and ipsilateral stimulation. Tone duration, 100 msec; repetition rate, 1/sec; 100 stimulus trials; frequency of tone 1762 Hz; period, 568 μsec; intensity to both ears, 20 dB SPL. Note that seldom was more than a single spike evoked for each of the 100 stimulus presentations. (From BRUGGE *et al.*, 1969)

The effect of interaural time difference on the responses to fast transients has not yet been studied adequately. In one series of experiments, 30 units which responded to clicks were studied. Click pairs were presented at variable interaural time delays from 0.5 msec left ear leading to 0.5 msec right ear leading. For only one of the units was the response clearly dependent on interaural delay. (ABELES, unpublished data.) One should note that interaural differences are not the only cues for the localization of sound source. (See the chapter localization of sound in this Handbook.)

C. Coding of Clicks

The impulse is the standard excitation for obtaining the transient response of a linear system. Acoustic impulses are called clicks. The auditory nervous system is certainly not a linear system, yet responses to clicks have been studied extensively, especially in tracing pathways in anesthetized animals. The gross electrode technique yields a characteristic click-evoked response from AI, and, to some extent, from other auditory cortical regions. Surprisingly enough, when unit studies were first done, it was observed that clicks were relatively ineffective stimuli as compared to tones (Erulkar *et al.*, 1956). Similar results were found in studies of unanesthetized, muscle-relaxed cats (Goldstein *et al.*, 1968).

For units in AI, there are two main types of responses to single clicks, excitatory and suppressive (de Ribaupierre *et al.*, 1972b). In the excitatory response, one or two spikes are evoked with invariant latency, usually from about 9–11 msec according to the unit. In the suppressive response, there is a suppression of spontaneous activity for about 50 msec, followed by a return of activity often with a rebound. When suppression stops, the latency to the first spike is variable from response to response.

In single unit studies of cortical neurons, it is usual to use metal microelectrodes for extracellular recording of nerve spikes. Units may be studied for hours. In the experiments of de Ribaupierre *et al.* (1972a) hyperfine pipettes were used to obtain intracellular recordings; these pipettes have the advantage of displaying post-synaptic potentials (PSPs) along with nerve spikes. Intracellular recording is a difficult technique because of the small size of the neurons in AI; stable recordings usually can be maintained only for minutes. Figure 7 shows average intracellular potentials for the two types of responses described, illustrating the post-synaptic events underlying the patterns of spike activity described. Figure 7A is from a cell that showed a suppressive pattern of spike activity, and Fig. 7C is from a cell that showed an excitatory pattern of spike activity. More complex patterns of responses to clicks have also been obtained (Gerstein and Kiang, 1964; Goldstein *et al.*, 1968).

As indicated in Section II A, there is a class of cortical units characterized by thin spikes, usually having high rates of spontaneous activity. When these units responded to clicks, there was only an excitatory response consisting of a burst of spikes with latency to the first spike showing little variability for repeated presentations of the click stimulus. In general, the thin spike units seem to differ from regular spike units in lacking inhibitory (suppressive) drive (de Ribaupierre *et al.*, 1972b).

In a series of experiments (Goldstein *et al.*, 1971; de Ribaupierre *et al.*, 1972b), trains of repetitive clicks have been used as stimuli. Units were classified in four categories according to their responses to click trains. About 40% were classified as *lockers*; these cells had spike discharges precisely time-locked to the individual clicks in the train. A limiting rate of locking was defined for each cell in the locker category. This was the highest rate of click repetition at which the spikes were phase-locked with the individual clicks in the train. The distribution of this limiting rate for the locker cells was broad, extending from 10/sec to 1000/sec with a median between 50/sec and 100/sec. Roughly one-fourth of the cells were *unresponsive* to repetitive click stimulation. The remainder of the population was evenly

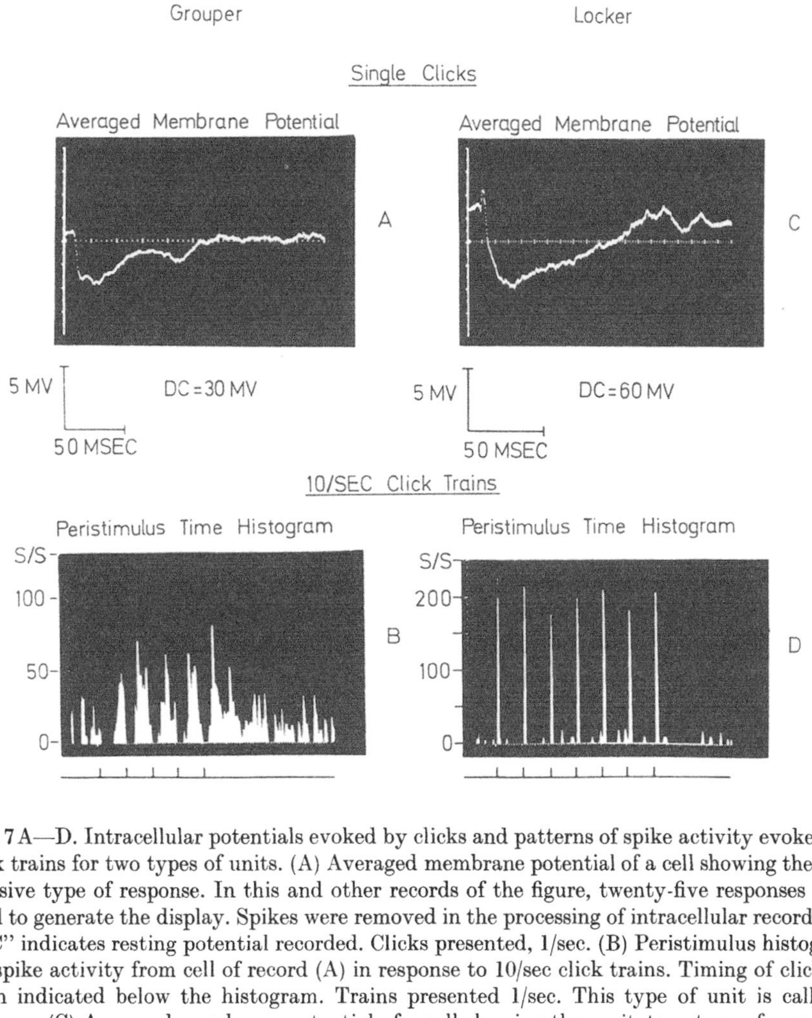

Fig. 7 A—D. Intracellular potentials evoked by clicks and patterns of spike activity evoked by click trains for two types of units. (A) Averaged membrane potential of a cell showing the suppressive type of response. In this and other records of the figure, twenty-five responses were used to generate the display. Spikes were removed in the processing of intracellular recordings. "DC" indicates resting potential recorded. Clicks presented, 1/sec. (B) Peristimulus histogram for spike activity from cell of record (A) in response to 10/sec click trains. Timing of clicks in train indicated below the histogram. Trains presented 1/sec. This type of unit is called a grouper. (C) Averaged membrane potential of a cell showing the excitatory type of response. Stimuli, etc., as in (A). (D) Peristimulus histogram for spike activity from cell of record (B) in response to 10/sec click trains. This type of unit is called a locker. (From GOLDSTEIN *et al.*, 1971)

divided between *special responders* and *groupers*. Cells with special response properties showed a change in response pattern to the click train as a function of click rate but showed no synchrony of discharge to the individual clicks in the train. Groupers had spike responses loosely synchronized to the individual clicks for low repetition rates. Thin-spike units were never in the grouper category, but were present in the other three categories. For regular-spike units, lockers showed excitatory responses for single clicks, and groupers showed suppressive responses to single clicks. Figures 7B and 7D show the pattern of spike activity in response to click trains for two cells, one of which had a suppressive response for single clicks (see Fig. 7A), the other had an excitatory response for single clicks (see Fig. 7C). Thus, Fig. 7B illustrates the spike pattern for a grouper, and Fig. 7D illustrates the spike pattern for a locker.

D. Coding of other Stimuli

Single tones and clicks are favorite stimuli of auditory scientists because they are compact in their spectral (tones) or temporal (clicks) dimension. Tones and clicks, however, are not necessarily good elementary waveforms for analysis of auditory responses to sounds of interest in communication. Furthermore, some cortical units do not respond to tones or clicks. Other stimuli have been used in an attempt to find whether these unresponsive units respond to any acoustic stimulus and also whether there is a class of stimuli more "appropriate" than others for the population of cortical units (see Section II A).

White or wide-band noise, like the click, has a broad distribution of energy in the frequency spectrum. Noise signals can be presented in bursts with quite the same envelope shape as used for tones. In one series of experiments (GOLDSTEIN *et al.*, 1968), cortical units exhibited a similar diversity of response patterns for noise bursts as for tone bursts. Generally, the predictability of the pattern of response to tone bursts from that to noise bursts (and vice versa) was poor (GOLDSTEIN *et al.*, 1968). An exception was the class of regular-spike units which had short latency excitatory responses to clicks. They usually showed *on* responses for both tone bursts and noise bursts (DE RIBAUPIERRE *et al.*, 1972b). In terms of effectiveness in stimulating cortical units, wide-band noise bursts occupy an intermediate position between tone bursts, which are more effective, and clicks, which are less effective. None of these "simple" stimuli is a good candidate for an appropriate type of stimulus.

There has been some study of cortical unit responses to frequency-modulated tones. The most extensive studies have been those of WHITFIELD and EVANS (1965) who used periodic sinusoidal and intermittent linear-ramp modulation in recording from single units of the cortex in cats. They found that many units respond to frequency modulated signals and do so more securely than when steady tones are used as stimuli. About 10% of their population, which responded to frequency changes, did not respond to steady tones. There was considerable diversity in the response patterns of the units. Some units responded in a way that could be predicted from their responses to single steady-tone stimulation, but, in most cases,

---

Fig. 8A—J. Dot displays and peristimulus time histograms for cells I and II. Stimuli: (A) Spontaneous activity; (B) Isolation Peep; (C) Err; (D) Twitter; (E) Chuck; (F) Tone bursts; (G) Rough Cackle; (H) Shriek; (I) Peep; (J) Vit. The stimulus is represented beneath each dot display. Vertical bars represent three times as many spikes in Cell I as in Cell II. Left-most bar in each histogram and vertical row of dots in each dot display represent onset of sweep. Time scale (horizontal rows of dots in A is 60 msec between dots. (From WOLLBERG and NEWMAN 1972)

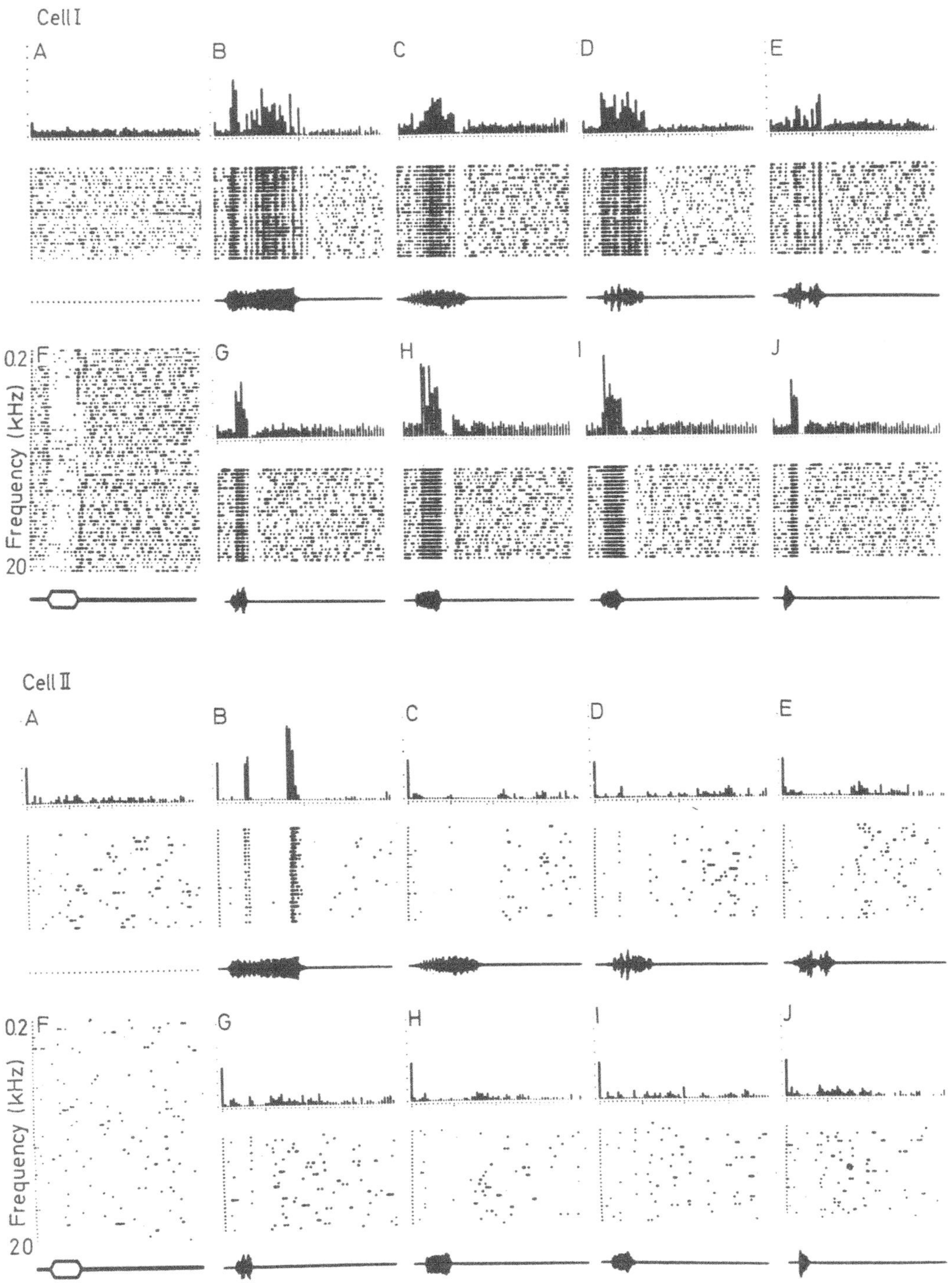
Cell I
A
B
C
D
E
F
G
H
I
J
0.2
Frequency (kHz)
20
Cell II
A
B
C
D
E
F
G
H
I
J
0.2
Frequency (kHz)
20

Fig. 8

such prediction was not possible. Many units responded when the frequency was rising but not when it was falling, or vice versa.

In our studies, we used a "swept" tone, gliding back and forth every four seconds between 500 Hz and 50 kHz, as the electrode was advanced through the cortex. Such a procedure facilitates detection of units when the electrode passes near them. This "search" stimulus has the advantage over other stimuli in that most units respond to it and, unlike tone bursts the cat's frequency range can be explored quickly. We have found that responses to tone-bursts are usually not easily predicted from responses to the ramp modulated FM signal that we used. It seems that although the latter signals are quite effective stimuli for auditory cortical units, they do not, in our view, meet the requirements of an appropriate stimulus class, in that, for most units, responses to FM stimuli are neither stronger nor present at lower thresholds than responses to some other stimuli. The search for an appropriate stimulus for the cat's auditory cortical neurons has not been exhaustive, but thus far there has been a great diversity of response properties of the units that have been studied, and no class of stimuli has been convincingly shown to be appropriate.

In some bats, FM signals are of special interest as they are used by the animals for echo-location. SUGA (1965a) has found units in the bat's auditory cortex that were more sensitive to FM than to pure tone pulses. The response properties of the units showed considerable diversity with regard to tuning curves, sensitivity to direction of frequency modulation, etc. (SUGA, 1965b). Neural models were described that allow prediction of the different types of response to FM stimuli from excitatory and inhibitory tuning curves.

Vocalizations of the species under study may be an interesting stimulus to use in studies of cortical units. No systematic study of this sort has been done in cats. One group (FUNKENSTEIN *et al.*, 1971; FUNKENSTEIN and WINTER, 1973; WOLLBERG and NEWMAN, 1972) has studied responses from neurons in the superior temporal cortex of awake squirrel monkeys to species-specific vocalizations. The squirrel monkey has a quite stereotyped and well studied vocal repertoire (WINTER *et al.*, 1966). Cell I of Fig. 8 responded with temporally complex patterns to many vocalizations and with a suppression response to tone-bursts over a broad frequency range. Cell II of Fig. 8 responded with an on-off pattern to one call and was relatively unresponsive to other calls and to tone bursts. More than 80% of the population studied by WOLLBERG and NEWMAN (1972) responded to tape-recorded vocalizations, most having properties between the extremes illustrated in Fig. 8. These investigators conclude that, in the squirrel monkey, the auditory cortex is concerned with analysis of complex auditory signals and suggest that it plays an important role in the coding and interpretation of sounds used in species-specific communication.

III. Functional Architecture

In the somesthetic and visual cortex, researchers have found correlates between response properties of units and cortical morphology (MOUNTCASTLE, 1957; HUBEL and WIESEL, 1962). The relationship of sensory stimulus representation to structural and morphological features of the cortex is usually called functional architecture. The features for which correlates have been sought are depth (the layers of cells observed anatomically), surface position (topical organization), and columns.

Studies of functional architecture of the auditory cortex have almost exclusively employed tonal stimuli. In one study in which tone bursts, clicks, and noise bursts were used, the only evidence of functional organization related to structure and morphology was for tone burst stimulation (ABELES and GOLDSTEIN, 1970).

A. Organization According to Depth

The three shapes of tuning curves of cortical units, sharp-peaked and narrow, flat and broad, and multi-peaked, were mentioned in the section on neural coding. OONISHI and KATSUKI (1965) stated that units with flat tuning curves were most superficial in the cortex; units with multi-peaked tuning curves were intermediate in depth; and units with sharp-peaked tuning curves were deepest.

In another study (ABELES and GOLDSTEIN, 1970), a statistical analysis was used to ask whether there was a dependence of certain response properties on depth in the cortex. The properties examined were responsiveness to clicks, to noise bursts and to tone bursts, broad versus narrow tuning, monaural versus binaural influence. For all of the properties, the hypothesis that functional properties were independent of cortical depth could not be rejected. This finding for narrow versus broad tuning fails to support the schema of organization according to depth that has been proposed by OONISHI and KATSUKI (1965). The two studies differed in anesthetic state of the cats, manner of presenting the acoustic stimuli (free field versus earphones), and in method of data analysis.

B. Tonotopic Organization

For subcortical loci of the auditory nervous system of the cat, the most striking relationship between structure and response properties of units is tonotopic organization. As an electrode is advanced through these structures in the appropriate direction, the units encountered exhibit an orderly and monotonic sequence of best frequencies (best frequency is the minimum of the unit's tuning curve). This feature of the auditory system is described in detail in other chapters of this Handbook.

Three groups (HIND *et al.*, 1960; EVANS *et al.*, 1965; GOLDSTEIN *et al.*, 1970) have reported systematic studies of the tonotopic arrangement of single units in the cat's primary auditory cortex (AI). The reports are in general agreement and show a tonotopic organization with responsiveness to low frequencies posterior and responsiveness to high frequencies anterior in AI. Cortical organization is qualitatively different from the very orderly arrangement reported for subcortical locations. The anterior posterior gradient is clear in terms of the average of best frequencies, but there is often considerable fluctuation from unit to unit. Technical difficulties are such that, to date, there are no reports of experiments in which an electrode has been advanced more or less parallel to the cortical surface and best frequencies of sequentially encountered units examined. Such data would be most directly comparable with data from subcortical locations.

C. Columnar Organization

The idea that there are functional columns in the sensory cortical areas was first advanced by LORENTE DE NÓ (1938) on the basis of observations that cortical cells are arranged in vertical (radial) chains. This idea received electrophysiological confirmation in the work of MOUNTCASTLE (1957). Units in the somatic cortex were classified as activated by movement of hairs, by pressure upon the skin, or by mechanical deformation of deep tissues. Neurons encountered in a radial traverse of the cortex were activated from almost identical peripheral receptive fields and were modality-specific. In off-radial tracks, a sequence of units of one modality would be followed by a sequence of another modality, and so on. A comparable columnar organization was found in the visual cortex (HUBEL and WIESEL, 1962). Units in a functional column had receptive fields in the same general retinal region, and all had the same receptive-field axis orientation (orientation of slits or edges producing maximal responses). As the electrode passed from one column to another in off-radial tracks, there were discontinuities in the axis-orientation of the units recorded.

There has been no convincing demonstration of discrete functional columns in the auditory cortex analogous to those found in somatic and visual cortical areas.

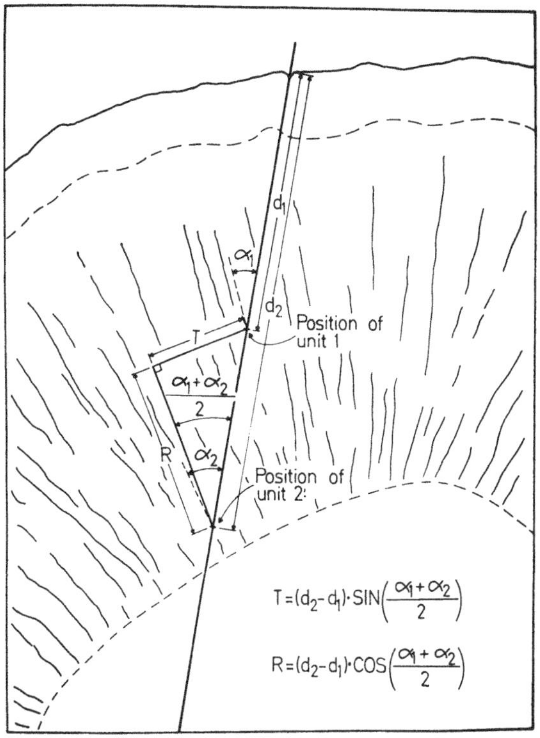

Fig. 9. Measurement of tangential and radial distance between units. The distance along the track, *d*, and the angle between the track and radial chains of cells, α, were assigned to each unit. For each pair of narrowly tuned units in the same track, the tangential separation, *T*, and the radial separation, *R*, were calculated using the indicated formulas. (From ABELES and GOLDSTEIN, 1970)

In such a demonstration, off-radial electrode tracks would have to result in sequences of units having a common response property with abrupt transitions between the sequences. ABELES and GOLDSTEIN (1970) demonstrated an organization which is very likely related to the chains of neurons considered by LORENTE DE NÓ (1938). The neural response property for which the relationship was demonstrated was best frequency of narrowly tuned units.

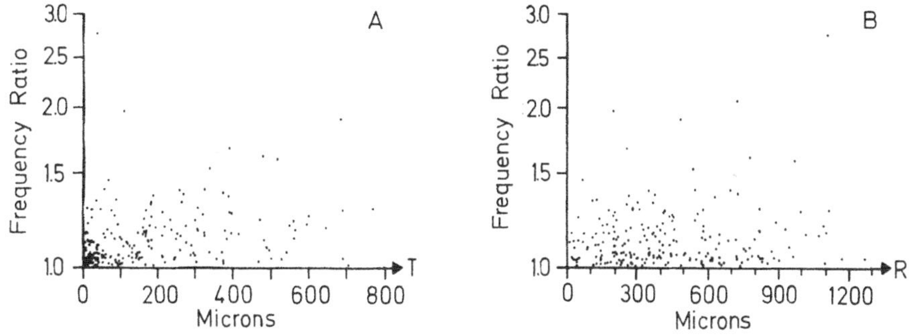

Fig. 10A, B. Closeness of best frequencies of narrowly tuned units as a function of their tangential and radial separation. (A) scatter plot. The abscissa of each point is the tangential separation, *T*, between a pair of narrowly tuned units, and the ordinate is the ratio between best frequencies of the pair. The two units in each pair are from the same electrode track. (B) scatter plot. As in (A) for radial separation, R. (From ABELES and GOLDSTEIN, 1970)

Narrowly tuned units comprise about half of those units in the primary auditory cortex which respond to tone bursts. A statistical analysis showed that narrowly tuned units in close radial alignment are very likely to have their response ranges in registration and, thus, to have nearly equal best frequencies. Figure 9 illustrates how the tangential (*T*) and radial (*R*) separation of pairs of narrowly tuned units in an electrode track were measured. For all such pairs (there were 200 in the experimental series), the ratio of best frequencies was calculated (higher best frequency/lower best frequency); results are shown in Fig. 10. It is clear, in Fig. 10A, that there is a high concentration of pairs with close best frequencies for small tangential separations, but no such concentration is seen when the best frequency ratios are plotted versus radial separation (Fig. 10B). Statistical analysis showed the experimental results to be consistent with the hypothesis that registration of best frequencies of units is dependent on the units' alignment with the radial chains and independent of radial separation in the cortex.

The above results led us to ask whether the narrowly tuned units were to be found in columns. The data indicated that if there are such columns in the auditory cortex, they are small (ca. 100 μ diameter) and at the limit of resolution of the techniques employed. Another question asked was whether binaural representation has a columnar organization. There was some tentative evidence for such organization, but certainly the question needs further study.

Summarizing, there is convincing evidence of a relationship between response ranges of narrowly tuned units and their alignment with the radial chains of neurons observed by anatomical techniques. As yet, there has been no convincing demonstration of the sort of "hard walled" columns observed in somesthetic and visual cortex. The more statistical picture of organization seen in the auditory cortex appears to be consistent with the anatomical structure typical of sensory cortex (VON BONIN and MEHLER, 1971).

IV. Conclusion

What general statements can we make about the auditory cortex as a result of knowledge gained by unit studies? What will be the directions of further work? These questions occupy our attention in the present section.

The scientist studying single units has been compared to the collector of beautiful butterflies. Attention is focussed on the colorful, dramatic, or unusual. It may be too easy to think of the cortex as populated by cells with functional properties best suited to study by our technique. Thus, strongly responding units with repeatable responses to acoustic signals dominate our mental image of the auditory cortex. But we should not forget the other neural units. There are the glial cells, silent and spikeless[1]. Probably more important functionally are the cells which do not respond to our stimuli and the ones that have such labile response properties that it is not possible to study them systematically. These labile units are very interesting. It is frustrating that they are so difficult to study without wading into the morass of such phenomena as "habituation", "attention", and "novelty". Perhaps they can be studied in a controlled way in behaving animals. It is intriguing that they are found intermixed with the 80% of the units that have quite stimulus-bound response properties. CHOW (1970) has speculated on the functional significance of a system of plastic cells in the neural axis parallel to the cells having rigid input and output circuits in the visual pathways. Such intermixing of neurons with functionally different properties could lead to revision of rigid concepts of "projection areas" and "association areas" in the cerebral cortex.

One obvious topic for further work is the systematic investigation of single units in cortical areas other than the primary auditory cortex. Areas other than AI are well delineated in the cat, yet rather little unit exploration of them has been reported. Unit studies of auditory cortex of animals other than the cat are very much in order. Monkeys are currently receiving attention in several laboratories (BRUGGE and MERZENICH, 1971; WOLLBERG and NEWMAN, 1972; MILLER, 1971).

A new experimental approach which is now being used in several laboratories (EVARTS, 1968; WURTZ, 1969; MILLER, 1971) is the combining of behavioral studies with unit recordings. Such studies may shed light on the questions of effect of previous experience and of the significance of the stimulus on unit responses. This approach may allow systematic studies of the mechanisms of short term memory of acoustic stimuli. These questions are of utmost importance for understanding the ever changing behavior of an animal. Associated with the problem of plasticity of unit properties in the mature brain is the problem of cortical development in the newborn and young animal. Scientists working on the

[1] In one study using intracellular techniques (DE RIBAUPIERRE *et al.*, 1972a) "idle cells" which are very likely glia were encountered. Indications were that their very stable resting potential is virtually unmoved during acoustic stimulation.

visual system have done interesting and clinically important work using the single unit technique (WIESEL and HUBEL, 1965; BLAKEMORE and COOPER, 1970), yet, little comparable work has been done on the auditory cortex.

Very little is known today about the neural mechanisms responsible for unit coding properties. Intracellular recordings indicate that most auditory cortical units (thin-spike units are an exception) are "held down" by inhibition and respond only under circumstances in which the inhibitory drive is removed and/or overcome by an excitatory drive (DE RIBAUPIERRE *et al.*, 1972a). More extensive studies of this question are clearly needed. Such studies could be combined with histological studies in which the same cells are stained and identified. At present, we have very little information about possible correlation between cell morphology and coding properties.

In this chapter and elsewhere, we have stressed the diversity of the neural coding properties of the units in the auditory cortex. This diversity makes the cortex a difficult region to study and makes it especially unattractive to those who like their science in neat packages. Let us hope that new studies, new techniques, and new findings will move us out of what will someday be called the early phases (or even the dark ages) of neuroscientific study of the cortex.

Acknowledgement

This chapter was written at Hadassah Medical School of the Hebrew University during Dr. GOLDSTEIN's Guggenheim Fellowship.

References

ABELES, M., GOLDSTEIN, M. H., JR.: Functional architecture in cat primary auditory cortex: Columnar organization and organization according to depth. J. Neurophysiol. **33**, 172—187 (1970).

ABELES, M., GOLDSTEIN, M. H., JR.: Responses of single units in the primary auditory cortex of the cat to tone and tone pairs. Brain Res. **42**, 337—352 (1972).

ADES, H. W.: Central auditory mechanisms. In: FIELD, J. (Ed.): Handbook of physiology Sect. I, Vol. I, Ch. 24. Amer. Physiol. Soc., Washington 1959.

BLAKEMORE, C., COOPER, G. F.: Development of the brain depends on the visual environment. Nature (Lond.) **228**, 477—478 (1970).

BONIN, G. VON, MEHLER, W. R.: On columnar arrangement of nerve cells in cerebral cortex. Brain Res. **27**, 1—9 (1971).

BRUGGE, J. F., DUBROVSKY, N. A., AITKIN, L. M., ANDERSON, D. J.: Sensitivity of single neurons in the auditory cortex of cat to binaural tone stimulation; effects of varying interaural time and intensity. J. Neurophysiol. **32**, 1005—1023 (1969).

BRUGGE, J. F., MERZENICH, M. M.: Representation of frequency in auditory cortex in the macaque monkey. In: SACHS, M. B. (Ed.): The physiology of the auditory system. Baltimore: National Educational Consultants 1971.

CHOW, K. L.: Integrative functions of the thalamocortical visual system of cat. In: PRIBRAM, K. H. (Ed.): Biology of memory, p. 273—292. New York: Academic Press 1970.

ERULKAR, S. D., ROSE, J. E., DAVIES, P. W.: Single unit activity in the auditory cortex of the cat. Bull. Johns Hopk. Hosp. **99**, 55—86 (1956).

EVANS, E. F., ROSS, H. F., WHITFIELD, I. C.: The spatial distribution of unit characteristic frequency in the primary auditory cortex of the cat. J. Physiol. (Lond.) **171**, 476—493 (1965).

EVANS, E. F., WHITFIELD, I. C.: Classification of unit responses in the auditory cortex of the unanesthetized and unrestrained cat. J. Physiol. (Lond.) **171**, 476—493 (1964).

EVARTS, E. V.: Relation of pyramidal tract activity to force exerted during voluntary movement. J. Neurophysiol. **31**, 14—27 (1968).

FUNKENSTEIN, H. H., NELSON, P. G., WINTER, P., WOLLBERG, Z., NEWMAN, J. D.: Unit responses in auditory cortex of awake squirrel monkeys to vocal stimulation. In: SACHS, M. B. (Ed.): The physiology of the auditory system. Baltimore: National Educational Consultants 1971.
FUNKENSTEIN, H. H., WINTER, P.: Responses to acoustic stimuli of units in the auditory cortex of awake squirrel monkeys. Exp. Brain Res. 18, 464—488 (1973).
GERSTEIN, G. L., KIANG, N. Y-S.: Responses of single units in the auditory cortex. Exp. Neurol. **10**, 1—18 (1964).
GOLDSTEIN, M. H., JR., ABELES, M., DALY, R. L., MCINTOSH, J.: Functional architecture in cat primary auditory cortex: Tonotopic organization. J. Neurophysiol. **33**, 188—197 (1970).
GOLDSTEIN, M. H., JR., HALL, J. L. II, BUTTERFIELD, B. O.: Single-unit activity in the primary auditory cortex of unanesthetized cats. J. acoust. Soc. Amer. **43**, 444—455 (1968).
GOLDSTEIN, M. H., JR., RIBAUPIERRE, F. DE, YENI-KOMSHIAN, G.: Cortical coding of periodicity pitch. In: SACHS, M. B., (Ed.): The physiology of the auditory system. Baltimore: National Educational Consultants 1971.
HALL, J. L. II, GOLDSTEIN, M. H., JR.: Representation of binaural stimuli by single units in primary auditory cortex of unanesthetized cats. J. acoust. Soc. Amer. **43**, 456—461 (1968).
HIND, J. E., ROSE, J. E., DAVIES, P. W., WOOLSEY, C. N., BENJAMIN, R. M., WELKER, W. I., THOMPSON, R. F.: Unit activity in the auditory cortex. In: RASMUSSEN, G. L., WINDLE, W. (Eds.): Neural mechanisms of the auditory and vestibular systems, p. 201—210. Springfield (Ill.): Ch. C. Thomas 1960.
HUBEL, D. H., HENSON, C. O., RUPERT, A., GALAMBOS, R.: "Attention" units in the auditory cortex. Science **129**, 1279—1280 (1959).
HUBEL, D. H., WIESEL, T. N.: Receptive fields, binocular interaction and functional architecture in the cat's visual cortex. J. Physiol. (Lond.) **160**, 106—154 (1962).
LORENTE DE NÓ, R.: The cerebral cortex: Architecture, intracortical connections and motor projection. In: FULTON, J. F. (Ed.): Physiology of the nervous system, Chapter 15. London: Oxford University Press 1938.
MILLER, JOSEF M.: Single unit discharges in behaving monkeys. In: SACHS, M. B. (Ed.): The physiology of the auditory system. Baltimore: National Educational Consultants 1971.
MOUNTCASTLE, V. B.: Modality and topographic properties of single neurons in cat's somatic sensory cortex. J. Neurophysiol. **20**, 408—434 (1957).
OONISHI, S., KATSUKI, Y.: Functional organization and integrative mechanism in the auditory cortex of the cat. Jap. J. Physiol. **15**, 342—365 (1965).
ROSE, J. E., BRUGGE, J. F., ANDERSON, D. J., HIND, J. E.: Phase-locked response to low-frequency tones in single auditory nerve fibers of the squirrel monkey. J. Neurophysiol. **30**, 769—793 (1967).
RIBAUPIERRE, F. DE, GOLDSTEIN, M. H., JR., YENI-KOMSHIAN, G.: Intracellular study of the cat's primary auditory cortex. Brain Res. **48**, 185—204 (1972a).
RIBAUPIERRE, F. DE, GOLDSTEIN, M. H., JR., YENI-KOMSHIAN, G.: Cortical coding of repetitive acoustic pulses. Brain Res. **48**, 205—225 (1972b).
ROSE, J. E., GROSS, N. B., GEISLER, C. D., HIND, J. E.: Some neural mechanisms in the inferior colliculus of the cat which may be relevant to localization of a sound source. J. Neurophysiol. **29**, 288—314 (1966).
SUGA, N.: Responses of cortical auditory neurones to frequency modulated sounds in echo-locating bats. Nature (Lond.) **206**, 890—891 (1965a).
SUGA, N.: Functional properties of auditory neurones in the cortex of echo-locating bats. J. Physiol. (Lond.) **181**, 671—700 (1965b).
WHITFIELD, I. C., EVANS, E. F.: Responses of auditory cortical neurons to stimuli of changing frequency. J. Neurophysiol. **28**, 655—672 (1965).
WIESEL, T. N., HUBEL, D. H.: Binocular interaction in striate cortex of kittens reared with artificial squint. J. Neurophysiol. **28**, 1041—1059 (1965).
WINTER, P., PLOOG, D., LATTA, J.: Vocal repertoire of the squirrel monkey, its analysis and significance. Exp. Brain Res. **1**, 359—384 (1966).
WOLLBERG, Z., NEWMAN, J. D.: Auditory cortex of squirrel monkey: Response patterns of single cells to species-specific vocalizations. Science **175**, 212—214 (1972).
WURTZ, R. H.: Visual receptive fields of striate neurons in awake monkeys. J. Neurophysiol. **32**, 727—742 (1969).

Chapter 5

Physiological Studies of the Efferent Recurrent Auditory System

JOHN E. DESMEDT, Brussels (Belgium)*

With 11 Figures

Contents

I. Introduction 219
II. Experiments on Conscious Animals 221
III. Reticular Versus Extra-Reticular Influences on Acoustic Input 224
IV. The Olivo-Cochlear System of Mammals 225
A. Crossed and Uncrossed OCB 225
B. Experimental Conditions for OCB Stimulation 227
C. Titration of OCB Effects 228
D. Parameters of OCB Effects 231
E. Effects of COCB Stimulations on the Acoustic Pathway 233
F. Ionic Mechanisms of Efferent COCB in the Inner Ear 234
G. Neurochemical Mechanism of OCB Axons 234
H. Mode of Action of OCB Axons in the Inner Ear 236
I. Psychophysiological Significance of the OCB System 238
V. CERACS 238
A. CERACS Driving of OCB Neurons 239
B. Efferent Effects on Cochlear Nucleus Neurons 240
C. The Functional Significance of CERACS 241
References 242

I. Introduction

Physiological studies of the auditory system have been mostly concerned with the analysis of responses to sound stimulation in the afferent pathway from the cochlea to the temporal cortex. However, scattered anatomical data have pointed for many years to the presence of neurons with centrifugal axons which run in caudal direction to terminate in nuclei of the classical auditory pathway (HELD, 1893; COLLIER and BUZZARD, 1901; LORENTE DE NÓ, 1933; METTLER, 1935; TASIRO, 1940). The problem raised by such centrifugal axons unrelated to the motor systems and presumably subserving some (unknown) controlling function in the processing of sensory information was dramatized when GRANT RASMUSSEN

* The research reported in this paper has been supported by the National Institutes of Health, by the Office of Naval Research, by the Fonds de la Recherche Scientifique Médicale and by the Fonds National de la Recherche Scientifique.

(1946, 1953) presented clear anatomical evidence of the sizeable olivo-cochlear system whose axons go out centrifugally in the eighth nerve and reach to the base of hair cells in the organ of Corti (see volume V/1, Chapters 8 and 11). The significance of these axons remained obscure for several years until GALAMBOS (1956) showed that faradization of the efferent bundle reduced the sound-evoked activity in the auditory nerve. That a genuine inhibitory effect was indeed exerted on the acoustic input was further established by a detailed parametric study (DESMEDT, 1962) and by the recording from single auditory axons (FEX, 1962; WIEDERHOLD and KIANG, 1970). The olivo-cochlear inhibition was also found to be reduced or suppressed by small doses of the alcaloids strychnine and brucine (DESMEDT and MONACO, 1961, 1962), which are now considered as specific antagonists of the glycinergic post-synaptic inhibitions of the mammalian spinal cord (ECCLES, 1964; CURTIS, DUGGAN, and JOHNSTON, 1971; TEN BRUGGENCATE and SONNHOF, 1972). The olivo-cochlear system offers a rather rare example of an homogeneous bundle of long inhibitory axons which can be selectively stimulated electrically, destroyed or brought under control in the mammalian central nervous system and there are no interneurons interposed between these axons and the first auditory synapse in the inner ear. Indeed the Rasmussen olivo-cochlear neurons served as a paradigm of centrifugal neurons in sensory systems, and their disclosure encouraged the search for other centrifugal mechanisms in the various sensory modalities. It must however be recognized at the outset that, in spite of current studies (see below), the actual significance of these neurons for auditory perceptual processes is still far from clear.

Although the best known part of the recurrent auditory efferent system, the olivo-cochlear bundle tells only a small part of the story. The investigation of the other descending neurons related to the auditory pathway on the brainstem is far less advanced and many problems remain unanswered. Anatomical studies with the Nauta and other techniques have identified at least two sets of corticofugal axons in experiments involving lesions of the temporal cortex. One set originates mainly from the primary auditory cortex and the other arises mainly from the ventral temporo-insular cortex (in the cat) or the superior temporal gyrus (in the monkey) (DESMEDT and MECHELSE, 1959a; WALTHER and RASMUSSEN, 1960; DESMEDT, 1960; RASMUSSEN and DESMEDT, unpublished). These fibres run along the classical ascending auditory pathway and appear rather loosely organized and mixed with other fibre systems in the lateral brainstem. Furthermore the corticofugal axons do not reach more caudal than the inferior colliculus and the dorsal nucleus of the lateral lemniscus. At the thalamic and midbrain levels, they synapse with other neurons whose axons descend to the level of the cochlear nuclei and superior olivary complex after one or more synapses. Some of the centrifugal axons terminate on the olivo-cochlear neurons (RASMUSSEN, 1964) which can thus be considered as the outermost effector link of the centrifugal system (DESMEDT, 1962, 1968). Both anatomical (see this volume, Chapter K) and electro-anatomical (see below) studies indicate that these systems are quite specifically organized with respect to the auditory modality. They are indeed considered as distinct from MAGOUN'S non-specific reticular system and for this reason they were given the comprehensive label of Centrifugal Extra-Reticular Auditory Control System or CERACS (DESMEDT, 1962). The detailed analysis of the neuron circuits of CERACS has been progressing slowly and the difficulties involved in the detailed investigation of this complex poly-synaptic system should be appreciated.

The disclosure of central pathways terminating not in motor nuclei, but in subcortical relays of sensory systems led to the general idea that sensory information could somehow be controlled before it reaches the cerebral cortex; this aroused vivid interest among psychologists in the mid-fifties (*cf.* for example BRUNER, 1958). Bold speculations, based largely on inferential evidence, were triggered by the further suggestion that the non-specific reticular formation of the brainstem could actually be involved in the control of auditory input to the brain (HERNANDEZ-PEON, SCHERRER and JOUVET, 1956; HERNANDEZ-PEON, 1961; MAGOUN, 1958). These views have been extensively quoted in textbook for more than a decade although they proved quite disappointing and were not substantiated by more critical experiments (DESMEDT and MECHELSE, 1958; DESMEDT, 1960, 1968). The initial observations of HERNANDEZ-PEON and his colleagues failed to be confirmed (WORDEN and MARSH, 1963; WORDEN, 1966). It is now clear that the hypothesis of HERNANDEZ-PEON as to reticular system effects offers no valid short-cut to the problems of the neural substrates of selective perception; the latter should indeed be approached by more realistic, alternative models based on adequately controlled experimental data. The following sections do not survey the published data comprehensively but, instead, attempt a critical discussion of a few current issues.

II. Experiments on Conscious Animals

There is a limited span for perception at any one time and only relatively few of the sensory stimuli received from environment are perceived consciously or elicit a behavioural response. The phenomena of selective perception for auditory cues can be studied consistently in psychological experiments (see for example BROADBENT, 1958; TREISMAN, 1964) but they appear rather elusive when discussed in relation to the neural mechanisms likely to be involved. The models so far presented in this respect appear to have little relation to the actual hardware of the brain. An obvious question frequently discussed is whether messages entering the auditory pathway are dealt with in a different way when the subject's attention shifts between sensory modalities in a simple experimental paradigm. One answer to such a question was sought by implanting gross steel electrodes chronically in the cochlear nuclei and other brain structures in cats and by subsequently recording the electrical responses to sound stimuli, generally clicks, while an attempt was made to manipulate the significance of these stimuli for the conscious animal (HERNANDEZ-PEON *et al.*, 1956; GALAMBOS, SHEATZ, and VERNIER 1956). Thus, when the cat was distracted by the sight of a live mouse or by the smell of fish, the auditory responses were said to decrease for a short interval of time. When the clicks were repeated at regular intervals for several hours, a very progressive reduction in the size of the responses was noticed; when the same clicks were then associated with an unpleasant electric shock to the cat's feet in a conditioning sequence, the auditory responses rapidly increased in size (HERNAN-

DEZ-PEON, 1961). These pioneering observations can be repeated when using the same conditions (*e.g.* DESMEDT, 1960, Fig. 111; BACH-Y-RITA, BRUST-CARMONA, PENALOZO-ROJAS, and HERNANDEZ-PEON, 1961; MARSH, MCCARTHY, SHEATZ, and GALAMBOS, 1961) but their significance now appears quite different from what was thought initially.

It may be appropriate at this stage to consider the assumptions in the interpretations made by the authors cited above. The gross electrodes record the summed potentials from a population of neurons and it is considered that a response of larger (or smaller) voltage reflects an increase (or a decrease) of the neural activity at the recorded site. The latter statement roughly simplifies the actual situation and may not apply as such under certain conditions, for example, when a change in the background excitatory drive shifts the dynamic range of the intensity function for the recorded response. However when recording with due precautions at the first two or three auditory relays, clear-cut changes in the gross responses can generally be interpreted in this straightforward way provided the electrodes are carefully placed in relation to current sources and sinks in the relay (*cf.* DESMEDT, 1962, p. 1495) (see electro-anatomical evidence, below). It must be recognized that most of the changes in the records illustrated by HERNANDEZ-PEON and co-workers are frequently rather small in voltage and they appear unconvincing when the spontaneous fluctuations of such responses are taken into account (HORN, 1965).

A second point is that HERNANDEZ-PEON emphasized throughout his reports that his observations are to be interpreted in terms of variations in the amount of centrifugal blocking of the sensory input at peripheral levels of the sensory pathway during shifts of attention. A word of caution is in order here. While the observation of a definite reduction in the summed potentials of a neuron population might indeed suggest the calling into action of an inhibitory process, this could not safely be construed into a reduction of the coded auditory information actually transmitted through the recorded relay nucleus (*cf.* WORDEN, 1966; DESMEDT, 1968).

The main issue raised by the data on conscious animals is that, contrary to initial interpretations, the recorded changes in auditory potentials result, not from a neural centrifugal gating process, but rather from uncontrolled changes in the sound stimulus impinging on the basilar membrane (Table 1). The initial Hernandez-Peon experiments had been carried out with no sound-proofing of the rooms and the clicks had been delivered by an overhead speaker. Under such conditions, room acoustics result in marked variations in the sound pressure of the stimulus for different positions in the room (MARSH, WORDEN and HICKS, 1962). Furthermore, the cat with implanted electrodes assumes, during the learning session, different body and head positions which further emphasize the stimulus distortions and changes in intensity (MARSH *et al.*, 1962; WORDEN, MARSH, ABRAHAM, and WHITTLESEY, 1964; SIMMONS and BEATTY, 1964a). Thus, when attending to the clicks, the animals sit up and orient head and pinna towards the overhead speaker which increases the actual sound pressure at the eardrum. By contrast, during acoustic habituation runs, the animals tend to move progressively into a quiet lying position with drooping head and pinna. As pointed out by WORDEN (1966), these differences in head and ear positions, which had been regarded as indications of the behavioral state of the animal, were in fact producing by themselves large changes in the actual sound intensity at the ear-drum although the stimulus delivered by the speaker remained constant.

Table 1. Factors affecting the size of an auditory stimulus acting at the inner ear

1. Room acoustics
2. Head shadow and pinna orientation
3. Masking by extraneous noise
4. Masking by self-generated noise
5. Middle ear muscles contractions (-acoustic); (- non-acoustic)

These difficulties can be removed by delivering the sound through small earphones inserted into the external auditory meatus. It is then found that most of the changes in amplitude of cochlear nucleus potentials are eliminated and the responses remain fairly stable in spite of marked changes in apparent attentiveness of the animal (WORDEN and MARSH, 1963; WORDEN, 1966). Furthermore, a good case can be made for relating the remaining variations in acoustic responses to the state of contraction of the middle ear muscles (M.E.M.). The disturbing observation that potentials recorded with chronically implanted electrodes were quite sensitive to variables little related to perception, such as general body movements or other motor activities, was carefully studied by CARMEL and STARR (1963). They recorded cochlear microphonic potentials through chronically implanted electrodes in cats equipped with small earphones, with or without surgical section of the M.E.M. The microphonic potentials allow evaluation of actual sound stimulation of the inner ear and they are reduced by M.E.M. contractions. Such studies using unanesthetized animals, later extended to man, (see volume V/1, Chapter 15) disclosed a remarkably wide repertoire of activities of M.E.M., whereas the earlier studies of anesthetized animals had given the impression that only loud noises would elicit M.E.M. contractions.

Changes in cochlear microphonics occur not only because M.E.M. reflexes are elicited by sound stimuli, but they are also induced by non-acoustic factors such as yawning, swallowing, blinking, gross bodily movements, stimulation of the skin of the face and pinna, etc.... The middle ear muscle contractions elicited by other than acoustic stimuli persist in experimentally deafened animals, and they appear developmentally in kittens before the acoustic reflex (CARMEL and STARR, 1963; SIMMONS and BEATTY, 1964b). M.E.M. contractions have also been recorded before first vocalizations. The significance of such non-acoustically triggered M.E.M. contractions may be to stabilize mechanically the ossicular chain in the middle ear and to minimize self-produced sounds (VON BÉKÉSY, 1962). It is of interest that such M.E.M. responses also occur in man, allowing interesting psychophysiological experiments to be made (SALOMON and STARR, 1963; DJUPESLAND, 1964, 1965).

In contrast to extrinsic attenuations of sound outside the head, the (neural controlled) attenuations resulting from M.E.M. contractions associated, for example, with teeth clenching or bodily movements do not produce a subjective change in the auditory perception of the human subject for sounds of moderate intensity. Such perceptual stability implies the operation of some sort of central mechanism integrated with the efferent motor commands to M. E. M., a mechanism which must be of a kind similar to the one maintaining stability of visual perception during saccadic movements of the eyeball. When considering sounds of near-threshold intensity, the situation is somewhat different and M.E.M. contractions, by increasing the middle ear impedance, elevate the acoustic threshold and make such sounds inaudible. It can be suggested that we tend to remain motionless when listening to faint auditory cues just in order to keep the M.E.M. relaxed and thereby to improve our ability to detect the near-threshold sounds.

The above discussion can be summarized by saying that there is no evidence for suggesting that selective attention influences the size of gross electrical responses in the auditory nerve or cochlear nuclei. The earlier claims of HERNANDEZ-PEON and his group were based on spurious experiments which can be accounted for by variations of the acoustic stimulus arriving at the eardrum (room acoustics effects; head shadowing; pinna orientation) and by variations in the state of contraction of the middle ear muscles in relation to both acoustic and non-acoustic factors (Table 1). Animals stimulated through small earphones and with middle ear muscles cut do not exhibit changes in gross electrical responses which could be related consistently to behavioral attentiveness.

III. Reticular Versus Extra-Reticular Influences on Acoustic Input

In the elaboration of his views, HERNANDEZ-PEON emphasized that the non-specific reticular arousal system of the brainstem was responsible for sensory control (HERNANDEZ-PEON, 1961; MAGOUN, 1958). However, his evidence in relation to the auditory pathway is largely inferential and unconvincing, such as the observation that the cochlear nucleus responses to clicks in cats would increase after an injection of Pentobarbital or after extensive coagulation in the mesencephalon (HERNANDEZ-PEON, JOUVET, and SCHERRER, 1957). It can reasonably be asked whether such procedures did nothing but "reduce a tonic reticular influence" in the acoustic pathway.

In better controlled acute experiments on the cat, systematic bipolar stereotaxic stimulation elicited reductions of the click-evoked responses of the cochlear nucleus only when stimulating discrete well-localized points of the lateral brainstem (DESMEDT and MECHELSE, 1957, 1958; DESMEDT, 1960). Such points organize themselves in a centrifugal sequence which is remarkably consistent with the anatomical evidence of RASMUSSEN (1964) and of DESMEDT and MECHELSE (1959a). Moreover, in the same experiments, an identical stimulus delivered more medially in the reticular core (and producing the expected reticular arousal effects) failed to elicit any significant reduction in the auditory responses. Such experiments are only valid when carefully controlled for stimulus conditions (earphones; clear path between earphone and eardrum), after immobilization of the animal by a curare drug (to exclude masking by movement) and after surgical section of the middle ear muscles (M.E.M.). When these muscles are left intact, a strong reticular faradization will elicit contractions of M.E.M. with associated reductions in sound evoked responses (HUGELIN, DUMONT, and PAILLAS, 1960). It is not certain whether *moderate* changes in the level of vigilance would also elicit changes in middle ear impedance through M.E.M. contractions. The potential significance of such reticular effects for sensory processing appears in any case limited when considering the wide repertoire of non-acoustic M.E.M. activities which can go on without changing the perceived sounds.

In any case the M.E.M. effects elicited by strong reticular activations should be distinguished from genuine centrifugal neural gating involving, not the sound transmission in the middle ear, but the coded signals at the auditory nerve, cochlear nucleus, or other levels of the pathways (DESMEDT, 1968).

It is currently recognized that the brainstem reticular system is essential to maintain vigilance and critical reactivity (DELAFRESNAYE, 1954). While this arousal system thus regulates the overall level of responsiveness of the nervous system, including that of the cerebral cortex (BREMER and STOUPEL, 1959; DUMONT and DELL, 1960; DESMEDT, 1960, p. 159), this by no means implies that it would also control the flow of information in the specific sensory pathways (DESMEDT, 1968). It should also be made clear that the possibility of some discrete interactions between reticular neurons and neurons belonging to CERACS should not be excluded a priori and actually deserve investigation. The point argued in the above discussion is that the non-specific reticular arousal system as a whole does not elicit any significant centrifugal inhibitory effect at auditory relays. The problem whether the reticular arousal system may trigger centrifugal mechanisms in other non-auditory sensory modalities is not discussed in this paper.

IV. The Olivo-Cochlear System of Mammals

Physiological studies of the Rasmussen olivo-cochlear bundle (OCB) have been carried out mostly on the cat for which extensive morphological data are available (RASMUSSEN, 1946, 1960; SPOENDLIN and GACEK, 1963; SPOENDLIN, 1966, 1972). The OCB effects on the ear can be conveniently demonstrated by electrical stimulation of the bundle in the medulla and by recording the changes thereby induced in the sound evoked responses (Fig. 1A).

A. Crossed and Uncrossed OCB

RASMUSSEN (1960) showed the OCB to contain about 500 myelinated axons of 3–4 μ diameter which originate from two separate groups. The cell bodies of the crossed OCB neurones (COCB) (about $^3/_4$ of the OCB neurons) are located close to the medial accessory superior olive on the opposite side; their axons course dorsalwards and decussate between the facial nerve genua, close to the floor of the fourth ventricle where carefully placed bipolar needle electrodes will activate them rather selectively without spurious stimulation of second-order auditory axons or other parts of the auditory pathway (*cf.* DESMEDT, 1962, p. 1490). The COCB axons then traverse the facial nerve root and join the vestibular nerve en route to the cochlea (Fig. 1E). Stimulation of COCB (floor of 4th ventricle) on the midline elicits a simultaneous reduction of the auditory nerve responses elicited by clicks presented to both ears (Fig. 1 C). Cell bodies of the uncrossed OCB neurons (UOCB) are located close to the main S-shaped nucleus of the superior olive on the same side and their axons move dorsalwards to join the COCB axons at the ventral part of the vestibular nucleus (Fig. 1 E). Electrical stimulation at this point elicits a

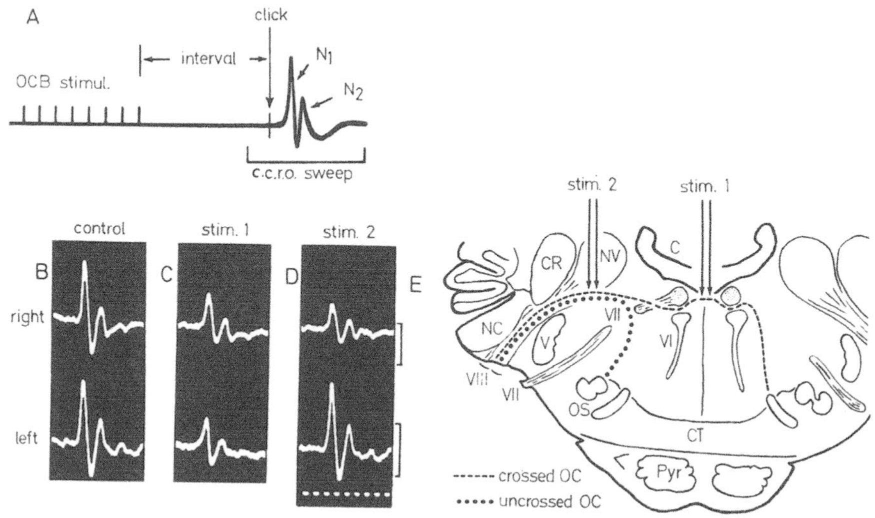

Fig. 1 A—E. The electric activation of the olivo-cochlear system in cat. (A) Sketch of the stimulation program; cathode ray oscilloscope discloses the responses to the sound which is either preceded or not by a train of electrical pulses to the OCB. (B) control to binaural click responses simultaneously recorded from the right and left round window membranes. (C) same with preceding stimulation by 40 pulses at 400/sec at point 1 in E. (D) same with stimulation at point 2. Calibrations, 200 μV and 0.5 msec. (E) drawing based on several frontal sections through the brainstem showing the approximate course of the COCB and UOCB according to Rasmussen's evidence. Abbreviations: *CN* cochlear nucleus; *CR* corpus restiforme; *NV* vestibular nucleus; *C* cerebellum; *OS* superior olive; *CT* trapezoid body. (From Desmedt and LaGrutta, 1963, Fig. 1)

stronger but unilateral reduction of the neural responses elicited by click stimulation of the ear (Fig. 1 D). Whereas it is easy to activate COCB axons selectively, the same is not true for UOCB axons which have been much less frequently studied for this reason. To study the effects of electrical stimulation of UOCB, one must cut the COCB at the decussation in a prior surgical procedure (with sterile precautions) and then make the UOCB experiment proper 4—10 weeks later when the COCB axons have had time to degenerate (Desmedt and LaGrutta, 1963). Stimulation on the midline is then ineffective while lateral stimulation through appropriately placed electrodes causes ipsilateral reduction of the click elicited neural responses; the reduction is about $^1/_4$ that produced by stimulation of COCB. It is of interest that UOCB effects do not appear to be associated with an increase of the cochlear microphonic (CM) potentials (see below) and that they are reduced or eliminated by strychnine and brucine, but not by other convulsants nor by the anticholinergic hyoscine or dihydro-β-erythroidine (Desmedt and LaGrutta, 1963). Such evidence suggests the neurochemical identity of the crossed and uncrossed OCB components and raises the question as to why the OCB system should have a dual anatomical organization in the medulla. As for the mode of termination it has been suggested that the UOCB may be related to the row of inner hair cells in the rat and chinchilla (IURATO, 1964; IURATO, SMITH, ELDREDGE, and HENDERSON, 1968) but the segregation of terminals of the COCB and UOCB respectively in the inner ear is apparently not a complete one (SMITH and TAKASAKA, 1971). There are also data for the cat (SPOENDLIN, 1966). The interesting question of the place of termination of COCB and UCOB fibers is still unsettled.

B. Experimental Conditions for OCB Stimulation

Experiments are generally carried out with cats or guinea pigs anesthetized with Pentobarbital or α-chloralose, but similar results are obtained in unanesthetized cats with a spinal transection at the level of C_2 (DESMEDT, 1962). Well-placed bipolar steel electrodes can elicit maximal OCB effects with electrical pulses of 30–50 μsec duration and about 100 μA strength. Stereotaxic orientation of the electrodes inserted through a small skull opening leaving the dura intact is less traumatic and more reliable than the surgical removal of the cerebellum and actual exposure of the floor of the fourth ventricle. The animals should be kept immobilized by curarizing drugs (gallamine or flaxedil, D-tubocurarine or decamethonium) which do not affect the results with usual dosages. The M.E.M. should

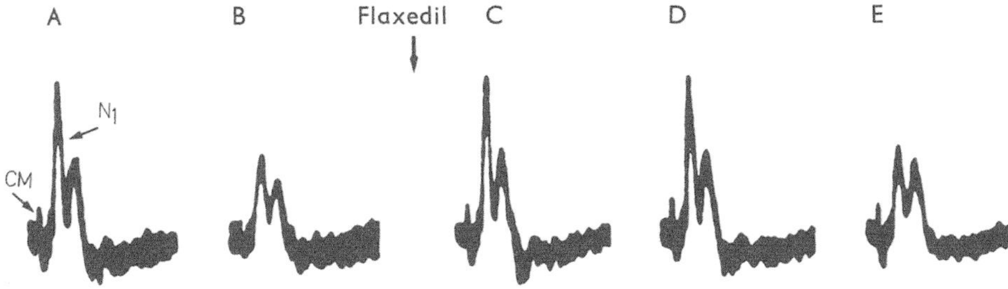

Fig. 2A—E. Effects of electrical stimulation at the midline COCB decussation before and after curarization in a cat with intact middle ear muscles. Electrical responses to clicks recorded from the round window membrane. (A) control. (B) same with click preceded by a single shock (stapedius muscle contraction effect). Before the next control in (C), a full curarizing dose of flaxedil (gallamine) was injected intravenously. (D) the single shock before the click no longer changes the response. (E) repetitive electrical stimulation at the same point preceding the click elicits a potentiation of CM and reduction of neural potentials

be cut surgically to exclude any change in middle ear impedance during the course of the experiment when the level of curarization may fluctuate. Stimulation of the COCB on the midline can easily diffuse to the nearby facial nerve genua and elicit a contraction of the stapedius muscle. With intact M.E.M. and before curarization, a single electric pulse can reduce the response to click (Fig. 2A, B) because the stapedius contraction interferes with the delivery of the sound stimulus, but after curarization this effect is removed (D) while repetitive stimulation at the same point elicits a pure COCB effect (E) with no reduction, but actually an increase, in the size of the CM potential.

C. Titration of OCB Effects

Maximal activation of OCB (for example, with 40 electric pulses delivered at 400/sec) elicits both an increase in the voltage of CM and a decrease in the voltage of the N_1 and N_2, the compound potentials that may be recorded from the auditory nerve in response to a click (Fig. 3A, B). When the click intensity is no more than about 30 dB above threshold, the neural response is considerably reduced or even eliminated. For stronger clicks, the percentage reduction is much less. Consistent titration of the OCB effects can be achieved when plotting the intensity function of N_1 with and without OCB activation; the latter simply shifts the intensity function by a fairly constant horizontal distance (DESMEDT, 1962) (Fig. 3D). The COCB effect is then estimated not as a variable percentage reduction of neural responses depending on the click intensity which happens to be used for testing but as a consistent inhibitory effect of so many equivalent dB which can be matched by a roughly constant dB reduction of sound intensity. The method is valid within a range of about 40–50 dB above threshold (DESMEDT, 1962; WIEDERHOLD and PEAKE, 1966; WIEDERHOLD, 1970). The maximal COCB inhibition recorded for gross auditory nerve responses in the cat ranges from a −12 to a −26 equivalent dB (DESMEDT, 1962; WIEDERHOLD and PEAKE, 1966). When recording the spike discharges of single auditory axons to pure tones at their characteristic frequency for the axon, COCB activation elicits only pure inhibitory effects, without any evidence that auditory axons would be activated by COCB (FEX, 1962; WIEDERHOLD and KIANG, 1970). The COCB inhibitory effect on single axons can be quite strong; it ranges from −1 to −25 equivalent dB, with a maximum potency for axons with characteristic frequency between 3 and 10 kc (Fig. 4); the maximum COCB inhibitory effect does not appear to be correlated with the axon acoustic threshold (WIEDERHOLD, 1970). A similar range of equivalent dB reductions of neural responses was recorded with COCB activation in the guinea pig, both for N_1 (KONISHI and SLEPIAN, 1971) and for single auditory axons activated by pure tones (TEAS, KONISHI, and NIELSEN, 1972).

The smaller COCB effects on the gross N_1 response at high click levels (DESMEDT, 1962; SOHMER, 1965; WIEDERHOLD and PEAKE, 1966) can be explained by the apparent reduction in inhibitory effect noticed by WIEDERHOLD (1970) for single auditory axons when the acoustic stimulus is increased to such an extent that the axon discharge rate has saturated.

The repetitive electrical activation of the uncrossed OCB in cats only yields maximum reductions of N_1 equivalent to −6 or −7 equivalent dB (DESMEDT and LAGRUTTA, 1963).

The equivalent dB method provides a consistent titration of the strength of efferent gating and it is also meaningful since the gross inhibitory effect is expressed in terms of an equivalent change in acoustic input. The use of such a method does of course not imply that the natural function of efferent pathways is only, if at all, to elicit gross shifts in the intensity function of sound evoked responses by massive discharges such as are elicited by the faradization of the

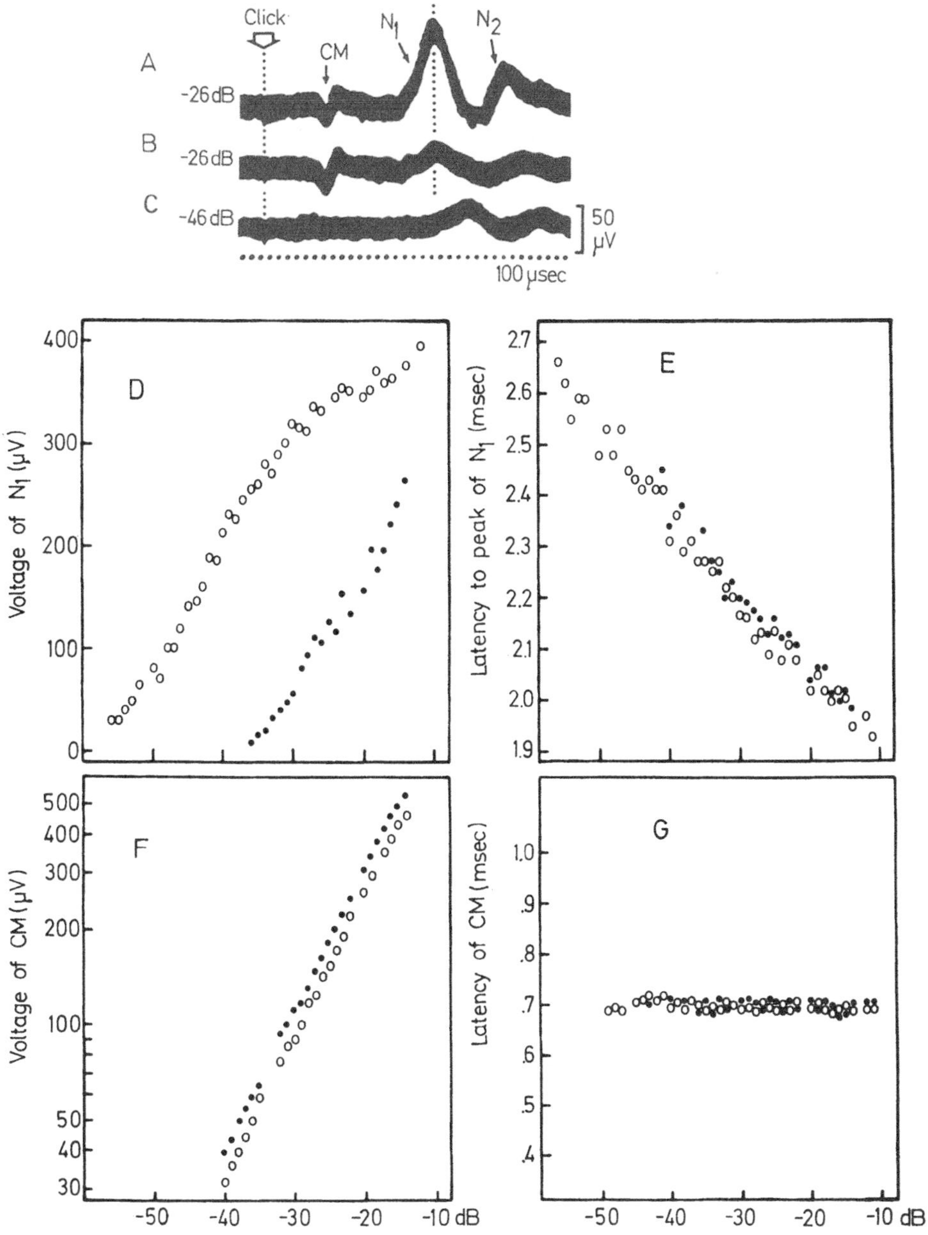

Fig. 3A—G. Effects of COCB repetitive stimulation in the cat paralyzed with flaxedil (gallamine) and with middle ear muscles cut. (A—C) electrical responses to click recorded from the round window membrane, showing the CM and neural components. (B) a maximal COCB stimulation preceding the click potentiates CM and decreases the neural potentials without increasing the peak latency of N_1. (C) response to a click attenuated by 20 dB without COCB stimulation which matches N_1 in (B) for size, but presents an increased latency. (D—G) diagram of the voltage (D, F) of latency (E, G) of N_1 (D, E) or CM (F, G) responses to clicks of relative intensity indicated on the abscissa in a single experiment on cat. Circles, without COCB activation. Dots, with preceding maximal COCB activation. The COCB effects in equivalent dB can be titrated by noticing the horizontal shift of the intensity functions in (D) and (F). The latencies are not significantly affected by COCB activation (E, G) (From DESMEDT, LAGRUTTA, and LAGRUTTA, 1971, Fig. 1 and 2)

bundle. The method allows the experimenter to evaluate pharmacological and other effects and also to compare data in different preparations and/or different animal species; this would not be possible by estimating percentage changes (DESMEDT, 1962). It should be stressed that the method only considers one feature at a time and that the matching of an OCB-conditioned response to the response to sound alone by no means implies that their respective response characteristics

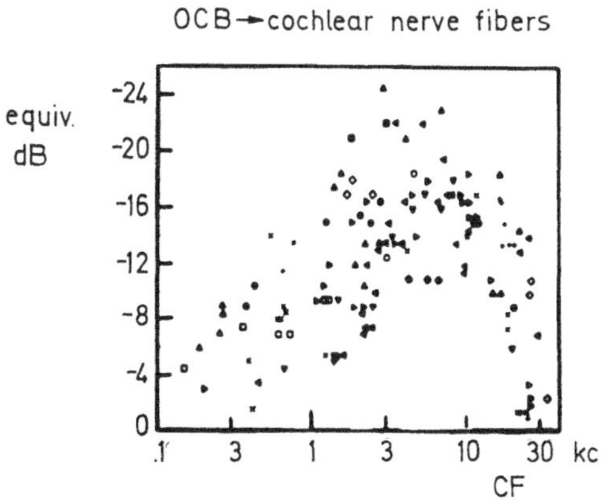

Fig. 4. Scatter plot of COCB inhibitory effect expressed in equivalent dB (ordinate) and the auditory nerve fibre characteristic frequency (abscissa) for a number of cats studied by WIEDERHOLD. (From WIEDERHOLD, 1970, Fig. 12A)

would be identical. In fact, they are not. For example, the response to a click attenuated by −20 dB, in Fig. 3C can be matched for the voltage of N_1 with the COCB-conditioned response in B, but it differs in two significant features: CM is very much smaller and the peak latency of N_1 is definitely longer in C. In any matching procedure, one should thus consider successively the different features of a response which can each be titrated separately by the equivalent dB method (Fig. 3D, F).

The virtual lack of latency increase of N_1 during COCB inhibition (Fig. 3B) is further documented in Fig. 3E for a series of click intensities; this may be significant for acoustic coding. N_1 represents the modal activity of high C.F. (characteristic frequency) units from the basal turn of the cochlea (KIANG, WATANABE, THOMAS, and CLARK, 1965). Since the ability to localize a brief sound in space involves interaction of inputs from the two ears, probably at neurons of the medial superior olive (GALAMBOS, SCHWARTZKOPFF and RUPERT, 1959; HALL, 1965; RUPERT, MOUSHEGIAN, and WHITCOMB, 1966) it is interesting that COCB activity can reduce the probability of firing in high C.F. auditory axons without significantly increasing their latency of discharge thereby preserving the binaural time interaction and perhaps allowing OCB-induced gating to be differentiated by the nervous system from external or M.E.M. attenuations of the sound (DESMEDT, LAGRUTTA and LAGRUTTA, 1971).

The increase in CM by COCB activation is a quite consistent finding, both for the CM evoked by click (Fig. 3B) and by tone pips (FEX, 1959; DESMEDT and

MONACO, 1961; FEX, 1962; DESMEDT, 1962; WIEDERHOLD and PEAKE, 1966). When estimated from the horizontal displacement of the CM intensity function, the effect is rather small and it does not exceed +3 equivalent dB (Fig. 3F). Observations have been made in the cat (DESMEDT and MONACO, 1961) and in the guinea pig (KONISHI and SLEPIAN, 1971).

When recording with a micropipette inserted into the scala media of the basal turn of the cochlea, the positive endocochlear potential of +70 to +90 mV is recorded (TASAKI, 1957; VON BÉKÉSY, 1960). A maximal COCB activation elicits a negative shift of up to 3 mV of this endocochlear potential, both in the cat (FEX, 1967; DESMEDT and ROBERTSON, 1974) and in the guinea pig (KONISHI and SLEPIAN, 1971). The time course of this negative shift is quite consistent with that recorded for COCB effects on either the neural N_1 response to click of the CM potential. The negative shift of the cochlear potential has a latency of about 10 msec, it increases with the number of electrical pulses delivered to the COCB up to about 40, and it dissipates thereafter with a time course corresponding rather closely to that of the dissipation of N_1 inhibition after a COCB activation (DESMEDT and ROBERTSON, 1975). The negative shift in the scala media must be related to an increased current flow through the outer hair cells when the membrane conductance at their base is increased by transmitter released from active COCB terminals (see below).

D. Parameters of OCB Effects

When using N_1 or CM responses as a test, single electric pulses to the OCB do not produce any detectable change. Minimal effects appear when at least 3 shocks are delivered at a frequency above 40/sec and maximal effects are recorded for 20–40 shocks at 300–400/sec (cat; DESMEDT, 1962) (Fig. 5). Frequencies below 40/sec have little effect and, for a given number of shocks, the COCB effects increase

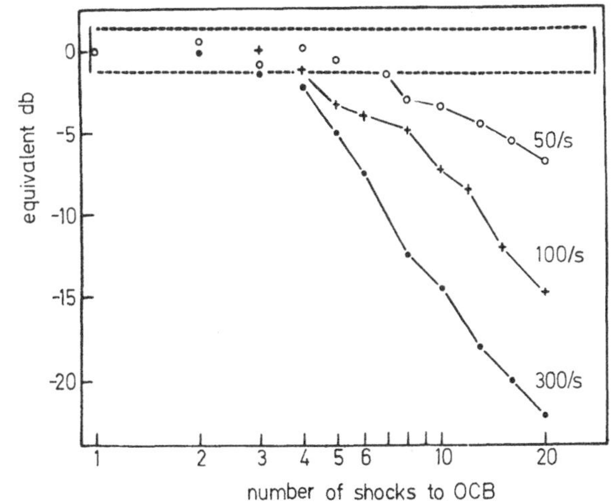

Fig. 5. Effect of frequency and number of electrical pulses delivered to COCB on the N_1 inhibition achieved in the cat. The latter is expressed in equivalent dB from the corresponding intensity function. (From DESMEDT, 1962, Fig. 7B)

with the frequency from 40/sec to 400/sec. Frequencies above 600/sec are less effective. In the guinea pig, the optimal frequency is also 300–400/sec (KONISHI and SLEPIAN, 1971). When recording from single auditory axons activated at their C.F. by pure tones, the maximal COCB inhibitory effects are also observed for stimulations at 300–400/sec. It is possible to average a number of PST histograms synchronized with the individual COCB shocks in the brain (WIEDERHOLD and KIANG, 1970). For COCB activations at 30–60/sec, each shock is then seen to elicit a consistent transient reduction in the number of auditory axon spikes per bin; this effect lasts somewhat longer than 10 msec and it is thus not surprising that the effects of each successive shock merge together for stimulations at 100/sec or more. The averaging of PST histograms affords a very sensitive method for detecting minimal COCB effects and it indicates that single COCB shocks can significantly depress the auditory axon response to sound. These important experiments also confirm that strong centrifugal gating requires the summation of repetitive COCB volleys at frequencies well above 100/sec (WIEDERHOLD and KIANG, 1970).

When the activation of COCB is prolonged, the inhibitory effects can be maintained at a fairly steady level for many minutes (WIEDERHOLD and KIANG, 1970).

When the activation lasts for about 20–100 sec, the effects dissipate as a first order reaction whose exponential time constant varies between 90 and 200 msec and there is an indication that the time constant of dissipation increases with the number of shocks delivered (DESMEDT, 1962) (Fig. 6). This phenomenon could result from accumulation of inhibitory transmitter released at active COCB terminals by the increased number of volleys. In the cat, the time course of dissi-

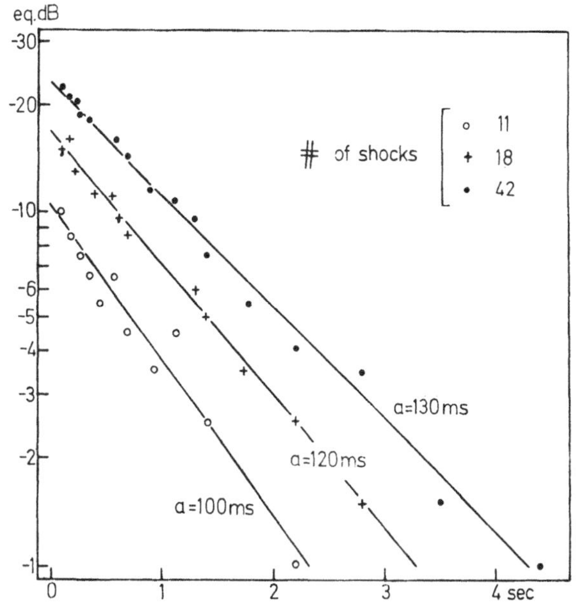

Fig. 6. Dissipation of COCB inhibition of N_1 after delivering a train of 11 (circles), 18 (crosses) or 42 (dots) electrical pulses to COCB before the testing click. Abscissa, interval between end of train and click. Ordinate, inhibition in equivalent dB, on a log scale. The calculated exponential constants of the regression lines are indicated (From DESMEDT, 1962, Fig. 9A)

pation of COCB effects on N_1 and CM respectively is consistent in any one experiment (DESMEDT and MONACO, 1961).

E. Effects of COCB Stimulations on the Acoustic Pathway

All the acoustic information reaching the nervous system is funneled through the auditory axons. The centrifugal inhibition achieved by OCB should thus regulate admission of signals to the brain provided it is genuinely efficient and wide-

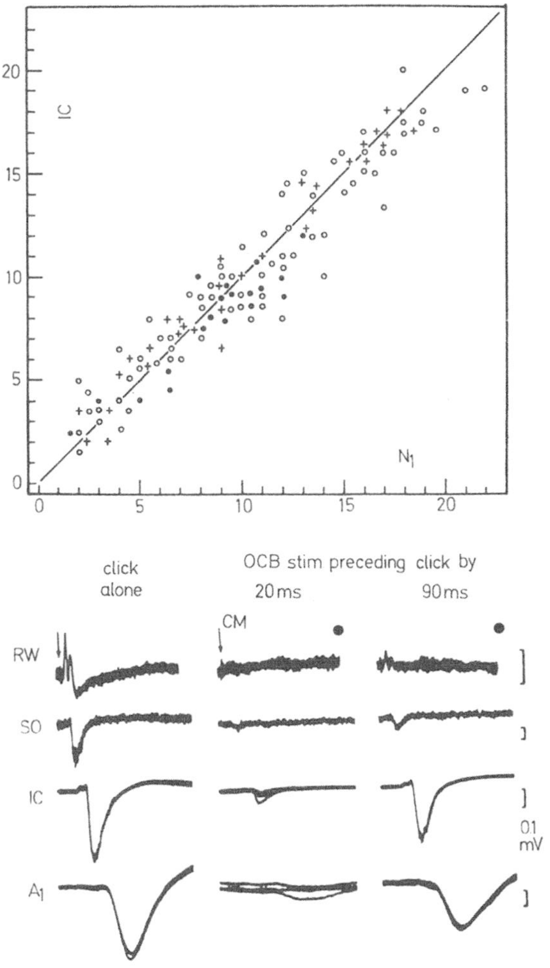

Fig. 7. Effect of COCB activation on the responses to click in the afferent auditory pathway. Responses are recorded simultaneously at the round window (*RW*), principal S-shaped superior olive (*SO*), inferior colliculus (*IC*) and primary auditory cortex (A_1). The COCB stimulation (42 shocks at 400/sec) preceded the testing click by 20 or 90 msec in the two examples shown. The diagram on the right side related the simultaneously recorded reductions in equivalent dB of N_1 (abscissa) and inferior colliculus (ordinate) responses for various COCB activations in cats. Results are pooled from experiments performed either under pentobarbital (crosses), α-chloralose (circles), or without anesthesia but high spinal transection (dots). The relation is fairly proportional, along the 45° line (From DESMEDT. 1962, Fig. 13 and 16B)

spread among the auditory axons. This problem was investigated for the COCB component which can be activated selectively at its decussation without spurious stimulation of the auditory pathway (DESMEDT, 1962). The gross responses elicited by sound in different relays were simultaneously recorded and the changes induced by COCB activations with various parametric sets were estimated in equivalent dB from the corresponding intensity functions of each relay. The findings can be summarized by saying that, for any activation of COCB, the reduction in equivalent dB at the cochlear nucleus superior olive, inferior colliculus, and primary auditory cortex was similar to that of the auditory nerve N_1 response. Figure 7 illustrates sample records of click responses simultaneously recorded at 4 levels of the pathway; the diagram also compares equivalent dB changes at the inferior colliculus and auditory nerve. These data substantiate the view that the COCB can effectively control the acoustic input. This of course tells only part of the story since the signals entering the auditory pathway will meet with other gates also operated by CERACS and moreover, the CERACS operations are presumably more differentiated than the massive volleys triggered by the electrical stimulations used in these experiments.

F. Ionic Mechanism of Efferent COCB in the Inner Ear

The presence of a synaptic space (with no membrane fusion or nexus) between the OCB terminals and the auditory dendrites or the base of hair cells points to a neurochemical synapse. In recent experiments the scala tympani of the cat's cochlea was perfused with low-chloride solutions. When the chloride was replaced by the larger-diameter gluconate or sulphate anions, the COCB-induced inhibition of neural responses was reduced or eliminated reversibly, provided that the rate of perfusion was adequate to reduce the chloride concentration below about 80 mM (as estimated by a chloride sensitive electrode inserted into the basal turn of the scala tympani) (DESMEDT and ROBERTSON, 1973, 1975). Replacement of the perilymph chloride with the small diameter bromide anion in similar experiments did not interfere with the COCB effects. These results indicate that the chemical transmitter released from COCB terminals elicits an increase in conductance to small anions (normally to chloride) in the post-synaptic membranes. It is of interest that the low chloride perfusions, adequate to remove the COCB inhibitory effect on the N_1 response, only reduce to a smaller extent the COCB-induced potentiation of CM and negative shift of the endocochlear potential (DESMEDT and ROBERTSON, 1975).

G. Neurochemical Mechanism of OCB Axons

Histochemical staining for acetylcholinesterase (AChE) indicates that the efferent OCB axons and their terminals contain AChE (SCHUKNECHT, CHURCHILL, and DORAN, 1959; ROSSI, 1961; SHUTE and LEWIS, 1965; see review in IURATO, LUCIANO, PANNESE, and REALE, 1971). The presence of choline-acetyltransferase has recently been reported (JASSER and GUTH, 1973).

Such evidence is not sufficient to identify the OCB axons as cholinergic (*cf.* ROBINSON and BELL, 1967) but the working hypothesis of an efferent cholinergic mechanism has been tested in several studies and deserves careful consideration. Intravenous injections of D-tubocurarine, gallamine (flaxedil), atropine, or hyoscine do not reduce the OCB effects, and the same is true for dihydro-β-erythroidine which is a liposoluble curarizing agent (DESMEDT and MONACO, 1961, 1962). Intra-cochlear injections of cholinomimetic compounds have various effects on responses to sound (GISSELSON, 1960; TANAKA and KATSUKI, 1968) while the effects of COCB activation appear either not influenced (TANAKA and KATSUKI, 1966) or antagonized (GALLEY, KLINKE, PAUSE, and STORCH, 1971; KONISHI, 1972) by the intra-cochlear injection of the anticholinergic drugs D-tubocurarine, flaxedil or atropine. The injection of hemicholinium HC-3 which blocks acetylcholine synthesis (BIRKS and MACINTOSH, 1961) has also been reported to be followed by some depression of COCB inhibition if the bundle is stimulated repetitively (GUTH and AMARO, 1969).

On the other hand, it is certain that the inhibitory COCB and UOCB axons are blocked specifically within about 2 min by small intravenous doses of the alcaloids, strychnine or brucine, while other convulsants such as picrotoxin, metrazol and thiosemicarbazide are ineffective even when injected in convulsive doses (DESMEDT and MONACO, 1961, 1962; DESMEDT and LAGRUTTA, 1963). Strychnine effects can nicely be graded by injecting, for example, 50 and 100 μg per kg in succession (Fig. 8). Topical application of a drop of 1/400 solution of

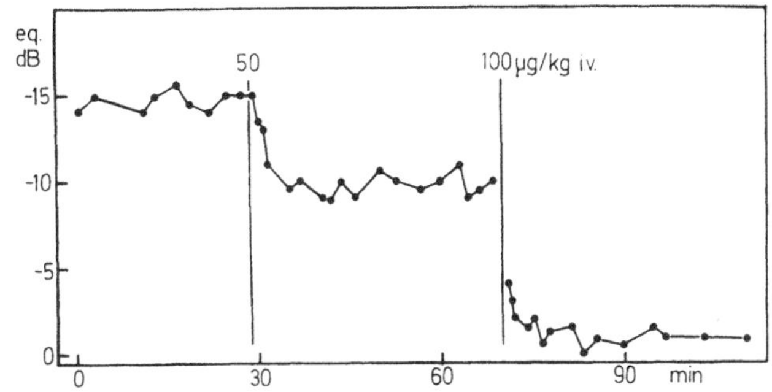

Fig. 8. Effect of small doses of intravenous strychnine on the COCB inhibition on N_1 in the cat. Abscissa, time in minutes. Ordinate, inhibition in equivalent dB. Fifty and 100 micrograms/kg of strychnine sulphate were injected at the vertical lines. The reductions of the COCB inhibitory effects develop quickly and can be graded

strychnine to the intact round window membrane blocks the COCB effects on that side only. Strychnine is now considered as a rather specific antagonist of postsynaptic glycinergic inhibitions in the spinal cord and brainstem of mammals (ECCLES, 1964; CURTIS, 1969; CURTIS, DUGGAN, and JOHNSTON, 1971; TEN BRUGGENCATE and SONNHOF, 1972) and it is surprising that glycine has apparently never been directly tested as a possible inhibitory transmitter in the inner ear. The question as to whether the sometimes suggestive effects observed by intracochlear perfusion of cholinomimetic drugs (BOBBIN and KONISHI, 1971) do result from membrane conductance changes similar to those elicited by transmitter release from COCB terminals remains unanswered and the specific antagonist to OCB, strychnine, was not tested in these experiments. In my opinion, there is no compelling evidence for the genuine cholinergic nature of OCB axons and it should be pointed out that strychnine does not block known cholinergic central synapses, including the centrifugal excitatory cholinergic pathway to the cochlear nucleus (COMIS and WHITFIELD, 1968). More experiments are required to elucidate the identity of the chemical transmitter (glycine or acetylcholine being the most likely candidates) released by olivo-cochlear axons in the inner ear.

H. Mode of Action of OCB Axons in the Inner Ear

In mammals, two sites of action have been considered for the OCB chemical transmitter which increases the membrane conductance for small anions in the post-synaptic membranes (DESMEDT and ROBERTSON, 1973). It appears premature to propose a simplistic view of the pattern of efferent terminals from COCB and UOCB on the inner and outer hair cells and on the auditory dendrites related to either of these hair cells. Extensive synaptic contacts are found on the short radial cochlear dendrites below inner hair cells in mammals while the presence of efferent synapses on the base of the inner hair cells is doubtful (SMITH and SJÖSTRAND, 1961; SMITH, 1961; SMITH and RASMUSSEN, 1963; WERSÄLL *et al.*, 1965; SPOENDLIN, 1966, 1969; SMITH and TAKASAKA, 1971). On the other hand the efferent terminals are quite prominent on the base of the outer hair cells (SMITH and RASMUSSEN, 1963; SPOENDLIN and GACEK, 1963; SPOENDLIN, 1966, 1969, 1972). The presence of axo-dendritic synapses on the auditory dendrites underneath the outer hair cells is less well documented, possibly because of technical difficulties: such synapses have been described in the guinea pig (SMITH and SJÖSTRAND, 1961; SMITH and TAKASAKA, 1971) and in the chinchilla (SMITH and RASMUSSEN, 1963), but SPOENDLIN (1966, 1969) and ECHANDIA (1967) consider them to be rare in the cat. In view of the fine diameter and quite extensive branching of the efferent olivo-cochlear axons in the inner ear, caution must be exercized in discussing these issues.

Activation of COCB elicits an increase of CM response to sound and a negative shift of the endocochlear potential: both effects can be related primarily to the increased anion conductance of the membrane at the base of the outer hair cells (few if any efferent synapses at the inner hair cells have been described). Such a reduced membrane resistance would increase leakage currents flowing through the hair cells from scala media to scala tympani. Furthermore the CM generator potential elicited by sound at the hair-bearing end of the outer hair cells will then result in a larger I-R drop in the external circuit and, therefore, in a larger *recorded* CM (for recording electrodes placed in scala tympani or on the round window membrane). At the same time the electrotonic potential at the basal hair cell membrane must be reduced by the local drop in membrane resistance induced by the activation of chloride ionophores. As a result the release of the chemical excitatory transmitter from hair cell base to auditory dendrite should decrease. Thus we consider that the local depolarization actually triggering the release of the excitatory transmitter (through an electro-secretory process presumably similar to that operating at the neuromuscular synapse: see KATZ, 1969) from outer hair cell is reduced by COCB activity, even though the CM potential recorded in the external circuit appears to be "paradoxically" (DESMEDT and

MONACO, 1961) increased (see detailed discussion in DESMEDT and ROBERTSON, 1975). In the mechanism just discussed, the inhibitory effect is to be considered as presynaptic since it involves the outer hair cell membrane where sound activation elicits the release of the excitatory transmitter.

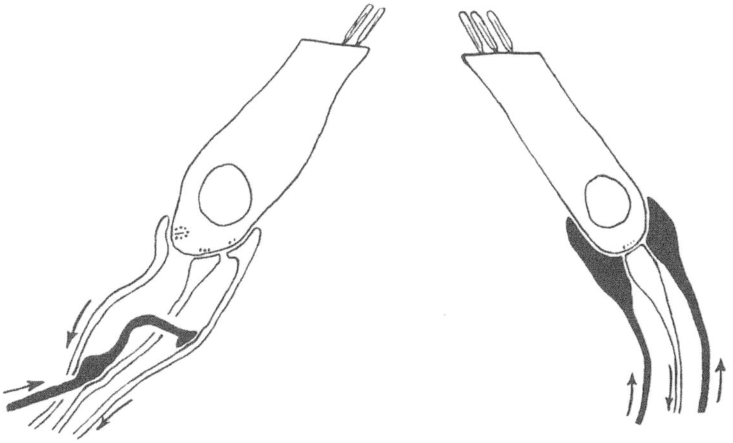

Fig. 9. Highly simplified sketch to indicate patterns of synaptic termination of efferent OCB axons (in black) either onto the auditory dendrite (postsynaptic inhibition) or on the base of hair cells (presynaptic inhibition) (based on references quoted in the text)

The OCB axons must also exert a post-synaptic inhibitory effect at the axo-dendritic synapses; this can be explained in a straightforward manner since the increase of the chloride conductance in the auditory dendrite membrane will produce a hyperpolarizing IPSP and/or a clamping of the membrane potential close to the equilibrium potential for chloride, as disclosed in cat's motoneurons (*cf.* ECCLES, 1964). The difficulty remaining is to decide which of these two mechanisms in the inner ear is preponderant when considering the function of gating the acoustic input to the brain. The data available do not allow pressing the argument towards either mechanism and it is likely that either the presynaptic or the postsynaptic efferent mechanisms may turn out to be more significant for gating auditory input when considering COCB versus UOCB axons or indeed when considering different animal species.

For example the efferent cochlear bundle of the pigeon (BOORD, 1961) is organized differently from the OCB system of mammals. Maximal electrical stimulation of the pigeon efferents, even under the best experimental conditions, does not reduce the neural responses by more than 2–5 equivalent dB while the recorded CM increase is definitely larger (up to 7 equivalent dB) than in the cat (DESMEDT and DELWAIDE, 1965). The dissipation of the efferent effects is slower for CM than for N_1 but both are selectively eliminated by strychnine (DESMEDT and DELWAIDE, 1963, 1965). These consistent features have been analyzed in detail and they raise interesting questions besides that of the behavioral significance, if any, of such a poorly efficient gating system. The efferent terminals in the pigeon cochlea are related exclusively to the base of the hair cells and no axo-dendritic synapses have yet been described (CORDIER, 1964; SMITH, 1967, 1968). The pigeon evidence, in so far as it could be used for arguing the case of other animal species, suggests that a strong efferent action restricted to the hair cells (presynaptic inhibition) is rather less efficient to suppress auditory nerve responses to sound (DESMEDT and DELWAIDE, 1965).

I. Psychophysiological Significance of the OCB System

Behavioral studies have compared the auditory performance of cats and monkeys before and after surgical section of the COCB. Inconsistencies or negative results might have been related to the incomplete inactivation of the OCB system (the UOCB being spared) or to the chosen testing procedures not involving the relevant function of these efferents. Thus the ability to discriminate a sound signal in the presence of noise was found to be reduced after COCB section in monkeys (DEWSON, 1968) and in cats (TRAHIOTIS and ELLIOTT, 1970). The performance of cats in a frequency discrimination task appeared to deteriorate (CAPPS and ADES, 1968). However, the pure tone auditory threshold and the perceptual signal-to-noise ratio presented no deficit in another series of experiments on cats after COCB section (IGARASHI, ALFORD, NAKAI, and GORDON, 1972).

V. CERACS

The main anatomical features of this polysynaptic centrifugal system have been sketched at the beginning of this chapter (see also Section II,K). Electro-anatomical studies have provided results which are consistent with anatomical data (DESMEDT, 1960; RASMUSSEN, 1964, 1967; RASMUSSEN and DESMEDT, unpublished). In these experiments gross electrical responses to clicks were recorded from both sides in various divisions of the cochlear nucleus complex and also at the round window membrane in cats paralyzed with curarizing drugs and in whom the middle ear muscles had been cut. Stereotaxic stimulations were delivered systematically in the brainstem and their effects on the responses analyzed. This type of experiment does not disclose the microphysiological organization of neurons and synapses but serves to identify the presence of functional interactions between different levels of the centrifugal systems as well as the anatomical location of the descending axons in the brainstem. The method is suitable for disclosing polysynaptic pathways which are more difficult to investigate with anatomical methods.

Activation of different portions of CERACS elicit consistent reductions in sound-evoked potentials. The positive points are close to, but not in, the main ascending auditory pathway: mesial to the lateral lemniscus, ventral and antero-lateral to the inferior colliculus, mesial to the brachium of the inferior colliculus and to the medial geniculate nucleus (DESMEDT and MECHELSE, 1957, 1958; DESMEDT, 1960, 1968). As mentioned above electrodes in the mesencephalic activating reticular formation do not elicit genuine neural gating effects.

The stimulation of CERACS must throw into action many descending axons but, with careful positioning of the bipolar electrodes, it is possible to sometimes achieve the differential activation of several subsets of descending axons. In the experiment illustrated in Fig. 10 both the antero-ventral cochlear nucleus and the auditory nerve responses to click were recorded on one side. An identical stimulus delivered to two nearby locations in the lateral brainstem elicited in B a reduction of the cochlear nucleus response without any significant change in the N_1 auditory nerve potential while in D, there was a reduction of both cochlear

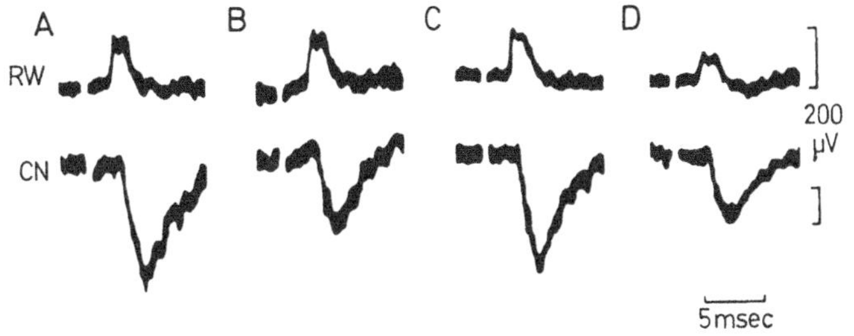

Fig. 10. Electrical activation of CERACS and the resulting effects on the responses to click recorded at the round window (upper trace) and in the anteroventral cochlear nucleus. Identical stimulations are delivered to two nearby points in (B) and in (D) (see text). (A and B, from Desmedt, 1960; Fig. 112)

nucleus and auditory nerve responses. In the latter instance the reduction in ventral cochlear nucleus response must result at least in part from the reduced input received via the auditory axons when the CERACS stimulation involves descending axons with an excitatory effect on the OCB inhibitory neurons in the medulla. This statement is based on the facts that CERACS can indeed activate OCB neurons (Fig. 11) and that OCB axons provide the only efferent pathway for eliciting a reduction in auditory nerve response after the middle ear muscles have been cut. On the other hand the observation of Fig. 10B implies that the CERACS axons and/or neurons activated by this particular electrode position would produce a significant change, not in OCB neurons themselves, but in other interneurons acting directly on the cochlear nucleus.

A. CERACS Driving of OCB Neurons

The activity of OCB axons can be recorded with microelectrodes in the eighth nerve (Fex, 1962) and the OCB neurons can be recorded from directly in the medulla (Rupert, Moushegian, and Whitcomb, 1968; Desmedt, 1968). Several physiological criteria (besides the final anatomical control) can be used to identify the latter nucleus (Galambos, Schwartzkopff, and Rupert, 1959; Clark and Dunlop, 1968) and COCB neurons present response characteristics which are clearly different from those of primary auditory afferent axons: the auditory threshold of COCB neurons is rather high and their discharge is more regular and at a low frequency (about 50/sec) (Fex, 1962; Rupert *et al.*, 1968).

The apparent discrepancy between the frequency of faradization of OCB (200–400/sec) necessary to achieve maximum efferent gating in artificial stimulations (Fig. 5) (DESMEDT, 1962; WIEDERHOLD and KIANG, 1970; KONISHI and SLEPIAN, 1971) and the recorded low frequency of firing of OCB neurons activated by sound led to some skeptical comments about the possibility for OCB to exert significant effects under physiological conditions of activation (PFALZ, 1969). The efficiency of OCB as an inhibitory mechanism would indeed appear rather poor in the cat or guinea pig when considering only the case of sound activation

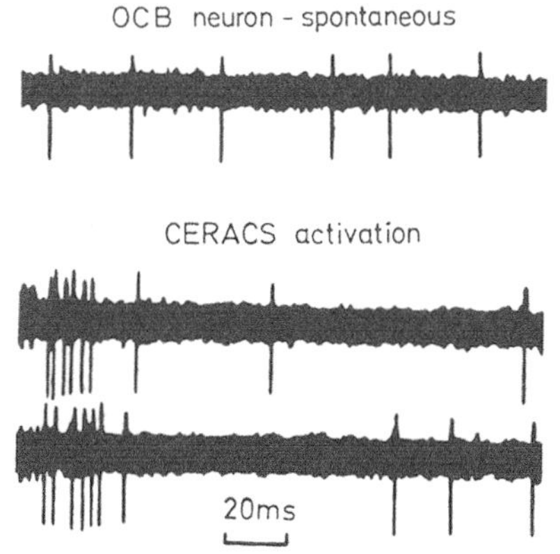

Fig. 11. Recording with platinum microelectrode from a COCB neuron in the cat's medulla. The activation of CERACS with single shocks elicits a burst of rather high frequency spikes. (From DESMEDT, 1968, unpublished)

of OCB neurons. However this is only one part of the story, since it was found that activation of CERACS even with single electric pulses can elicit bursts of spikes at much higher frequency in single COCB neurons (Fig. 11) (DESMEDT, 1968, 1971). Stimulation of the ventral temporo-insular cortex which sends corticofugal axons into CERACS is also effective. This means that OCB function should not be discussed with a model involving only their "segmental" connections with auditory input because the most significant excitatory drive to OCB neurons is received rather from higher levels of the brain through CERACS (DESMEDT, 1968).

B. Efferent Effects on Cochlear Nucleus Neurons

The cochlear nucleus has several subdivisions and many different synaptic patterns (*cf.* Section III, D) which should be considered when discussing efferent effects. Several types of efferent axons have been identified (RASMUSSEN, 1960, 1967). The cochlear nucleus neurons receive a variety of synaptic terminals, some with flattened synaptic vesicles and others with round vesicles, which stain for acetylcholinesterase; the respective origin of these terminals is not entirely clear (OSEN and ROTH, 1969; McDONALD and RASMUSSEN, 1971). Collaterals of COCB axons have been found to terminate in the cochlear nucleus (RASMUSSEN, 1967) and COCB electrical stimulation can either depress or activate the spike discharges of cochlear nucleus neurons even after cochlear destruction (STARR and WERNICK, 1968). COMIS and WHITFIELD (1968) observed that electrical stimulation in the medial portion of the principal superior olive elicits after a rather long latency an excitatory effect in neurons of the anterior ventral cochlear nucleus. There is good evidence that these efferent axons are cholinergic and their effects can be blocked by locally applied atropine, gallamine or dihydro-β-erythroidine, but not by strychnine (COMIS and WHITFIELD, 1968; COMIS and DAVIES, 1969). Electrical stimulation in the dorso-lateral part of the principal superior olive elicits in cochlear nucleus neurons inhibitory effects which are not blocked by strychnine (COMIS, 1970). Evaluation of these interesting data requires additional microphysiological studies.

C. The Functional Significance of CERACS

The limited data available do not make it possible to elaborate detailed models of the possible CERACS operations. There is evidence that in conscious bats (Vespertilionidae) a neural gating acoustic signals can be achieved in the auditory pathway (without any reduction at the first synapse in the inner ear) in relation to emitted sound in echolocation (SUGA and SCHLEGEL, 1972). Other studies have involved chronic lesions in the auditory cortex. It is known that ventral temporal regions receive auditory and/or other sensory input. For example cross modal integration of auditory and visual input has been found in the auditory IV area of the anterior sylvian gyrus (DESMEDT and MECHELSE, 1959b; DESMEDT, 1960; LOE and BENEVENTO, 1969; AVANZINI, MANCIA, and PELLICIOLI, 1969), Auditory responses have also been recorded in the ventral sylvian gyrus (SINDBERG and THOMPSON, 1962). These cortical areas represent presumably only one part of the neural systems which may be involved in the control of CERACS and which could somehow "program" differential centrifugal gating in the auditory relay nuclei receiving CERACS axon terminals (DESMEDT, 1968, 1971). Such statements are rather speculative although there is consistent evidence that chronic surgical lesions of the ventral temporo-insular cortex (in cats and dogs) or of the superior temporal gyrus (in monkeys) disrupt the discrimination of auditory patterns (sequences of sounds) (GOLDBERG, DIAMOND, and NEFF, 1957; STEPIEN, CORDEAU, and RASMUSSEN, 1960; GOLDBERG and NEFF, 1961; NEFF, 1962; CHORAZYNA and STEPIEN, 1963; DEWSON, COWEY, and WEISKRANTZ, 1970). It is not unreasonable to speculate that current cognitive experiences elaborated in auditory associative cortex might influence through CERACS the structuring and the selection of the subroutines which process the incoming auditory signals in the nuclei of the classical auditory pathway, in active behaviors concerned with the search of anticipated auditory cues and with the identification of sound patterns. CERACS could then be viewed as the descending limb of a long circuit going from cochlea to cortex and back down (DESMEDT, 1960, 1971). It must be stressed that simplistic views of such a loop should be avoided since a number of results would rather suggest CERACS to include differentiated subsystems of neurons which operate several different terminal links of which only a few (COCB, UOCB, efferents to cochlear nucleus) have been identified physiologically with some detail.

References

Avanzini, G., Mancia, D., Pelliccioli, G. Ascending and descending connections of the insular cortex of the cat. Arch. ital. Biol. **107**, 696—714 (1969).

Bach-y-Rita, G., Brust-Carmona, H., Penalozo-Rojas, J., Hernandez-Peon, R.: Absence of para-auditory influences on the cochlear nucleus during distraction and habituation. Acta neurol. lat.-amer. **7**, 73—81 (1961).

Békésy, G. von: Experiments in hearing. London: McGraw-Hill 1960.

Békésy, G. von: The gap between the hearing of external and internal sounds, p. 267—288. In: Biological receptors mechanisms. Cambridge: University Press 1962.

Birks, R. H. E., MacIntosh, F. C.: Acetylcholine metabolism of a sympathetic ganglion. Canad. J. Biochem. **39**, 787—827 (1962).

Bobbin, R. P., Konishi, T.: Acetylcholine mimics crossed olivo-cochlear bundle stimulation. Nature (Lond.) New Biol. **231**, 222—223 (1971).

Boord, R. L.: The efferent bundle in the caiman and pigeon. Exp. Neurol. **3**, 225—239 (1951).

Bremer, F., Stoupel, N.: Facilitation et inhibition des potentials évoqués corticaux dans l'éveil cérébral. Arch. int. Physiol. **67**, 240—275 (1959).

Broadbent, D. E.: Perception and communication. London: Pergamon 1958.

Bruggencate, G. ten, Sonnhof, U.: Effects of glycine and GABA, and blocking actions of strychnine and picrotoxin in the hypoglossus nucleus. Pflügers Arch. **334**, 240—252 (1972).

Bruner, J. S.: Neural mechanisms in perception. Res. Publ. Ass. nerv. ment. Dis. **36**, 118—143 (1958).

Capps, M. J., Ades, H. W.: Auditory frequency discrimination after transection of the olivo-cochlear bundle in squirrel monkeys. Exp. Neurol. **21**, 147—158 (1968).

Carmel, P. W., Starr, A.: Acoustic and nonacoustic factors modifying middle ear muscle activity in waking cats. J. Neurophysiol. **26**, 598—616 (1963).

Chorazyna, H., Stepien, L.: Effect of bilateral sylvian gyrus ablations on auditory conditioning in dogs. Bull. Acad. pol. Sci. **11**, 43—45 (1963).

Clark, G. M., Dunlop, C. W.: Field potentials in the cat medial superior olivary nucleus. Exp. Neurol. **20**, 31—42 (1968).

Collier, J., Buzzard, F.: Descending mesencephalic tracts in cat, monkey and man; Monakow's bundle; the dorsal longitudinal bundle; the ventral longitudinal bundle; the ponto-spinal tracts lateral and ventral; the vestibulo-spinal tract; the central tegmental tract (centrale Haubenbahn); descending fibres of the fillet. Brain **24**, 177—221 (1901).

Comis, S. D.: Centrifugal inhibitory process affecting neurones in the cat cochlear nucleus. J. Physiol. (Lond.) **210**, 751—760 (1970).

Comis, S. D., Davies, W. E.: Acetylcholine as a transmitter in the cat auditory system. J. Neurochem. **16**, 423—429 (1969).

Comis, S. D., Whitfield, I. C.: Influence of centrifugal pathways on unit activity in the cochlear nucleus. J. Neurophysiol. **31**, 62—68 (1968).

Cordier, R.: L'ultrastructure des synapses au niveau des récepteurs de la sensibilité somatique spéciale, et plus particulièrement dans l'organe de Corti du pigeon. Bull. Acad. Med. Belg. **4**, 729—753 (1964).

Curtis, D. R.: The pharmacology of spinal postsynaptic inhibition. Progr. Brain Res. **31**, 171—189. Amsterdam: Elsevier 1969.

Curtis, D. R., Duggan, A. W., Johnston, G. A. R.: The specificity of strychnine as a glycine antagonist in the mammalian spinal cord. Exp. Brain Res. **12**, 547—565 (1971).

Delafresnaye, J. F. (Ed.): Brain mechanisms and conciousness. Oxford: Blackwell 1954.

Desmedt, J. E.: Neurophysiological mechanisms controlling acoustic input, In: Rasmussen, Windle (Eds.): Neural mechanisms of the auditory and vestibular systems, Chapter 2 p. 152—164. Springfield/Ill.: Ch. Thomas 1960.

Desmedt, J. E.: Auditory evoked potentials from cochlea to cortex as influenced by activation of efferent olivo-cochlear bundle. J. acoust. Soc. Amer. **34**, 1478—1496 (1962).

Desmedt, J. E.: Feedback control in sensory systems. In: 24th Int. Congress of Physiol. Sci. **6**, 264. Washington 1968.

Desmedt, J. E.: The centrifugal auditory system, 8, 209—210. In: Symposium Neurophysiology of Hearing 25th Int. Congr. Physiol. Sci. Munich 1971.

DESMEDT, J. E., DELWAIDE, P. J.: Neural inhibition in a bird: effect of strychnine and picrotoxin. Nature (Lond.) **200**, 583—585 (1963).

DESMEDT, J. E., DELWAIDE, P. J.: Functional properties of the efferent cochlear bundle of the pigeon revealed by stereotaxic stimulation. Exp. Neurol. **11**, 1—26 (1965).

DESMEDT, J. E., LAGRUTTA, V.: Function of the uncrossed efferent olivo-cochlear fibres in the cat. Nature (Lond.) **200**, 472—474 (1963).

DESMEDT, J. E., LAGRUTTA, V., LAGRUTTA, G.: Contrasting effects of centrifugal olivo-cochlear inhibition and of middle ear muscle contraction on the response characteristics of the cat's auditory nerve. Brain Res. **30**, 375—384 (1971).

DESMEDT, J. E., MECHELSE, K.: Sur un phénomène d'inhibition centrifuge dans la voie acoustique centrale chez le chat. C.R. Soc. Biol. (Paris) **151**, 2209—2212 (1957).

DESMEDT, J. E., MECHELSE, K.: Suppression of acoustic input by thalamic stimulation. Proc. Soc. exp. Biol. (N.Y.) **99**, 772—775 (1958).

DESMEDT, J. E., MELCHESE, K.: Corticofugal projections from temporal lobe in cat and their possible role in acoustic discrimination J. Physiol. (Lond.) **147**, 17—18 (1959a).

DESMEDT, J. E., MELCHESE, K.: Mise en évidence d'une quatrième aire de projection acoustique dans l'écorce cérébrale du chat. J. Physiol. (Paris) **51**, 448—449 (1959b).

DESMEDT, J. E., MONACO, P.: Mode of action of the efferent olivo-cochlear bundle on the inner ear. Nature (Lond.) **192**, 1263—1265 (1961).

DESMEDT, J. E., MONACO, P.: The pharmacology of a centrifugal inhibitory pathway in the cat's acoustic system. Proc. First Int. Pharmacol. Meeting, Stockholm. 8, p. 183—188. Oxford: Pergamon 1962.

DESMEDT, J. E., ROBERTSON, D.: The perfusion of the scala tympani of the cat's cochlea with low-chloride solutions: effect on olivo-cochlear inhibition. Arch. int. Physiol. Biochim. **81**, 553—554 (1973).

DESMEDT, J. E., ROBERTSON, D.: Ionic mechanism of the efferent olivo-cochlear inhibition studied by cochlear perfusion in the cat. J. Physiol. (Lond.) **247**, 407—428 (1975).

DEWSON, J. H.: Efferent olivo-cochlear bundle: some relationships to stimulus discrimination in noise. J. Neurophysiol. **31**, 122—130 (1968).

DEWSON, J. H., COWEY, A., WEISKRANTZ, L.: Disruption of auditory sequence discrimination by unilateral and bilateral cortical ablations of superior temporal gyrus in the monkey. Exp. Neurol. **28**, 529—548 (1970).

DJUPESLAND, G.: Middle ear muscle reflexes elicited by acoustic and nonacoustic stimulation. Acta oto-laryng. (Stockh.) Suppl. **188**, 287—292 (1964).

DJUPESLAND, G.: Electromyography of the tympanic muscles in man. Int. Audiol. **4**, 34—41 (1965).

DUMONT, S., DELL, P.: Facilitation réticulaire des mécanismes visuels corticaux. Electroenceph. clin. Neurophysiol. **12**, 769—796 (1960).

ECCLES, J. C.: The physiology of synapses, Berlin-Göttingen-Heidelberg-New York: Springer 1964.

ECHANDIA, E., RODRIGUEZ, L.: An electron microscopic study on the cochlear innervation. I. The recepto-neural junctions at the outer hair cells. Z. Zellforsch. **78**, 30—46 (1967).

FEX, J.: Augmentation of the cochlear microphonics by stimulation of efferent fibres to cochlea. Acta oto-laryng. (Stockh.) **50**, 540—541 (1959).

FEX, J.: Auditory activity in centrifugal and centripetal cochlear fibres in cat. A study of a feedback system. Acta physiol. scand. Suppl. **189**, 1—68 (1962).

FEX, J.: Efferent inhibition in the cochlea related to hair-cell dc activity: study of postsynaptic activity of the crossed olico-cochlear fibres in the cat. J. acoust. Soc. Amer. **41**, 666—675 (1967).

GALAMBOS, R.: Suppression of auditory nerve activity by stimulation of efferent fibers to cochlea. J. Neurophysiol. **19**, 244—427 (1956).

GALAMBOS, R., SCHWARTZKOPFF, J., RUPERT, A.: Microelectrode study of superior olivary nuclei. Amer. J. Physiol. **197**, 527—536 (1959).

GALAMBOS, R., SHEATZ, G., VERNIER, V. G.: Electrophysiological correlates of a conditioned response in cats. Science **123**, 376—377 (1956).

GALLEY, N., KLINKE, R., PAUSE, M., STORCH, W.-H.: The effect of flaxedil (Gallaminetriethiodid) on the efferent endings in the cochlea. Pflügers Arch. **230**, 1—4 (1971).

GISSELSSON, L.: Effect on microphonics of acetylcholine injected into the endolymphatic space. Acta oto-laryng. (Stockh.) **51**, 636—638 (1960).
GOLDBERG, J.M., DIAMOND, I.T., NEFF, W.D.: Auditory discrimination after ablation of temporal and insular cortex in cat. Fed. Proc. **16**, 47—48 (1957).
GOLDBERG, J.M., NEFF, W.D.: Frequency discrimination after bilateral ablation of cortical auditory areas. J. Neurophysiol. **24**, 119—128 (1961).
GUTH, P.S., AMARO, J.: A possible cholinergic link in olivo-cochlear inhibition. Int. J. Neuropharmac. **8**, 49—53 (1969).
HALL, J.L.: Binaural interaction in the accessory superior-olivary nucleus of the cat. J. acoust. Soc. Amer. **37**, 814—823 (1965).
HELD, H.: Die centrale Gehörleitung. Z. Anat. Entw. Gesch. (Leipzig) 201—248 (1893).
HERNANDEZ-PEON, R.: Reticular mechanisms of sensory control. In: Rosenblith (Ed.): Sensory communication, p. 497—530. New York: Wiley 1961.
HERNANDEZ-PEON, R., SCHERRER, H., JOUVET, M.: Modification of electrical activity in cochlear nucleus during "attention" in unanesthetized cats. Science **123**, 331—332 (1956).
HERNANDEZ-PEON, R., JOUVET, M., SCHERRER, H.: Auditory potentials at cochlear nucleus during acoustic habituation. Acta neurol. lat.-amer. **3**, 144—156 (1957).
HORN, G.: Physiological and Psychological aspects of selective perception. In: Advances in the study of behavior, I, p. 155—215. New York: Academic Press 1965.
HUGELIN, A., DUMONT, S., PAILLAS, N.: Formation réticulaire et transmission des informations auditives au niveau de l'oreille moyenne et des voies acoustiques centrales. Electroenceph. clin. Neurophysiol. **12**, 797—818 (1960).
IGARASHI, M., ALFORD, B.R., NAKAI, Y., GORDON, W.P.: Behavioral auditory function after transection of crossed olivo-cochlear bundle in the cat. I. Pure tone threshold and perceptual signal-to-noise ratio. Acta oto-laryng. (Stockh.) **73**, 455—466 (1972).
Iurato, S.: Fibre efferenti dirette e crociate alle cellule acustiche del'organo del Corti. Atti Soc. ital. Anat. **72**, 60—63 (1964).
IURATO, S.: Submicroscopic structure of the inner ear. Oxford: Pergamon 1967.
IURATO, S., SMITH, C.A., ELDREDGE, D., HENDERSON, D.: Electronmicroscopic observations and cochlear potentials after sections of the crossed olivocochlear tract in the chinchilla. Electron Microsc. Proc. 4th Eur. Reg. Conf. Rome p. 561 (1968).
IURATO, S., LUCIANO, L., PANNESE, E., REALE, E.: Histochemical localization of acetylcholinesterase (AChE) activity in the inner ear. Acta oto-laryng. (Stockh.) Suppl. **279**, 1—50 (1971).
JASSER, A., GUTH, P.S.: The synthesis of acetylcholine olivo-cochlear bundle. J. Neurochem. **20**, 45—53 (1973).
KATZ, B.: The release of neural transmitter substances. Liverpool: Liverpool University Press 1969.
KIANG, N.Y.S., WATANABE, T., THOMAS, E.C., CLARK, L.F.: Discharge patterns of single fibers in the cat's auditory nerve. Cambridge (Mass.) MIT Press 1965.
KONISHI, T.: Action of tubocurarine and atropine on the crossed olivo-cochlear bundles. Acta oto-laryng. (Stockh. **74**, 252—264 (1972).
KONISHI, T., SLEPIAN, J.Z.: Effects of the electrical stimulation of the crossed olivo-cochlear bundle on cochlear potentials recorded with intracochlear electrodes in guinea pigs. J. acoust. Soc. Amer. **49**, 1762—1769 (1971).
LOE, P.R., BENEVENTO, L.A.: Auditory-visual interaction in single units in the orbito-insular cortex of the cat. Electroenceph. clin. Neurophysiol. **26**, 395—398 (1969).
LORENTE DE NÓ, R.: Anatomy of the eighth nerve. III. General plan of structure of the primary cochlear nuclei. Laryngoscope (St. Louis) **43**, 327—350 (1933).
MAGOUN, H.W.: The waking brain. Springfield (Ill.) Charles Thomas 1958.
MARSH, J.T., MCCARTHY, D.A., SHEATZ, G., GALAMBOS, R.: Amplitude changes in evoked auditory potentials during habituation and conditioning. Electroenceph. clin. Neurophysiol. **13**, 224—234 (1961).
MARSH, J.T., WORDEN, F.G., HICKS, L.: Some effects of room acoustics on evoked auditory potentials. Science **137**, 280—282 (1962).
MCDONALD, D.M., RASMUSSEN, G.L.: Ultrastructural characteristics of synaptic endings in the cochlear nucleus having acetylcholinesterase activity. Brain Res. **28**, 1—18 (1971).

METTLER, F. A.: Cortifugal fiber connections of the cortex of Macaca mulatta. The temporal region. J. comp. Neurol. **63**, 25—47 (1935).

NEFF, W. D.: Neural structures concerned in localization of sound in space. Psychologische Beiträge **6**, 492—500 (1962).

OSEN, K. K., ROTH, K.: Histochemical localization of cholinesterase in the cochlear nuclei of the cat, with notes on the origin of acetylcholinesterase-positive afferents and the superior olive. Brain Res. **16**, 165—185 (1969).

PFLALZ, R. K. J.: Absence of a function for the crossed olivocochlear bundle under physiological conditions. Arch. klin. exp. Ohr.-Nas. u. Kehlk.-Heilk. **193**, 89—100 (1969).

RASMUSSEN, G. L.: The olivary peduncle and other fiber projections of the superior olivary complex. J. comp. Neurol. **84**, 141—219 (1946).

RASMUSSEN, G. L.: Further observations of the efferent cochlear bundle. J. comp. Neurol. **99**, 61—74 (1953).

RASMUSSEN, G. L.: Efferent fibers of the cochlear nerve and cochlear nucleus. In: Rasmussen, Windle (Eds.): Neural mechanisms of the auditory and vestibular systems, p. 105—115. Springfield (Ill.): Charles Thomas 1960.

RASMUSSEN, G. L.: Anatomic relationship of the ascending and descending auditory systems. In: Neurological aspects of auditory and vestibular disorders, p. 1—19. Springfield (Ill.): Charles Thomas 1964.

RASMUSSEN, G. L.: Efferent connections of the cochlear nucleus. In: Sensorineural hearing processes and disorders. p. 61—75. Boston: Little, Brown and Company 1967.

ROBINSON, P. M., BELL, C.: The localization of acetylcholinesterase at autonomic neuromuscular junction. J. Cell Biol. **33**, 93—102 (1967).

ROSSI, G.: L'acétylcholinesterase au cours du développement de l'oreille interne du cobaye. Acta oto-laryng. (Stockh.) Suppl. **170**, 1—91 (1961).

RUPERT, A., MOUSHEGIAN, G., WHITCOMB, M. A.: Superior-olivary response patterns to monoaural and binaural clicks. J. acoust. Soc. Amer. **39**, 1069—1076 (1966).

RUPERT, A., MOUSHEGIAN, G.: Olivo-cochlear neuronal responses in medulla of cat. Exp. Neurol. **20**, 575—584 (1968).

SALOMON, G., STARR, A.: Electromyography of middle ear muscles in man during motor activities. Acta neurol. scand. **39**, 161—168 (1963).

SCHUKNECHT, H. F., CHURCHILL, J. A., DORAN, R.: The localization of acetylcholinesterase in the cochlea. Arch. Otolaryng. (Chic.) **69**, 549—559 (1959).

SHUTE, C. C. D., LEWIS, P. R.: Cholinesterase-containing pathways of the hindbrain: afferent cerebellar and centrifugal cochlear fibres. Nature (Lond.) **205**, 242—246 (1965).

SIMMONS, F. B., BEATTY, D. L.: Habituation (adaptation) in the middle-ear muscle reflexes of the cat. Acta oto-laryng. (Stockh.) **57**, 89—96 (1974a).

SIMMONS, F. B., BEATTY, D. L.: Sound field variability: some measurements from the cat. Electroenceph. clin. Neurophysiol. 332—334 (1964b).

SINDBERG, R. M., THOMPSON, R. F.: Auditory response field in ventral, temporal and insular cortex of cat. J. Neurophysiol. **25**, 21—28 (1962).

SMITH, C. A.: Innervation pattern of the cochlea. The internal hair cell. Ann. Otol. Rhinol. Laryng. **70**, 504—527 (1961).

SMITH, C. A.: Innervation of the cochlea. Rev. Panam. Otorinolaringol. Broncoesofagol. **1**, 77—92 (1967).

SMITH, C. A.: Morphological features of axo-dendritic relationship between efferent and cochlear nerves in the cochlea of mammals and pigeon. In: Structure and functions of inhibitory neuronal mechanisms, p. 141—155. New York: MacMillan (Pergamon) 1968.

SMITH, C. A., RASMUSSEN, G. L.: Recent observations on the olivo-cochlear bundle. Ann. Otol. Rhinol. Laryngol. **72**, 489—506 (1963).

SMITH, C. A., SJÖSTRAND, F. S.: A synaptic structure in the hair cells of the guinea pig cochlea. J. Ultrastruct. Res. **5**, 184—192 (1961).

SMITH, C. A., TAKASAKA, T.: Auditory receptor organs of reptiles, birds and mammals. In: Contributions to sensory physiology, Vol. 5, p. 129—178. London: Academic Press 1971.

SOHMER, H.: The effect of contralateral olivo-cochlear bundle stimulation on the cochlear potentials evoked by acoustic stimuli of various frequencies and intensities. Acta oto-laryng. (Stockh.) **60**, 59—70 (1965).

SPOENDLIN, H.: The organization of the cochlear receptor. In: Advances in Oto-Rhino-Laryngology, Vol. 13, p. 23—93. Basel: S. Karger 1966.

SPOENDLIN, H.: Innervation patterns in the organ of Corti of the cat. Acta oto-laryng. (Stockh.) **67**, 239—254 (1969).

SPOENDLIN, H.: Innervation densities of the cochlea. Acta oto-laryng. (Stockh.) **73**, 235—248 (1972).

SPOENDLIN, H., GACEK, R.: Electronmicroscopic study of the efferent and afferent innervation of the organ of Corti in the cat. Ann. Otol. **72**, 660—686 (1963).

STARR, A., WERNICK, J. S.: Olivo-cochlear bundle stimulation: effects on spontaneous and tone-evoked activities of single units in cat cochlear nucleus. J. Neurophysiol. **31**, 549—564 (1968).

STEPIEN, L. S., CORDEAU, J. P., RASMUSSEN, T.: The effect of temporal lobe and hippocampal lesions on auditory and visual recent memory in monkeys. Brain **83**, 470—489 (1960).

SUGA, N., SCHLEGEL, P.: Neural attenuation of responses to emitted sounds in echolocating bats. Science **177**, 82—84 (1972).

TANAKA, Y., KATSUKI, Y.: Pharmacological investigations of cochlear responses and of olivo-cochlear inhibition. J. Neurophysiol. **29**, 94—108 (1966).

TASAKI, I.: Hearing. Ann. Rev. Physiol. **19**, 417—438 (1957).

TASIRO, S.: Experimentell-anatomische Untersuchung über die efferenten Bahnen aus den Vierhügeln der Katze. Z. mikr. anat. Forsch. **47**, 1—32 (1940).

TEAS, D. C., KONISHI, T., NIELSEN, D. W.: Electrophysiological studies on the spatial distribution of the crossed olivo-cochlear bundle along the guinea pig cochlea. J. acoust. Soc. Amer. **51**, 1256—1264 (1972).

TRAHIOTIS, C., ELLIOTT, D. N.: Behavioral investigation of some possible effects of sectioning the crossed olivo-cochlear bundle. J. acoust. Soc. Amer. **47**, 592—596 (1970).

TREISMAN, A. M.: Selective attention in man. Brit. med. Bull. **20**, 12—16 (1964).

WALTHER, J. B., RASMUSSEN, G. L.: Descending connections of auditory cortex and thalamus of the cat. Fed. Proc. **19**, 291 (1960).

WERSÄLL, J., FLOCK, A., LUNDQVIST, P. G.: Structural basis for directional sensitivity in cochlear and vestibular sensory receptors. Cold Spr. Harb. Symp. quant. Biol. **30**, 115—132 (1965).

WIEDERHOLD, M.: Variations in the effects of electric stimulations of the crossed olivo-cochlear bundle on cat single auditory nerve-fiber-responses to tone burst. J. acoust. Soc. Amer. **48**, 966—977 (1970).

WIEDERHOLD, M. L., KIANG, N. Y. S.: Effects of electric stimulation of the crossed olivo-cochlear bundle on single auditory-nerve-fibers in the cat. J. acoust. Soc. Amer. **48**, 950—965 (1970).

WIEDERHOLD, M. L., PEAKE, W. T.: Efferent inhibition of auditory-nerve responses: dependence on acoustic-stimulus parameters. J. acoust. Soc. Amer. **40**, 1427—1430 (1966).

WORDEN, F. G.: Attention and auditory electrophysiology. Progr. Physiol. Psychol. **1**, 45—116 (1966).

WORDEN, F. G., MARSH, J. T.: Amplitude changes of auditory potentials evoked at cochlear nucleus during acoustic habituation. Electroenceph. clin. Neurophysiol. **15**, 866—881 (1963).

WORDEN, F. G., MARSH, J. T., ABRAHAM, F. D., WHITTLESEY, J. R. B.: Variability of evoked auditory potentials and acoustic input control. Electroenceph. clin. Neurophysiol. **17**, 524—530 (1964).

Chapter 6

The Influence of the External and Middle Ear on Auditory Discriminations

ROBERT A. BUTLER, Chicago, Illinois (USA)

With 4 Figures

Contents

I. The External Ear 247
A. Introduction 247
B. The Pinna's Influence on the Localization of Sound 249
1. Monaural Localization of Sound in Space 249
2. Localization in the Median Sagittal Plane 251
3. Binaural Localization of Sound in the Horizontal Plane 254
4. Theoretical Considerations of the Pinna's Role in Sound Localization 255
II. The Middle Ear 256
References 259

I. The External Ear

A. Introduction

In evaluating the role of the outer ear, HENSEN (1879) concluded that the pinnae are not completely useless for hearing acuity and auditory perception, but their contributions are minor. Similar statements have been made repeatedly in the past by those working in audition. Contributors to present-day editions of physiology handbooks express the same view if, indeed, they refer to the pinnae at all. Generally speaking, they are correct. But, the extent to which the outer ear does influence hearing requires elaboration. In the course of doing this, it will be seen that in some situations the outer ear's contribution is critical for accurate auditory discrimination.

Without question, considerably more interest in the role of the outer ear was expressed by 19th century investigators of hearing than by those thus far in the 20th century. What, it was asked, is the purpose of the convolutions? To MÜLLER (1840), the complex configurations seemed most suitable for conducting, or guiding, sound waves into the ear canal. Granted that this may be so, the question remained whether the pinna actually improved hearing acuity. Several writers remarked on the observation that hard-of-hearing persons cup their hand behind the pinna which seemed to result in better acuity. The implication was that the cupped hand essentially served to enlarge the pinna, thereby collecting more sound. Were this the case, then hearing should be more acute with the pinna than without it. The otologist TOYNBEE (1860) failed to confirm this expectation. He examined a patient whose left auricle was missing almost entirely. The right one was intact. Careful testing showed no measurable difference in acuity between ears. DARWIN (1907) cited TOYNBEE in support of his argument that man's pinnae are vestigial. He held that the various folds and prominences within the human pinna were merely rudiments which in lower mammals provided structural support for holding the pinna erect[1].

The pinna, however, does have an important function in auditory perception, and the origin of the experimental approach to this problem can be traced to the work of KESSEL (1882). He moved a tuning fork slowly past the pinna from front to rear and from above to below. A sudden increase in loudness was detected when the sound source, proceeding toward the rear, passed by the tragus. Loudness decreased when the tuning fork reached the region behind the pinna. Changes in loudness were perceived also when the tuning fork was moved vertically. It was soon realized that what appeared as a loudness change when simple stimuli were presented at various locations became a change in timbre when listening to complex stimuli. And it is this perceptual event which reveals the important function of the outer ear: The consequence of its interaction with the sound field provides a cue for location of the stimulus.

BLOCH (1893) performed the basic experiments, but many others besides KESSEL preceded him in pointing to the role that the pinna must play in directional hearing. MACH (1875), perhaps the most influential, claimed that the pinna served as a resonator for high frequency sounds and, in doing so, affected the perceived timbre of the stimulus. Still, it remained for BLOCH to make the systematic empirical observations. He studied the proficiency of observers in discriminating between the location of complex sounds originating in the horizontal and frontal planes under binaural and monaural listening conditions. The fact that his listener demonstrated some ability to locate sounds monaurally in the horizontal and frontal planes eliminated binaural cues as critical for the task. He reported also that listeners could locate sounds in the median sagittal plane and hypothesized that the pinnae must have furnished the localization cue. To test this idea, he filled the upper part of the pinnae with wadding and taped the lobules to the skull. Spatial discrimination between sounds in the sagittal plane was severely impaired. BLOCH concluded that, whereas the pinnae exerted only a secondary influence on localization of sounds originating in the horizontal and frontal planes under normal listening conditions, they were of primary importance for distinguishing between locations of sounds emanating in the median sagittal plane. RAYLEIGH (1907) concurred. He stated that the only possible explanation for distinguishing between sounds originating in front from those behind is "an alteration of quality due to the external ears" (p. 231). Actually, WEBER suggested this role for the pinna as early as 1851 (cited by BORING, 1942, p. 383). And, it is probably a fair appraisal of the situation to say that this idea has not been challenged seriously throughout the ensuing decades.

---

[1] No one appeared to question the pinna's importance for hearing in many other mammals. The pinna of the horse, for example, which is controlled by no less than 17 muscles, was often chosen to emphasize the degree to which the pinna could be utilized. More recently, the pinna of the bat (the sub-order, Microchiroptera) has attracted attention for similar reasons.

Since BLOCH's original study, our understanding of the pinna's function in sound localization has not advanced significantly, but not because of a lack of interest in localization of sound in space. What happened was that a testing technique was adopted which by-passed the pinnae as a price for greater precision of measurement. Starting with KLEMM (1918) and HORNBOSTEL and WERTHEIMER (1920), interest burgeoned in the relative importance of time and intensity differences for locating sound. Equipment for generating sound and controlling its presentation had improved vastly. Through the use of headphones, time of stimulus arrival and stimulus intensity could be varied at each ear independently, and a most sophisticated series of experiments has been conducted during the past 40—50 years. The important point with respect to the present discussion is that in most studies of auditory space perception between the years 1918 and say, 1965, the pinnae were essentially inoperative. They were covered by headphones.

B. The Pinna's Influence on the Localization of Sound

Although the idea that localization of sound in space could be accounted for by binaural cues gained wide acceptance, there were, and continue to be, a few who find this explanation inadequate. They stand on firm ground, for how can one explain the proficiency of sound localization that takes place without the benefit of binaural differences in time and intensity? One should, however, first ask how proficient is localization of sound when listening monaurally or when sounds are positioned in the median sagittal plane. Both situations are devoid of binaural cues. With regard to monaural listening, the consensus is that persons locate sounds reasonably well when they originate ipsilaterally to the functioning ear. Nearly all of the numerous investigators, beginning with POLITZER (1876), have mentioned, however, that sounds originating in the horizontal plane tend to be displaced toward the ear that is functioning normally. This bias obviously decreases accuracy in localizing sounds as compared to binaural performance. Most investigators also report that localization of sounds originating contralaterally to the functioning ear are impaired; in some cases, severely so. The ability of man to locate sounds arising in the median sagittal plane has been demonstrated repeatedly. Spatial discrimination can be quite precise. Noisebursts, for instance, positioned vertically and separated only by 6 deg can be located significantly more accurately than expected by chance (BUTLER, 1969). To localize sound in the median sagittal plane requires that the sounds be complex and that they contain frequencies of 7000 Hz and above (ROFFLER and BUTLER, 1968). Presumably, these prerequisites also apply to monaural localization.

1. Monaural Localization of Sound in Space

GILSE and ROELOFS (1930), serving as their own experimental subjects, located sounds generated by POLITZER's Hörmesser[2]. Adjacent sound sources were separated by 20 deg and extended from 0—80 deg azimuth. First, the subjects listened binaurally, then monaurally (right ear), and finally monaural (right ear) with the pinna filled completely with Plasticene. Monaural performance was

[2] This instrument, frequently employed by 19th and early 20th century investigators, generated an impulse sound resembling the ticking of a watch.

inferior to binaural performance; sounds appeared to originate from a more restricted region of the horizontal plane. This was most noticeable when the pinna was filled. All stimuli, regardless of their actual origin, appeared from the same narrow sector of the arc.

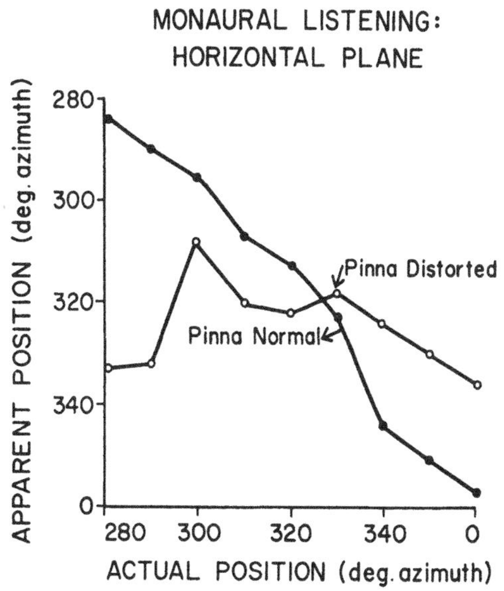

Fig. 1. Monaural localization of sound in the horizontal plane with and without distortion of the pinna

The present writer has confirmatory data in a form that is more suitable for illustration. The listener, a 21 year old medical student who had been deaf in his right ear since childhood, was asked to identify which of 9 loudspeakers generated bursts of white noise. Head movements were restricted. After a series of 180 judgments, the pinna of the normal ear was folded forward and taped to the cheek. The test was repeated. Figure 1 summarizes the data. It is clear that the listener was unable to locate the sounds accurately when his pinna was distorted. In general agreement with Gilse and Roelofs' report, there was a tendency to perceive the stimuli as originating from a restricted segment of the arc. Since the sounds were presented ipsilaterally to the normal ear, the effect of the head shadow on location judgments was eliminated and one can reasonably infer that the normal pinna contributed to the accuracy of the performance — that its distortion brought about a deterioration of performance[3].

[3] Measures to rule out the effects of the head shadow when studying the pinna's function in monaural listening have not always been taken. Angell and Fite (1901), testing the hypothesis that the pinna was responsible for proficient monaural localization performance, filled it with oily putty. They compared sound localization accuracy with that recorded when the pinna was left in its normal state. No differences were found. Unfortunately, in this series of tests, the sound sources were separated by 90 deg so timbre cues provided by the head's sound shadow must have been operating. A few years later, Hocart and McDougall, 1908, made the same error. Hence, the oft suspected critical role played by the pinna in monaural localization was never demonstrated unequivocally in these early studies.

Brunzlow (1925) collected data on the ability to distinguish between the locations of two successively presented sounds when listening monaurally. His stimuli were either impulse sounds or longer duration samples of noise. The observed performances together with the introspective reports led him to divide each semicircular arc, 0-90-180 deg and 0-270-180 deg, into two zones of hearing: a central and a peripheral zone. The former extended from the azimuth location where the forehead casts a sound shadow on the external meatus entrance to the azimuth location where the posterior border of the pinna also shadows the same area. The peripheral zone occupied the remaining segments of each semicircle. Spatial differentiation between sound sources was most refined when the locations of the two successive sounds spanned either the anterior or the posterior border of the central zone. Brunzlow interpreted his data to mean that at the border separating zones of hearing, the qualitative differences between the stimuli were most apparent. He also pointed out that his results were in essential agreement with those of Bloch despite the fact that the auditory stimuli differed in each case. Actually, the important features of the stimuli were common to both experiments — the sounds were complex and contained high frequency components.

2. Localization in the Median Sagittal Plane

Perekalin (1929) requested subjects to locate sounds when rubber tubes were inserted in each ear canal or, at other times, when the pinnae were bent fully forward. Performance in the localizing task suffered; front-rear and above-below confusions increased appreciably. Gilse and Roelofs (1930) filled portions of the concha selectively with plastic material. In nearly every instance performance was poorer with partially filled pinnae, but no systematic relation was apparent between performance decrements and that particular region of the pinna filled. What seems to be important is the extent to which the pinnae cavities are filled. Gardner and Gardner (1973), serving as their own subjects, located broadband noise originating from 1 of 9 loudspeakers positioned in the anterior segment of the median sagittal plane. The task was exacting as the loudspeakers were placed only 4.5 deg apart extending from −18 to +18 deg *re* ear level. Localization tests were carried out under a variety of conditions which differed from one another with respect to the degree to which the pinnae cavities were occluded. A specially prepared rubber composition was used as the occluder. Figure 2 depicts the range of experimental conditions and the localization performances associated with each. Proficiency varied inversely with the degree of pinna cavity occlusion. And, in agreement with Gilse and Roelofs, no one region of the pinna cavity was more critical to the localization task than any other. The experiment, however, was not designed to study systematically this particular point. The authors merely stated that all regions are of importance. Jongkees and Groen (1946) placed glass tubes in the ear canals, hence, by-passing completely the pinna. They reported that localization of sounds was impossible. The ability to locate noise-bursts positioned vertically in the medial sagittal plane with and without the pinna's contribution was studied also by Roffler and Butler (1968). The loudspeakers were straight ahead of the listeners and were separated from one another by 15 deg. Listeners made few errors under normal conditions, but performed at chance level of accuracy when their pinnae were flattened against their heads.

Several investigators have claimed that head movements are mainly responsible for the front-rear distinction; that the pinna's role is negligible. Certainly head movements can contribute to the consistency of correct front-rear discriminations, but they are not essential. Burger's (1958) experiment makes this

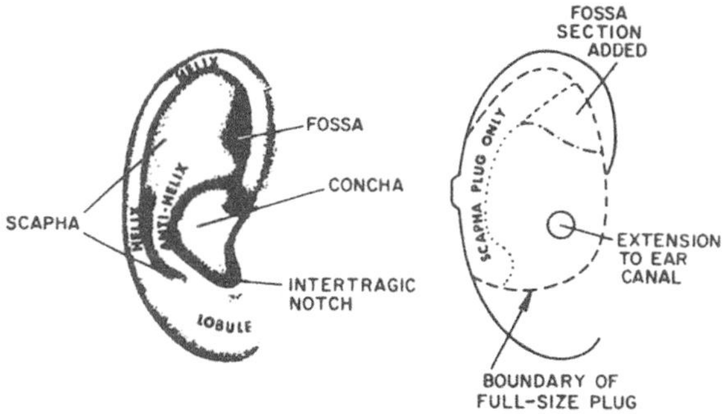

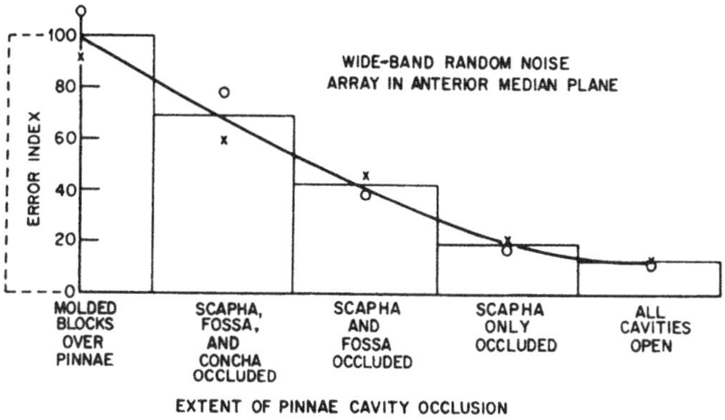

Fig. 2. Localization proficiency as a function of extent of pinnae cavity occlusion. An error index of "*O*" represents errorless performance; that of 100 represents a chance level of accuracy. The symbols "*o*" and "*x*" represent the data points for the two listeners

point. Listeners were requested to tell whether the sounds originated from in front or behind. Stimuli were generated by loudspeakers displaced 15 deg to the right of the median sagittal plane. Burger reasoned that with the sound source removed somewhat from the midline, slight head movements would not cause the sound to appear at one time to the listener's left and at another time to his right. Without this precaution, what was intended to be a front-rear judgment could turn out to be a left-right judgment without the listener's awareness. The auditory stimuli were filtered octave bands of noise presented at three different intensity levels. The same stimulus presented in front was not presented from behind on the succeeding trial and vice versa. This served to prevent a direct comparison between successive stimuli with respect to their timbre and/or loudness. Sometimes the head was fixed. Other times it was free to move, but listeners were asked to refrain from excessive head movements. Performance was superior when the head was not fixed. Nevertheless, even when the head was clamped, accuracy of front-rear discriminations clearly exceeded chance level. Moreover, when the head was free, but the pinnae's influence was counteracted by covering them with muffs, the percentage of correct front-rear discriminations fell markedly. It approached chance expectancy.

Further experimental evidence that the pinna is critical for front-rear discrimination of sound sources was contributed by KIETZ (1953). He by-passed the pinnae by inserting 3 cm length tubes in each ear canal. Then, he tested subjects under three listening conditions: 1) a model of the pinna was attached to the distal end of each tube, 2) a small cup, comparable in size to a "cupped" hand was attached to the distal end of each tube, and 3) the distal ends of the tubes remained free. Noise stimuli were presented in front and behind the listener. Only when listening with the artificial pinnae could front-rear distinctions be made with reasonable consistency; performance was inferior to that when listening normally, but clearly superior to the other experimental conditions.

The most recent, as well as the most sophisticated, study on front-rear discrimination of sound sources is that of BLAUERT'S (1969/1970). While he did not subject the pinna to experimental manipulations, his experiments were designed with the view that the pinna is important for this type of localization task. BLAUERT presented one-third octave bands of noise whose center frequencies ranged from 125–16000 Hz. Different stimulus intensities were employed and the sounds came from either in front of or behind the listener. Irrespective of the actual location of the sounds, narrow bands of noise, centered at approximately 250, 500, 4000, and 6000 Hz, appeared to originate in front; that noise band centered around 8000 Hz seemed to come from above, and those centered around 1000 and 12000 Hz were perceived as originating from behind. Next, he recorded sound pressure level (SPL) at the ear drum when broadband noise emanated from loudspeakers in front of or behind and, in one case, above the listener. A boost in SPL at specific frequencies coincided with the location judgment of those same frequencies. For example, the SPL at approximately 4000 and 6000 Hz as measured in the ear canal was greater when the loudspeaker was in front; 1000 Hz was boosted when the loudspeaker was behind the subject. If the distinction of front from rear positioned sounds is based on spectral differences, then a sound which has the spectral composition of that originating in front should appear to come from in front even though it may be emanating from a loudspeaker placed behind the listener. BLAUERT placed the tip of a probe tube, connected to a microphone, at the entrance of the ear canal. Sound (pink noise) was presented from in front of or behind the listener. The sound pressure waves were picked up by the probe microphone, compensated electronically for the spectral changes caused by the pinnae and the head, and recorded on magnetic tape. Then he played back the sounds to the subjects. His findings were, indeed, interesting. Listeners perceived the sound as originating from the location at which it was when originally recorded despite the fact that at times it was coming from the opposite direction.

3. Binaural Localization of Sound in the Horizontal Plane

The binaural cues of time and intensity differences are so dominant that it is difficult to detect the influence of the pinna when locating sounds binaurally in the horizontal plane. But, it exists. Employing techniques already mentioned, the pinnae have been either by-passed or modified. Perekalin (1929) placed 10-cm long tubes in the ear canals and compared localization performance with that observed under normal conditions. Unquestionably, the subjects performed less accurately when listening through the tubes. Aside from the front-rear reversals, they made many more localization errors when the stimuli originated from the sides. An objection to this study is that the interaural time differences were also changed because of the excessive length of the tubing. Freedman and Fisher (1968) claimed, however, that extending the interaural axis does not cause performance decrements provided an artificial pinna is attached to the distal end of the tubes. They placed metal tubes as long as 19.7 cm in each ear canal and found that accuracy of locating noise-bursts, separated by at least 22.5 deg, was not significantly reduced from that exhibited under normal conditions. If no artificial pinnae were attached to the ends of the tubes, the localization proficiency was impaired severely. These data, together with those on monaural localization, have been marshalled by Freedman and Fisher to support the idea that binaural difference cues are neither necessary nor sufficient for adequate localization of sound in space; that the role of the pinna is a significant, if not dominant, factor in the localizing task.

Rather than by passing the pinnae, Gilse and Roelofs (1930) filled them with Plasticene. Localization proficiency was somewhat degraded. Many of the sound sources originating along the arc extending from 270 through 0—90 deg azimuth appeared to be displaced laterally.

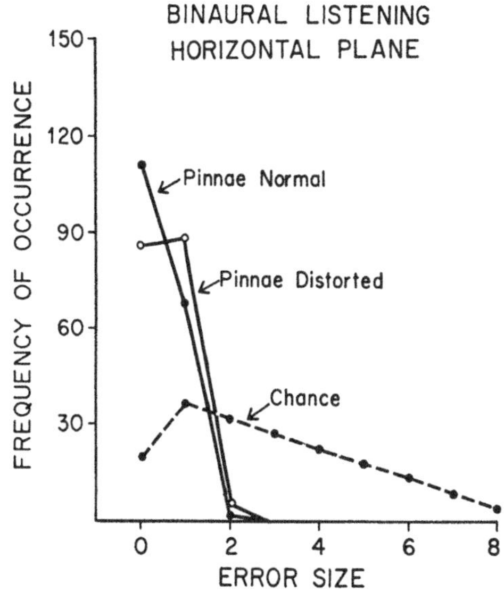

Fig. 3. Distribution of errors when locating sounds binaurally in the horizontal plane with and without distortion of the pinnae

BUTLER (unpublished data) tested 4 listeners on a sound localization task under normal conditions as well as when the pinnae were bent forward and taped to the cheek or folded downward. The loudspeakers were 5 deg apart, center to center, and covered an arc from 340 deg to 20 deg azimuth. This is a region in which proficiency of binaural localization excels. Figure 3 summarizes the data in terms of error scores. A correct response was assigned an error score of 0; an error of 5 deg was assigned a score of 1; an error score of 2 reflected a 10 deg error, etc. It is evident that listeners with distorted pinnae performed well, but not as well as when the pinnae were left in their normal position.

4. Theoretical Considerations of the Pinna's Role in Sound Localization

Frequent mention has been made of the assertion that the pinna's presence in the sound field brings about a change in the quality or timbre of the sound. The so-called Klang or timbre theory of sound localization has had many supporters during the past century. THOMPSON (1882, p. 415) provided one of the earliest expressions of this theory:

> The ear has been trained from childhood to associate certain differences in the quality of sounds (arising from differences in the relative intensities of some of the partial tones that may be present) with definite direction, and relying on these associated experiences, judgments are drawn concerning sensations of sounds whose direction is otherwise unknown. The quality of the sounds being influenced by both the head and the pinna.

Several instances have been given in this discussion where the pinna alone seems primarily responsible for the localization cue.

Spectral, or its equivalent in psychological terms, timbre, differences may indeed be the cue for localization resulting from the presence of the pinna. And association of these differences with direction of the sound source may well be established through learning. There exist, however, other interpretations of the pinna's role. BATTEAU (1967) proposed that the configurations of the pinnae provide multiple delay paths for sound waves directed toward the ear drum; that the pattern of delay is characteristic of the particular azimuth and elevation of the sound source. Somewhat earlier, PETRI (1932) published a paper in which he described how the manifold curvatures of man's pinna form an array of intersecting parabolic surfaces which conceivably could serve to reflect sound waves toward the ear drum with built-in delays[4]. BATTEAU supported his argument primarily with electro-acoustical data collected from models of the pinna. The electrical response to a click stimulus was picked up from a microphone positioned in place of an artificial ear drum and was registered on an oscilloscope screen. The delay of the response onset was a function of the azimuth and elevation of the click. Delays of from 15—80 μsec were observed when the sound source was moved from 90—0 deg azimuth. In the frontal plane, delays of approximately 100—300 μsec in response onset were observed when the sound source was moved from the above position through the lateral position to the below position. BATTEAU suggested that the pinna performs a transformation of the sound pres-

[4] The assumption that the pinna could serve as a reflector of sounds was considered by MACH (1875) as untenable. He pointed out that the wave length of audible sounds are too long to be reflected by surface areas as small as those comprising the configurations of the pinna.

sure pattern. To extract information about the location of the sound source, the auditory nervous system must perform an inverse transform.

It is clearly evident that most of the behavioral research on man's outer ear during the past seven to eight decades has done little more than to confirm the findings of Bloch. The direction taken by Blauert, in which the perceived front-rear position of a sound is correlated with its fine structure as measured in the ear canal, does, however, promise to advance our knowledge of the pinna's contribution to sound localization. Without doubt, this technique will be applied also to the monaural localization task. Whether the auditory system utilizes the presumed temporal differences resulting from multiple delay paths provided by the pinna is problematical.

II. The Middle Ear

The transfer function of acoustic energy and the resonance properties of the middle ear have been discussed elsewhere in this volume. Relevant to the present discussion is that any diminution in the efficiency of energy transfer has an adverse effect on hearing. Otologists, in their concern with the diseased middle ear, have developed a variety of surgical procedures which can result in a marked improvement in the efficiency of acoustic energy transmission. What is probably an almost continuous fluctuation in the efficiency of energy transfer across the middle ear during wakefulness is brought about by the action of the middle ear muscles. Physical, physiological, and behavioral indices of reflex activity have been employed. The first two are discussed by other contributors to this volume. Behavior denoting a change in hearing as a consequence of the middle ear reflex will be treated briefly in the remaining part of this chapter.

The aural reflex resulting in a change in middle ear impedance can be elicited by non-acoustic as well as by acoustic stimulation, but the vast majority of investigators have employed acoustic stimuli when studying this reflex. It appears that the contraction of the stapedius muscle is primarily responsible for initiating events leading to impedance changes. One function frequently attributed to the reflex is that it protects the inner ear from acoustic trauma. The question arises: How much protection is provided? Although the answer is beset with qualifications, the following statements furnish a reasonable appraisal of the situation. First, despite the fact that thousands of persons are without one or both stapedius muscles as a consequence of middle ear surgery, no data are available on the relative incidence of permanent hearing loss in these people following exposure to high intensity noise. What has been measured is the amount of temporary threshold shift (TTS) in normal hearing persons caused by strong acoustic stimulation with and without-activation of the middle ear reflex. Ward (1962b), for example, presented high intensity noise-bursts (133 dB and above) once each 2.4 sec for 2 min either alone or preceded by a 1000 Hz tone-burst at 100 dB SPL. The rationale for the experimental design was that a less intense stimulus, yet one capable of eliciting the reflex, should protect the inner ear from the traumatic effect of a more intense stimulus. It did, provided the interval between the two stimuli was sufficiently long to permit the muscle contraction to reach its full strength. Specifically, the TTS was reduced only 1 dB (median value) when the two stimuli were separated by 25 msec, but a 13 dB (median value) reduction in TTS was measured when the tone-burst preceded the noise-burst by 100 msec. WARD suggested that a 150 msec interval between the reflex activating stimulus and the more intense stimulus would have served better to protect the inner ear from the deleterious effects of the latter sound. In fact, this was the interval chosen by LOEB and FLETCHER (1961) to protect the inner ear from the acoustic trauma caused by 30 caliber machine gun shots; TTS for 6000 and 8000 Hz were lessened by 20 dB. The requirement that the reflex must be elicited *before* the onset of a potentially damaging sound, in order to protect against it, all but eliminates the reflex as an effective mechanism for conserving hearing. Some protection, however, may accrue in the presence of a rapidly repeating impulse stimulus. Here, the muscles remain in a state of partial contraction over a period of at least several minutes (WARD *et al.*, 1961).

Another behavioral index of middle ear activity is the amount of contralateral threshold shift (CTS) brought about by intense ipsilateral stimulation. Conceptually, this is an elegant experimental technique. The reflex, being consensual, permits the tracing of threshold to a sound in one ear while the opposite ear is being exposed to a high intensity sound of a different frequency composition. With the proper correction for the effect of contralateral masking on threshold, LOEB and FLETCHER (1961) showed that the aural reflex elicited by a train of 100 dB clicks elevated the threshold in the opposite ear to a 500 Hz tone by approximately 20 dB. Adaptation of the reflex can be demonstrated clearly by the CTS technique as illustrated by the data of Fig. 4.

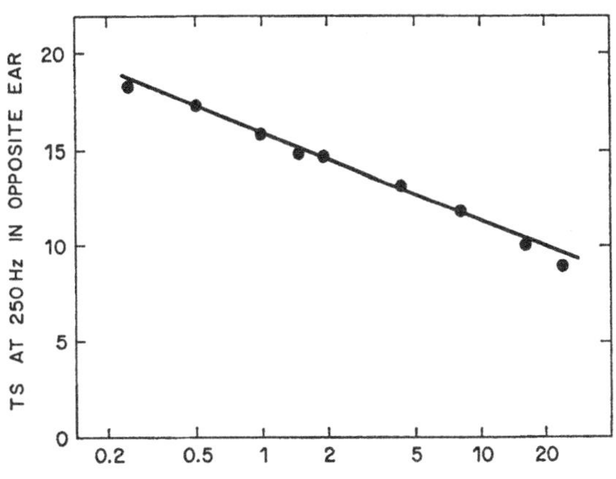

Fig. 4. Adaptation of the middle ear reflex as indicated by a progressive decrease in the threshold shift (*TS*) for a 250 Hz tone. The reflex was elicited by a narrow band of noise (120 dB SPL) presented continously to one ear. (After WARD, 1962a)

It follows that if the middle ear reflex elevated threshold, it should bring about a decrease in the apparent loudness of suprathreshold sounds. Conversely, if the reflex is inoperative, intense sounds should appear even louder than normally.

PERLMAN (1938) published a clinical report on a patient suffering from a facial paralysis on the right side which also involved the stapedius muscle. Loud sounds experienced in day-to-day urban living caused considerable discomfort in the patient's right ear. Careful testing in the laboratory indicated a loudness imbalance between ears when the sound level exceeded 70 dB. For instance, a 512 Hz tone presented to the right ear at 85 dB was equivalent in loudness to a 95 dB tone delivered to the left ear. This disturbed loudness function returned to normal upon recovery from paralysis. These data imply that the loudness increase of sounds presented to the affected ear were due to the absence of the stapedius reflex. According to MELNICK (1968), many stapedectomized patients are hyper-irritable to relatively intense sounds. Since they have no intact stapedius muscle, it is reasonable to assume that the absence of a normal middle ear reflex may contribute to their irritability; *i.e.*, sounds which appear loud to persons with normal functioning ears may be too loud for the stapedectomized patient. Reasoning from the fact that the middle ear reflex by far has its greatest effect on the transmission of low frequency sinusoids — a fact whose behavioral manifestation was unequivocally demonstrated by REGER (1960) — MELNICK calculated, from a series of loudness judgments, the intensity levels at which a low frequency tone (250 Hz) was subjectively equal in loudness to a middle frequency tone (1 000 Hz) delivered over a wide range of stimulus intensities. All stimuli were presented to the ear deprived of an intact stapedius muscle. A normal hearing group of listeners was tested for purposes of comparison. Instead of finding a more rapid growth of loudness in the impaired ears for the 250 Hz tone, both the pathological and normal groups gave essentially the same results. Cognizant of the possibility that these negative results may have been due to the requirements peculiar to the testing technique, MELNICK employed the method of magnitude estimation devised by STEVENS (1956) and suitable for estimating the growth of loudness. Again, no appreciable differences in the loudness function occurred between the impaired and normal ears. The author was forced to conclude that some factor other than an abnormal loudness function must be responsible for the stapedectomized patient's hyperirritability to sounds. He suggested that without the stapedius muscle, the normal tolerance of ossicular displacement may be exceeded when driven by relatively intense sounds. Hence, it may be cutaneous and perhaps mild pain sensations which could accompany the sensation of loudness that give rise to the patient's discomfort.

LOEB and RIOPELLE (1960) carried out an experiment on loudness judgments as influenced by the middle ear reflex in which only normal hearing listeners were employed. A brief 2200 Hz tone, presented at 105 dB SPL to the left ear, served as the reflex activating stimulus. 200 msec later a 500 Hz tone-burst (standard stimulus) was presented to the right ear. One second later another 500 Hz tone-burst (comparison stimulus) was given to the right ear, and the subject was required to say whether the comparison stimulus was softer or louder than the standard. As the intensity of the standard stimulus was increased progressively over the 70 dB SPL reference level, the comparison stimulus required less increments in intensity to appear equally loud. As an illustration, when the intensity of the standard stimulus reached 100 dB SPL, the comparison stimulus appeared just as loud when presented at approximately 93 dB SPL (median value). The reflex, therefore, must have reduced the apparent loudness of the standard stimulus.

In summary, these studies imply that the middle ear muscle activity is reflected by changes in apparent loudness. Behavioral methods, however, are a poor choice for gaining information in the state of the middle ear when more direct and simpler techniques are available (see Vol. V/1, Chapter 15 and 16). Nevertheless, these data on the influence of the aural reflex on hearing thresholds are important to psychoacousticians, particularly, when interpreting the results of auditory masking experiments, an area in which a great deal of research is in progress.

The functional significance of the aural reflex in man is, indeed, obscure. It obviously cannot furnish protection against the onslaught of the intense noise characteristic of much of our present environment. A moment's reflection on the probable low noise levels in the environment throughout the phylogenetic development of our species is all that is necessary to convince us that the reflex could not have been naturally selected to protect the inner ear from damage. What then are other possible functions? LAWRENCE (1965) proposed a fascinating idea. He argued from the data collected in animal experiments that the middle ear muscles can influence the intensity and phase of even low intensity sounds and that the activity in one ear is, in part, independent of that in the other ear. Since interaural phase and intensity differences are of prime importance in sound localization, LAWRENCE suggested that the joint action of the two sets of middle ear muscles may serve to direct and alternate attention to specific sound sources originating among many sound sources occurring simultaneously. SIMMONS (1964), in his provocative paper on the aural reflex, suggested another interesting role of the middle ear muscle activity. Based on the fact that it is the low frequency sounds which are attenuated by the reflex and that middle ear muscle activity accompanies chewing and head movements, he proposed that the differential attenuation of the internal low frequency noise generated by eating prevents them from masking higher frequencies originating in the external environment. This then would better enable an animal, while eating or grazing, to detect auditorily the approach of a predator. Other speculations exist. Further research will enable us to identify those less tenable and, hence, eliminate them from serious consideration.

References

ANGELL, J. R., FITE, W.: The monaural localization of sound. Psychol. Rev. 8, 225—246 (1901).

BATTEAU, D. W.: The role of the pinna in human localization. Proc. roy. Soc. **168**, Series B, 158—180 (1967).

BLAUERT, J.: Sound localization in the median plane. Acustica **22**, 205—213 (1969/1970).

BLOCH, E.: Das binaurale Hören. Z. Ohrenheilk. **24**, 25—85 (1893).

BORING, E. G.: Sensation and perception in the history of experimental psychology. New York: Appleton-Century 1942.

BRUNZLOW, D.: Über die Fähigkeit der Schallokalisation und ihre Bedingtheit durch die Schallqualitäten und die Gestalt der Ohrmuschel. Z. Sinnesphysiol. **56**, 326—363 (1925).

BURGER, J. F.: Front-back discrimination of the hearing system. Acustica 8, 301—302 (1958).

BUTLER, R. A.: On the relative usefulness of monaural and binaural cues in locating sound in space. Psychon. Sci. **17**, 245—246 (1969).

DARWIN, C.: The descent of man, 2nd Ed. New York: Appleton 1907.

Freedman, S. J., Fisher, H. G.: The role of the pinna in auditory localization. In: Freedman, S. J. (Ed.): The neuropsychology of spatially oriented behavior. Homewood, Ill.: Dorsey Press 1968.

Gardner, M. B., Gardner, R. S.: Problem of localization in the median plane: effect of pinnae cavity occlusion. J. Acoust. Soc. Amer. **53**, 400—408 (1973).

Gilse, v., Roelofs, O.: Untersuchungen über die Schallokalisation. Acta oto-laryng. **14**, 1—20 (1930).

Hensen, E.: Die Functionen des äußeren und mittleren Ohrs. In: Hermann, L. (Hrsg.): Handbuch der Physiologie, Bd. III, Tl. 2. Leipzig: Vogel 1879.

Hocart, A. M., McDougall, W.: Some data for a theory of the auditory perception of direction. Brit. J. Psychol. **2**, 386—405 (1908).

Hornbostel, E. M. v., Wertheimer, M.: Über die Wahrnehmung der Schallrichtung. Sitzungsber. Akad. Wiss. Berlin **15**, 388—396 (1920).

Jongkees, L. B. W., Groen, J. J.: On directional hearing. J. Laryngol. Otol. **61**, 494—504 (1946).

Kessel, J.: Über die Funktion der Ohrmuschel bei den Raumwahrnehmungen. Arch. Ohrenheilk. **18**, 120—129 (1882).

Kietz, H. v.: Das räumliche Hören. Acustica **3**, 73—86 (1953).

Klemm, O.: Untersuchungen über die Lokalisation von Schallreizen. Arch. ges. Psychol. **38**, 71—114 (1918).

Lawrence, M.: Middle ear muscle influence on binaural hearing. Arch. Otolaryngol. **82**, 478—482 (1965).

Loeb, M., Riopelle, A. J.: Influence of loud contralateral stimulation on the threshold and perceived loudness of low-frequency tones. J. Acoust. Soc. Amer. **32**, 602—610 (1960).

Loeb, M., Fletcher, J. L.: Contralateral threshold shift and reduction in temporary threshold shift as indices of acoustic reflex action. J. Acoust. Soc. Amer. **33**, 1558—1560 (1961).

Mach, E.: Bemerkungen über die Funktion der Ohrmuschel. Arch. Ohrenheilk. **9**, 72—76 (1875).

Melnick, W.: Loudness post-stapedectomy. Ann. Otol. Rhinol. Laryngol. **77**, 305—316 (1968).

Müller, J.: Von den Formen und akustischen Eigenschaften der Gehörwerkzeuge. In: Handbuch der Physiologie des Menschen, Bd. II. Koblenz: Hölscher 1840.

Perekalin, W. E.: Über die akustische Orientierung. Z. Hals-, Nas.- u. Ohrenheilk. **25**, 442—461 (1929).

Perlman, H. B.: Hyperacusis. Ann. Otol. Rhinol. Laryngol. **47**, 947—953 (1938).

Petri, J.: Über Aufbau und Leistung der Ohrmuschel. Z. Hals-, Nas.- u. Ohrenheilk. **30**, 605—608 (1932).

Politzer, A.: Studien über die *Paracusis loci*. Arch. Ohrenheilk. **11**, 231—236 (1876).

Rayleigh, O. M.: On our perception of sound directions. Phil. Mag. **13**, Series 6, 214—232 (1907).

Reger, S. N.: Effect of middle ear muscle action on certain psychophysical measurements. Ann. Otol. Rhinol. Laryngol. **69**, 1179—1198 (1960).

Roffler, S. K., Butler, R. A.: Factors that influence the localization of sound in the vertical plane. J. Acoust. Soc. Amer. **43**, 1255—1259 (1968).

Simmons, F. B.: Perceptual theories of middle ear muscle function. Ann. Otol. Rhinol. Laryngol. **73**, 724—739 (1964).

Stevens, S. S.: The direct estimation of sensory magnitudes-loudness. Amer. J. Psychol. **69**, 1—25 (1956).

Thompson, S. P.: On the function of the two ears in the perception of space. Phil. Mag. **13**, Series 5, 406—416 (1882).

Toynbee, J.: Diseases of the ear. Philadelphia: Blanchard and Lea 1860.

Ward, W. D.: Studies on the aural reflex. II. Reduction of temporary threshold shift from intermittent noise by reflex activity; implications for damage-risk criteria. J. Acoust. Soc. Amer. **34**, 234—241 (1962a).

Ward, W. D.: Studies on the aural reflex. III. Reflex latency as inferred from reduction of temporary threshold shift from impulses. J. Acoust. Soc. Amer. **34**, 1132—1137 (1962b).

Ward, W. D., Selters, W., Glorig, A.: Exploratory studies on temporary threshold shift from impulses. J. Acoust. Soc. Amer. **33**, 781—793 (1961).

Chapter 7

Behavioral Tests of Hearing and Inner Ear Damage

GÖRAN BREDBERG* and IVAN M. HUNTER-DUVAR**, Uppsala (Sweden)

With 16 Figures

Contents

I. Introduction 261
II. Review of Literature 262
A. Animal Experiments 262
B. Human Studies 270
III. Discussion of Literature 277
A. General Considerations 277
B. Anatomical Frequency Scale 278
C. Recruitment and Hair Cell Populations 279
D. Ganglion Cell Population and Speech Discrimination 281
E. Histology 282
IV. Current Perspectives 287
A. Animal Experiments 287
B. Human Studies 293
C. Histology 297
D. Comment 301
References 302

I. Introduction

Results of recent animal experiments (WARD and DUVALL, 1971; HUNTER-DUVAR and ELLIOTT, 1972) and human studies (BREDBERG, 1968) cannot help but puzzle the investigator who has accepted an orderly "anatomical frequency scale" cochlea and an obvious relationship between hair cell loss and hearing impairment. A better understanding of the current status of "behavior studies and inner ear damage" may be facilitated by a review of the studies and reports which contain audiometric behavioral data correlated with inner ear histology. Theorists and model builders have, of course, combined behavioral data with anatomical changes, as well as with mechanical and electrophysiological measurements, to fashion

* This work was supported by the Swedish Medical Research Council (Project No. 3542).
** Postdoctoral Fellow. Medical Research Council of Canada.

their products. In this chapter, we shall look only at studies that report both behavioral and histological results in an effort to determine which of our current beliefs about inner ear function have foundation in these data. The literature can generally be separated into reports on human patients where histories, audiograms, and histology are available, and experimental studies on trained animals with audiograms before and after experimental manipulation of the hearing mechanism.

II. Review of Literature

A. Animal Experiments

Pavlov (1927) very early established the practicality of using conditioned animals in auditory experimentation. He reports on an experiment wherein Andreev used a classically conditioned dog to test the resonance theory of Helmholtz. After one cochlea had been surgically destroyed, a fine needle was used to destroy the apical portion of the other experimental cochlea. The animal showed a permanent behavioral hearing loss for tones lower than 310 Hz. Wittmaack (1928) reported the histology for the experimental ear of this animal and found the organ of Corti degenerated from the middle of the basal coil to the apex. Kemp's (1935) review of the experimental literature describes only one study (Davis *et al.*, 1935) in which conditioned behavior was used in conjunction with an histological evaluation of the cochlea. In that study 5 of 14 guinea pigs, exposed to pure tone stimuli, showed behavioral losses as measured by a conditioned respiratory reflex. Stimuli were of low frequency and intensity and exposure was for long duration. Of the 5 animals one was excluded because of otitis media and a second because it died early in testing. One of the remaining three animals showed a behavioral loss of unspecified degree from 800 Hz to 2 kHz; no histological damage was found. The second guinea pig showed a behavioral loss of 30 dB at the exposure frequency of 600 Hz with no histological damage. The remaining animal showed a 10–20 dB behavioral loss from 4–8 kHz; the histological report indicated scattered missing hair cells in the first outer row of the first, second and third turns. These inconclusive and diverse results are a precursor for many of the subsequent studies. The diversity, however, is generally not reflected in the conclusions of the experimenters.

Davis *et al.* (1937) conditioned cats to respond to pure tone stimuli by opening a food box. Lesions were made in the middle and inner ear with a dental drill and/or cautery. Hearing losses were measured and compared to histological lesions. Five animals were included in the report. The particular damage done to individual animals was not clearly specified. The conclusion of a satisfactory correlation between behavior and histology is questionable.

Neff's (1947) study on the effects of partial section of the auditory nerve introduced avoidance conditioned cats to auditory experimentation. For the following 20 years the cat and the avoidance conditioning procedure were almost exclusively used in behavioral research involving the inner ear. This 1947 experiment was the first of a series of behavioral studies to come out of the University of Chicago. In these studies, one of the cat's cochleas was surgically destroyed in order to confine stimulation and hearing tests to a single ear. Lindquist (1949,

1954) subjected cats to multiple temporary hearing loss exposures and then to impulse noise from a 0.32-caliber starting pistol. Seven of the eight animals showed permanent behavioral losses. The extent of the losses for individual animals is not given. If any cochlear damage was caused by the temporary threshold shifts, it is unfortunately confounded with damage from the permanent losses. In two animals with 30–40 dB permanent losses, no cochlear damage was found. In five animals, hair cells and ganglion cells appeared normal, but minor damage to supporting structures was reported. SCHUKNECHT *et al.* (1951) compared behavioral losses and cochlear damage in 9 of 10 cats that had been subjected to severe head blows. Two of the animals had sharply defined 20 dB hearing losses at 4000 Hz which generally was the frequency of maximum loss. One of the two had no visible inner ear damage and the other had some abnormal appearing inner and outer hair cells confined to 1 mm of the upper basal coil. For the remaining animals, increases in permanent hearing losses were generally accompanied by reports of increased damage to hair cells, nerve fibers, and ganglion cells.

In their next experiment, SCHUKNECHT and NEFF (1952) followed the early example of ANDREEV and made lesions in the apical end of the cochlea with a fine needle. Six trained cats were used. Two animals with organ of Corti damage confined to the apical turn showed 20–30 dB losses for all test frequencies. Two animals with damage to apical and middle turns had behavioral losses of 20–50 dB for frequencies below 8 kHz and normal thresholds above that frequency. Two animals with damage throughout the cochlea had threshold losses of 60–80 dB below 4 kHz and complete hearing loss above that frequency. Similar needle lesions to the basal coil of the cochlea of six cats (SCHUKNECHT and SUTTON, 1953) gave generally high frequency behavioral losses with relatively sharp cut offs. Histological correlates were much better for this group than for the apical lesion group but still were not always well defined. As an example, one animal showed a normal hair cell population with a 40–70 dB hearing loss at 8 kHz and above. The correlate for this loss was a complete ganglion cell loss for less than 1 mm extending to a 50% loss for less than 3 mm. SCHUKNECHT and WOELLNER (1953) experimented with partial section of the cochlear nerve in 5 cats. Three of the operations proved successful. The other two animals had cochlear blood supply complications. Eighty per cent losses in spiral ganglion cells gave audiometric shifts that varied from 0 dB in one of the animals to 40–60 dB below 8 kHz in the other two animals.

SCHUKNECHT (1953a) selected the data from 16 of the cats used in the experiments discussed above and proposed the anatomical frequency scale for the cat cochlea shown in Fig. 1. In addition, the overall evidence from the experimentation on avoidance conditioned cats was considered and a number of conclusions were presented (SCHUKNECHT, 1953b). Several of the suggestions were in line with those previously made by other experimenters. It was suggested that outer hair cells were more susceptible to injury than inner hair cells (except in drug lesions). Inner hair cells were calculated to be at least 50 dB less sensitive to acoustic stimuli than outer hair cells. Hair cells could be lost without spiral ganglion degeneration, and primary degeneration of the spiral ganglion produced no change in Corti's organ. Up to 50% of the spiral ganglion cells could be degenerated in any area of the cochlea with no elevation of pure tone threshold. Degeneration of pillar cells gave the best correlate with spiral ganglion lesions. How these early conclusions have fared over the years will be considered below.

SCHUKNECHT and WOELLNER (1955) presented additional data for eight cats with partial section of the cochlear nerve. Four animals with extensive ganglion cell losses and normal hair cell population are reported. Two of these animals showed no

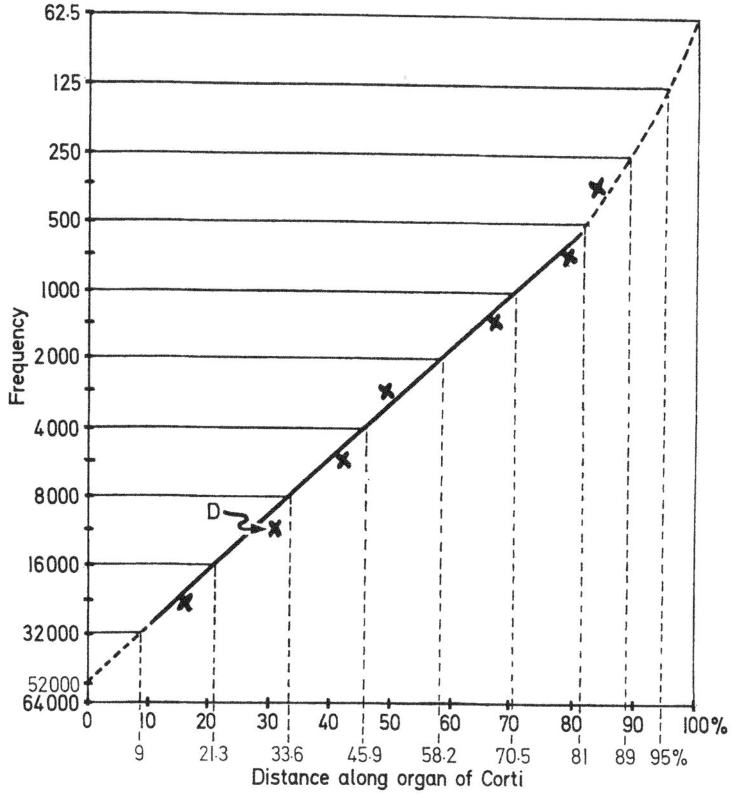

Fig. 1. Anatomical frequency scale for the cat cochlea. Distance along cochlear duct in percentage is plotted as function of frequency. Frequencies are represented along ordinate on logarithmic scale. Mean percentage distances for various frequency ranges (table) have been placed in proper locations. Thus point D representing 8000—16000 cps range is located on line extending from point midway between 8000 and 16000 cps and 31% of way along the cochlea. (From SCHUKNECHT, 1953a)

hearing loss; the third had a 30 dB loss at 500 Hz while the fourth had a loss of 60 dB at 2000 Hz and 30–40 dB loss for 4000 Hz and higher frequencies. Correlation between hearing losses and inner ear pathology for these animals was difficult to make because of the irregularity in ganglion cell losses. In the other four animals almost complete ganglion cell losses are accompanied by severe hair cell losses and cochleas showed severe degeneration attributed to blood supply complications. Hearing losses ranged from severe to total loss.

ELLIOTT (1961) brought a new dimension to auditory behavioral experimentation by measuring changes in pitch discrimination of avoidance conditioned cats after cochlear insult. Two of four cats, that received cochlear nerve lesions, had extensive localized ganglion cell losses (90%). Behaviorally these two animals showed no change in pure tone thresholds and no deviation from normal pitch discrimination thresholds. The other two animals with extensive ganglion cell losses throughout the cochlea exhibited normal pitch discrimination but had pure tone threshold deficits which varied from slight to moderate. In this same study, four other cats were exposed to high frequency, high intensity, pure tone stimulation. Three of these animals had severe pure tone losses (80 dB) at 8 kHz and above. Pitch discrimination and thresholds for 4 kHz and lower frequencies were normal. Histology on two of the animals showed extensive hair cell and ganglion cell losses generally corresponding to locations that would be predicted from SCHUKNECHT's anatomical frequency scale. The third animal had no ganglion cell losses but extensive hair cell loss in the basal coil, again correlating well with the Schuknecht scale. The fourth sound exposed animal had normal thresholds for frequencies of 2 kHz and below with a greater than 80 dB loss at 8 kHz and higher frequencies. This animal had deficits in pitch discrimination at 2 and 4 kHz; histological examination of the inner ear surprisingly showed no hair cell or ganglion cell damage.

MCGEE and OLSZEWSKI (1962) used avoidance conditioned cats in the first animal behavioral study of the effects of ototoxic drugs on the inner ear. Animals were given selected doses of streptomycin and dihydrostreptomycin for different periods of time and were allowed long survival time after drug treatment. Only representative audiograms and histology were presented in the report. Permanent losses were reported to correlate with severe and moderate hair cell and/or ganglion cell damage; temporary threshold shifts with slight damage. This experiment is unique in the literature on animal behavioral tests in relation to inner ear damage in that a histological control group was included. In view of data on normal cat cochleas presented below (see Fig. 7), we find the report of no pathology in the seven control animals surprising. MCGEE *et al.* (1963) exposed cats to ultrasonic stimulation. The basal turn of the cochlea was thinned and irradiated. The report gives audiograms and histology for only 3 of 7 experimental animals. In the three animals shown, high frequency losses and location of cochlear damage appear related; however, this result is not seen for the low frequency behavioral losses.

MILLER *et al.* (1963) looked at the effects of noise on the behavioral audiograms and on damage to the organ of Corti of cats whose hearing was tested by avoidance conditioning. Forty-two trained animals were exposed in the study. Inner ear damage was rated on a nine point scale. Among other results, the experimenters found that an increase in duration of exposure to continuous broad band noise correlated with an increase in amount and extent of permanent hearing loss. The location of cochlear damage was generally confined to the upper part of the basal turn and the lower and middle part of the second turn. This area extends from 9–16 mm as measured from the round window. Greatest behavioral losses occurred for the frequencies 1, 2, and 4 kHz with the point of maximum loss varying among these frequencies for different animals. The histological ratings used make comparison with data from other studies difficult. According to these ratings, only occasional missing hair cells were seen for losses of less than 60 dB although hair cells, supporting cells, and other structures may have shown marked changes.

Elliott and McGee (1965) added intensity discrimination thresholds to the behavioral measurements made on cats with inner ear lesions. Cochlear lesions were made in eleven animals using high intensity pure tone stimuli that ranged in frequency from 1 through 26 kHz. In a twelfth animal, the lesion was made with an ultrasonic stimulus. Four of the animals had hearing losses so severe that discrimination testing was not possible. Post exposure intensity discrimination thresholds did not vary from the normal range in the testable animals. The evaluation of the prediction of a decrease in intensity discrimination thresholds with hair cell dysfunction was complicated by the occurence of neural degeneration. Hair cell lesions were in generall considerably smaller than might be predicted from the extent of behavioral losses.

In an experiment designed to determine the effects of salicylates on the auditory system, Myers and Bernstein (1965) introduced avoidance conditioned squirrel monkeys to inner ear research. Elevated pure tone thresholds of up to 40 dB, resulting from acute salicylate intoxication, were found to be fully reversible when the drug was withdrawn. No cochlear lesions that could be attributed to the treatment were found when cochlear sections were viewed by light and by transmission electron microscopy. Cummings (1968) tested the ototoxicity of nitrogen mustard on four conditioned cats. Hearing losses ranged from 15–20 dB at 125 Hz to 70 dB at 500 Hz; all animals had complete loss for test frequencies above 500 Hz. Surprisingly, there was no change in thresholds from the first day after treatment until the animals were sacrified two weeks later. Histological examination showed complete hair cell loss except for the apical 4 or 5 mm of the cochleas where 20–60% of the hair cells remained. Ganglion cells were reported normal in all animals.

Stebbins *et al.* (1969) used monkeys (macaques), conditioned with food reinforcement, in a study of hearing losses and cochlear pathology produced by ototoxic drugs. Animals received daily doses of kanamycin or neomycin and permanent shifts in pure tone thresholds were followed. Treatment was stopped at different times for different animals. Audiograms showed a change from normal thresholds to losses of 60–90 dB (or complete loss) in less than an octave. Losses progressed over time; the highest test frequency was first affected followed by losses for lower frequencies. This experiment was the first behavioral study to use phase microscopy of surface preparations (Engström *et al.*, 1966) in the post mortem examination of cochleas. Histology showed a transition from all hair cells present to all hair cells missing over a distance of 1–2 mm. On the basis of definitive data from four of the five animals, the authors suggested locations for three frequencies on the macaque cochlea. Figure 2 shows these locations as well as that for a fourth frequency added in a later paper (Stebbins *et al.*, 1971). Dolan *et al.* (1970) reported audiograms of cats that had been exposed to pure tones at high intensities for different durations of time. Animals for which results were given in this preliminary report had been subjected to tones of 1, 2, or 4 kHz. Post mortem examinations of the cochleas made use of surface preparations with phase microscopy. Sizable apical lesions without corresponding audiogram deficits were seen in some animals. Extent of hair cell losses were generally smaller than might be expected from the hearing losses. Audiograms and cochleographs for these animals are included in Fig. 10.

In reference to damage risk criteria, ELDREDGE and MILLER (1969) discussed results from an avoidance conditioned chinchilla with a 20 dB hearing loss from 750–3000 Hz and an extensive hair cell loss in the region 6–12 mm from the basal end. This introduced a new species to this field of research. ELDREDGE *et al.* (1971) reported on four chinchillas that exhibited slower than normal recovery when exposed to an octave band noise of moderate intensity and long duration. None of

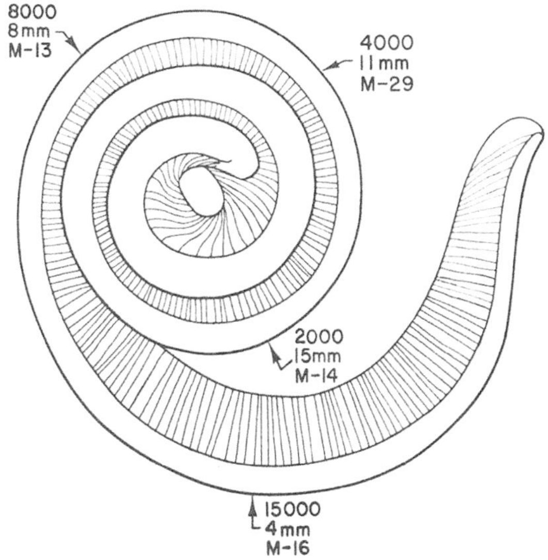

Fig. 2. Cochlear locations of the regions for threshold responses at 15, 8, 4, and 2 kHz in the monkey as determined in experiments with kanamycin and neomycin. (From STEBBINS, CLARK, PEARSON, and WEILAND, 1971)

the animals had significant audiogram changes when tested 3 months after exposure. Subsequent histology showed moderate localized hair cell losses in all animals. Histological examination involved a new procedure (BOHNE, 1972), which combined the embedding techniques used for transmission electron microscopy with surface preparation and phase microscope techniques. Although animals in this study exhibited normal thresholds with hair cell losses, there is some possibility that the losses seen were a product of the selection procedure rather than the stimulation. Thus, Fig. 3 (HUNTER-DUVAR, 1971) shows data from a squirrel monkey which had normal thresholds but considerable hair cell loss in both exposed and control cochleas. The experimenter reports that this animal always recovered very slowly from temporary threshold shift exposures. The unexposed ear prohibits attributing the hair cell loss to the fatiguing stimulus in this case.

HUNTER-DUVAR and ELLIOTT (1972) interrupted the ossicular chain in one ear of each of fourteen squirrel monkeys so that the cochlea might be used as an histological control. The experiment involved relatively small (20 dB) permanent and temporary hearing losses. The animals were avoidance conditioned and two, used as normal and operation controls, received no traumatic exposure. The experimental ears of six of the animals were exposed to pure tone stimulation at either 1 or 2 kHz (120 dB, SPL) for sufficient durations to create permanent threshold shifts (PTS) of 15–20 dB. The remaining six animals received multiple pure tone exposures at either 1 or 2 kHz (120 dB, SPL) each of sufficient duration to create temporary threshold shifts (TTS) of 15–25 dB measured 30 min after exposure. Subsequent histology done by phase microscopy of surface preparations

Cochlea

	Row	Apex							Base	Total
Left	I	4	4	4	1	4	1	4	6	28
	O_1	13	22	25	44	65	102	87	48	406
	O_2	12	7	9	6	6	1	2	3	46
	O_3	23	22	18	25	3	4	4	8	107
Right	I	3	1	4	0	1	1	1	1	12
	O_1	10	21	31	46	66	123	102	53	452
	O_2	8	8	19	14	4	6	6	10	75
	O_3	31	26	30	22	12	9	1	16	147

Missing hair cells

Fig. 3. Cochleographs for exposed (left) and unexposed ears of a squirrel monkey. Audiograms were normal before and after TTS exposures however abnormally long recovery times were required for this animal. Cochleographs extend from 1—17 mm from the apex. Each square represents 2 mm of cochlea and contains approximately 210 hair cells for inner rows and 260 hair cells for outer rows. (From HUNTER-DUVAR and ELLIOTT, 1972)

showed no difference in hair cell losses between eperimental and control cochleas of exposed animals with either temporary or permanent hearing losses. There was no difference in hair cell losses between exposed and unexposed animals nor was there any difference between the operated and unoperated ears of the unexposed animals. Scattered hair cell losses were present in all cochleas.

WARD and DUVALL (1971) exposed avoidance conditioned chinchillas to octave band noise of different intensities and durations. In the first group of animals, permanent losses ranged from 20 dB at 250 Hz to 70 dB at 2 kHz with a mean loss of 50–60 dB from 4–16 kHz. Postmortem examination of cochleas made use of light microscopy of serial sections supplemented by transmission electron microscopy. The organ of Corti was destroyed in the basal two turns (75%) of the cochlea. A second group of animals, with 30–40 dB means losses from 2–16 kHz, also showed an absence of hair cells in the basal and middle turns and a degenerated organ of Corti. Surprisingly, two animals with essentially normal thresholds were missing all outer hair cells in the basal turn as well as in the greater part of the middle turn. Inner hair cells appeared normal except for some fused stereocilia. WARD (1971) reports that he has additional chinchillas exhibiting normal thresholds with all outer hair cells missing in the basal and middle turns. One animal showed no behavioral loss, and histological examination showed both inner and outer hair cells absent in the basal turn. These unusual findings will be further discussed below.

ERNSTSON (1972), using a conditioning technique originated by ANDERSSON and WEDENBERG (1965), obtained audiograms for normal guinea pigs and for a

Table 1. Animal behavioral studies with inner ear histology

Subject	Stimulus						
	Airborne sound	Ultrasound	Head blows	Mechanical insult	Drugs	Nerve lesion	Genetic
Cat	Lindquist *et al.* (49) Elliott (61) Miller *et al.* (63) Elliott and McGee (65) Dolan *et al.* (70)	McGee *et al.* (63)	Schuknecht *et al.* (51)	Davis *et al.* (37) Schuknecht and Neff (52) Schuknecht and Sutton (53)	McGee and Olszewski (62) Cummings (68)	Neff (47) Schuknecht and Woellner (53) Schuknecht and Woellner (55)	
Chinchilla	Eldredge and Miller (69) Eldredge *et al.* (71) Ward and Duvall (71)						
Dog				Pavlov (27)			
Guinea pig	Davis *et al.* (35)						Ernstson (72)
Monkey (macaque)					Stebbins *et al.* (69) Stebbins *et al.* (71)		
Monkey (Saimiri sciureus)	Hunter-Duvar and Elliott (72)				Myers and Bernstein (65)		

strain of waltzing guinea pigs. Thresholds for the waltzing guinea pigs, initially the same as those of the normals, declined to total hearing losses over a period of approximately 40 days. Pure tone audiograms were monitored and hair cell counts done for 5 animals – one normal and 4 waltzers. Normal thresholds as reported for frequencies from 1–4 kHz are suspect because of the failure to test at intensities lower than 0 dB when the modal response for these frequencies was at 0 dB. Although there may be some increase in missing hair cells with an overall increase in hearing loss we are not able to relate the patterns of hair cell losses to the hearing losses shown in the audiograms.

Table 1 is a summary of the animal experimental literature discussed above. During the last half century approximately 25 studies have been reported in which audiograms were obtained by behavioral methods and were related to inner ear damage. The studies have used six different species of animals and a wide variety of stimuli. Results have often been diverse and sometimes even contradictory. Some of the possible reasons for the diversity will be discussed below.

B. Human Studies

The first systematic studies relating human temporal bone pathology to results of hearing tests were done at Johns Hopkins Hospital in Baltimore in the beginning of the thirties. Temporal bones had been collected before 1930 in Europe, but the approach was not systematic and pure tone audiometry was not used in hearing tests. Guild (1921) published a design for the graphic reconstruction of the guinea pig cochlea from serial sections. This technique was adopted at Johns Hopkins Hospital in the late 1920's for the study of the human cochlea. A large number of temporal bones, for which hearing tests taken prior to death were available, were collected and studied with the new histological technique. The specimens were grouped according to specific patterns of hearing loss. Crowe *et al.* (1934) reported in detail the findings for 94 temporal bones from 58 cases with audiograms. Seventy-nine of these temporal bones were associated with a "high tone hearing loss". This study is exceptional in that it reports in detail the histology of a control group of 15 ears from 8 cases with "normal" hearing, that is, 20 dB or less hearing loss up to 8192 Hz. In this control group, nerve atrophy, averaging as much as 50% in the lower 4 mm of the basal turn and 20% in the next 3 mm, was frequently observed. The organ of Corti was generally undamaged at all levels. An occasional exception occurred in the last 2 mm at the basal end of the cochlea. In several cases, the stria vascularis displayed atrophic areas where it was less than half its usual width. Of 46 ears with a "gradually increasing" hearing loss for tones above 2000 Hz, in 24, partial atrophy of nerves in the basal 10–15 mm was a characteristic histopathological finding. In 17 of the 46 ears, no lesion correlate was found for a hearing loss that in some cases reached 55 dB in the 4–8 kHz range. In 24 of 33 ears with an "abrupt" high tone hearing loss, the typical finding was a degeneration of nerves and of the organ of Corti in the basal coil. In 10 of these 24 ears, the degeneration extended throughout the entire length of the basal coil. In the other 14, degeneration was described as "patchy". In 5 of the remaining 9 ears, nerve degeneration was observed, while, in the other 4, no pathology explaining the hearing losses was found.

ODA (1938) extended the study of the Johns Hopkins Hospital temporal bone material with a report of 35 ears from 19 cases with impaired hearing for both low and high tones. In a search for a morphological correlate to the low frequency hearing loss, he compared these cases with others having only a high frequency hearing loss from the CROWE, GUILD, and POLVOGT, 1934 study. Changes in the cochlea involved the middle and apical coils in only seven of the ears. These ears had extensive partial atrophy of the nerve. In three, it was continuous and, in four, it was more or less localized. In all seven ears, the nerve atrophy extended beyond the upper basal turn, up to or nearly up to the apex. Only one cochlea showed extensive degeneration of the organ of Corti; there was complete atrophy in the basal 17 mm. The organ of Corti appeared normal above 21,5 mm. The hearing loss in this ear was 40–45 dB at 64, 128 and 256 Hz; 70 dB at 500 Hz and more than 100 dB above 500 Hz. ODA concluded that for frequencies of 8192, 4096, and 2048 Hz, the evidence for tonal localization along the basilar membrane was in complete agreement with the data presented by CROWE *et al.* (1934). However, ODA also stated: "The evidence as to localization of lower tones on the contrary, does not support the theoretical views that there is even a moderably restricted area of response to these frequencies". CIOCCO (1934) analyzed the data of CROWE *et al.* (1934) and concluded that there was evidence for frequency localization in the cochlea. He suggested that the area of most importance for 8192 Hz was the middle of the lower basal turn; for 4096 Hz, the upper end of the lower basal turn; and for 2048 Hz, the lower part of the upper basal turn.

The term presbyacusis (presbycusis) has been used to describe a gradual decline with age in pure tone threshold sensitivity, affecting the high frequencies more than the low. Several population studies of hearing at different ages have been carried out (*e.g.*, GLORIG *et al.*, 1957, 1962; HINCHCLIFFE, 1959; JATHO and HECK, 1959). SPOOR (1967) analyzed eight different studies and presented graphs of presbyacusis. A number of temporal bone studies claim to describe presbyacusis; however, in most cases, the audiograms do not meet the audiological criteria of presbyacusis as given by population studies (SPOOR, 1967). The studies, instead, describe temporal bone pathology in elderly hard of hearing people. The most constantly described histological finding in presbyacusis is a loss of ganglion cells and nerves in the basal end of the cochlea (CROWE *et al.*, 1934; FLEISCHER, 1956; SCHUKNECHT, 1964a; MAKISHIMA, 1967). CROWE *et al.* (1934) discussed several cases where no correlate to the hearing loss could be found. SCHUKNECHT (1967) has also stated that no obvious histophathological correlate has been found for the typical audiometric curve that shows a high frequency loss that gradually becomes more severe and involves the next lower frequencies as age increases. However, BREDBERG (1968), applying the surface specimen technique to the study of the human cochlea, demonstrated a loss of sensory cells with increasing age. The inner hair cells showed a degeneration confined to the basal region, whereas the outer hair cells showed an over-all degeneration accentuated at both the basal and the apical ends of the cochlea. HANSEN and RESKE-NIELSEN (1965) described the central auditory pathways and the temporal bones of 12 elderly subjects. All audiograms showed a high tone loss and different amounts of low frequency hearing loss. They described a moderate degeneration of the ganglion cells in the basal coil and a flattening of the organ of Corti throughout all coils. The latter finding was regarded as due mainly to preparation artefacts. The authors felt that the high tone loss could be explained by the cochlear findings, but they assumed that the hearing loss for low tones was exclusively central in origin. Schuknecht (1964) and Schuknecht and Igarashi (1964) described three cases of hearing loss of unknown etiology with "flat" audiometric curves at about 40–60 dB. The only pathology found was a partial atrophy of the stria vascularis in the upper parts of the cochlea. Schuknecht (1964) classifies this as a type of presbyacusis.

An early symptom of Ménière's disease is an irregulary fluctuating hearing loss for low tones. In more advanced cases, the hearing loss includes all frequencies and the fluctuations are less pronounced and less frequent. The final stage is a flat loss of 50–70 dB associated with poor speech discrimination. A characteristic feature throughout the disease is the presence of loudness recruitment. Since the first temporal bone reports of Hallpike and Cairns (1938) and Yamakawa (1939), studies have generally reported late stages of the disease with audiograms revealing a flat loss. A finding almost invariably reported has been different degrees of hydrops of the endolymphatic spaces. The hydrops may be found in the cochlear duct alone, most markedly in the apical coil, or it may extend to include the saccule and utricle. Ruptures of Reissner's membrane and ruptures or herniations of the walls of the saccule and utricle have also been described. In spite of a pronounced flat loss (40–70 dB) and recruitment, the organ of Corti has been reported to be morphologically normal in many cases of Ménière's disease (Lindsay, 1944, 1960; Schuknecht, 1953, 1968; Schuknecht *et al.*, 1962; Hansen and Reske-Nielson, 1964; Lindsay *et al.*, 1967). In other advanced cases, the organ of Corti has been described as pathological, mostly as more or less compressed (Hallpike and Cairns, 1938; Dix *et al.*, 1948; Lindsay and von Schultess, 1958; Kristensen, 1961; Lindsay *et al.*, 1967). In a few cases, atrophy of the spiral ganglion and of the nerves in the osseous spiral lamina has been reported (Lindsay and von Schultess, 1958; Kristensen, 1961). In Ménière's disease, Schuknecht (1968) considers a disturbance of the biochemical and metabolic activity and not a morphological change in the cochlea as the cause for the loss in threshold sensitivity and the presence of loudness recruitment.

Several drugs have been shown to have toxic effects of the inner ear. There exist reports of human temporal bones where the hearing loss has been attributed to such toxic effects. Several antibiotics of the streptomyces group, which are of clinical value, have the unfortunate side effect of being ototoxic. Graf (1951) described degeneration of sensory cells in the basal coil of the cochlea in patients who became deaf under streptomycin therapy for tuberculous meningitis. Lindsay *et al.* (1960) reported on the temporal bones of a patient with a profound deafness attributed to neomycin. Histology showed an almost complete absence of inner hair cells and 60–100% irregular degeneration of the outer hair cells. No damage was seen in the peripheral neurons and apical ganglion. Benitez *et al.* (1962) described two cases of kanamycin induced hearing loss. In one case, only 3 g of the drug were given as the drug was discontinued once a high tone hearing loss developed. Despite the cessation of the therapy, the hearing loss progressed and resulted in high tone loss involving 2000 Hz and higher frequencies. One of the ears showed an abrupt hearing loss whereas the loss in the other was more gradual. However, the two ears showed almost identical lesions with an almost total loss of hair cells in the basal 15 mm of the cochlea. Some inner hair cells remained between 12 and 15 mm from the end. The possibility exists that the audiogram, obtained three weeks before death, did not represent the actual hearing loss at time of death. The other case reported had a profound hearing loss after 28 g of kanamycin. An audiogram was not obtained. The damage to the cochlea consisted of a severe loss of hair cells extending upwards 27 mm from the basal end. JØRGENSEN and SCHMIDT (1962) reported the histology of a case of complete deafness attributed to 11 g of kanamycin. There was a total loss of inner and outer hair cells in the basal and middle turns with a small number of hair cells present in the apical turn. MATZ *et al.* (1965) described a case of a 73 year old patient who developed a complete deafness after treatment with kanamycin. The damage consisted of a severe destruction of outer hair cells and only infrequent loss of inner hair cells. A case of hearing loss attributed to ethacrynic acid, a potent diuretic, was reported by MATZ *et al.* (1969). The audiogram of the left ear showed a gradually falling curve from 40 dB at 250 Hz to 70 dB at 8000 Hz. The preservation of the cochlea appeared excellent with the outer hair cells completely degenerated in the basal half of the cochlea and 20–50% missing in the apical half. The inner hair cells, the ganglion cells, and all other structures appeared normal in both ears. The audiogram of the right ear revealed a gradually increasing loss in threshold sensitivity for the lower frequencies reaching 25 dB at 1000 Hz, a rapid fall to 60 dB at 2000 Hz, and again a gradual fall to 70 dB at 8000 Hz. The outer hair cells were completely gone in the most basal 5 mm of the cochlea. From 5–15 mm, about two thirds of the outer hair cells were gone, and all cells appeared normal in the rest of the cochlea. SCHUKNECHT (1964a) described a case in which a hearing loss occured after the administration of nitrogen mustard for the treatment of Hodgkin's disease. There was a 10–20 dB loss for 250 Hz with descending audiometric curves reaching a maximum of about 50 dB for 4000 and 8000 Hz. The organ of Corti appeared shrunken. Although the compression of the organ of Corti very much resembled an artefact, "the compression phenomenon" (FERNANDEZ, 1958), the author attributed it, in part, to the effects of chemical ototoxicity. The change was very severe in the basal coil and progressively less severe toward the apex. Salicylate intoxication produces a "flat" hearing loss which is completely reversible (MYERS *et al.*, 1965; MYERS and BERNSTEIN, 1965; McCABE and DEY, 1965; and GOLD and WILPIZESKI, 1966). BERNSTEIN and WEISS (1967) reported no inner ear changes in two subjects having severe hearing losses due to chronic salicylate intoxication.

A characteristic finding in subjects exposed to intense acoustic stimuli is a loss in threshold sensitivity for pure tones of frequencies in the neighborhood of 4000 Hz. More severe injury extends the spectrum of hearing loss to include higher and lower frequencies, but the maximum loss is usually in the 4000 Hz region. Our knowledge of structural changes in the human cochlea related to acoustic trauma is very sparse. The first report to be found in the literature was made by HABERMANN (1890) who described the cochleas of a blacksmith, whose occupation had exposed him to high intensity noise during his working life. The most extensive damage was found in the upper part of the lower basal coil where the organ of Corti, the nerves in the osseous spiral lamina, and the spiral ganglion were all completely degenerated. IGARASHI *et al.* (1964) reported two cases with audiograms displaying abrupt high tone losses with a maximum of 75 dB at 4 kHz in one and

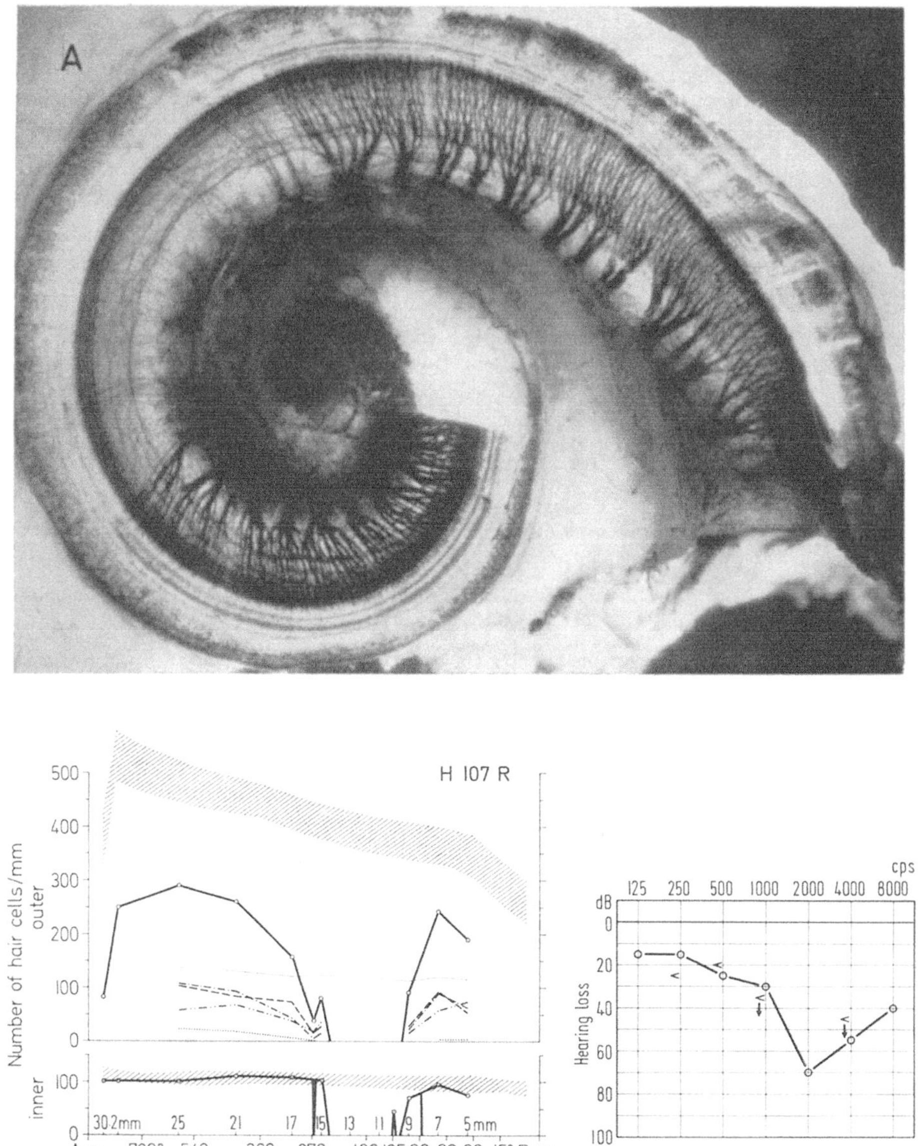

Fig. 4A

Fig. 4A, B. Right (A) and left (B) cochleas from a 71 year old man exposed for many years to high intensity noise in a saw mill. The right cochlea shows an almost complete degeneration of radial nerves in the osseous spiral lamina as well as a corresponding degeneration of the organ of Corti in the region between 135° and 270° (10.5—14 mm). The hair cell population and the audiogram are shown below the cochlea. The left cochlea is associated with an audiogram which reveals a hearing loss similar to that in the right ear. Note the striking difference between the two ears in the appearance of the nerve bundles in the osseous spiral lamina. The population of outer hair cells shows a pronounced degeneration in the region corresponding to that of total degeneration in the right ear. Note that the inner hair cells show almost no degeneration in this area of the organ of Corti. (From Bredberg, 1968)

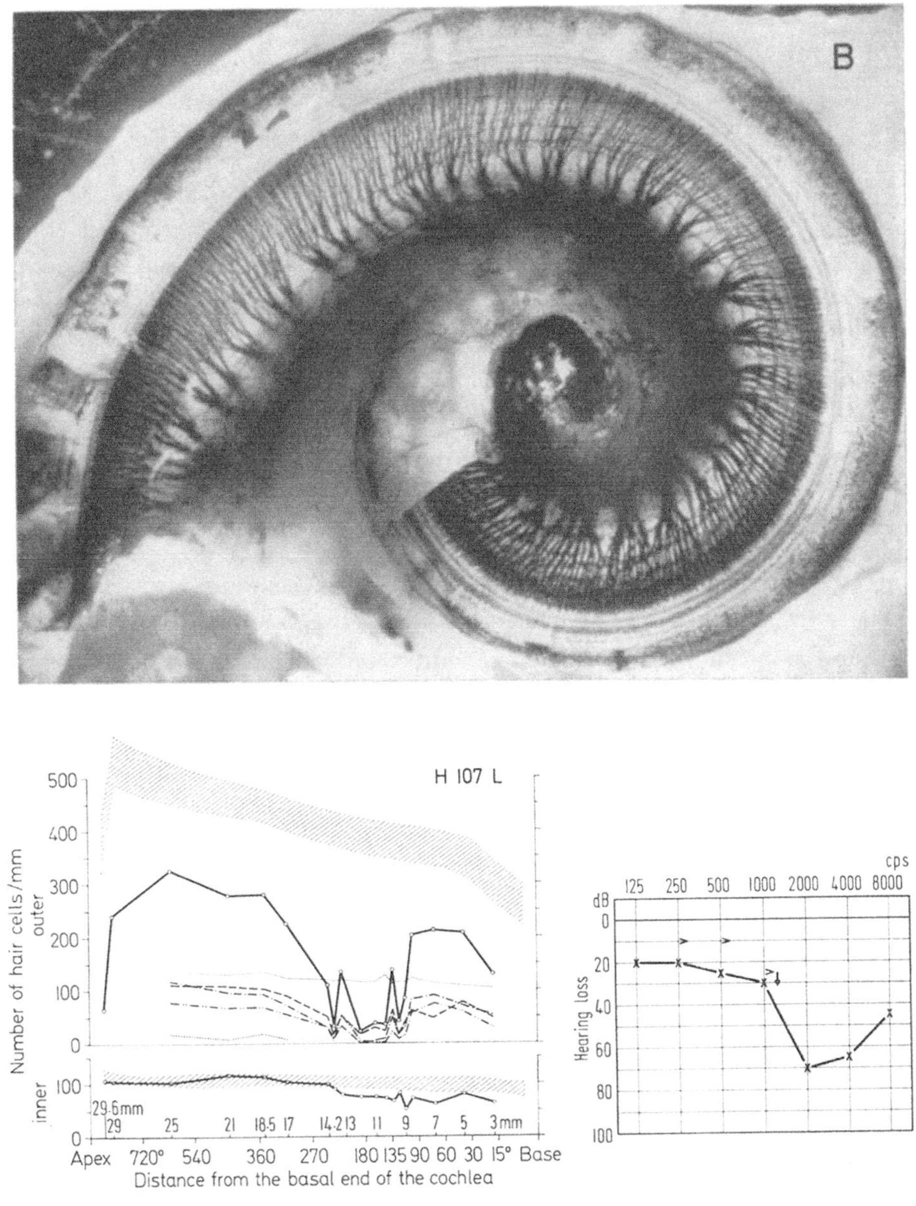

Fig. 4B

60 dB at the same frequency in the other. In the first case, noise exposure was the probable cause. In the second, a skull trauma could not be ruled out as causing the hearing loss. The first case showed a partial loss of hair cells in a circumscribed region of the midportion of the basal coil extending from the 5 mm to the 13 mm region. In the second, there was a similar lesion between 9 mm and 15 mm. The spiral ganglion and all other structures were reported normal. BREDBERG (1968), using the surface specimen technique, reported 28 ears from 16 subjects for whom

hearing tests had been carried out. Of these, 5 cases had histories of noise exposure. A 51 year old man had an audiogram with a typical 4000 Hz dip of 45 dB. In both ears, there was a moderate overall degeneration of outer hair cells, typical for the age. The loss was slightly more pronounced in the region 11—13 mm from the basal end. Inner hair cells and nerves in the osseous spiral lamina were normal. A 71 year old man, noise exposed in a saw-mill for many years, had a wide "dip" in his audiogram, with a maximum loss of 70 dB at 2 kHz and a loss of 55—65 dB at 4 kHz (see Fig. 4). In the left ear, there was an almost total loss of outer hair cells from 9 mm to 14.2 mm and an overall loss of about 30%. The inner hair cells and nerves showed slight degeneration in the basal turn. The right ear showed the same pattern of outer hair cell absence with the important difference that all inner hair cells and nerves in the osseous spiral lamina were degenerated in the same area where the outer hair cells were completely lost; that is from 9 mm to almost 15 mm from the base. In two subjects with histories of noise exposure, no correlate to the hearing loss was found in the sensory cell population nor in the nerve supply in the osseous spiral lamina. In one of these cases, the audiogram revealed a "dip" of 55—65 dB at 6 kHz; the other had an abrupt high tone hearing loss above 2 kHz. A very different pattern was seen in a 62 year old man who had been exposed for many years to the noise in the engine room of a diesel motor ship. The audiogram showed a "flat" 20—30 dB hearing loss with degeneration of about 50% of the outer hair cells in the middle and apical coils. In addition, a moderate overall reduction of outer hair cells was seen.

It appears that a history of noise exposure and/or a change in threshold has no simple relationship to a loss of sensory cells. Several studies report cases without a history of noise exposure but with audiograms similar to those discussed above. Guild *et al.* (1931) described a case in which the audiogram showed a narrow "4000 Hz dip" of 45 dB. Histological examination revealed a circumscribed partial degeneration in the region 6.8—11 mm from the basal end of the cochlea. In the middle third of the lesion, outer hair cells were almost totally degenerated. The inner hair cells, the nerves in the osseous spiral lamina, and the spiral ganglion were all normal throughout this region. Wever (1942) studied 51 temporal bones of the Johns Hopkins Hospital temporal bone collection, all associated with audiograms displaying a "tonal dip". The dips were from 30 dB to a maximum of 85 dB. Despite these rather extensive hearing losses, Wever was unable to correlate the dips with any localized structural change in the cochlea. He concluded: "The results showed a somewhat higher degree of pathology of inner ear structures in cases in which large dips were present, but no region was identified as specifically responsible for the local depressions". Bredberg (1968) described one ear of a 63 year old man without any known history of noise exposure but with an audiogram showing the maximum hearing loss of 75 dB at 4 kHz. In the cochlea, there was a complete loss of the organ of Corti in the region 7—9 mm from the basal end and a loss of nerves in the corresponding portion of the osseous spiral lamina. The outer hair cells showed an overall degeneration of 30—40%.

A few reports exist of patients with a tumor of the cerebellopontine angle in which the hearing was assessed and the temporal bones were studied. Dix *et al.* (1948) described a patient with a high degree of deafness resulting from neurofibroma of the eighth nerve. There was a gross reduction of nerve fibers and almost total loss of spiral ganglion cells. The organ of Corti was essentially normal. In a case reported by JØRGENSEN (1962), a small intracochlear neurinoma was found in one ear. The audiogram showed a flat bilateral loss of about 50 dB. The neurinoma was not regarded as the cause of the hearing loss although, no other abnormality was found. CITRON *et al.* (1963) reported a normal pure tone audiogram in a case where a tumor had caused severe degeneration of the spiral ganglion and associated neurons. Approximately 20% of the normal cell population was present. The organ of Corti and the other cell elements within the scala media were reported as well preserved, including the stria vascularis. SCHUKNECHT (1964) described a case of profound deafness attributed to a meningioma resulting in complete degeneration of the cochlear nerve and ganglion cells. In another case (SCHUKNECHT, 1966) an acoustic neurinoma resulted in a different lesion. The hearing loss increased from 50 dB at 125 Hz to 70–80 dB at 4 kHz and 8 kHz. Complete loudness recruitment was present, and speech discrimination was 90%. Histological examination showed a loss of nerve fibers only in the 12–16 mm region of the cochlea. There was an absence of the organ of Corti in the basal 4 mm, partial loss of hair cells from 4–12 mm, and severe shrinking of the organ of Corti in the region 14–21 mm from the base. BREDBERG (1968) reported a case of degeneration of approximately two thirds of the nerve fibers throughout the cochlea. Still the speech discrimination scores were 92% for one ear and 84% for the other. The audiogram revealed a moderate high tone loss (see Fig. 6).

In otosclerosis, the typical hearing loss is conductive due to fixation of the stapedius footplate. In advanced cases, especially those where the otosclerotic lesions involve the cochlear capsule, a profound conductive and sensorineural hearing loss is often found. Because of problems in evaluating the sensorineural hearing loss, these cases are not reviewed here. The most extensive reports on the subject have been published by BENITEZ and SCHUKNECHT (1962), ALTMANN (1966), NAGER (1966), RÜEDI and SPOENDLIN (1966), SCHUKNECHT and GROSS (1966), and SANDO *et al.* (1968).

III. Discussion of Literature

A. General Considerations

With both humans and animals, the ototoxic stimulus that may result in a severe PTS for one subject gives no threshold change in another. The search for the reason behind this fact goes hand in hand with the study of the effects of different ototoxic stimuli on inner ear anatomy. In the meantime this high between subject variability seriously restricts the usefulness of statistical measures of central tendency and contributes much to the diversity seen when the results of different studies are compared.

Additional complicating factors are seen in the light of improved technology. Earlier studies used methods for testing hearing that are no longer considered reliable; behavioral testing methods were often inadequate and conditioning of some animals (*e.g.* guinea pig) is very difficult and gives questionable results; different species may differ greatly in their susceptibility to the same ototoxic stimuli; histological techniques differ in the studies, as do the structures that are examined. Some of these factors will be considered in greater detail in appropriate sections.

One additional factor is not discussed further but may be important. The studies reported here have generally assumed that the primary location for damage associated with permanent hearing losses resulting from ototoxic stimuli is the cochlea. Foundations for this assumption and the possibility of other locations deserve consideration.

B. Anatomical Frequency Scale

The first serious attempts to construct an anatomical frequency scale based on the correlation between pure tone hearing loss and inner ear damage were made from studies of human temporal bones collected at the Johns Hopkins Hospital in Baltimore. From selected cases, Crowe *et al.* (1934) and Ciocco (1934) concluded that normal hearing thresholds for frequencies of 8192, 4096 and 2048 Hz depended upon the intact state of specific regions of the organ of Corti. From examination of similar data for cases having low frequency losses, Oda (1938) concluded that there was no support for postulating a relationship between condition of even moderately restricted regions of the organ of Corti and threshold hearing for frequencies below 2048 Hz. Most later studies in which human temporal bones have been examined postmortem have not been concerned primarily with this problem of frequency localization in the cochlea; however, there are some data that are relevant. Bredberg (1968) plotted threshold sensitivity for pure tones versus density of outer hair cells at the cochlear locations predicted to correspond to certain frequencies (von Békésy, 1960; Greenwood, 1961). Losses of outer hair cells (up to 40%) at a location 30 mm from the basal end of the cochlea were accompanied by hearing losses of only 15 dB or less for a 250 Hz tone. Several cochleas had 50–75% of the outer hair cells missing in this area, yet the most severe hearing loss at 250 Hz was 40 dB. Conversely, a less severe loss of outer hair cells 7 mm from the basal end of the cochlea was accompanied by a pronounced hearing loss for an 8000 Hz pure tone. In the 7 mm location, a 25% loss of outer hair cells corresponded to a 20–30 dB threshold shift, and a 50% loss of hair cells was accompanied by a 50–75 dB hearing deficit. The relation between outer hair cell loss and hearing loss that was found for these locations (250 Hz – 30 mm and 8000 Hz – 7 mm) is shown in Fig. 5. Between these extremes a graded transition was found in the relationship.

The early formulation of an anatomical frequency scale in the cat (Schuknecht, 1953a) has received support from later behavioral studies where extensive threshold shifts for high frequencies have been found. The low frequency end of the scale has not received substantial support from subsequent behavioral studies. Cats with large losses of hair cells in the apex (Elliott, 1961; McGee *et al.*, 1963; Elliott and McGee, 1965; Dolan *et al.*, 1970) have shown normal low frequency thresholds. The better fit of high frequency losses and hair cell lesions in the basal end of the cochlea is not surprising if we remember that locations for frequencies above 2 kHz, which involve less than 60% of the cochlea, were initially made on the basis of data from 14 animals. Unfortunately, mapping of frequency location in the apical 40% of the cochlea was done using data from only 4 selected animals.

Although the CUMMINGS (1968) ototoxicity study on the effects of nitrogen mustard presents data which strongly supports the location of low apical frequency function suggested by SCHUKNECHT, there is considerable evidence that apical lesions or their absence do not generally correspond to particular audiometric findings (see Figs. 8, 10, 11). The behavioral experiments of STEBBINS *et al.* (1969, 1971) present strong evidence for frequency location above 2 kHz in the monkey cochlea. The progressive effect of the ototoxic drugs used suggests a method for testing frequency localization along the basilar membrane in other species.

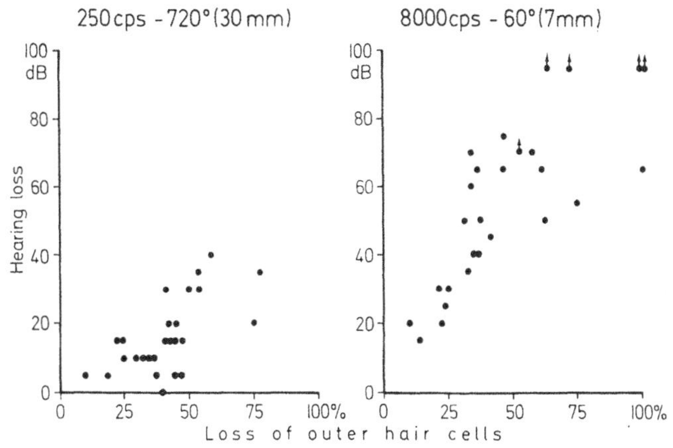

Fig. 5. Relationship between hearing loss and outer hair cell loss for 250 Hz — 720° (30 mm)and for 8000 Hz — 60° (7 mm). (From BREDBERG, 1968)

C. Recruitment and Hair Cell Populations

Recruitment has been defined as an abnormally rapid rise in loudness with increasing stimulus intensity (DAVIS and SILVERMAN, 1970), and it has been suggested that this phenomenon occurs in cases with certain kinds of end organ pathology (DIX *et al.*, 1948; JERGER, 1961). As yet, the suspected correlation between hair cell damage or other cochlear lesion and recruitment has not received support from studies relating behavioral tests of hearing with changes in the inner ear. Recruitment is a typical finding in cases diagnosed as Ménière's disease. Although initial reports described end organ hair cell changes in Ménière's disease (HALLPIKE and CAIRNS, 1938; DIX, 1948; LINDSAY and v. SCHULTESS, 1958; KRISTENSEN, 1961) the observation has not been supported by later studies (LINDSAY, 1960; SCHUKNECHT, 1968; SCHUKNECHT *et al.*, 1962; HANSEN and RESKE-NIELSEN, 1964; LINDSAY *et al.*, 1967). BREDBERG (1968) described a case with diminished intensity difference limen in which sensory cell population was normal for the subject's age. Audiometric data and histology for this case are shown in Fig. 6. Other cases with pronounced outer hair cell losses showed no decrease in intensity difference limen. ELLIOTT and MCGEE (1965) found no change in the intensity discrimination of cats with hearing and hair cell losses created by exposure to high intensity pure tones.

SCHUKNECHT'S (1953b) suggestion that a 50 dB or less hearing loss corresponds to a total or partial loss of outer hair cells and that greater than 50 dB hearing

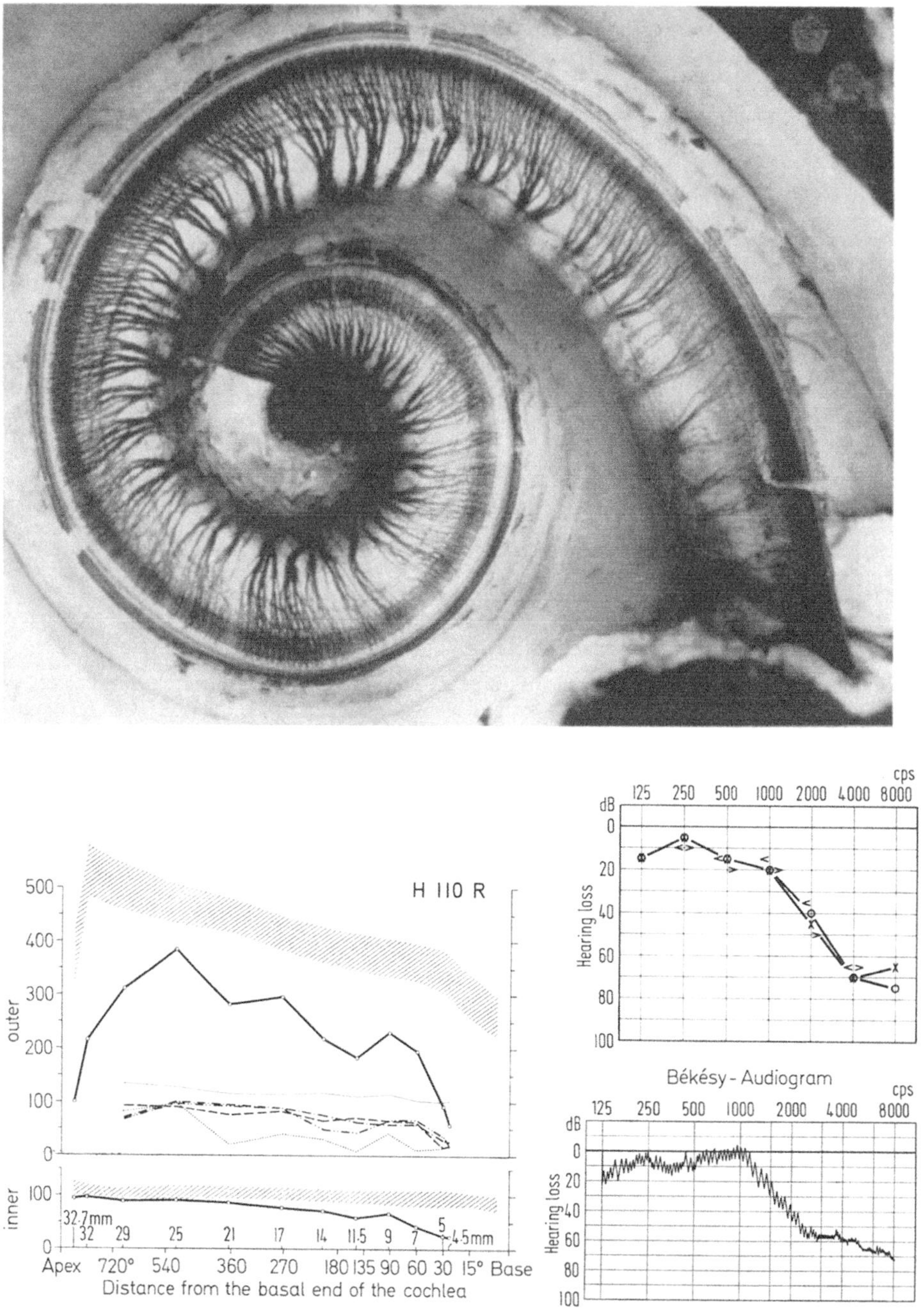

Fig. 6. Right cochlea from a man aged 81 years. No history of noise exposure or ear disease. Both cochleas (only one is shown here) show a pronounced patchy degeneration of nerve fibres throughout all coils. Approximately one third are present. The sensory cell population is well preserved for the age and does not show a similar patchy degeneration. The pure tone audiogram for both ears is shown. The von Békésy audiogram for the right ear was similar to that for the left shown above. There is a diminished difference limen for the frequencies of 3000Hz and higher. The speech discrimination score was 92% for the right ear and 84% for the left. Speech thresholds were 23 and 25 dB respectively. (From BREDBERG, 1968)

losses involve inner hair cell damage survived for over fifteen years. The failure of many experimenters to differentiate between inner and outer hair cell losses when reporting damage has made assessment of the above conclusion difficult. A review of the animal studies which do report inner and outer hair cell damage separately (McGee *et al.*, 1963; Elliott and McGee, 1965; Stebbins *et al.*, 1969; Dolan *et al.*, 1970) illustrates the difficulty in finding behavioral material that satisfies the assumption as originally stated. Behavioral data from the Ward and Duvall (1971) experiment are in sharp conflict with the conclusion. Schuknecht (1971) has questioned his original assumption on the basis of the subsequent electrophysiological data of Kiang. Since Spoendlin (1966, 1972) presented data indicating that 90–95% of the afferent innervation is from inner hair cells, the number of outer hair cells from which electrical recordings have been obtained by investigators using microelectrodes at the level of the cochlear nerve becomes questionable. In view of the variability in lesions found in the few definitive behavioral cases reported, differentiation between inner and outer hair cell function must wait for more experimental results. In this direction, cat "Goldie" shown in Fig. 11 becomes theoretically interesting. Almost all inner hair cells were missing or severely damaged for a distance of 2 mm in the lower middle turn of the cochlea of this animal, the outer hair cell population in the same area was intact. The audiogram shows a 45 dB hearing loss at 3 kHz. On no other occasion have we observed this type of a lesion. The cochleas shown in Fig. 4 are interesting in relation to the sensitivity of the inner hair cells. One ear shows a complete degeneration of the organ of Corti and associated nerves in the osseous spiral lamina in one sector, whereas the other ear shows only a loss of outer hair cells in the corresponding region. The hearing loss was essentially the same in both ears, 55–70 dB for the frequencies involved (2000–4000 Hz).

D. Ganglion Cell Population and Speech Discrimination

On the basis of earlier studies Schuknecht (1953a) suggested that 50% losses of ganglion cells could be tolerated without pure tone threshold changes. Spiral ganglion cell loss and its relationship to pure tone thresholds has become more confusing with subsequent studies (Schuknecht and Woellner, 1955; Elliott, 1961). In these studies, some animals had threshold elevations with ganglion cell losses considerably less than 50%. In other animals, normal thresholds were seen where histology indicated ganglion cell losses as great as 90%. Similar findings were reported for a human case by Citron *et al.* (1963). Although it has been suggested that a considerably higher percentage of ganglion cells is essential for good speech discrimination (Schuknecht and Woellner, 1955; Jerger, 1960) Bredberg (1968) presented a case with normal speech discrimination scores where only one third of the nerves were present in the osseous spiral lamina (see Fig. 6). There is some evidence that speech discrimination may be affected by lesions at all levels of the auditory system from cochlea to cortex (Lidén, 1967). In both Ménière's disease and acoustic trauma, deficits in speech discrimination scores may be predictable from pure tone audiograms (Lidén, 1967). Conversely, in cases of eighth nerve tumor, severe speech discrimination losses may be accompanied by normal pure tone audiograms. Here, once again, actual behavioral data with histological followup are limited.

E. Histology

One possible source of the diverse results discussed above could be the histological technique used. Since the sensory regions within the labyrinth are enclosed in bone, the histological preparation of these structures presents special problems. The majority of both human and animal studies have used light microscopy of serially sectioned specimens. In recent years, a few studies have used the surface specimen technique, and transmission electron microscopy (TEM) has supplemented light microscopy in at least two studies. In animal experimentation, it is possible to make histological preparation under optimal conditions whereas this is not possible with humans. A very important factor in human studies is the degree of postmortem autolysis. When autolysis is pronounced, it makes a specimen impossible to evaluate. The crucial factor is time between death and fixation. Bredberg (1968) determined that the maximum time lapse for satisfactory preservation is approximately 10 h for light microscopy of surface specimens. For transmission electron microscopy Kimura *et al.* (1964) suggested a corresponding time of 4–5 h.

The preparation for serial sectioning of a temporal bone takes several months and involves many steps. The bone must be fixed, decalcified, and embedded before it can be sectioned and stained. Any of these steps may produce artefacts which may be difficult to evalute. Since the early part of the century, when Wittmaak described "hypotonic degeneration" of the inner ear, similar descriptions have appeared in the literature. This change consists of a shrinkage and flattening of the entire organ of Corti. Fernandez (1968) regarded this change as a preparation artefact and called it "the compression phenomenon". The "angiosclerotic degeneration" described by von Fieandt and Saxén (1937) as a type of presbyacusis was similary regarded by Bredberg (1968). A striking example of this type of artefact is the "absence" of the organ of Corti combined with almost normal hearing reported by Bunch and Wolff (1935). Wever (1942) was well aware of the possibilities of preparation artefacts when evaluating the inner ear in cases with a "dip" in the audiogram. He presented a case of an apparent artefact located in the region of the cochlea where the lesion may have been located. This made it impossible to evaluate the relation between the lesion and the hearing loss. The possibility exists that a lesion in a given region makes the sensory organ more fragile and sensitive to some preparation procedures. Schuknecht (1964) has described a temporal bone where it was impossible to tell whether the changes observed were preparation artefacts or actual pathology. One of the advantages of animal experimentation is that the factors discussed above may be reduced to a minimum.

In serial sectioning technique, generally every 10th section is stained and examined. The thickness of each section is usually 16–20 μ. Hence, in a cross section of the organ of Corti, there may be included either 3 or 4 outer hair cells in each row. This creates the possibility of at least a 25% error in the evaluation of each section. Problems of evaluating lesions are even greater in the parts of the sensory organ which are cut tangentially due to difficulties in orientation. These inaccuracies are reflected in serial section reports which state that there is no loss of sensory cells in some persons 50–70 years of age. The same problem may have occurred in experimental animal studies where no mention of sensory cell losses is made in untreated ears. Evidence that there may be cell damage in supposedly normal ears is given in Fig. 7. One advantage of the serial sectioning technique is that it allows examination not only of the hair cells but also of the spiral ganglion, the nerves in the osseous spiral lamina, the stria vascularis and other associated structures. Other advantages are that the sections may be reviewed for additional information during later studies; and that, with different staining techniques, histochemical information may be obtained about cochlear structure.

In the surface specimen technique (Engström *et al.*, 1966) used in recent studies, the cochlea is opened directly after fixation and the different parts of the organ of Corti (or other structures) are dissected out. The specimens are studied by phase contrast or by ordinary light microscopy. Specimens are placed so that the surface facing the tectorial membrane is "upward". It is thereby possible to examine every single sensory cell. Only a few reported studies have used this technique. Two studies have used the principle of surface preparation on material which has been embedded for TEM (Eldredge *et al.*, 1971; Ernstson, 1972). Myers and Bernstein (1965) and Ward and Duvall (1971) have evaluated cochlea material from behaviorally trained animals with the transmission electron microscope. The gains in resolution and the possibility of seeing changes which cannot be seen in light microscopy are accompanied by the problem of surveying the whole organ of Corti. To sample enough areas to be representative is a difficult and time consuming task.

To date, no behavioral studies using the surface specimen technique have looked at the spiral ganglion. This is also true of other structures such as the stria vascularis, the tectorial membrane, and Reissner's membrane. In order to study these structures the procedure used has to be directed towards such a purpose; this is possible. Present work in that direction will be discussed below.

The failure of experimenters to report histological controls in many studies creates a major problem in interpretation of results. Figure 7 shows data for both ears of 16 untrained cats (Bredberg and Ades, 1970). Hair cell evaluation was by phase microscopy of surface preparations. Close inspection of this figure reveals the type of error that may occur when the term "normal ear" is used synonymously with an interpretation of "no hair cell loss". Most histological reports fail to show the apical hair cell lesions so often found in exposed and unexposed cats using the surface preparation techniques (see Figs. 7, 10 and 11). The cats with middle ear problems could be eliminated in training as would animals with large hair cell losses like that shown for the right ear of cat number 29 (Fig. 7). Many of the other animals could possibly exhibit pure tone thresholds within the limits of "normal" variability. The etiology of the losses seen in both ears or in the unexposed ears of these cats is unknown. These are, however, a random selection of the "normal" type of cat that has been used in the behavioral studies reported above. Only one of the studies using cats (McGee and Olszewski, 1962) reported histological control. In the McGee and Olszewski study, a control group of seven cats were reported as "normal". Histological examination in the study used light microscopy of

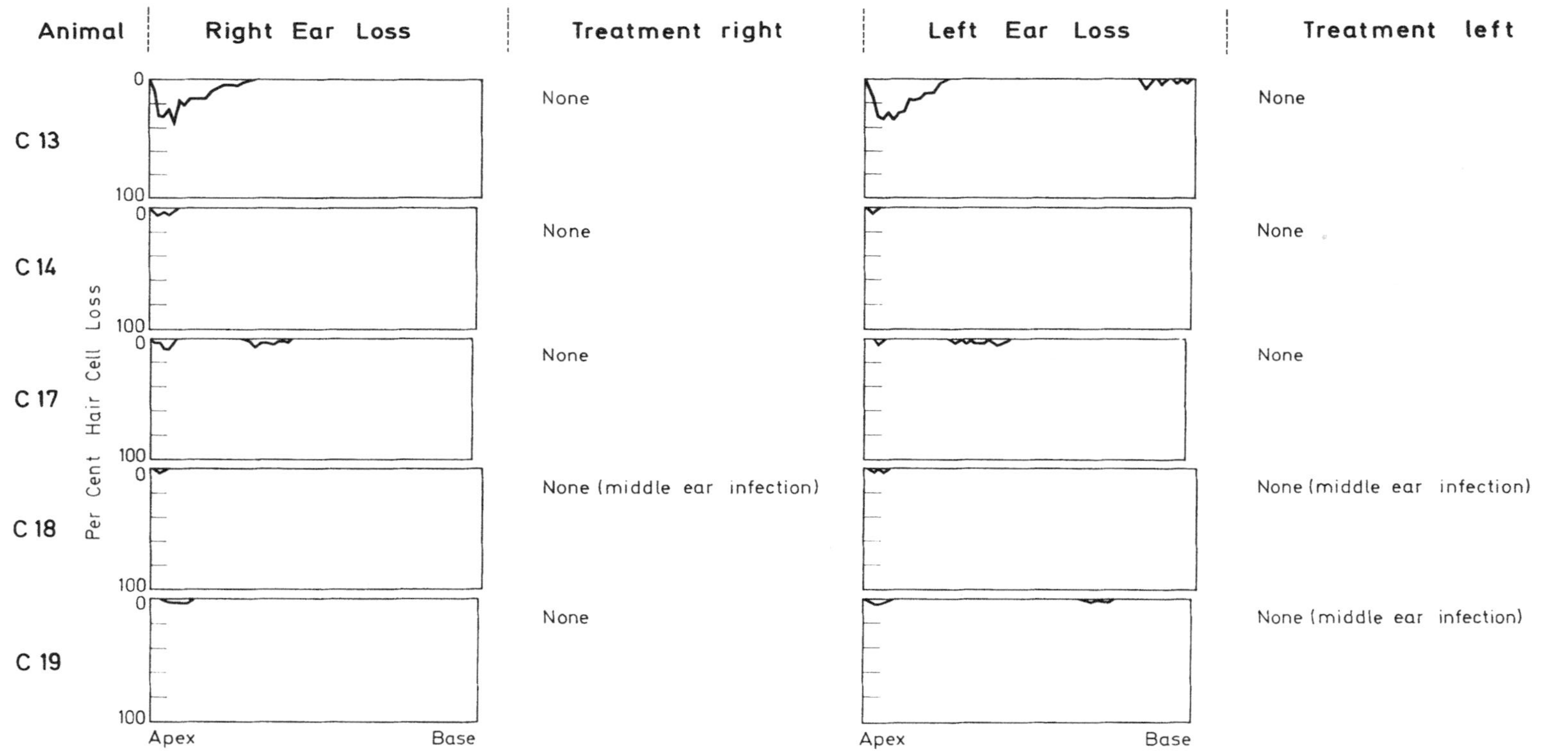

Fig. 7a

Fig. 7a—c (pp. 284—286). Cochleographs showing hair cell losses found in 16 binaural cats. In six of the animals *one* ear received treatment with either an ototoxic drug or a high intensity pure tone stimulus. Two of the animals were sick and in poor condition. The remaining nine animals were healthy "normal" cats with no known history of noise exposure

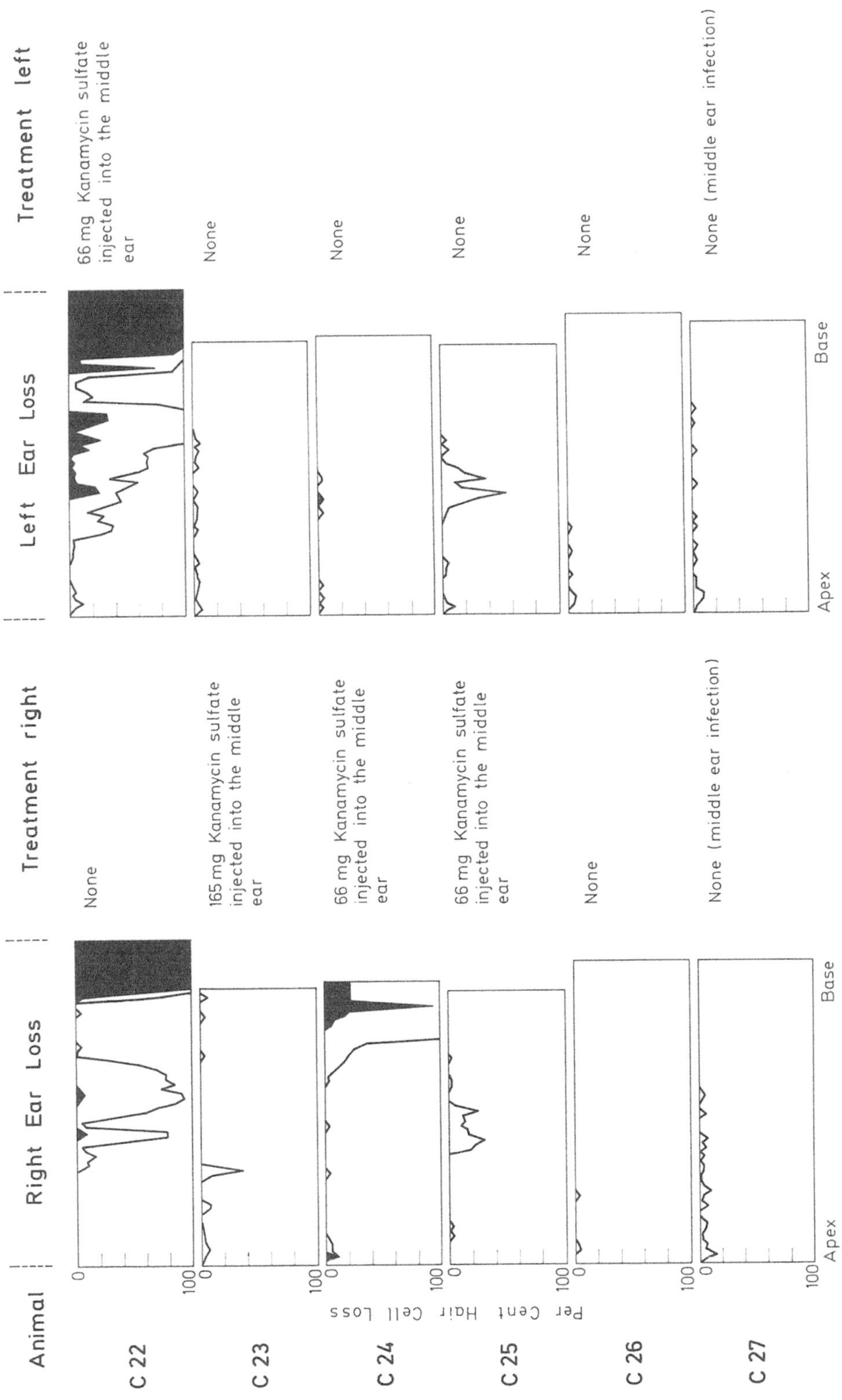
Animal
C 22
C 23
C 24
C 25
C 26
C 27
Right Ear Loss
Treatment right
None
165 mg Kanamycin sulfate injected into the middle ear
66 mg Kanamycin sulfate injected into the middle ear
66 mg Kanamycin sulfate injected into the middle ear
None
None (middle ear infection)
Left Ear Loss
Treatment left
66 mg Kanamycin sulfate injected into the middle ear
None
None
None
None
None (middle ear infection)
Per Cent Hair Cell Loss
0
100
Apex
Base

Fig. 7b

Animal
Right Ear Loss
Treatment right
Left Ear Loss
Treatment left
Per Cent Hair Cell Loss
0
100
Apex
Base
C 29
None. Sick cat in poor condition. Loosing weight (middle ear infection.)
None. Sick cat in poor condition. Loosing weight (middle ear infection.)
C 31
None (some fibrous tissue in the middle ear.)
None
C 40
None
4000 Hz, 130 dB, 1 hour (exposed through a tube fitting the ear canal, pinna removed.)
C 50
2000 Hz, 130 dB, 1 hour (exposed through a tube fitting the ear canal, pinna removed.)
None
C 52
None. Sick cat
None. Sick cat (middle ear infection.)

Fig. 7c

serial sections so that the finding most probably represents a generalization based upon examination of 10–20% of the hair cells and adjacent structures. Inspection of unexposed ears, of two monkeys and control ears with an incus removed of 14 squirrel monkeys (HUNTER-DUVAR, 1971) showed scattered hair cell losses ranging from less than 1–33%. No significant difference in pure tone thresholds was seen between an animal with a 30% hair cell loss in one row and an animal with a 1% loss in the same location. Experimenters designing behavioral studies would be wise to incorporate sufficient histological control material to minimize the possibility of random lesions being reported as a product of the treatment.

IV. Current Perspectives

A. Animal Experiments

Some recent animal experiments where hair cell losses have been accurately assessed with surface preparation techniques differ in results as to the relation between hair cell loss and hearing loss. Figure 8 shows audiograms and corresponding cochleographs for 12 chinchillas (MILLER, ELDREDGE, and BREDBERG; unpublished data). One may note the large hair cell losses for animals in the group exposed to octave band, low frequency noise. This hair cell loss appears to be accompanied by a slight shift in threshold. In contrast, Fig. 9 shows audiograms and corresponding hair cell losses for 4 squirrel monkeys exposed to high intensity pure tone stimulation (HUNTER-DUVAR and ELLIOTT, 1972, 1973). Ears with 15–45 dB losses have no accompanying hair cell losses when compared to control ears. The difference in the findings was further investigated (HUNTER-DUVAR and BREDBERG, 1973) by exposing 4 trained chinchillas to a pure tone stimulus (1000 Hz, 120 dB). Results showed that 12 min exposure to this stimulus resulted in the same or greater hearing loss in the chinchilla as was obtained by a 12 h exposure to the same stimulus in the squirrel monkey. Hearing losses in the chinchillas were accompanied by discrete lesions of approximately 1 mm which included inner and outer hair cells and nerve fibers.

The WARD and DUVALL (1971) study reported normal thresholds in chinchillas with outer hair cells completely missing in the basal and middle turns of the cochlea. Such results were quite different from the findings of sizable hearing losses in monkeys (HUNTER-DUVAR and ELLIOTT, 1972, 1973) and in humans (BREDBERG, 1968) with no corresponding hair cell losses. As a consequence of these different results, 4 trained chinchillas were exposed to the same stimulus (HUNTER-DUVAR and BREDBERG, 1973) reported in the Ward and Duvall paper (octave band noise, 123 dB, 15 min). Five weeks after exposure all 4 animals still showed PTS's. Hearing losses were 15–20 dB for 3 of the animals and greater than 40 dB for all frequencies measured (0.5–8 kHz) for the fourth animal. Histological examination of the cochleas from the exposed ears showed no hair cell loss for the animal with the smallest hearing loss; a few localized missing hair cells for one of the PTS animals and a small (< 1 mm) lesion of inner hair cells, outer hair cells and nerve fibers for the other. The animal with the PTS > 40 dB had inner and outer hair cells completely missing for the basal and middle turns.

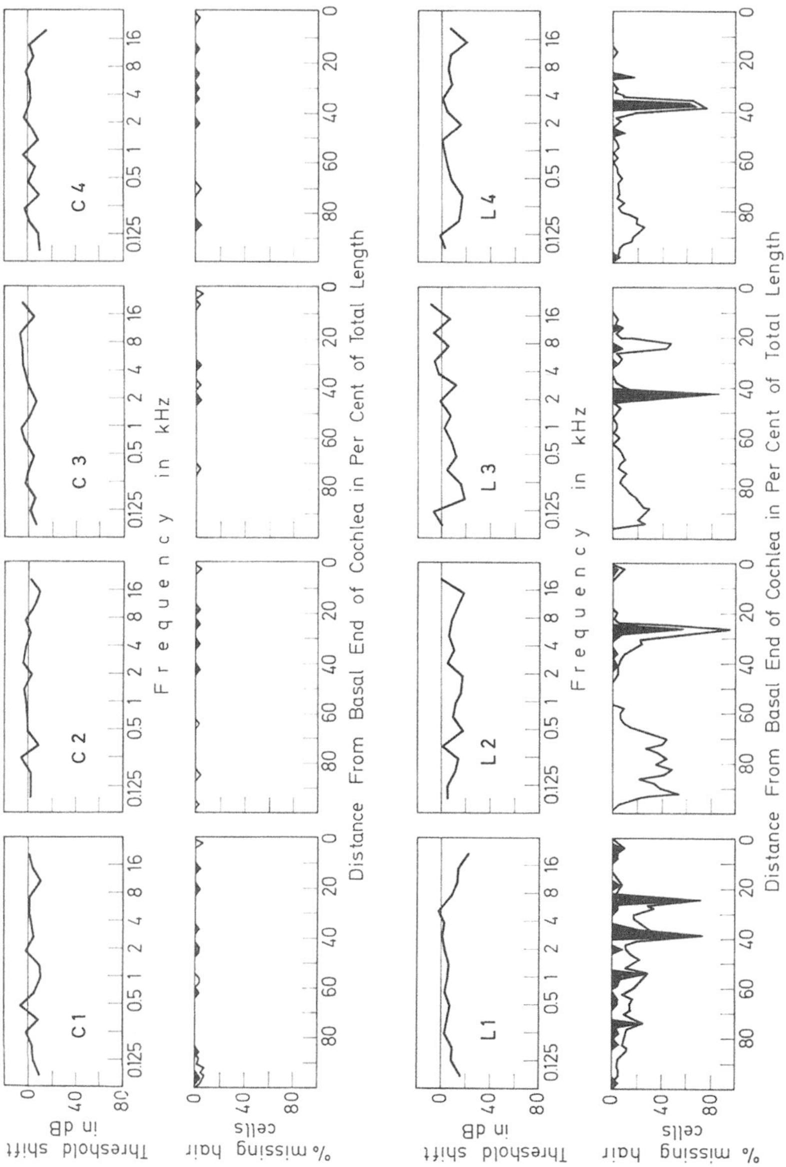

Fig. 8. Behavioral audiograms and cochleographs showing hair cell losses for a group of 8 chinchillas. Unexposed animals (*C1*—*C4*) each have a littermate in the exposed PTS group (*i.e. C1* is a littermate to *L1* etc.). Stimulus was an octave band noise centered at 500 Hz and exposure was for 9 days at an intensity of 103 dB. Shaded areas denote missing inner hair cells and the unbroken line represents missing outer hair cells. (From MILLER, ELDREDGE, and BREDBERG, unpublished data)

Figure 10 shows audiograms and cochleographs for 17 cats exposed to pure tones of different frequencies, intensities and durations (DOLAN *et al.*, 1970 and unpublished data). Figure 11 (FRAZIER *et al.*, 1972) shows audiograms and corresponding hair cell losses for 11 cats exposed to a pure tone of 2880 Hz at different intensities and for different durations. The diversity in resulting hearing losses and hair cell lesions is indicative of the problems involved in determining correlations and localization. It is probably wise to refer to the cochleographs of unexposed

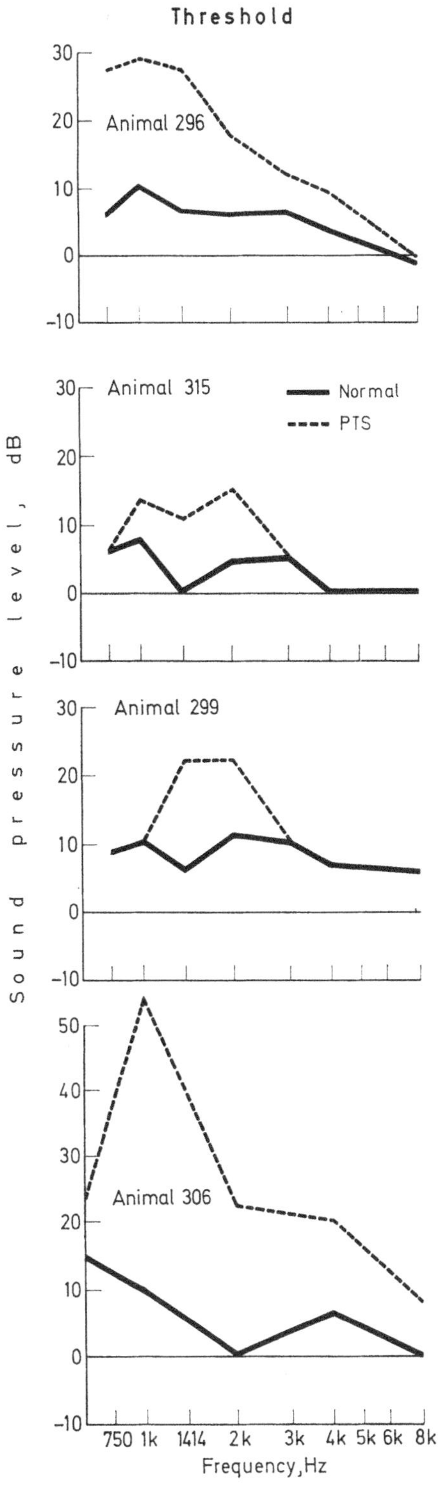

Cochlea

Animal 296, Left

Row									Total
I	3	1	1	2	2	1	2	3	15
0_1	8	9	4	3	7	2	10	7	50
0_2	8	10	4	2	2	3	5	6	40
0_3	5	10	10	3	3	8	15	4	58

Animal 296, Right

Row									Total
I	2	1	1	0	1	0	0	2	7
0_1	10	12	5	4	7	2	9	8	57
0_2	10	8	4	0	8	2	3	6	41
0_3	11	14	11	4	7	7	12	8	74

Animal 315, Left

Row									Total
I	2	1	1	1	0	0	0	0	5
0_1	1	2	1	3	1	2	0	0	10
0_2	1	2	0	0	0	2	2	0	7
0_3	6	3	7	1	2	0	1	1	21

Animal 315, Right

Row									Total
I	1	0	0	0	0	0	1	0	2
0_1	2	0	0	0	1	0	1	1	5
0_2	3	1	1	1	0	0	1	1	8
0_3	0	0	2	4	3	0	0	1	10

Animal 299, Left

Row									Total
I	2	1	1	1	0	1	4	2	12
0_1	9	15	10	1	4	2	5	6	52
0_2	8	15	7	4	3	3	0	6	46
0_3	14	18	17	5	11	3	5	10	83

Animal 299, Right

Row									Total
I	2	2	1	1	3	3	5	4	21
0_1	4	6	5	4	5	9	8	17	58
0_2	10	10	8	8	5	8	2	11	62
0_3	21	12	13	6	5	6	13	10	86

Animal 306, Left

Row									Total
I	1	1	3	0	1	0	0	0	6
0_1	2	1	21	1	1	1	2	0	29
0_2	4	6	5	0	0	1	2	1	19
0_3	3	5	4	4	0	0	0	0	16

Animal 306, Right

Row	Apex							Base	Total
I	2	1	1	1	1	1	0	0	7
0_1	6	3	1	1	3	1	1	2	18
0_2	2	2	1	1	2	2	0	1	11
0_3	2	4	2	1	0	1	1	1	12

Missing hair cells

Fig. 9. PTS audiograms and hair cell loss cochleographs for exposed (left) und unexposed (right) ears of 4 squirrel monkeys. Stimulus was a 1 kHz pure tone of varied intensity and duration. Cochleographs extend from 1—17 mm from the apex. Each square represents 2 mm of cochlea and contains approximately 210 hair cells for inner rows and 260 hair cells for outer rows. (From HUNTER-DUVAR, and ELLIOTT, 1972, 1973)

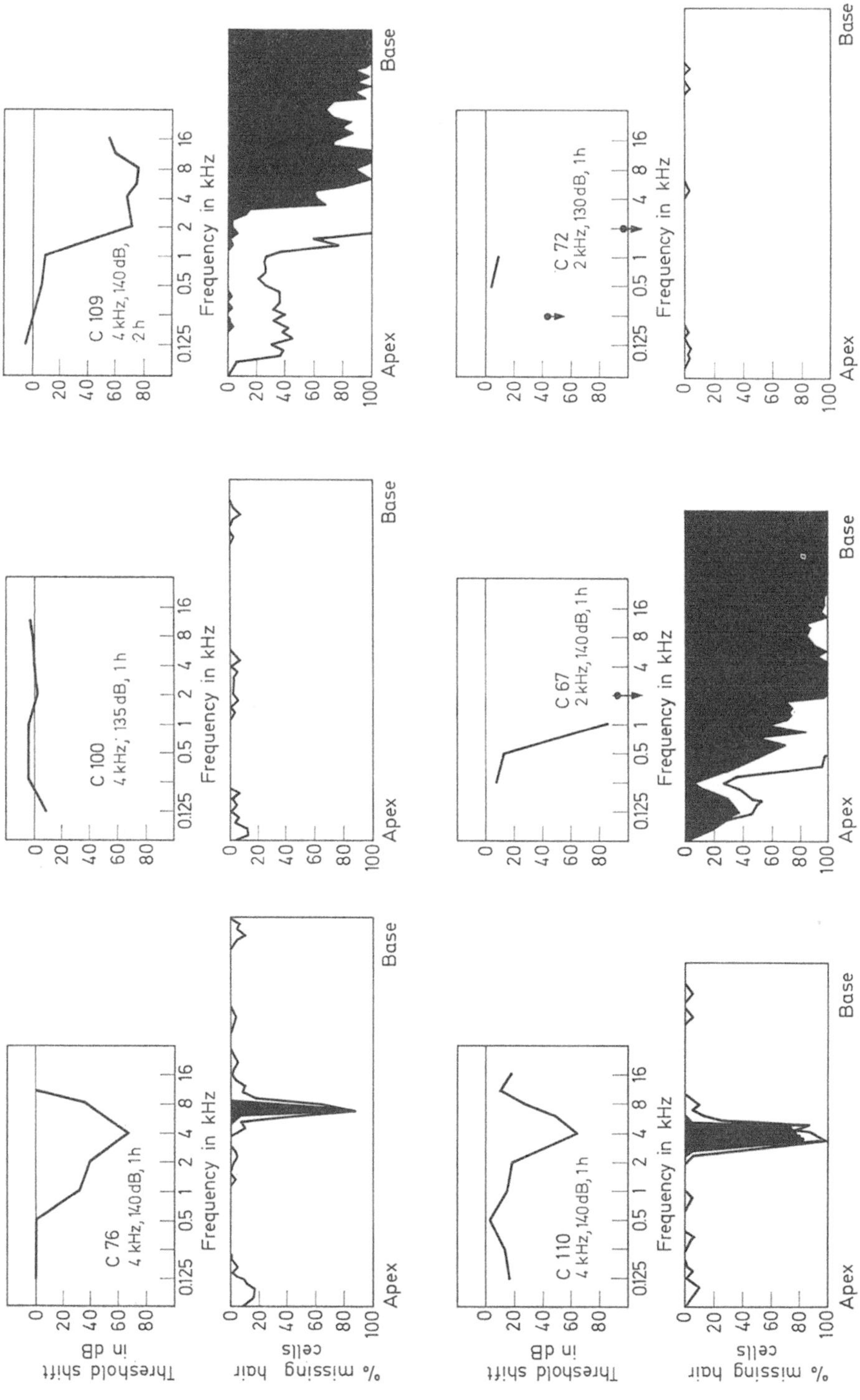
Threshold shift in dB
% missing hair cells
Frequency in kHz
Apex
Base
C 76
4 kHz, 140 dB, 1 h
C 100
4 kHz, 135 dB, 1 h
C 109
4 kHz, 140 dB, 2 h
C 110
4 kHz, 140 dB, 1 h
C 67
2 kHz, 140 dB, 1 h
C 72
2 kHz, 130 dB, 1 h

Fig. 10a

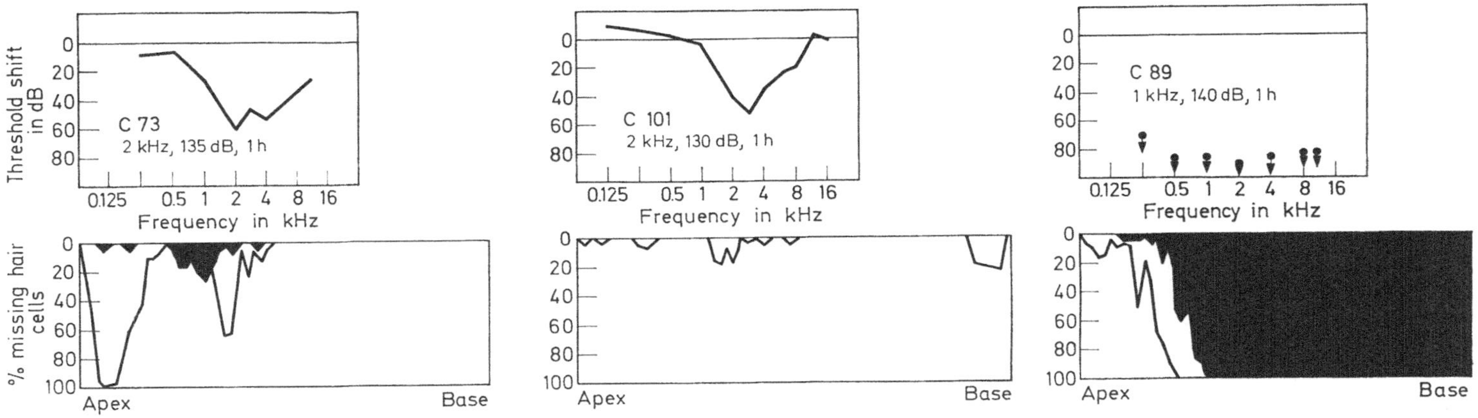

Fig. 10b

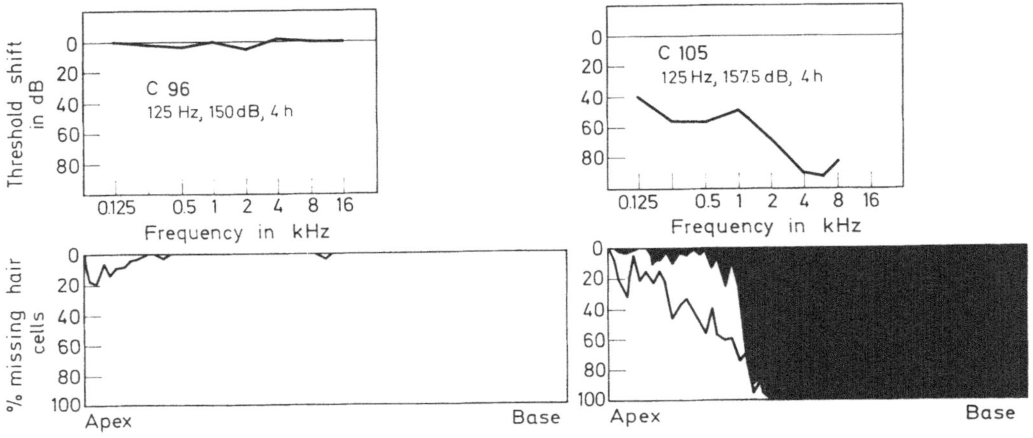

Fig. 10c

Fig. 10a—d (pp. 290—292). Behavioral audiograms and cochleographs showing hair cell losses for a group of 17 cats. Animals were exposed to high intensity pure tone stimulation and audiograms show the resulting PTS. Shaded areas denote missing inner hair cells and the unbroken line represents missing outer hair cells. (From DOLAN, BREDBERG, ADES, and NEFF, 1970, and unpublished data)

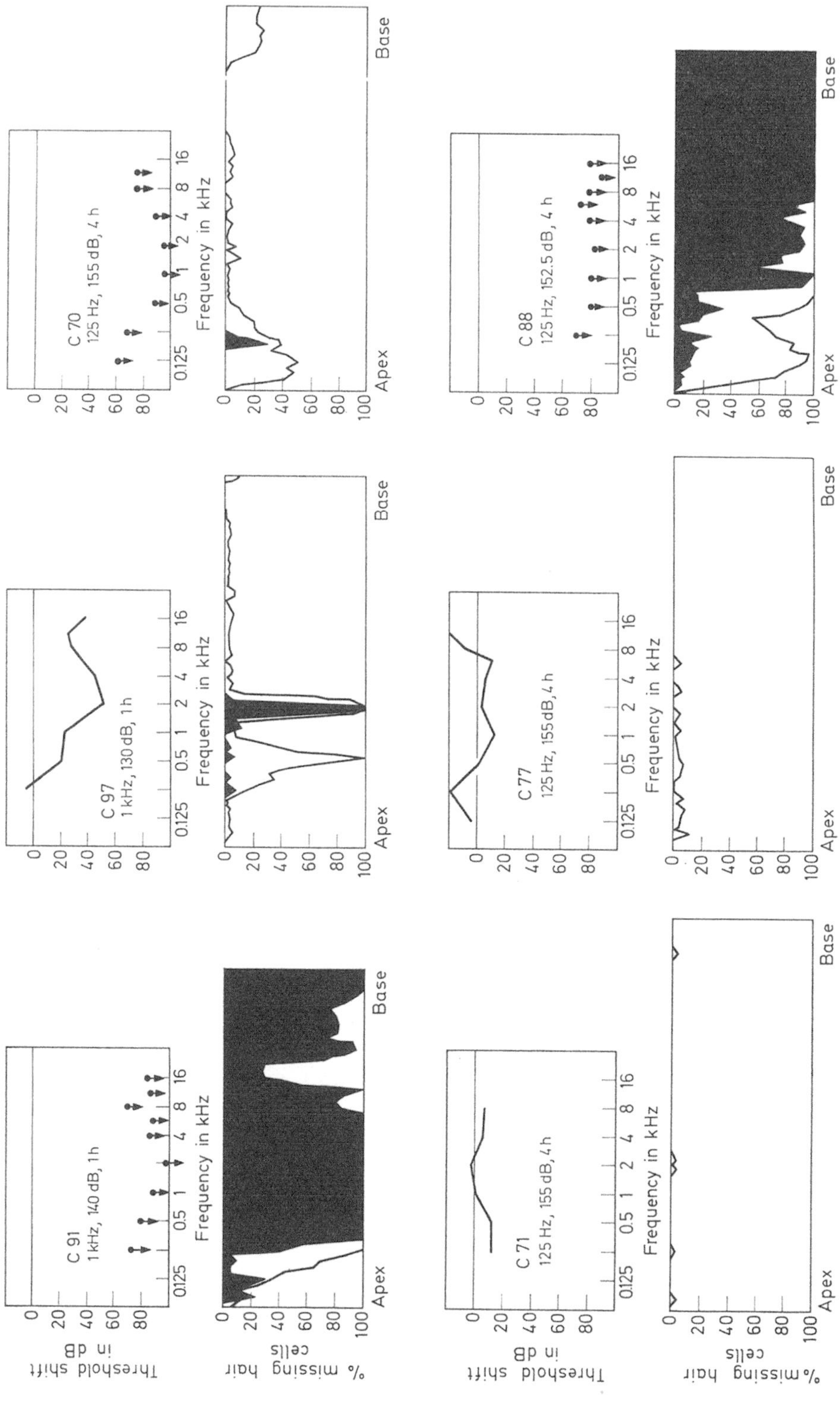

Threshold shift in dB
% missing hair cells
C 91
1 kHz, 140 dB, 1 h
C 97
1 kHz, 130 dB, 1 h
C 70
125 Hz, 155 dB, 4 h
C 71
125 Hz, 155 dB, 4 h
C 77
125 Hz, 155 dB, 4 h
C 88
125 Hz, 152.5 dB, 4 h
Frequency in kHz
0.125 0.5 1 2 4 8 16
0 20 40 60 80 100
Apex
Base

Fig. 10d

cats (Fig. 7) when considering the data shown in Fig. 10 and 11. Interpretation of results in terms of hearing loss as related to inner ear damage must be made with some caution until additional evidence is available.

In all of the studies reviewed above, only hair cell losses have been throughly investigated with the presumption that hair cells are the structures whose change is most likely to be associated with changes in the audiogram. It now appears that this assumption is not completely warranted; future histology should include all cochlear structures. The lack of definition of pathology in the different cochlear structures becomes an important consideration. At present, there are not sufficient data on the range of "normality" of cochlear anatomy in any species to draw conclusions of normal or abnormal except in cases of gross variance. Consequently, histological control for comparison is important in behavioral studies. The experimental question should determine if this requirement is better met by a control group or a control ear in the same animal. In an attempt to have animals function as their own controls, the authors are experimenting with interruption of the ossicular chain and suturing of the external meatus in chinchillas. HUNTER-DUVAR (1971) has found that 35–45 dB attenuation may be obtained with squirrel monkeys by interrupting the ossicular chain and cementing a rubber plug in the external meatus. Interruption of the ossicular chain in the chinchilla results in a threshold shift of 40–50 dB. This procedure should provide sufficient protection from sound trauma of moderate intensity; however, testing problems can arise in animals with hearing losses greater than 40 dB. One possible solution is to deafen the control ear by sectioning the cochlear nerve. This has the disadvantage of producing retrograde degeneration of nerves and degeneration in the spiral cochlear ganglion and if the arterial blood supply is accidentally partly or wholly interrupted degeneration of hair cells.

B. Human Studies

Although the surface specimen technique has been applied to the study of the human inner ear (BREDBERG *et al.*, 1965; JOHNSSON and HAWKINS, 1967), only one extensive study has related hearing function to cochlear structure using this technique (BREDBERG, 1968). Figures 4 and 6 show data from this study which illustrates how the inner ear may be mapped. In several instances, it appears that the accurate evaluation of the sensory cell population does not give a satisfactory explanation of the audiogram. As in animal experimentation the need for more refined histological techniques is apparent. In a few human specimens, we have applied scanning electron microscopy (SEM) in order to evaluate the structures under high power. Figure 12 compares hair cells with stereocilia from a 19 year old

---

Fig. 11 (pp. 294, 295). PTS audiograms and cochleographs for 11 monaural cats exposed to a 2880 Hz pure tone of varied intensity and duration. Cochleographs extend from 1—19 mm from the apex. Each square represents 2 mm of cochlea and contains approximately 210 hair cells for inner rows and 260 hair cells for outer rows. A cochleograph for one unexposed cat (Comber) from the same laboratory is included in the figure. (From FRAZIER, ELLIOTT, and HUNTER-DUVAR, unpublished data)

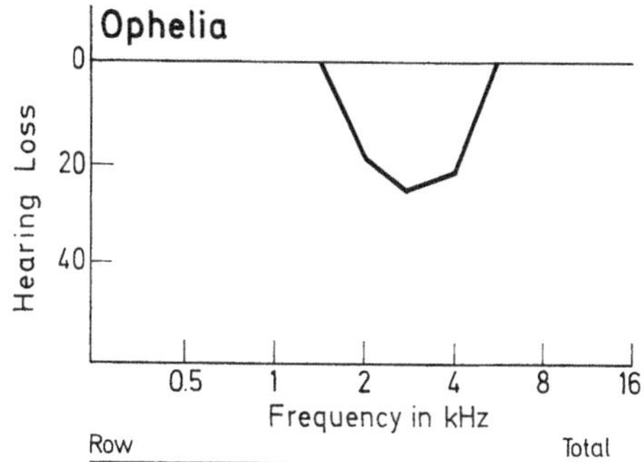

Cochlea

Row										Total
I	3	1	0	3	0	0	0	1	0	8
O_1	17	2	4	6	16	6	2	1	1	55
O_2	46	5	10	5	11	5	3	1	3	89
O_3	65	33	15	7	20	23	10	3	19	195

Apex Hair Cell Loss Base

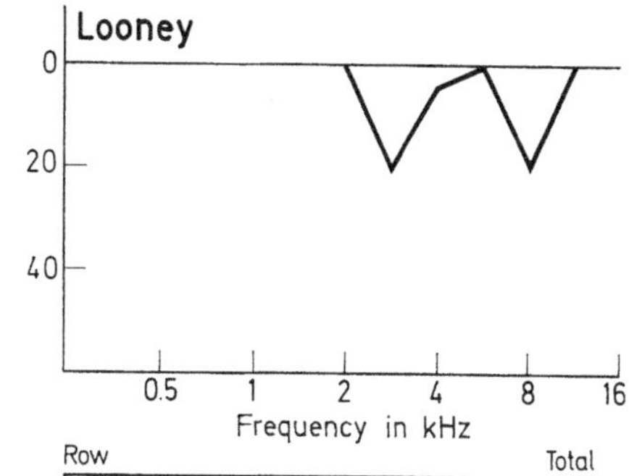

Row										Total
I	5	4	5	13	5	5	1	1	0	39
O_1	22	42	16	9	6	7	17	62	25	206
O_2	31	19	15	9	12	9	33	87	42	257
O_3	78	47	28	20	23	36	50	65	101	448

Apex Hair Cell Loss Base

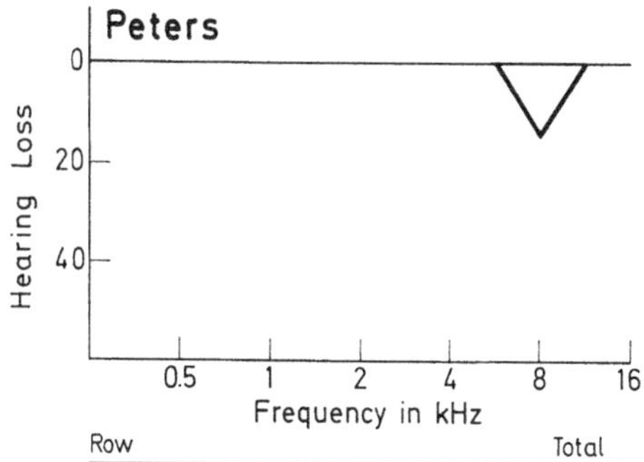

Cochlea

Row										Total
I	2	1	2	0	10	11	0	1	2	29
O_1	5	3	8	7	1	5	5	6	2	42
O_2	4	5	14	11	2	2	2	2	1	43
O_3	12	14	16	8	1	0	4	6	6	67

Apex Hair Cell Loss Base

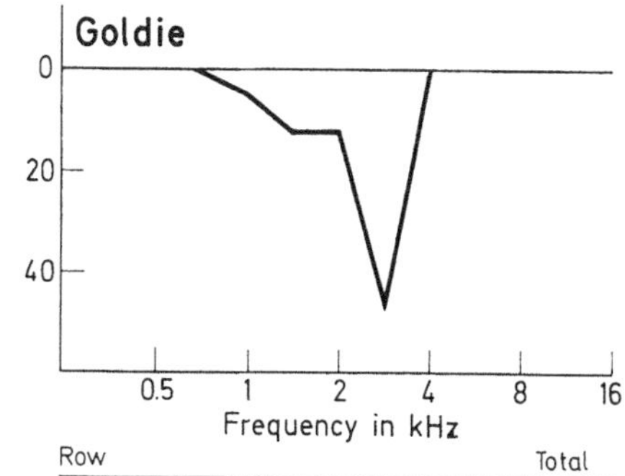

Row										Total
I	70	3	2	75	81	0	0	0	6	237
O_1	108	1	6	7	7	0	8	2	6	145
O_2	116	10	4	5	2	1	3	0	3	144
O_3	124	17	9	15	6	13	6	5	4	199

Apex Hair Cell Loss Base

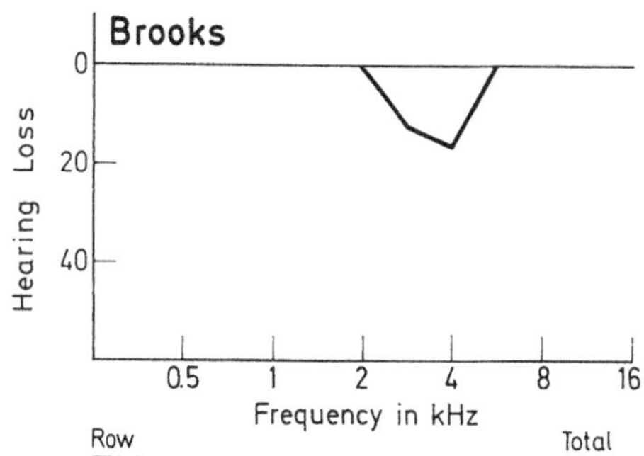

Cochlea

Row										Total
I	5	4	9	5	1	1	1	1	1	28
O_1	21	5	8	5	2	3	3	3	5	55
O_2	40	3	4	2	3	2	2	1	1	58
O_3	94	17	13	6	4	8	7	3	5	157

Apex Hair Cell Loss Base

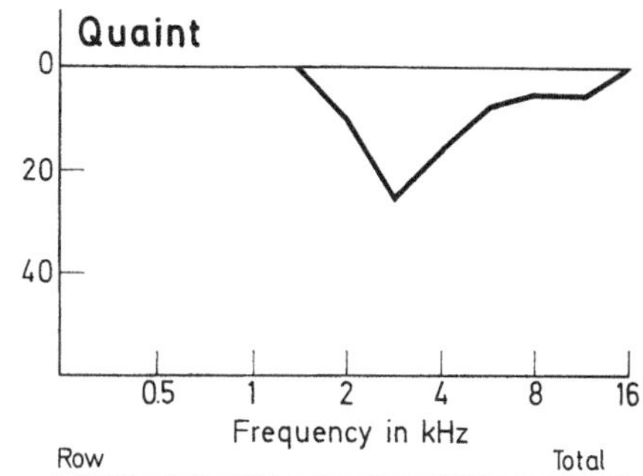

Row										Total
I	0	0	0	0	2	2	0	1	2	7
O_1	0	1	1	7	2	8	1	0	2	21
O_2	6	1	3	12	14	9	3	2	7	57
O_3	11	6	1	7	53	49	24	28	8	187

Apex Hair Cell Loss Base

Fig. 11a (Legend see page 293)

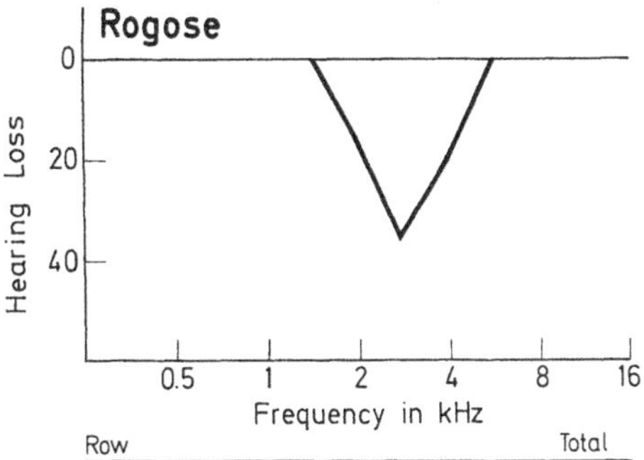

Cochlea

Row										Total
I	0	3	3	22	2	1	0	0	0	31
O_1	7	1	4	46	53	25	1	1	6	144
O_2	15	5	3	24	75	29	2	0	9	162
O_3	36	5	3	23	72	32	1	1	8	181

Apex Hair Cell Loss Base

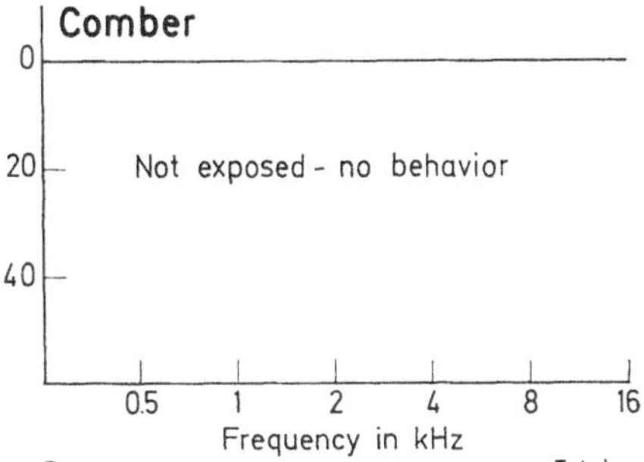

Row										Total
I	2	3	2	2	2	0	0	0	0	11
O_1	16	6	4	0	0	3	3	0	4	36
O_2	22	4	5	2	2	1	3	4	5	48
O_3	41	20	21	10	5	6	3	5	7	118

Apex Hair Cell Loss Base

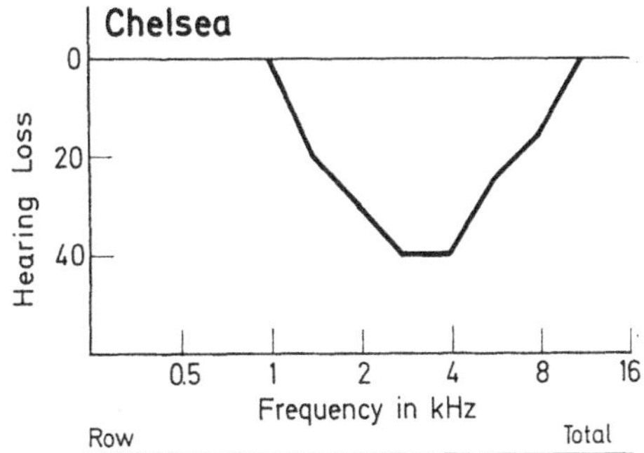

Cochlea

Row										Total
I	11	1	8	5	4	2	0	1	1	33
O_1	7	4	9	69	32	15	2	4	4	146
O_2	27	9	6	30	46	19	3	3	3	146
O_3	36	31	21	59	60	30	4	2	8	249

Apex Hair Cell Loss Base

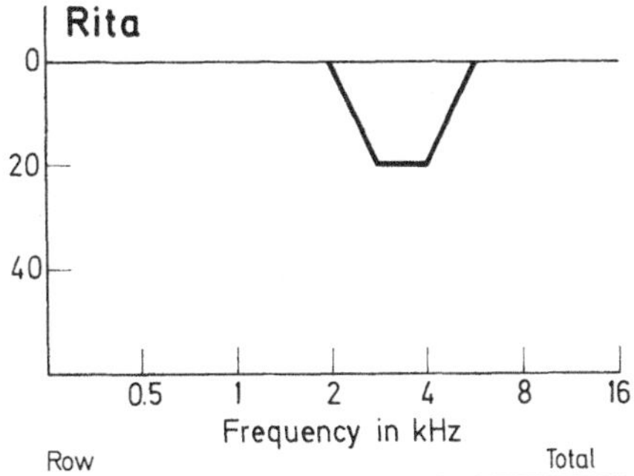

Row										Total
I	1	1	0	1	0	1	1	0	1	6
O_1	4	4	2	5	6	0	5	15	16	57
O_2	9	4	0	3	3	1	1	4	5	30
O_3	23	13	5	8	7	2	2	6	5	71

Apex Hair Cell Loss Base

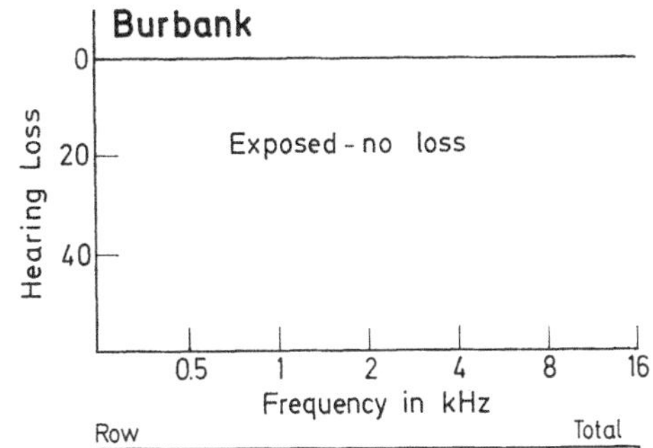

Cochlea

Row										Total
I	3	9	7	7	4	2	2	3	5	42
O_1	4	15	20	34	23	10	5	6	10	127
O_2	18	13	20	20	7	8	6	5	14	112
O_3	20	53	49	22	13	13	9	7	15	201

Apex Hair Cell Loss Base

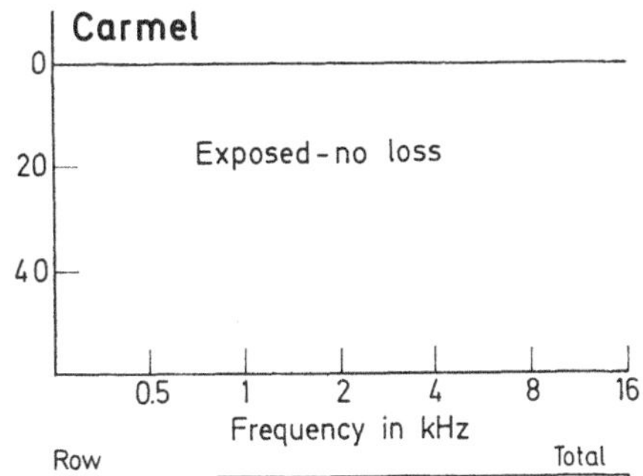

Row										Total
I	2	0	0	0	0	2	0	0	0	4
O_1	15	7	2	0	5	1	1	0	4	35
O_2	29	6	1	3	5	2	1	0	8	55
O_3	68	12	5	4	4	6	4	8	9	120

Apex Hair Cell Loss Base

Fig. 11b

man displaying no sensorineural hearing loss with hair cells from the corresponding region of a man with a history of noise exposure where the corresponding audiogram revealed a "dip" of 55–65 dB for 6000 Hz. The sensory cell population revealed no loss in the area where the lesion might be expected (von Békésy, 1960; Greenwood, 1961). No apparent changes of the stereocilia or of the surface of the organ of Corti were found. These pictures are from specimens included in Bredberg's (1968) study; the specimens have since been stored on slides in glycerol. The comparison of the quality of specimens from animal (Fig. 15) and human cochleas (Fig. 12) illustrates problems such as postmortem autolysis previously discussed.

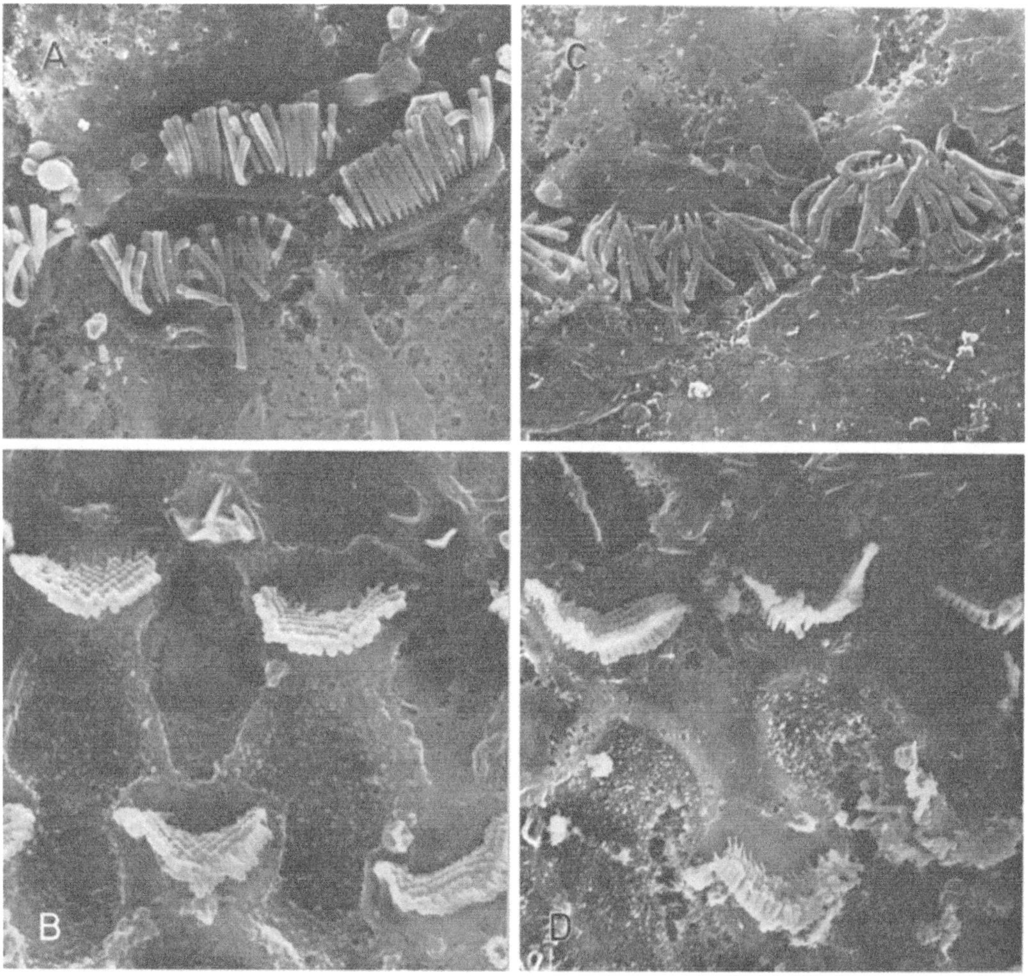

Fig. 12A—D. Scanning electron micrographs from corresponding areas of two human cochleas. A and B show cilia on inner and outer hair cells located 11 mm from the basal end of the cochlea of a 19 year old man. The audiogram indicated no sensorineural hearing loss. C and D show cilia on inner and outer hair cells from the corresponding area of the cochlea of a subject with a history of noise exposure. The audiogram showed a 60 dB loss for 6000 Hz. No alterations are seen in the cilia in the area where a lesion might theoretically be expected. Magnification A 3900 ×, B 4400 ×, C 3400 ×, D 3800 ×

The careful selection of cases were autolysis is at a minimum may yield valuable information if the surface specimens are first viewed with the inverted microscope then selectively used for SEM and TEM. KIMURA *et al.* (1964) showed the practicality of studying the human organ of Corti with TEM.

The need for histological controls in animal experiments is no less important in human studies. Since so little SEM and TEM have been done on human material, almost no ultrastructure anatomy is available for reference.

C. Histology

The introduction of phase microscopy of surface preparations (ENGSTRÖM *et al.*, 1966) presented an excellent method for accurately assessing hair cell populations. It now appears that this technique has demonstrated that, if initial hearing loss correlates are to be found, more refined techniques are required to examine ultra structural damage in the cochlea. SEM presents an excellent tool for the examination of surface structure (LIM and LANE, 1969; BREDBERG *et al.*, 1970). Structural changes such as giant cilia which are at the limits of resolution of the phase microscope can be easily inspected with SEM (Fig. 13). Using SEM, LINDEMAN and BREDBERG (1972) have found an increase in giant cilia in the apical portion of the cochleas of some exposed cats. However, preliminary investigation has shown no correlation between permanent hearing losses not exceeding 45 dB and major changes in the cilia of sound exposed monkeys. Figure 14 shows cilia on hair cells from areas where lesions might be expected (VON BÉKÉSY, 1960; GREENWOOD, 1961; STEBBINS *et al.*, 1969). Exposed and control ears show no differences. The possibility of preparation artefacts must be kept in mind until more experience with the new methods has been acquired. The specimens in Fig. 14 have been stored on slides in glycerol under cover glasses for over a year. This may account for the apparent artefactual deterioration of the supporting cells not seen in the specimens in Fig. 15 which were dissected, examined, and then stored in 70% alcohol.

The time and effort involved in obtaining behavioral data bring increased importance to the handling of the histological material. BOHNE (1972) has described a technique for handling chinchilla cochleas which by all appearances preserves the entire structure after the whole cochlea has been embedded. Preliminary inspection of the organ of Corti is possible from the basal side with the phase microscope. Selected areas may then be reembedded and cut in thick sections for survey or in thin sections for TEM.

In an effort to obtain accurate cochleograms, and to then examine selected material by TEM or SEM, use may be made of an inverted phase microscope. Specimens can be dissected and hair cell counts made while the organ of Corti is still in buffer, saline or alcohol. After cochleograms have been completed, portions to be studied under higher power can be either dried for SEM or embedded for TEM. In this way, localized or suspected damage may be accurately assessed with the tool considered most appropriate.

Material that has been on slides in glycerol for periods of longer than one year can also be surveyed with TEM. Unfortunately, there is some loss in ultrastructure definition. Figure 16 shows sample sections from sound exposed squirrel monkey

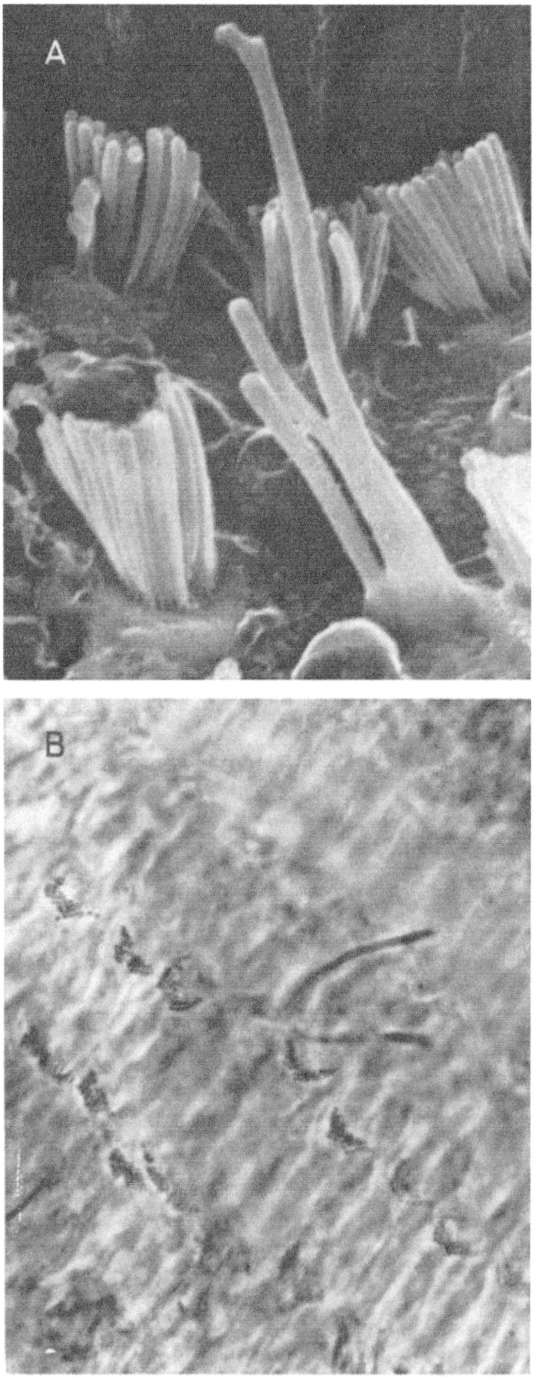

Fig. 13A, B. Giant cilia as resolved by scanning electron (A) and phase microscopes (B) Squirrel monkey. Magnification A 4200 ×, B 1400 ×

and demonstrates a slight loss of quality probably associated with the glycerol treatment. The lack of definition of a "normal" range of minor ultrastructural changes also makes the inclusion of control material for comparison mandatory when the transmission electron microscope is the research tool in inner ear studies.

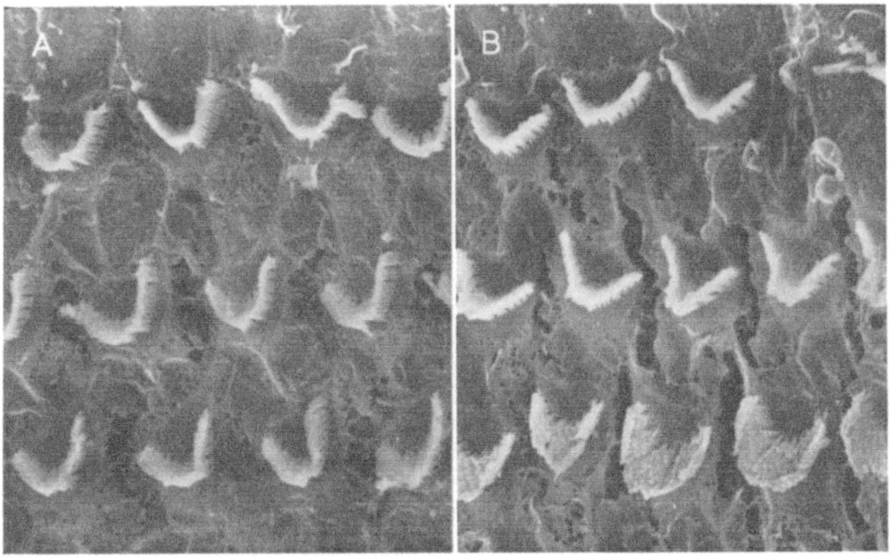

Fig. 14. Scanning electron micrographs of cilia from unexposed (A) and exposed (B) ears of a squirrel monkey. Micrographs show corresponding locations from upper basal turns of left and right ears of monkey 309 from the Hunter-Duvar and Elliott study (1972). Specimens were freeze dried after being stored on slides in glycerol for one year. Magnification A & B 2500 ×

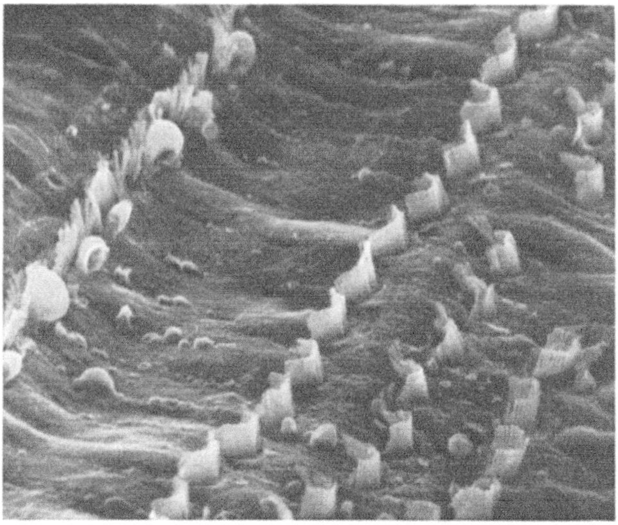

Fig. 15. Cilia middle turn squirrel monkey. Specimens stored in 70% alcohol and freeze dried after cochleograming. Magnification 830 ×

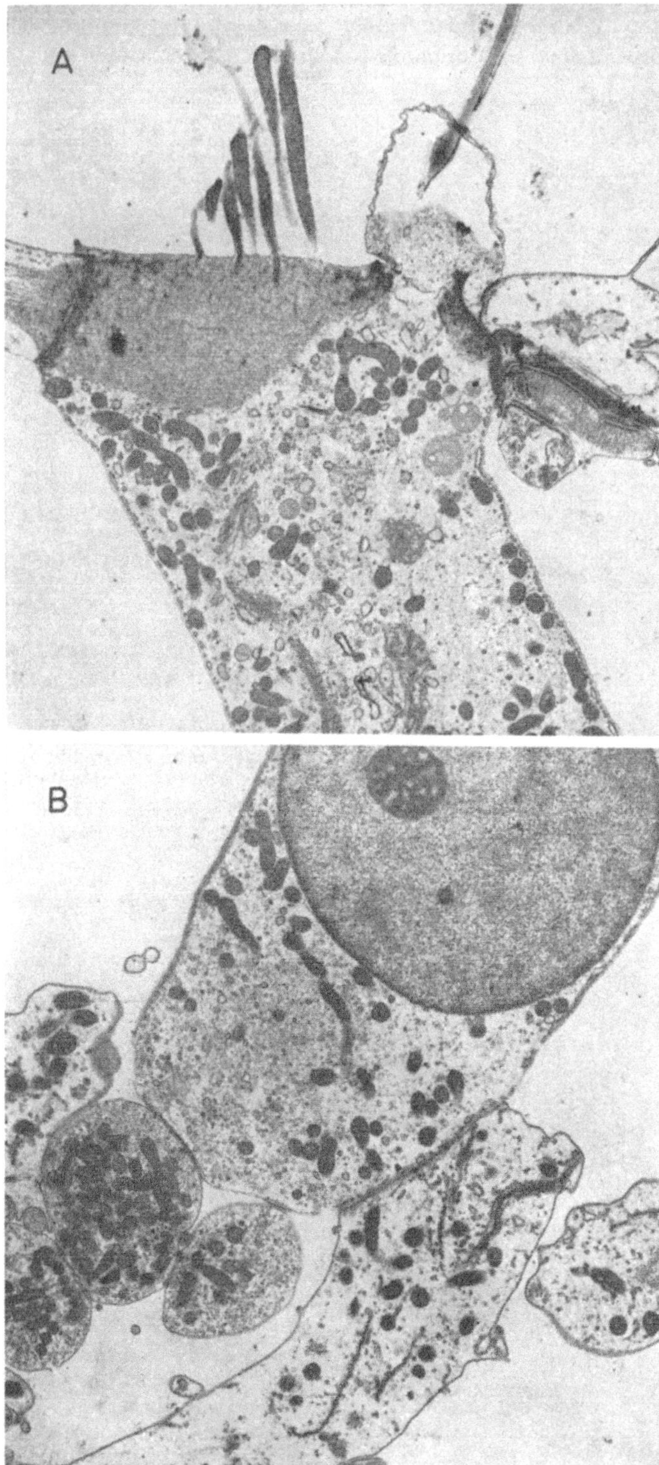

Fig. 16 A u. B

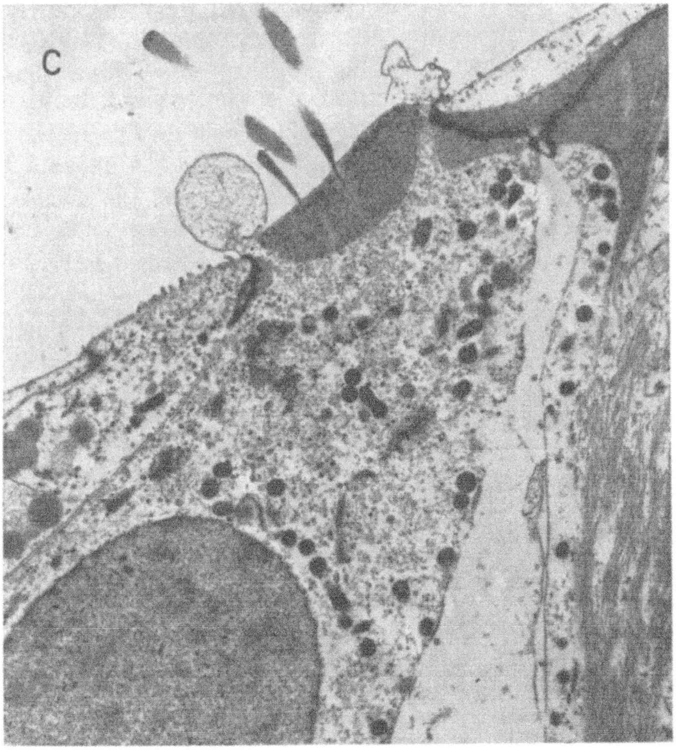

Fig. 16 C

Fig. 16 A—C. Transmission electron micrographs of hair cells from squirrel monkey. This material had been stored on slides in glycerol for approximately one year before being embedded. Although some definition is lost sufficient detail is present for damage survey purposes. (A) and (B) are the top and base of outer hair cells from the middle turn of monkey 298 from the Hunter-Duvar and Elliott study (1972). (C) is an inner hair cell from the middle turn of monkey 315 (see Fig. 9) Magnification A and B 9500 ×; C 8100 ×

D. Comment

Behavioral tests of auditory function provide information about the state of the human auditory system not obtainable by any other means. Behavioral studies of auditory function in animals provide a method for observing the effects of stimuli on the auditory system. If future studies are to add reliable information to our knowledge of the auditory system and its function certain considerations are important.

Training and testing of animals for auditory discrimination tasks requires the use of sophisticated procedures to ensure that thresholds recorded are those of the tested animals rather than those of the testing humans. Improved procedures can come from a knowledge of the animal learning literature.

Future investigators like their predecessors will use stimuli which are of immediate interest to them. Experimental designs which allow the exploration of stimulus boundaries should be used whenever possible. For some stimuli, large differences in individual susceptibility suggest that the same treatment should be given to several subjects to ensure that subject variability is not mistaken for differential treatment effects.

Collection of behavioral data is both time consuming and arduous. Selection of the appropriate methods for assessing anatomical damage is extremely important. A knowledge of the advantages and disadvantages of the different histological techniques available for the postmortem examination of the cochlea will prove expedient and should minimize artifacts and material losses.

The importance of procedural and histological controls has been sufficiently stressed above.

These are a few of the considerations necessary to ensure productive studies. Each experiment will bring with it new problems. An understanding of past problems may assist in the solutions.

References

Altman, F.: Sensorineural deafness in otosclerosis. Ann. Otol. (St. Louis) **75**, 469 (1966).

Anderson, H., Wedenberg, E.: A new method for hearing tests in the guinea pig. Acta oto-laryng. (Stockh.) **60**, 375 (1965).

Békésy, G. von: Experiments in Hearing (E. G. Wever, Ed.). New York: McGraw-Hill 1960.

Benitez, J. T., Schuknecht, H. F.: Otosclerosis: a temporal bone report. Laryngoscope (St. Louis) **72**, 1 (1962).

Benitez, J. T., Schuknecht, H. F., Brandenburg, J. H.: Pathological changes in human ear after kanamycin. Arch. Otolaryng. **75**, 192 (1962).

Bernstein, J. M., Weiss, A. D.: Further observations on salicylate ototoxicity. J. Laryng. **81**, 915 (1967).

Bredberg, G.: Cellular pattern and nerve supply of the human organ of Corti. Acta oto-laryng. (Stockh.) Suppl. 236 (1968).

Bredberg, G., Ades, H. W.: Unpublished data 1970.

Bredberg, G., Engström, H., Ades, H. W.: Cellular pattern and nerve supply of the human organ of Corti. A preliminary report. Arch. Otolaryng. **82**, 462 (1965).

Bredberg, G., Lindeman, H. H., Ades, H. W., West, R., Engström, H.: Scanning electron microscopy of the organ of Corti. Science **170**, 861 (1970).

Bohne, B. A.: Location of small cochlear lesions by phase contrast microscopy prior to thin sectioning. Laryngoscope (St. Louis) **82**, 1—16, (1972).

Bunch, C. C., Wolff, D.: The absence of the organ of Corti: An audiometric and histologic study. Ann. Otol. (St. Louis) **44**, 754 (1935).

Ciocco, A.: A statistical approach to the problem of tone localization in the human cochlea. Hum. Biol. **6**, 714 (1934).

Citron, L., Dix, M. R., Hallpike, C. S., Hood, J. D.: A recent clinicopathological study of cochlear nerve degeneration resulting from tumor pressure and disseminated sclerosis, with particular reference to the finding of normal threshold sensitivity for pure tones. Acta oto-laryng. (Stockh.) **56**, 330 (1963).

Crowe, S. J., Guild, S. R., Polvogt, L. M.: Observations on the pathology of high-tone deafness. Bull. Johns Hopk. Hosp. **54**, 315 (1934).

Cummings, C. W.: Experimental observations on the ototoxicity of nitrogen mustard. Laryngo-scope (St. Louis) **78**, 530 (1968).

Davis, H., Derbyshire, A. J., Kemp, E. H., Lurie, M. H., Upton, M.: Functional and histo-logical changes in the cochlea of the guinea pig resulting from prolonged stimulation. J. gen. Physiol. **12**, 251 (1935).

Davis, H., Dworin, S., Lurie, M. H., Katzman, J.: Behavioral electrical, and anatomical studies of abnormal ears. Laryngoscope (St. Louis) **47**, 435 (1937).

Davis, H., Silverman, S. R.: Hearing and deafness. New York: Holt, Rinehart and Winston 1970.

DIX, M. R., HALLPIKE, C. S., HOOD, J. D.: Observations upon the loudness recruitment phenomenon. Proc. roy. Soc. Med. **41**, 516 (1948).

DOLAN, T. R., BREDBERG, G., ADES, H. W., NEFF, W. D.: The behavioral and histological effects of auditory overexposure on the cat. Paper read at the eightieth meeting of the Acoustical Society of America, Houston/Texas 1970. J. acoust. Soc. Amer. **49**, 92 (1971).

ELDREDGE, D. H., MILLER, J. D.: Acceptable noise exposures. Damage risk criteria. Amer. Speech Hearing Assoc., Rpt. number **4**, 110 (1969).

ELDREDGE, D. H., MILLS, J. H., BOHNE, B. A.: Anatomical behavioral and electrophysiological observations on chinchillas after long exposures to noise. International Symposium on Otophysiology. Ann. Arbor/Michigan 1971. Adv. Oto-Rhino-Laryng. **20**, 64 (1973).

ELLIOTT, D. N.: The effects of sensorineural lesions on pitch discrimination in cats. Ann. Otol. (St. Louis) **70**, 582 (1961).

ELLIOTT, D. N., MCGEE, T. M.: Effects of cochlear lesions upon audiograms and intensity discrimination in cats. Ann. Otol. (St. Louis) **74**, 386 (1965).

ENGSTRÖM, H., ADES, H. W., ANDERSSON, A.: Structural pattern of the organ of Corti. Stockholm: Almqvist and Wiksell 1966.

ERNSTSON, S.: Cochlear physiology and hair cell population in a strain of the waltzing guinea pig. Acta oto-laryng. (Stockh.) Suppl. 297 (1972).

FERNANDEZ, C.: Postmortem changes and artefacts in human temporal bones. Laryngoscope (St. Louis) **68**, 1586 (1958).

FIEANDT, H. VON, SAXEN, A.: Pathologie und Klinik der Alterschwerhörigkeit. Acta oto-laryng. (Stockh.) Suppl. 23 (1937).

FLEISCHER, K.: Histologische und audiometrische Studie über den altersbedingten Struktur- und Funktionswandel des Innernohres. Arch. Ohr.-, Nas.- u. Kehlk.-Heilk. **170**, 142 (1956).

FRAZIER, L., ELLIOTT, D. N., HUNTER-DUVAR, I. M.: Unpublished data (1972).

GLORIG, A., NIXON, J.: Hearing loss as a function of age. Laryngoscope (St. Louis) **72**, 1596 (1962).

GLORIG, A., WHEELER, D., QUIGGLE, R., GRINGS, W., SUMMERFIELD, A.: Wisconsin State Fair hearing survey 1954. Los Angeles: Amer. Acad. of Ophthal. and Otolaryng. (1957).

GOLD, A., WILPIZESKI, C. R.: Some effects of sodium salicylate on evoked auditory potentials in cats. Laryngoscope (St. Louis) **76**, 674 (1966).

GRAF, K.: Histologische Veränderungen des Innerohres nach Behandlung der Meningitis tuberculosa. Acta oto-laryng. (Stockh.) **39**, 121 (1951).

GREENWOOD, D. D.: Critical bandwidth and the frequency coordinates of the basilar membrane. J. acoust. Soc. Amer. **33**, 344 (1961).

GUILD, S. R.: A graphic reconstruction method for the study of the organ of Corti. Anat. Rec. **22**, 141 (1921).

GUILD, S. R., CROWE, S. J., BUNCH, C. C., POLVOGT, L. M.: Correlations of differences in the density of innervation of the organ of Corti with differences in the acuity of hearing, including evidence as to the location in the human cochlea of the receptor for certain tones. Acta oto-laryng. (Stockh.) **15**, 269 (1931).

HABERMANN, J.: Über die Schwerhörigkeit der Kesselschmiede. Arch. Ohrenheilk. **30**, 1 (1890).

HALLPIKE, C. S., CAIRNS, H.: Observations on the pathology of Ménière's syndrome. Proc. roy. Soc. Med. **31**, 1317 (1938).

HANSEN, C. C., RESKE-NIELSEN, E.: Pathological studies in perceptive deafness. A patient with hydrops of the labyrinth. Acta oto-laryng. (Stockh.) Suppl. **188**, 162 (1964).

HANSEN, C. C., RESKE-NIELSEN, E.: Pathological studies in presbyacusis. Arch. Otolaryng. **82**, 115 (1965).

HINCHLIFFE, R.: Threshold of hearing as a function of age. Acustica **9**, 303 (1959).

HUNTER-DUVAR, I. M.: Effects of intense auditory stimulation: Hearing losses and inner ear changes in the squirrel monkey. Ph. D. thesis. Wayne State University, Detroit/Michigan 1971.

HUNTER-DUVAR, I. M., ELLIOTT, D. N.: Effects of intense auditory stimulation: Hearing losses and inner ear changes in the squirrel monkey. J. acoust. Soc. Amer. **52**, 1181 (1972).

HUNTER-DUVAR, I. M., ELLIOTT, D. N.: Effects of intense auditory stimulation: Hearing losses and inner ear changes in the squirrel monkey II. J. acoust. Soc. Amer. **54**, 1179 (1973).

HUNTER-DUVAR, I. M., BREDBERG, G.: Effects of intense auditory stimulation: Hearing losses and inner ear changes in the chinchilla. J. acoust. Soc. Amer. **55**: 795 (1974).

Igarashi, M., SCHUKNECHT, H. F., MYERS, E. N.: Cochlear pathology in humans with stimulation deafness. J. Laryng. **78**, 115 (1964).

JATHO, K., HECK, K. H.: Schwellenaudiometrische Untersuchungen über die Progredienz und Charakteristik der Alterschwerhörigkeit in den verschiedenen Lebensabschnitten. Z. Laryng. Rhinol. **38**, 72 (1959).

JERGER, J. F.: Audiological manifestations of lesions in the auditory nervous system. Laryngoscope (St. Louis) **70**, 417 (1960).

JERGER, J. F.: Recruitment and allied phenomena in differential diagnosis. J. Aud. Res. **2**, 145 (1961).

JOHNSSON, L.-G., HAWKINS, J. E., JR.: A direct approach to cochlear anatomy and pathology in man. Arch. Otolaryng. **85**, 599 (1967).

JØRGENSEN, M. B.: Intracochlear neurinoma. Acta oto-laryng. (Stockh.) **54**, 227 (1962).

JØRGENSEN, M. B., SCHMIDT, R.: The ototoxic effect of kanamycin. Acta oto-laryng. (Stockh.) **55**, 537 (1962).

KEMP, E. H.: A critical review of experiments on the problem of stimulation deafness. Psychol. Bull. **32**, 325 (1935).

KIMURA, R. S., SCHUKNECHT, H. F., SANDO, I.: Fine morphology of the sensory cells in the organ of Corti in man. Acta oto-laryng. (Stockh.) **58**, 390 (1964).

KRISTENSEN, H. K.: Histopathology in Méniêr's disease. Acta oto-laryng. (Stockh.) **53**, 237 (1961).

LIDÉN, G.: Undistorted speech audiometry. In: Sensorineural hearing processes and disorders. Henry Ford Hospital International Symposium 1965. Boston: Little, Brown. 1967.

LIM, D. J., LANE, W. C.: Three dimensional observation of the inner ear with the scanning electron microscope. Tr. Amer. Acad. Ophthal. Otolaryng. **73**, 842 (1969).

LINDEMAN, H. H., BREDBERG, G.: Scanning electron microscopy of the organ of Corti after intense auditory stimulation: Effects on stereocilia and cuticular surface of hair cells. Arch. Ohr.-, Nas.- u. Kehlk.-Heilk. **203**, 1 (1972).

LINDQUIST, S. E.: Stimulation deafness: A study of temporary and permanent hearing losses resulting from exposure to noise and to blast impulses. Ph. D. thesis, University of Chicago, Chicago/Illinois 1949.

LINDQUIST, S. E., NEFF, W. D., SCHUKNECHT, H. F.: Stimulation deafness: A study of hearing losses resulting from exposure to noise of blast impulses. J. comp. Physiol. Psychol. **47**, 406 (1954).

LINDSAY, J. R.: Ménière's disease. Histopathologic observations. Arch. Otolaryng. **39**, 313 (1944).

LINDSAY, J. R.: Hydrops of the labyrinth. Arch. Otolaryng. **71**, 500 (1960).

LINDSAY, J. R., KOHUT, R. I., SCIARRA, P. A.: Ménière's disease: Pathology and manifestations. Ann. Otol. (St. Louis) **76**, 5 (1967).

LINDSAY, J. R., PROCTOR, L. R., WORK, W. P.: Histopathologic inner ear changes in deafness due to neomycin. Laryngoscope (St. Louis) **70**, 382 (1960).

LINDSAY, J. R., SCHULTESS, G. VON: An unusual case of labyrinthine hydrops. Acta oto-laryng. (Stockh.) **49**, 315 (1958).

MAKISHIMA, K.: Histopathological studies in presbycusis. J. Otorhinolaryng. Soc. Jap. **70**, 63 (1967).

MATZ, G. J., BEAL, D. D., KRAMES, L.: Ototoxicity of ethacrynic acid. Arch. Otolaryng. **90**, 152 (1969).

MATZ, G. J., WALLACE, T. H., WARD, P. H.: The ototoxicity of kanamycin. Laryngoscope (St. Louis) **75**, 1960 (1965).

MCCABE, P. A., DEY, F. L.: The effect of aspirin upon auditory sensitivity. Ann. Otol. (St. Louis) **74**, 312 (1965).

McGee, T. M., Olszewski, J.: Streptomycin sulphate and dihydrostreptomycin toxicity. Arch. Otolaryng. **75**, 295 (1962).

McGee, T. M., Cole, G. G., Van Den Ende, H.: Effects of quantified ultrasonic sound on labyrinthine structures in animals. Laryngoscope (St. Louis) **73**, 683 (1963).

Miller, J. D., Watson, C. S., Covell, W. P.: Deafening effects of noise on the cat. Acta otolaryng. (Stockh.) Suppl. 176 (1963).

Myers, E. N., Bernstein, J. M.: Salicylate ototoxicity. Arch. Otolaryng. **82**, 483 (1965).

Myers, E. N., Bernstein, J. M., Fostiropolous, G.: Salicylate ototoxitity. A clinical study. New Engl. J. Med. **273**, 587 (1965).

Nager, G. T.: Sensorineural deafness and otosclerosis. Ann. Otol. **75**, 481 (1966).

Neff, W. D.: The effects of partial section of the auditory nerve. J. comp. Physiol. Psychol. **40**, 2031 (1947).

Oda, D.: Observations on the pathology of impaired hearing for low tones. Laryngoscope **48**, 765 (1938).

Pavlov, I. P.: Conditioned Reflexes. (G. V. Anrep, Ed.) N. Y. Dover Publications Inc. 1972.

Rüedi, L. and Spoendlin, H.: Pathogenesis of sensorineural deafness in otosclerosis. Ann. Otol. **75**, 525 (1966).

Sando, I., Hemenway, W. G., Hilyard, V. H., English, G. M.: Cochlear otosclerosis: A human temporal bone report. Ann. Otol. **77**, 23 (1968).

Schuknecht, H. F.: Techniques for study of cochlear function and pathology in experimental animals: Development of the anatomical frequency scale for the cat. Arch. Otolaryng. **58**, 377 (1953a).

Schuknecht, H. F.: Lesions of the organ of Corti. Tr. Amer. Acad. Ophthal. Otolaryng. **157**, 366 (1953b).

Schuknecht, H. F.: Further observations on the pathology of presbycusis. Arch. Otolaryng. **80**, 369 (1964a).

Schuknecht, H. F.: The pathology of several disorders of the inner ear which cause vertigo. Southern Med. J. **57**, 1161 (1964b).

Schuknecht, H. F.: The pathophysiology of angle tumors. In: The vestibular system and its diseases (R. J. Wolfson, Ed.) Philadelphia 1966.

Schuknecht, H. F.: The effect of aging on the cochlea. In: Sensorineural Hearing Processes and Disorders. Henry Ford Hospital International Symposium. P 393. Boston: Little, Brown 1967.

Schuknecht, H. F.: Pathology of Ménière's disease, Otolaryng. Clin. N. Amer. 331 (1968).

Schuknecht, H. F.: Discussion of Ward and Duvall paper. Trans. Amer. otol. Soc. **59**, 133 (1971).

Schuknecht, H. F., Benitez, J. T., Beekhuis, J.: Further observations on the pathology of Ménière's disease. Ann. Otol. (St. Louis) **71**, 1039 (1962).

Schuknecht, H. F., Gross, C. W.: Otosclerosis and the inner ear. Ann. Otol. (St. Louis) **75**, 423 (1966).

Schuknecht, H. F., Igarashi, M.: Pathology of slowly progressive sensorineural deafness. Trans. Amer. Acad. Ophthal. Otolaryng. **68**, 222 (1964).

Schuknecht, H. F., Neff, W. D.: Hearing losses after apical lesions in the cochlea. Acta otolaryng. (Stockh.) **42**, 263 (1952).

Schuknecht, H. F., Neff, W. D., Perlman, H. B.: An experimental study of auditory damage following blows to the head. Ann. Otol. (St. Louis) **60**, 273 (1951).

Schuknecht, H. F., Sutton, S.: Hearing losses after experimental lesions in the basal coil of the cochlea. Arch. Otolaryng. **57**, 129 (1953).

Schuknecht, H. F., Woellner, R. C.: Hearing losses following partial section of the cochlear nerve. Laryngoscope (St. Louis) **63**, 441 (1953).

Schuknecht, H. F., Woellner, R. C.: An experimental and clinical study of deafness from lesions of the cochlear nerve. J. Laryng. **69**, 75 (1955).

Spoendlin, H.: The organization of the cochlear receptor. Adv. Oto-Rhino-Laryn 9. **13**, 1 (1966).

Spoendlin, H.: Innervation densities of the cochlea. Acta oto-laryng. (Stockh.) **73**, 235 (1972).

Spoor, A.: Presbycusis values in relation to noise induced hearing loss. Int. Audiology **6**, 48 (1967).

Stebbins, W. C., Clark, W. W., Pearson, R. D., Weiland, M. G.: Noise and drug induced hearing loss in monkeys. International Symposium on Otophysiology, Ann Arbor/Michigan 1971. Adv. Oto-Rhino-Laryng. **20**, 42 (1973).

Stebbins, W. C., Miller, J. M., Johnsson, L-G., Hawkins, J. E., Jr.: Ototoxic hearing loss and cochlear pathology in the monkey. Ann. Otol. (St. Louis) **78**, 1007 (1969).

Ward, W. D.: Personal communication 1971.

Ward, W. D., Duvall, A. J.: Behavioral and ultrastructural correlates of acoustic trauma. Ann. Otol. (St. Louis) **80**, 881 (1971).

Wever, E. G.: Problem of tonal dip. Laryngoscope (St. Louis) **52**, 169 (1942).

Wittmaack, K.: Demonstration zur Resonanz-Theorie. Acta oto-laryng. (Stockh.) **12**, 32 (1928).

Yamakawa, K.: Über pathologische Veränderungen bei einem Ménièrekranken. Zbl. Hals.-, Nas.- u. Ohrenheilk. **32**, 697 (1939).

Chapter 8

Behavioral Studies of Auditory Discrimination: Central Nervous System*

WILLIAM D. NEFF, Bloomington, Indiana (USA), IRVING T. DIAMOND, Durham, North Carolina (USA), and JOHN H. CASSEDAY, Durham, North Carolina (USA)

With 25 Figures

Contents

I. Historical Review . . . 309
A. 1860—1900 . . . 309
B. 1900—1930 . . . 310
C. 1930—1940 . . . 310
D. 1940—1975 . . . 311
II. Relation of Structure and Function . . . 311
A. Subdividing the Auditory System . . . 311
1. The Ascending System . . . 311
2. The Descending System . . . 314
3. Tonotopic Organization . . . 315
4. Limitations of Behavior-Ablation Method . . . 315
5. Conclusions . . . 315
B. Definition of Auditory Cortex . . . 316
1. Retrograde Degeneration Studies . . . 317
2. Anterorgrade Degeneration Studies . . . 318
C. Species Differences . . . 320
III. Research in Modern Era . . . 323
A. Ablation of Auditory Cortex . . . 323
1. Discrimination of Onset of Sound and Absolute Thresholds . . . 324
a) Cat . . . 324
b) Monkey . . . 325
c) Other Animals . . . 325
d) Man . . . 326
e) Retention of Learned Discriminations: One-Stage vs. Two-Stage Ablations 330
f) Effects of Decortication . . . 331
g) Summary . . . 331
2. Discrimination of Changes in Intensity . . . 332
a) Cat . . . 333
b) Monkey . . . 334
c) Man . . . 334
d) Summary . . . 335

* Research of the authors of this chapter has been supported in part by the National Science Foundation (W.D.N.) (J.H.C.), the National Aeronautics and Space Administration (W.D.N.), the National Institute of Mental Health (I.T.D.) and the National Institutes of Health (J.H.C.).

3. Discrimination of Changes in Frequency . . . 335
a) Cat . . . 335
b) Monkey . . . 339
c) Other Animals . . . 339
d) Man . . . 340
e) Summary . . . 340
4. Pattern Discrimination . . . 341
a) Cat . . . 341
b) Monkey . . . 344
c) Man . . . 344
d) Summary . . . 345
5. Discrimination of Temporal Order . . . 346
a) Man . . . 346
b) Summary . . . 347
6. Recent or Short Term Memory . . . 347
a) Dog . . . 347
b) Monkey . . . 348
c) Summary . . . 348
7. Discrimination of Changes in Duration . . . 349
a) Cat . . . 349
b) Summary . . . 350
8. Localization and Lateralization of Sound . . . 350
a) Cat . . . 351
b) Monkey . . . 353
c) Other Animals . . . 353
d) Man . . . 354
e) Reflex Responses . . . 356
f) Summary . . . 357
9. Other Binaural Discriminations . . . 358
a) Cat . . . 358
b) Man . . . 361
c) Summary . . . 362
10. Sounds with Complex Spectra . . . 362
a) Cat . . . 362
b) Monkey . . . 362
c) Man . . . 362
d) Summary . . . 363
11. Summary and Conclusions . . . 363
B. Ablation of Subcortical Centers and Transection of Subcortical Pathways . . . 364
1. Discrimination of Onset of Sound and Absolute Thresholds . . . 365
a) Experimental Animals . . . 365
b) Man . . . 366
c) Summary . . . 368
2. Discrimination of Changes in Intensity . . . 368
a) Experimental Animals . . . 368
b) Man . . . 369
c) Summary . . . 369
3. Discrimination of Changes in Frequency . . . 370
a) Experimental Animals . . . 370
b) Man . . . 371
c) Summary . . . 371
4. Attention to Auditory Stimuli . . . 371
5. Localization and Lateralization of Sound . . . 372
a) Experimental Animals . . . 372
b) Man . . . 376
c) Summary . . . 377

6. Echolocation . . . 378
7. Pattern Discrimination . . . 379
8. Summary and Conclusions . . . 380
C. Descending Auditory Pathways . . . 381
1. Absolute Thresholds . . . 381
2. Discrimination of Changes in Intensity or Frequency . . . 381
3. Localization of Sound in Space . . . 382
4. Masking . . . 382
5. Summary . . . 384
IV. Discussion . . . 384
A. Relation of Cortex to Subcortex . . . 384
B. Evolution of Sensory and Association Cortex . . . 386
C. Relation between Cat and Other Species . . . 387
D. Conclusions . . . 388
List of Abbreviations . . . 389
References . . . 390

I. Historical Review

For convenience of review, the history of research that has used behavioral tests before and after ablation of auditory centers or transection of neural pathways in order to understand the neural basis of auditory discrimination[1] may be divided into four periods; each period represents a time during which new methods were being developed that led to new information: (A) 1860–1900; (B) 1900–1930; (C) 1930–1940; (D) 1940–1975. Throughout all of these periods, advance in knowledge depended upon progress along a number of lines, principally the following:

(1) refinement of methods used to trace neural pathways and identify neural centers,

(2) development of adequate methods by which to measure auditory discriminations,

(3) improvement of surgical techniques.

A. 1860-1900

By the last half of the 19th century, neuroanatomical techniques and tools had reached the stage of development so that some investigators undertook to distinguish parts of the nervous system having different functions. Peripheral pathways of sensory systems could be traced and attempts could be made to

[1] The definition of sensory discrimination as used in this chapter is as follows: "The living organism makes a sensory discrimination when it shows by its behavior that it has made a response to a change in a physical stimulus applied to one of its sense organs. It may respond to the presence or absence of a particular stimulus or may make a choice between two or more stimuli. When man is used as a subject in an experiment on sensory discrimination, he may be instructed to report verbally when a stimulus change occurs; often he is asked to respond by pushing a button or by making some other nonverbal response. When a lower animal is used as an experimental subject, it is usually trained to indicate the stimulus change by making a movement, such as flexing a leg, pressing a lever with its foot, or moving from one position to another in the test apparatus. In each case, the discrimination consists simply of a motor response to a change in some aspect of a physical stimulus which acts upon the sensory end organs" (NEFF, 1960).

define centers in the cortex. Differential staining techniques were used to trace degenerating pathways and to identify differences in cellular organization of nuclei. Methods used to define auditory cortex included electrical stimulation to elicit movement of the ears and turning the eyes and head as if towards a sound source (FERRIER, 1886) and mapping of areas in which early myelination of cortical elements occurred (VOGT, 1898).

Behavioral studies during this early period consisted of observations of changes in behavior before and after central nervous system lesions produced experimentally in animals or occurring because of accident or disease in human patients. No particular training and testing methods were used with animals, and careful psychophysical tests were not made in human patients. Consequently, interpretations of results led to controversy more often than agreement among investigators (see, for example, FERRIER, 1886, 1889, 1890; BROWN and SCHÄFER, 1888; MUNK, 1881; LARIONOW, 1899; MOTT, 1907).

Experimental surgery prior to 1875 (approximately) was done without asepsis. Observations on experimental animals could be made for only a short time post-surgery and never after complete recovery of general physical condition. Aseptic techniques in human neurosurgery were not introduced until after 1865.

B. 1900-1930

Some new information about structure but very little about function of the auditory system was produced during the period, 1900–1930. But this was a time when new methods were being developed that would later make possible rapid advance in knowledge in all areas of sensory physiology. Many investigators recognized the potential value of recording electrical responses of sense organs and of sensory pathways and centers of the nervous system; therefore, techniques and instrumentation were steadily improved.

The interest of psychologists and physiologists in studying learned behavior in experimental animals led to the development of methods that were readily adaptable to the study of sensory capacities. PAVLOV, BECHTEREV, WATSON, THORNDIKE, and YERKES, and their students are the names most often associated with the development of effective behavioral methods for training and testing experimental animals.

Aseptic procedures were now universally used in animal experiments as well as in the clinic; thus, behavior after surgery could be observed for months or years.

C. 1930-1940

From 1930–1940, progress at an accelerating rate was made in methodology in all three lines of investigation noted above; new anatomical methods were introduced, most importantly the use of electrophysiological techniques of tracing neural pathways; procedures for training and testing hearing of experimental animals were improved. Surgical methods were also improved through more precise techniques of ablation, use of sulfa drugs, and use of careful aseptic procedures. Furthermore, in a number of laboratories, programs of research on the auditory system were planned and carried out over a period of several years.

D. 1940-1975

By 1940, there was fair agreement among investigators as to the principal pathways and centers of the auditory system in mammals. Information was based mainly on experimental studies of the cat, dog, and monkey. Limited information was available for man.

Since 1942, there has been a steady flow of published reports in which the *experimental surgery-behavior* method has been used to study the auditory system. The need for more precise definition of the pathways and centers and of subdivisions within these structures stimulated anatomical research. Findings of the anatomical research, in turn, made possible refinement of the ablation technique.

The review of research results that follows in Section III will be confined almost entirely to studies reported since 1940. Reference will be made to early studies only when the results had significance that was not appreciated or not correctly interpreted but which has been clarified by later findings.

Before reviewing the results of experiments and of clinical studies done in the modern era, we shall consider the possibilities and the limitations in interpretation of results obtained by the procedure in which auditory discrimination behavior is examined before and after ablation or injury of auditory centers or before and after experimental or accidental transection of auditory pathways.

II. Relation of Structure and Function

In the intact animal, all parts of the auditory system are probably involved to some extent in all auditory discriminations, but to say that the auditory system works as a whole need not imply that the contribution of each part is the same. For example, the role of the superior olive in sound localization is not the same as the role of the medial geniculate body. The goal of behavioral studies of the auditory system is to identify the special contribution of each part to the whole. The ablation-behavior method is complicated by the fact that it relies on anatomical and electrophysiological methods to define subdivisions of the central auditory system. It is beyond our purpose to review in detail the anatomy of the ascending and descending pathways, but some of the problems of subdividing the auditory system need to be discussed before the results obtained from ablation can be evaluated.

A. Subdividing the Auditory System

1. The Ascending System

The starting point for subdivision is to define the ascending pathway from ear to cortex. By 1930, it had been shown by studies using the Marchi method for tracing degenerated myelinated fibers that the cochlear nucleus (CN) is the first center, the superior olivary complex (SOC) the second, the inferior colliculus (IC) the third, the medial geniculate body (GM) the fourth and the auditory cortex itself the fifth and final center (Fig. 1). To stop at this description of the system would be an oversimplification. Several qualifications are necessary, and several issues remain unresolved.

First, there is the problem of defining precisely the borders of an auditory center. In the case of the cochlear nucleus, there is hardly any question about the borders, but at the level of the thalamus, it is another matter. The thalamus consists of a number of cytoarchitectonic subdivisions, and one of these, GM, was shown at the turn of the century to receive fibers from IC and to project

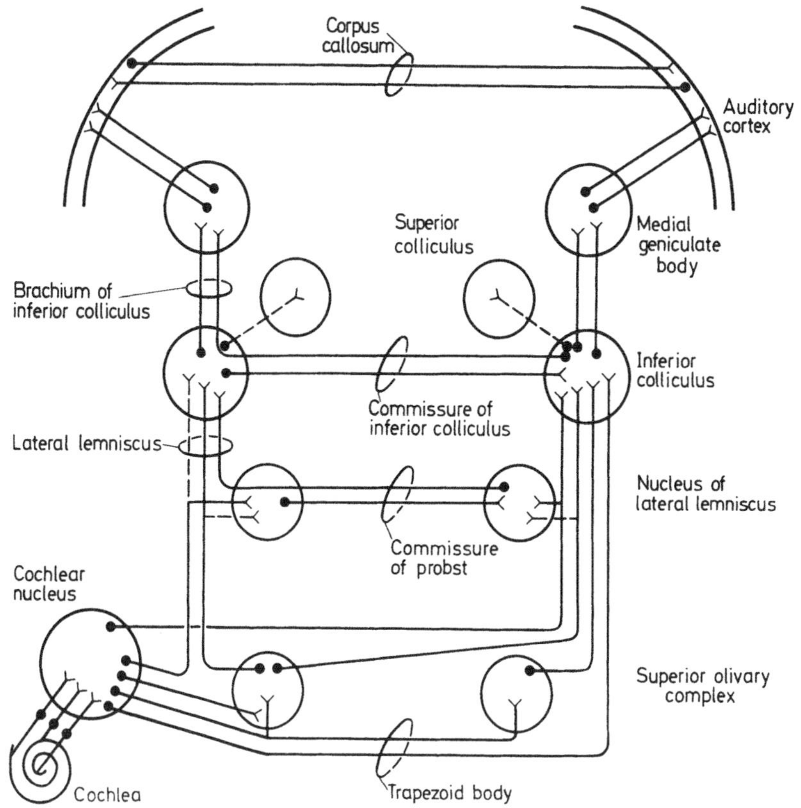

Fig. 1. Diagram of the major subdivisions and connections in the ascending auditory system. Dashed lines represent connections for which experimental evidence is questionable or incomplete. The diagram is intended to summarize anatomical evidence indicating that there are several pathways, in parallel, from cochlear nucleus to cortex. The connections for one cochlea only are shown

to neocortex (Fig. 2). The pattern of innervation of GM by fibers from the brachium of the inferior colliculus (BIC) and the pattern of projection to the cortex was not known clearly at that time. But GM is divisible into a lateral sector densely populated with small cells, and a medial sector with larger cells and more sparsely populated. The division of fibers from BIC between these two sectors and their projection to cortical areas were not clear. In advanced mammals and in the cat, in particular, the principal division of GM is divisible into several sectors (Morest, 1964), and it is almost certainly the case that not all of these sectors receive direct neural input from BIC. To further complicate the picture the posterior nuclei (Po), dorsomedial to GM receive auditory input not only from BIC (MOORE and GOLDBERG, 1963) but also probably from the lateral tegmental area (MOREST, 1965). At least Po projects to cortex which seems to have auditory function (ROSE and WOOLSEY, 1958), and cells in Po are responsive to auditory stimuli (POGGIO and MOUNTCASTLE, 1960).

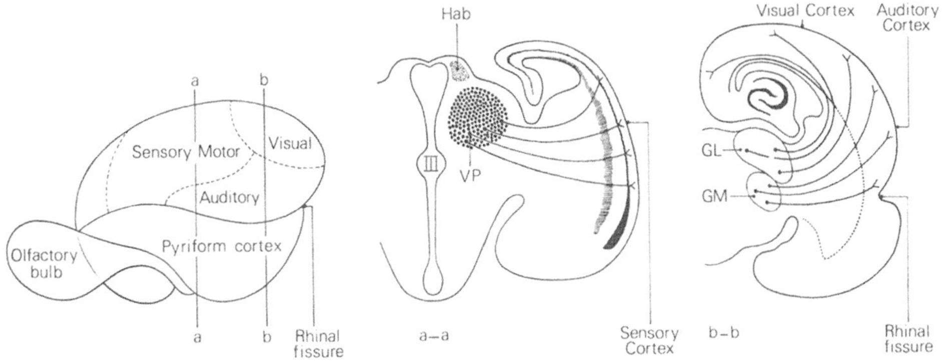

Fig. 2. To show the auditory cortex in a primitive animal, this drawing was adapted from ELLIOT SMITH (1910). The figure shows that the position of the auditory cortex as well as the visual and somatic cortex depends on thalamic fibers from the different relay nuclei taking the shortest and most direct route. This figure also serves in making the point that the concept of thalamic relay nuclei was well established at the turn of the century

The problem of defining the limits of auditory cortex poses special difficulties because, in an obvious sense, the whole cortex is a single organ; auditory cortex is more like visual cortex than it is like the auditory thalamus or the auditory midbrain. Our present conception of the limits of both auditory cortex and thalamus depends on attempts to collate the results of different anatomical methods. In addition to the methods of tracing fibers from center to center, the techniques of electrophysiology have proved essential in defining what should be included under the concept of auditory thalamus and auditory cortex. Furthermore, results from both anatomical degeneration and electrophysiological studies must be viewed in the light of cortical and thalamic subdivisions that are based upon cytoarchitecture.

Below, we shall try to reach a working concept of the mammalian auditory cortex for the purpose of reviewing the literature. First, there are some issues about the concept of a subdivision which must be raised. Within each level, including the cochlear nucleus and SO complex as well as GM and cortex, there are cytoarchitectonic subdivisions, which have different connections; this structural arrangement supports the idea that the subdivisions have functional significance. The resulting picture of the auditory system is much more complicated than one which shows the system as a series of obligatory relay stations. From one part of the cochlear nucleus, which corresponds to a cytoarchitectonic unit, fibers project to SOC; from another part, fibers bypass SOC and project directly to IC. In this way, there are parallel and separate pathways ascending to IC.

We cannot say whether, in IC, separate ascending pathways converge on neurons or remain separate. There is good evidence that some pathways ascend in parallel without convergence all the way to the cortex. Thus, only the ventral nucleus of the principal division of the GM receives fibers from the central nucleus of IC and projects to the auditory koniocortex. The auditory thalamic belt receives fibers from the lateral tegmental area and projects to the belt of cortex adjacent to the core auditory cortex. These facts raise a central issue for the functional analysis of the auditory system. Is the fundamental unit or part "the level" or "the path" – *e.g.*, do the cochlear nuclei serve one common function and the inferior colliculi another, or alternatively, is the fundamental unit a particular path in parallel with other paths? As an example of the second alternative, the SOC may be a critical center in its own pathway, and it may be misleading to try and assign it a level above cochlear nucleus and below IC. To continue the example, we might imagine that SOC is for sound localization and that there is a parallel path to IC for frequency discrimination. Both pathways could, of course, converge in the cortex.

The notion that there is a common function associated with each "level" is most reasonable when we think of the cortex. As we noted earlier, the cortex seems to be a single organ; it is the last part of the path to develop in vertebrate history; and it reaches its greatest development in the higher primates and man. Such considerations make it tempting to think of the cortex as the site of the highest level of integration and to seek some psychological function common to the whole cortex. Early investigators who were interested in the central nervous system proposed that the cortex was the basis of sensation, and, therefore, that the auditory cortex was necessary for auditory sensation, the visual cortex for visual sensation and so on.

To summarize, the idea that the auditory nervous system is a pathway from ear to cortex is useful as a first approximation, but, on closer study, the single path can be divided into two or more parallel paths. Some "levels" such as SOC may be centers for only some of the parallel paths. "Levels", such as the thalamus, may be divided into parts which serve as nuclei for different parallel pathways.

2. The Descending System

While it makes sense to designate the cortex as the culmination of the auditory pathways, neural impulses, originating at the ear, do not end in the cortex but are relayed back to the centers of the ascending pathway itself as well as to other cortical and subcortical centers. The auditory cortex projects to GM and to IC, and, thus, the integration at these centers can be considered to involve a higher level than would at first appear to be the case. In considering the relation between the ascending and descending paths, it is important to distinguish reciprocal interconnections between levels from nonreciprocal interconnections. As an example of a reciprocal interconnection, we cite the anatomical finding that the part of GM that projects to a given sector of auditory cortex receives fibers from the same sector (DIAMOND *et al.*, 1969). Perhaps the very cell whose axon terminates in a cortical column of cells receives fibers from that column. Not every descending path projects in this reciprocal fashion; for example, the relation between IC and the auditory cortex is not reciprocal. Only the central nucleus of IC receives fibers of the lateral lemniscus (LL) and projects to the cortex via GM; but the external nucleus, not the central, is one of the recipients of corticotectal efferents (DIAMOND *et al.*, 1969). In principle, then, it is possible to separate the afferent and efferent systems at IC; to do so in practice, *e.g.*, by ablation, is extremely difficult if not impossible. There is, at least, one place where the descending system is separated clearly from the ascending system and that is in the olivocochlear bundle. Later, we shall describe the effects on behavior of transecting this tract. We shall also describe an ingenious way of separating the ascending and descending systems by taking advantage of the fact that, in the cochlear nucleus, one of them is adrenergic and one is cholinergic. Certain drugs can selectively block the synapses from one system without affecting synaptic transmission from the other.

3. Tonotopic Organization

Centers at each level of the auditory system may be subdivided not only on the basis of structural organization but also according to topographic (tonotopic) projection of the basilar membrane upon the auditory centers. We refer to the fact that each neuron can be designated by its "best frequency" (*e.g.*, the frequency which most readily evokes an increase in firing rate) with the result that an orderly representation of the basilar membrane emerges at different levels of the system. If, at one level, there is *more than one* representation of the basilar membrane, or if one representation does not include all of the responsive cells at a given level, then some further subdivision is in order. It is probably fair to conclude that the cytoarchitectonic and electrophysiological subdivisions within each level will correspond to differences in connections.

4. Limitations of Behavior-Ablation Method

The behavior-ablation method must not only be content with an incomplete picture of the auditory system; it is also limited by the fact that a deficit after ablation does not reflect simply a function of the ablated part. On the contrary, we see, in an animal deprived of a part, the functioning of the remaining system, and we recognize that each of the remaining parts is itself more or less disturbed by the ablation. To give an extreme example, the effects of completely removing the cochlear nuclei cannot reflect the function of the cochlear nuclei, because every other center in the path is deprived of auditory input. Finally, we come to a physical limitation in the ablation method. In any nucleus, it may be impossible to destroy only one subdivision without concomitant damage to adjacent subdivisions.

5. Conclusions

The foregoing account has indicated that the anatomical definition of a part and the behavioral study of that part's function are aspects of the same inquiry. The description of a relationship between structure and function does not provide a complete account of the relations between anatomy, physiology, and behavior. The aim of neurophysiological investigations of the auditory system is, of course, much more than the definition of the borders of a center or the subdivision of a center into one or more tonotopic maps. A major goal that is especially important for behavioral study is to identify features of the stimulus that excite different neural units. For example, the discovery that cells in SO respond to small differences in the time of arrival of sounds at the two ears suggests functions that can be tested by behavioral methods. Anatomy also has this double significance of defining the limits of organs and suggesting functional relationships between organs. Just as the valves of the heart are clues to the circulation of the blood, so is the arrangement of some cells in SO a sign that their function may be to relate impulses originating from the two ears. Functional studies must consider not only gross connections but also the arrangement of cell bodies and of their dendrites and axons.

At the organ level, anatomical, electrophysiological and behavioral studies are all aspects of a *single* structure-function inquiry. The cellular level is of great significance for the organ level in two ways: first, it provides clues about function; second, the long-run hope is that function at the organ level can be explained by events occurring at the cellular level. For the present, the organ level of inquiry looks to the cellular level for neural models that may be useful in explaining results of behavioral experiments.

B. Definition of Auditory Cortex

The definition of the limits of auditory cortex is a special problem for our chapter, not only for the general reasons just described, but because a good portion of the studies we shall review are about the cortex, and it will be necessary to define some terms for different subdivisions so these studies can be compared. In other words, for each study we must try to assess whether the auditory cortex was completely ablated or what subdivisions were removed.

In this section, we shall emphasize studies in which the cat was used as experimental subject because the cat is the animal for which most anatomical as well as behavioral work has been done. Whether the cat can be taken as representative of all mammals is a question to be considered later.

There are several methods of defining the limits of auditory cortex and its subdivisions; each method yields results that are helpful to studies employing another method. Our final picture of auditory cortex depends on attempting to collate results of all methods. It is convenient to begin with the cytoarchitectural and evoked potential studies of ROSE (1949) and ROSE and WOOLSEY (1949). ROSE defined a core zone in the middle ectosylvian gyrus, AI, which has dense supra-granular layers and a complex fifth layer, one band of which is very light. While the characteristics of the core do not fit all of the criteria for koniocortex – *e.g.*, as seen in area 17 of primates – it is closer to koniocortex than is any part of the surrounding belt of cortex. ROSE subdivided the belt into several zones, SF, AII, and Ep. Thus, cytoarchitecture led to the notion of a core and a surrounding belt.

ROSE was aware of the studies of his colleague and collaborator, WOOLSEY, who had used the evoked potential method to define auditory cortex. One complete representation of the basilar membrane was located in the middle ectosylvian gyrus and corresponds to cytoarchitectonic area, Auditory I. A second responsive area was identified below AI and was called AII. Later evidence showed that the insular and temporal areas and somatic area II were connected to auditory thalamus. We then have the following divisions that, according to one or another criteria, are auditory; AI, AII, SF, SII, Ep, I, and T. These are depicted in Figs. 3–5.

1. Retrograde Degeneration Studies

Since auditory cortex is primarily the cortex that receives fibers from the auditory thalamus (see Fig. 3), it is important for our purpose to review the studies of thalamo-cortical projections. One of the classical methods for studying thalamo-cortical projections is the observation of cell change, atrophy, and death in the thalamus after cortical ablation, which necessarily, transects the ascending axons. Earlier work, before 1940, established auditory cortex in monkey (WALKER), rat (LASHLEY), as well as in cat. The flourishing of ablation studies in the 1950's gave impetus to this approach in defining auditory cortex since every case with cortical ablation could provide evidence about thalamic degeneration. By 1960, it was established that, in the cat, the rostral one-half of the principal division of GM degenerates after removal of AI (ROSE and WOOLSEY, 1958; NEFF and DIAMOND, 1958). This degenerated region corresponds to MOREST'S GMv as shown

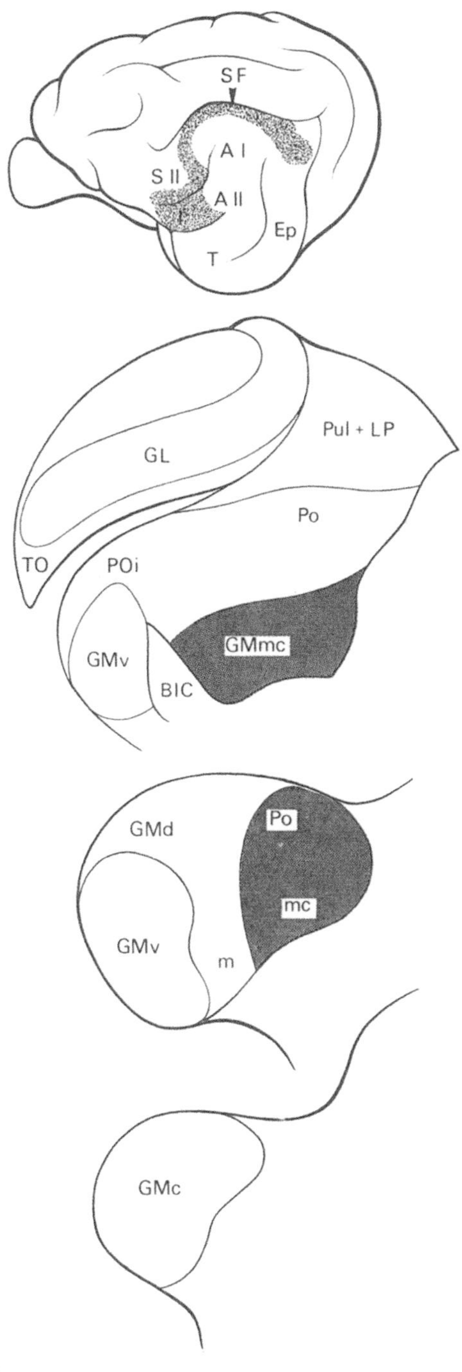

Fig. 3. Summary diagram showing the relationships in the cat of the auditory thalamus and cortex. The chief point of the diagram is to show the distinction between a core and a belt in both thalamus and cortex. AI is the cortical core and receives fibers from GMv. Both areas are shown in the darker stippling. The belt around GMv, which includes GMd, mc, m, and Po, projects to the belt around AI which includes SII, I, T, Ep, and SF

in Figs. 3–5. As the ablation extended ventrally, more and more of the caudal half of GM showed degeneration, which led to the conclusion that the caudal extremity of GM projects to areas below AI, including I and T (Butler et al., 1957; Diamond and Neff, 1957; Rose and Woolsey, 1958; Neff and Diamond, 1958; Diamond et al., 1958). If AI is spared, but an extensive belt of cortex is removed, including SII, I, T, and Ep, degeneration is found in caudal GM (principal division) and in the magnocellular division (GMmc) (see Fig. 4). Small lesions in the cortical belt around AI do not produce any discernible thalamic degeneration. These findings gave rise to the concept of "sustaining projection" (Rose and Woolsey, 1958). As defined, a sustaining projection is one in which two or more collaterals project to two or more cortical subdivisions; each collateral can sustain the thalamic cell body, so, in order to produce retrograde degeneration, several cortical subdivisions must be removed. Thus, GMmc does not degenerate after a small ablation confined to one division but does show degeneration after removal of AI, AII, and Ep.

2. Anterograde Degeneration Studies

The early anterograde studies using the Marchi method reached conclusions surprisingly close to our present conception of auditory cortex in the cat. Both Woollard and Harpman (1939) and Ades (1941) identified the middle ectosylvian gyrus as the target of GM,

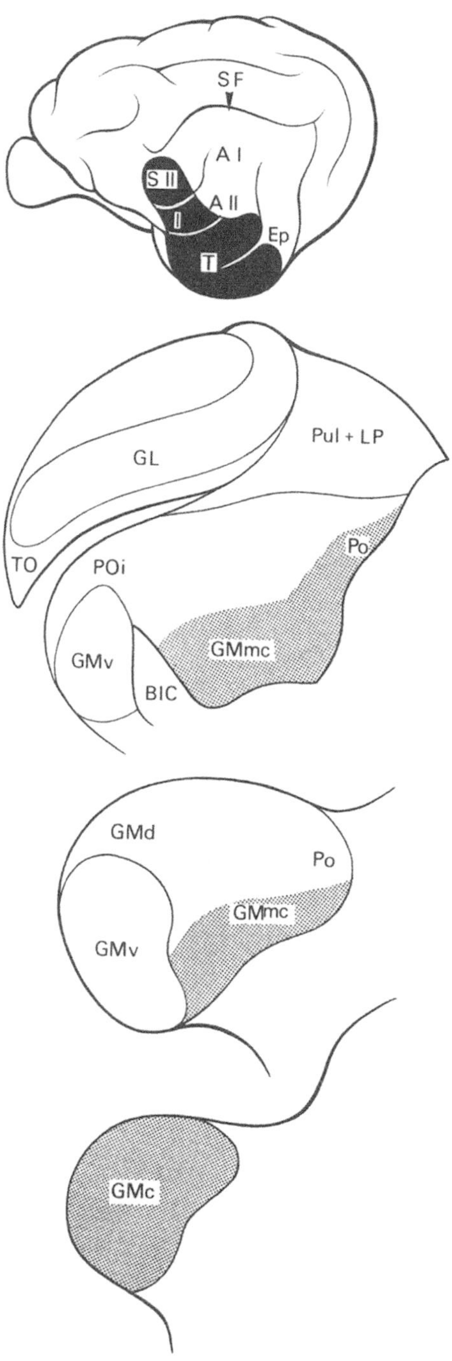

Fig. 4. Diagram showing retrograde degeneration in the auditory thalamus after a cortical lesion of SII, portions of AII and the insular temporal areas. (From Diamond et al. 1958; Diamond and Chow, 1962). The cortical lesion is shown in black and the retrograde degeneration in GMmc, GM, and Po is shown in stipple

but the ventral border of auditory cortex was closer to the rhinal fissure in WOOLLARD and HARPMAN's map (Fig. 6). Their lesions probably extended into the caudal half of GM. With the advent of the NAUTA method for detecting degeneration of unmyelinated fibers, the prospects of defining auditory cortex by anterograde degeneration were improved, and a number of investigators have made lesions in GM and Po and have traced patterns of projection to the cortex. In at least one major respect, the results of anterograde degeneration and retrograde degeneration agree: the ventral division of GM projects to cortical area AI. There is some disagreement about the projection of the auditory thalamic belt to auditory cortical belt. According to evidence ob tained by observing retrograde degeneration, GMmc sends sustaining projections to AI, AII, and Ep, and to the insular-temporal zone. HEATH and JONES (1971) report that GMmc as well as Po projects just to the insular area and to the fringe area, SF. This conclusion is shown in Fig. 5. Since SF is almost always damaged after ablations of AI, it is difficult, on the basis of published retrograde studies, to distinguish the contributions of SF and AI to thalamic degeneration. HEATH and JONES also interpret their findings as indicating that dorsal and caudal GM project to Ep and to the temporal area; this interpretation is in fair agreement

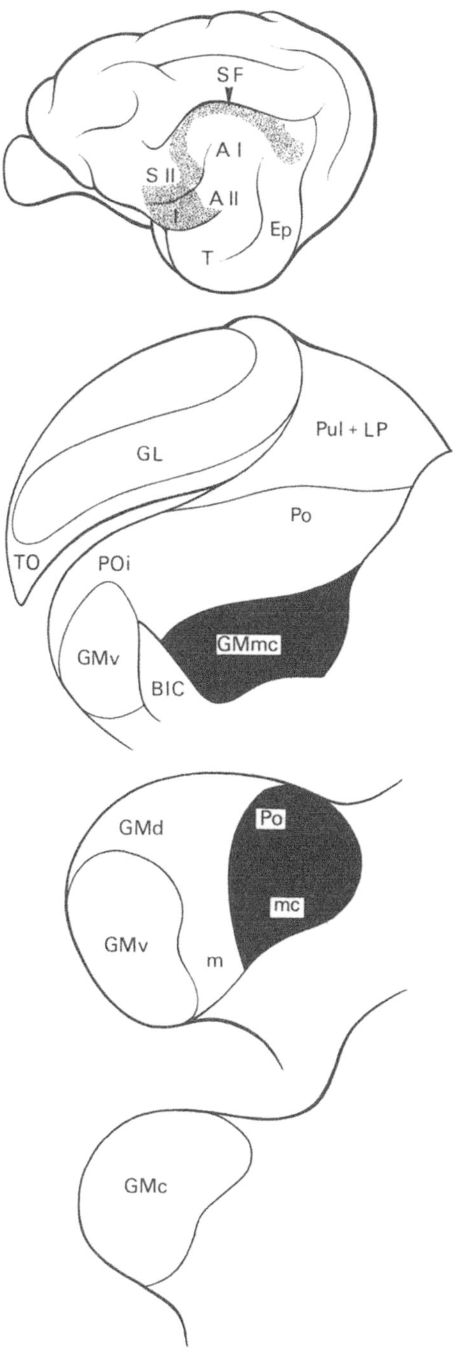

Fig. 5. Diagram showing the cortical target of the magnocellular division of GM and a portion of the posterior nuclear group (based upon HEATH and JONES, 1971). The thalamic lesion is shown in black and the extent of the cortical target as determined by the Nauta method for tracing anterograde degeneration is shown in the cortex by stippling. According to HEATH and JONES, the target of GMmc and Po includes the insular area and the suprasylvian fringe. There is no projection to AI

with the conclusions from retrograde studies. SOUSA-PINTO (1973), from his experiments, reports that each of MOREST'S subdivisions projects to a different cortical zone.

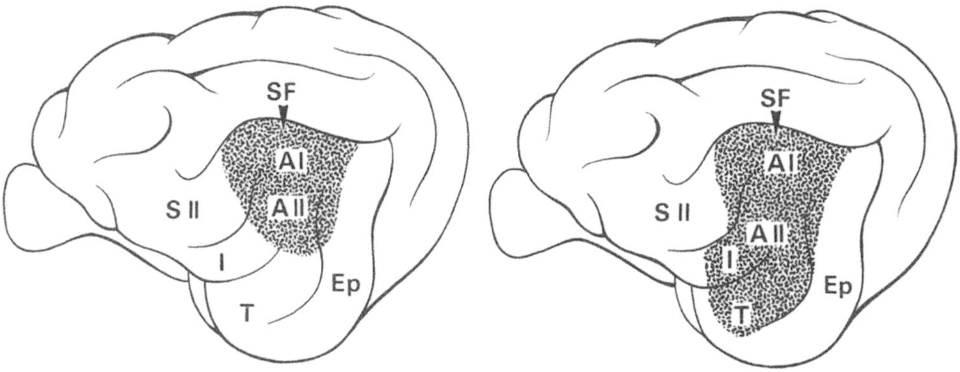

Fig. 6. Auditory area in the cat as defined by anterograde degeneration studies. On the left is the auditory area defined by ADES (1941) and on the right is the auditory area defined by WOOLLARD and HARPMANN (1939). The difference in the ventral boundary in the experiments might be explained by the more caudal extent of the thalamic lesion in the experiment by WOOLLARD and HARPMAN

All of the results – from experimental anatomy, from neurophysiology and from architectonic analysis – can be brought under one scheme if we divide both the thalamus and the cortex into a central core and a peripheral belt (Fig. 3). Then, it is possible to say, first, that all evidence points to a precise topographic projection from the thalamic core GMv to the cortical core AI. Second, this thalamo-cortical projection seems to be reciprocated by a cortico-thalamic projection in a precise and organized way (DIAMOND *et al.*, 1969). Third, questions remain as to the projection of the thalamic belt to the cortical belt, the point to point precision of this projection, and the projection of each thalamic division to one or to more than one cortical division.

Whatever the final answers may be, present evidence suggests that the general pattern of thalamo-cortical relationships holds for all mammals; this we shall discuss in the next section.

C. Species Differences

When we speak in this chapter of the auditory system we mean the auditory system of mammals. This implies that we can identify a plan of organization common to mammals and that we can identify species differences as orderly departures from this plan. We expect that bats, in some ways, may be unique because of the role of discrimination of high frequencies in guiding their behavior, and we note that small animals differ from large animals and that animals living in water differ from animals living on land because of the effects of head size and speed of sound conduction in different media. Differences among animals as to sound cues used in auditory localization are reflected in the relative size of the medial and lateral nuclei of SOC (MSO and LSO).

In any case, species differences in the relative size of brainstem centers, such as MSO, LSO or IC, do not obscure the organization common to all mammals; these differences can be used to advantage in behavioral ablation studies. Thus, MASTERTON and DIAMOND (1967) and HARRISON and IRVING (1966) have used comparative anatomy as a way of approaching the behavioral significance of SOC. There are further species differences that may be more difficult to describe as variations on a central theme. For example, the auditory cortex of the cat is much larger and more complicated than the auditory cortex of simpler and more generalized mammals; this difference seems to be related to what DARWIN called "the advancement of organization". First, mammals differ more in respect of cortical than of brainstem structural organization. This is another way of saying that, in the various lineages, the cortex has evolved at a greater pace than the brainstem and that the size and development of neocortex reflects behavioral traits most recently acquired in evolution; the cortex has not expanded in several lines simply to increase the range of frequencies that can be detected. The search for function of auditory cortex in higher mammals such as cat and monkey has been guided by attempts to imagine in what way the perception of the auditory world has changed. For example, the task of distinguishing complex differences in the temporal pattern of sounds seems to be beyond the capacity of lower mammals such as an opossum.

A second factor that makes the comparative study of auditory cortex more difficult is that the borders of auditory cortex and subdivisions within auditory cortex are more difficult to establish in some animals than in others. In the cat, the cortical target of projections from GM and Po includes a number of divisions: AI, SF, AII, Ep, SII, I, and T. If all of these are auditory, in the sense of receiving auditory input from the thalamus, a problem arises because it seems unlikely that all of these divisions are present in a lower mammal.

Partly in anticipation of this issue, we have decided to divide the cortex of the cat into core and belt; a core area, approximating koniocortex in its cytoarchitecture and receiving fibers from GMv, can be identified in such diverse mammals that it is reasonable to assume that the core area was present in the common ancestor of mammals. Figure 7 shows auditory thalamus and cortex of tree shrew, an insectivore which may resemble the ancestor of primates. Finally, there is recent evidence that the core area in all mammals corresponds to AI of the cat. Recording from single cells, MERZENICH and his colleagues have defined a core area in the cortex of the cat (MERZENICH *et al.*, 1975), the squirrel (MERZENICH, 1975), and the monkey (MERZENICH and BRUGGE, 1973).

We conclude that, despite great differences in the size and complexity of the auditory thalamus and cortex of mammals, we can, in every species, identify the thalamic core which corresponds to GMv. This core projects to a cortical sector which more or less approximates koniocortex. Surrounding is a belt of cortex that SANIDES (1970) calls pro-koniocortex; the belt area receives fibers from the thalamic region dorsal and medial to GMv. This general arrangement of core and belt areas can be regarded as the mammalian plan.

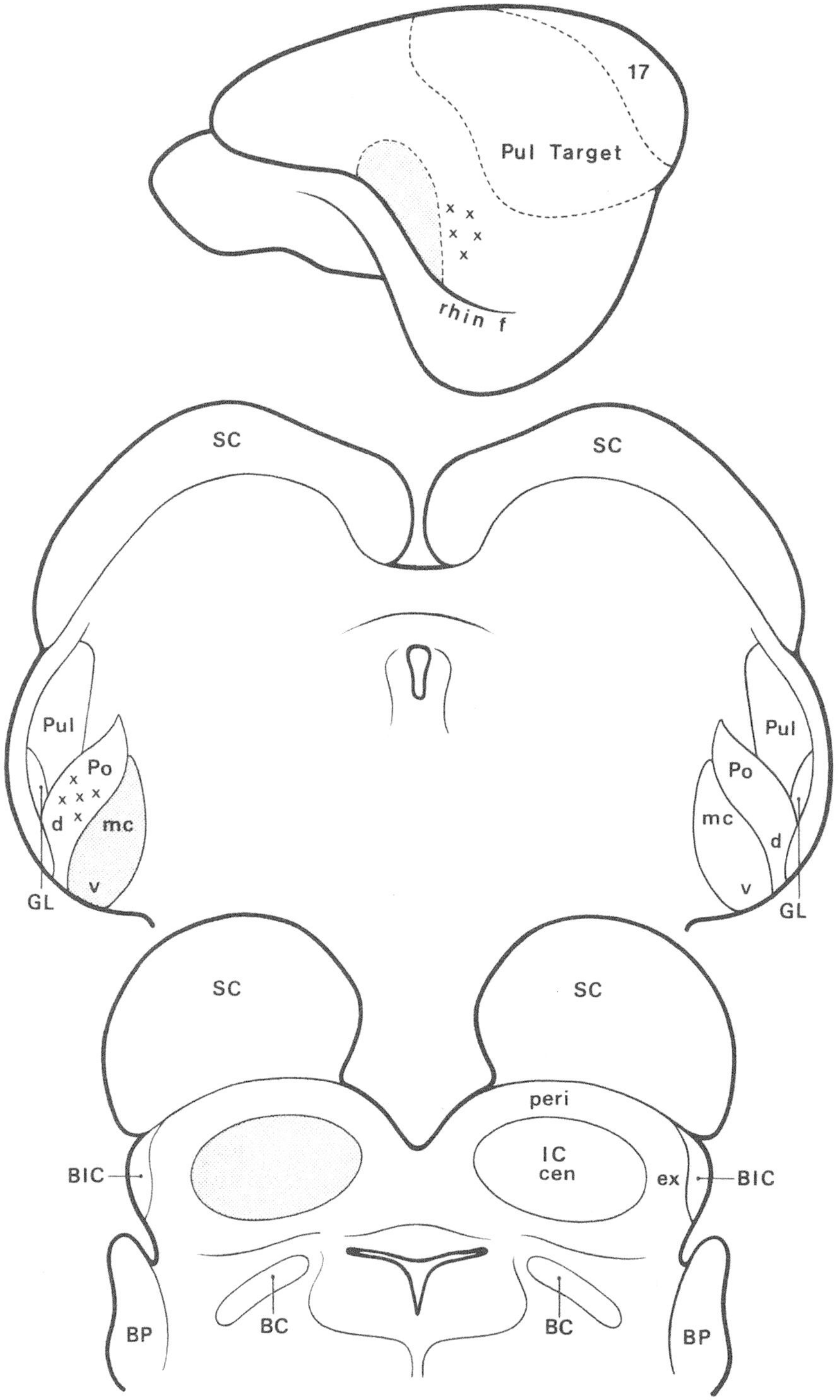

Fig. 7. Diagram to show that a distinction between two parallel paths can be drawn for the tree shrew, a mammal more primitive than the cat. The diagram shows that the central nucleus of IC projects to GMv and GMmc which in turn project to the auditory core. This path is shown by the stippled areas. A second path from GMd and Po to the cortex behind the auditory core is indicated by X's (CASSEDAY *et al.*, 1975; OLIVER and HALL, 1975)

III. Research in Modern Era

The review of research that follows will be divided into two main sections: ablation of auditory cortex and ablation of subcortical centers or transection of pathways of the auditory nervous system. Each main section will be subdivided according to different kinds of auditory discrimination.

For most discriminations, evidence is available from experimental studies of the cat and monkey and from clinical studies of human patients. The dog and rat have been used in a number of experimental studies, and, quite recently, investigators interested in the comparative approach have developed methods of testing mammals such as the hedgehog, opossum, tree shrew, and bush-baby.

A. Ablation of Auditory Cortex

Auditory projection areas of the cerebral cortex of mammals that have been used in numerous behavioral studies are shown in Figs. 3 and 7–9. The kinds of evidence upon which these maps are based has been discussed above (see II. Relation of Structure and Function).

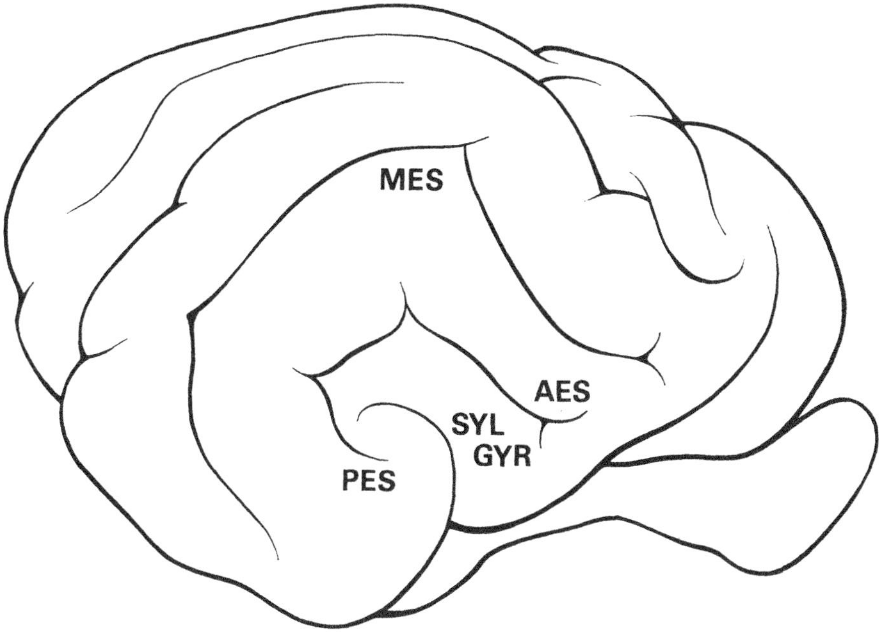

Fig. 8. Surface view of the cortex in the dog showing the auditory areas: the middle ectosylvian (MES), anterior ectosylvian (AES), sylvian (SYL), and posterior ectosylvian gyri (PES)

It will become apparent in the following review of research that the delineation of areas that constitute the auditory cortex has developed over a period of years from evidence provided by anatomical, electrophysiological, and behavioral investigations. In evaluating results of behavioral studies, the state of knowledge at the time experiments were done and the consequent accuracy of cortical ablations need to be considered.

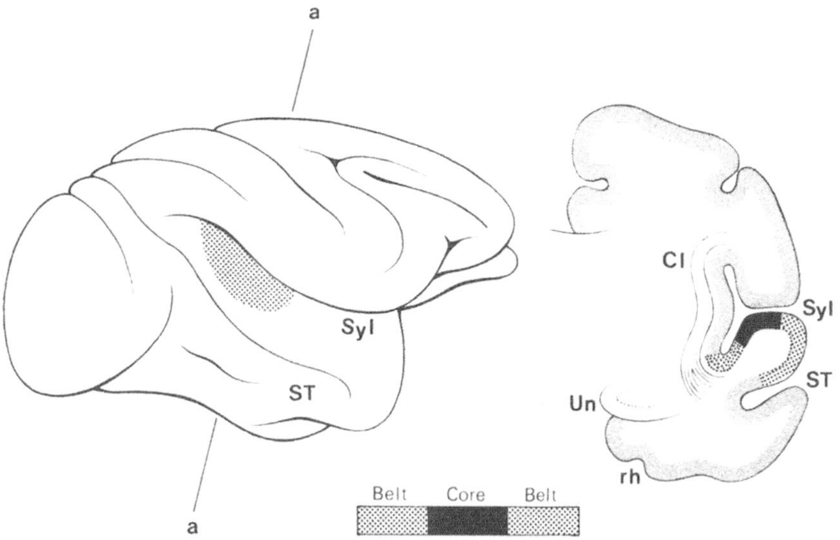

Fig. 9. Lateral view of the cortical surface in rhesus monkeys to show locus of the auditory area between the sylvian (Syl) sulcus and superior temporal sulcus (ST). The frontal section in plane labelled a-a shows that the core area is entirely within the sylvian sulcus and is surrounded on both sides by a belt. Further abbreviations are as follows: Cl, claustrum; rh, rhinal area; Un, uncus. These figures were taken from SANIDES (1970)

1. Discrimination of Onset of Sound and Absolute Thresholds

With the introduction of methods by which experimental animals could be trained to make learned responses to auditory signals, logical steps in exploring the function of auditory cortex were, first, to test an animal's ability to learn a response to the onset of a suprathreshold sound after ablation of auditory cortex, and, second, to measure the absolute intensity thresholds for tonal stimuli at different frequencies. The test of pure tone thresholds was already a clinical tool. Not knowing the range of frequencies audible to experimental animals, the first experiments made tests at approximately the same frequencies usually tested in measuring human audiograms, namely, octave steps from 125 through 8000 Hz. Most clinical audiometers tested octave steps from 128 through 8192 Hz, in correspondence with musical scales.

a) Cat. KRYTER and ADES in 1943 reported that, after bilateral ablation of auditory cortex in the cat (Fig. 3), the experimental animals could relearn to make a conditioned response (move forward in a rotating cage) in order to avoid electric shock applied to their foot-pads through the bars that formed the floor of the cage. The acoustic signals used as conditioned stimuli were tones of 125, 1000, and 8000 Hz. Furthermore, when relearning had taken place, absolute thresholds measured after cortical removal were not significantly different from the thresholds that had been determined before surgery; the animals without auditory cortex could still detect the same weak acoustic signals. The cortical ablations, in at least some of the animals in the KRYTER and ADES study, included most of areas AI, AII, and Ep and invaded I–T and SII.

STOUGHTON (1954) reported no change in absolute thresholds of four animals that had bilateral ablation of cortical areas AI, AII, EP, and SII. Threshold measurements were obtained for the frequencies of 70, 250, 1000, 4000, and 16000 Hz. The amount of insular-temporal region removed and the extent of thalamic degeneration were not reported.

WEGENER (1954) also found no significant changes in absolute thresholds of cats tested before and after bilateral ablation of AI, AII, Ep, SII, and most of I–T or of AI, AII, Ep, SI, and SII. Thalamic degeneration was not reported.

MARUYAMA and KANNO (1961) reported small increases in absolute thresholds of cats after bilateral ablation of auditory cortex and larger increases (up to 65 dB) after ablation of the sigmoid gyrus as well as auditory cortex. From the brief report given of the behavioral tests, it appears that the apparent hearing losses may have been due to testing some animals too soon after surgery and to general behavioral upset such that appropriate avoidance responses could not be made in the rotating cage used for threshold tests.

b) Monkey. ELLIOTT tested the ability of rhesus monkeys to perform a number of different auditory discriminations after bilateral ablations of auditory cortex (Fig. 9). As a control, absolute thresholds (audiograms) for the frequency range 250–16000 Hz were obtained before and after the cortical ablations. Ten animals had bilateral ablations of the primary auditory cortex (core cortex, Fig. 9) located on the superior surface of the superior temporal gyrus, posterior to the central sulcus. Two animals had larger ablations which included not only the primary area but cortex of the superior temporal surface anterior to the central sulcus (belt cortex, Fig. 9). In one animal, only the anterior area was ablated. ELLIOTT reported that none of the animals showed more than a 5 dB shift in its audiogram at any frequency. In a later study, ELLIOTT reported increase in thresholds ranging from 11 to 34 dB for frequencies of 500, 1000, and 2000 Hz after bilateral ablations that included all of the cortex of the superior surface of the superior temporal convolutions.

c) Other Animals. GIRDEN (1943) found temporary increases in pure tone thresholds but recovery to near normal in dogs after bilateral ablations that included most of auditory cortex.

As part of experiments to study localization of sound in space, RAVIZZA and MASTERTON (1972) measured pure-tone thresholds of the opossum before and after removal of not only auditory cortex but all cortex except a few remnants outside the auditory areas. For frequencies of 500, 2000, 8000, and 32000 Hz, the one decorticate opossum tested post-operatively differed little from normal animals.

For a number of years, GERSUNI and BARU, and their colleagues have been studying the effects of signal duration on detection of an auditory signal or of a change in one of its attributes, such as, frequency or intensity. [See GERSUNI

(1971) and BARU and KARASEVA (1972) for summaries.] In experiments, using dogs as subjects, BARU (1966, 1967) has reported that absolute thresholds for pure tones or for noise of brief durations are increased (decrement in hearing) after unilateral ablation of auditory cortex when the sound is presented to the ear contralateral to the cortical lesion. Bilateral cortical ablations affect detection of sounds presented to either ear. In normal animals, the effect of signal duration on sound pressure level necessary for detection of an acoustic signal is in the range, 100–200 msec. In animals with auditory cortex ablated, the absolute thresholds are likewise increased for durations below 100–200 msec, but an additional deficit appears for durations shorter than 10–15 msec (Fig. 10).

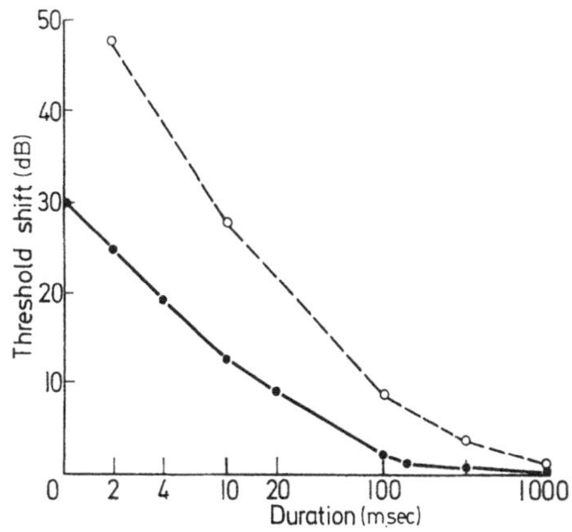

Fig. 10. Mean threshold shifts as a function of duration of 1000 Hz tone. Solid line shows mean values for a group of normal dogs. The dashed line shows the results for a dog with unilateral ablation of auditory cortex when stimulus is presented to the ear contralateral to side of cortical ablation. (From GERSUNI, 1971)

d) Man. The primary auditory projection area in man is shown in Fig. 11. Evoked response, (CELESIA and PULETTI, 1969), cytoarchitectural, and electrical stimulation studies agree that this region probably corresponds to the core or primary auditory cortex of other mammals. In most cases of temporal lobe surgery, the primary projection area is spared if at all possible, particularly in surgery of the left temporal lobe.

A number of reports have been made of human cases in which audiograms were obtained before and after hemispherectomy for infantile hemiplegia. In such cases, it is difficult or impossible to assess the functional state of the auditory cortex of the hemisphere that was removed. The effects of unilateral ablation of auditory cortex in such cases should not be considered to be the same as unilateral ablation in a normal brain. Studies of infantile hemiplegia indicate that many of the functions of the damaged cerebral hemisphere are taken over by the intact side.

In general, investigators, who have measured pure-tone thresholds in cases of infantile hemiplegia, have reported normal or near normal hearing for both ears and little change in the audiograms after hemispherectomy (GOLDSTEIN *et al.*,

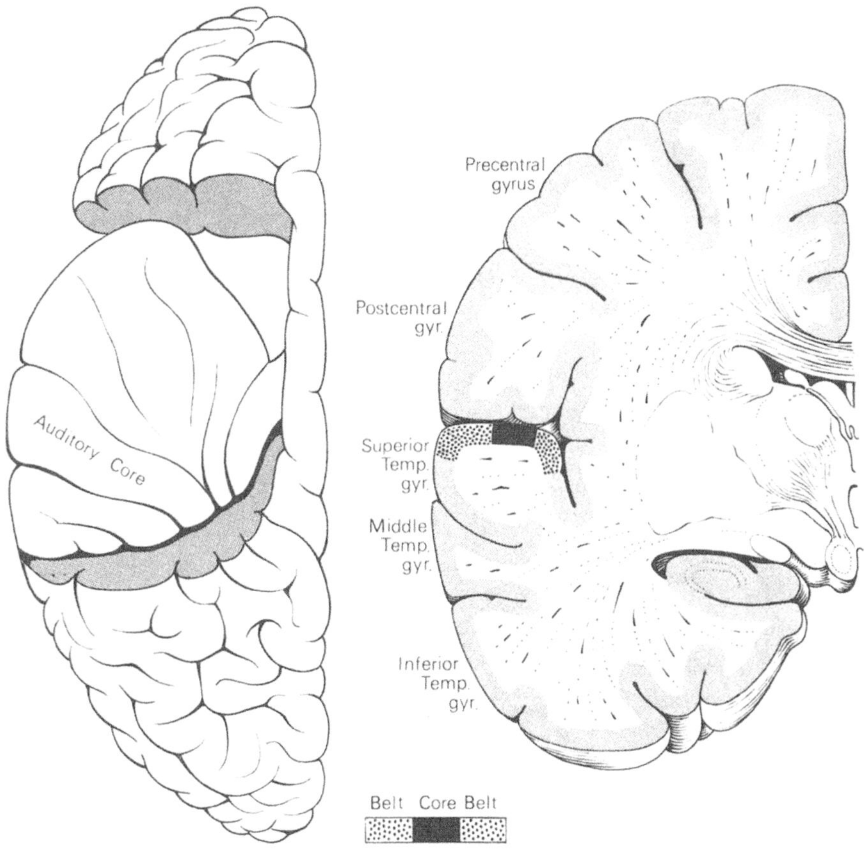

Fig. 11. To show the auditory cortex in man it is necessary to cut away the post central gyrus which lies above the superior temporal sulcus. Such a cut-away is shown in the diagram on the left which is a dorsal view of the hemisphere. On the right is a second diagram of a frontal section through the auditory cortex showing that the core is buried in the superior temporal sulcus and is surrounded by a belt. [Adapted from figures in a paper by W. Penfield and P. Perot (1963)]

1956; Goldstein, 1961; Hodgson, 1967; Damasio *et al.*, 1975). Karp *et al.* (1969), making measurements at only 500 Hz, reported that, in a comparison of 19 left hemiplegic patients with a group of 19 control patients without brain damage, 10 of the 19 hemiplegics showed slightly raised thresholds for the ear opposite to the side of the brain damage.

Landau *et al.* (1960) have described a case of congenital aphasia in which the brain was obtained and examined after death of the child at age 10. There appeared to be not only severe damage of tissue in sections through the region of the sylvian fissure but also complete or nearly complete degeneration of the medial geniculate bodies bilaterally. Pure-tone thresholds had been difficult to measure in this case, but there was evidence of losses no greater than 25 dB throughout the range 125–8000 Hz. Thresholds for a noise stimulus were within the normal range.

GERSUNI *et al.* (1967, 1971) and KARASEVA (1972) have measured absolute thresholds for tones and noise as a function of signal duration in patients with temporal lobe damage. As in experimental animals, in normal human subjects and in patients with cortical damage other than in the temporal lobe, thresholds increase as signal duration is decreased below 100–200 msec. In patients with unilateral damage of the temporal lobe that includes the primary projection area (core area) of the auditory system, there is a further increase in absolute thresholds for signals having a duration less than about 20 msec when the signals are presented to the ear contralateral to the lesion. GERSUNI *et al.* have related these behavioral results to results of electrophysiological experiments (GERSUNI, 1963, 1971).

Cases of bilateral temporal lobe damage involving the auditory areas of the cortex occur infrequently. A few cases have been reported of patients in whom bilateral damage to auditory cortex occurred as the result of disease or circulatory disorder. For all the cases described prior to 1963, evaluation of results is difficult because of deficiencies in the audiological testing, in the neurological examinations, or in both (see JERGER *et al.* 1969, pp. 7–8).

Recently, two cases of bilateral temporal lobe damage have been described by JERGER *et al.* (1969, 1972). In both cases, cite of lesion was accurately determined, and a comprehensive battery of hearing tests was administered. It is important, for comparison with data from animal experiments, that a number of non-verbal tests were given. Results of these tests will be described in appropriate sections below. Here, we shall consider only the tests of pure-tone thresholds.

The first case (J-1) described by JERGER *et al.* (1969) was a 20 year old white male, a maintenance worker in the Air Force. The patient first suffered a cerebral vascular accident (CVA) affecting branches of the left middle cerebral artery. A similar CVA occurred in the right hemisphere five months later. Pure-tone thresholds had been measured for the patient before his induction into military service, two years before the first CVA. No audiograms were made after the first CVA, but whispered voice, Schwaback, Rinne, and Weber tests indicated normal hearing.

After the second CVA, pure-tone thresholds were measured immediately on the first day of admission to the hospital and at intervals, thereafter. Results of these tests are shown in Fig. 12. On the first test given immediately after the second CVA, there was a hearing loss of 60–85 dB at all frequencies in both ears. There was gradual recovery and, less than three months later, hearing at 1000 and 2000 Hz was within the normal range. Severe loss at high frequencies continued in the right ear.

A second case (J-2) examined by JERGER *et al.* (1972) also suffered successive cerebral vascular accidents involving the middle cerebral arteries. This patient was a 62 year old male for whom pure-tone audiograms, made five years before his first CVA, were available. Immediately after the second CVA, the patient had hearing difficulties but pure-tone thresholds were not measured until two months later; these are shown and can be compared with the pre-accident audiograms (Fig. 13). How much of the hearing change was the result of the cerebral vascular accidents and how much was due to other factors during the six year period cannot be determined. In any event, the change in threshold levels was not great. The brain of the patient was obtained at time of death, four months after the last hearing tests. The results of post-mortem examination of the brain have been described by KANSHEPOLSKY *et al.* (1973). The examination revealed

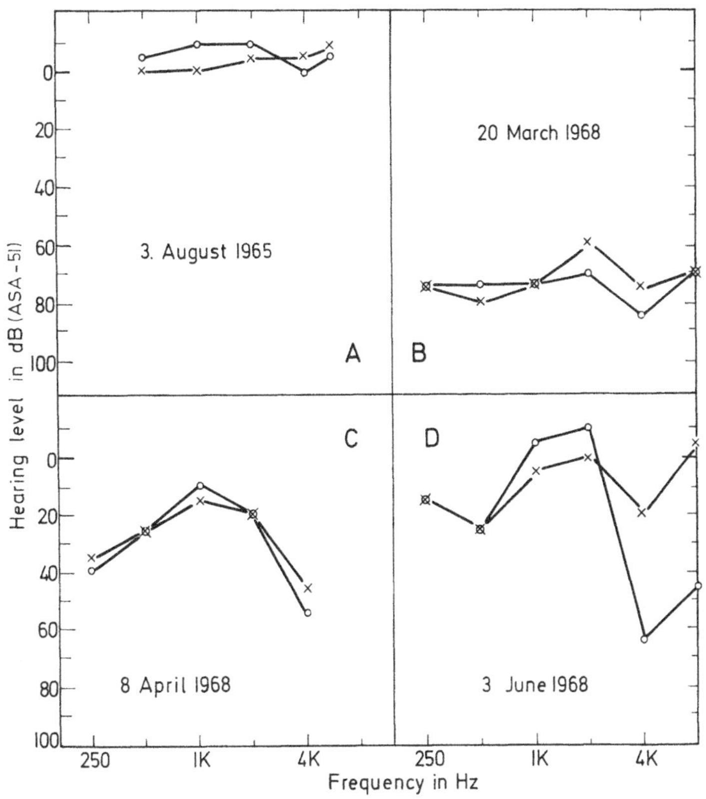

Fig. 12A—D. Pure tone air conduction audiograms for patient, J-1. Audiograms measured before the cerebral vascular accidents are shown in (A) (○ — ○ right ear, × — × left ear). (B) shows audiograms measured on the day of admission to the hospital immediately after the second cerebral vascular accident. (C) and (D) show partial recovery of pure tone thresholds at later dates. (From JERGER *et al.*, 1969)

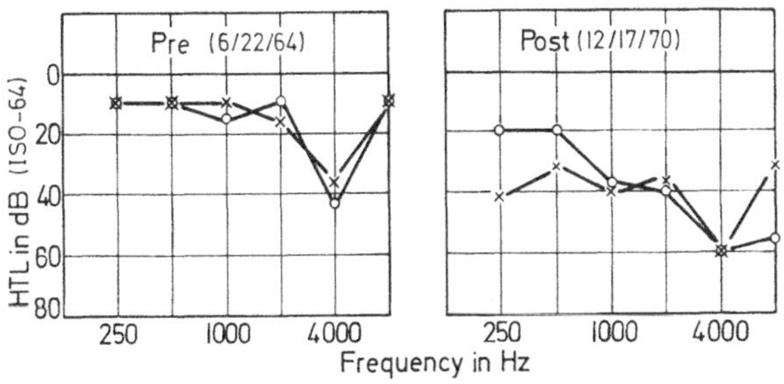

Fig. 13. Pure tone air conduction audiograms for JERGER patient J-2. The audiograms on the left were obtained six years before the second cerebral vascular accident; those on the right, three months after (From JERGER *et al.*, 1972)

bilateral lesions of the superior surface of the superior temporal gyri. The lesions extended from the posterior tip of the sylvian fissure to a point about two centimeters in an anterior direction. There were also smaller lesions in the right middle temporal gyrus and in the left basal ganglia, the left internal capsule, and in the superior lateral medulla. From the descriptions given, it would appear that most of the primary auditory cortex, as shown in Fig. 11, was destroyed. The report contains no description of thalamic degeneration.

LHERMITTE *et al.* (1971) have reported the results of examination of hearing for three patients with bilateral temporal lobe lesions caused by cerebral vascular accidents. In two of the patients, audiograms were obtained at intervals after the last CVA. As in the first case described by JERGER, a severe loss occurred throughout the frequency range on first tests given soon after the CVA. There was considerable recovery so that hearing was in the normal or near normal range for most frequencies two to four months later. The results of post-mortem examination of the brains were available for only one of the two cases; in this case, there was bilateral damage to the auditory cortex, probably incomplete on the right side. In a third case, with bilateral damage of auditory areas confirmed post-mortem, audiograms were not obtained immediately after the accidental lesions. For those recorded later, after time for recovery, hearing was in the normal or near normal range for frequencies below 2000 Hz. Severe losses that occurred at 4000 Hz and 8000 Hz were probably due primarily to peripheral damage rather than to the cortical lesions.

e) Retention of Learned Discriminations; One-Stage vs. Two-Stage Ablations. For most training and test procedures that have been used with experimental animals in studies of auditory discriminations, including response to onset of sound, there is amnesia for the learned habit after one-stage bilateral ablations of all or nearly all of auditory core cortex. For many discriminations, relearning is possible. Typically, post-operative behavior of the animal indicates that it does retain ability to respond appropriately to non-auditory cues in the test situation. For example, in avoidance conditioning, a prompt escape response to the unconditioned stimulus is retained when the avoidance reponse to the auditory signal is lost and must be relearned.

STEWART and ADES (1951), STOUGHTON and NEFF (1950), and STOUGHTON (1954) have compared the effects of one-stage bilateral and two-stage successive unilateral ablations of auditory cortex on the retention of a conditioned avoidance response to sound stimuli. STEWART and ADES trained monkeys to respond to a buzzer signal in a double-grill box. After one-stage bilateral ablation of auditory cortex, there was loss of the learned response but rapid relearning. In two-stage ablations, animals showed amnesia for the learned response when the period between the successive unilateral ablations was six days or less. For two-stage ablations with interoperative periods of seven or more days, there was retention of the conditioned response.

In the STOUGHTON and NEFF experiment, cats were trained to make an avoidance response in a rotating cage to pure-tone stimuli. Results obtained were similar to those reported by STEWART and ADES. After one-stage bilateral ablation of areas AI, AII, Ep, and part of I–T, animals showed amnesia for the learned response but relearned rapidly. After two-stage surgery with 12–14 days between successive unilateral ablations, animals retained the conditioned response. Results were not affected by over-training nor by training to a wide range of frequencies and intensities. A further finding of STOUGHTON and NEFF was amnesia for the conditioned response in two-stage as well as one-stage cortical removal when the cortical areas ablated included SII as well as AI, AII, Ep, and part of I–T. No such loss was found after control ablations of SI and SII, SII alone, or large parts of the suprasylvian gyrus.

f) Effects of Decortication. After complete or near complete ablation of all cortex, CULLER and METTLER (1934) reported that a dog could learn to make a diffuse response (struggling) to an auditory signal at a level well above threshold; it could not learn to make a discrete response (flexing fore-leg) to avoid shock.

BROMILEY (1948) reported successful establishment in the dog of a more specific response (flexing of hind-leg) in a conditioned avoidance situation in which the conditioned stimulus was sound and the unconditioned stimulus, shock to foot-pad.

BARD and RIOCH (1937), in a study of the decorticate cat, reported that animals with almost all of the cortex removed appeared to localize sound or, at least, to orient in the direction of the sound source. No special training to localize sound in space was given. Reactions of animals to familiar sounds in their laboratory environment were observed. Responses were less prompt than in normal animals. The tests made did not completely eliminate the possibility that cues other than auditory guided the experimental animals.

g) Summary. When measured carefully over a period of time after surgery, there is little or no deficit in pure-tone thresholds of the experimental animals that have been tested after bilateral ablation of the primary auditory projection areas of the cortex or after bilateral ablations that include not only primary cortex but also the belt areas (I–T and SII in cat). It should be noted, however, that in the experiments that have been cited, there may have been small remnants of insular-temporal cortex in the cats tested and minimal amounts of belt cortex in the monkeys. In the cat studies, thresholds were not measured for frequencies above 16000 cps although it is now known that cats hear frequencies up to approximately 60000 Hz (NEFF and HIND, 1955; ELLIOTT *et al.*, 1960; MILLER *et al.*, 1963).

The limited evidence from human cases with bilateral lesions of auditory cortex suggests that, as in experimental animals, pure-tone thresholds are not severely affected. The temporary losses that have been described in some cases may be due to other factors than destruction of cortical tissue in the auditory areas.

Because of the difficulty of fitting headphones on experimental animals, monaural tests have not often been made, although, recently, a number of investigators have used head phones in experiments involving lateralization of sounds. Evidence is lacking for monaural tests of pure tone thresholds. In human subjects with unilateral temporal lobe damage, no large shifts in threshold have been reported; in some studies, small increases have been found for the ear contralateral to side of cortical lesion, and the deleterious effect of short signal durations has likewise been noted for the contralateral ear.

Although it is difficult to evaluate the results of some of the early studies in which conditioning methods were used to measure pure-tone thresholds, the report of deficits after cortical ablation may be related to the kind of temporary deficit seen in one of the cases described by JERGER and in two of the cases in the study of LHERMITTE *et al.* In the early animal experiments, testing was started within a few days after surgery and, because of danger of infection, often completed within two weeks (for example, see METTLER *et al.*, 1934). GIRDEN (1943) did report large initial losses after bilateral ablation of auditory cortex in dogs, but, when tests were continued over a period of time, there was almost complete recovery.

2. Discrimination of Changes in Intensity

In training experimental animals to respond to a change in auditory signal and to obtain measures of difference thresholds (difference limens: DL's), a method of stimulus presentation and conditioning introduced by KAPPAUF (1943) has been used in many of the studies that will be outlined below. Typical stimulus sequences are shown in Fig. 14. A principal advantage of presenting acoustic

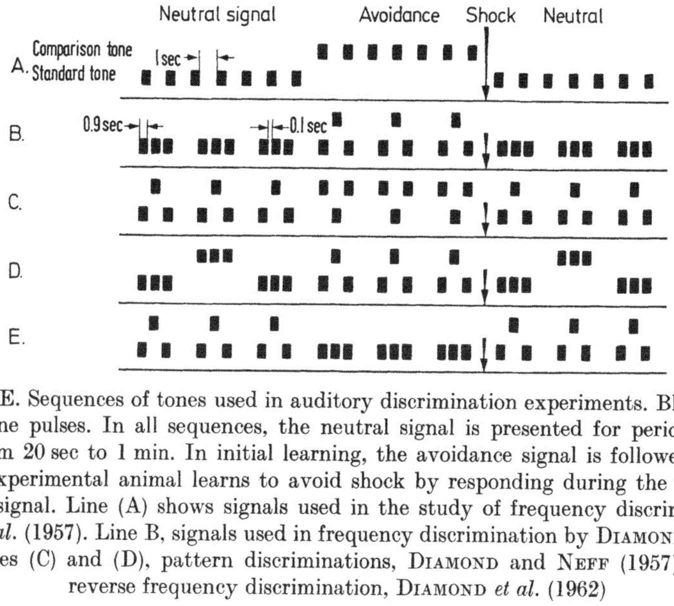

Fig. 14A—E. Sequences of tones used in auditory discrimination experiments. Black squares indicate tone pulses. In all sequences, the neutral signal is presented for periods of time, usually from 20 sec to 1 min. In initial learning, the avoidance signal is followed by shock until the experimental animal learns to avoid shock by responding during the time of the avoidance signal. Line (A) shows signals used in the study of frequency discrimination by BUTLER *et al.* (1957). Line B, signals used in frequency discrimination by DIAMOND and NEFF (1957). Lines (C) and (D), pattern discriminations, DIAMOND and NEFF (1957). Line (E), reverse frequency discrimination, DIAMOND *et al.* (1962)

stimuli, particularly tones, as regularly spaced pulses, is that switching from one signal to another can be done during silent intervals, thus, avoiding introduction of switching artifacts. The tone pulses may be made of sufficient duration that they can be keyed on and off at rates that introduce minimal distortion.

The top line of Fig. 14 illustrates sequences of stimuli that may be used in testing discriminations such as change in intensity, change in frequency, and change in duration.

a) Cat. RAAB and ADES (1946) and ROSENZWEIG (1946) reported that, after bilateral ablation of auditory cortex, cats were able to relearn a conditioned response to change in intensity of pure tones (125, 1000, and 8000 Hz) and that there was little or no change in the difference limens. In the RAAB and ADES study, auditory areas AI, AII, and Ep appeared to be completely removed, and, in some cases, the lesions invaded I–T, SII, and SS. Retrograde degeneration in the medial geniculate bodies was reported as complete or near complete. ROSENZWEIG reported complete bilateral removal of cortex in one of two animals tested before and after surgery. It should be kept in mind that, at the time of the RAAB and ADES and ROSENZWEIG studies, auditory cortex in the cat was considered to be the areas AI, AII, and Ep of Fig. 3. Both RAAB and ADES and ROSENZWEIG made use of the conditioned avoidance procedure in a rotating cage.

Fig. 15. Double-grill apparatus used in avoidance conditioning

OESTERREICH *et al.* (1971) have repeated and extended the earlier experiments; a conditioned avoidance procedure in a double-grill box (Fig. 15) was used. The cortical areas ablated and the DL's before surgery and after unilateral and bilateral ablations are shown in Table 1. There was little or no change in the DL's, the maximum (3.3 dB) occurring in cat OE-844. Post-mortem examination of the thalamus of the experimental animals revealed complete degeneration of the principal division of the medial geniculate body (GMv) except for a small region at the caudal tip. From other experiments, it is known that this caudal tip projects to insular-temporal cortex; apparently some of the latter, adjacent to the rhinal fissure, was spared. Severe degeneration was also present in the magnocellular division of the medial geniculate and in the posterior group of thalamic nuclei.

Table 1. Difference thresholds for change in intensity measured before (pre-op) and after successive unilateral ablations of auditory cortex (*L* = left auditory cortex; R = right auditory cortex). (OESTERREICH *et al.*, 1971)

Cat	Lesion	Differential intensity limen in decibels
OE-734	Pre-op	1.2
	L. AI AII Ep I—T	1.4
	R. AI AII Ep I—T	1.2
OE-805	Pre-op	2.0
	L. AI AII Ep I—T SS SII	2.0
	R. AI AII Ep I—T SS	4.2
OE-788	Pre-op	1.3
	L. AI AII Ep I—T SS	1.4
	R. AI AII Ep I—T SS	1.7
OE-844	Pre-op	2.0
	L. AI AII Ep I—T SS	1.2
	R. AI AII Ep I—T SS SII	5.3

b) Monkey. ELLIOTT has reported some deficits in intensity discrimination in monkeys in which the primary auditory projection areas of the cortex or those areas plus the surface cortex anterior to the primary area were ablated. The results of his experiments are difficult to compare with those of other studies because animals were tested on a battery of 17 auditory discrimination tasks and the training procedure used was complex.

c) Man. MILNER (1962), using the Loudness test from the Seashore Measures of Musical Talents, found a small but significant increase in the difference limen (poorer discrimination of changes in intensity of a tone of 440 Hz) in patients with right temporal lobectomy; no increase was found in patients with left temporal lobectomy. The test was presented through a loudspeaker so both ears received the stimuli.

JERGER *et al.* (1969, 1972) measured loudness discrimination in two patients who had bilateral lesions of the temporal lobe. The quantal psychophysical method (STEVENS *et al.*, 1941) was used in the loudness tests. In both patients, a decrement in ability to discriminate intensity changes was found.

LHERMITTE *et al.* (1971) reported good discrimination of change in intensity of tonal signals by three patients with bilateral lesions of auditory areas of the temporal lobe.

In contrast to the results of the studies of MILNER and of JERGER, SWISHER (1967) reported improved intensity discrimination in patients with left temporal lobectomies that included the transverse gyri of Heschl. No significant changes were found in patients with left temporal lobectomies that spared Heschl's gyri, in patients with right temporal lobectomies that included Heschl's gyri, or in control patients without brain damage. Measurements that showed hypersensitivity in the one group of left temporal lobectomy patients were made approximately three weeks after surgery. Tests of patients in whom similar surgery had been done a year earlier showed no hypersensitivity. Likewise, one patient, tested at three weeks and again at one year showed increased sensitivity only on the first test.

In his examination of a patient with left hemispherectomy, HODGSON (1967) found that SISI scores for tones of high intensity levels were increased when stimuli were presented to the right ear, *i.e.*, the ear contralateral to the brain lesion. This finding indicates abnormal intensity discrimination (hypersensitivity).

d) Summary. In all animals that have been studied, including man, discriminations of change in intensity can be learned or relearned after bilateral ablation of or damage to the primary and most of the belt auditory areas of the cerebral cortex. Furthermore, there is little effect of the ablation on sensitivity to small changes in sound pressure levels.

3. Discrimination of Changes in Frequency

The earliest studies cited in the above sections had failed to shown any severe deficits in the capacity of experimental animals to make learned auditory discriminations after bilateral ablation of auditory areas of the cerebral cortex. Because of the tonotopic organization of the auditory cortex as shown by the experiments of WOOLSEY and WALZL (1942), TUNTURI (1944–1946), LICKLIDER and KRYTER (1942), HIND (1953), KENNEDY (1955), and KENNEDY *et al.* (1954), the hypothesis that discrimination of changes in frequency might be affected by cortical ablation led several investigators to undertake experiments in which they tested frequency discrimination before and after cortical ablations. We know now that tonotopic organization is characteristic not only of subdivisions of the auditory cortex but of subdivisions of the auditory system from cochlea to cortex. While this had not been shown through electrophysiological experiments when the experiments described below were initiated, an orderly projection of the cochlea on the primary cochlear nucleus (CN) was known (LORENTE DE NO, 1933; LEWY and KOBRAK, 1936).

a) Cat. Experiments in which frequency discrimination was measured before and after bilateral ablation of auditory cortex were reported by BUTLER and NEFF (1950), MEYER and WOOLSEY (1952), and BUTLER *et al.* (1957). Both groups of investigators used a conditioned avoidance procedure: BUTLER *et al.* a double grill box (Fig. 15) and MEYER and WOOLSEY, a rotating cage. The manner in which tonal stimuli were presented in the two studies is shown in Fig. 16, lines A and B. Results of the two studies agreed in part, namely, that frequency discrimination can be relearned after bilateral ablation of auditory cortical areas AI, AII, and Ep. Results differed, however, in that MEYER and WOOLSEY reported that animals with SII ablated, in addition to AI, AII, and Ep, could not relearn the frequency discrimination but could learn to discriminate a change in intensity of a tonal signal. BUTLER *et al.*, on the other hand, were able to retrain animals to make a frequency discrimination after bilateral ablations that included SII. Furthermore, in the experiments of BUTLER *et al.*, the experimental animals were trained to respond to small changes in frequency (near the frequency DL's for the cat), and the bilateral ablations had very little effect on this capacity.

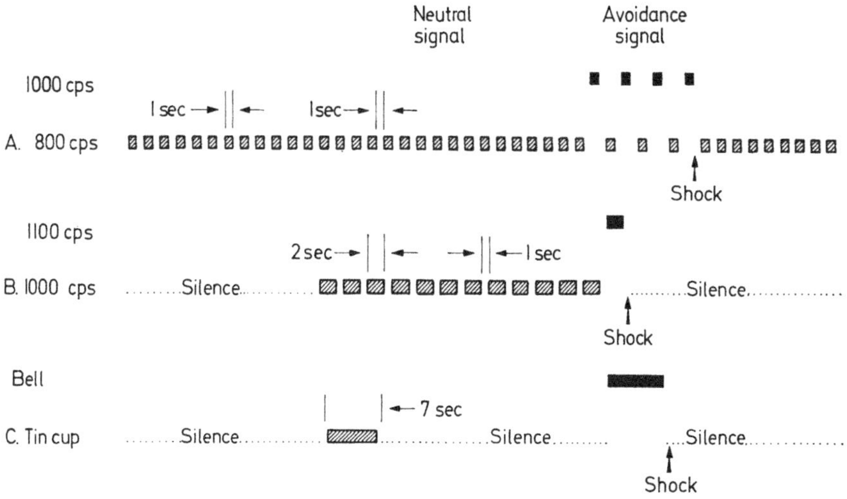

Fig. 16A and B. Sequences of sound signals used in auditory discrimination experiments: (A) by BUTLER *et al.* (1957) and GOLDBERG and NEFF (1961a); (B) by MEYER and WOOLSEY (1952; (C) by ALLEN (1945)

The difference in the results of the experiment by MEYER and WOOLSEY and that by BUTLER *et al.* was later explained in studies by THOMPSON (1960) and by GOLDBERG and NEFF (1961a). THOMPSON showed that, after bilateral ablation of auditory cortex, cats could relearn a frequency discrimination when the positive signal was a new frequency replacing a background signal (repetitive pulses) as shown in line A of Fig. 16. The same animals were unable to relearn when the stimuli were presented as shown in line B of Fig. 16, a situation in which a silent interval is followed sometimes by a neutral signal to which no response is made, and then the neutral signal is replaced by the avoidance signal. THOMPSON offered the explanation that "removal of auditory cortex appears to interfere with the ability to inhibit response to negative stimuli in frequency discrimination, rather than to interfere with frequency discrimination as such". GOLDBERG and NEFF (1961a) and DIAMOND *et al.* (1962) have offered a different explanation to account for the results of different kinds of stimulus presentation.

The suggested explanation of GOLDBERG and NEFF and of DIAMOND *et al.* is an extension of a model of neural events proposed by NEFF (1960b, 1961a). The model initially proposed that the difference between auditory discriminations that can be learned after bilateral ablation of auditory cortex and those that cannot is that, for the former, new neural elements are excited when a background signal is changed to a positive signal, but for the latter, no new units are excited, since the change from neutral to positive signal involves only a change in temporal events in the same neural units. GOLDBERG and NEFF and DIAMOND *et al.* added to the model by pointing out that when a neutral signal is repeated for a given time, some habituation of neural activity takes place. A change of the signal to a new frequency (warning signal) elicits a larger neural response to which the animal is able to respond after ablation of auditory cortex.

Brown *et al.* (1967) have shown that after bilateral ablation of auditory areas AI, AII, Ep, SII, and most of I–T, cats can relearn to make an avoidance response when a pulsing background tone (1000 Hz) is changed to a pulsing tone of another frequency (1200 Hz) and to refrain from responding when the change is from 1000 Hz to 800 Hz. The authors conclude that the evidence of their experiments supports the explanation of Goldberg and Neff and of Diamond *et al.* rather than that of Thompson.

Ehmer (1953) has also reported successful training of cats to make a frequency discrimination after bilateral ablation of auditory cortex; he used a conditioned avoidance situation and presented the stimuli as in the Butler *et al.* study. None of the animals in his study had complete ablation of auditory cortex as we now define it.

In the report by Butler *et al.*, it was noted that there was incomplete degeneration in the caudal end of the medial geniculate body after even the largest cortical ablations. It was also noted that the further the cortical lesion extended into the insular-temporal region, the further degeneration of the geniculate occurred in a caudal direction. This observation plus similar findings in other experiments (Neff *et al.*, 1956; Rose and Woolsey, 1958; Neff and Diamond, 1958; Diamond *et al.*, 1958) led to additional studies of frequency discrimination by Goldberg and Neff (1961a) and by Thompson (1964).

Goldberg and Neff (1961a) trained cats in a conditioned avoidance situation to discriminate changes in frequency (signals presented as in line A, Fig. 16) before and after bilateral cortical ablations that varied in extent: AI, AII, Ep in the animal with the smallest lesion; AI, AII, Ep, I–T, SII, and SS in the animal with the largest lesion. For the largest lesion, there was complete degeneration of the principal division of the medial geniculate and complete degeneration of the anterior part of the magnocellular division with only a few large cells remaining in the posterior part. There was also complete degeneration of cells in the rostral part of the posterior group of nuclei. The cortical ablations, thus, included all of the cortical areas that receive projection from the thalamic nuclei known to form part of the primary ascending pathways of the auditory system. The results of the experiment are given in Table 2 which shows the areas

Table 2. Effects of bilateral ablation of auditory areas of cortex on frequency discrimination. The columns on the right show the number of trials to reach a criterion of 9 correct responses in 10 trials on three successive days. The auditory areas ablated are indicated for each animal (From Goldberg and Neff, 1961)

Cat no.	Lesion	Days to criterion	
		Pre-op	Post-op
G 673	AI, AII, Ep	19	10
G 657	AI, AII, Ep, SII	16	9
G 615	AI, AII, Ep, I—T	19	11
G 617	AI, AII, Ep, I—T	25	14
Na 616	AI, AII, Ep, I—T, SII	16	19
G 672	AI, AII, Ep, I—T, SII, Anterior SS	28	33
NS 680	AI, AII, Ep, I—T, SII, SS	23	25

ablated in each animal and the number of trials for it to learn the frequency discrimination (800 Hz vs. 1000 Hz) before and after cortical ablation. All animals were able to relearn; the three with the largest lesions required slightly more postoperative than preoperative training.

THOMPSON (1964), using similar methods of stimulus presentation and avoidance conditioning (rotating cage rather than double-grill box used by GOLDBERG and NEFF), trained animals on a frequency discrimination before and after ablations of different extent: the smallest included AI, AII, Ep, SII, and I–T; the largest, the above areas plus the suprasylvian and lateral gyri. Three animals with the largest lesions failed to relearn the frequency discrimination although given many more training trials than was required for preoperative learning. Thalamic degeneration in the experimental animals was not reported.

In a further experiment, THOMPSON and SMITH (1967) tested frequency discrimination of animals with bilateral ablations confined to suprasylvian and lateral gyri (areas from which evoked responses of long latency may be obtained to acoustic stimuli). Stimuli were presented in the two sequences as shown in lines A and B of Fig. 16. Experimental animals were not trained prior to the cortical ablations. None of the four animals with ablation of association areas was able to reach a criterion of 90% correct responses in 20 trials on one day when a stimulus was used that consisted of a pulsing 250 Hz neural tone replaced at intervals by a 2000 Hz positive signal (line A, Fig. 16); all animals performed at a better than chance level. When the sequence of stimuli consisted of silence followed by either the neutral or positive signal, three of the four experimental animals reached the criterion level of learning. The one that failed was later found to have some undercutting of auditory cortex as well as ablation of the suprasylvian and lateral gyri.

In considering the results of the above experiments on frequency discrimination in the cat, it is important to emphasize differences in experimental procedures and the effect of the differences on results and interpretation of results. As we have pointed out, one difference among the experiments was the manner of stimulus presentation and reinforcement. The go, no-go situation in which the experimental animal has to respond to a warning signal and refrain from a similar response to a neutral signal leads to results that differ from the situation in which a repeating pulsing tone forms the background or neutral signal, and response is made to a change in frequency of the pulsing tone (warning signal). This appears to be the most simple form in which a behavioral response to frequency change can be taught to an experimental animal.

The experimental apparatus and the kind of response required are also important factors that affect learning, especially in animals with cortical ablation. This is illustrated by the difference between results in which a rotating cage or a double-grill box was used for avoidance conditioning. In the rotating cage, giving shock for a false response to a neutral stimulus as well as for failure to respond to a positive stimulus produces confusion and consequent behavior in which the animal either responds or fails to respond to both neutral and positive stimuli. In the double-grill box, this confusion does not exist as is illustrated by the success of BROWN et al. (1967) in training animals to respond when the frequency of a background stimulus was changed in one direction (higher) and to refrain from response when the frequency was changed in the opposite direction (lower). In the double-grill box, false responses to the negative stimulus can be punished without disrupting correct responses to the positive stimulus.

b) Monkey. EVARTS (1952) trained monkeys *(Macaca mulatta)* in a food reward situation in which the experimental animal could lift the cover of a food well and obtain food reward when a 350 Hz signal was given but had to refrain from response when a 3500 Hz signal was sounded. Two animals were trained and tested before surgical ablation of auditory cortex; two others were trained only after surgery. All four animals were able to make the frequency discrimination after bilateral ablations that included cortex of the middle and posterior parts of the superior surface of the superior temporal gyrus. EVARTS noted that retrograde degeneration did not occur in the caudal part of the medial geniculate body.

In 1953, JERISON and NEFF reported results of a study in which frequency and pattern discriminations were tested in monkeys *(Macaca mulatta)* before and after bilateral ablations of auditory areas of the cortex (see also JERISON, 1954; NEFF, 1961b). The conditioned avoidance procedure in a double-grill box was used in the training and testing of animals (Fig. 15). Acoustic stimuli were presented as shown in line B, Fig. 14. Bilateral ablations varied in extent but, in all cases, included the core projection area on the superior surface of the superior convolution of the temporal lobe. Some parietal cortex was also ablated. The belt cortex of the superior temporal lobe anterior to the central sulcus was not removed completely in any animal. It is now known that complete retrograde degeneration in the medial geniculate occurs only when this anterior portion of the superior surface is removed (AKERT *et al.*, 1959). This region, like insular-temporal cortex in the cat, receives projection from the caudal end of the medial geniculate body. Animals that were trained and tested on a frequency discrimination before cortical ablation were able to relearn the discrimination postoperatively.

MASSOPUST *et al.* in a series of studies (1965–1967, 1970, 1971, Summary, 1971) have tested frequency discriminations in monkeys *(Macaca mulatta)* after bilateral ablations of different regions of the cortex of the temporal and parietal lobes. Animals with the largest ablations failed to relearn the frequency discriminations; other groups showed smaller deficits. It is impossible to compare the results of the experiments of MASSOPUST *et al.* with results of other experiments, cited above, because of important differences in training and testing procedures and in definition of areas that constitute auditory cortex in the monkey. The authors point out "a frequency discrimination task was designed which not only involved the conditioned-avoidance procedures but also recall (memory) and simple conceptualization".

c) Other Animals. ALLEN (1944, 1945) tested the ability of dogs to discriminate between sound of bell and sound of tapping on a tin cup (Fig. 16). The fundamental frequencies of the two stimuli were 2600 Hz and 72 Hz. While there were other cues available, the animals' discriminations were probably based upon the frequency difference of the two sounds. After bilateral ablation of the then known auditory cortex of the dog, the experimental animals did not retain the learned discrimination habit. As in the THOMPSON experiment, where the positive and negative stimuli were separated by a silent period (go, no-go situation), ALLEN's dogs were unable to refrain from making a response to the negative stimulus.

TUNTURI (1955) has reported that dogs can make a frequency discrimination after bilateral ablations of anterior, middle and posterior ectosylvian gyrus.

d) Man. Few tests of frequency discrimination have been made for patients with temporal lobectomies. In those cases where some kind of frequency discrimination has been measured, little or no change has been noted when test procedures involved simply the discrimination between two tones of one or more seconds duration presented either monaurally or binaurally. For example, MILNER (1962) found no significant change in frequency discrimination when using the Seashore Pitch test to examine patients before and after unilateral temporal lobectomy.

Good frequency discrimination was reported by LHERMITTE *et al.* (1971) for three patients with bilateral lesions of the temporal lobe.

An exception to the above generalization is the finding by BARU, KARASEVA, GERSUNI, and others (see BARU and KARASEVA, 1972) that duration of the acoustical stimulus becomes a critical factor in comparing discrimination of normal animals or human subjects and animals or patients with temporal lobe lesions. Absolute and differential thresholds for signals of short duration are affected by the lesions.

e) Summary. Frequency discrimination, like intensity discrimination and discrimination of onset of acoustic signals, may be learned or relearned after bilateral lesions involving all or nearly all of primary auditory cortex in animals such as the cat and monkey and in human patients. As noted in the introduction of this section, the early results of electrophysiological studies showed a tonotopic organization of auditory cortex and led to the hypothesis that the cortex, therefore, might be essential for accurate discrimination of change in frequency. That this is not the case is not surprising since we now know that this tonotopic organization is present at all levels of the auditory system. Furthermore, tonotopic organization, although suggesting a "place" analysis by the auditory system, need not be the necessary basis for frequency discrimination. For tones in the lower range of frequencies, at least, other neural cues may provide the mechanism for such discriminations.

DIAMOND (1967), in discussing the functional significance of topographic organization of sensory neocortex (tonotopic for auditory system), has pointed out that there is no direct way of finding out what sensory experiences the cat has in response to tonal stimuli of different frequencies and whether sensory experience is changed by ablation of cortex. He suggests that "one way of investigating an alteration of sensory experience is to show a change in the equivalence of stimuli". He further assumes that "the gradients of generalization described by GUTTMAN (1963) can be interpreted as reflecting the animals sensory organization". In line with this assumption, GUTTMAN and DIAMOND (1963) trained animals to make multiple lever pushing responses to a 1000 Hz tone when it replaced a background tone of 5000 Hz. When the response had been learned, generalization tests were given, substituting tones of other frequencies for the 1000 Hz avoidance tone. Tests trials were given both with the 5000 Hz background or neutral tones present between trials and with a silent background. Generalization curves for normal animals and for animals with bilateral ablation of auditory cortex were strikingly similar, even to an octave peak effect in the curves when a 2000 Hz tone was given. GUTTMAN and DIAMOND concluded: "We know of no way of penetrating the sensory experience of cats more deeply and are satisfied that tonotopic organization at the cortical level is not necessary for the perception of tones."

4. Pattern Discrimination

Two of the authors of this chapter were involved in the first experimental study that tested discrimination of change in temporal patterns of tones after bilateral ablation of auditory cortex. With limited confidence in our memories, we can explain in slightly more detail, than given in the original research report, the considerations that led to the study.

At the time the pattern experiments were undertaken, other investigations, most with the cat as subject, had shown that after bilateral ablation of auditory cortex, experimental animals could learn to respond to tone signals at threshold levels and that they could discriminate changes in intensity and frequency at threshold or near threshold levels. Some loss in ability to localize sound in space had been found.

Review of experimental and clinical literature that dealt with the effects of cortical ablations or damage on discriminations in other sensory modalities, particularly vision and to a lesser extent somesthesis, indicated that the kinds of discriminations affected were those involving stimulus patterns. For both vision and somesthesis, spatial patterns had been used in the investigations although, considering how the stimuli act upon sense organs and the resulting initiation of neural activity, it certainly could be argued that, in most instances, the sensory input had both spatial and temporal characteristics.

For the auditory system, it was convenient to use temporal patterns of sound and, of course, it has often been pointed out that the auditory system deals primarily with such patterns as contrasted to the visual system for which spatial patterns appear to be more important. That this is a critical difference between the two is open to question, but, in any event, we chose to use change in the temporal sequences of tonal stimuli in the first experiments that will be described below.

a) Cat. DIAMOND and NEFF (1957) examined the ability of cats to respond to a change in the temporal pattern of tonal pulses before and after bilateral ablation of auditory cortex. The conditioned avoidance situation, double-grill box (Fig. 15), was used in training and testing the experimental animals. Two sequences of background stimuli were used (lines C and D of Fig. 14). In both instances, the same frequencies were present in the background and avoidance stimuli; the only difference which the animal could use as a cue for response was a change in the temporal arrangement of the tonal pulses.

The size of the cortical ablations in the experimental animals was intentionally varied, although the final differences, because of damage to blood supply or inaccuracies in estimates at time of surgery, were not known until results of electrophysiological tests at time of sacrifice and post-mortem examination of cortical lesions and thalamic degeneration had been completed. On the basis of postoperative behavioral tests as related to cortical damage, the animals could be divided into three groups: (1) retention of the learned discrimination in animals that had lesions of AI and possibly some of AII but with some remnants of AI remaining; (2) loss of the learned discrimination but relearning in animals with lesions confined to AI, AII, and Ep; and (3) loss and inability to relearn the pattern discrimination in animals with complete ablations of AI, AII, Ep, and part of I–T. Experimental animals that were unable to relearn the pattern discrimination could learn a frequency discrimination. The ablations in this experiment were made before the projection of caudal GM on I–T was known. It was one of the experiments that led to further study of thalamic degeneration after ablation of auditory areas of the cortex.

SCHARLOCK *et al.* (1963) has reported that cats, in whom auditory cortex was ablated bilaterally in infancy (7–10 days after birth), were able to learn a pattern discrimination when training was given at six months of age, whereas animals, in whom auditory cortex was ablated at the age of six or more months, failed.

As a further test of the importance of auditory cortex for discriminations that involve differences in temporal organization of stimuli, KELLY and WHITFIELD (1971) trained and tested cats to discriminate between a continuous rising tone, e.g. from 3780–4220 Hz and a similar falling tone, from 4220–3780 Hz. These workers used a training procedure similar to that used by DIAMOND and NEFF; *i.e.*, well-trained cats remained at rest in a double-grill box during the repeated presentation of a neutral stimulus and moved across a barrier at presentation of a positive stimulus. After bilateral ablations of auditory areas AI, AII, Ep, and I–T, animals were able to relearn the discrimination but with considerable difficulty. This ability to reacquire the discrimination is in contrast to loss and inability to relearn other sequences such as low-high-low vs. high-low-high or low-high vs. high-low (see KELLY, 1973). As suggested by the authors, the explanation of the difference in discriminating the two sets of stimuli may "correlate with the occurrence of some differential neural activity at brain stem level". Electrophysiological studies have shown that, at several levels of the auditory system, cells may be found that are selectively responsive to direction of change of frequency modulated tones (WHITFIELD and EVANS, 1965; NELSON *et al.*, 1966; ERULKAR *et al.*, 1968a, b; EVANS and NELSON, 1966; WHITFIELD, 1969).

In 1957, GOLDBERG *et al.* reported that after bilateral ablation of insular-temporal cortex (I–T) in the cat, the experimental animals had a severe deficit in ability to make a learned pattern discrimination (high-low-high vs. low-high-low). The learned habit was not retained after removal of I–T and could not be relearned to preoperative criterion level by any animal, and, for some animals, postoperative performance never exceeded the chance level. All animals were able to learn a frequency discrimination (line A, Fig. 14). Examination of the brains of the experimental animals revealed degeneration of the caudal part of the medial geniculate body (GMp) after the I–T lesions. This evidence confirmed results of earlier experiments which had shown that cells in the caudal tip of GMp did not degenerate after ablation of AI, AII, and Ep, and that the further a lesion of these areas was extended ventrally into I–T, the further degeneration extended caudally in GMp (NEFF *et al.*, 1956; BUTLER *et al.*, 1957; DIAMOND and NEFF, 1957; NEFF and DIAMOND, 1958; ROSE and WOOLSEY, 1958).

DIAMOND *et al.* (1958), in a study of retrograde degeneration after lesions of I–T, further clarified the relationship between insular-temporal cortex and the medial geniculate; their results indicated that caudal GMp had both essential and sustaining projection to the I–T region.

Since the 1957 study, NEFF *et al.* (1974) have trained and tested cats on pattern discrimination before and after I–T ablations of different locus and extent. Ablations were varied in size and place within the insular-temporal and adjacent regions. Preliminary evaluation of the data indicates that: (1) pattern discrimination is lost and is not relearned after bilateral ablations that include most of both insular and temporal cortex; (2) a similar result, loss and failure to relearn, occurs after ablations of temporal cortex plus cortex in the area ventral to Ep; (3) ablations of insular, temporal, or insular plus cortex immediately anterior to the insular area do not result in lasting deficits. Results need to be re-evaluated after examination of thalamic degeneration. The results are of special interest because of recent evidence that different divisions of the auditory thalamus project selectively to insular and temporal areas (HEATH and JONES, 1971; SOUSA-PINTO, 1973).

A number of other investigators have also shown that, in the cat, a severe deficit in temporal pattern discrimination occurs after I–T ablations (CORNWELL, 1967; KELLY, 1973; COLAVITA, 1974). The stimuli in the CORNWELL experiment were pairs of sounds; two presentations of the same stimulus were a positive signal, two different stimuli were negative (buzzer-buzzer or bell-bell, positive; buzzer-bell or bell-buzzer, negative). This is the same sort of discrimination suggested by KONORSKI (1959) as a test for recent memory (see p. 347 below). CORNWELL noted that animals with I–T lesions made errors on both the positive (go) trials and on the negative (no-go) trials. This finding is of interest in interpreting the results of other experiments such as that of THOMPSON (1960) reviewed above (p. 336).

KELLY (1973) trained cats to make a shock-avoidance response to FM modulated rising vs. falling tones and to patterns of tone pulses in the same frequency range, low-high vs. high-low. After bilateral ablation of insular-temporal cortex, cats were able to relearn the discrimination of rising vs. falling tones although they did so with great difficulty and could not reach preoperative level of performance; they were unable to relearn the two-tone sequences.

A possible explanation of the difference between the frequency modulated tones and tone pairs has been noted above (p. 342) in discussing the KELLY and WHITFIELD (1971) experiment. KELLY (1974) later used the same pairs of tonal stimuli to test the effects of bilateral ablation of non-auditory cortex in the cat. Large lesions of suprasylvian gyri, anterior lateral gyri, and pericruciate cortex did not severely affect either type of pattern discrimination. After the largest lesions, in which total amount of cortex removed was approximately equivalent to that of AI, AII, Ep, and I–T, there was evidence of some retention of the pattern discriminations and rapid relearning to preoperative performance levels.

The method of stimulus presentation and of conditioning used by COLAVITA *et al.* (1974) were essentially the same as those of DIAMOND and NEFF (1957) except that, for auditory patterns, he used two intensities of an 800 Hz tone to produce the sequences (loud-soft-loud vs. soft-loud-soft). Animals with bilateral ablations of I–T were unable to relearn the discrimination. In additional tests made with pairs of tones, COLAVITA found that the experimental animals could relearn the discriminations of loud-soft-vs. soft-loud or loud-silence-loud vs. silence-loud-silence, components of the original triads of tones, but, even, after learning these discriminations, they could not learn loud-soft-loud vs. soft-loud-soft although the cues were present to which they were able to respond when some components of the triad had been removed.

COLAVITA (1972) has also tested cats in discriminations of temporal patterns of visual (dim-bright-dim vs. bright-dim-bright) and vibrotactile (weak-strong-weak vs. strong-weak-strong) stimuli (1974). He reports that both kinds of discrimination, like temporal patterns of sound, are lost after bilateral ablation of I–T and are not relearned after long postoperative retraining.

COLAVITA concludes that the insular-temporal region in the cat is concerned with temporal pattern discriminations of several sensory modalities and not of auditory patterns alone. If this is the case, further experiments are needed to find out if some parts of I–T and adjacent cortex are critical for auditory pattern discrimination and other parts for visual and tactile pattern discrimination. In the monkey, it has been shown that ablation of the anterior part of the superior surface of the superior convolution of the temporal lobe leads to a deficit in discrimination of auditory patterns without affecting discrimination of visual patterns; the converse is true for ablation of inferotemporal cortex (NEFF, 1961b).

b) Monkey. JERISON and NEFF (JERISON, 1954; NEFF 1961b) tested discrimination of changes in temporal patterns of tones in monkeys using the same experimental procedures as in the DIAMOND and NEFF (1957) study of cats. The results were similar to those of the latter study. Animals with the largest ablations (core plus most of belt areas) failed to relearn pattern discrimination but could learn a frequency discrimination. One animal with a lesser lesion was able to relearn the pattern discrimination.

NEFF (1961a) reported results of tests of temporal pattern discrimination in monkeys *(Macaca mulatta)* after bilateral ablation of cortex of the superior surface of the superior temporal gyrus. The lesions included most of the belt area and spared part of the core area of auditory cortex. Animals had a severe postoperative deficit and required as many or more trials to relearn than had been necessary for preoperative learning of the discrimination.

c) Man. MILNER (1962), taking cognizance of the results of the animal experiments, made use of the Seashore Test of Tonal Memory in examining human patients in whom part of one temporal lobe had been removed by surgery. In the Tonal Memory Test, two sequences of tones are presented in close succession. One tone in the second sequence is different from that in the first and must be identified by the subject being tested. The test increases in difficulty from sequences of 3 tones each to sequences of 5 tones. MILNER found that patients with right temporal lobectomies made more errors (a significant increase) than patients with left temporal lobectomies and were also significantly different from the normal subjects used in standardizing the Seashore tests.

SHANKWEILER (1966), using a test involving unfamiliar melodies presented dichotically, and MILNER *et al.* (1965), a test of recorded birdsongs presented binaurally, found that patients with temporal lobectomies made poorer scores than normal subjects or patients with damage outside of the temporal lobes. Furthermore, patients with right temporal lobectomies scored lower than patients with left lobectomies. On the dichotic tests, SHANKWEILER found that the greatest deficit occurred for material presented to the left ear in patients with right temporal lobectomies.

SCHULOFF and GOODGLASS (1969) used dichotic tests made up of both verbal and non-verbal signals to test and compare results for patients with left or right hemisphere lesions. Non-verbal tests included tonal sequences and trains of clicks presented dichotically. The results indicated that "injury to the hemisphere dominant for a given material produced bilateral deficits specific to that material". For digits, left hemisphere was dominant; for tone sequences, right hemisphere; for clicks, neither hemisphere. A further finding was that the greatest deficit occurred for the ear opposite the hemisphere dominant for a given stimulus.

LHERMITTE *et al.* (1971) have also made use of non-verbal tests in examination of three patients with bilateral temporal lobe damage. The patients were unable to make temporal pattern discriminations, but they could discriminate differences in intensity and frequency of pure tones.

KARASEVA (1972) reported that patients with unilateral damage to auditory projection areas of the temporal lobe had difficulty in discriminating rhythmic patterns of clicks when the signals were given to the ear contralateral to the cortical lesion. KARASEVA concluded that "the human auditory cortex plays an important role in the estimation of time relationships between sounds, as well as in the perception of short sound signals".

In a case report of a patient with a thalamic tumor that affected the auditory projections to the right temporal lobe, ROESER and DALY (1974) noted that the patient stated that music sounded "fuzzy and blurred" when listened to with both ears or with the left ear but sounded "clear" when listened to primarily with the right ear. Verbal stimuli presented to the left ear were perceived as normal.

d) Summary. Investigation of discriminations involving change in the temporal patterns of sounds calls attention to an important role of the auditory cortex in behavior initiated and guided by acoustic stimuli. As we shall also see in the following sections on recent or short term memory, duration, and localization of sound in space, auditory cortex appears to be essential for behavioral discriminations in which the effects of neural activity set off by acoustic stimuli must be maintained in the brain in some form for short intervals (milliseconds to a few seconds). For sequences of tones such as used in the experiments described above, there must be interaction of neural events elicited by the components of a triad of tones in order that the triad is recognized as one of two patterns that form the basis of a discrimination. Furthermore, the first pattern must be represented in some manner in the nervous system for a short time in order to allow comparison with a new second pattern.

5. Discrimination of Temporal Order

Discriminations involving the temporal order of acoustic signals presented in sequences have been investigated in normal human subjects and have been discussed by Hirsh (1959) and by Hirsh and Sherrick (1961). Although identical tests have not been used in animal experiments, temporal pattern discriminations that have been studied in lower animals are similar to temporal order discrimination and may involve similar neural processes.

a) Man. Efron (1963) reported that aphasic patients had difficulty in making judgments about the temporal order of acoustic signals. Noting Efron's report, Jerger *et al.* (1969) used tests of temporal order in examining a patient with bilateral lesions of the temporal lobe (see p. 328 above). One of the tests that he used was similar to that of Efron; another was similar to one that Hirsh and Sherrick had employed in experiments with normal human subjects (Fig. 17). The time,

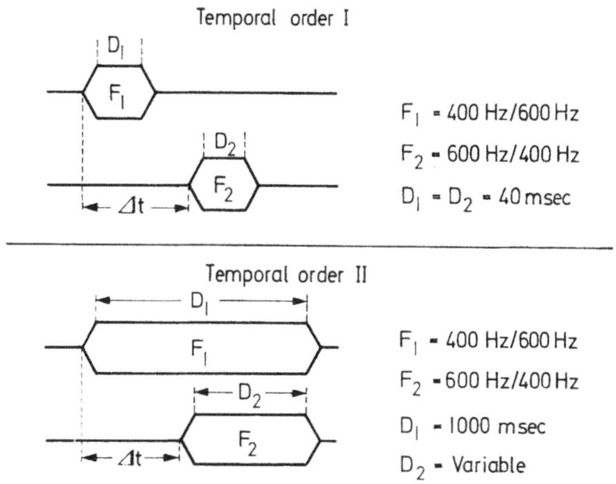

Fig. 17. Temporal order tests used by Jerger (1969)

Δt, between onsets of the two signals, as shown in Fig. 17, was the critical measurement made in the tests. In order to make correct discriminations of the temporal pair (stating whether the first tone was the one of high pitch or low pitch in the pair), Δt for the temporal lobe patient had to be in the range of 300–600 msec. A normal control subject made correct discriminations when Δt was 10–25 msec. When given a visual test, also requiring temporal order judgments, the patient performed as well as the control subject.

SWISHER and HIRSH (1972) examined brain damaged patients with auditory and visual tests of temporal order. On the auditory tests, patients with both right and left brain damage required longer times between signal onsets in order to make correct judgments. Fluent aphasics made the poorest scores. Patients with right brain damage were most affected when successive tonal stimuli were presented to separate ears. In some cases, the quality of two tones appeared to change when they were presented in rapid sequence although they could be identified correctly when each tone was presented alone.

b) Summary. Temporal order tests have not been used in animal experiments. The time intervals between components of the patterns and between patterns in the discriminations that have been studied in lower animals were of a different order of magnitude from the critical time intervals of the temporal order tests.

The deficit of patients with temporal lobe damage, when given the temporal order tests, suggests that auditory cortex may be important in organizing and maintaining a sequence of neural events so that they represent appropriately a sequence of acoustic events.

6. Recent or Short Term Memory

In a short paper published in 1959, KONORSKI pointed out that few neurophysiological investigations had attempted to study what he termed recent or dynamic memory. He proposed a method of presenting sensory stimuli to experimental animals so that recent memory would be measured. The discrimination required by his proposal consisted of a negative or neutral signal composed of two different stimuli, AB or BA (two acoustic, two visual, or two stimuli of any other sensory modality). The positive signal to which an animal would be trained to respond consisted of two identical signals, AA or BB. With such an arrangement of stimuli, a correct response cannot be made until the second component of the second pair appears. By increasing the interval between pairs, the time over which recent memory has to operate can be increased.

a) Dog. A series of experiments to measure the effects of ablations of neural centers on recent memory have been reported by KONORSKI and his associates. Dogs were used as experimental subjects in experiments by CHORAZYNA (1959) and CHORAZYNA and STEPIEN (1961a, 1963). The results may be summarized as follows:

In all experiments with dogs, a conditioned response (lifting right foreleg and placing it on a food-tray in front of the animal) was required when the positive signal was presented. Food reward was given for a correct response. When the negative signal was presented, the animal had to refrain from the foreleg movement. In most instances, the stimuli consisted of pairs of tones, usually of two seconds duration each, with three seconds between the two components of a pair. In some cases, other sounds were used, such as bell and train of clicks. When the stimuli were tones, the pair, composed of two different components, typically differed in frequency, the second being higher than the first by 500–1000 Hz. Differences in intensity were also used in some tests.

In the earliest experiments reported from the Nencki Institute (CHORAZYNA, 1959; CHORAZYNA and STEPIEN, 1961a,b) the negative stimuli were always unlike sounds, for example, 110 Hz and 1100 Hz; the positive stimuli were two tones of the same frequency. Frequencies of both negative and positive stimuli were changed from trial to trial so that a correct response could not be made on an absolute basis, *i. e.*, by recognition of one particular frequency. In a later study (CHORAZYNA and STEPIEN, 1963), one group of animals was trained as above (two like tones, positive; two different tones, negative); and, for a second group, two like tones were negative and two different tones, positive.

Only normal animals were tested in the 1959 study. In the later studies, different parts of the auditory cortex were ablated bilaterally.

Bilateral ablations of parts or all of the ectosylvian gyri did not impair performance on the recent memory test (two unlike tones, negative; two like tones, positive), (CHORAZYNA and STEPIEN, 1961a). After bilateral ablations of the sylvian gyri, animals failed completely to make the discrimination and were unable to relearn although tested over a long period. Animals tested after unilateral ablations showed little or no impairment of performance.

The results of the 1961a experiments were confirmed and extended by CHORAZYNA and STEPIEN (1961b). After bilateral ablation of the sylvian gyri, animals trained to tonal signals showed a complete inability to make the discrimination. Animals trained with stimulus pairs composed of combinations of bell and train of clicks or tones of different intensity suffered severe deficits in performance and, after long retraining, were still unable to reach the stable performance shown in preoperative tests. Animals tested after unilateral ablations showed little impairment for the discrimination.

In a third study (CHORAZYNA and STEPIEN, 1963), the results of the earlier experiments were confirmed, in that it was again found that loss of the learned response and failure to relearn occurred after bilateral ablation of the sylvian gyrus in animals trained on the discrimination: two like tones, positive; two unlike tones, negative. In contrast, a second group of animals, for which the signal, two like tones, was negative and two unlike tones, positive, showed no impairment of performance after similar bilateral ablations. The difference in these two kinds of discrimination is of particular interest in considering the neural model suggested by NEFF *et al.* (see NEFF 1960b; 1961a; 1967; and p. 336 of this chapter).

b) Monkey. The same arrangement of stimulus pairs that was used in the dog experiments was also used in an experiment with monkeys (African green) as subjects (STEPIEN *et al.*, 1960). After bilateral ablation of the first and second temporal convolutions, anterior to the core auditory area, animals failed and were unable to relearn the auditory discrimination. They were able to perform correctly when pairs of visual stimuli were used instead of acoustic signals. In contrast, after bilateral infero-temporal lesions, monkeys could not relearn the visual discrimination but were able to make the appropriate responses to the auditory signals.

c) Summary. When the negative stimuli are two unlike sounds and the positive, two like sounds, the recent memory tests employed by KONORSKI and his colleagues resemble the reverse frequency tests of DIAMOND *et al.* (1962). In the latter, the neutral signal was a triad made up of tones of two different frequencies (low-high low); the positive signal was a triad of three like tones (low-low-low)

(Fig. 14, line E). Cats with bilateral ablation of auditory cortex were unable to learn the reverse frequency discrimination, just as dogs in the experiments of the Nencki Institute group could not learn low-high vs. low-low, when the latter was the positive signal. In both the experiments of DIAMOND *et al.* and of the Nencki group, animals without auditory cortex could learn a discrimination in which the neutral or negative stimuli were like tones and the positive signal contained a tone of new frequency, *e. g.*, low-low-low (neutral) and low-high-low (positive) or low-low (neutral) and low-high (positive).

As KONORSKI and his associates have pointed out, the tests he devised require the maintenance of some kind of neural "sign" produced by one set of stimuli (negative) so that it may be compared with the neural events produced by a new set of stimuli (positive) This requirement is similar to that noted for discrimination of temporal patterns (DIAMOND and NEFF *et al*, above). As KONORSKI has noted, in one kind of pattern discrimination test used by DIAMOND and NEFF (1957), the positive signal (high-low-high) begins with a different frequency than the background stimulus (low-high-low). Assuming that the triad sequences are heard as groups, the change in the first component of a group can provide the cue for a correct discrimination. However, in the DIAMOND and NEFF study, the background stimuli in some cases were (low-low-low, high-high-high, – etc.) and the positive signal (high-low-high) (Fig. 14, line D). The effect of cortical ablation on this discrimination, in which the first tone of the positive signal cannot serve as the cue for a correct discrimination, was the same as for the low-high-low vs. high-low-high stimuli.

7. Discrimination of Changes in Duration

The development of a neural model to explain the difference between discriminations that could be made by experimental animals after cortical ablations and discriminations that could not (NEFF, 1960b, 1961a, b; GOLDBERG and NEFF, 1961a; DIAMOND *et al.*, 1962) led to the study of duration discrimination by SCHARLOCK *et al.* (1965). The neural model stated that, after bilateral ablation of auditory cortex, discriminations could be relearned if new neural units or more neural units (the latter would, of course, imply new units) were activated by a change from background (including silence) to a positive signal; discriminations could not be relearned if no new units or no additional units were activated. The model explained, for example, the difference between discriminations involving response to onset of a tone, to change in intensity, or to change in frequency and discriminations of change in temporal patterns. It was assumed that a discrimination of change in duration of a tone would be a relatively simple case of a change in neural activity confined to the same neural units; *i. e.*, no new or additional units activated.

a) Cat. Results of the duration experiment of SCHARLOCK *et al.* are shown in Table 3. After the largest bilateral ablations, which included area SII in addition to other auditory areas, animals failed to relearn the discrimination, which required an avoidance response to a change from a pulsing 800 Hz background tone of four seconds duration to a one second pulsing signal of the same frequency and intensity.

Table 3. Preoperative learning and postoperative relearning by cats with bilateral ablations of different auditory areas of cortex (From SCHARLOCK *et al.*,1965).

Cat no.	Lesion	Days to criterion	
		Pre-op	Post-op
668[a]	AI, AII, Ep	30	27
723	AI, AII, Ep	24	2
113	AI, AII, Ep, I—T	26	14
721	I—T	32	26
671[a]	AI, AII, Ep, I—T, SII	34	No relearning[b]
774	AI, AII, Ep, I—T, SII	38	No relearning
739	AI, AII, Ep, I—T, SII, Anterior SS	31	No relearning

[a] Animals 668 and 671 received 20 trials per day; all others received 10 trials.
[b] Performance not above chance during 900 postoperative trials.

NEFF and CASSEDAY (1975) have repeated the experiment of SCHARLOCK *et al.* with a reversal of the background and avoidance signals (a background tone of one second duration and avoidance signal, tone of 4 sec duration). Again, the inclusion of area SII as part of the total ablation appeared to make the critical difference between relearning and failure to relearn the duration discrimination.

SCHARLOCK *et al.* (1963) have compared the effects of auditory cortex ablation in infants (7–10 days after birth) and in adult cats on discrimination of changes in duration. When tested at age of six months, the animals in which the cortex was removed in infancy learned the duration discrimination at approximately the same rate as unoperated control animals. Animals in which the auditory cortical areas were ablated after they were six months or more of age showed deficits in performance on the discrimination.

b) Summary. As predicted by the proposed neural model of NEFF *et al.*, discrimination involving response to a change in duration of a tonal signal was affected by bilateral ablation of auditory cortex. The finding that the inclusion of SII in the ablation was critical to the deficit in postoperative performance was a third piece of evidence that this area is an important part of auditory cortex. Earlier, MEYER and WOOLSEY (1952) had reported that ablations that included SII as well as other auditory projection areas produced deficits in postoperative learning of a frequency discrimination, and STOUGHTON (1954) found failure to retain a learned response to onset of tones in a two-stage surgical ablation that included SII in addition to AI, AII, Ep, and part of I–T.

8. Localization and Lateralization of Sound

For most mammals, the ability to localize sound in space plays an important role in the detection and tracking of prey and the avoidance of predators. The cues that an animal can use in localizing will depend upon the frequency range of its hearing, the size of its head, configuration and motility of external ears, as well as upon the acoustic spectrum of the sound source.

In its most primitive form, localization may be only a reflex response involving orientation of eyes and head towards the spatial position of the sound source.

This response can bring to bear the visual sense to guide further attack, retreat, or freeze behavior.

For mammals that hunt or are hunted at night or for those whose prey or predator may be difficult to see because of size or coloration, simple reflex orientation is not enough; the ability to track a source of sound by hearing and to move towards or away from it becomes critical.

If we assume that increase in the ability to utilize information about the position of a sound source was important in the evolutionary process, then we might except to find a higher level of localizing behavior in animals with a greater development of the central nervous system. From this standpoint, examination of localizing behavior appears to be a logical step in the study of function as related to cortical structure.

a) Cat. One of the first kinds of discrimination to reveal a deficit in performance after bilateral ablation of auditory cortex was a discrimination that required the experimental animal to approach the place from which an acoustic signal had been given in order to obtain food reward (NEFF *et al.*, 1950, 1956). Figure 18 shows apparatus typical of that used in a number of experiments to be described.

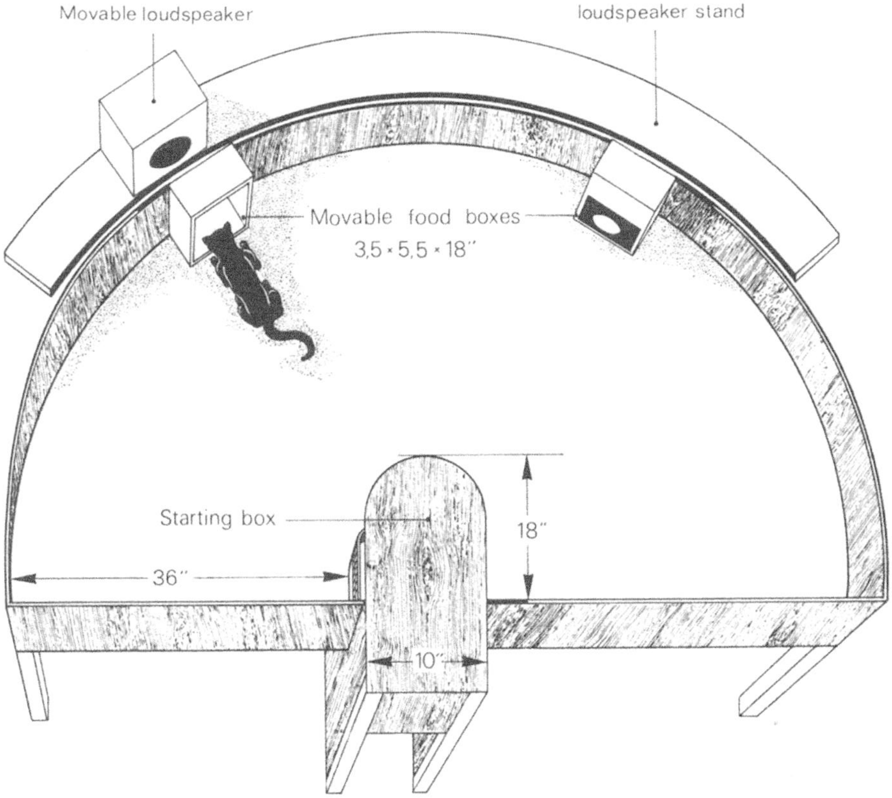

Fig. 18. Apparatus used in testing localization of sound in space

In a series of studies, NEFF and his associates have found:

(1) After bilateral ablations of cortical areas AI, AII, Ep, I–T, SII, and a large part of SS, cats failed to relearn to localize sound; performance at wide angles was at chance level (NEFF 1968a; STROMINGER, 1969a). The signal used was a loud buzzer; it was sounded a number of times but turned off before animals were released from the starting cage.

(2) After bilateral ablations of AI, AII, Ep, I–T, and SII with all or most of SS spared, animals were able to perform only slightly better than chance even when the sound signal on successive presentations was separated by an angle of 90° (NEFF, 1968a; STROMINGER, 1969a).

(3) After ablations confined to AI, AII, Ep, and part of I–T, animals could localize but reached the 75% level of correct performance (chance = 50%) at angles of 40° as compared to 5° before ablation (NEFF *et al.*, 1950; ARNOTT, 1953; see NEFF, 1962, 1968a).

(4) After bilateral ablations confined to AI, AII, and Ep, localization behavior for wide angles was better than chance, but there was wide fluctuation in performance from day to day (NEFF *et al.*, 1956).

(5) Cats with one ear destroyed have been tested before and after unilateral ablation of auditory cortex (core and belt) ipsilateral or contralateral to the intact ear. A clear cut difference between the effects of the two ablations has been found (NEFF and CASSEDAY, 1972). After ablation of cortex contralateral to the intact ear, animals were completely unable to localize the sound source and approach it even when the sound signal (pulses of broad band noise) was continued after release of the animal from the starting cage; that is, the animals could not track the sound. After ablation of cortex ipsilateral to the intact ear, animals were able to relearn to localize and could make correct responses when the sound was turned off before release from the starting cage.

Acoustic stimuli were presented to the ears of cats through headphones by MASTERTON and DIAMOND (1964) in experiments designed to measure the effect of cortical and subcortical lesions on lateralization of sound. Experimental animals were first trained (avoidance response in double-grill box) to respond when short trains of clickes presented to one ear were switched to the opposite ear (Fig. 19). After this discrimination had been learned, tests were made with each click in the sequence first stimulating one followed after 0.5 msec by a click at the opposite ear, *e.g.* R–L in Fig. 19. When the temporal order of clicks presented to the two ears was reversed so that after R–L, each click at the left ear preceded a click at the right ear (L–R), it was found that the experimental animals, having learned R vs. L, now responded immediately to R–L vs. L–R. To the human subject, with the time intervals used, R–L sounds like clicks in right ear only and L–R like clicks in left ear only. The clicks are fused and lateralized to the side in which the clicks first appear.

After bilateral ablation of auditory areas of the cortex, AI, AII, Ep, I–T, animals were unable to discriminate correctly between R–L and L–R although they could still respond to the change R vs. L.

The R–L vs. L–R discrimination was relearned when training was given with the R stimuli of greater intensity in R–L and the L stimuli of greater intensity in L–R. As the R and L stimuli were made equal in gradual steps, animals continued to make correct responses but never reached their preoperative level of performance.

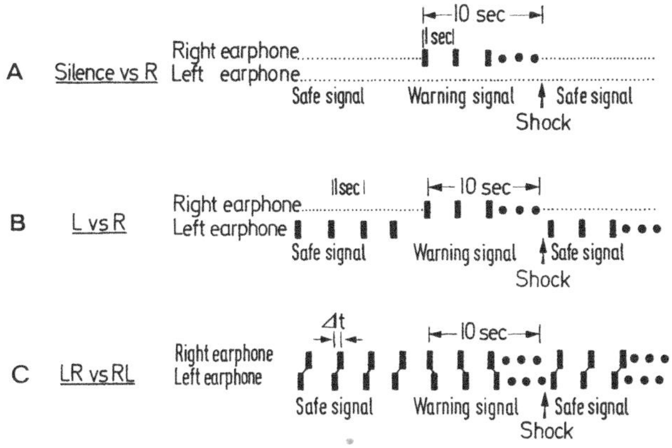

Fig. 19A—C. Schematic representation of temporal sequence of clicks presented to the left (L) or right (R) ear. (A) Silence vs clicks to R. (B) Clicks to L vs. clicks to R. (C) Binaural clicks presented so that each click to L precedes by 0.5 msec a click to R (LR); in the warning signal, the temporal order of clicks presented to the two ears is reversed (RL). (From MASTERTON *et al.*, 1967)

b) Monkey. WEGENER (1964) has reported that, after bilateral ablation of primary auditory cortex, all of nine monkeys tested showed gross deficits in performance at all angles of separation of successive positions of a buzzer signal (angles from 5° to 180°). No animal made more than 60% correct responses at an angle of 90°. After surgery, the discrimination had to be relearned, and more trials were required than in initial preoperative learning.

Experiments by HEFFNER and MASTERTON (1975) have shown that the monkey with bilateral ablation of auditory cortex can localize a sound in a free-field situation if the response is one in which the animal has only to indicate right or left by pressing a lever immediately in front of it in order to obtain a reward. The same animals were unable to make correct localizations when they were required to approach the place where the sound had been given (sound off during approach) in order to obtain reward.

c) Other animals. RAVIZZA and MASTERTON (1972) tested localization of sound in space in the opossum by using a training procedure in which the experimental animal, water deprived, learned to lick a water spout placed in front of its cage in order to obtain water reward. This response kept the animal's head in a position facing two loud-speakers, one on the right and one on the left. When brief noise bursts were sounded through one speaker (*e. g.*, safe signal, on the right), water reward was available. When the sound shifted to the left (warning signal), shock to the feet was given; the shock caused the animal to stop licking the water spout. Experimental animals such as the opossum are able to learn to drink only when the safe signal is present and to cease at the onset of the warning signal. After ablations of all but remnants of the neocortex, opossums were able to discriminate between sounds from the left and sounds from the right. The accuracy of their discrimination for horizontal shift in the direction of the sound source was less than that of normal animals (24.1° vs. 4.6°). The decorticate animals were also able to make a response to an absolute cue, *i. e.*, to sound from left when it was not preceded by sound from the right. The results of this study led to the experiments by Ravizza and Diamond (1974) described below and to an analysis of the several components involved in localizing behavior.

Localization of sound in space before and after ablations of auditory cortex has been measured in the hedgehog *(Paraechinus hypomelas)* and bushbaby *(Galago senegalensis)* by Ravizza and Diamond (1974). Animals were trained to approach a sound source in order to obtain food reward. After bilateral ablations of auditory cortex, the animals were tested under three stimulus conditions: (C) stimulus (white noise) on until animal reached one of six goal boxes, the one in front of the speaker from which the sound had emerged on a given trial; (H) sound turned off just as the animal was released from the starting cage; (0) sound turned off before release. After auditory cortex ablation, performance of the bushbabies was better than that of the hedgehogs under all three stimulus conditions, particularly under condition 0. The authors suggest that the difference in the performance of the hedgehogs and the bushbabies was probably due to the bushbabies' greater natural ability to use body orientation as an aid for localization, although the possibility of a difference in cortical lesions could not be ruled out. In discussing the results of the experiment, Ravizza and Diamond point out that "the requirements of the task with the sound 'off' before release seem to be divisible into three parts: identify the locus of the sound source, 'store' this spatial information, and move toward the sound source". As Ravizza and Masterton (1972) observed in their study of localizing behavior in the decorticate opossum, an animal can make a differential response to sound on left versus sound on right but not necessarily be able to move toward it. The results of the hedgehog and bushbaby study indicate that both animals, after removal of auditory cortex, can identify the direction from which a sound comes and that they have sufficient short-term storage of spatial information to enable them to move to the source when there is no delay between signal offset and start of movement towards the source. The greatest deficit in performance occurs when the animal must store information as to place of sound source for a longer interval in order to guide movement toward the source. In this case, the ability of the animal to orient head or body towards the sound source and to maintain this orientation may be a critical factor. Ravizza and Diamond conclude with the speculation that "spatial organization implies knowing not only where an object or source of stimulation lies relative to the observer, but it also implies the ability to remember the locus and to locomote to the locus. In other words, the intact mammal *knows* where its body is relative to the objects of the environment, *knows* the position of the objects relative to each other, and the meaning of "knows" implies that the organism can manipulate these objects including changing its own position relative to them. It may be just this high level of integration that is disrupted by removal of auditory cortex".

d) Man. The ability to localize or lateralize sound has been tested in numerous patients after partial or total *unilateral* temporal lobectomy (McNally in Pen-

FIELD and EVANS, 1934; GREENE, 1929; WORTIS and PFEFFER, 1948; SANCHEZ-LONGO *et al.*, 1957; SANCHEZ-LONGO and FORSTER, 1958; STROMINGER and NEFF, 1967). In general, there is agreement that a small deficit in localizing in the horizontal plane is found when sounds are presented in the auditory field opposite the side of cortical damage. A patient tested by STROMINGER and NEFF (1967; NEFF, 1968) had a right hemispherectomy. The test situation required the patient to place by remote control a spot of light on a semi-circular screen in order to indicate the position from which he heard a sound signal (broad-band noise). With this test arrangement, very accurate localizations could be made. Table 4 gives the results for the hemispherectomy patient when the noise signal was presented in the left, center, and right auditory fields. The patient was tested on two occasions; he showed improvement on the second test. His largest errors occurred in the left auditory field, the field opposite the side of brain lesion.

Table 4. Scores on localization test for male patient with right hemispherectomy. Scores are errors in degrees between center of loud-speaker and position of spot of light that patient placed where he localized sound source. Tests 1 and 2 were given on different days. Note that largest errors were made when the position of the sound source was in auditory field opposite the brain lesion. Average scores for 9 normal control subjects ranged from 1.69—3.18 degrees. (From STROMINGER and NEFF, 1967).

	Auditory field			
	Left	Center	Right	Average
Test 1	8.56°	2.58	3.63	4.92
Test 2	4.41°	2.91	2.67	3.33

BOSATRA and VERONESE (1965) have reported abnormalities in lateralization tests of patients after hemispherectomy; performance improved when sound signals increased from 0.5–10 sec. It appears likely that their patient had a deficit similar to that of JERGER's patient J-1 whose performance on localization and lateralization tests is described below.

MATZKER (1959) has described a lateralization test in which tonal signals, presented binaurally, were interrupted eight times per second. By means of a delay network, Δt's of 180–650 μsec were introduced at each interruption of the signal. Unfortunately, click artifacts at the onset of the signal were not eliminated. In spite of these clicks, frequency related differences in lateralization were found in some patients with brain damage. Interaural Δt's were adjusted until the patient lateralized the signal at intracranial midline. Tests were usually first conducted with the signal leading in the right ear, then with the signal leading in the left ear. In four cases of unilateral cortical damage, lateralization judgments deviated to the side contralateral to the cortical lesion. One case of particular interest was that of a 16 year old girl whose audiogram indicated normal hearing in both ears. A large temporal meningioma on the left hemisphere was found at surgery. In lateralization tests, she performed normally at all frequencies tested except 1000 Hz. At this frequency, all Δt's were lateralized to the right.

Careful tests of localization and lateralization in patients with bilateral temporal lobe damage were made in the two cases recently reported by JERGER *et al.* (1969; 1972). One of these cases (J-1) is of special interest because a peculiarity of the initial results and information volunteered by the patient led to a further analysis of factors affecting the localization response.

Patient J-1, examined by JERGER in 1969, is the same patient for whom tests of absolute thresholds and DL's for frequency and intensity have been reported in earlier sections of this chapter. To measure localization, the subject was placed blindfolded in a chair with a head-rest in an anechoic room. A loudspeaker on a boom could be moved in an arc of 180° in front of the subject. Sounds at 60 dB SPL and 10 sec in duration were presented through the loudspeaker. The patient was asked to indicate the direction from which the sound came either by a verbal response or by pointing with a yardstick. Sound stimuli used were: train of clicks, sawtooth noise, and tones of 400, 1000, and 4000 Hz. Patient J-1 made large errors for all stimuli on this localization task. In the course of the tests, in which he was instructed to point to the sound source, the patient surprised the experimenters by asking, "Should I point to where the sound begins or where it ends?" When given lateralization tests with headphones, it was discovered that the sound at onset was perceived by the patient as coming from one position, but as the sound continued, it appeared to move often from left to right but sometimes from right to left or in both directions successively.

In further tests, JERGER *et al.* found that the patient's threshold for pure tones as a function of on-time of the signal was abnormal as compared to results for a control subject. As the on-time was decreased from 1000 msec to 20 msec, the sound pressure level necessary for the tone to be heard by the patient needed to be increased by as much as 30–50 dB at the shortest durations. The threshold change was much greater for the right than for the left ear. These results, suggest an explanation of the patients's difficulty in localization and lateralization. At onset, the tone is heard in the left ear only or is louder in that ear and is lateralized to that side. As the tone continues, it reaches levels above threshold or becomes louder for the right ear and is perceived as moving from left to right.

In the second patient (J-2) examined by JERGER *et al.* (1972) no serious deficit in localizing behavior was found. Asymmetry between the two ears in the on-time detection test was not found in this patient.

e) Reflex Responses. As described above, NEFF *et al.* (1956) reported that, after bilateral ablation of auditory cortex (incomplete), the localization of sound in space by cats was considerably less accurate than in normal animals. Earlier, TEN CATE (1934) and BARD and RIOCH (1937) had reported that decorticate cats could react to the direction of sound; no special training and testing procedure was used by TEN CATE nor by BARD and RIOCH. The apparent contradiction of the results of the latter studies and that of NEFF *et al.* led RISS (1959) to test the ability of cats to orient to familiar sounds before and after bilateral ablations of auditory cortex (AI, AII, Ep). He found that normal animals responded promptly and accurately to a short sound signal such as dropping a food pellet on a metal plate; furthermore, they moved quickly to the sound source in order to obtain the food. Animals without auditory cortex did not orient to brief sounds and showed only partial success in searching and finding the position of a continuing sound.

THOMPSON and WELKER (1963) also tested the reflex orientation of cats after bilateral cortical ablation of AI, AII, Ep, I–T, and part of SS. They reported that, compared to normal animals, the responses of animals with cortical ablations were less frequent, less accurate, and less consistent, and had longer latencies.

CHAN (1975) has observed and measured reflex startle and orientation responses after unilateral and bilateral ablations of AI, AII, Ep, I–T, and part of SS. There was little or no effect upon startle responses of short latency (less than 15 msec) after unilateral or bilateral ablations. After unilateral ablations, there was some impairment of orientation; animals were less responsive and did not orient as accurately as they had before cortical removal. After bilateral ablations, there was a further small increase in failures to respond and decrease in accuracy of responses as measured by orientation of the head in relation to source of sound stimulus.

f) Summary. A hierarchy of responses based upon localization and lateralization of sound has been examined in a fairly wide range of animals before and after ablation of auditory cortex. Similar tests have been given to human patients with unilateral or bilateral temporal lobe damage.

Reflex orientation to sound is only moderately disturbed by unilateral ablation of auditory cortex in the cat; it is not severely affected by bilateral ablations although there is some impairment in both accuracy and consistency of the response. There are also reports of some degree of reflex orientation in decorticate cats.

In a test situation not requiring approach or tracking, the decorticate opossum can learn to discriminate between sound on the left and sound on the right although not as accurately as a normal animal.

After bilateral ablation of auditory cortex the bushbaby and hedgehog can learn to approach the source of a sound if the sound is continued during approach or at least until the animal has started to move in the direction of the sound source. The hedgehog fails if the sound is terminated before the animal starts its approach; the bushbaby performs somewhat better under this condition, apparently because of better body orientation.

After complete bilateral ablation of auditory cortex, the cat and monkey perform at chance or near chance levels when required to approach the position of a sound source when the sound has been discontinued before the animal starts its approach. The monkey with auditory cortex ablated can discriminate between sound on the right and sound on the left when it has only to push a lever on the right or left and is not required to locate the sound by moving toward it.

Human patients with unilateral damage to the temporal lobe often have trouble in localizing sound in the auditory field opposite the side of brain lesion; the deficit is usually slight. After bilateral temporal lobe damage (probably incomplete lesions of auditory cortex) accuracy of localizing or lateralizing may be quite severely affected. The deficit appears to be dependent upon asymmetry of the bilateral lesion.

The nature of the behavioral response that is required is an important factor in considering the effects of auditory cortex ablation on localizing and lateralizing behavior. From the evidence available, it is reasonable to assume that auditory cortex is essential for behavior in which appropriate behavior must be guided by an acoustic cue that is no longer present. Perhaps, in the absence of auditory cortex, organization of a "spatial world" based upon acoustic information is no longer possible.

9. Other Binaural Discriminations

a) Cat. In experiments, designed in part to follow up the effects of auditory cortex ablation as related to manner of stimulus presentation (change from neutral background signal to avoidance signal and change from silence to a positive or negative signal), AXELROD and DIAMOND (1965) trained cats to make an avoidance response in a double-grill box to two stimulus conditions. Stimuli were presented to the cats through headphones held in place by a leather helmet. Animals first learned to make the avoidance response when a train of one per second clicks being presented to one ear was switched to the opposite ear (detection-situation). After learning this discrimination, the same animals were trained to make an avoidance response when clicks were presented to the right ear after a silent period and to refrain from responding when clicks were presented to the left ear after silence (identification-situation). After bilateral ablation of auditory areas AI, AII, Ep, and I–T and, in some cases, invasion of SII, animals were able to relearn to make the avoidance response in the detection situation, but they showed a severe deficit in performance in the identification situation. AXELROD and DIAMOND reject the explanation of results in terms of failure to inhibit the response to a negative stimulus and support an explanation in terms of habituation during the neutral signal in the detection situation (see p. 33b, above).

Using similar training and stimulus presentation methods as in the AXELROD and DIAMOND experiment, KAAS *et al.* (1967) trained cats to respond to a change in binaurally presented stimuli shown in Fig. 20. There were two possible cues that the animal could use to avoid shock by responding to the warning signal. If it "listened" only to the signal in one ear (attentive side in Fig. 20), it needed to respond when a high tone appeared (frequency discrimination). If it integrated the sequences as presented in temporal order to the two ears, then the discrimination required was a change in temporal pattern (LLLH vs. LLHH). The human observer can detect either change by attending to the signals in one ear only or in both. In the KAAS *et al.* experiments, when tests (Fig. 21) were given to find

	Neutral signal	Warning signal
attending ear	L – L – L – L –	L – H – L – H –
ignoring ear	– L – H – L – H	– L – H – L – H
binaural pattern	L L L H L L L H	L L H H L L H H

Fig. 20. Sequences of signals used in initial training of animals in experiment of KAAS *et al.* (1967). Tonal pulses, 0.9 sec. in duration were used; the interval between pulses in the binaural pattern was 0.1 sec. L = 800 Hz; H = 1000 Hz

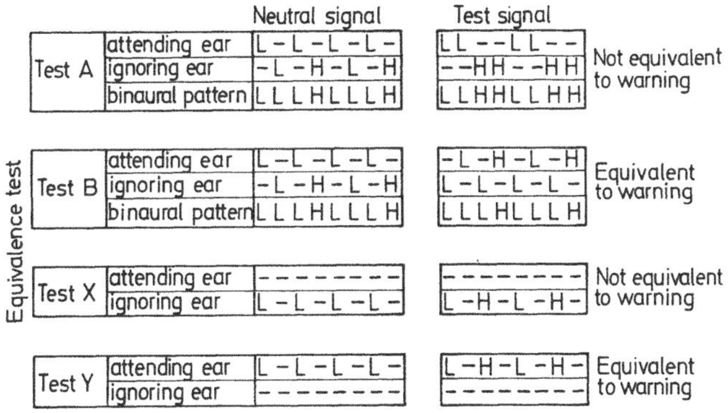

Fig. 21. Sequences of signals used in equivalence tests (KAAS *et al.*, 1967). Signals were tonal pulses, 0.9 sec. in duration; interval between pulses in binaural patterns was 0.1 sec. L = 800 Hz; H = 1000 Hz

out what cues the animals were using after they had learned to respond correctly in at least nine of ten trials on two successive days, all animals immediately made 90% or better correct responses on test B (Table 5). As shown in the table, the experimental animals also responded correctly on test Y but failed completely on tests A and X. The results indicate that normal animals learned to attend to one ear, primarily, the ear which provided a relatively simple cue for correct response. That the signal to the opposite ear also had some effect is indicated by the lack of transfer in test X. For example, animals trained to respond to LLLL vs. LHLH presented to the left ear and no signal to the right ear showed immediate transfer when the neutral and warning signals were presented to the right ear only.

Continuing their experiments, KAAS *et al.* next ablated auditory cortex (AI, AII, Ep, I–T, SII) contralateral to the attending ear in five animals and ipsilateral to the attending ear in three others. A severe initial deficit with slow relearning was found in the group receiving contralateral ablations; a smaller initial deficit and rapid relearning in the ipsilateral group. After retraining to a 90% performance criterion, the remaining auditory cortex was ablated in six animals.

Table 5. Ratio of conditioned responses to presentation of four types of equivalence tests by trained normal cats. See Fig. 21 for tests used. (From KAAS *et al.*, 1967)

	Tests			
Cat no.	A	B	X	Y
430	0/10	9/10	1/10	9/10
458	0/10	10/10	0/10	10/10
424	0/10	10/10	0/10	10/10
490	0/10	9/10	1/10	8/10
489	1/10	10/10	0/10	7/10
499	0/10	9/10	1/10	8/10
488	0/10	10/10	0/10	9/10
515	0/10	9/10	0/10	9/10
Total all cats	1/80	76/80	3/80	70/80

All showed severe initial loss of the discrimination and gradual recovery with retraining. The animals in which cortex contralateral to the attending ear was ablated in the second operation performed more poorly than those in which ipsilateral cortex was removed.

KAAS *et al.* made use of the concept of habituation to an ongoing signal and response to increase in neural activity set off by a new signal to explain the results of their experiments.

For human subjects, if acoustic signals are presented from two sound sources separated in space with only a few milliseconds separating the onsets of the signals, a single fused sound is heard, and it is localized in the position of the source of the signal sounded first. This perceptual phenomenon has been called, the "precedence effect". The effects of different stimulus parameters on the precedence effect have been studied in numerous psychophysical experiments.

In a series of experiments with several combinations of stimuli presented from sound sources on the right and left (90° apart) in a Y maze, WHITFIELD *et al.* (1972) (also CRANFORD *et al.*, 1971, preliminary report) found that unilateral ablation of auditory areas AI, AII, Ep, I–T, and SII in the cat disrupted its ability to make a discrimination based upon localization of the sound source or sources. Normal animals, having learned to respond appropriately to a signal on the right (approach sound source to obtain food reward) or on the left, immediately made correct responses when the signal on the right was followed, after 5 msec, by a signal on the left. This is the precedence effect as described above. Apparently the cat, as man, hears the combined signal as fused and as coming from the direction of the first signal.

After unilateral ablation of auditory cortex, there was no longer a precedence effect for the two successive signals when the first signal was presented in the auditory field opposite the side of cortical ablation. Animals with unilateral lesions also showed deficits in a discrimination in which they went to the left when a single sound was presented on the left and to the right when two signals were presented, first on the left and then on the right, but separated in time so that they were heard as two separate sounds by a human observer.

CRANFORD (1975) tested contralateral masking in cats, using avoidance conditioning (double-grill box) and head phones for presenting stimuli separately to the two ears. The neutral stimulus was broad-band noise pulses presented to one ear; the positive stimulus, a pulsing 1000 Hz tone delivered to the opposite ear. Tests were given with and without noise pulses in the first ear while the tones were presented to the second ear. In normal animals, a mean contralateral masking effect of 5.4 dB was found for five animals tested. After unilateral ablations of auditory cortex (AI, AII, Ep, I–T, and SII), an asymmetry in the masking effect occurred, depending on the ear masked in relation to the side on which cortex was removed. When noise was presented to the ear ipsilateral to the side of the ablation and tones to the contralateral ear, the mean masking effect was increased to 10.9 dB. When noise was presented to the contralateral ear and tones to the ipsilateral, the mean masking effect was only 5 dB. When the remaining auditory cortex was ablated, resulting in a bilateral lesion, contralateral masking effects for all conditions of presentation were the same and amount of masking was approximately equal to that found before the first unilateral ablation.

b) Man. The effects of cortical or subcortical lesions on speech discrimination will not be reviewed in this chapter. There is a large literature concerning speech discrimination and lesions of the central nervous system, the results of which cannot be compared readily with experiments on lower animals. Furthermore, construction of speech discrimination tests has often had the primary purpose of providing a diagnostic tool rather than a means of revealing neural mechanisms of hearing.

Brief mention is made below of speech tests only as an introduction to other studies which replaced speech, presented dichotically, with non-speech acoustic signals. The reader interested in dichotic tests of speech is referred to a number of reviews of dichotic listening with emphasis on speech tests and central nervous system function which have appeared in recent years (Studdert-Kennedy and Shankweiler, 1970; Berlin and Lowe, 1972; Berlin and McNeil, 1975).

Dichotic tests, verbal and nonverbal, have proved of value in examination of patients with lesions of the temporal lobe. The use of such tests was suggested by the psychophysical experiments of Broadbent (1956); these initial experiments tested normal subjects with no known disorders of hearing or damage to the central system. Kimura (1961) was the first to report results of giving similar dichotic tests to patients with temporal lobe damage, although Jerger and Mier (1960) had examined temporal lobe patients with a dichotic test patterned after tests used in experiments by Cherry (1953) and others with normal subjects. Jerger and Mier found a deficit in speech perception for the ear contralateral to temporal lobe damage when the ipsilateral ear was simultaneously exposed to different verbal stimuli.

In the study reported by Kimura in 1961, spoken digits were recorded on a two-channel tape recorder in such fashion that the two ears could be stimulated by two different digits simultaneously, alternately, or all of a group could first be presented to one ear. Scores on the tests were based on total number of digits reported correctly for each ear. Kimura compared test scores before and after surgery for patients with partial ablations of the left temporal lobe, right temporal lobe, or of one frontal lobe. The average score for patients with left temporal lobe lesions was lower than that of any other group and was significantly different from the average score for right temporal lobe patients.

Kimura (1964) has also constructed a two-channel test with which she can present two different melodies simultaneously to the two ears. She has found that normal subjects make a significantly better score for those melodies presented to the left ear.

Using Kimura's melodies test, Shankweiler (1966) found that lower scores were made by patients with right temporal lobe damage than by those with left temporal lobe damage. Furthermore, the effects of surgery in which Heschl's gyri were spared in some cases and removed in others suggest that the cortex of the superior surface of the superior temporal lobe anterior to Heschl's gyri may be of particular importance for recognition of temporal patterns; a similar region in the monkey has been shown to be important for pattern discriminations (NEFF, 1961a).

c) Summary. Discriminations dependent upon binaural interaction are obviously appropriate tests of central nervous system function in hearing. It is not surprising, therefore, that after damage to auditory cortex, deterioration of performance has been found for a number of different binaural tests.

10. Sounds with Complex Spectra

Discrimination of changes in sounds of complex spectra and recognition of such sounds that may be identified with a source have been tested in numerous studies of human patients but only in a few animal experiments.

a) Cat. DEWSON (1964) used operant conditioning methods to train cats to discriminate between the speech sounds [*u*] and [*i*] as well as between differences in frequency and intensity of tones. After bilateral ablation of insular-temporal cortex, the animals retained the ability to discriminate between the differences in the tonal stimuli but did not retain and could not reacquire the ability to discriminate between speech sounds.

b) Monkey. DEWSON *et al.* (1969) have also tested the monkey's *(Macaca mulatta)* ability to discriminate differences between speech sounds before and after ablation of different regions of the temporal lobe. After bilateral ablation of the posterior part of the superior surface of the superior convolution of the temporal lobe (core area of auditory cortex), experimental animals lost the ability to discriminate between the speech sounds [*i*] and [*u*] and were unable to relearn the discrimination. Severe but reversible deficits occurred in animals with bilateral ablation of midtemporal cortex. The speech sound discrimination was retained after bilateral ablation of infero-temporal cortex.

SYMMES (1966) reported a severe and lasting deficit in the ability of monkeys *(Macaca speciosa* and *Macaca mulatta)* to discriminate between intermittent and steady state noise after ablations of primary auditory cortex.

c) Man. For examination of patients diagnosed as having lesions of either the left or right hemisphere, FAGLIONI *et al.* (1969; see also SPINNLER and VIGNOLO, 1966; VIGNOLO, 1969) employed two kinds of tests, both requiring discrimination between sounds of complex spectra. One test consisted of pairs of meaningless sounds (noises of different complex spectra). Subjects given the test were asked to report whether a pair of sounds, recorded on tape and presented binaurally, were the same or different. A second test required the subject to identify a picture that corresponded to the source of a meaningful sound. To make the identification, the subject was asked to indicate which of four pictures corresponded to the source of the sound presented. The four pictures included one that would be the natural source of the sound, two pictures representing sources belonging to the same semantic category as the real source but not capable of producing the sound, and a fourth picture of a source that was completely unrelated to the given sound. For example, when the sound of ringing bells was presented to the subject, the four pictures from which he had to make a choice were: (1) ringing bells (real source), (2) people praying in a church (semantic category), (3) a doorbell (semantic category), and (4) a farm tractor in motion (odd or absurd category). In tests of 41 patients with right and 60 with left hemisphere lesions, it was found that the mean score on the meaningless sounds test was significantly lower for the right hemisphere patients, and the mean score on the meaningful sounds test was significantly lower for the left hemisphere patients who were aphasic.

d) Summary. For discriminations that involve sounds of complex spectra, it is not possible on the basis of present neurophysiological knowledge to make meaningful inferences about neural events elicited by the sounds. Results of experiments, such as those of DEWSON, are of interest because they show that not only non-human primates but also carnivores have the auditory capacity to distinguish between components of speech and that this capacity is disrupted by ablation of auditory areas of the cortex.

The results of the meaningful and meaningless sound tests are important in that they provide evidence of interhemispheric differences for sounds other than speech.

11. Summary and Conclusions

From experimental and clinical investigations, evidence has been obtained that severe deficits occur for the following auditory discriminations after bilateral ablation of or damage to auditory areas of the cerebral cortex:

1. Localization of sound in space.
2. Lateralization of dichotic signals.
3. Change in frequency in which a frequency is dropped from a sequence.
4. Change in temporal pattern of sounds.
5. Change in duration of sounds.
6. Comparison of signals involving recent or short term memory.
7. Identification of temporal order.
8. Detection of signal (absolute threshold) as a function of duration.
9. Change in attribute of signal (*e.g.* frequency or intensity DL) as a function of duration.
10. Change in sounds of complex spectra.

After unilateral or partial cortical ablations, some loss in ability to make the above discriminations may occur. In such cases, the loss may be seen only for sound signals presented to the ear contralateral to the cortical lesion.

Discriminations that can be made by experimental animals after complete bilateral ablations of core and belt areas of auditory cortex include: response to onset of sound, to change in intensity of tones, and to change in frequency of tones.

A common characteristic of those discriminations that are adversely affected by ablation of auditory cortex is a requirement for integration of neural events for short periods of time, in many instances for a few seconds. In temporal pattern discrimination and in the recent memory type of discrimination used by the Nencki Institute group, some neural representation of a first sequence of acoustic signals had to be maintained for comparison with neural events elicited by a second sequence of sounds. In many localization of sound in space experiments, the locus of a sound source no longer present had to be represented in the nervous system in order to guide the required approach to the position of the sound source.

During the years that the studies reviewed above were carried out, new information as to auditory pathways and centers was being added continually through advances in anatomical and physiological methods. We still lack complete knowledge as to the cortical areas that receive input when an acoustic stimulus excites nerve endings in the cochlea. There is considerable evidence that "primary" or "classical" auditory pathways and centers are not the only route by which nerve impulses reach the cortex. It is a futile exercise to ablate all cortical areas that receive some auditory input from subcortical centers, areas that give evoked responses of long latency as well as short latency and areas that receive input from subcortical systems that have collateral neural connections with the primary auditory system and with other systems as well. If such ablations were made, many cortical functions, sensory, motor, and integrative would necessarily be affected, and it would be impossible to parcel out deficits due only to disruption of the auditory input. It may well be that, in the experimental animals that have been most carefully studied, we have already reached the point where further extension of ablations will tell us little more about the role of the cortex in receiving and processing auditory information. An approach that at present appears more likely to yield new insights is to apply the behavior-ablation method to studies of other animals, studies that will enable us to relate differences in structural development with differences in discriminatory capacity.

Devising more elaborate and complicated tests of auditory discrimination may have value in diagnosis of damage to the auditory nervous system in man. Such tests are not likely to help us in understanding the basic functions or capabilities of the primary auditory nervous system. To understand the latter, the important tests are those for which we can make reasonable hypotheses about events in the auditory paths and centers (see Neff and Neff, 1972).

B. Ablation of Subcortical Centers and Transection of Subcortical Pathways

In Section III A, we have seen that many auditory discriminations survive ablation of auditory cortex. An animal with all auditory areas of the cortex removed can not only respond to the onset of sound, but it can learn to discriminate between two sounds that differ in intensity or in frequency. Such findings have made it apparent that auditory centers of the brainstem are more than relay stations or centers serving merely reflex functions.

The focus of early behavioral studies of subcortical auditory centers was to determine which of these centers might be important in mediating a conditioned response to sound. For example, in 1943, Kryter and Ades on the basis of behavioral experiments suggested that in the absence of auditory cortex, the tectum might provide the neural organization and connections essential for learning a response to the onset of sound. Later investigators were less concerned with the question, at what level of the nervous system is it possible for an animal to learn an auditory discrimination? Attention was more often directed at discovering how the function of the auditory system was disrupted by ablation of subcortical centers or transection of subcortical pathways.

In addition to reviewing results of animal experiments, we shall cite a few examples of cases of brainstem lesions in man from the clinical literature. These studies were selected for their potential significance in understanding the central auditory system, not for their clinical or diagnostic worth. Bearing in mind that any brainstem lesion may affect eighth nerve function, or even cochlear function through occlusion of the vascular supply, we have tried to choose cases in which there was reasonable evidence that the principal damage to the auditory pathway was at a point central to the eighth nerve. Unlike lesions made in the brainstem of experimental animals, lesions in the brainstem of man seldom affect only one part of the central auditory pathway, with the result that auditory disorders after brainstem damage in man are highly variable and difficult to interpret.

Behavioral changes in auditory discrimination have been examined most often after transection of parts of the auditory pathway that lie close to the surface of the brainstem and, thus, can be exposed with relative ease. In the cat, this method has been used successfully for transection of the brachium of the inferior colliculus (BIC), the commissure of the inferior colliculus (CIC), the lateral lemniscus(LL) and the trapezoid body (TB) (see Fig. 1). All of these pathways can be transected without causing substantial damage to surrounding neural tissue. Stereotaxic techniques have made it possible to destroy auditory structures in less accessible parts of the brainstem, such as SOC, but this method has seldom been used in behavioral studies.

1. Discrimination of Onset of Sound and Absolute Thresholds

a) Experimental Animals. Having found no change in pure tone thresholds after bilateral ablation of auditory cortex in cats, KRYTER and ADES (1943) went on to examine the effects of bilateral transection of the brachium of the inferior colliculus (BIC), bilateral ablation of the inferior colliculus (IC), bilateral transection of the lateral lemniscus (LL), and in some cases, a combination of cortical and subcortical lesions. After bilateral lesions limited to BIC, only small losses were found, 6 dB or less. Slightly larger losses occurred after bilateral ablation of IC and the largest threshold shifts, a range from 30–50 dB, were measured after deep lesions of BIC plus partial ablation of IC or after transection of LL.

In more recent experiments, the extent to which bilateral lesions of LL or BIC have transected pathways to auditory cortex has been tested by electrophysiological methods after completion of behavioral testing.

Cats in which bilateral transection of BIC or LL eliminated evoked responses as recorded from auditory cortex under sodium pentobarbital anesthetic, were still able to respond to the onset of sound (CASSEDAY and NEFF, 1975). DIXON (1973) measured pure tone thresholds of cats for the frequencies 250, 1000, 4000, 16000, and 32000 Hz before and after transection at the lateral lemniscal level of the crossed auditory pathway, the uncrossed pathway, and both. Tests of the crossed pathway were made by destroying one cochlea and transecting the lateral lemniscus on the opposite side. Tests of the uncrossed pathway were made by destroying one cochlea and transecting the lateral lemniscus on the same side.

At time of sacrifice, the cortex was explored for click-evoked responses. Chloralose anesthetic was used to increase the possibility of obtaining evoked responses. In animals with unilateral LL transections, including those with complete transection of crossed or uncrossed auditory pathways, no significant changes in pure tone thresholds were found. In animals with incomplete bilateral transection of LL, moderate hearing losses (30 dB or less) were measured. In two animals with complete bilateral transection of LL as indicated by both anatomical examination and absence of evoked responses from cortical areas, severe hearing losses (80 dB or more) occurred. Intensity of tones used had to be increased and rise time of the tone pulses shortened in order to retrain the two animals with the largest bilateral lesions. With tones of such high intensity, the animals may have been responding to non-auditory cues, *e.g.* tactual cues from movement of the eardrum.

In the DIXON study, the largest lesions of LL extended medially to the border of the peri-aqueductal gray. It appears likely that the ascending auditory system includes pathways outside of the lateral lemniscus as the latter is usually defined.

JANE *et al.* (1965) measured the absolute intensity threshold for one frequency, 300 Hz, in one cat after transection of the trapezoid body (TB), and of the commissures of HELD, VON MONAKOW, and PROBST. The threshold did not differ from that of normal animals tested under the same conditions. CASSEDAY (1970) found normal pure tone thresholds for frequencies, 250, 1000, and 8000 Hz, after complete transection of TB and the other auditory commissures of the medulla.

b) Man. It is difficult to correlate damage to auditory structures of the brain stem of man and changes in hearing because the brain stem lesions typically damage parts of more than one auditory center or tract as well as non-auditory structures. A lesion or a tumor, even if confined to a given region, may interfere with the blood supply of other regions.

Changes in pure tone thresholds have been reported by a number of investigators for clinical cases with confirmed damage to auditory structures of the brain stem. Elevated thresholds, with greater deficit for high than for low frequencies, have been found in a number of studies, for example, in cases of brainstem tumor (SLOAN *et al.*, 1943; BENITEZ *et al.*, 1966) and in patients with kernicterus (CARHART, 1967; DIX and HOOD, 1973) as well as in other cases having suspected damage at the brain stem level. As JERGER (1960a) has pointed out, high frequency hearing loss may have a variety of etiologies so that this symptom has limited value for clinical diagnosis.

Threshold shift affecting all frequencies or, infrequently, primarily the low frequencies, have also been reported, for example, for patients with multiple sclerosis (NOFFSINGER *et al.*, 1972) and for patients with other degenerative diseases of the nervous system.

Several lines of evidence have indicated that lesions in the brainstem of man can result in increased absolute thresholds for stimuli of long duration or for continuous stimuli, while thresholds for interrupted sounds or sounds of short duration show relatively less increase. BARU and KARASEVA (1972) have compared thresholds for a 1000 Hz tone of two durations (1 msec and 1200 msec) for 9 patients with lesions in the brainstem. While thresholds for stimuli of both durations were higher than those of normal subjects, the greatest increase was found for the 1200 msec tone. This finding is in contrast to the threshold increase for short duration tones found in patients with damage to auditory cortex. In two cases, damage to auditory structures in the brainstem was verified. In one case, a 24 year old man, autopsy showed that a tumor had resulted in nearly complete bilateral degeneration of IC and LL; in the second case, a 24 year old woman, surgery revealed that a tumor on the floor of the fourth ventricle had damaged the cochlear nuclei. The increase in threshold for the long duration stimuli was greatest in the second case. BARU and KARASEVA suggested that thresholds for long duration stimuli are increased more by damage to the cochlear nuclei than by damage to the inferior colliculi.

An increase in threshold during sustained auditory stimulation (adaptation) that is abnormally large is a common finding in lesions of the eighth nerve (JERGER, 1960b, c); this abnormal adaptation can sometimes be seen in cases with brainstem lesions. One method of testing for abnormal adaptation utilizes the Békésy self-recording audiometer. The subject is instructed to press a key as long as he hears the signal and to release the key when the signal disappears. The key controls the direction of a motor driven attenuator that automatically decreases or increases the intensity of the signal. A continuous graph recording of the attenuator values shows oscillations about the subject's threshold. Thresholds for continuous presentation of a tone can be compared with thresholds for the same tone when it is periodically interrupted. In subjects with normal hearing, the Békésy tracing will indicate that the same average sound intensity is required to detect the signal when it is presented continuously and when it is presented intermittently (JERGER, 1960b). For cases in which adaptation is severe, the response to the intermittent signal will remain at one intensity level, as in normal subjects, but with continuous stimulation, the sound must be more and more intense for detection. Adaptation, as measured in this way, results in a Békésy tracing which has been termed type III (JERGER, 1960b).

CARHART (1957) has devised a threshold tone decay test for determining unusual adaptation. The subject is presented with a pure tone at his threshold. If at the end of one minute the tone is no longer detectable, its intensity is raised 5 dB, and the test continues for another minute. Increments of 5 dB are then added after each minute until the subject reports hearing the tone for a full minute. Total adaptation refers to cases in which the subject cannot continue to detect the tone for a full minute at the highest intensity available from the audiometer (about 100 dB re 0.0002 dynes/cm^2). Abnormal adaptation as measured by the tone decay test is often seen in patients with brain stem damage as well as in cases with 8th nerve damage (BENITEZ *et al.*, 1966; PARKER *et al.*, 1962). PARKER *et al.* (1968) found abnormal adaptation (tone decay test) to be present but less severe in cases with lesions in the pons than in cases with eighth nerve lesions; hearing loss was less severe also in the cases with pontine lesions. It was suggested that abnormal adaptation in the presence of other symptoms, auditory and non-auditory, might characterize pontine lesions. Autopsy findings for one of the patients examined by PARKER *et al.* revealed a glioblastoma multiforme which extended into the medulla and pons. The patient had complete adaptation in one ear for all frequencies tested (250–8000 Hz, in octave steps) and complete adaptation in the other ear for 8000 Hz. Pure tone thresholds were not markedly increased.

Severe adaptation (Békésy audiometry) without threshold loss for interrupted pure tones has been reported in a case of pinealoma; the adaptation was not evident after the tumor had been removed (Kos, 1955).

MORALES-GARCIA and HOOD (1972) found that 11 of 27 patients with brainstem lesions, which apparently did not involve the eighth nerve, had a pronounced slow elevation in threshold when presented with sustained pure tones (tone decay test). In half of the patients that showed this adaptation, thresholds for short pulses of pure tones were normal. In several of the patients, abnormal adaptation was found at only one frequency. JERGER and JERGER (1974) failed to find abnormal adaptation to pure tones, as tested by Békésy audiometry, in 16 patients with brainstem lesions not involving the eighth nerve.

c) Summary. In experimental animals: (1) after major portions of the auditory system have been destroyed bilaterally at or below the level of the tectum, substantial increases in absolute thresholds have been measured; (2) small increases in absolute thresholds are found after transection of BIC; (3) little or no threshold increase is found after complete transection of auditory commissures of the medulla; (4) cats can learn to respond to the onset of sound after large bilateral lesions of LL; (5) after bilateral lesions that completely transect LL and invade adjacent tracts, learned responses can be made only to sounds of high intensity; the possibility that the responses are made to non-auditory cues cannot be ruled out.

In human patients with damage to auditory structures of the brain stem, pure tone thresholds are often increased (hearing deficit). In many cases, the greatest loss occurs for the higher frequencies. Because of the organization of nerve fibers in the auditory nerve, with fibers from the base on the outside and entering the cochlear nucleus in a dorsal position, high frequency hearing loss might be expected for lesions affecting the cochlear nucleus or the 8th nerve as it enters the nucleus (see DANDY, 1934; NEFF, 1947).

Abnormal adaptation to continuous tonal stimuli and increase in threshold for tones of brief duration have been noted in patients with brain stem damage.

2. Discrimination of Changes in Intensity

a) Experimental Animals. In their experiments reported in 1946, RAAB and ADES measured DL's for intensity not only before and after cortical ablations but before and after bilateral ablations of (1) inferior colliculus, (2) superior colliculus, (3) auditory cortex and inferior colliculus, (4) auditory cortex and superior colliculus. The experimental animals (cats) were trained and tested using avoidance conditioning in a rotating cage. The learned discrimination was retained after bilateral ablation of IC or SC. The discrimination was not retained after bilateral cortical lesions; relearning was possible. DL's for intensity were not changed after bilateral ablations of IC or SC. DL's were increased by 5–10 dB after bilateral ablation of auditory cortex and IC or auditory cortex and SC.

Ablations of auditory cortex as we now know it were incomplete in the cats used in the RAAB and ADES experiments. Some cortex in the I–T region was probably spared in all animals.

OESTERREICH *et al.* (1971) repeated the RAAB and ADES experiments and added new information by making more complete ablations of auditory cortex and by including animals with bilateral transection of BIC. The avoidance conditioning procedure in a double-grill box was used. In Section III.A, Table 1, the results of cortical ablations are shown; there was little or no change in DL's for intensity. After bilateral ablations of auditory cortex and IC, DL's were increased by 5–7 dB. The largest increase in DL's, approximately 10 dB, occurred after bilateral transection of BIC and adjacent tracts. Retraining was required after the combined ablation of cortex and IC and after bilateral transection of BIC.

b) Man. A clinical test that is directly related to experimental measures of intensity discrimination is the short incremental sensitivity index (SISI) (JERGER *et al.*, 1959). A tone is presented to the subject at 20 dB above his threshold. At 5 sec intervals, the intensity of the tone is increased 1 dB for 200 msec. The subject is asked to respond when he hears an increase in loudness. A high score on this test (hypersensitivity) is found in cases of cochlear dysfunction such as in Meniere's disease, whereas in retrocochlear lesions, low scores are found (JERGER *et al.*, 1959). In tests of patients with multiple sclerosis, NOFFSINGER *et al.* (1972) found a high incidence (about 50%) of decreased sensitivity for short intensity increments. These authors point out that in most groups of subjects, chosen for neurological symptoms, the occurrence of low SISI scores is 50–60%. They suggested that such results indicate that lesions anywhere in the central auditory pathway may produce deficits that result in low SISI scores.

Recruitment, the abnormally steep growth in loudness as sound intensity is increased, is usually considered to result from disorders of the inner ear but is occasionally seen in patients with lesions of the brainstem (*e.g.*, GOODMAN, 1957; FLOWER and VIEHWEG, 1961; DIX and HOOD, 1973). DIX and HOOD (1953) suggested that tumors which occlude the vascular supply to the cochlea may account for the occurrence of recruitment; they later (DIX and HOOD, 1973) concluded that loudness recruitment may occur in cases of brainstem damage, such as patients with kernicterus, in whom the cochlea is unlikely to be damaged.

c) Summary. Experimental studies of the effects of subcortical lesions on intensity discrimination have been reported only for the cat. Amnesia for a learned discrimination of change in intensity occurs after bilateral transection of BIC but not after bilateral ablations of IC unless the latter is combined with bilateral ablation of auditory cortex.

DL's for intensity are increased (5–7 dB) by combined bilateral ablations of cortex and IC and slightly more (10 dB) by BIC transection.

With the largest BIC transections and with the combined IC and auditory cortex ablations, very little information concerning auditory stimuli is reaching centers rostral to the medulla or tectum. With BIC transection, the functional state of IC is questionable since axons of most IC cells have been transected. Furthermore, input from the descending auditory system has probably been eliminated. Nevertheless, animals with bilateral transection of BIC or bilateral combined ablation of IC and auditory cortex can learn to respond to change in intensity of tones and difference limens are only moderately increased.

3. Discrimination of Changes in Frequency

a) Experimental animals. In an experiment by Goldberg and Neff (1961b), cats were trained to discriminate 800 from 1000 Hz by avoidance conditioning in a double-grill box (signals presented as in line A, Fig. 16). After transection of BIC, none of the animals retained the learned discrimination but some relearned. At the time of sacrifice, auditory cortex was explored under pentobarbital sodium anesthesia for evoked responses to click stimuli. On the basis of anatomical and electrophysiological findings, animals could be divided into three groups: (1) animals with incomplete transection of BIC; evoked responses were recorded from auditory cortex; (2) animals with complete transection of BIC; evoked responses recorded from auditory cortex; (3) animals with complete transection of BIC plus adjacent pathways; no evoked responses recorded from auditory cortex. The loss in ability to make the frequency discrimination was related to the extent of the lesion at the level of BIC. Animals in Group 1, above, were able to relearn the frequency discrimination, but three of four animals required many more trials than for initial preoperative learning. Only one of the animals in Group 2 was able to relearn the discrimination to criterion in 100 days of retraining. The two other animals in Group 2 did not reach criterion but performed well above chance level. Animals in Group 3 were unable to relearn even when the frequency difference was increased. All animals in Groups 2 and 3 were able to learn to respond to onset of a buzzer or tonal stimulus. In Section III.A, p. 336, it was reported that frequency discrimination as tested in the Goldberg and Neff (1961a) study can be relearned after ablation of auditory cortex and that, with the largest ablations (AI, AII, I–T, SII, and SS, and some undercutting of adjacent cortical areas), only slightly more postoperative than preoperative trials were required to reach the learning criterion. The significance of the finding that a severe deficit occurs in ability to learn an auditory discrimination after BIC transection whereas the same discrimination can be relearned after cortical ablation will be discussed at the end of Section III.B (p. 379).

Effects of lesions in the inferior colliculus on frequency discrimination have been investigated by Goreva and Kalinina (1967). Cats were tested in an instrumental task in which food reinforcement was given for approach to the place where a 1000 Hz tone was sounded. No reinforcement was given when another frequency was presented from the same place. The animals were tested postoperatively only; their performance was compared with that of normal animals. Animals with the most complete bilateral destruction of IC were not able to discriminate 750 or 800 Hz from 1000 Hz; discrimination of 500 from 1000 Hz was possible in these and in other animals that had less severe damage to the inferior colliculus. In all animals with lesions of the inferior colliculus, acquisition of the discrimination occurred more slowly than in normal animals. Control tests to eliminate the possibility that animals may have made use of intensity cues were not reported.

Testing frequency discrimination has not been the primary purpose of experiments in which auditory pathways of the lower brainstem have been transected. In a study principally concerned with auditory localization, cats that failed to relearn to localize after transection of the trapezoid body were able to learn frequency discrimination, 800 vs. 1200 Hz, in approximately the same number of trials required by normal animals (CASSEDAY and NEFF, 1975).

b) Man. Measures of frequency discrimination have not often been reported in clinical investigations. SALTZMAN (1952) described a case of a 29 year old woman in whom bilateral lesions had been placed in the midbrain for the relief of intractable pain. After several operations, the patient suffered tinnitus and reported that music lacked tonality, sounding like noise, and that children's voices were unintelligible, sounding shrill. Improvement in discrimination of music and speech occurred over a period of 8 months.

Instances of diplacusis have been seen in patients with lesions of the cerebellopontine angle (GOODMAN, 1957), but it is likely that such lesions affected the eighth nerve as well as centers of the medulla.

c) Summary. The cat is unable to relearn a frequency discrimination after complete transection of BIC plus adjacent pathways. In contrast, animals with bilateral ablations of all auditory areas of the cortex readily relearn frequency discrimination when trained with the same procedures. Limited evidence indicates that transection of all auditory commissures in the medulla has no effect on learning a frequency discrimination.

There are few published reports of frequency discrimination tested in patients with brainstem lesions.

4. Attention to Auditory Stimuli

A prepotent stimulus is a stimulus to which an animal responds in the presence of competing stimuli; prepotency has been used as an operational definition of attention for purposes of behavioral testing of experimental animals. JANE *et al.* (1965) tested the prepotency of auditory over visual stimuli in cats that had lesions at several levels of the auditory system. Animals were first trained in a double-grill box to respond to a compound stimulus which consisted of a bright flashing light and a 300 Hz tone, the latter set at 25 dB above the threshold of normal animals. After animals had learned to respond to the sound-light combination, unreinforced test trials were given; only the light or only the sound was presented. Normal cats responded predominantly to the auditory stimulus. This prepotency of sound over light was also seen in animals in which the trapezoid body was transected or in which the auditory cortex was ablated bilaterally prior to training and testing. Animals, in which BIC was transected bilaterally or in which the dorsal part of IC was ablated bilaterally, responded predominantly to the light rather than to the sound. Other behavioral tests indicated that the animals with dorsal IC lesions had normal absolute thresholds at 300 Hz and performed normally in a behavioral task requiring discrimination of small differences in the time of arrival of sound at the two ears.

5. Localization and Lateralization of Sound

Localization of sound in space and lateralization of sounds presented to the ears through headphones are dependent on differences in time, intensity, and frequency spectra of acoustic signals arriving at the two ears. Sounds can be localized by animals or human subjects with hearing in only one ear, but this kind of localization is based upon changes in sound as the head is moved.

In planning behavioral experiments to discover the neural processes involved in localization and lateralization, obvious steps were to examine the effects of cortical ablation (see Section III.A, 8), to transect commissural pathways of the auditory nervous system in order to determine what level or levels were important for interaction of nervous impulses from the two ears, and to measure accuracy of performance and test for acoustic cues used after unilateral lesions at different levels.

a) Experimental Animals. The experimental apparatus shown in Fig. 18 was used in many of the localization experiments reviewed below. In lateralization experiments, avoidance conditioning in a double-grill box (Fig. 15) was used. BLAU *et al.* were the first to report experiments in which localization of sound in space was tested before and after transection of commissures of the auditory system (BLAU, 1953; NEFF, 1954, 1962; NEFF and DIAMOND, 1958). From results of tests of a small number of animals (cats), the above investigators reported no change in localization ability after transection of the commissure of the inferior colliculus (CIC), but a severe deficit in one animal with complete or near complete transection of the trapezoid body (TB).

More extensive investigations of localization behavior of the cat after transection of commissures of the auditory system and other lesions of subcortical structures have been made by MOORE *et al.* (1974) and by CASSEDAY and NEFF (1975).

MOORE *et al.* found that transection of the corpus callosum (CC), CIC, or both had little or no effect on the cat's ability to localize a buzzer sound source. The test apparatus shown in Fig. 18 was used. Lesions of TB, none of which completely transected the tract, produced deficits in accuracy of localization. Deficits in performance on the localization task also occurred after lesions that involved the superior olivary complex (SOC) or the lateral lemniscus (LL).

CASSEDAY and NEFF (1975) were successful in making more complete transections of TB and the commissures of Held and of von Monakow with minimal damage to other structures of the medulla. One animal with complete transection of all three commissures (Fig. 22) was unable to localize under the usual test conditions in which the sound ceased at the time the animal was released from the starting cage. Moreover, the animal could not be trained to find the sound source when the pulsing noise signal was continued after the animal's release. Analysis of behavioral and post-mortem anatomical findings brought out an additional important result, namely, that localization was not affected for animals in which the lesion spared the anterodorsal part of TB but that there was complete loss in ability to localize and no relearning in animals in which the anterodorsal part of TB was transected. WARR (1966) has reported that, in the cat, fibers from the cochlear nucleus to the medial superior olive (MSO) travel in the anterodorsal part of TB. The behavioral results thus support results of anatomical and electrophysiological (GALAMBOS *et al.*, 1959) experiments in indicating that, in the cat at least, MSO may be the center for the initial binaural interaction that is essential for accurate localization of sound in space.

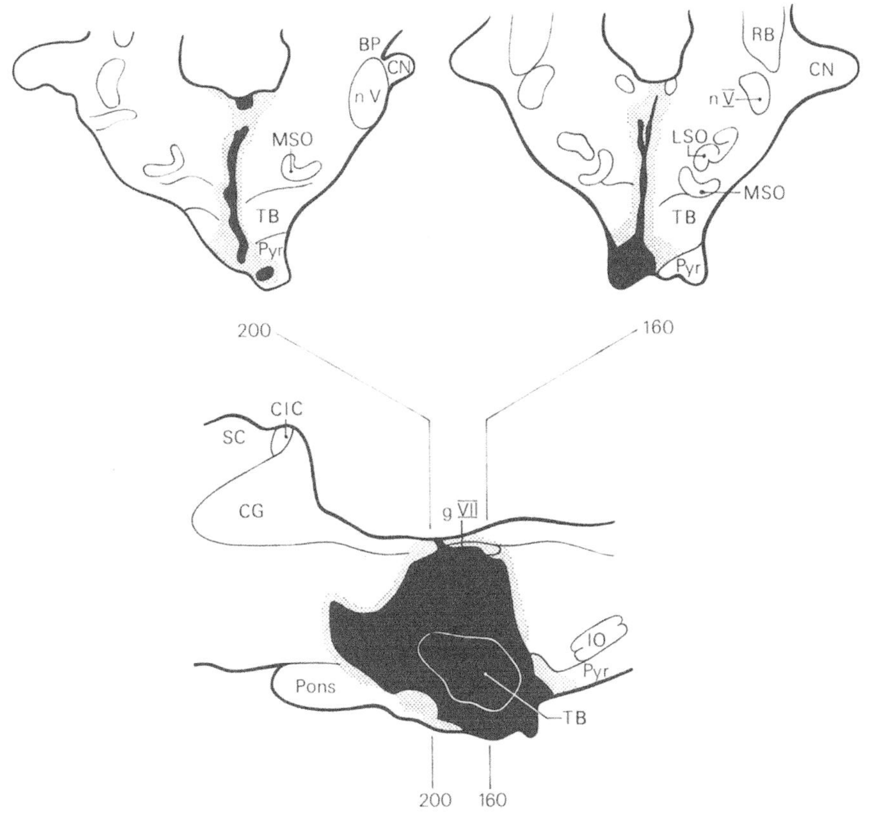

Fig. 22. Drawings and reconstructions of the lower brainstem of a cat that was unable to relearn to localize after transection of TB and other auditory commissures in the medulla. At the top of the figure are drawings of frontal sections through the area of the lesion. At the bottom of the figure a parasagittal view of the brainstem and of the lesion has been reconstructed from frontal sections of the brain. Blackened portions represent complete degeneration; stippling, incomplete degeneration. For explanation of abbreviations, see List of Abbreviations (p. 389). (From CASSEDAY and NEFF, 1975)

Behavioral evidence that the superior olive is involved in encoding binaural cues for lateralization has come from experiments by MASTERTON *et al.* (1967). Cats were trained in a double-grill box to make three successive discriminations to signals presented through headphones. The first was a discrimination of the onset of a train of clicks presented to the right ear followed by a silent intertrial interval (silence vs. R in Fig. 19). The second was a discrimination between a train of clicks in the left ear (the safe signal) and a train of clicks in the right ear (L vs. R in Fig. 19). The third was a discrimination of differences in the time of arrival (Δt) of sound at the two ears (LR vs. RL in Fig. 19). Each click in the train to the left ear was followed, 0.5 msec later, by a click to the right ear (LR) for the safe signal; the temporal order of clicks presented to the two ears was reversed (RL) for the warning signal. MASTERTON *et al.* found that normal cats, having learned the L vs. R discrimination, needed no training to discriminate correctly LR vs. RL. After lesions that completely transected TB, animals readily learned silence vs. R and L vs. R. But the same animals did not transfer to LR vs. RL and were unable to learn the LR vs. RL discrimination with a Δt of 0.5 msec after extensive retraining. In the LR vs. RL condition, time differences between the left and right clicks could be varied so that thresholds for discrimination of Δt could be established. Animals with transection of TB were able to discriminate binaural time differences when Δt was increased to 1.0 msec. The magnitude of the deficit in Δt discrimination can be seen by comparing the performances of a normal animal and of an animal with complete transection of TB. Figure 23 A shows performance on Δt threshold tests for an

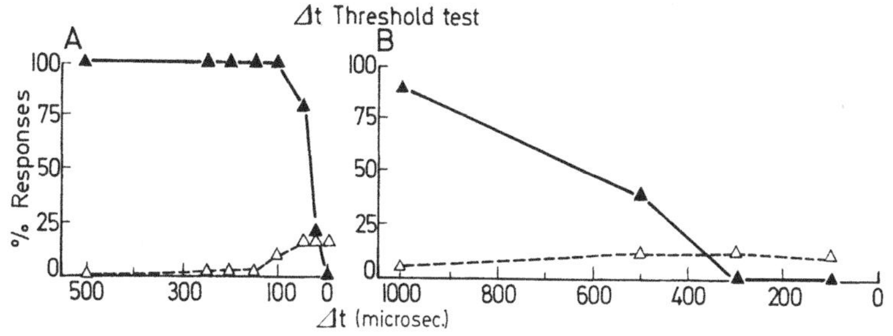

Fig. 23A and B. Performance (percent correct responses) as a function of binaural time differences (Δt) for 2 cats. Closed symbols indicate response levels during warning signal (RL); open symbols, response level during safe signal (LR). (A) Performance in discriminating Δt by an animal with an intact auditory system. (B) Performance in discriminating Δt by an animal with destruction of TB and left SOC. (From MASTERTON *et al.*, 1967)

animal in which no auditory structures were damaged. Figure 23 B gives the results of Δt threshold tests for an animal in which destruction of the trapezoid body was complete. Animals with complete TB transection were unable to discriminate binaural time differences that were within the range naturally available to the animal, as determined by the time it would take a sound in space to travel from one ear to the other. Animals with partial lesions of TB transferred from L vs. R to LR vs. RL like normal animals, but an increase in Δt threshold was found that was related to the extent of the lesion.

MASTERTON *et al.* (1969) have shown that among different species there is a negative correlation between the highest audible frequency and interaural distance. As the acoustic shadow which results in interaural intensity differences is greater for high frequencies, MASTERTON *et al.* suggested that high frequency hearing is an adaptation for localizing sound in primitive mammals that have small heads and close-set ears.

Inferences about the role of SOC in sound localization can be made from comparative studies of the subdivisions of SOC, the MSO and LSO. The relative size of these two divisions differs from species to species, and the ratio of MSO to LSO correlates highly with head size. The significance of this correlation is that animals with small heads are at a disadvantage if they have to rely on Δt, so such animals are more dependent on Δi cues. Since Δi cues are greater for high frequencies than for low frequencies, we might expect that animals with small heads would be very sensitive to high frequencies. MASTERTON and his colleagues (MASTERTON *et al.*, 1975) have tested this hypothesis in an experiment in which the performance of several species of animals was tested in a task requiring localization of pure tone pips. The species were selected for variation in the size of MSO and, in order of decreasing size of MSO, were: cat, tree shrew, rat and hedgehog. The apparatus used to test localization was similar to that shown in Fig. 18; localization tests were made with sound sources separated by 60°. Animals of each species could localize high frequencies, 8–32 kHz, but their ability to localize middle and low frequencies seemed to be related to the size of MSO. The most interesting comparison was that between hedgehogs and tree shrews: the head width of these two animals is similar; the tree shrew has a clearly distinguishable MSO, while MSO appears to be absent in the hedgehog. Hedgehogs did not perform above chance in localizing tone pips below 8000 Hz; tree shrews were above chance at all frequencies. The authors suggest that the encoding of binaural time cues in MSO is important for localization of low frequency sounds. Another possible function of MSO has been suggested by HARRISON and his colleagues (HARRISON and IRVING, 1966; IRVING and HARRISON, 1967; HARRISON and FELDMAN, 1970). On the basis of a correlation between the size of MSO and indices of visual function, it was suggested that MSO participated in some way in interactions between the auditory and visual systems (see MASTERTON and DIAMOND, 1967; HARRISON and HOWE, 1974, for further discussion of this idea).

After unilateral transection of LL or BIC, cats relearn to localize sound rather quickly although accuracy of localization is less (STROMINGER and OESTERREICH, 1970; CASSEDAY and NEFF, 1975). Animals with unilateral transection of BIC tend to make more errors in localizing sounds in the auditory field contralateral to the site of the lesion as compared to sounds in the ipsilateral field (STROMINGER and OESTERREICH, 1970). This difference between localizing in the ipsi- and contralateral fields has also been observed in animals and man after unilateral lesions of auditory cortex.

MASTERTON *et al.* (1967) compared the effects of unilateral transection of LL with the effects of TB transection in the lateralization discrimination described above. It was found that the effects of these two lesions differed in the following respect. After transection of LL, it is the failure to transfer from L vs. R to LR vs. RL that is the outstanding symptom of the deficit. The deficit after TB lesions has to do with the magnitude of Δt that is discriminable. The lack of transfer from L vs. R to LR vs. RL after LL transection was not the result of inability to discriminate binaural time differences; when an animal with LL lesion had relearned the lateralization based upon time differences, its threshold for Δt was the same as that for normal animals.

After complete bilateral transection of BIC and adjacent tracts, STROMINGER and OESTERREICH found that animals were unable to relearn to localize even when sound sources were separated by wide angles. Animals with more limited transections were able to relearn to localize sound, but lengthy retraining was necessary and the accuracy of localization was impaired (CASSEDAY and NEFF, 1975). The results of BIC transection is two-fold: first, transection of primary BIC, which presumably is that part of the path which contains IC axons projecting to GMv and ultimately to AI, considerably disturbs the capacity for sound localization. Second, the deep pathways in or adjacent to BIC, which include the central acoustic tract and which probably project to the belt cortex, are sufficient to enable an animal to learn to localize sound albeit with great difficulty and with less accuracy.

GOREVA and KALININA (1967) have reported that cats with bilateral lesions in IC require more trials to learn to localize sound than do normal animals. Anatomical details of the lesions were not given.

In the lateralization tests described earlier, MASTERTON *et al.* (1968) found that after bilateral transection of BIC, in which auditory tracts in the lateral tegmentum were spared, the LR vs. RL discrimination could not be learned, and the L vs. R discrimination was impaired. After large bilateral ablations of IC, the silence vs. R task was quickly learned, but the R vs. L task could not be learned in several times the number of trials required by normal animals. In contrast, animals with bilateral ablation of the dorsal pericentral nucleus and the dorsal portion of the central nucleus of IC performed essentially normally on L vs. R and LR vs. RL discriminations. These results suggest that: (1) auditory pathways outside of BIC, in the lateral tegmental area, are capable of maintaining simple lateralization where only one ear is stimulated (L vs. R) but not lateralization based on binaural time differences to the two ears (LR vs. RL); (2) ascending pathways from and in the region of the ventral portion of IC are essential for lateralization based on time or intensity cues; (3) all but the ventral portion of IC is unessential for lateralization.

b) Man. WALSH (1957) tested the ability of three patients with brainstem damage to localize sound in both the horizontal and vertical planes. Patient No. 1 had a brainstem thrombosis; Patient No. 2, a tumor in the upper brainstem including the posterior thalamus on one side; Patient No. 3 disseminated sclerosis in the brainstem. Patients Nos. 2 and 3 had some difficulty localizing in the horizontal plane; localization in the vertical plane was severely deficient for all three patients.

LIDEN and KORSAN-BENGTSEN (1973) reported decreased ability to localize sound for all of 12 patients who had pontine tumors in the region of the 4th ventricle. Speech discrimination and absolute thresholds for tones were reported to be normal or near normal in all of the patients. Details of the localization tests were not given.

WALSH (1957) reported thresholds for discriminating binaural time differences for three patients. Two of the patients, with midbrain damage, could discriminate Δt's of less than 200 μsec which was not greatly different from the Δt threshold for normal subjects tested in the study. A third patient for whom the locus of brainstem lesion was not accurately determined could discriminate Δt's of 2.0 and 1.0 msec, but not 200 μsec.

MATZKER (1959) reported tests of lateralization for a patient with a brainstem lesion (see p. 355 above for description of test). The patient died a few days after testing and was described as having a large glioblastoma on the left side of the brainstem. Audiograms obtained during testing showed a 20–30 dB hearing loss in both ears for frequencies up to 3000 Hz and larger losses at frequencies above 3000 Hz. The patient was unable to lateralize when presented with frequencies from 250–1000 Hz or 3000 Hz, the highest frequency tested. At 2000 Hz, lateralization judgments were made to the left, the side of the lesion, for all binaural time differences. As mentioned earlier, MATZKER reported that with cortical damage, lateralization judgments deviated toward the side contralateral to the affected cortex.

NOFFSINGER *et al.* (1972) tested lateralization based upon interaural intensity differences for 60 patients with multiple sclerosis. Twelve of the patients either did not report the apparent image of the sound to be at intracranial midline for any binaural difference in intensity or required unequal intensities binaurally in order to judge the image at the midline.

c) Summary. From behavioral experiments in which localization or lateralization of sound was tested before and after brainstem lesions in experimental animals (in most cases, the cat), the following generalizations may be made:

1. Total inability to localize sound in space is found after transection of all auditory commissures of the medulla.
2. The most critical band of commissural fibers for localization are those of the anterodorsal part of the trapezoid body. In the cat, this part of TB includes fibers passing from the cochlear nucleus to the medial superior olive.
3. Lateralization of sound is also severely affected by TB transection. Δt, the difference in time of arrival of sound at the two ears, is increased.
4. There is evidence from comparative studies that differences in ability of different species to localize frequencies in the middle and low frequency range is correlated with the size of MSO; animals with large MSO's perform better than those with small MSO's.
5. Unilateral transection of BIC or LL in the cat leads to amnesia for a learned localization discrimination but relearning is rapid. Some deficit in accuracy of localization is present after relearning.
6. Cats cannot learn to localize sound in space after complete bilateral transections of BIC and adjacent tracts or of LL and adjacent tracts.
7. Unilateral transection of LL has a detrimental effect on lateralization of sound, but after relearning, Δt is in the normal range.
8. The ventral part of IC is of critical importance for lateralization based upon either time or intensity differences of sounds presented at the two ears.
9. Evidence from clinical studies indicates that deficits in performance on localization and lateralization tests often occur after brainstem lesions. In most cases that have been examined, the exact locus of the lesion was not known.

It is of particular interest to note the differences in localization behavior between animals with one ear destroyed and animals with transection of TB.

The former can learn to localize by using cues not used when both ears are intact; by head movements it can scan the sound field and locate the position where sound is louder and, for complex sounds, the position where high frequencies are best heard. In contrast, the animal with trapezoid transection is apparently unable to find new cues by which he can become oriented in the sound field. Head movements bring conflicting cues, *e.g.* sound becoming louder in one ear and, at the same time, weaker in the opposite ear. Commissural pathways above the medulla do not provide the binaural interaction essential for localization.

6. Echolocation

A study has been reported by SUGA (1969) on changes in the bat's ability to utilize the reflected sound of its own vocalizations as cues in locating and avoiding objects in space. The ability of bats *(Myotis yumanensis)* to avoid obstacles, consisting of wire strands or rods varying in diameter from 1.9–3.7 mm, was tested after lesions of the inferior colliculus. Preoperatively, the animals had no difficulty in avoiding wires of the smallest diameter.

Transection of the commissure of IC, bilateral ablation of the dorsal half of IC, or unilateral destruction of the ventral part of IC did not affect performance on the task. Bilateral lesions that invaded the ventral region of IC resulted in a deficit in avoiding the smallest diameter obstacle; the rate of sound emission by the bat during avoidance was changed. In animals in which all or nearly all of ventral IC was destroyed bilaterally, complete inability to avoid even the largest obstacles was observed. Auditory evoked potentials recorded at the cortex were greatly diminished but were present in at least one hemisphere, indicating that not all ascending auditory connections to the forebrain were severed.

It is not clear whether the neural mechanisms of echolocation and sound localization are the same or different. It is possible that both are variants of one fundamental mechanism, but that possibility is by no means a foregone conclusion. In either case, the results of the effects of brainstem lesions are consistent with the idea of one mechanism, because lesions of the dorsal part of IC have no effect on the cat's ability to lateralize or localize sound or on the bat's ability to echolocate, while lesions that invade ventral IC have a profound effect on localization and echolocation.

7. Pattern Discrimination

Systematic studies of pattern discrimination have not been made in experimental animals with subcortical lesions. Incidental to tests of localization of sound in space, two cats that failed to localize after transection of TB in one case and all auditory commissures of the medulla in the second were able to learn a pattern discrimination (800–1200–800 Hz vs. 1200–800–1200 Hz). The number of trials for learning was in the range of that for normal animals (CASSEDAY and NEFF, 1975).

8. Summary and Conclusions (Section III. B)

Research results reviewed in Section III.A showed that many discriminations survive ablation of auditory cortex, a finding that directed attention to the role of subcortical centers and pathways. The role of the subcortical structures has been studied by the behavioral-ablation method with many of the same tests used in the experimental and clinical investigations of cortical function. In considering the effects of subcortical lesions, it makes a difference whether the capacity for the discrimination in question is one that survives or one that does not survive ablation of auditory cortex. Investigations employing discriminations that survive cortical ablations have first been directed at finding the subcortical center or level, the ablation of which eliminates the capacity for the discrimination in question. If cortical ablation impairs the capacity for a given discrimination, as in the case of sound localization, the goal of experiments involving subcortical lesions is to find out which of the several parallel pathways from cochlea to cortex are essential in order that the necessary information be encoded and transmitted to auditory cortex.

Effects of brainstem lesions on discriminations that survive cortical ablation. Learned responses to the onset of sound, to weak acoustic signals, to change in intensity, and to change in frequency, can all be relearned after complete removal of auditory cortex.

Animals are still capable of responding to the onset of sound after large bilateral lesions of LL, lesions that have retrograde effects on SOC and CN. The evidence is not conclusive that animals can be trained to acoustic signals after complete bilateral transection of LL and adjacent tracts. As noted above (p. 365) two animals in the experiments of DIXON were able to learn an avoidance response to sound only when the sound pressure level was 90–100 dB above their preoperative thresholds and, after learning, failed to respond to signals less than 80 dB above threshold. At such high sound pressure levels, the animals may have been responding to non-auditory sensory cues, *e.g.* tactual cues from movement of the ear-drum.

While response to sound is still possible after deep bilateral transections of BIC and large bilateral lesions of LL, absolute thresholds for intensity are greatly increased after these transections. Sensitivity to weak acoustic signals apparently depends on IC and the integrity of it's projections to GM.

Discrimination of differences in frequency and in intensity are both affected by bilateral transection of BIC and adjacent auditory pathways. Thresholds for discrimination of differences in intensity are increased after such lesions, and discrimination of even large differences in frequency appears to be impossible. We lack conclusive evidence to explain the difference between the effects of cortical lesions and of deep BIC transections, especially the difference in the effects on frequency discrimination. In some experiments, the cortical ablations certainly included all of core and belt auditory areas so it cannot be argued that transection of BIC is simply a more effective way of eliminating the function of auditory cortex than is cortical ablation. The difference may be that cortical ablation has no retrograde effect on IC while transection of BIC does direct damage to the axons of IC neurons. Although cutting of the axons whose cell bodies are in IC is not reflected in retrograde atrophy of the cell bodies, the cells can hardly function normally.

Effects of brainstem lesions on discriminations that are severely affected by ablation of auditory cortex. Of the auditory discriminations that are severely affected by ablation of auditory cortex there are only two for which we have a reasonable amount of information from experimental and clinical investigations; the two are pattern discrimination and localization or lateralization of sound.

Pattern discrimination can be learned after transection of all auditory commissures in the medulla. The necessary neural information for the discrimination is carried by ipsilateral pathways at least to the tectal level. It is also known that cats with one ear destroyed can learn a pattern discrimination and that it is retained or relearned quickly after ablation of auditory cortex either ipsilateral or contralateral to the intact ear (NEFF and MUNCHEL, 1975). There are alternate pathways, therefore, by which the necessary information for pattern discrimination can be transmitted from cochlea to cortex. Interaction of nerve impulses from the two ears is not necessary and there is no evidence that brain stem centers rostral to the primary cochlear nucleus play a critical role in pattern discrimination other than as relay centers. Information is lacking as to what parts of CN may be involved in encoding and as to what parts of ascending auditory pathways are critical.

When we consider localization or lateralization of sound, we have a quite different picture. Interaction of nerve impulses from the two ears is the first requirement for localization or lateralization based upon binaural cues. We know from psychophysical studies of human subjects that the most accurate localization or lateralization only takes place when binaural cues are available. While sounds can be localized when only one ear receives the signals, the discrimination is highly inaccurate if the head is held in a fixed position, and, although accuracy improves somewhat with head movements to scan the soundfield for differences in intensity and in frequency spectra (complex sounds), it is a completely different task for the listener than that when binaural cues are present.

Experimental animals, having learned to localize sound with both ears intact, are at a complete loss when one ear is destroyed; relearning is slow. The auditory cortex contralateral to the intact ear is essential for relearning (see p. 352).

In contrast to animals with one ear destroyed, animals with transection of the trapezoid body fail to relearn to localize sound. It may be that the failure is due to transection of neural tracts that by-pass SOC and go from CN to contralateral IC. Another possible explanation is that the animal with transection of TB receives conflicting cues when it scans the sound field by head movements. Whatever the explanation, a comparison of the localization or lateralization of the one-eared animal and the animal with TB transection is helpful in understanding the relationship between auditory cortex and auditory centers of the brainstem. The cortex appears to be necessary to achieve an integrated picture of space based upon sound cues; lower brainstem centers are necessary for analysis of the neural correlates of the differences in time and intensity of sounds stimulating the two ears.

C. Descending Auditory Pathways

In describing the effects of lesions in the auditory pathway, we have not discussed possible differences between the influence of descending and ascending systems. Ablation of auditory centers and transection of auditory pathways at all levels of the auditory nervous system will probably involve both the ascending and descending components of the system. WHITFIELD (1967) has pointed out that this possibility has not been taken into account in studies of auditory discrimination. Except at the level of the medulla, techniques have not been developed that allow the two systems to be studied separately in behavioral experiments.

At the medulla, portions of the efferent system diverge from the ascending system; it is possible to transect the olivocochlear bundle (OCB) at the floor of the fourth ventricle without causing damage to the afferent auditory system.

1. Absolute Thresholds

Results of behavioral experiments which have measured the effects of transection of OCB on absolute thresholds have been uniformly negative. GALAMBOS (1960) reported no change in absolute thresholds for frequencies of 300, 1500, and 5000 Hz for two cats tested before and after transection of OCB. This finding was extended by TRAHIOTIS and ELLIOTT (1970) who found no loss in absolute thresholds of cats tested at octave intervals from 250 Hz to 15 kHz. Both of these studies used avoidance techniques in a double-grill box. IGARASHI *et al.* (1972) have also confirmed the results of the earlier experiments.

In the study by TRAHIOTIS and ELLIOTT measures of temporary threshold shift after exposure to intense sound were made for the animals with OCB transection and for a control group of normal animals; no differences were found.

2. Discrimination of Differences in Frequency or in Intensity

GALAMBOS (1960) reported that DL's for frequency and intensity in one cat tested after transection of OCB were in the range of those obtained for another animal that was subjected to a sham operation. In their experiments, CASSEDAY and NEFF (1975) transected OCB as well as the commissures in the medulla of the ascending auditory system in one animal; the animal was able to learn to discriminate 800 vs. 1200 Hz. DL's for frequency were not measured. CAPPS and ADES (1968) trained four squirrel monkeys to make frequency discriminations for food reward in a Wisconsin General Test Apparatus modified for use in auditory experiments. Each of two frequencies was presented through one of two loudspeakers on any given trial. The animal's task was to select the food box in front of the loudspeaker that produced the tone of lower frequency. After transection of OCB, frequency difference thresholds increased from preoperative levels by about 450 Hz at 1000 Hz and 600 Hz at 4000 Hz. The preoperative thresholds for discriminating frequency differences were large, 150 Hz at 1000 Hz and slightly over 300 Hz at 4000 Hz. The authors interpreted the large DL's as due to the unreliable performance of squirrel monkeys when they are required to make difficult discriminations.

3. Localization of Sound in Space

While localization tests were not systematically explored in the study by CAPPS and ADES, the initial training was essentially a localization task (discrim inating from which of the two loudspeakers a tone was presented). Postoperatively, none of the animals showed evidence of having any difficulty in making this discrimination.

4. Masking

Discrimination of signals masked by noise after disruption of the efferent pathways in the medulla have been investigated in four experiments. In three of these the OCB was transected; in the fourth, the efferent system was blocked at the cochlear nucleus by pharmacological methods.

DEWSON (1968) used operant conditioning techniques to train monkeys *(Macaca mulatta)* to discriminate between the human speech sounds [*i*] and [*u*] in the presence of masking noise with a low pass filter at 2400 Hz. The intensity of the noise was varied according to the animal's performance from trial to trial. After transection of the crossed OCB, all three monkeys showed deficits in discriminating the vowel sounds at low signal to noise ratios. Figure 24 shows the

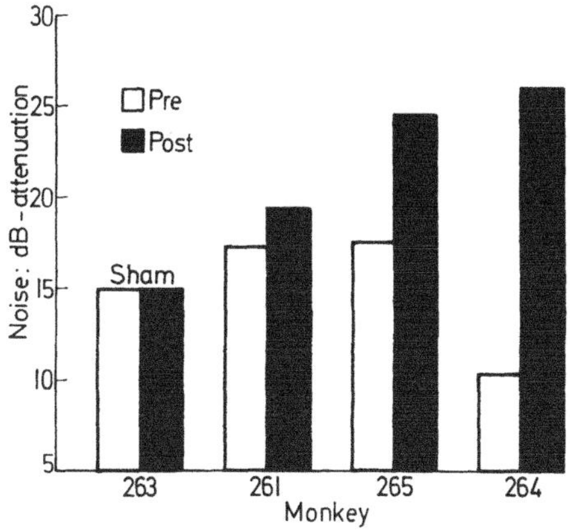

Fig. 24. Discrimination of speech sounds in noise before (open bars) and after (solid bars) transection of OCB. The most intense noise is at 5 dB attenuation. The height of the bars represents the mean level of attenuation in order for performance to be at a specified level (63.5%) above chance. Preoperative means of the noise levels were significantly different from postoperative means ($p < 0.002$, 2-tailed *t*-test) for the animals (**261**, **265**, **264**) with OCB transection. Monkey 263 had a sham operation. (From DEWSON, 1968)

amount by which the noise had to be attenuated for each animal's performance to be at a specified point above chance on its psychometric function. For the three animals that had transection of OCB, this performance level was achieved only with greater attenuation of the noise (*i.e.* higher signal to noise ratios) than was necessary preoperatively. It should be noted that two of the three monkeys with OCB transection had previously undergone bilateral ablation of superior temporal cortex or inferotemporal cortex. The animal with intact cortex performed at the lowest signal to noise ratio preoperatively, and the highest postoperatively. DEWSON suggested that the deficits might have been greater for two of the animals had they not received cortical ablation prior to testing.

TRAHIOTIS and ELLIOTT (1970) measured the amount of masking by noise of tonal signals, 500, 1000, and 2000 Hz, after transection of OCB. Postoperatively, performance of three animals (cats) indicated that slightly greater masking was found at 1000 and 2000 Hz than was found for a control group of three animals that underwent a sham operation. The differences were not statistically significant. TRAHIOTIS and ELLIOTT speculated that if transection of the olivocochlear bundle has an effect on masking, it might be either the result of an increase in the critical band or a degradation of the tonal stimulus such that it becomes "noisier". Measures of temporary threshold shift after exposure to intense sound did not reveal differences between animals in the experimental and control groups.

IGARASHI *et al.* (1972) measured amount of masking in four cats after transection of OCB. Avoidance conditioning was used in a rotating cage. No masking differences were found in comparing animals with OCB transection and sham-operated animals that served as controls.

PICKLES and COMIS (1973) have investigated the role of another component of the efferent pathway. They took advantage of the fact that the pathway, which travels from the superior olivary complex to the cochlear nucleus via the intermediate acoustic stria has cholinergic properties, whereas the afferent system does not (COMIS and WHITFIELD, 1968). The efferent system was blocked by applying atropine in awake animals to the surface of the cochlear nucleus via an implanted cannula. Immediately before and a few minutes after injections, thresholds for 1000 Hz tone pips were measured in the quiet and in the presence

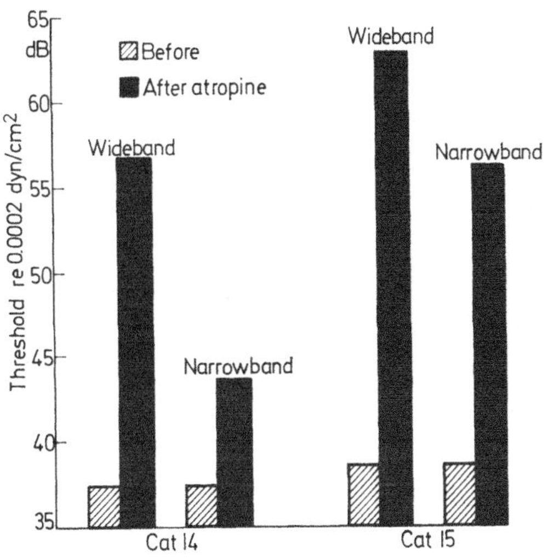

Fig. 25. Changes in thresholds for detecting a pure tone, masked by wide-band or narrow-band noise, after injection of atropine. The bandwidth of the narrow-band noise was 200 Hz, centered on the tonal stimulus. Before injection, intensities of the masking noises were adjusted so that thresholds for the tone were the same in both narrow-band and broad-band masking noises. The mean difference between the deficits with wide-band vs. those with narrow-band masking noises, was significant ($p < 0.01$, 2-tailed t-test). (From PICKLES and COMIS, 1973)

of several levels of masking noise. Avoidance conditioning in a double grill box was used for training and testing. After injection of atropine, absolute thresholds were slightly increased, but a greater increase was found for the masked thresholds. Control injections of sodium pentobarbital indicated that atropine was not simply acting as a general depressant. Pickles and Comis compared masked thresholds for a wide-band and a narrow-band noise; the latter had a band width of 200 Hz centered at 1000 Hz. Before injection of atropine, masked thresholds were equal for the wide and narrow-bands of noise (Fig. 25). After injection of atropine, the masked thresholds increased under both masking conditions, but the masked threshold for wide-band noise was significantly greater than that for narrow-band noise (Fig. 25). From the results, Pickles and Comis concluded that blocking of the cochlear nucleus component of the efferent system interferred with the mechanism of masking. Specifically, it was suggested that the width of the critical band was increased so that frequencies further from the tonal stimulus contributed to masking.

5. Summary

Behavioral effects of transecting or blocking the efferent auditory system have been investigated only at the level of the medulla. The efferent pathways have been transected at higher levels of the brain stem and their function has undoubtedly been affected by cortical ablations. Except at the level of the medulla, no methods have been devised by which the role of afferent and efferent systems can be separated in behavioral experiments.

Results to date have not given any clear picture of the function of the efferent system at any level.

IV. Discussion

The modern era of ablation-behavior research has been dominated by a desire to take advantage of improved methods in neurophysiology, neuroanatomy, and behavioral testing. Most experiments were undertaken to answer a particular question rather than to test some general theory about brain function. To find an overall view of the relation of brain and behavior it is necessary to go back to the period before Lashley's important studies of the 1920's and 1930's. For the purpose of our discussion, it is worthwhile to recall the accepted views of brain function and brain evolution 50 years ago and in particular the ideas of Sherrington and Pavlov. Then we can assess how the older traditional views must be modified in the face of the results accumulated since 1940.

A. The Relation of Cortex to Subcortex

The traditional view of the sensory systems in mammals sharply distinguished between the function of sensory cortex and the function of the rest of the sensory path from end-organ to cortex. For example, Sherrington felt that a decorticate dog is an automaton without intent or conscious sensation. It follows from

SHERRINGTON's general conclusions that the auditory centers of the brainstem might serve as reflex centers in a manner analogous to the spinal cord; thus, he might predict that a dog without auditory cortex would prick up its ears in response to a sound just as a spinal dog withdraws its leg in response to a noxious stimulus presented to the leg. SHERRINGTON was aware of the contention that it is difficult or impossible to test for conscious sensation in animals other than man, but he argued that if a brain damaged dog showed appropriate behavior with normal affect and obvious intent – in other words, if it did not act like an automaton – it is likely to have sensation (SHERRINGTON, 1906).

While PAVLOV's position appears to be antagonistic to SHERRINGTON in regard to conscious sensation, the views of the two are similar in that both considered the cortex qualitatively different from subcortical centers. PAVLOV held that the purpose of physiology is best served by considering all behavior as reflexive, but he distinguished between unconditioned and conditioned reflexes. The pathway for conditioning, in his opinion, required that the "primary" sensory areas of the cortex and cortico-cortical fibers from the sensory areas to the association and motor areas be intact. Thus, both PAVLOV and SHERRINGTON regarded the brainstem as an extension of the spinal cord in which cranial nerves "replaced", so to speak, spinal nerves to provide access to the periphery. Only when the nerve impulse attained the level of the cortex were the functions studied by psychology – learning and sensation – achieved. From this historical perspective, a major contribution of the ablation-behavior method in the study of hearing was the discovery that a cat without auditory cortex can *learn* to sit quietly at the presentation of one sound, *e.g.*, a tone of a given frequency, and can *learn* to move in response to a second sound, a tone of another frequency. This behavior would seem to require conscious sensation, as SHERRINGTON defined the term, but, in any event, the behavior is clearly learned and does not resemble the reflexes of a spinal dog.

The failure to find a deficit in frequency or intensity discrimination cannot be attributed to sparing a part of the auditory cortex. The lesion made in some animals by GOLDBERG and NEFF (1961a) not only removed all of the core and belt auditory cortex but undercut or removed the caudal two-thirds of the rest of the neocortex. In view of this evidence, it is difficult to interpret the failure of cats to respond to a change in frequency after transections of BIC and adjacent pathways. We have argued that the transection of BIC injures the neurons of IC whose axons have been cut even though this injury does not seem to be reflected in conspicuous retrograde degeneration. Whatever interpretation is given to this difference between the effects of cortical and deep BIC lesions, it is important that after either lesion all of the animals can *learn* to respond to onset of sound. Therefore, we can conclude that subcortical centers are not merely reflex centers or relay stations in a path to cortex; they can serve in learned auditory discrimination. It must be conceded that present evidence does not rule out the possibility that nerve impulses elicited by auditory stimuli may reach sensory-motor cortex via pathways outside LL and BIC, perhaps by way of collaterals to the reticular formation. As noted earlier, to make larger cortical lesions that include cortex beyond that ablated in past experiments becomes meaningless as such ablations would interfere with normal locomotion and responses to non-auditory cues.

That different pathways from cochlea to cortex can serve for different auditory discriminations is clearly illustrated by comparing the effects of TB transection on localization of sound in space and on frequency, intensity and pattern discrimination. Localization is completely disrupted; the other discriminations are not.

Finally, it should be emphasized that in a discrimination such as localization of sound in space, auditory cortex is essential in order that neural events set off by binaural interaction in lower centers can be used to guide behavior after cessation of peripheral stimuli. It appears likely that the auditory cortex is necessary for the perception of objects in space based upon auditory cues and the perception of the organism's position in relation to external objects.

B. Evolution of Sensory and Association Cortex

Returning again to the period before 1930, the traditional view of neocortex sharply distinguished between the structure, function and evolution of sensory and of association cortex. Sensory cortex was defined as cortex receiving sensory elicited neural impulses from nuclei extrinsic to the cortex, and in particular, from the relay nuclei of the thalamus. Association cortex was defined as those areas receiving sensory impulses only from other cortical areas, for example, from sensory cortex. While it is beyond our purpose to discuss the rationale underlying the distinction between simple sensation and complex perception, the function of sensory cortex was identified with "simple" discriminatory dimensions such as the wave length of light or sound and the function of association cortex was thought to be more complex and to depend on the function of sensory cortex. The function of sensory cortex was regarded as essentially the same in all mammals. Species comparisons further suggested that the relative size of the sensory cortex was stable in mammalian evolution. On the other hand, species differences in the extent and development of association cortex were conspicuous, reflecting the role of association cortex in the "higher" processes of learning, memory and thinking. Indeed, the phyletic level of the species could be assessed by the ratio of association to sensory cortex. In lower mammals, the ratio was thought to be very low while in higher mammals such as primates, the ratio was thought to be very high.

Recent studies require a revision in many of these old ideas. The primary auditory cortex of the cat is surrounded by a belt of cortex, but that belt cannot be called association cortex in the sense of receiving intrinsic fibers only. As we have shown, the belt receives fibers from a group of auditory nuclei surrounding GMv. Furthermore, there is evidence that the pathway (or pathways) to the belt auditory cortex is parallel to primary or core path from IC. For the most part cortico-cortical projections between core and belt are reciprocal so on these grounds too, there is no special reason to define one part as sensory and another as association.

There is considerable evidence that similar general principles apply to the other neocortical sensory systems. We know that there is a visual tecto-pulvinar path (GRAYBIEL, 1972a, b; PALMER *et al.*, 1972; ROSENQUIST *et al.*, 1974) parallel to the primary geniculo-striate system. So the visual cortex too can be regarded as a core and belt. In any case, the visual cortex adjoins the auditory cortex, since the so called CLARE-BISHOP area lies adjacent to SF, with the result that auditory cortex is almost surrounded by sensory cortex – somatic at one border, olfactory and visual at the other borders. In the cat, there is little or no association cortex in the traditional sense of the term.

It is appropriate to ask whether the functional relation between core and belt cortex bears any resemblance to the expectations of the traditional theory of a sensory cortex and an association surround. There is no evidence that the belt area serves more complex functions than the core. On the other hand, experimental results show that deficits in discrimination are usually more severe when both core and belt are removed rather than either area alone. This is to be expected if the core and belt represent targets of parallel paths; by the same token the behavioral results contradict the idea that the belt is dependent on the core. Of course, the traditional idea was that association cortex entirely depended on sensory cortex for sensory input in which case the deficit from removal of the sensory area could not be exacerbated by removal of the dependent association cortex.

There is too little evidence available to speculate on whether or not a dissociation of deficits might be achieved by separate removal of belt and core. Removal of the belt alone seems to have a profound effect on pattern discrimination (GOLDBERG *et al.*, 1957) and removal of AI alone appears to have an effect in certain localization situations (WHITFIELD and DIAMOND, 1975).

C. Relation between Cat and Other Species

The failure to identify extensive association areas in the cat casts a doubt about the traditional view that sensory areas are stable in evolution. The cortex of the cat represents a significant advance over the level represented by basal insectivores. Therefore, it would appear that sensory cortex itself expanded in the evolution of carnivores. Clearly, significant species differences cannot be restricted just to differences in the expansion of association cortex. We can ask first whether some function of auditory cortex can be identified that is *common* to diverse mammals. If so, this function might be attributed to the common ancestor. Such a common function might provide clues to the reason why cortex evolved in early mammals. Then, we can ask what is the functional significance of the development of auditory cortex in carnivores. Finally, we can ask did auditory cortex expand similarly in other lineages, notably the primates and if so, were the same or different behaviors achieved by the development.

There is evidence that in widely different mammals, such as insectivore and primate, removal of the auditory cortex affects sound localization (RAVIZZA and DIAMOND, 1974). It is tempting to speculate that the ability to localize brief sounds in space, an ability that is so clearly adaptive for both prey and predator, was improved by the origin of neocortex in early mammals and that this structure-function relation was maintained in all mammalian lines of descent. We can speculate further that one purpose served by the development of auditory cortex was improvement not just in localization but in the perception of auditory space and in the locomotion appropriate to the spatial location of sounds. As we have pointed out, the simple laboratory measure of sound localization, such as requiring the animal to approach the source of sound after the sound ceases, is quite complex and can be analyzed into several components. The integration of sensory and motor components is a subtle achievement and may even contribute to the animal's concept of a three dimensional environment. In this light, it is not hard to picture selective pressures that would result in the further development of cortex.

Whether development of structure and function has been the same in different lines – say primates and carnivores – cannot be assessed. We can say that the anatomical organization of the auditory thalamus appears to have undergone parallel or even convergent evolution in carnivores and primates. In both groups the cytoarchitectonic differences between dorsal, magnocellular, and ventral divisions of GM has widened, and Po has expanded and subdivided. It is not clear whether the various thalamic divisions project to homologous cortical subdivisions in both groups.

D. Conclusions

In conclusion, the present era of behavioral study of the auditory nervous system was initiated by a series of experiments in which auditory discriminations were measured before and after ablation of auditory cortex; the results of the first experiments contradicted the expectations of the traditional view that auditory sensory cortex was necessary for learned auditory discriminations. The next step was an attempt to find a deficit after auditory cortex removal; this search was not guided by any general theory of the function and evolution of neocortex, and preconceptions about the relationship between sensory cortex and association cortex were abandoned. Now we can say that the deficits uncovered make sense. The capacity to localize sounds in space has adaptive value for prey and predator alike. We can imagine that auditory neocortex conferred upon the mammalian ancestor the advantage of accurate sound localization and that this function evolved in several mammalian lineages. In an advanced mammal such as the cat, the locus of a source of a brief sound is not only identified quickly, but at the same time the animal has the ability to move toward or away from the locus, and it perceives the spatial relation between the source and other objects including the position of its own body. Temporal pattern discrimination likewise has adaptive value. Complex sound information, except for locus, is in most instances conveyed over time and takes the form of a temporal pattern. The elements of the pattern typically have different frequency spectra. In fact, the simplest speech sounds can be regarded as temporal patterns. There is little doubt that as auditory cortex evolved, so did the ability to interpret the significance of temporal patterns, thus facilitating communication between members of the same species and recognition of sounds made by members of other species.

Acknowledgement

The authors wish to thank Claudia Lemon, Nada Staddon, Gail Stillwagon, and Sandra Weaver for their assistance in preparation of the manuscript and illustrations for this chapter.

List of Abbreviations

AES	anterior ectosylvian gyrus or sulcus
AI, AII	auditory areas of cortex in the cat
BC	brachium conjunctivum
BIC	brachium of inferior colliculus
BP	brachium pontis
CC	corpus callosum
CG	central gray
CIC	commissure of inferior colliculus
Cl	claustrum
CN	cochlear nucleus
DL	difference limen or difference threshold
Ep	posterior ectosylvian gyrus; auditory area in the cat
GL	lateral geniculate body
GM	medial geniculate body
GMc	caudal division of medial geniculate body
GMd	dorsal division of medial geniculate body
GMm	medial division of medial geniculate body
GMmc	magnocellular division of medial geniculate body
GMp	principal division of medial geniculate body
GMv	ventral division of medial geniculate body
g VII	genu of seventh cranial nerve
I	insular cortex; auditory area in cat
IC	inferior colliculus
ICcen	central nucleus of inferior colliculus
ICperi	pericentral nucleus of inferior colliculus
ICex	external nucleus of inferior colliculus
IO	inferior olive
I—T	insular and temporal cortical areas; auditory areas in the cat
LL	lateral lemniscus
LP	lateral posterior nucleus of thalamus
MES	middle ectosylvian gyrus
MSO	medial superior olivary nucleus
LSO	lateral superior olivary nucleus
nV	fifth cranial nerve
OCB	olivocochlear bundle
PES	posterior ectosylvian gyrus or sulcus
Po	posterior group of nuclei of thalamus
POi	intermediate Po
Pul	pulvinar
Pyr	pyramidal tract
RB	restiform body
rh	rhinal area
rhin f	rhinal fissure
SC	superior colliculus
SF	sylvian fringe; auditory area in the cat
SOC, SO	superior olivary complex
SS	suprasylvian gyrus or sulcus
ST	superior temporal sulcus
SYL	sylvian gyrus or sulcus
SII	somatic area II of cerebral cortex
T	temporal cortex; auditory area in the cat
TB	trapezoid body
Un	uncus
17	area 17 of cerebral cortex
Δi	binaural difference in intensity of a sound
Δt	binaural difference in time of arrival of sound

References

ADES, H. W.: Connections of the medial geniculate body in the cat. Arch. Neurol. Psychiat. **45**, 138—144 (1941).

AKERT, K., WOOLSEY, C. N., DIAMOND, I. T., NEFF, W. D.: The cortical projection area of the posterior pole of the medial geniculate body in *Macaca mulatta*. Anat. Rec. **134**, 242 (1959).

ALLEN, W. F.: A comparison of the dogs ability to respond correctly to conditioned differential tests with auditory, olfactory, general cutaneous, and optic analysers. J. gen. Psychol. **31**, 37—52 (1944).

ALLEN, W. F.: Effect of destroying three localized cerebral cortical areas for sound on correct conditioned differential responses of the dog's foreleg. Amer. J. Physiol. **144**, 415—428 (1945).

ARNOTT, G. P.: Impairment following ablation of the primary and secondary areas of the auditory cortex. Unpublished Ph. D. dissertation, Univ. of Chicago (1953).

AXELROD, S., DIAMOND, I. T.: Effects of auditory cortex ablations on ability to discriminate between stimuli presented to the two ears. J. comp. physiol. Psychol. **59**, 79—89 (1965).

BARD, P., RIOCH, D. McK.: Study of four cats deprived of neocortex and additional portions of the forebrain. Johns Hopkins Hosp. Bull. **60**, 73—147 (1937).

BARU, A. V.: Role of the temporal cortex in the detection of acoustic stimuli of different duration. Zh. Vyssh. Nerv. Deyat. **16**, 655—665 (1966).

BARU, A. V.: Differential thresholds of frequency in relation to duration of tonal series in animals after ablation of the auditory cortex. In: Mechanisms of hearing. Leningrad: Nauka 1967.

BARU, A. V., KARASEVA, T. A.: The brain and hearing (translated by B. Haigh). New York: Consultants Bureau 1972.

BENITEZ, J. T., CORVERA, J., NOVOA, V.: Auditory manifestations of cochlear and retrocochlear lesions in humans. Ann. Otol. Rhinol. Laryngol. **75**, 149—161 (1966).

BERLIN, C. I., LOWE, S. S.: Temporal and dichotic factors in central auditory testing. In: KATZ, J. (Ed.): Handbook of clinical audiology. Baltimore: Williams and Wilkins 1972.

BERLIN, C. I., MCNEIL, M. R.: Dichotic listening. In: LASS, N. J. (Ed.): Contemporary issues in experimental phonetics. Springfield, Ill.: Charles C. Thomas 1975.

BLAU, W.: Localization of sound in space: the role of midbrain commissural pathways. Master's thesis, Univ. of Chicago (1953).

BOSATRA, A. B., VERONESE, A.: Clinical observations on some aspects of binaural hearing under physiological and pathological conditions. Intern. Audiol. **4**,49—51 (1965).

BROADBENT, D. E.: Successive responses to simultaneous stimuli. Quart. J. exp. Psychol. 8, 145—162 (1956).

BROMILEY, R. B.: Conditioned responses in a dog after removal of neocortex. J. comp. physiol. Psychol. **41**, 102—110 (1948).

BROWN, S. M., SCHÄFER, E. A.: An investigation into the function of the occipital and temporal lobes of the monkey's brain. Phil. Trans. roy. Soc. **179**B, 303—327 (1888).

BROWN, T. S., GEDVILAS, G. R., MARKO, L. A.: Effect of auditory cortical lesions on a test of frequency discrimination in the cat. Proc. Amer. Psychol. Ass. **1967**, 101—102.

BUTLER, R. A., NEFF, W. D.: Role of the auditory cortex in the discrimination of changes in frequency. Amer. Psychol. **5**, 474 (1950).

BUTLER, R. A., DIAMOND, I. T., NEFF, W. D.: Role of auditory cortex in discrimination of changes in frequency. J. Neurophysiol. **20**, 108—120 (1957).

CAPPS, J. J., ADES, H. W.: Auditory frequency discrimination after transection of the olivocochlear bundle in squirrel monkeys. Exp. Neurol. **21**, 147—158 (1968).

CARHART, R.: Clinical determination of abnormal auditory adaptation. Arch. Otolaryngol. **65**, 32—39 (1957).

CARHART, R.: Probable mechanisms underlying kernicteric hearing loss. Acta Oto-laryngol. Suppl. **221**, 41 (1967).

CASSEDAY, J. H.: Auditory localization: the role of the brainstem auditory pathways of the cat. Unpublished doctoral dissertation, 71 pp. Indiana University (1970).

CASSEDAY, J. H., DIAMOND, I. T., HARTING, J. K.: Unpublished observations (1975).

CASSEDAY, J. H., NEFF, W. D.: Auditory localization: role of auditory pathways in brainstem of the cat. J. Neurophysiol. **38**, 842—858 (1975).

CELESIA, G. G., PULETTI, F.: Auditory cortical areas of man. Neurology **19**, 211—220 (1969).

CHAN, J.: Unpublished results. Center for Neural Sciences. Indiana University (1975).

CHERRY, E. C.: Some experiments upon the recognition of speech, with one and with two ears. J. acoust. Soc. Amer. **25**, 975—979 (1953).

CHORAZYNA, H.: Investigation of recent memory of acoustic stimuli in normal dogs. Bull. Acad. Polon. Sci. Ser. Sci. Biol. **7**, 119—121 (1959).

CHORAZYNA, H., STEPIEN, L.: Impairment of auditory recent memory produced by cortical lesions in dogs. Acta Biol. Exp. **21**, 177—187 (1961a).

CHORAZYNA, H., STEPIEN, L.: Impairment of recent memory of the auditory stimuli after bilateral ablations of sylvian gyri in dogs. Bull. Acad. Polon. Sci., Ser. Sci. biol. **9**, 117—120 (1961b).

CHORAZYNA, H., STEPIEN, L.: Effect of bilateral sylvian gyrus ablations on auditory conditioning in dogs. Bull. Acad. Polon. Sci. Ser. Sci. Biol. **11**, 43—45 (1963).

COLAVITA, F. B.: Auditory cortical lesions and visual pattern discrimination in cat. Brain Res. **39**, 437—447 (1972).

COLAVITA, F. B.: Insular-temporal lesions and vibrotactile temporal pattern discrimination in cats. Psychol. Behav. **12**, 215—218 (1974).

COLAVITA, F. B., SZELIGO, F. V., ZIMMER, S. D.: Temporal pattern discrimination in cats with insular-temporal lesions. Brain Res. **79**, 153—156 (1974).

COMIS, S. D., WHITFIELD, I. C.: Influence of centrifugal pathways on unit activity in the cochlear nucleus. J. Neurophysiol. **31**, 62—68 (1968).

CORNWELL, P.: Loss of auditory pattern discrimination following insular-temporal lesions in cats. J. comp. physiol. Psychol. **63**, 165—168 (1967).

COWAN, W. M.: Anterograde and retrograde transneuronal degeneration in the central and peripheral nervous system. In: NAUTA, W.J.H., EBBESSON, S.O.E. (Eds.): Contemporary research methods in neuroanatomy. New York-Heidelberg-Berlin: Springer 1970.

CRANFORD, J.: Personal communication (1975).

CRANFORD, J., RAVIZZA, R., DIAMOND, I. T., WHITFIELD, I. C.: Unilateral ablation of the auditory cortex in the cat impairs complex sound localization. Science **172**, 286—288 (1971).

CULLER, E. A., METTLER, F. A.: Conditioned behavior in a decorticate dog. J. comp. Psychol. 18, 291—303 (1934).

DAMASIO, A. R., LIMA, A., DAMASIO, H.: Nervous function after right hemispherectomy. Neurology **25**, 89—93 (1975).

DANDY, W. E.: Effects on hearing after subtotal section of the cochlear branch of the auditory nerve. Bull. Johns Hopkins Hosp. **55**, 240—243 (1934).

DEWSON, J. H.: Speech sound discrimination by cats. Science **144**, 555—556 (1964).

DEWSON, J. H.: Efferent olivocochlear bundle: some relationships to stimulus discrimination in noise. J. Neurophysiol. **31**, 122—130 (1968).

DEWSON, J. H. III, PRIBRAM, K. H., LYNCH, J. C.: Effects of ablations of temporal cortex upon speech sound discrimination in the monkey. Exp. Neurol. **24**, 579—591 (1969).

DIAMOND, I. T.: The sensory neocortex. In: NEFF, W. D. (Ed.): Contributions to sensory physiology, Vol. 2, pp. 51—100. New York: Academic Press 1967.

DIAMOND, I. T., CHOW, K. L.: Biological psychology. In: KOCH, S. (Ed.): Psychology: a study of a science, Study II, Vol. 4. New York: McGraw-Hill 1962.

Diamond, I. T., Chow, K. L., Neff, W. D.: Degeneration of caudal medial geniculate body following cortical lesion ventral to auditory area II in cat. J. comp. Neurol. **109**, 349—362 (1958).

Diamond, I. T., Goldberg, J. M., Neff, W. D.: Tonal discrimination after ablation of auditory cortex. J. Neurophysiol. **25**, 223—235 (1962).

Diamond, I. T., Jones, E. G., Powell, T. P. S.: The projection of the auditory cortex upon the diencephalon and brainstem in the cat. Brain Res. **15**, 305—340 (1969).

Diamond, I. T., Neff, W. D.: Ablation of temporal cortex and discrimination of auditory patterns. J. Neurophysiol. **20**, 300—315 (1957).

Dix, M. R., Hood, J. D.: Modern developments in pure tone audiometry and their application to the clinical diagnosis of end-organ deafness. J. Laryngol. Otol. **67**, 343—357 (1953).

Dix, M. R., Hood, J. D.: Symmetrical hearing loss in brainstem lesions. Acta Otolaryngol. **75**, 165—177 (1973).

Dixon, N. T.: Tonotopic organization of the inferior colliculus and the effects of lateral lemniscus lesions on hearing sensitivity in the cat. Doctoral dissertation, Indiana University (1973).

Efron, R.: Temporal perception, aphasia and *deja vu*. Brain **86**, 403—424 (1963).

Ehmer, R. H.: The role of the auditory areas of the cortex in a conditioned frequency discrimination. Doctoral dissertation, University of Rochester, New York (1953).

Elliott, D. N.: Personal communication.

Elliott, D. N., Stein, L., Harrison, M. J.: Determination of absolute intensity thresholds and frequency-difference thresholds in cats. J. acoust. Soc. Amer. **32**, 380—384 (1960).

Elliot Smith, G.: Some problems relating to the evolution of the brain. Lancet **1910** I, 1—6, 147—153, 221—227.

Erulkar, S. D., Butler, R. A., Gerstein, G. L.: Excitation and inhibition in cochlear nucleus. II. Frequency modulated tones. J. Neurophysiol. **31**, 537—548 (1968a).

Erulkar, S. D., Nelson, P. G., Bryan, J. S.: Experimental and theoretical approaches to neural processing in the central auditory pathway. In: Neff, W. D. (Ed.): Contributions to sensory physiology. New York: Academic Press 1968b.

Evans, E. F., Nelson, R. A.: Behavior of neurons in cochlear nucleus under steady and modulated tonal stimulations. Fed. Proc. **25**, 463 (1966).

Evarts, E. V.: Effect of auditory cortex ablation on frequency discrimination in monkey. J. Neurophysiol. **15**, 443—448 (1952).

Faglioni, P., Spinnler, H., Vignolo, L.: Contrasting behavior of right and left hemisphere-damaged patients on a discriminative and a semantic task of auditory recognition. Cortex **5**, 366—389 (1969).

Ferrier, D.: Functions of the Brain. New York: G. P. Putnam's Sons 1886.

Ferrier, D.: Schäfer on the temporal and occipital lobes. Brain **11**, 7—30 (1889).

Ferrier, D.: The croonian lectures on cerebral localization. London: Smith, Elder, and Co. 1890.

Fletcher, H.: Auditory patterns. Rev. Mod. Phys. **12**, 47—65 (1940).

Flower, R. M., Viehweg, R.: A review of audiologic findings among patients with cerebellopontine angle tumors. Laryngoscope **71**, 1105—1126 (1961).

Galambos, R.: Studies of the auditory system with implanted electrodes. In: Rasmussen, G. L., Windle, W. F. (Eds.): Neural mechanisms of the auditory and vestibular system, pp. 137—151. Springfield, Ill.: Thomas 1960.

Galambos, R., Schwartzkopff, J., Rupert, A.: Microelectrode study of superior olivary nuclei. Amer. J. Physiol. **197**, 527—536 (1959).

Gersuni, G. V.: Evoked potentials and the mechanism of discrimination of an external stimulus. Zh. Vyssh. Vervn. Deyat. **13**, 882—890 (1963).

Gersuni, G. V.: Sensory processes at the neuronal and behavioral levels (translated by J. E. Rose). New York: Academic Press 1971.

GERSUNI, G.V., BARU, A.V., KARASEVA, T.A. On the role of the auditory cortical projection zone in discriminating acoustic signals. J. Higher Nervous Activity (USSR) **17**, 932—946 (1967) (in Russian).

GERSUNI, G.V., BARU, A.V., KARASEVA, T.A., TONKONOGII, I.M.: Effects of temporal lobe lesions on perception of sounds of short duration, Chapter 16, Sensory processes at the neuronal and behavioral levels (Ed. GERSUNI, G.V., trans., ROSE, J.). New York: Academic Press 1971.

GIRDEN, E.: The acoustic mechanism of the cerebral cortex. Amer. J. Psychol. **55**, 518—527 (1943).

GOLDBERG, J.M., DIAMOND, I.T., NEFF, W.D.: Auditory discrimination after ablation of temporal and insular cortex in cat. Fed. Proc. **16**, 204 (1957).

GOLDBERG, J.M., MOORE, R.Y.: Ascending projections of the lateral lemniscus in the cat and the monkey. J. comp. Neurol. **129**, 143—155 (1967).

GOLDBERG, J.M., NEFF, W.D.: Frequency discrimination after bilateral ablation of cortical auditory areas. J. Neurophysiol. **24**, 119—128 (1961a).

GOLDBERG, J.M., NEFF, W.D.: Frequency discrimination after bilateral section of the brachium of the inferior colliculus. J. comp. Neurol. **116**, 265—290 (1961b).

GOLDSTEIN, R.: Hearing and speech in follow-up of left hemispherectomy. J. Spch. and Hearing Dis. **26**, 126—129 (1961).

GOLDSTEIN, R., GOODMAN, A., KING, R.B.: Hearing and speech in infantile hemiplegia before and after left hemispherectomy. Neurology **6**, 869—875 (1956).

GOODMAN, A.C.: Some relations between auditory function and intracranial lesions with particular reference to lesions of the cerebellopontine angle. Laryngoscope **67**, 987—1010 (1957).

GOREVA, O.E., KALININA, T.E.: Role of inferior colliculus in analysis of auditory stimuli. Neurosci. Translations **1**, 47—52 (1967—1968).

GRAYBIEL, A.M.: Some ascending connections of the pulvinar and nucleus lateralis posterior of the thalamus in the cat. Brain Res. **44**, 99—125 (1972a).

GRAYBIEL, A.M.: Some extrageniculate visual pathways in the cat. Invest. Ophthal. **11**, 322—332 (1972b).

GREENE, T.C.: The ability to localize sound: a study of binaural hearing in patients with tumor of the brain. Arch. Surgery **18**, 1825—1851 (1929).

GUTTMAN, N.: Laws of behavior and facts of perception. In: KOCH, S. (Ed.): Psychology: a study of a science, Vol. 5. New York: McGraw Hill 1963.

GUTTMAN, N., DIAMOND, I.T.: Unpublished experiments (1963).

HARRISON, J.M., FELDMAN, M.M.: Anatomical aspects of the cochlear nucleus and superior olivary complex. In: NEFF, W.D. (Ed.): Contributions to sensory physiology, Vol. 4. New York: Academic Press 1970.

HARRISON, J.M., HOWE, M.E.: Anatomy of the afferent auditory nervous system of mammals. In: KEIDEL, W.D., NEFF, W.D. (Eds.): Handbook of sensory physiology, Vol. V/1, Chapter 9. Berlin-Heidelberg-New York: Springer 1974.

HARRISON, J.M., IRVING, R.: Visual and non-visual auditory systems in mammals. Science **154**, 738—742 (1966).

HEATH, C.J., JONES, E.G.: An experimental study of ascending connections from the posterior group of thalamic nuclei in the cat. J. comp. Neurol. **141**, 397—426 (1971).

HEFFNER, H.E., MASTERTON, R.B.: The contributions of auditory cortex to sound localization in the monkey. J. Neurophysiol. (in press).

HIND, J.E.: An electrophysiological determination of tonotopic organization in auditory cortex of cat. J. Neurophysiol. **16**, 475—489 (1953).

HIRSH, I.J.: Auditory perception of temporal order. J. acoust. Soc. Amer. **31**, 759—767 (1959).

HIRSH, I.J., SHERRICK, C.E.: Perceived order in different sense modalities. J. exp. Psychol. **62**, 423—432 (1961).

HODGSON, W. R.: Audiological report of a patient with left hemispherectomy. J. Speech Hearing Dis. **32**, 39—45 (1967).

IGARASHI, M., ALFORD, B. R., NAKAI, Y., GORDON, W. P.: Behavioral auditory function after transection of crossed olivocochlear bundle in the cat. Acta Oto-laryngol. **73**, 455—466 (1972).

IRVING, R., HARRISON, J. M.: Superior olivary complex and audition: a comparative study. J. comp. Neurol. **130**, 77—86 (1967).

JANE, J. A., MASTERTON, R. B., DIAMOND, I. T.: The function of the tectum for attention to auditory stimuli in the cat. J. comp. Neurol. **125**, 165—192 (1965).

JERGER, J. E.: Observations on auditory behavior in lesions of the central auditory pathways. Arch. Otolaryngol. **71**, 747—806 (1960a).

JERGER, J.: Bekesy audiometry in analysis of auditory disorders. J. Speech, Hear. Res. **3**, 175—187 (1960b).

JERGER, J.: Audiological manifestations of lesions in the auditory nervous system. Laryngoscope **70**, 417—425 (1960c).

JERGER, J., JERGER, S.: Auditory findings in brainstem disorders. Arch Otolaryngol. **99**, 342—350 (1974).

JERGER, J., LOVERING, L., WERTZ, M.: Auditory disorder following bilateral temporal lobe insult: report of a case. J. Speech Hear. Disorders **37**, 523—553 (1972).

JERGER, J. F., MIER, M.: The effect of brain stem lesions on auditory responses of humans. Paper presented Psychonomic Society, University of Chicago (1960).

JERGER, J., SHEDD, J. L., HARFORD, E.: On the detection of extremely small changes in sound intensity. Arch. Otolaryngol. **69**, 200—211 (1959).

JERGER, J., WEIKERS, N. J., SHARBROUGH, F. W. III., JERGER, S.: Bilateral lesions of the temporal lobe. Acta Otolaryng. Suppl. **258**, 1—51 (1969).

JERISON, H. J.: Effect of auditory cortex ablation on tone pattern discrimination in the rhesus monkey. Doctoral dissertation, University of Chicago (1954).

JERISON, H. J., NEFF, W. D.: Effect of cortical ablation in the monkey on discrimination of auditory patterns. Fed. Proc. **12**, 237 (1953).

KAAS, J., AXELROD, S., DIAMOND, I. T.: An ablation study of the auditory cortex in the cat using binaural tonal patterns. J. Neurophysiol. **30**, 710—724 (1967).

KANSHEPOLSKY, J., KELLEY, J. J., WAGGENER, J. D.: A cortical auditory disorder. Neurology **23**, 699—705 (1973).

KAPPAUF, W. E.: The application of conditioning methods to the study of discrimination and the measurement of differential thresholds in animals. J. Psychol. **15**, 120—135 (1943).

KARASEVA, T. A.: The role of the temporal lobe in human auditory perception. Neuropsychology **10**, 227—231 (1972).

KARP, E., BELMONT, I., BIRCH, H.: Unilateral hearing loss in hemiplegic patients. J. Nerv. Ment. Dis. **148**, 83—86 (1969).

KELLY, J. B.: The effects of insular and temporal lesions in cats on two types of auditory pattern discrimination. Brain Res. **62**, 71—87 (1973).

KELLY, J. B.: Polysensory cortical lesions and auditory temporal pattern discriminations in the cat. Brain Res. **80**, 317—327 (1974).

KELLY, J. B., WHITFIELD, I. C.: Effects of auditory cortical lesions on discriminations of rising and falling frequency-modulated tones. J. Neurophysiol. **34**, 802—816 (1971).

KENNEDY, T. T.: An electrophysiological study of the auditory projection areas of the cortex in monkey *(Macaca mulatta)*. Doctoral dissertation, University of Chicago (1955).

KENNEDY, T., KIANG, N., SUTTON, S.: Auditory projection areas of the cortex in monkey *(Macaca mulatta)*. Amer. J. Physiol. **183**, 634 (1954).

KIMURA, D.: Some effects of temporal-lobe damage on auditory perception. Canad. J. Psychol. **15**, 156—169 (1961).

KIMURA, D.: Left-right differences in the perception of melodies. J. exp. Psychol. **16**, 355—358 (1964).

KONORSKI, J.: A new method of physiological investigation of recent memory in animals. Bull. Acad. Polon. Sci., Ser. sci. biol. **7**, 115—117 (1959).

KOS, C. M.: Auditory function as related to the complaint of dizziness. Laryngoscope **65**, 711—721 (1955).

KRYTER, K. D., ADES, H. W.: Studies on the function of the higher acoustic nervous centers in the cat. Amer. J. Psychol. **56**, 501—536 (1943).

LANDAU, W. M., GOLDSTEIN, R., KLEFFNER, F. R.: Congenital aphasia: a clinicopathologic study. Neurology. **10**, 915—921 (1960).

LARIONOW, W.: Über die musikalischen Zentren des Gehirns. Pflug. Arch. ges. Physiol. **76**, 608—625 (1899).

LASHLEY, K. S.: Thalamo-cortical connections of the rat's brain. J. Comp. Neurol. **75**, 67—121 (1941).

LEWY, F. H., KOBRAK, H.: The neural projections of the cochlear spirals on the primary acoustic centers. Arch. Neurol. Psychiat. **35**, 839—852 (1936).

LHERMITTE, F., CHAIN, F., ESCOUROLLE, R., DUCARNE, B., PILLON, B., CHEDRU, G.: Etude des troubles perceptifs auditifs dans les lesions temporales bilaterales. Rev. Neurol. **124**, 329—351 (1971).

LICKLIDER, J. C. R., KRYTER, K. D.: Frequency localization in the auditory cortex of the monkey. Fed. Proc. **1**, 51 (1942).

LIDEN, G., KORSAN-BENGTSEN, M.: Audiometric manifestations of retrocochlear lesions. Adv. Otorhinolaryngol. **20**, 271—287 (1973).

LORENTE DE NO, R.: Anatomy of the eighth nerve. The central projection of the nerve endings of the inner ear. Laryngoscope (St. Louis) **43**, 1—38 (1933).

MARUYAMA, N., KANNO, Y.: Experimental study on functional compensation after bilateral removal of auditory cortex in cats. J. Neurophysiol. **24**, 193—202 (1961).

MASSOPUST, L. C., Jr., BARNES, H. W., VERDURA, J.: Auditory frequency discrimination in cortically ablated monkeys. J. Aud. Res. **5**, 85—93 (1965).

MASSOPUST, L. C., Jr., BARNES, H. W., MEDER, J., MEDER, R.: Deficits in auditory frequency discrimination following temporal and frontal lobe lesions in monkeys. J. Aud. Res. **6**, 261—276 (1966).

MASSOPUST, L. C., Jr., WOLIN, L. R., FROST, V.: Changes in auditory frequency discrimination thresholds after temporal cortex ablations. Exp. Neurol. **19**, 245—255 (1967).

MASSOPUST, L. C., Jr., WOLIN, L. R., FROST, V.: Increases in auditory middle frequency discrimination threshold after cortical ablations. Exp. Neurol. **28**, 299—305 (1970).

MASSOPUST, L. C., Jr., WOLIN, L. R., FROST, V.: Frequency discrimination thresholds following auditory cortex ablations in the monkey. J. Aud. Res. **11**, 227—233 (1971).

MASTERTON, R. B., DIAMOND, I. T.: Effects of auditory cortex ablation on discrimination of small binaural time differences. J. Neurophysiol. **27**, 15—36 (1964).

MASTERTON, B., DIAMOND, I. T.: The medial superior olive and sound localization. Science **155**, 1696—1697 (1967).

MASTERTON, B., HEFFNER, H., RAVIZZA, R.: The evolution of human hearing. J. acoust. Soc. Amer. **45**, 966—985 (1969).

MASTERTON, B., JANE, J. A., DIAMOND, I. T.: Role of brainstem auditory structures in sound localization. I. Trapezoid body, superior olive, and lateral lemniscus. J. Neurophysiol. **30**, 341—359 (1967).

MASTERTON, R. B., JANE, J. A., DIAMOND, I. T.: Role of brainstem auditory structures in sound localization. II. Inferior colliculus and its brachium. J. Neurophysiol. **31**, 96—108 (1968).

MASTERTON, B., THOMPSON, G. C., BECHTOLD, J. K., ROBARDS, J.: Neuroanatomical basis of binaural phase-difference analysis for sound localization: a comparative study. J. comp. physiol. Psychol. **89**, 379—386 (1975).

MATZKER, J.: Two new methods for the assessment of central auditory function in cases of brain disease. Ann. Otol. Rhinol. Laryngol. **68**, 1185—1197 (1959).

MERZENICH, M. M.: Personal communication (1975).

MERZENICH, M. M., BRUGGE, J. F.: Representation of the cochlear partition on the superior temporal plane of the macaque monkey. Brain Res. **50**, 275—296 (1973).

MERZENICH, M. M., KNIGHT, P. L., ROTH, G. L.: Representation of cochlea within primary auditory cortex in the cat. J. Neurophysiol. **38**, 231—249 (1975).

METTLER, F. A., FINCH, G., GIRDEN, E., CULLER, E. A.: Acoustic value of the several components of the auditory pathway. Brain **57**, 475—484 (1934).

MEYER, D. R., WOOLSEY, C. N.: Effects of localized cortical destruction on auditory discriminative conditioning in cat. J. Neurophysiol. **16**, 149—162 (1952).

MILLER, J. D., WATSON, C. S., COVELL, W. P.: Deafening effects of noise on the cat. Acta Oto-Laryngol. Suppl. **176**, 1—91 (1963).

MILNER, B.: Laterality effects in audition. In: MOUNTCASTLE, V. B. (Ed.): Interhemispheric relations and cerebral dominance. Baltimore, Md.: Johns Hopkins Press 1962.

MILNER, B., KIMURA, D., TAYLOR, L. B.: Noverbal auditory learning after frontal or temporal lobectomy in man. Paper presented at Eastern Psychological Assoc. (1965).

MOORE, R. Y., GOLDBERG, J. M.: Ascending projections of the inferior colliculus in the cat. J. Comp. Neurol. **121**, 109—136 (1963).

MOORE, C. N., CASSEDAY, J. H., NEFF, W. D.: Sound localization: the role of the commissural pathways of the auditory system of the cat. Brain Res. **82**, 13—26 (1974).

MORALES-GARCIA, C., HOOD, J. D.: Tone decay test in neuro-otological diagnosis. Arch. Otolaryngol. **96**, 231—247 (1972).

MOREST, D. K.: The neuronal architecture of the medial geniculate body of the cat. J. Anat. (London) **98**, 611—630 (1964).

MOREST, D. K.: The lateral tegmental system of the midbrain and the medial geniculate body: study with Golgi and Nauta methods in cat. J. Anat. (Lond.) **99**, 611—634 (1965).

MOTT, F. W.: Bilateral lesion of the auditory cortical centre. Brit. Med. J. 1—15, 10 Aug. (1907).

MUNK, H.: Über die Funktionen der Großhirnrinde: gesammelte Mitteilungen aus den Jahren 1877—1880. Berlin: Hirschwald 1881.

NEFF, W. D.: The effects of partial section of the auditory nerve. J. Comp. Physiol. Psychol. **40**, No. 4, 203—215 (1947).

NEFF, W. D.: Hearing theory. Paper presented at 48th meeting of Acoustical Soc. of America (1954).

NEFF, W. D.: Behavioral studies of auditory discrimination. Ann. Otol. **66**, 506—514 (1957).

NEFF, W. D.: Functions of the auditory cortex. Laryngoscope **68**, 413—419 (1958).

NEFF, W. D.: Sensory discrimination. Chapter 60, pp. 1447—1470. In: FIELD, J., MAGOUN, H. W., HALL, V. E. (Eds.): Handbook of physiology, Section 1, Neurophysiology, Volume 3. Washington, D.C.: American Physiological Society 1960a.

NEFF, W. D.: Role of the auditory cortex in sound discrimination. In: RASMUSSEN, G. L., WINDLE, W. F. (Eds.): Neural mechanisms of the auditory and vestibular systems. Springfield, Ill.: C. C. Thomas 1960b.

NEFF, W. D.: Neural mechanisms of auditory discrimination. Chapter 15, pp. 259—278. In: ROSENBLITH, W. A. (Ed.): Sensory communication. New York: John Wiley 1961a.

NEFF, W. D.: Discriminatory capacity of different divisions of the auditory system. In: BRAZIER, M. A. B. (Ed.): Brain, Volume 1, pp. 205—224. Washington, D.C. AIBS. 1961b.

NEFF, W. D.: Neural structures concerned in localization of sound in space. Psychol. Beit. **6**, 492—500 (1962).

NEFF, W. D.: In: DE REUCH, A. V. S., O'CONNOR, M. (Eds.): Ciba Foundation symposium on disorders of language, pp. 183—142. London: J. and A. Churchill Ltd. 1964a.

NEFF, W. D.: Physiological aspects of binaural hearing. Int. Audiol. **3**, 170—173 (1964b).

NEFF, W. D.: Auditory discriminations affected by cortical ablations. In: GRAHAM, A. B. (Ed.): Sensorineural hearing processes and disorders, pp. 201—210. Boston: Little, Brown 1967.

NEFF, W. D.: Localization and lateralization of sound in space. In: DE REUCH, A. V. S., KNIGHT, J. (Eds.): Ciba Foundation Symposium on Hearing Mechanisms in Vertebrates, pp. 207—231. London: J. and A. Churchill Ltd. 1968a.

NEFF, W. D.: Experimental studies of auditory discrimination. Int. Audiol. **7**, 12—15 (1968b).

NEFF, W. D., ARNOTT, G. P., FISHER, J. D.: Function of auditory cortex: localization of sound in space. Amer. J. Physiol. **163**, 738 (1950).

NEFF, W. D., CASSEDAY, J. H.: Unpublished results (1972).

NEFF, W. D., CASSEDAY, J. H., NEFF, P. L.: Unpublished results (1974).

NEFF, W. D., DIAMOND, I. T.: The neural basis of auditory discrimination. In: HARLOW, H. F., WOOLSEY, C. N. (Eds.): Biological and biochemical bases of behavior, pp. 101—126. Univ. of Wisconsin Press 1958.

NEFF, W. D., FISHER, J. F., DIAMOND, I. T., YELA, M.: Role of auditory cortex in discrimination requiring localization of sound in space. J. Neurophysiol. **19**, 500—512 (1956).

NEFF, W. D., HIND, J. E.: Auditory thresholds of the cat. J. acoust. Soc. Amer. **27**, 480—483 (1955).

NEFF, W. D., MUNCHEL, M.: Unpublished results (1975).

NEFF, W. D., NEFF, P. A.: The temporal lobe and auditory discrimination. In: RAMPP, D. L. (Ed.): Proc. first annual symposium on auditory processing and learning disabilities. Memphis Tenn.: Memphis State Press 1972.

NELSON, P. G., ERULKAR, S. D., BRYAN, J. S.: Responses of units of the inferior colliculus to time-varying acoustic stimuli. J. Neurophysiol. **29**, 834—860 (1966).

NOFFSINGER, D., OLSEN, W. O., CARHART, R., HART, C. W., SAHGAL, V.: Auditory and vestibular aberrations in multiple sclerosis. Acta Oto-laryngol. Suppl. **303**, 1-63 (1972).

NIIMI, K., NAITO, F.: Cortical projections of the medial geniculate body in the cat. Exp. Brain Res. **19**, 326—342 (1974).

OESTERREICH, R. E., STROMINGER, N. L., NEFF, W. D.: Neural structures mediating sound intensity discrimination in the cat. Brain Res. **27**, 251—270 (1971).

OLIVER, D. L., HALL, W. C.: Subdivisions of the medial geniculate body in the tree shrew *(Tupaia glis)*. Brain Res. **86**, 217—227 (1975).

PALMER, L. A., ROSENQUIST, A. C., SPRAGUE, J. M.: Corticotectal systems in the cat; their structure and function. In: FRIGYESI, T., RINVIK, E., YAHR, M. D. (Eds.): Corticothalamic projections and sensorimotor activities. New York: Raven Press 1972.

PANDYA, D. N., SANIDES, F.: Architectonic parcellation of the temporal operculum in rhesus monkey and its projection pattern. Z. Anat. Entwickl.-Gesch. **139**, 127—161 (1973).

PARKER, W., DECKER, R. L., GARDNER, W. H.: Auditory function and intracranial lesions. Arch. Otolaryngol. **76**, 425—435 (1962).

PARKER, W., DECKER, R. L., RICHARDS, N. G.: Auditory function and lesions of the pons. Arch. Otolaryngol. **87**, 228—240 (1968).

PAVLOV, I. P.: Conditioned reflexes. New York: Dover 1960.

PENFIELD, W., EVANS, J.: Functional defects produced by cerebral lobectomies. Res. Publ. Ass. Res. Nerv. Ment. Dis. **13**, 352 (1934).

PENFIELD, W., PEROT, P.: The brain's record of auditory and visual experience. Brain **86**, 595—697 (1963).

PICKLES, J. O., COMIS, S. D.: Role of centrifugal pathways to the cochlear nucleus in detection of signals in noise. J. Neurophysiol. **36**, 1131—1137 (1973).

POGGIO, G. F., MOUNTCASTLE, V. B.: A study of the functional contribution of the lemniscal and spinothalamic systems to somatic sensibility: central nervous mechanisms in pain. Bull. Johns-Hopkins Hosp. **106**, 266—316 (1960).

RAAB, D. H., ADES, H. W.: Cortical and mid-brain mediation of a conditioned discrimination of acoustic intensities. Amer. J. Psychol. **59**, 59—83 (1946).

RANDALL, W. L.: Generalization after frequency discrimination in cats with central nervous system lesions. In: MOSTOFSKY, D. J. (Ed.): Stimulus generalization. Stanford: Stanford University Press 1965.

RAVIZZA, R., DIAMOND, I. T.: Role of auditory cortex in sound localization: a comparative ablation study of hedgehog and bushbaby. Fed. Proc. **33**, 1917—1919 (1974).

RAVIZZA, R. J., MASTERTON, B.: The habituation of auditory reflexes in decorticate opossum *(Didelphis virginiana)*. Physiol. Behav. **6**, 717—722 (1971).

RAVIZZA, R. J., MASTERTON, R. B.: Contribution of neocortex to sound localization in the opossum *(Didelphis virginiana)*. J. Neurophysiol. **35**, 344—356 (1972).

RISS, W.: Effect of bilateral cortical ablation on discrimination of sound direction. J. Neurophysiol. **22**, 374—384 (1959).

ROESER, R. J., DALY, D. D.: Auditory cortex disconnection associated with thalamic tumor: A case report. Neurology **24**, 555—559 (1974).

ROSE, J. E.: The cellular structure of the auditory region of the cat. J. comp. Neurol. **91**, 409—440 (1949).

ROSE, J. E., WOOLSEY, C. N.: The relations of thalamic connections, cellular structure, and evocable electrical activity in the auditory region of cat. J. comp. Neurol. **91**, 441—466 (1949).

ROSE, J. E., WOOLSEY, C. N.: Cortical connections and functional organization of thalamic auditory system of cat. In: HARLOW, H. F., WOOLSEY, C. N. (Eds.): Biological and biochemical bases of behavior. Madison: Univ. Wisconsin Press 1958.

ROSENZWEIG, M.: Discrimination of auditory intensities in the cat. Amer. J. Psychol. **59**, 127—136 (1946).

ROSENQUIST, A. C., EDWARDS, S. B., PALMER, L. A.: An autoradiographic study of the projections of the dorsal lateral geniculate nucleus in the cat. Brain Res. **80**, 71—93 (1974).

SALTZMAN, M.: Audiometric studies following mesencephalotomy and thalamotomy. Arch. Otolaryngol. **56**, 194—203 (1952).

SANIDES, F.: Functional architecture of motor and sensory cortices in primates in the light of a new concept of neocortex evolution. In: NOBACK, C. R., MONTAGNA, W. (Eds.): The primate brain. New York: Appleton-Century-Crofts 1970.

SANCHEZ-LONGO, L. P., FORSTER, F. M., AUTH, T. L.: A clinical test for sound localization and its applications. Neurology **7**, 655—663 (1957).

SANCHEZ-LONGO, L. P., FORSTER, F. M.: Clinical significance of impairment of sound localization. Neurology **8**, 119—125 (1958).

SCHARLOCK, D. P., NEFF, W. D., STROMINGER, N. L.: Discrimination of tone duration after bilateral ablation of cortical auditory areas. J. Neurophysiol. **28**, 673—681 (1965).

SCHARLOCK, D. P., TUCKER, T. J., STROMINGER, N. L.: Auditory discrimination by the cat after neonatal ablation of temporal cortex. Science **141**, 1197—1198 (1963).

SCHULHOFF, C., GOODGLASS, H.: Dichotic listening, side of brain injury and cerebral dominance. Neuropsychology. **7**, 149—160 (1969).

SHANKWEILER, D.: Effects of temporal lobe damage on perception of dichotically presented melodies. J. comp. physiol. Psychol. **62**, 115—119 (1966).

SHERRINGTON, C. S.: The integrative action of the nervous system. New Haven: Yale University Press 1906.

SLOANE, P., PERSKY, A., SALTZMAN, M.: Midbrain deafness: tumor of the midbrain producing sudden and complete deafness. Arch. Neurol. Psychiat. **49**, 237—243 (1943).

SOUSA-PINTO, A.: Cortical projections of the medial geniculate body in the cat. Adv. Anat. Embryol. Cell Biol. **48**, No. 2, 1—42 (1973).

SPINNLER, H., VIGNOLO, L.: Impaired recognition of meaningful sounds in aphasia. Cortex **2**, 337—348 (1966).

STEPIEN, L. S., CORDEAU, J. P., RASMUSSEN, T.: The effect of temporal lobe and hippocampal lesions on auditory and visual recent memory. Brain **83**, 470—489 (1960).

STEVENS, S. S., MORGAN, C. T., VOLKMANN, J.: Theory of the neural quantum in the discrimination of loudness and pitch. Amer. J. Psychol. **54**, 315 (1941).

STEWART, J. W., ADES, H. W.: The time factor in reintegration of a learned habit lost after temporal lobe lesions in the monkey. J. Comp. Physiol. Psychol. **44**, 474—486 (1951).

STOUGHTON, G. S.: The role of the cortex in retention of a conditioned response to auditory signals. Doctoral dissertation. University of Chicago (1954).

STOUGHTON, G. S., NEFF, W. D.: Function of the auditory cortex: the effects of one-stage vs. two-stage ablation. Amer. Psychol. **5**, 474 (1950).

STROMINGER, N. L.: Localization of sound in space after unilateral and bilateral ablation of auditory cortex. Exp. Neurol. **25**, 521—533 (1969a).

STROMINGER, N. L.: Subdivisions of auditory cortex and their role in localization of sound in space. Exp. Neurol. **24**, 348—362 (1969b).

STROMINGER, N. L., NEFF, W. D.: Sound localization in temporal operated patients. Anat. Rec. (Abstract, 1967).

STROMINGER, N. L., OESTERREICH, R. E.: Localization of sound after section of the brachium of the inferior colliculus. J. comp. Neurol. **138**, 1—18 (1970).

STUDDERT-KENNEDY, M., SHANKWEILER, D.: Hemispheric specialization for speech perception. J. acoust. Soc. Amer. **48**, 579—594 (1970).

SUGA, N.: Echo-location and evoked potentials of bats after ablation of inferior colliculus. J. Physiol. (Lond.) **203**, 707—728 (1969).

SWISHER, L. P.: Auditory intensity discrimination in patients with temporal lobe damage. Cortex **11**, 179—193 (1967).

SWISHER, L., HIRSH, I. J.: Brain damage and the ordering of two temporally successive stimuli. Neuropsychologia (Basel) **10**, 137—152 (1972).

SYMMES, D.: Discrimination of intermittent noise by macaques following lesions of the temporal lobe. Exp. Neurol. **16**, 201—214 (1966).

TEN CATE, J.: Können die bedingten Reaktionen sich auch außerhalb der Großhirnrinde bilden? Arch. Neerl. Physiol. **19**, 469—481 (1934).

THOMPSON, R. F.: Function of auditory cortex of cat in frequency discrimination. J. Neurophysiol. **23**, 321—334 (1960).

THOMPSON, R. F.: Role of cortical association fields in auditory frequency discrimination. J. comp. physiol. Psychol. **57**, 335—339 (1964).

THOMPSON, R. F., SMITH, H. E.: Effects of association area lesions on auditory frequency discrimination in cat. Psych. Sci. **8**, 123—124 (1967).

THOMPSON, R. F., WELKER, W. I.: Role of auditory cortex in reflex head orientation by cats to auditory stimuli. J. comp. physiol. Psychol. **56**, 996—1002 (1963).

TRAHIOTIS, C., ELLIOTT, D. N.: Behavioral investigation of some possible effects of sectioning the crossed olivocochlear bundle. J. acoust. Soc. Amer. **47**, 592—596 (1970).

TUNTURI, A. R.: Audio frequency localization in the acoustic cortex of the dog. Amer. J. Physiol. **141**, 397—403 (1944).

TUNTURI, A. R.: Further afferent connections to the acoustic cortex of the dog. Amer. J. Physiol. **144**, 389—394 (1945).

TUNTURI, A. R.: A study on the pathway from the medial geniculate body to the acoustic cortex of the dog. Amer. J. Physiol. **147**, 311—319 (1946).

TUNTURI, A. R.: Physiological determination of the boundary of the acoustic area in the cerebral cortex of the dog. Amer. J. Physiol. **160**, 395—401 (1950).

TUNTURI, A. R.: Effects of lesions of the auditory and adjacent cortex on conditioned reflexes. Amer. J. Physiol. **181**, 225—229 (1955).

VIGNOLO, L. A.: Auditory agnosia: a review and report of recent evidence. In: BENTON, A. L. (Ed.): Contributions to clinical neuropsychology, pp. 172—208. Chicago: Aldine 1969.

VOGT, O.: Sur la myelinisation de l'hemisphere cerebral du chat. C.R. Soc. Biol., Paris, 10 ser. **5**, 528—536 (1898).

Walker, A. E.: The projection of the medial geniculate body to the cerebral cortex in the macaque monkey. J. Anat. **71**, 319—331 (1937).

Walker, A. E.: The primate thalamus. Chicago: Univ. Chicago Press 1938.

Walsh, E. G.: An investigation of sound localization in patients with neurological abnormalities. Brain **80**, 222—250 (1957).

Warr, W. B.: Fiber degeneration following lesions in the anterior ventral cochlear nucleus of the cat. Exp. Neurol. **14**, 453—474 (1966).

Wegener, J. G.: The role of special cerebral and cerebellar systems in sensory discrimination. Doctoral dissertation. Univ. of Chicago (1954).

Wegener, J. G.: The effect of cortical lesions on auditory and visual discrimination behavior in monkeys (personal communication).

Wegener, J. G.: Auditory discrimination behavior of brain-damaged monkeys. J. Aud. Res. **4**, 227—254 (1964).

Whitfield, I. C.: The auditory pathway. London: Arnold 1967.

Whitfield, I. C.: Response of the auditory nervous system to simple time-dependent acoustic stimuli. Ann. N.Y. Acad. Sci. **156**, 671—677 (1969).

Whitfield, I. C., Cranford, J., Ravizza, R., Diamond, I. T.: Effects of unilateral ablation of auditory cortex in cat on complex sound localization. J. Neurophysiol. **35**, 718—731 (1972).

Whitfield, I. C., Diamond, I. T.: Unpublished results.

Whitfield, I. C., Evans, E. F.: Responses of auditory cortical neurons to stimuli of changing frequency. J. Neurophysiol. **28**, 655—672 (1965).

Woollard, H. H., Harpman, J. A.: Marchi degeneration study of medial geniculate body on the cortex in cat. Projections to sylvian, middle ectosylvian, and dorsal part of anterior ectosylvian gyri. Amer. J. Neurol. Psychiat. **2**, 35—44 (1939).

Woolsey, C. N., Walzl, E. M.: Topical projection of nerve fibers from local regions of the cochlea to the cerebral cortex of the cat. Johns Hopkins Hosp. Bull. **71**, 315—344 (1942).

Wortis, S. B., Pfeffer, A. Z.: Unilateral auditory-spatial agnosia. J. Nerv, Ment. Dis. **108**, 181—186 (1948).

Chapter 9

Scaling

E. Zwicker, München, Germany

With 40 Figures

Contents

I. Methods 401
II. Threshold and Masked Threshold (Simultaneous Masking) 402
A. Steady State Condition 402
B. Temporal Effects 406
III. Critical Bands, Critical Band Rate and Psychoacoustical Excitation 409
A. Steady State 409
B. Temporal Effects 414
IV. Just Notable Differences 415
A. In Amplitude 415
B. In Frequency 416
C. Discussion of JNDs by Means of the Excitation Pattern 417
V. Loudness and its Calculation 419
A. Steady State 419
B. A Model of Loudness Formation 422
C. Loudness Calculation 425
D. Temporal Effects 426
VI. Pitch 428
A. The Pitch of Pure Tones 428
B. The Pitch of Complex Sounds 430
VII. Roughness 430
VIII. Timbre and Sharpness 433
A. Timbre 433
B. Sharpness 434
IX. Subjective Duration 436
X. Influence of Phase Changes on the Auditory Sensation 439
XI. Nonlinear Distortion 441
References 444

I. Methods

From experiments in psycho-acoustics several basic questions relating stimulation and sensation can be answered. These questions or the problems the questions are dealing with can be grouped into the following 4 categories:

1. Threshold (absolute or differential — also called just-noticeable difference, JND).
2. Equality of some aspect of perceived sounds.
3. Order or similarity,
4. Equality of intervals or ratios.

These problems can be related to fundamental scales; S. S. STEVENS (1951) named four: the nominal, the ordinal, the interval, and the ratio scale. STEVENS also described seven empirical operations to derive these psychophysical scales. Three of the methods are more often used than the other four.

The *method of adjustment* can be used for all scales except that of order. With this method the observer adjusts the stimulus until he has subjectively reached the criterion he was asked for. The method of adjustment was extended by VON BÉKÉSY (1947) to the method of tracking. The observer operates a switch alternately in such a way that the stimulus changes automatically, through motor-driven equipment, from a situation below the criterion to a situation above the criterion. Controlled by the observer's switch, the stimulus oscillates very slowly around the criterion. If a printed recording is used, a track is the result of the observer's operation. The average of this track represents the value of the stimulus for the criterion.

The *method of constant stimuli* can be used for similar problems. With this method, comparison stimuli are paired randomly with a fixed standard. The observer indicates whether the comparison stimulus is greater or less than the standard with regard to the criterion.

The *method of paired comparison* is similar to the method of constant stimuli but more general. The stimuli are again presented in pairs, but each stimulus is paired with every other stimulus. Thereby, the number of presentations increases considerably with increasing number of stimuli. After each comparison the observer indicates which pair is similar to or different from a given stimulus attribute. This method is primarily used for problems of order or similarity.

Since subjective measurements depend not only on the individual differences of observers but also on the "condition" of the observer himself, the results are always samples of data which have to be averaged. The normal mean or the geometrical mean is primarily used; but the median with interquartile ranges should be preferred, especially in averaging data to compose a scale of unknown characteristic, because it is independent of nonlinear transformation.

II. Treshold and Masked Threshold (Simultaneous Masking)

A. Steady State Condition

The threshold in quiet is the sound pressure level, L_T, of a sinusoidal tone as a function of its frequency, f, at which the tone is just audible. It can be measured using the constant stimulus method, but the method of tracking with a BÉKÉSY-audiometer is faster and easier. Thus, the threshold in quiet is the boundary between the area in which the tone can be heard and that in which it cannot be heard. The variations of different observers with normal hearing are small (about ± 7 dB), but typical. For young observers (less than 25 years of age) with normal hearing, the average threshold in quiet for the free field condition is about 70 dB at 20 Hz, drops first quickly then more slowly and reaches 0 dB at about 2 kHz (see ROBINSON and DADSON, 1957; ISO Rec. 226; ZWICKER and FELDTKELLER, 1967). The most sensitive range exists between 2 kHz and 5 kHz, where the threshold drops below 0 dB and, for some observers, reaches -10 dB to -15 dB. Above 5 kHz, the threshold rises again and, for frequencies higher than 12 kHz the threshold increases rapidly to reach the frequency limit for hearing at 16 kHz to 18 kHz (see Fig. 1, dashed curve).

The *masked threshold* proceeds from the threshold in quiet by adding a constant audible sound which masks, at least in a certain frequency range, the sinusoidal tone for which the threshold is measured. In this way, three parts of the hearing area (sound pressure level, L_T, of a sinusoidal tone versus its frequency, f) can be distinguished: One, in which only the masker is heard; a second in which the masker and the sinusoidal test tone are heard; and a third, which is the difference between the second and the threshold in quiet. The last indicates the effect of masking.

The threshold, L_T, of sinusoidal tones masked by white noise of different density level, l_{WN}, is shown in Fig. 1. White noise is a special noise which has no pitch and no rhythm. If the noise is separated into equidistant segments of frequency or into segments of time, none of the segments can be distinguished from the others.

The masked threshold in Fig. 1 depends not much on frequency in the range below 500 Hz but it is clearly dependent for frequencies above 500 Hz, although the density level, l_{WN}, of the masking white noise is independent of frequency. The masked threshold rises at approximately 10 dB per decade of frequency in the range above 500 Hz. On the other hand, a 10 dB increase of the density level,

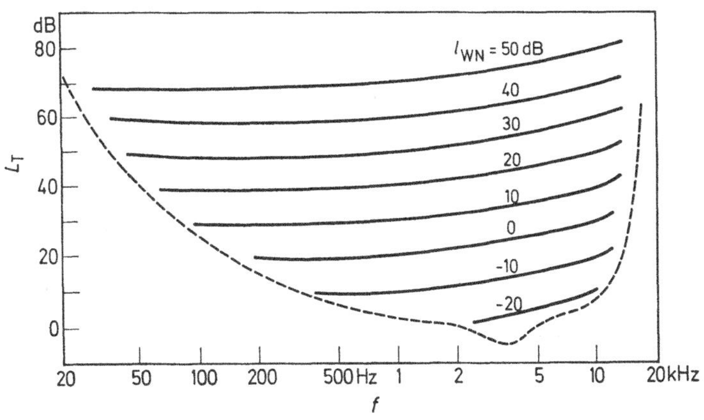

Fig. 1. Level, L_T, of a sinusoidal tone as a function of its frequency, f, at threshold in quiet (dashed curve) and at threshold masked by white noise of different density levels, l_{WN}, (solid curves)

l_{WN}, of the masking white noise results in an equal increase of 10 dB for the masked threshold. This holds for low (except near threshold in quiet) as well as for high levels throughout the audible frequency range.

The threshold, L_T, of pure tones masked by sinusoidal tones can be measured with simple equipment. Unfortunately, the results are not so simple and need explanation (Wegel and Lane, 1924; Zwicker and Feldtkeller, 1967; Greenwood, 1971). The threshold L_T masked by a 1 kHz-tone with a level of 80 dB is shown in Fig. 2 together with the threshold in quiet. For low frequencies, the 1 kHz-masker-tone does not influence the threshold of the test-tone. Above 500 Hz the masking effect starts and the threshold rises. At about 950 Hz, beats become audible and the threshold of the test-tone as such does not exist anymore until frequencies higher than 1050 Hz are reached.

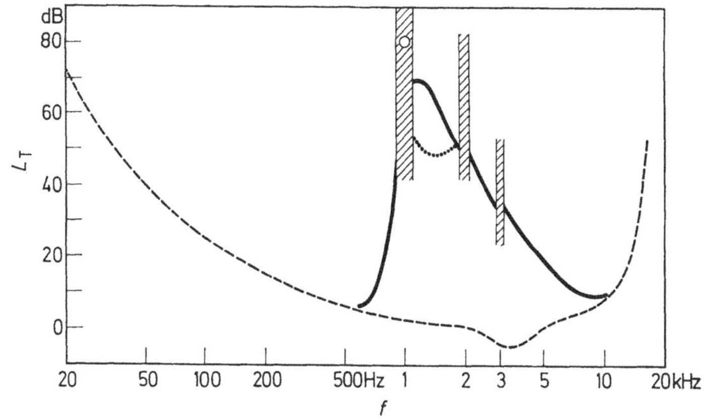

Fig. 2. Level, L_T, of a sinusoidal tone as a function of its frequency, f, at threshold in quiet (dashed curve) and at threshold masked by a 80 dB, 1 kHz masker-tone (solid curve). The hatched areas indicate beats. The dotted curve belongs to the threshold of difference tones

In the frequency range between about 1.1 kHz and 1.8 kHz, two types of threshold can be measured. When the test-tone level is increased, first, a threshold is reached which separates the area where the masker-tone alone is heard from the area in which the masker-tone and difference-tones are audible. The difference-tones are generated in the ear by nonlinearities; the masker tone and the test-tone are primaries (Greenwood, 1971). With further increase of the level of the test-tone, a second threshold is reached above which masker-tone and difference-tones are again audible, but in addition the test-tone can be heard. This threshold, which is the one we are looking for, is not at all prominent and can be measured only by well-experienced observers (Zwicker and Feldtkeller, 1967).

Around 2 kHz and 3 kHz, but sometimes at higher harmonics, beats are observed and again wipe out a well-defined threshold (as indicated in Fig. 2).

Changes in masker level as well as changes in masker frequency alter the masked threshold in a very complex way. An extensive study on many effects correlated with masked thresholds was done by GREENWOOD (1971). It shows that the nonlinearities of the ear have a strong influence on the shape of the masked threshold.

The threshold of pure tones masked by narrow bands of noise is not influenced by beats. Even though the apparatus, especially the filter, is not simple (the filter should have an almost rectangular frequency response), the results are quite remarkable because these masked thresholds give a good approximation of the frequency selectivity of the ear, not only over a wide range of levels but also over a large frequency range.

Figure 3 shows thresholds, L_T, of pure tones masked by narrow bands of noise (band width equal to one critical band) with center frequencies, f_c, of 250 Hz,

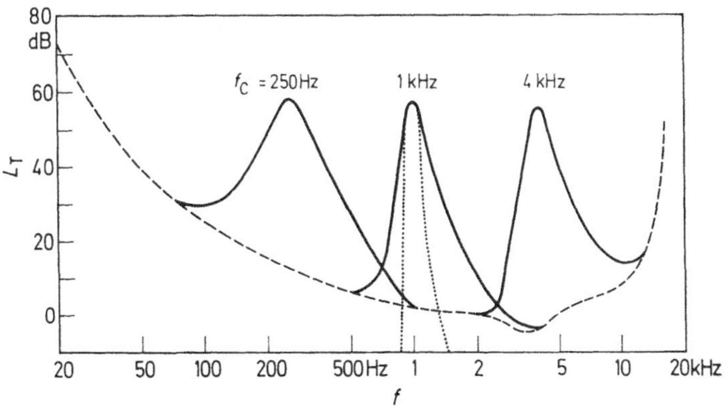

Fig. 3. Level, L_T, of a sinusoidal tone as a function of its frequency, f, as threshold in quiet (dashed curve) and at threshold masked by critical band wide noises with level of 60 dB, centered at 250 Hz, 1 kHz and 4 kHz (solid curves). The dotted curve indicates the characteristic of the filter used at 1 kHz

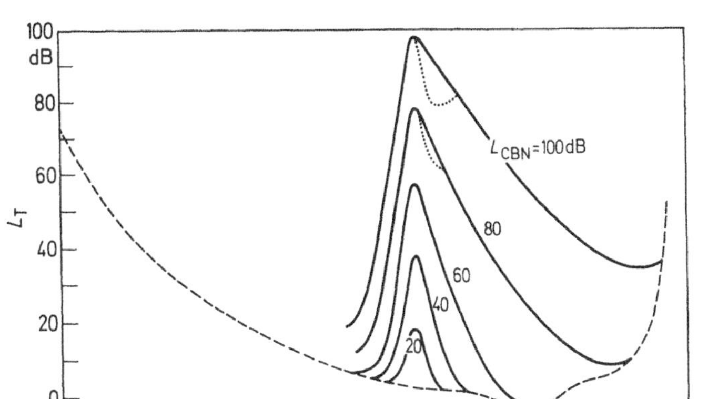

Fig. 4. Level, L_T, of a sinusoidal tone as a function of its frequency, f, at threshold in quiet (dashed curve) and at threshold masked by a critical band noise centered at 1 kHz with different levels, L_{CBN} (solid curves). The dotted parts indicate regions at which the threshold of difference noises is first reached

1 kHz, and 4 kHz and identical critical band levels, L_{CBN}, of 60 dB. The frequency response of the filter (dotted curve) is shaped almost rectangularly while the masked threshold shows a more trapezoidal form with steep slopes towards lower frequencies and flatter slopes towards higher frequencies.

For different levels, L_{CBN}, of the critical band noise, the masked threshold does not change its shape on the low frequency side but changes distinctly on the high frequency side. There, the slope gets flatter (as shown in Fig. 4 for a center frequency of 1 kHz) when the level of the narrow band noise is increased. This effect holds for all center frequencies. At very high levels, the nonlinearity of the ear again influences the threshold in the octave above the center frequency. Difference-noises occur which become audible before the test-tone becomes audible.

B. Temporal Effects

For continuous masker and gated test-sounds, two fundamental parameters can be changed (see Feldtkeller and Oetinger, 1956; Zwislocki, 1960; Zwicker, 1965a, b): the duration of single test-impulses or the repetition rate of test-impulses of certain (small) pulse-widths. Using tone-impulses or tone-bursts as test-sounds, the results can be represented by two simple graphs, which hold for the masked threshold as well as for the threshold in quiet. In the graph, the sound pressure level, L_T (T_T), of the test-tone at threshold can be shown, from which the impulse of different duration, T_T, is cut out. Then for different levels of the masker, many curves are necessary. Using as a reference the value L_T (∞), which represents the threshold of a long test-tone as described above, the level difference $\Delta L_T = L_T\,(T_T) - L_T\,(\infty)$ as a function of the duration (see Fig. 5) unfolds the effect totally: For long durations ($T_T > 500$ msec), the

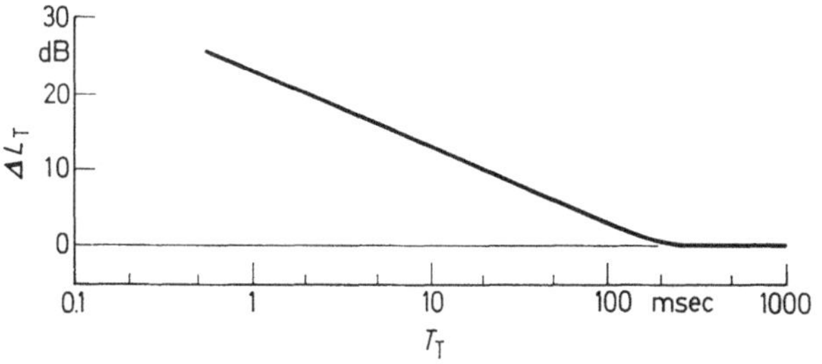

Fig. 5. Masked threshold of single test sound impulses as a function of their duration, T_T, expressed by the level difference, $\Delta L_T = L_T\,(T_T) - L_T(\infty) \cdot L_T\,(T_T)$ is the level of the test sound impulse with the duration T_T and $L_T(\infty)$ is the level of a long sound impulse as shown in Figs. 1—4

"normal" threshold is reached. When the duration, T_T, is decreased, the level L_T (T_T) and, accordingly, the level difference ΔL_T [L_T (∞) is constant for a certain masker condition] has to be enlarged in order to reach the threshold. For durations shorter than about 200 msec, ΔL_T rises 10 dB if the duration is shortened by a factor of 10. In this range the product of the sound intensity of the tone from which the impulse is cut out and the duration of the tone-impulse seems to be responsible for audibility. This holds for the masked threshold as well as for the threshold in quiet and for tone-impulses of different frequencies up to about 8 kHz as well as for narrow band noise-impulses of corresponding center frequencies. This general behaviour of the threshold of narrow band sounds (including tones) can be written in the form of the equation

$$\Delta L_{\mathrm{T}} = L_{\mathrm{T}}(T_{\mathrm{T}}) - L_{\mathrm{T}}(\infty) = 10 \lg \frac{1}{1 - e^{\frac{-T_{\mathrm{T}}}{200\,\mathrm{msec}}}} \,\mathrm{dB}$$

which can be simplified for short durations T_{T} into

$$\Delta L_{\mathrm{T}} = 10 \lg \frac{200\,\mathrm{msec}}{T_{\mathrm{T}}} \,\mathrm{dB}.$$

Instead of varying the duration of the test-tone with a fixed repetition rate (very small for single impulses), it is also possible to vary the repetition rate for constant (short) tone durations. In both cases, an extreme variation of the parameter results in a transition from single short impulses to the steady state condition. The extreme values must be the same but the shapes of the curves are quite different. For a constant duration of 5 msec, the threshold shift, $\Delta L_{\mathrm{T}} = L_{\mathrm{T}}(f_{\mathrm{r}}) - L_{\mathrm{T}}(\infty)$, is shown in Fig. 6 as a function of the repetition frequency f_{r}. Thereby, $L_{\mathrm{T}}(\infty)$ is

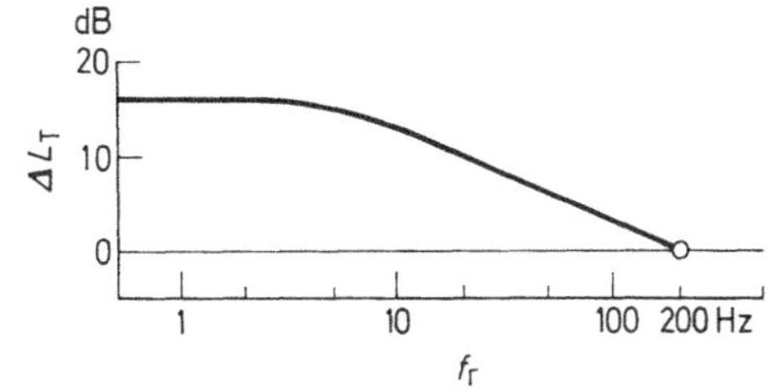

Fig. 6. Masked threshold of test sound pulses as a function of their repetition rate, f_{r}, expressed by the level difference, $\Delta L_{\mathrm{T}} = L_{\mathrm{T}}(f_{\mathrm{r}}) - L_{\mathrm{T}}(\infty)$. Since the duration, T_{T}, of the impulses is 5 msec, $L_{\mathrm{T}}(\infty)$ is equal to $L_{\mathrm{T}}(f_{\mathrm{r}} = 200\ \mathrm{Hz})$

equal to $L_{\mathrm{T}}(f_{\mathrm{r}} = 200\ \mathrm{Hz})$. The steady state condition is reached at a repetition rate of 200 Hz whereas the 5 msec impulses fuse into a continuous tone. Again, this curve is quite general and holds for all levels of the broad band masker as well as for the different frequencies of the test-sound as long as the spectrum remains a narrow band.

For broad band noises as test-sounds temporal effects can be described by similar curves with two modifications. The threshold of broad band noise masked by broad band noise rises for durations of the test-sound shorter than about 70 msec (not 200 msec). For the threshold in quiet the elevation starts at about 200 msec corresponding to the results for narrow band noise as test-sounds.

Many other temporal effects in masking can be measured, if not only the time pattern of the test-sound but also the time pattern of the masker is varied (Zwicker, 1973). The time sequence of the masker bursts and the test-sound impulses for such a pattern is given in Fig. 7. The masker is presented in bursts of duration, T_M, and level, L_M. The interval between the bursts is the time gap, T_G, *i.e.* the period, $T_p = T_M + T_G$. The test sound with duration, T_T, and level, L_T, is presented at a time delay, Δt, after the onset of the masker. To concentrate on the important facts, three out of many possible sound combinations may

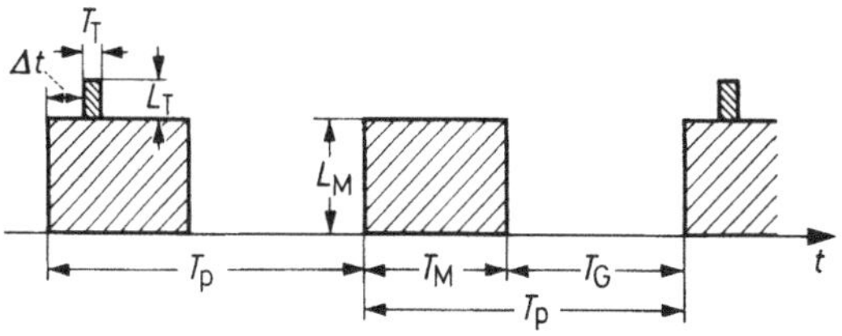

Fig. 7. Sequence of test signal and masker used in measurements of temporal effects in masking

be selected: Broad band masker and broad band test-sound; broad band masker and small band test-sound; small band masker and small band test-sound with a constant test-sound duration of 2 msec and a center frequency for small band sounds of 5 kHz. The masker duration, T_M, and the gap, T_G, are chosen at 600 msec each, in order to reach a steady state condition during the duration of the masker on the one hand and within the gap on the other hand. In such a condition, three ranges of masking can be differentiated: simultaneous masking, where the test-sound is presented simultaneously with the masker ($0 \leqq \Delta t \leqq T_M$); forward masking with a signal presented after the masker is switched off, up to the middle of the gap $\left(T_M \leqq \Delta t \leqq T_M + \frac{T_G}{2}\right)$; and backward masking for which the signal is presented at a time before the masker is switched on ($T_M + \frac{T_G}{2} = -\frac{T_G}{2} \leqq \Delta t \leqq 0$).

The complete masking effect for the 3 described conditions is shown in Fig. 8. Backward masking (see ELLIOTT, 1962; RAAB, 1963; ZWICKER and FELDTKELLER, 1967) depends mainly on the threshold in quiet ($\Delta t \approx -300$ msec) and the masked threshold in a steady state condition ($\Delta t \approx +300$ msec). The threshold rises in a steep ascent within the last 10 msec (for low frequencies) to 30 msec (for high frequencies of the test-tone) before the masker's onset. Backward masking does not last as long as forward masking, which needs about 120 msec to drop the threshold for the same amount of dB as backward masking increased it. Backward and forward masking depend only insignificantly on the frequency spectra of masker and signal, while simultaneous masking shows a distinct dependence. White-noise test-sounds masked by a white-noise burst and 5 kHz signals masked by a narrow band noise centered at 5 kHz show an almost constant simultaneous masking. However, a special "overshoot" is observed in the 5 kHz signal — white-noise masker condition (ZWICKER, 1965a, b; ELLIOTT, 1965; ZWICKER and FASTIL, 1972). The term "overshoot" is used to describe the phenomenon in which the masking effect upon short tone-impulses is stronger at the begining of the masker burst than later.

All the measured "overshoot" effects can be concisely summarized if small effects of less than 3 dB are ignored. The threshold of a short test-impulse, masked by a masker-burst, shows a transient with an "overshoot", provided that test-sound and masker have differently shaped frequency spectra. This "overshoot" does not occur if signal and masker have the same or similar frequency spectra.

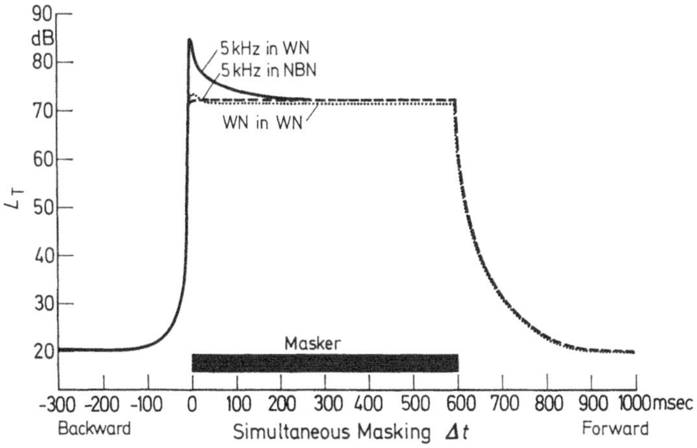

Fig. 8. Backward, simultaneous and foreward masking expressed by the level, L_T, of 2 msec test signals (white noise, WN, and 5 kHz, tone-impulse) as a function of the delay time, Δt, between onset of masker and onset of signal. The level of the masker (white-noise, WN, and narrow band noise, NBN) is chosen in such a way that steady state masking is the same for all types of masker

Most observers perform an "overshoot" only if the duration of the signal is less than 10 msec. The amount of the "overshoot" depends very little on the level of the masker but drops to zero near threshold in quiet. The threshold of short pulses with small time delays between onset of masker and onset of signal are not affected by the amount by which the masker outlasts the signal, except when the duration of the masker becomes shorter than 10 msec.

Thresholds of short tone-pulses masked by narrow band noise (Zwicker and Schütte, 1973; Fastl, 1974) or by frequency modulated tones (Zwicker, 1974a) are influenced by backward, simultaneous and foreward masking.

III. Critical Bands, Critical Band Rate and Psychoacoustical Excitation

A. Steady State

The hypothesis that there is a correlation between stimulus frequencies and specific places along the basilar membrane was introduced by von Helmholtz (1863). Von Békésy (1943) was able to show that a single tone produces traveling-wave displacements of the basilar membrane, not only at a certain point or in a small region but over a wide area of the inner ear, with a steeper slope toward the helicotrema and a flatter slope toward the oval window.

The corresponding psychoacoustical counterpart is represented by the concepts of critical band and psychoacoustical excitation: Experiments on the hearing threshold of complex sounds by GAESSLER (1954), on two-tone masking (ZWICKER, 1954), and on phase sensitivity at threshold of modulation (ZWICKER, 1952) have led to an approximation of the frequency selectivity of the ear: the critical band (ZWICKER *et al.*, 1957; ZWICKER, 1961). The critical band may be understood as a band filter with infinitely steep slopes and a frequency dependent width, Δf_G (see Fig. 9). Experiments to measure the critical band are described

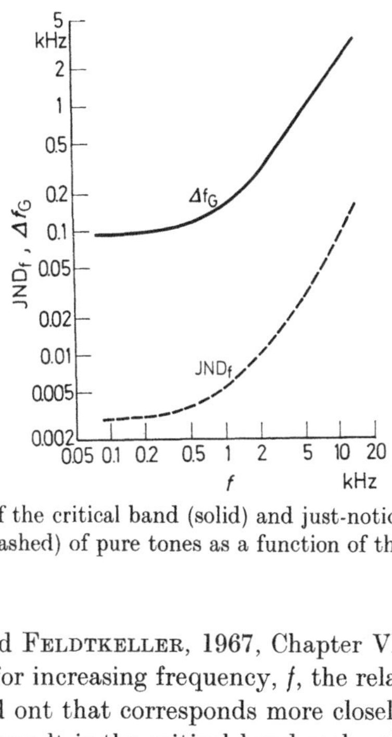

Fig. 9. Band width, Δf_G, of the critical band (solid) and just-noticeable difference, JND_f, for frequency (dashed) of pure tones as a function of the frequency, f

elsewhere (ZWICKER and FELDTKELLER, 1967, Chapter VI). When adjacent critical bands are ordered for increasing frequency, f, the relation between the physical frequency scale and ont that corresponds more closely to psychological data can be calculated. The result is the critical band scale, plotted in Fig. 10 as the critical band rate z in Bark (one Bark corresponds to one critical band) with linear and logarithmic frequency scales as abscissae. The relation between the frequency scale and the critical band scale seems to be very fundamental, because the critical band rate (Fig. 10b), the mel scale of pitch (Fig. 27) and the point of the maximum of the displacement along the basilar membrane (Fig. 10a) — all plotted versus frequency — result in similar functions (ZWICKER, 1956). In connection with threshold measurements, the critical band may be understood as the range of the z-scale or the distance along the basilar membrane over which the ear is able to integrate intensity. In this way the critical band represents only a very first approximation of the ear's frequency selectivity.

A better approximation is given by the "psychoacoustical excitation" which is derived from the "psychoacoustical incitation" and threshold patterns (see ZWICKER and FELDTKELLER, 1967, Chapter IX). To shorten the expressions, the adjective "psychoacoustical" is omitted in this chapter, nevertheless always "psychoacoustical excitation or incitation" is meant even in derived expressions like "excitation factor" or "excitation level". Furthermore, the attenuation, a_0, representing the frequency dependent sound transmission from the free field

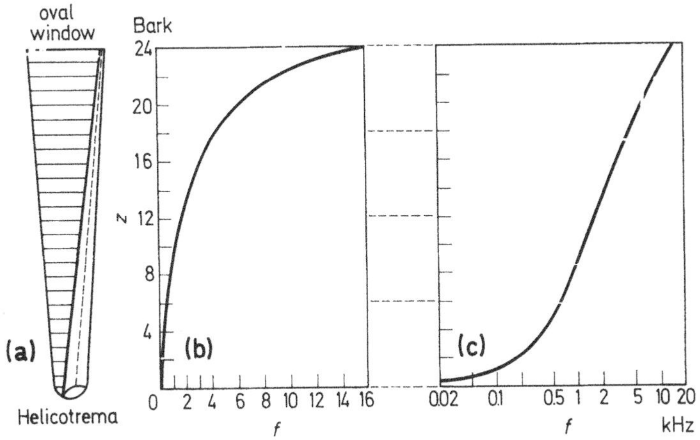

Fig. 10a—c. Critical band rate, z, as a function of the frequency, f, given on linear (b) and logarithmic scale (c). The drawing (a) indicates the relation to the locations along the basilar membrane from the oval window to the helicotrema

through the ear canal and the middle ear to the oval window, may be neglected in favor of a didactical simplification (for detailed description, see ZWICKER and FELDTKELLER, 1967, Chapter IX).

If filters with rectangular shape of attenuation, continuously shiftable along the frequency scale are assumed, the intensity within each critical band can be calculated. This intensity is called the "psychoacoustical incitation". Using the intensity density $\frac{dI}{df}$ and the width Δf_G of the critical band, the incitation A can be calculated for any sound by the equation

$$A(f_r) = \int_{f=f_r-1/2\Delta f_G(f_r)}^{f=f_r+1/2\Delta f_G(f_r)} \frac{dI}{dz}\, df$$

Since the critical band rate z depends in definite form on frequency (Fig. 10) the equation can be expressed as

$$A(z_r) = \int_{z=z_r-1/2\,\text{Bark}}^{z=z_r+1/2\,\text{Bark}} \frac{dI}{dz}\, dz.$$

To avoid problems with units the incitation factor $\frac{A\,(z)}{A_0}$ or even better the incitation level $L_A = 10 \lg \frac{A\,(z)}{A_0}$ dB is used where $A_0 = I_0 = 10^{-12} \frac{\text{W}}{\text{m}^2}$

So far, only critical bands are used as a first approximation, assuming filters with rectangular shape and infinitely steep filter slopes. As a second approximation, also embodying the form of the slopes of the ear's selectivity, the threshold patterns of pure tones masked by narrow bands of noise are used as filter shapes. They have the form of trapezoids with steep slopes towards lower frequencies and flat slopes towards higher frequencies (see Fig. 4) and are available for a wide range of levels as well as over a large frequency range. The threshold is raised even at frequencies well above the range in which the narrow band noise is physically effective. To describe this, in relation to the incitation, weaker selectivity, the "psycho-acoustical excitation", is introduced. This excitation is sometimes expressed as excitation factor $E_{(z)}/E_0$ but mostly as excitation level

$$L_{\text{E}} = 10 \lg \frac{E\,(z)}{E_0} \text{ dB}.$$

For narrow band noise with a center frequency f_{c} corresponding to the critical band rate z_{c}, the excitation level at z_{c} is equal to the incitation level and is called the main excitation level. Here, the masked threshold reaches its peak value. Furthermore, the excitation level is in general equal to the incitation level for a particular range of critical band rates in which the masked threshold is defined by the physically existing spectral density of the masking sound. This holds often for broad band noises. Outside of this range, there is accessory excitation, the level of which gradually drops down to the threshold in quiet. The accessory excitation is

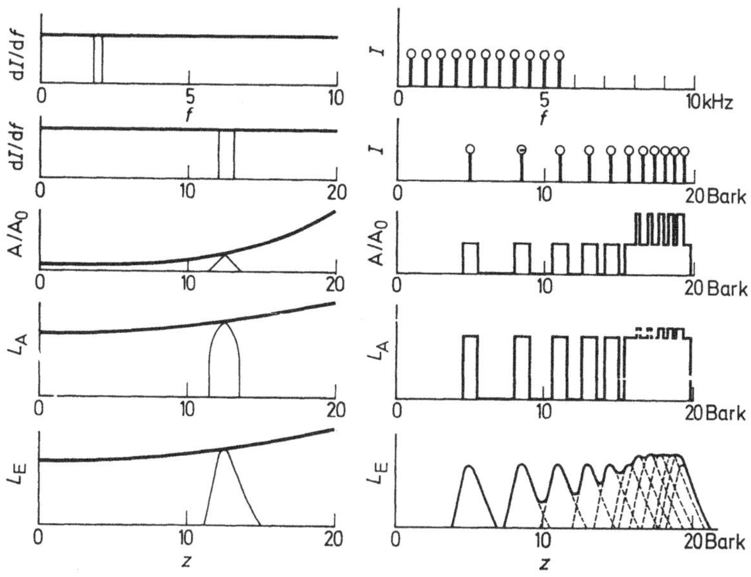

Fig. 11. Derivation of the excitation pattern for broad band noise and narrow band noise (left) as well as for a sound consisting of eleven harmonics (right)

obtained by shifting the masked threshold so far that main excitation and accessory excitation merge without a gap. In Fig. 11, the derivation of the excitation pattern is demonstrated for narrow band and for white-noise as well as for a sound consisting of eleven harmonics. Thereby the excitation pattern for tones is derived with the assumption that a noise, one critical band wide, and a tone of its center frequency produce the same excitation, if their sound pressure levels are equal (for more details see ZWICKER and FELDTKELLER, 1967).

One of the advantages of the excitation patterns (excitation level, L_E, versus critical band rate, z) in relation to the masking pattern (Fig. 3) becomes clear in Fig. 12. The excitation for critical band wide noises of different center frequencies

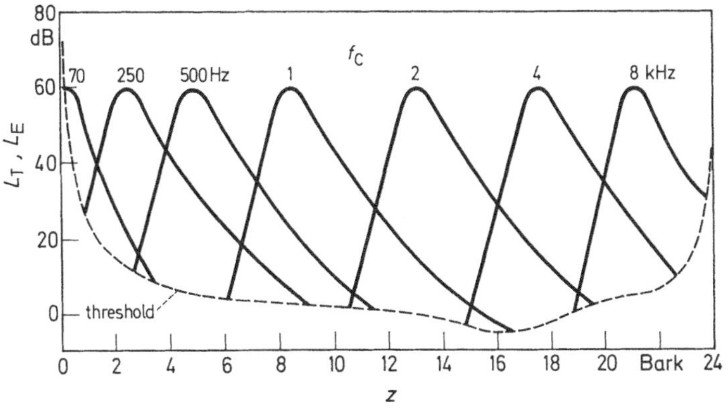

Fig. 12. Excitation pattern (excitation level, L_E, versus critical band rate, z) of critical band wide noises of different center frequencies, f_c, but equal sound pressure levels of 60 dB (solid). Level, L_T, of a pure tone at threshold as a function of the critical band rate, z, corresponding to its frequency (dashed)

but equal sound pressure levels is plotted there. The curves look very similar and may be derived from each other by shifting them up and down the z-scale. This means that, for different frequencies, the ear shows almost the same selectivity in location along the basilar membrane, when the frequency scale is transferred into the critical band scale.

On the other hand the ear's selectivity is somewhat nonlinaer. When the level of the masking narrow band noise is increased, the main excitation and the lower accessory excitation level increase by the same amount, whereas the slope of the upper accessory excitation decreases. This nonlinear effect, as illustrated in Fig. 13, is the reason for presenting the selectivity in graphic form rather than by equations. The excitation patterns in this form completely describe the selectivity of the ear for steady state conditions and are a much better approximation than the simple subdivision of the audible frequency range into critical bands. The usefulness of the excitation pattern will become more evident as soon as applications are discussed.

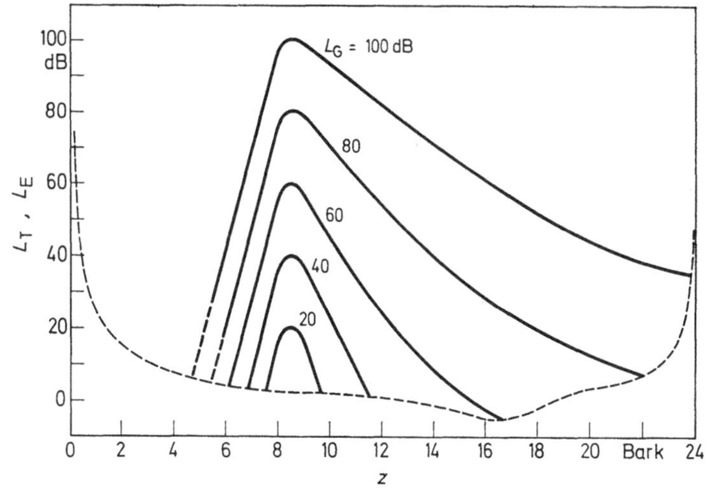

Fig. 13. Excitation pattern (excitation level, L_E, as a function of the critical band rate, z) of a critical band wide noise at different levels, L_G, with a constant center frequency of 1 kHz corresponding to 8.5 Bark (solid). Level, L_T, of a pure tone at threshold in quiet (dashed)

Recently the correlation between the psychoacoustical masking or excitation pattern and the neurophysiologically measured tuning curves has been described (ZWICKER, 1974b).

B. Temporal Effects

In many psychoacoustical sensations, the critical band plays an important role. But many sounds, especially when containing information like speech or music, have a strong time pattern, whereas steady state conditions with durations > 200 msec are rarely reached. Since foreward and backward as well as "overshoot" effects exist in masking, it is of great interest to know whether the critical band needs additional time to build up or only the rise time τ, which is correlated to the band width of the critical band by the equation, $\tau = 1/\Delta f$. Knowing this, it may be possible to construct time variable excitation-critical band rate pattern for any sound, which are even more important for understanding the creation of sensations than the excitation-critical band rate patterns in steady state conditions.

The discussion of the problem started with Scholl's (1962) experimental results. His results have meanwhile been confirmed (ZWICKER and FASTL, 1972). But the interpretation that the critical band mechanism seems to require a build up time of about 10 msec, was and still is disputed. Loudness measurements (PORT, 1963a, b; ZWICKER, 1965b) with short sound impulses of different band width led to the assumption that the critical bands are independent of duration. On the other hand some of the results of threshold measurements (ELLIOTT, 1967; GREEN, 1969) indicated the contrary. A recently published paper (ZWICKER and FASTL, 1972) settled most of the discrepancies and led to the explanation that not only the critical band but the whole excitation pattern builds up as quickly as forward and backward masking (Fig. 8) and the band width, respectively, permit, while most of the measured effects belong to the "overshoot" phenomenon.

IV. Just Noticeable Differences

A. In Amplitude

The just-noticeable-difference (JND) is the just audible change of a certain sound from one value of amplitude directly to another. If the change is interrupted by a pause in such a way that, for example, a tone-impulse followed by a quiet period is compared with another tone-impulse of different amplitude, then different results are obtained (see TERHARDT, 1968a). Summarizing we find that for loudness levels greater than 20 phon a difference in level larger than 1 dB is audible.

The direct change of an amplitude of a narrow band sound like a pure tone produces a click due to the short time spectrum of this change. To avoid this effect, which disturbs the measurement of JND, the amplitude is changed either with a smooth transient or sinusoidally. The latter is often preferred because of its well-defined frequency spectrum. The JND is measured then by the just-noticeable degree of slow (4 Hz) amplitude modulation, AM (RIESZ, 1928; ZWICKER, 1952; ZWICKER and FELDTKELLER, 1967). The relation between the degree m of AM and the corresponding level difference is given by

$$L = 10 \lg \frac{I_{max}}{I_{min}} \text{ dB} = 20 \lg \frac{1+m}{1-m} \text{ dB}.$$

For values of m smaller than 0.3, the approximation $\Delta L \approx 17.5$ m dB can be used. The just-noticeable difference in level (JND_L) can be calculated from results of measurements of the just-noticeable degree of AM. This is shown in Fig. 14 for a 1 kHz-tone as a function of the sound pressure level, L. At the low level of 15 dB the JND_L is about 2 dB and drops down with increasing level to 0.2 dB at about 90 dB. At high levels, the ear is very sensitive to changes of the level of pure tones. Similar curves for pure tones of different frequencies have been measured, which can be summarized into one single curve, if the loudness level of the tone is used as abscissa instead of the sound pressure level.

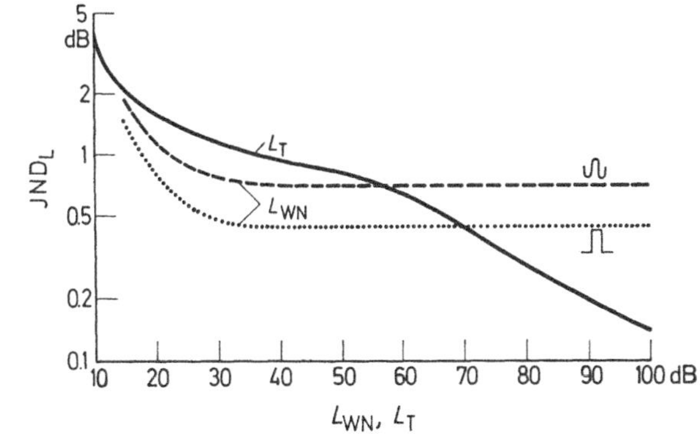

Fig. 14. Just-noticeable difference, JND_L, in level as a function of the level, L_T, for tones (solid) and as a function of the level L_{WN}, for white noise (dashed: sinusoidal modulation with 4 Hz; dotted: rectangular modulation)

The JND_L for white noise depends on level in a similar way for low levels up to 30 dB, but it then remains constant at a value of about 0.7 dB up to high levels. This means that the ear is less sensitive for changes in level for broad band noise than for tones. For such a broad band noise, the sinusoidal AM becomes needless because the short time spectra of the abrupt change are masked completely by the noise. Therefore, AM with rectangular shape can be used for JND_L measurements. The results show that the ear is more sensitive by almost a factor of two in this case. This difference holds for bands of noise with limited band width. However, if rectangular modulation is used, the just-noticeable change in level increases almost linearly with decreasing band width of the noise on a double logarithmic scale from 0.5 dB for 16 kHz band width to 5 dB for 10 Hz band width (see Fig. 15). For narrower band width the statistical amplitude changes of the noise become more and more audible and their influence on the audibility of changes in level increases.

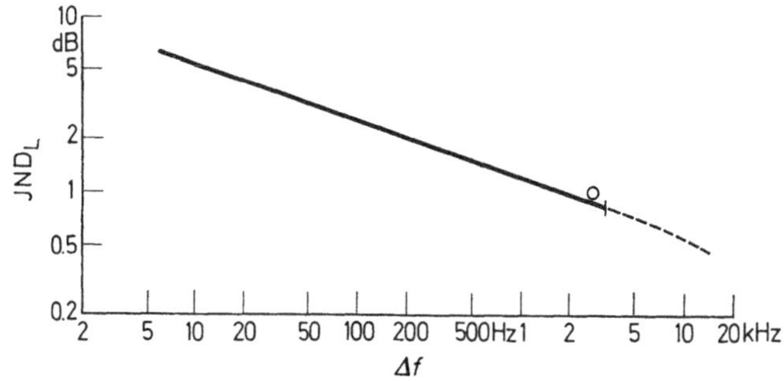

Fig. 15. Just-noticeable difference, JND_L, in level of bands of noise for rectangular modulation as a function of the bandwidth, Δf. The circle indicates 1 dB at a critical band width as large as possible

For the largest critical band width (about 3 kHz for a center frequency of 12 kHz) the noise sounds almost steady. In this case, the just-noticeable difference in level for rectangular changes reaches almost exactly 1 dB. This value is independent over a wide range of level and is assumed to be a basic value for the steady state condition (see Paragraph 4c).

The JND_L's for pure tones as well as for noises depend on the frequency of AM in such a way that for low frequencies near 4 Hz the ear is most sensitive, while for lower and especially for higher frequencies of modulation (repetition rate), the JND_L's increase by a factor of 10 at 300 Hz repetition rate.

B. In Frequency

The just-noticeable difference in frequency (JND_f) is measured in a similar way to that for amplitude, by using slow (4 Hz) frequency modulation (FM) to avoid clicks. In electrical engineering, the frequency deviation Δf of the FM is defined as the frequency change from the middle value f_m to the extreme values. Since the frequency passes the region from $f_m - \Delta f$ to $f_m + \Delta f$, the JND_f becomes equal to $2\Delta f$, if Δf is measured as the just-noticeable frequency deviation in sinusoidal FM as defined above.

The JND_f as a function of the level does not change much for levels higher than about 20 dB above threshold (ZWICKER and KAISER, 1952), while, for lower levels it increases. As a function of the frequency, the JND_f of pure tones stays almost absolutely (3.5 Hz) constant at low frequencies (< 500 Hz) and increases almost proportionally to frequency for higher frequencies, which means that the JND_f stays relatively constant at about 0.07, as presented earlier in Fig. 9. This means that the ear is most sensitive to absolute frequency shifts at low frequencies, while it is most sensitive to relative frequency shifts, such as in musical production or sound conserving systems, within the frequency range above 500 Hz. In Fig. 9, the quite similar dependence of JND_f and Δf_G on frequency is striking. This indicates a fundamental behaviour of the ear, which will be discussed later.

The JND_f can also be measured for noises, if, instead of the frequency of a pure tone, the low cut-off frequency of high-pass noises, the high cut-off frequency of low-pass noises or the center frequency of narrow band noises is modulated in frequency. For narrow band and high-pass noises, the JND_f is about 7 times larger than for pure tones but it does not increase as much for higher cut-off frequencies (ZWICKER, 1956; MAIWALD, 1967b). An interesting effect can be observed for high-pass noise with respect to low-pass noise. While the JND_f of the former stays constant with increasing level, as the JND_f of pure tones, the JND_f of the latter increases by a factor of about 5 when proceeding from 40 dB sound pressure level to 100 dB (MAIWALD, 1967c).

As a function of the modulation frequency (repetition rate) the JND_f has its lowest value around 4 Hz and rises, similar to the JND_L, somewhat towards lower and more towards higher frequencies of modulation. This holds for pure tones as well as for noises and means that the ear is most sensitive to frequency shifts at the rate where vibrato is performed, but also where most of the unwanted flutter takes place.

C. Discussion of JNDs by Means of the Excitation Pattern

The just-noticeable change in level of 1 dB for the steady sounding critical band wide noise at high center frequencies is used for the description of all JNDs. The basic assumption (ZWICKER, 1956) when using this value is very simple:

The ear is able to detect any change of a steady sound if the excitation level L_E is changed anywhere along the critical band scale z by the value $\Delta L_E \geqq 1$ dB.

MAIWALD (1967a—c) even extended the model to fluctuating sounds and found good agreement between the expected and the psychoacoustically measured results. For didactical rcasons only the model for steady sounds is discussed.

The JND_L for the level of a sinusoidal tone can be understood by taking into account that the tone produces not only a distinct frequency limited incitation pattern but an excitation pattern including lower and upper accessory excitations. As shown earlier, the upper accessory excitation increases nonlinearly with increasing level. This is the tool to describe the effect that the JND_L of 1 dB can be found psychoacoustically only for tones at loudness levels of 30 phon to 50 phon, while at higher levels the JND_L drops down to 0.2 dB. In Fig. 16a, the ΔL-increment is schematically shown. The nonlinearity is responsible for the fact that the upper accessory excitation increases much more than the main excitation when the level of the tone is increased. In Fig. 16 this magnification of the effect may only be a factor of 2; but the individual curves, often considerably different from each other, show ranges along the z-scale where the increment of the accessory excitation is 5—10 times larger than the increment of the main excitation, which corresponds to the increment of the tone. In such a case, the JND_L becomes 5—10 times smaller than 1 dB as it is measured for high levels of tones.

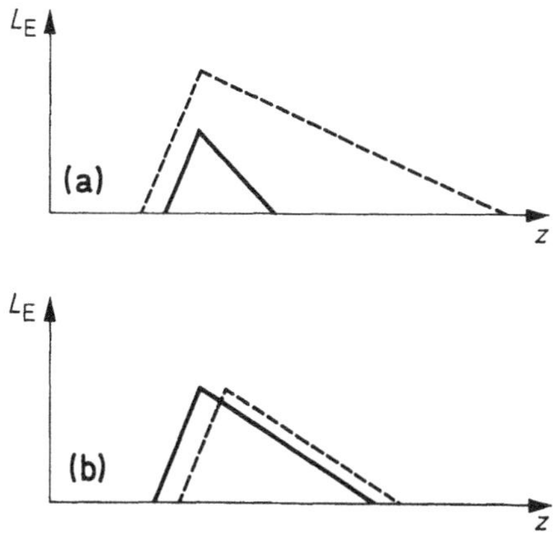

Fig. 16a and b. Schematic drawings of the change of the excitation pattern, $L_E(z)$, produced by a change in level (a) and produced by a change in frequency (b)

The fact that the upper accessory excitation influences the JND_L can be checked by adding a high-pass filtered noise to a tone periodically changed in level, in order to mask out the region of the upper accessory excitation. When doing so (ZWICKER, 1956), the JND_L increases to 1 dB almost independent of level because the main excitation has then to be changed by 1 dB to produce an audible effect.

The JND_f for frequency has to be interpreted in a similar way although the configuration shown in Fig. 16b is different. When changing the level, the excitation pattern shifts up and down with the nonlinear effect of the upper accessory excitation. For changes in frequency, the excitation pattern is shifted back and forth along the critical band scale (z-scale). No nonlinearity is involved in this case. The largest change of the excitation level during the shifting occurs at the lower accessory excitation with the steep slope. Since the steepness of this slope is almost independent of both the level and the frequency of the exciting tone, the JND_f should depend on frequency in the same way as the critical band scale z depends on frequency f. An average value for the slope of the lower accessory excitation is 27 dB/Bark (see MAIWALD, 1967a; ZWICKER, 1954). Since the unit 1 Bark of the critical band scale is correlated to the critical band width Δf_G of the frequency scale, the relation between $\mathrm{JND_f}$ and Δf_G can be calculated using again the assumption that $\Delta L_E = 1$ dB for JND, *i.e.* $\mathrm{JND_f} = 1\ \mathrm{dB}\ \frac{1\ \mathrm{Bark}}{27\ \mathrm{dB}} = \frac{1}{27}\Delta f_G$.

This relation fits very well with the psychoacoustical results as shown in Fig. 9. The $\mathrm{JND_f}$ and the critical band width Δf_G increase almost parallel as a function of frequency. This means, because both scales are logarithmic, that the curves are correlated through a frequency independent factor. This factor is 30 and is in good agreement with the predicted value of 27.

The model and its applications can further be checked by changing level and frequency at the same time. This can be realized most easily by simultaneous amplitude and frequency modulation with the same rate of modulation (4 Hz). The phase relation between these two modulations then determines the shift of the excitation pattern (ZWICKER, 1962a); configurations can be produced in which the upper accessory excitation is kept almost stationary whereas the lower moves strongly or, when changing the phase relation by 180°, the lower accessory excitation is kept almost stationary while the upper one moves very strongly. In the latter situation, a high-pass noise can be added, so that the higher accessory excitation is masked, whereas the almost stationary lower accessory excitation produces only a very faint audibility of the modulation. When appropriate values of modulation are chosen, the interesting effect can be produced, that simultaneous amplitude and frequency modulation reduce the audibility of modulation when compared with the audibility of the same but separate amplitude or frequency modulation, respectively (see ZWICKER, 1962; MAIWALD, 1967b).

V. Loudness and its Calculation

A. Steady State

Loudness depends, outside of other sound parameters like frequency, band width, and duration, mainly on the level. The sensation of loudness can be measured quantitatively either by using the loudness level, L_N (the sound pressure level of an equally loud 1 kHz-tone in the free field condition and frontally incident measured in "phon" using comparison methods), or by the loudness, N (a real subjective sensation value measured in "sone" using ratio estimation and production methods, where the number gives the factor of *how much* louder the sound in question is in relation to a 40 dB-1 kHz-tone in a free field). Both values are often used, sometimes even simultaneously.

The loudness function has been measured by many authors (*i.e.* STEVENS, 1957; ROBINSON, 1957; ZWICKER, 1958) and finally was standardized (ISO Rec. 131) in the form of the relation between loudness level, L_N, and loudness, N. For loudness levels of more than 40 phon the relation is given in the form

$$\frac{N}{\mathrm{sone}} = 2^{\left(\frac{\frac{L_N}{\mathrm{phon}} - 40}{10}\right)} = \frac{1}{16}\left(\frac{I_{1\,\mathrm{kHz}}}{I_0}\right)^{0.3} = \frac{1}{16}\left(\frac{p_{1\,\mathrm{kHz}}}{p_0}\right)^{0.6},$$

where I_0 is the reference intensity of $10^{-16}\,\frac{\mathrm{W}}{\mathrm{cm}^2}$ and p_0 is the reference sound pressure of $2 \cdot 10^{-4}\,\frac{\mathrm{dyn}}{\mathrm{cm}^2}$.

Figure 17 shows this relation as a straight line in a double logarithmic scale indicating a power function with a fixed point at 40 phon = 1 sone. For loudness levels below 40 phon, the loudness decreases more rapidly so that it reaches 0 sone at threshold.

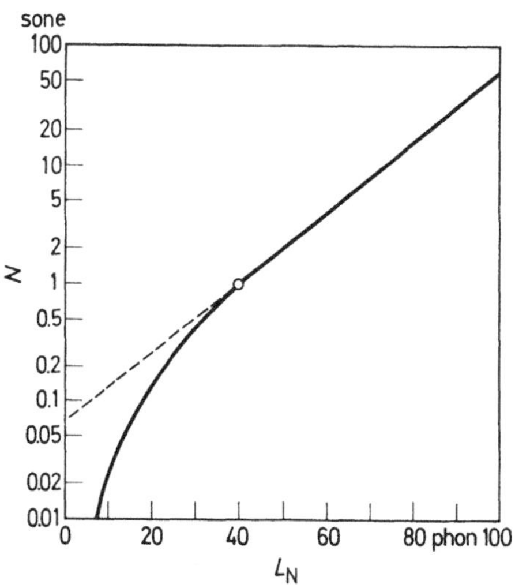

Fig. 17. Loudness, N, as function of the loudness level, L_N (solid). The dashed line indicates the power law with the exponent of 0.6 for the sound pressure

The loudness of tones and narrow band noises also depends on frequency and center frequency, respectively. This dependence is often given in graph form by the equal loudness contours which are also standardized (ISO Rec. 226). The sound pressure level, L, of tones is plotted as a function of their frequency and the points of equal loudness are connected to contours. Figure 18 shows such contours for the free field condition. The parameter can be given as loudness level, L_N, or, using the relation given in Fig. 17, as loudness, N. In Fig. 18 both expressions are used, indicating that for loudness levels above 40 phon an increment of 10 phon in loudness level, L_N, is equivalent to an increase of the loudness, N, by a factor of 2.

The loudness depends on band width in such a way that — keeping the total sound pressure level constant — it remains constant for band widths smaller than the critical band, while increases with increasing band width. This effect is strong at medium levels, but it becomes less pronounced for high levels or low levels. At very low levels just above threshold, the loudness even decreases with increasing band width. Figure 19 shows this effect and indicates that broad band noise can be 10 — 15 phon louder than a 1 kHz tone of the same sound pressure level, L, which is the parameter in Fig. 19. This is an important effect which

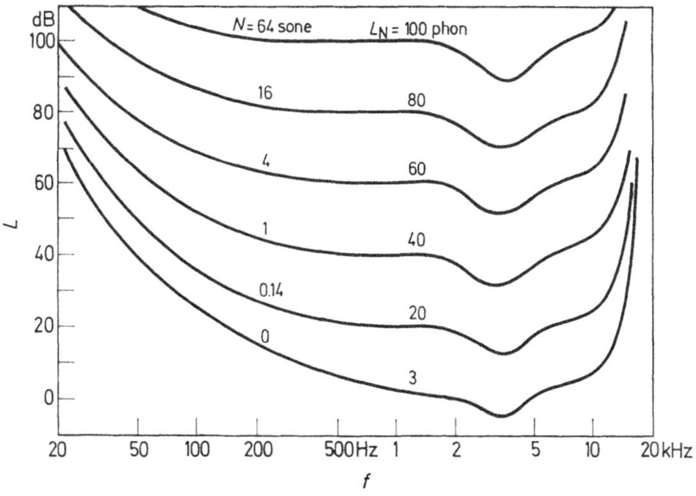

Fig. 18. Equal loudness contours for the free field condition. The level, L, of pure tones producing the same loudness, N, and loudness level, L_N, respectively is drawn as a function of the frequency, f, of the tones

has been measured by several authors (*i.e.* ZWICKER and FELDTKELLER, 1955; ZWICKER *et al.*, 1957; NIESE, 1958) and plays an essential part in loudness calculation. The loudness level, L_N, of uniform exciting noise 5 noise with equal intensity in each critical band — is shown as a function of its level, L_{UEN}, in Fig. 20. For all comparison measurements using the method of adjustment, two sets of data must be produced. In one experiment the noise is adjusted in such a way that it sounds as loud as the tone of fixed level; in the second experiment the tone is adjusted to sound as loud as the noise of fixed level. Using the medians and

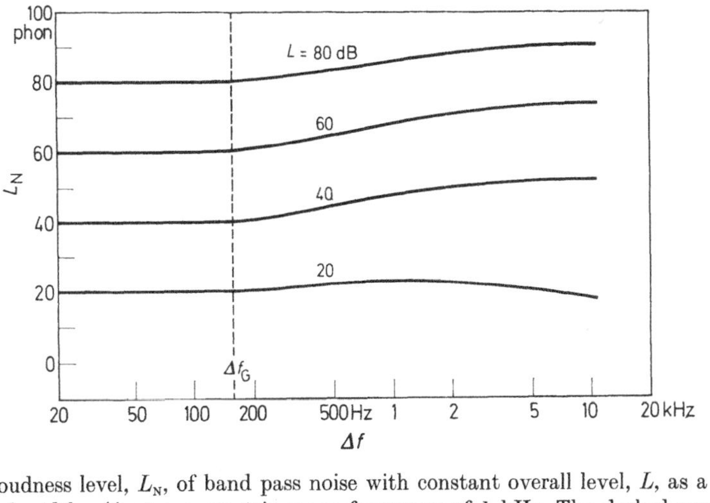

Fig. 19. Loudness level, L_N, of band pass noise with constant overall level, L, as a function of the band width, Δf, at a geometric mean frequency of 1 kHz. The dashed vertical line indicates the width, Δf_G, of the critical band

interquartile ranges to indicate the distribution of the results, the data of the first experiment are presented by vertical bars, data of the second adjustment procedure by horizontal bars. The difference between the results of the two experiments appears very clearly. It depends on level and in the range of lower and medium levels reaches a size of more than 10 dB. This effect is typical and makes it necessary to average the data of both experiments as represented by the dashed

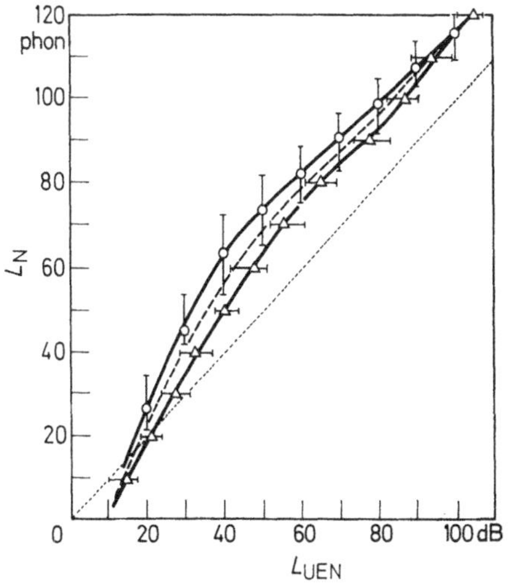

Fig. 20. Loudness level, L_N, of uniform exciting noise as a function of its level, L_{UEN}. Circles indicate results when adjusting the 1 kHztone, while triangles represent the adjustments of the noise. The interquartile ranges are also indicated. The dashed curve is the average of both results, whereas the dotted line indicates the equation $L_{UEN} = L_N$

line of Fig. 20. The difference between the average curve (dashed) and the dotted line (45°-line) indicates that a uniform exciting noise with, for example, a sound pressure level of 50 dB elicits a loudness level of 68 dB, which means that the noise is almost 4 times as loud as a tone at the same level (BRITTAIN, 1939; STEVENS, 1955; ZWICKER, 1958).

B. A Model of Loudness Formation

This model is based on the excitation pattern of the sound, on STEVENS' power law (1957), and on the assumption that the loudness of any sound is the integral of the specific loudness along the critical band scale. In this sense, every sound, even a pure tone, elicits loudness summation because the excitation and, in a similar way, the specific loudness of a pure tone spreads over a considerable portion of the critical band scale.

The excitation patterns of different sounds have been extensively described. In order to use this pattern for the derivation of loudness, it is necessary to know the relation between the specific loudness, N', and the excitation, E. A psychophysical equation expressing this relation has been formulated (ZWICKER, 1958, 1963). It is based on one form of STEVENS' power law which says that equal intensity ratios yield equal loudness ratios, or expressed in an equation:

$$\frac{\Delta N}{N} = C \frac{\Delta I}{I}.$$

In terms of the excitation model, we must assume that this law applies only to the excitation which, for a given sound, spreads over a large portion of the z-scale, and does not apply to the total intensity of the sound. The value by which the excitation pattern can be expressed in terms of the loudness is the "specific loudness, N'", with the unit $\frac{\text{sone}}{\text{Bark}}$. Replacing the intensity I by the excitation E and adding an excitation E_{gr} produced by a physiological background noise and a corresponding inaudible specific "loudness" N'_{gr}, STEVENS' formula proceeds to

$$\frac{\Delta N'}{N' + N'_{gr}} = k \cdot \frac{\Delta E}{E + E_{gr}}.$$

When this equation is treated as a differential equation it can be integrated. The just-noticeable increment in level for critical band noises — for medium center frequencies of about 1.8 dB (see Fig. 15) — can be used to determine the background noise; the exponent k can be determined when the postulate is observed that for uniform exciting noise as well as for a 1 kHz-tone the integral of the specific loudness along the z-scale has to be the total loudness, which gives $k = 0.23$. Finally, an arbitrary unit has to be chosen in such a way that the total loudness of a 40 dB — 1 kHz tone is exactly 1 sone. Using these data, the integrated equation can be expressed by the formula

$$\frac{N'}{\frac{\text{sone}_G}{\text{Bark}}} = 0.08 \left(\frac{E_{TH}}{E_0}\right)^{0.23} \left[\left(\frac{1}{2}\frac{E}{E_{TH}} + \frac{1}{2}\right)^{0.23} - 1\right],$$

where E_{TH} is the excitation at threshold and E_0 the excitation corresponding to the reference intensity I_0.

This equation is graphed in Fig. 21 where both E_{TH} and E are given as the levels L_{TH} and L_E. For simplification, the level difference a_0 is ignored describing the fact that at some frequencies more energy is transmitted from the free field to the stapes than at others (ZWICKER, 1958).

The formation of loudness due to the model is illustrated in Fig. 22. For two different sounds a 1 kHz-tone and uniform exciting noise, both with a sound pressure level of 60 dB, the distribution of the incitation factor, $\frac{A}{A_0}$, the excitation level, L_E, and the specific loudness along the critical band scale are shown (a_0 is ignored). Since both sounds have the same level, the areas between the incitation factor and the abscissa have to be the same. Because 24 adjacent critical bands (Bark) along the z-scale are involved, the size of the incitation factor of the noise is only $\frac{1}{24}$ of the incitation factor of the tone, whereas the main excitation level of the tone is $10 \log 24$ dB $= 14$ dB above the excitation level of the noise. The peak of the specific loudness contour of the tone reaches a value which is about $(24)^{0.23}$ times larger than the specific loudness of the noise. Nevertheless, the total loudness (equivalent to the areas below the curves) of the noise is about 3 times larger than that of the tone, corresponding to the level difference of about 16 dB in Fig. 17 for a noise with 60 dB sound pressure level. This example may be

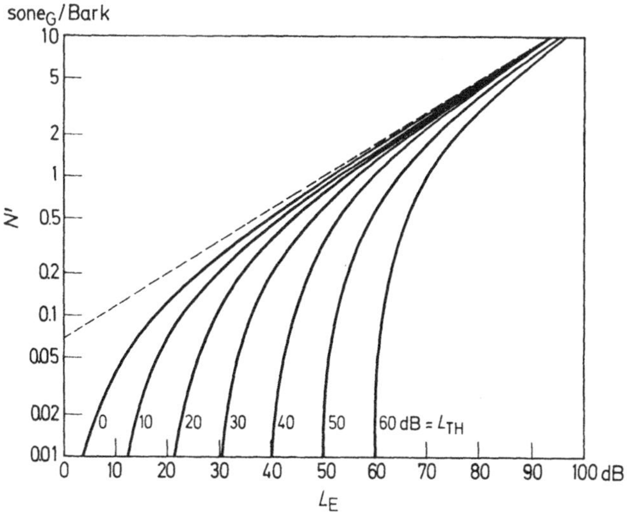

Fig. 21. Specific loudness, N', as a function of the excitation level, L_E, with the level at threshold L_{TH} as parameter. The dashed line indicates the power law $N' \sim \left(\frac{E}{E_0}\right)^{0.23}$

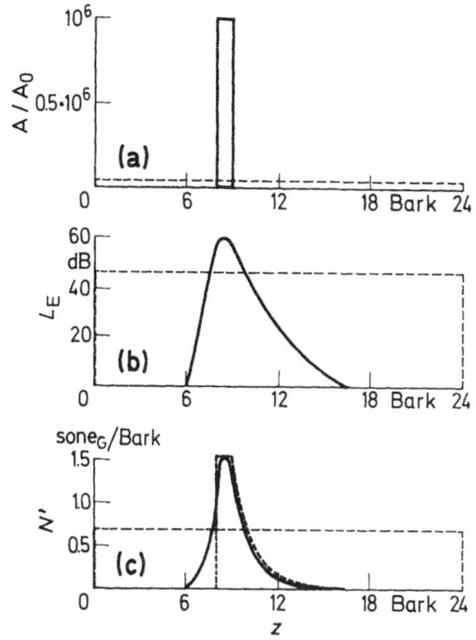

Fig. 22a—c. The incitation factor, $\frac{A}{A_0}$ (a), the excitation level, L_E (b), and the specific loudness, N' (c), as a function of the critical band rate, z. Schematic drawings are given for a 1 kHz-tone (solid) and for a uniform exciting noise (dashed) with the same level $L = 60$ dB. The dotted curve in (c) indicates the approximation used in the loudness calculation procedure

sufficient to demonstrate the application of the model. Many other examples are given in the literature (ZWICKER, 1958; ZWICKER and SCHARF, 1965; ZWICKER and FELDTKELLER, 1967).

The model of loudness formation has been extended to partially masked sounds (ZWICKER, 1963; SCHARF, 1964; ZWICKER and SCHARF, 1965) in such a way that, for example, the loudness of complex tones (composed of several pure tones) which are partially masked by a well audible background noise can be calculated.

The model and the extended model explain very precisely not only why loudness summates across frequency — as it does — but also why this effect depends on level. The application of the model can also explain why a 2 kHz-tone becomes less loud if a high-pass noise with a cut-off frequency of 2.6 kHz and a steep slope is added (ZWICKER, 1958). The excitation pattern transferred to the specific loudness pattern, which makes clear that even a pure tone produces a wide spread of activity along the critical band scale, turns out to be a useful tool for the understanding of many effects associated with loudness.

C. Loudness Calculation

To simplify the application of the loudness formation model a calculation procedure was developped (ZWICKER, 1958, 1959, 1960, 1963; ISO Rec. 532), for which sound level and spectral shape have to be presented as critical band levels or third octave band levels. The whole process of loudness formation from the excitation pattern is simplified by a graphical procedure approximating the specific loudness pattern of narrow band sounds (solid curve in Fig. 22c) by a curve with a vertical lower slope, a horizontal part of 1 Bark width and a slightly higher upper slope (dotted curve in Fig. 22 c). The dotted curve is fitted in such a way that the area below the (dotted) approximation — the total loudness — is the same as for the original curve. The procedure is standardized (ISO Rec. 532) and Fig. 23 illustrates an example of factory noise. All thin lines (solid, dashed and dotted) belong to the graphs, which are prepared for 5 different level ranges and 2 kinds of sound fields, the free and the diffuse field. The measured third octave band levels of the noise are displayed in the graph as thick horizontal lines between the corresponding cut-off frequencies of the third octave band using the ladders, the steps of which are marked in third octave band levels. At each left end of the horizontal thick lines, a vertical line towards the abscissa is added while at each right end a curve is connected, which has to be drawn parallel to the thin dashed curves travelling down smoothly to the abscissa. Thus an area is produced, which is limited by the abscissa, by the left and right end of the graph and by the upper most thick lines (see Fig. 23). This area corresponds very well to the total loudness of the noise. To calculate the area, it is transformed into a rectangular area by means of a line parallel to the abscissa from the left to the right end of the graph, the height of which is chosen by eye in such a way that the former area and the rectangular area are equal in size (thick dashed line in Fig. 23). Outside of the graph we find scales marked in loudness level as well as in loudness. The height of the rectangle is a measure for the size of the area and,

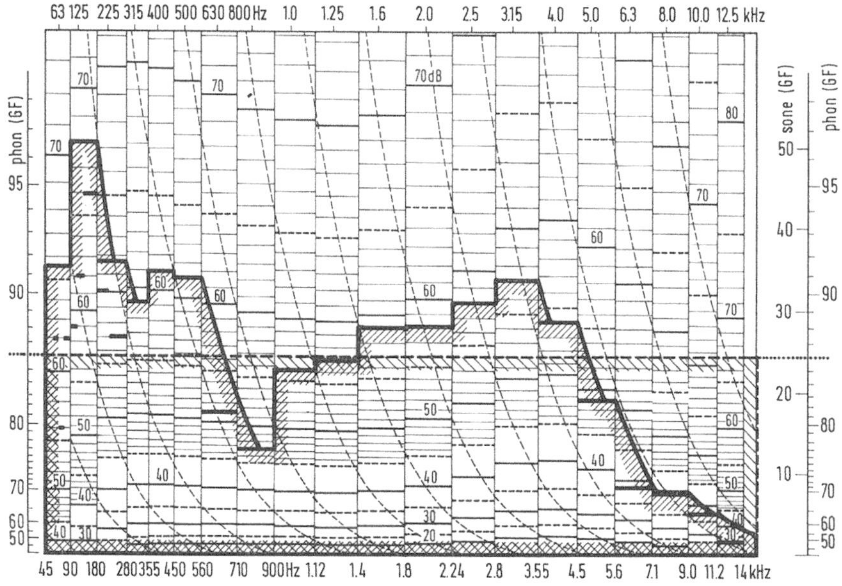

Fig. 23. Calculation of the loudness of factory noise from its third octave band levels using a graphical method based on the described model

therefore, a measure for the total loudness of the noise. In the case demonstrated in Fig. 23 the loudness reaches a value of 24 sone_{GF} corresponding to a loudness level of 86 phon_{GF}. The indices stand for critical bands (Frequenz*g*ruppen) and for *f*ree field. Such a graphical procedure, in which partial areas belong to parts of the loudness, has the advantage that not only the total loudness is calculated, but that also those spectral components of the noise which are responsible for the largest part of the loudness become clearly visible. On the other hand, the spectral parts which are masked by others are cancelled out completely because they lie below the upper most thick lines and are ignored.

D. Temporal Effects

The loudness, as well as the loudness level, depends on the temporal pattern of the sound. Effects were primarily measured then for single sound impulses as a function of the duration and for repeated short impulses as a function of the repetition rate (NIESE, 1956, 1958; PORT, 1959, 1963a, b; REICHARDT, 1965; STEVENS and HALL, 1966; WRIGHT, 1965; ZWICKER, 1966, 1969a; ZWISLOCKI, 1969). Figure 24 shows the loudness level L_N of single 1-kHz-impulses as a function of their duration T_T. The loudness level is constant for long durations above 200 msec. It decreases for shorter impulses, about 10 phon per factor 10 of shortening. For such an approximation, 100 msec is the limiting duration as indicated in Fig. 24 by the dashed line. At durations shorter than 5 msec the spectrum of the impulse exceeds the critical band drastically. Then, besides the temporal effect, a spectral effect is involved. For tones of higher frequency, the critical band is larger and measurements with shorter durations are possible. The results lead to very similar functions as shown in Fig. 24.

The loudness level increases with increasing repetition rate as shown in Fig. 25. 1 kHz impulses of 5 msec duration can be presented almost as single impulses (low repetition rate). On the other hand, the repetition rate can be increased to

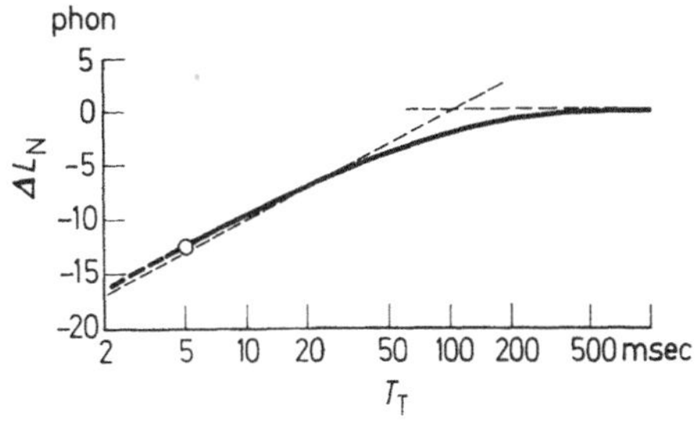

Fig. 24. Change, $\Delta L_N \equiv L_N(T_T) - L_N(T_T \to \infty)$, of the loudness level of a tone impulse as a function of its duration, T_T. The level from which the impulse is cut out is kept constant; the loudness level diminishes for decreasing duration. The dashed line indicates an approximation with a critical duration of 100 msec and a decrease of 10 phon when the duration is shortened by a factor of 10

such an extent that the end of an impulse concides with the begin of the next. In this case a continuous 1 kHz-tone is reached (for 5 msec impulses at a repetition rate of 200 Hz). This means that in Fig. 25 the endpoints or asymptotes of the curve must be the same as in Fig. 24; in both cases, the transition from single impulses to the steady state condition is measured.

Measurements with impulses cut out of broad band noise gave results similar to measurements with tone impulses (Port, 1963a, b; Niese, 1958; Zwicker, 1965b), however, there are smaller differences. Temporal effects in loudness are still discussed intensively especially because there are some discrepancies among the results of different authors (Reichardt, 1965; Zwicker, 1966). Therefore a Round Robin Test was initiated, of which only the first results are as yet available (Study Group B, ISO/TC43/SC1, 1971). Nevertheless, models

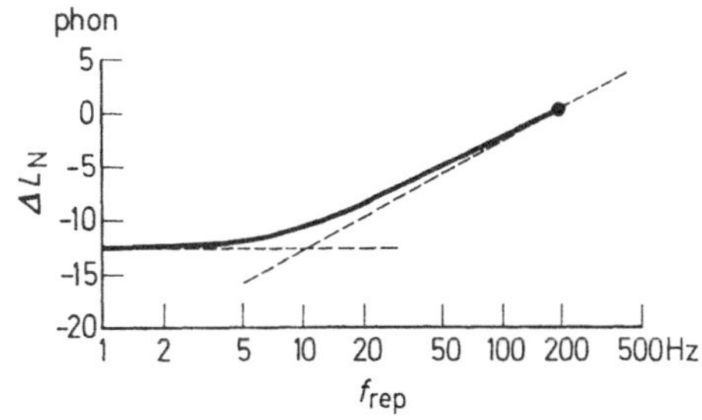

Fig. 25. Increment, $\Delta L_N = L_N(f_{rep}) - L_N(T_T \to \infty)$, of a tone burst consisting of 5 msec-1 kHz impulses as a function of their repetition rate, f_{rep}. At $f_{rep} = 200$ Hz, the impulses merge into a continuous tone. The dashed lines indicate useful approximations

describing temporal effects in loudness have been published (ZWICKER, 1968b, 1969a) and measurements on backward partial masking were undertaken (ZWICKER, 1965) to prove the influence of time structure of the excitation pattern on loudness.

VI. Pitch

Very often pitch of pure tones (STEVENS *et al.*, 1937; ZWICKER and FELDTKELLER, 1967) is meant if someone talks about the subjective magnitude pitch, although sounds composed of many harmonics (SCHOUTEN, 1938; RITSMA, 1962; WALLISER, 1969; TERHARDT, 1972a, b) or sounds as complex as bands of noise (FASTL, 1971) produce a sensation of pitch. All pitch sensations can be measured by comparing the unknown pitch of a sound with the pitch of a pure tone at a loudness level of 40 phon and variable frequency. The frequency of the pure tone is adjusted in such a way that both pitches have the same value. Thus, the pitch of the complex sound can be expressed by the frequency of the pure tone. Since the pitch of a pure tone seems to be the more basic sensation (TERHARDT, 1970, 1972a, b; 1974b) it has to be handled separately, with respect to the pitch of complex sounds.

A. The Pitch of Pure Tones

A pure tone produces a pitch sensation, which depends mainly on its frequency. If the loudness level is kept constant, the pitch of a pure tone can be measured by the method of halving and doubling the sensation by changing the frequency of the tone. The results of such measurements lead to a relation between the frequency and the pitch sensation which is well defined except for an arbitrary fixed point.

STEVENS *et al.* (1937) plotted their results using "mel" as unit for pitch sensation and chose as the fixed point, 1 kHz equal to 1000 mel (dotted line in Fig. 26).

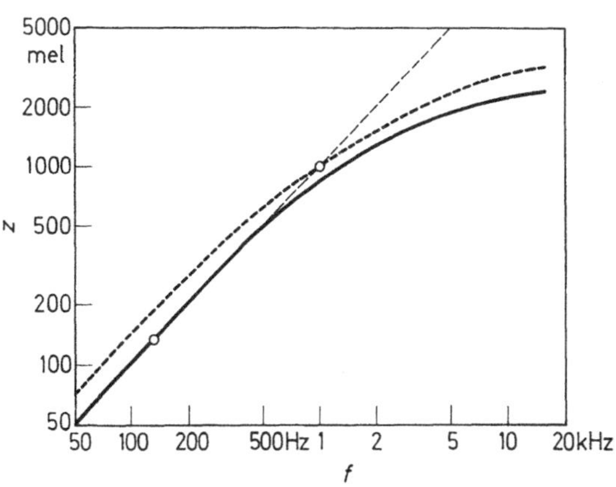

Fig. 26. Pitch, z, of a pure tone as a function of its frequency, f, at a constant loudness level of 40 phon. Fixed point 1000 Hz$\triangleq$1000 mel (dotted curve); fixed point 131 Hz$\triangleq$131 mel (solid curve). The dashed line indicates direct proportionality, $\frac{z}{\text{mel}} = \frac{f}{\text{Hz}}$

Another reasonable possibility for the fixed point was introduced by ZWICKER and FELDTKELLER (1967). They transferred 131 Hz (C_0 in the musical scale) into 131 mel; this has the advantage that frequency and pitch magnitude have the same number in the range of low frequencies, where pitch and frequency are proportional to each other, as may easily be seen in Fig. 26 (dashed line). This interesting effect disappears if the other fixed point is used. Nevertheless, the most important facts are shown by both curves: The pitch of pure tones is directly proportional to

(1) the JND_f-scale resulting from the integration of JND_fs

(2) the critical band-scale resulting from the integration of critical bands and

(3) the scale of distance of the maximum displacement on the basilar membrane from the helicotrema. If pitch and frequency are plotted on a linear scale, these facts become very clear, as shown in Fig. 27 using the fixed point 131 Hz = 131 mel.

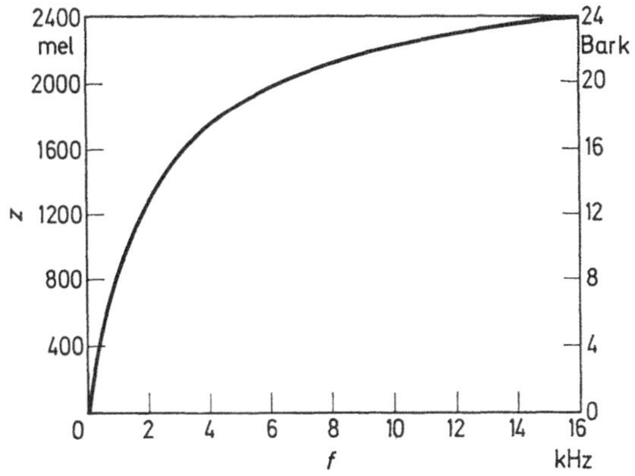

Fig. 27. Pitch, z, of a pure tone as a function of its frequency, f, on linear scales using 131 Hz ≙ 131 mel as a fixed point

The pitch sensation increases linearly with the frequency for frequencies below 500 Hz. For frequencies above 500 Hz, the pitch rises still further with increasing frequency but much less, so that, at very high frequencies around 16000 Hz, only 2400 mel are reached. This makes it understandable why high tones in the range of 8—16 kHz sound are much more equal in pitch than tones in the range of 400—800 Hz.

The pitch of pure tones does not only depend on frequency, but to a small extent also on level (S. S. STEVENS, 1935; WALLISER, 1969b; TERHARDT, 1974c). This effect is almost negligible at medium frequencies around 1 kHz, but increases for lower and higher frequencies. Low frequencies produce a pitch sensation which decreases with increasing level, while high frequencies produce a pitch which increases with increasing level. Although this effect is not very large it can easily be observed and, for a jump in loudness level from 50 phon to 90 phon, reaches a relative value of $\pm 6\%$ if tones of 150 Hz and 6 kHz, respectively, are used.

The pitch of pure tones also changes somewhat, if a partial masking sound is added. This effect was investigated by SCHUBERT (1950), WEBSTER *et al.* (1952),

WALLISER (1969b), and by TERHARDT and FASTL (1971). The latter were able to explain the diverging results by means of the assumption that the total pitch shift is composed of 2 different effects: the pitch change as a function of loudness level for different frequencies and the pitch change produced by the partial masking sound.

B. The Pitch of Complex Sounds

The discussion on the pitch of harmonic spectra started with a paper of SCHOUTEN [1938, 1940; a summary SCHOUTEN *et al.* (1962)] in which he pointed out that, besides the pitch sensation of pure tones, there exists another kind of pitch sensation which he called "Residuum". He proved that the pitch of a sound composed of many harmonic tones remains, if the fundamental tone or even several of the lowest tones are filtered out. Since that time many details of the "Residuum" have been investigated (DE BOER, 1956; FLANAGAN and GUTTMAN, 1960a, b; RITSMA, 1962, 1963, 1967; PLOMP, 1967; WALLISER, 1969c, d) for harmonic as well as for unharmonic spectra. WALLISER pointed out that the pitch of high-pass-filtered complex tones can be determined as a subharmonic of the lowest present partial tone. Recently TERHARDT (1970, 1972a, b) summarized not only the effects measured on the "Residuum" but also effects measured on pitch shifts of pure tones (S. S. STEVENS, 1935; WALLISER, 1969c; TERHARDT and FASTL, 1971) and of complex tones (LEWIS and COWAN, 1936; WALLISER, 1969a) as well as the effect of octave enlargement (WARD, 1954; WALLISER, 1969d; TERHARDT, 1971a). TERHARDT suggested the existence of 2 kinds of pitch sensation of different quality: (a) the "spectral pitch" correlated to the excitation pattern and to pure tones as a "primary sensation" and (b) the "virtual pitch" composed of the spectral pitches of dominant partials in a rather complex process. The knowledge of harmonic pitch intervals is stored in the hearing mechanism and plays a crucial part in processing "virtual pitch". From our early youth we are confronted with harmonic spectra like speech and learn to distinguish harmonic spectra from others. A generalized model according to this concept and based on results of psychoacoustic measurements (TERHARDT, 1972b) accounts for most of the phenomena known in pitch perception.

VII. Roughness

A sound with an amplitude changing periodically as a function of time with a repetition rate of about 50 Hz produces the sensation of roughness (HELMHOLTZ, 1863; WEVER, 1929; VON BÉKÉSY, 1935; MATTHES and MILLER, 1947). In the past, roughness was not studied very intensively. In the more extensive investigations of TERHARDT (1968a, b; 1974a), sinusoidal amplitude modulated tones were primarily used as sounds with a well-defined amplitude time structure. By means of the results of these measurements, one can extrapolate the roughness of sounds with almost any other periodic amplitude time structure.

Measurements of the roughness of amplitude modulated tones at different sound pressure levels indicated that roughness depends only very little on the level of the sound, but essentially on the relative amplitude fluctuation $m = \frac{\Delta p}{p}$ of the sound pressure. Comparing the roughness of a modulated tone with a sound pressure level, L, of 80 dB with the roughness of a modulated 60 dB-tone, both with $m = 1$, the difference in roughness is almost negligible, although the absolute change in sound pressure produced by the modulation differs by a factor of 10. Therefore, the relative amplitude modulation, m, is suggested to be the essential variable to describe roughness.

The relation between roughness, r, and the relative amplitude modulation, m, has been measured by TERHARDT (1968a). It is shown in Fig. 28 (the relative

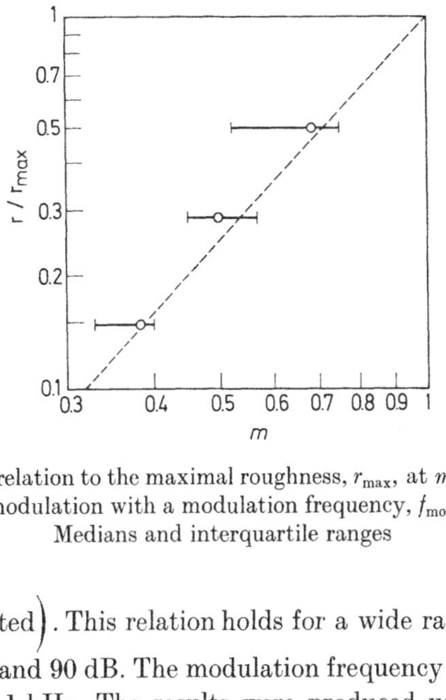

Fig. 28. Roughness, r, in relation to the maximal roughness, r_{max}, at $m = 1$ as a function of the degree, m, of sinusoidal modulation with a modulation frequency, $f_{mod} = 68$ Hz ($f_{car} = 1$ kHz). Medians and interquartile ranges

roughness $\frac{r}{r_{max}}$ is plotted). This relation holds for a wide range of sound pressure levels between 30 dB and 90 dB. The modulation frequency was 68 Hz, the center or carrier frequency, 1 kHz. The results were produced using the fractionation method (halving and doubling). Similar results have been found for other center or modulation frequencies as well as for other sounds like noises. The measured data can be described in good approximation by the power function, $r = \text{const} \cdot m^2$, which means that roughness rises with the square of the relative amplitude fluctuation. The dashed line in Fig. 28 indicates this simple relation.

The "const" in the equation mentioned above depends on center and modulation frequency. The relative roughness as a function of the modulation frequency, f_{mod}, with carrier frequency, f_{car}, as parameter is shown in Fig. 29 for a given L and m. At low modulation frequency no sensation of roughness is created although the tone fluctuates slowly in amplitude. Roughness seems to begin at about 20 Hz and rises very quickly for higher modulation frequencies. It reaches its peak value at modulation frequencies which depend on the carrier frequency.

For carrier frequencies below 2 kHz, the maximum is reached at lower modulation frequencies and is not as high as for carrier frequencies above 2 kHz where the modulation frequency for maximal roughness as well as for vanishing roughness does not depend on the carrier. In this case, the peak is reached at 70—80 Hz. For higher modulation frequency the roughness decreases quickly and reaches, at 250 Hz, $^1/_{10}$ of the peak value, which means that it almost vanishes.

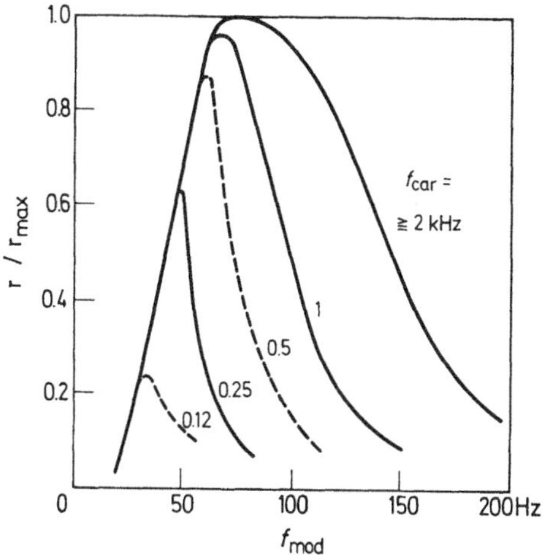

Fig. 29. Relative roughness, $\frac{r}{r_{max}}$, as a function of the modulation frequency, f_{mod}, for amplitude modulated pure tones with different carrier frequencies, f_{car}

The modulation frequency, $f_{mod\text{-}v}$, at which the sensation of roughness vanishes was studied more carefully in separate measurements as a function of the carrier frequency, f_{car}. The results are plotted in Fig. 30 in comparison to the band width of the critical band (dashed). For carrier frequencies below 1 kHz, the boundary for the existence of roughness agrees very well with the critical band width, while,

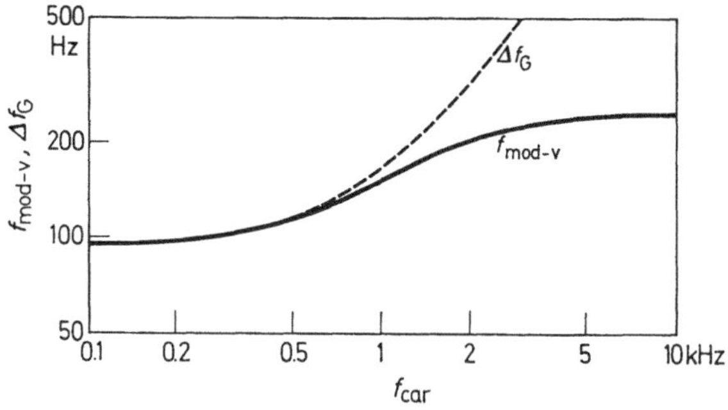

Fig. 30. Modulation frequency, $f_{mod\text{-}v}$, at which the roughness of amplitude modulated pure tones vanishes as a function of their carrier frequency, f_{car} (solid). The dashed curve indicates the width, Δf_G, of the critical band

for carrier frequencies above 2 kHz, this boundary remains constant, as already indicated in Fig. 29.

Since the critical band is a measure of the frequency selectivity of the ear, it is obvious that an amplitude modulation at high modulation frequencies can produce side tones which fall outside the critical band around the carrier. In this way, the critical band produces a boundary outside of which no overlapping of the excitations corresponding to the single lines of the modulation spectrum exists and, therefore, no roughness based on fluctuations of the excitation is produced. This spectral effect seems to be responsible for the fact that the critical band width, Δf_G, and the modulation frequency, f_{mod-v}, for vanishing roughness coincide for center frequencies below 1 kHz (Fig. 30). On the other hand, it becomes clear from Fig. 30 that roughness does not depend on critical band width for carrier frequencies above 2 kHz, where the roughness vanishes, if the modulation frequency is larger than 250 Hz. In this region, the ear is not able to follow very quick fluctuations even if its frequency selectivity allows for a strong interference between the partials of the spectrum. In this way, the shape of the curve shown in Fig. 29 for $f_{car} \geqq 2$ kHz may be interpreted as the characteristic of a kind of band pass which may be installed in the neural transmission of fluctuating excitations.

As of yet, we do not know much about the roughness produced by frequency modulated sounds, but it seems reasonable that it may be understood by means of the fluctuations of the excitation-critical band rate pattern they produce (VOGEL, 1974). Frequency modulation is thereby transferred into amplitude fluctuations of the excitation and can be treated as such. Again the described band pass characteristic is involved. The dependence of roughness on duration and other temporal effects of this sensation have not been studied.

VIII. Timbre and Sharpness

A. Timbre

Many papers have been written about timbre — see for example VON HELMHOLTZ (1863), STUMPF (1890), LICHTE (1941), LICKLIDER (1951), WELLEK (1963), S. S. STEVENS *et al.* (1965), RAHLFS (1966), PLOMP and STEENEKEN (1971), PLOMP (1970), VON BISMARCK (1971, 1974a, b) —; however, only few conclusions can be justified. One reason for this situation is the generally used definition of timbre (American Standard Association) as "that attribute of auditory sensation in terms of which a listener can judge that two sounds similarly presented and having the same loudness and pitch are dissimilar". This means that timbre should be a measure of all kinds of sensation produced by a sound except loudness and pitch. This is without doubt a very heavy (may be too heavy) load for an attribute of auditory sensation which should be measured quantitatively in number and units. Therefore, the psychophysicists tried to find a set of verbal attributes such as brightness, fullness, volume, density, sharpness, and roughness to describe timbre by splitting it into several sensations which are easier to measure quantitatively. The second reason is attributed to the fact that there are not only many verbal attributes which may be useful to describe timbre but also many physical parameters of the sound which can be changed to produce different timbre. Since only restricted samples of sounds can be used in quantitative experiments, many problems are left to be solved. Only two conclusions can be drawn without contradiction from most of the published papers: (a) the attribute timbre is multidimensional and (b) reliable measurements on verbal attributes can be obtained only for particular types of sounds.

These facts were taken into account by VON BISMARCK (1971, 1974a) when he extracted from the timbre perception those independent features, which can be described in terms of verbal attributes. VON BISMARCK used the methods of the semantic differential and factor analysis. Thereby, pairs of opposite attributes such as dark-bright or hard-soft characterized the endpoints of 30 scales. 35 different sounds (composed of line spectra and continuous spectra) were rated along those scales by two groups of subjects (musicians and non-musicians). Loudness and pitch of the sounds were equalized; only the spectral envelope was systematically changed. The results of the ratings of the two groups differed very little, but the factor analysis indicated that the 35 timbres can be described almost completely (90%) when rated only on 4 of the original 30 scales.

The factor carrying the largest part of the variance (44%) was represented by the scale dull-sharp. The scales representing the other factors appeared to be less suitable for the description of timbre in general. The portion of timbre not accounted for by sharpness seemed to lack a useful verbal psychophysical description.

Since many details of roughness had already been studied for the steady state condition, VON BISMARCK (1974b) started a second series of measurements to determine the sensation sharpness as a function of the parameters of a sound. The results of the first set of experiments suggested that sharpness is determined by the frequency location of the overall energy concentration of the spectrum.

B. Sharpness

The sharpness of steady state sounds depends mainly on the low and the high cut-off frequencies, which means on the center frequency for narrow band sounds and the spectral envelope for broad band sounds, respectively, and somewhat on level (VON BISMARCK, 1974a, b). Sharpness as an attribute of the auditory sensation can be doubled and halved or compared directly when sounds of different spectral envelopes are used. As shown in Fig. 31, sharpness of both noise and complex tones composed of many harmonics is doubled if the upper limiting frequency, f_{ul}, is shifted upward by about a factor of 3 or, expressed in terms of critical bands, by about 7 Bark. This relation can also be stated approximately as: sharpness is a power function of the upper limiting frequency with an exponent of 0.66.

Sharpness of narrow band noise depends on its center frequency. Results of experiments on halving and doubling the sharpness of narrow band noises using the method of constant stimuli (Fig. 31) indicate (Fig. 32) that sharpness can increase by more than a factor of 10 from low to high center frequencies.

A broad band noise as well as a harmonic complex tone sounds more sharp if the spectral envelope rises towards higher frequencies, instead of being flat. This effect is not as large as the effect of the upper limiting frequency but is clearly measurable. The data are given separately in Fig. 32 for noise and harmonic complex tones, using the relative sharpness S/S_0 on the ordinate, where S_{0B} and S_{0C} stand for the sharpness of a flat broad band noise and of a harmonic complex tone, respectively, both with an upper limiting frequency of 1 kHz. The shape of the spectral envelope is indicated by the slopes given in dB/octave.

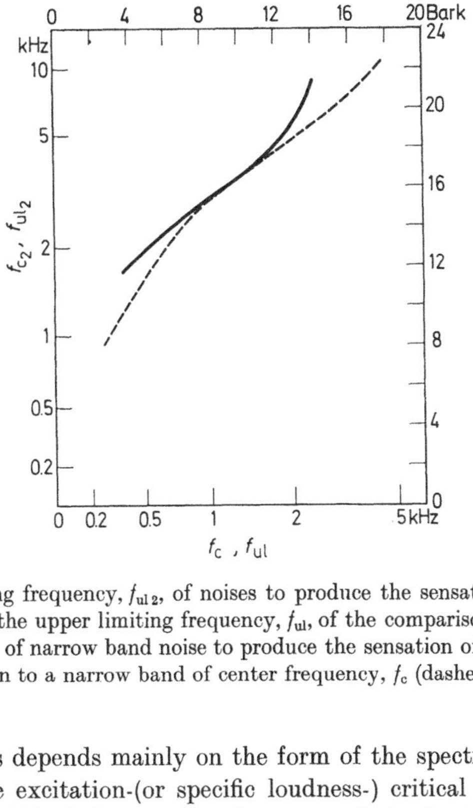

Fig. 31. Upper limiting frequency, $f_{ul\,2}$, of noises to produce the sensation of doubled sharpness as a function of the upper limiting frequency, f_{ul}, of the comparison sound (solid curve). Center frequency, f_{c2}, of narrow band noise to produce the sensation of doubled sharpness in relation to a narrow band of center frequency, f_c (dashed curve)

Since sharpness depends mainly on the form of the spectral envelope, which is correlated to the excitation-(or specific loudness-) critical band rate pattern, von Bismarck (1974b) tried to express sharpness of steady state sounds in terms of specific loudness, N', and critical band rate, z, and succeeded with the formula:

$$\frac{S}{S_0} = C \cdot \frac{\int_0^{24\,\text{Bark}} N'(z) \cdot g(z) \cdot dz}{\int_0^{24\,\text{Bark}} N' \, dz}$$

where $\int_0^{24\,\text{Bark}} N' dz$ is the total loudness, N, and $g(z)$ accounts for the fact that the sharpness of narrow band sounds like tones increases with frequency or critical band rate z, respectively, as indicated in Fig. 32 in the dashed curve. The measured sharpness and the sharpness calculated by this formula, which gives some kind of a first order momentum of the specific loudness-critical band rate pattern, are in good agreement. Even some special effects like the increment of sharpness when cutting off some low frequency components of the sound can be calculated.

Sharpness seems to be an attribute of the auditory sensation which is closely related to an attribute that was described by S. S. STEVENS (1934) as "density" and was examined by S. S. STEVENS and GUIRAO (1965).

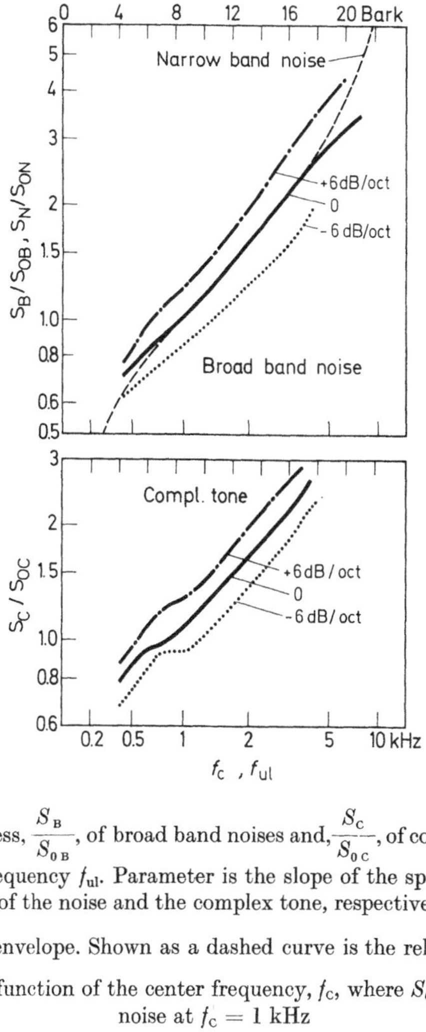

Fig. 32. Relative sharpness, $\frac{S_B}{S_{0B}}$, of broad band noises and, $\frac{S_C}{S_{0C}}$, of complex tones as a function of the upper limiting frequency f_{ul}. Parameter is the slope of the spectral envelope. S_{0B} and S_{0C} are the sharpnesses of the noise and the complex tone, respectively, with $f_{ul} = 1$ kHz and flat (0dB/oct) spectral envelope. Shown as a dashed curve is the relative sharpness, $\frac{S_N}{S_{0N}}$, of narrow band noise as a function of the center frequency, f_C, where S_{0N} is the sharpness of the noise at $f_C = 1$ kHz

IX. Subjective Duration

When listening to speech, we can concentrate on the subjective duration of parts of words such as vowels, consonants, or even pauses between them. Only two distinct changes of the parameters of the sound as a function of time are needed, to create a subjective impression of duration — the "subjective duration". Very simple examples are impulses (with a beginning and an end as the changes), pauses (quite similar), or two short clicks which mark a temporal distance that can also be judged by subjective duration. Besides loudness, pitch, and timbre, subjective duration belongs among the most important auditory attributes. Only very few studies on subjective duration have been published (Henry, 1948; Zwicker, 1969b; Burghardt, 1971; 1973a, b), perhaps because, for a long time, it was assumed that the subjective duration of a sound is equal to its physical duration. This happens only in special cases, for the most part this is not true. The difference reaches, as an extreme value, a factor of 5, which sounds unbelievable but is experimentally provable when impulses and pauses are compared, the latter in the range below 500 msec physical duration.

The just-noticeable differences in duration (JND_d) are on the order of 10% and are almost independent of level and duration. Only at low levels (<10 dB) and at very short durations (<10 msec) the JND_d increases (Henry, 1962; Zwicker, 1969b; Burghardt, 1971).

As mentioned above the auditory attribute subjective duration depends very little on level. A level difference of 40 dB from 65 dB to 25 dB, of a 1 kHz tone produces no change for long durations down to 100 msec. Only at short durations can an effect be measured so that, for example, a 15 msec - 25 dB - tone is matched to a 10 msec — 65 dB — tone (Burghardt, 1973a) for equal subjective duration.

The effect of frequency on subjective duration is not negligible, at least for durations of impulses shorter than 300 msec. Compared with a 3.2 kHz-impulse an impulse of lower or higher frequency must have a larger physical duration to produce equal subjective duration (Burghardt, 1973a). Low frequency tone impulses "last" longer compared with 3 kHz-impulses. If measurements are made for several frequencies and durations, a graph with frequency as abscissa and physical duration as ordinate can be drawn containing the contours of equal subjective duration for tone impulses (see Fig. 33), where 1 kHz is used as a

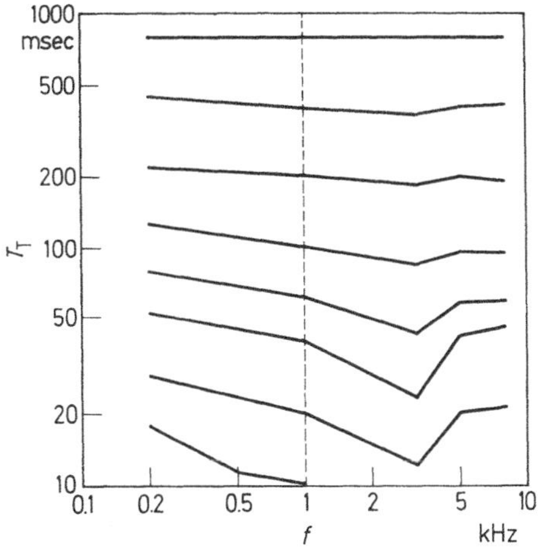

Fig. 33. Contours of equal duration. The physical duration, T_T, is plotted as a function of the frequency of the pure tones. The contours connect points of equal duration

reference frequency. The contours show a real dip at short durations for frequencies around 3 kHz. At long durations the curves are almost flat, *i.e.* the dependence on frequency disappears.

Pauses and impulses clearly differ in physical duration if they are matched to produce the same subjective duration. Figure 34 shows the results for 3 different sounds. The sequence of impulses and pauses used in the experiment is also shown in Fig. 34. Again two sets of experimental results have to be used to

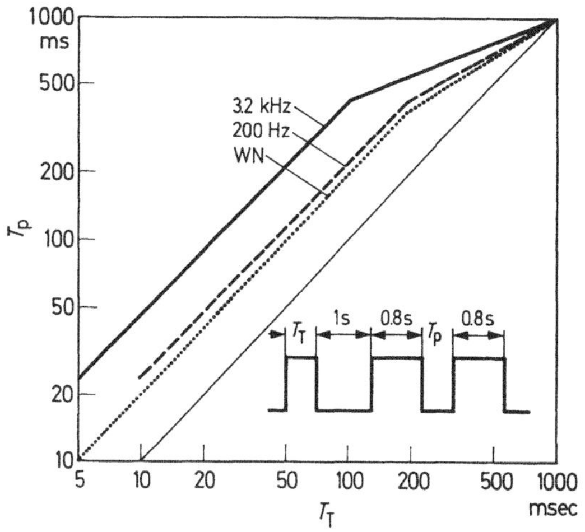

Fig. 34. Physical duration, T_P, of a pause producing the same subjective duration as an impulse as a function of the physical duration, T_T, of the impulse. Parameter is the kind of sound as indicated. The schematic drawing shows the sequence of T_T and T_P within one period of presentation

average errors in the results. The observers first adjust the duration, T_T, of the impulse in such a way that its subjective duration is equal to that of the pause with the duration, T_P. Then, T_P is varied to match the subjective duration of the impulse. For 200 Hz signals or white noise, the pauses have to be physically longer than the impulses by a factor of 2 or 3 in order to subjectively elicit the same duration. As expected from Fig. 33, this effect is largest for 3.2 kHz where it reaches a factor of 5. At pause durations larger than 500 msec, the effect decreases for all sounds and almost disappears for durations larger than 1 sec.

Measurements of halving and doubling can be arranged to obtain the relation between the stimulus' physical duration and the sensation's subjective duration (ZWICKER, 1969b; BURGHARDT, 1973). The results are shown in Fig. 35 for impulses with the reference frequency 1 kHz. The relative subjective duration, $\frac{D}{D_0}$, is drawn as a function of the impulse duration, T_T, where D_0 is the subjective duration of a 1 sec, 1 kHz tone impulse filtered by a narrow band filter to avoid clicks. For long duration both physical and subjective duration are proportional, *i.e.* the power function has an exponent of 1. For durations shorter than 100 msec, the exponent becomes smaller and the proportionality disappears, so that a physically short impulse lasts subjectively longer than expected.

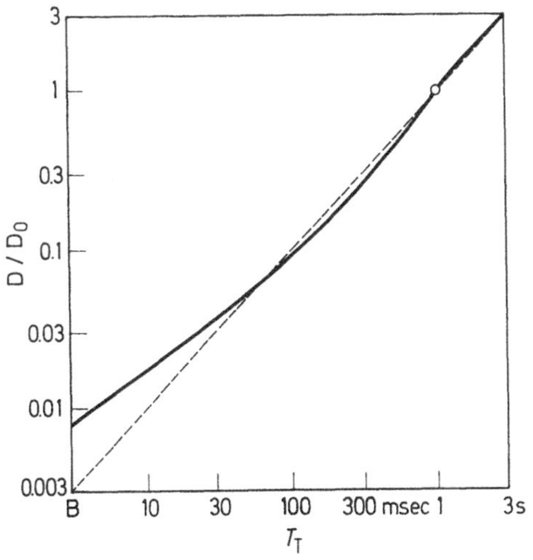

Fig. 35. Relative subjective duration, $\frac{D}{D_0}$, of a 1 kHz-tone-impulse as a function of its physical duration, T_T. As fixed point the subjective duration, D_0 at $T_T = 1$ sec, is chosen. The dashed line indicates proportionality

X. Influence of Phase Changes on the Auditory Sensation

Ohm's acoustical law (1843) can be reduced to the assumption that the phase relation between single components of the stimulus has no influence on the auditory sensation. This holds for very many stimuli and has been proved for JNDs of loudness, pitch, and timbre. All effects which disagree with OHM's law can be interpreted by the incomplete Fourier analysis of the ear which can be expressed most conveniently by the excitation pattern: If several stimuli produce separate excitation patterns without overlap, the phase condition between the stimuli cannot influence the sensation. As soon as an overlap occurs the sensation may be influenced by the phase condition.

This effect has been measured for the loudness of amplitude modulated tones (Bauch, 1956). In Fig. 36, the loudness level, L_N, of a 50% amplitude modulated 1 kHz tone is shown as a function of the modulation frequency, f_{mod}. For high modulation frequency, *i.e.* for separate excitation curves of the three spectral lines, the total loudness is the sum of the loudnesses of the single parts independent of phase. For modulation frequencies lower than half a critical band ($1/2\, \Delta f_G$), *i.e.* for total frequency separation of the spectral lines smaller than a critical band, the sound intensity of all spectral lines together is responsible for loudness, which is independent of the phase conditions. Only for very low frequencies of modulation, phase changes between the spectral lines influence the loudness in such a way that the peak value of the produced sound pressure becomes responsible for loudness. Since amplitude modulated tones can be transferred by changing phase conditions into quasi frequency modulated tones with almost constant amplitude, one can expect and finds that frequency modulated tones are less loud than amplitude modulated tones for which the peak value is responsible for loudness.

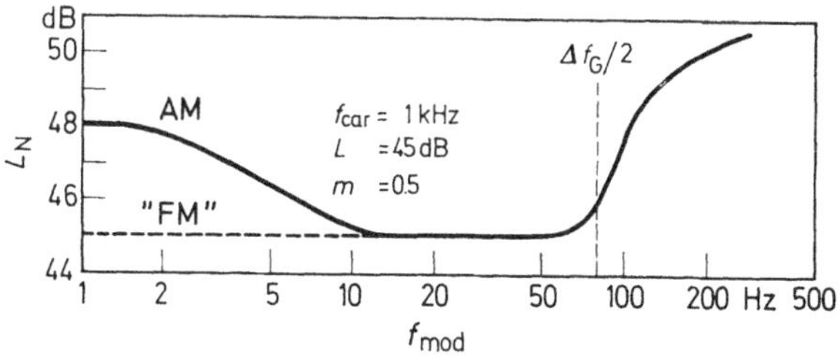

Fig. 36. Loudness level, L_N, of a 50% amplitude modulated 1 kHz tone with a level of 45 dB as a function of its modulation frequency, f_{mod} (AM). "FM" indicates the loudness level of a quasi frequency modulated tone with the same spectrum (dashed)

The results of pitch measurements diverge somewhat with respect to the question how much sensation of pitch is influenced by phase changes (RITSMA and ENGEL, 1964; PLOMP, 1964; WIGHTMAN, 1972).

The magnitude described varies from little to none; however, at low frequencies there seems to be an effect (TERHARDT and FASTL, 1971), which occurs if only two tones are presented in harmonic condition so that a peak in compression or a peak in rarefaction is realized.

Timbre is influenced by the phase relation between the partials (PLOMP and STEENEKEN, 1969). Only few measurements have been published so far, but it has been indicated (ZWICKER, 1970) that at least some of the measured effects can be described using the excitation-pattern model.

The JND_L and the JND_f can be measured as thresholds of amplitude modulation and as threshold of frequency modulation, respectively, both as functions of the repetition rate, *i.e.* the frequency of modulation, f_{mod}. As described above, amplitude modulation can be transferred into quasi frequency modulation and vice versa by changing phase conditions. The thresholds of modulation are expressed in the just-noticeable degree, m, of amplitude modulation and the just-noticeable change, Δf, of frequency modulation as a function of the frequency, f_{mod}, of modulation. If m for amplitude modulation and $\frac{\Delta f}{f_{mod}}$ for frequency modulation are plotted, as shown in Fig. 37, the spectral components of both kinds of modulation are the same if m and $\frac{\Delta f}{f_{mod}}$ have the same value and are smaller than 0.3. In such a presentation of the JND_L and the JND_f (ZWICKER, 1952), the difference between both kinds of just-noticeable modulation disappears for modulation frequencies larger than half a critical band. In this range, only the amplitude of the spectral lines is responsible for the threshold of modulation regardless of amplitude or frequency modulation, *i.e.* regardless of phase relation of the spectral lines. For lower frequencies of modulation, the two thresholds of modulation diverge, indicating that phase relation is important if the total frequency separation of the 3-line-spectrum becomes smaller than a critical band. As soon as 3 lines of a harmonic spectrum lie within one critical band, this situation may occur and phase relation may influence the sensation even high above threshold of modulation as demonstrated for roughness by TERHARDT (1968a).

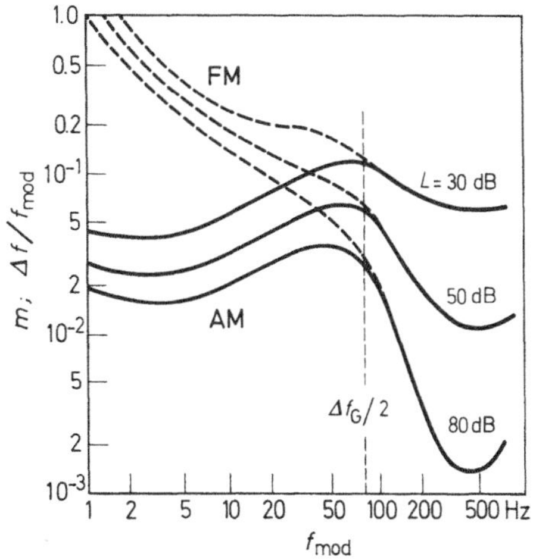

Fig. 37. Just-noticeable amplitude modulation (AM) expressed by the degree, m, of modulation and just-noticeable frequency modulation (FM) expressed by $\frac{\Delta f}{f_{\mathrm{mod}}}$ as a function of the modulation frequency, f_{mod}. Parameter is the level, L, of the modulated 1 kHz tone

XI. Nonlinear Distortion

The transmission characteristic of the ear is — as the characteristic of any transducer — somewhat nonlinear. Most prominent and important are the combination tones produced by this nonlinearity. They can easily be heard if, for example, relatively small intervals are played simultaneously on two flutes. Regarding all combination and harmonic tones produced by nonlinearity, the quadratic and the cubic difference tones are the dominant ones.

A method of compensation was introduced (ZWICKER, 1956) to measure quantitatively the magnitude of the difference tones produced by two pure tones. The possibility to compensate for a difference tone by an externally produced tone of the same frequency but variable amplitude and phase leads to the assumption that the nonlinearity is processed before the excitation of the inner ear is transferred into neural activity. The transmission system of the ear previous to the transformation into neural activity can be discussed as a system containing only smaller nonlinearity. In such a system, the input sound pressure, p_1, and the output sound pressure, p_2, are related by the equation

$$p_2 = p_1 + \alpha_2\, p_1^2 + \alpha_3\, p_1^3,$$

if only quadratic and cubic distortion are taken into account. If p_1 is composed of two tones (A and B) with the frequencies f_A and f_B (with $f_A < f_B$) and the levels L_A and L_B, the level L_{D2} of the quadratic difference tone calculated on the basis of the mentioned equation rises with L_A as well as with L_B. The measured results (ZWICKER, 1956) show exactly this behaviour. Therefore, the quadratic distortion of the ear can be assumed to behave "regularly" as expressed in the equation above. The coefficient, α_2, varies from person to person by about a factor of 10, and as a function of frequency by about the same factor.

The cubic difference tone behaves differently and does not follow the law for regular distortions as given in the equation above. The sound pressure, P_{D3B}, of the cubic difference tone with the frequency, $2f_1$-f_2, should rise with P_B (proportional), but with $P_A{}^2$ (quadratic). The measured results (ZWICKER, 1968a; HELLE, 1967, 1970) show this increase as a function of L_B (Fig. 38) only for levels $L_B < L_A$

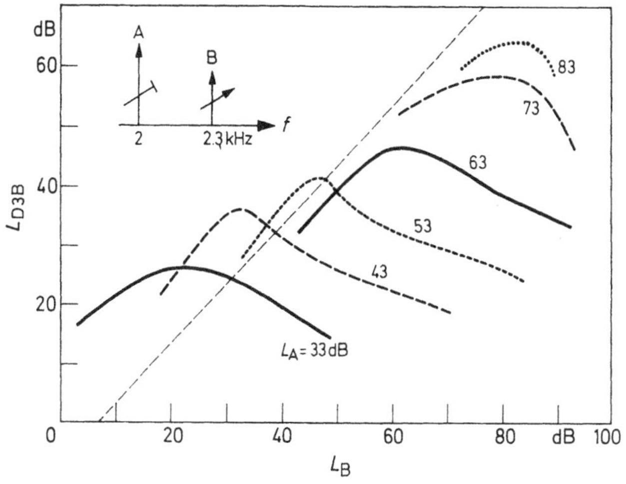

Fig. 38. Level, L_{D3B}, of the cubic difference tone produced by the primaries, A and B, (indicated) as a function of the level, L_B. Parameter is the level, L_A. The dashed line indicates the slope expected for regular nonlinear distortion

as a comparison with the dashed thin line (expected function) indicates. For levels $L_B > L_A$, the cubic difference tone does not at all follow the rules of regular distortion. As a function of L_A the slope of the L_{D3B}-curves should be twice as steep as shown in Fig. 38 (see dashed thin line in Fig. 39). The measured results (Fig. 39) show less agreement even for long levels: Instead of varying only the level L_A *or* L_B, the levels L_A *and* L_B can be changed simultaneously. The disagreement of the measured results in this condition with the expected shape of the dependence becomes clear in Fig. 40. The measured level, L_{D3B}, is shown as a function of $L_A = L_B + 10$ dB (ZWICKER, 1956). A decrease of L_A from 100 dB to 70 dB produces a decrease of L_{D3B} of only 10 dB although a decreeas of as much as 90 dB is expected as indicated by the dashed lines. To stress this effect several lines are drawn; between two adjacent lines the coefficient α_3 changes by a factor of 10.

This unexpected dependence of the level, L_{D3B} of the cubic difference tone was found for all subjects with only small differences in the shape of the curves. L_{D3B} depends somewhat on the frequency of f_A and f_B, respectively. It depends strongly

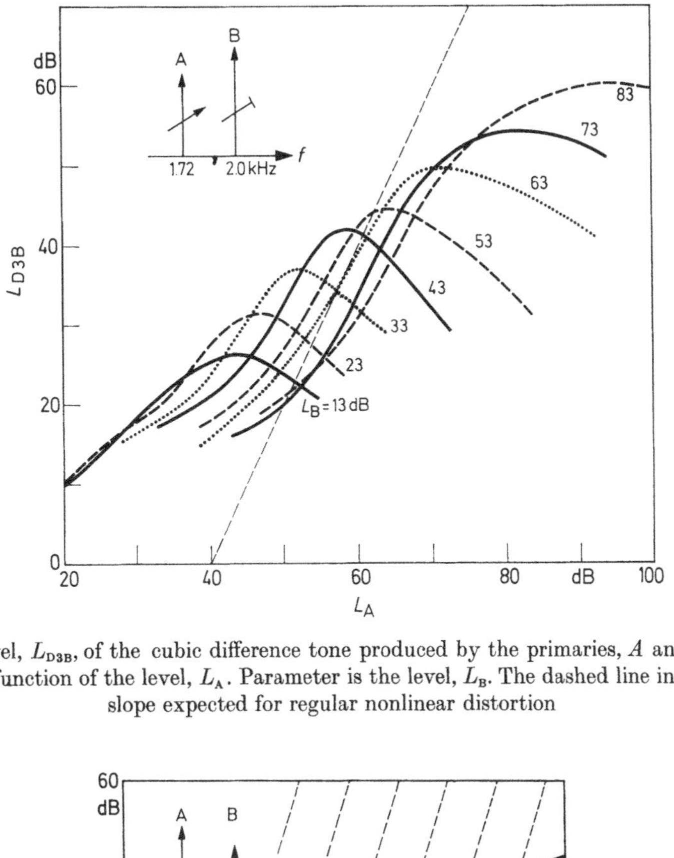

Fig. 39. Level, L_{D3B}, of the cubic difference tone produced by the primaries, A and B, (indicated) as a function of the level, L_A. Parameter is the level, L_B. The dashed line indicates the slope expected for regular nonlinear distortion

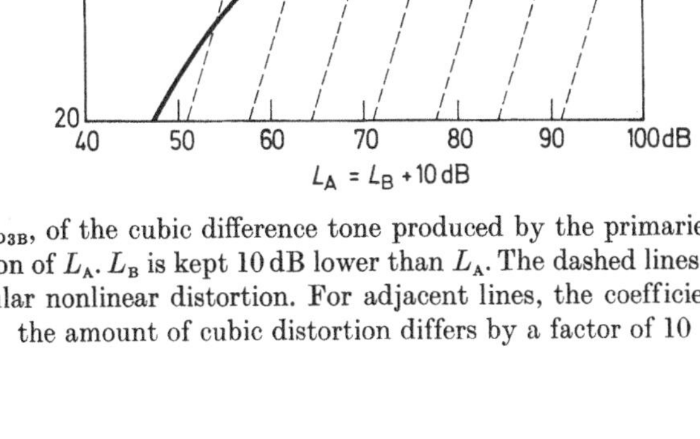

Fig. 40. Level, L_{D3B}, of the cubic difference tone produced by the primaries, A and B, (indicated) as a function of L_A. L_B is kept 10 dB lower than L_A. The dashed lines indicate the slope expected for regular nonlinear distortion. For adjacent lines, the coefficient responsible for the amount of cubic distortion differs by a factor of 10

on the frequency difference, $f_A - f_B$, in such a way that it stays constant for frequency differences smaller than the critical band and drops down quickly for larger differences. This phenomenon indicates some correlation between the spatial excitation pattern and the cubic distortion (ZWICKER, 1968). The great cubic distortion as a source for the cubic difference tone seems to take place only in the region where the excitation patterns produced by the tone, A, and the tone, B, respectively, overlap. Detailed studies of GOLDSTEIN (1967), HELLE (1969, 1970), GOLDSTEIN and KIANG (1968) confirmed earlier results.

GREENWOOD (1971) measured the threshold of pure tones masked by two tones or by a tone and a narrow band noise, respectively. The results indicate that the difference tone and difference noises produced by the cubic distortion, influence the masked threshold drastically in the frequency range, where they are expected (ZWICKER and FASTL, 1973).

Most authors assume that the cubic distortion originates in the inner ear prior to the transformation of hydroacoustical stimulation of the haircells into neural activity. Since the frequency selectivity of the distortion seems to be much smaller than the displacement of the basilar membrane, we may look for the source within the cochlear duct (ZWICKER, 1967; GOLDSTEIN, 1967).

References

BAUCH, H.: Die Bedeutung der Frequenzgruppe für die Lautheit von Klängen. Acustica **6**, 40—45 (1956).

BÉKÉSY, G. VON: Über akustische Rauhigkeit. Z. Techn. Phys. **16**, 276 (1935).

BÉKÉSY, G. VON: Über die Resonanzkurve und die Abhängigkeit der verschiedenen Stellen der Schneckentrennwand. Akust. Z. **8**, 66 (1943).

BÉKÉSY, G. VON: Über ein neues Audiometer. A. E. Ü. **1**, 13 (1947).

BISMARCK, G. VON: Timbre of steady sounds: scaling of sharpness. Proc. 7th Int. Congr. on Acoustics, Budapest, Vol. 3, pp. 637—640 (1971).

BISMARCK, G. VON: Timbre of steady sounds: a factorial investigation of its verbal attributes. Acustica **30**, 146 (1974).

BISMARCK, G. VON: Sharpness as an attribute of the timbre of steady sounds. Acustica **30**, 159 (1974).

BOER, E. DE: On the "residue" in hearing. Dissertation Universität Amsterdam (1956).

BRITTAIN, F. H.: The loudness of continuous spectrum noise and its application to loudness measurements. J. acoust. Soc. Amer. **11**, 113 (1939).

BURGHARDT, H.: Subjective duration of sinusoidal tones. Proc. 7th Int. Congr. on Acoustics, Budapest, Vol. 3, pp. 353—356 (1971).

BURGHARDT, H.: Die subjektive Dauer schmalbandiger Schalle bei verschiedenen Frequenzlagen. Acustica (1973a).

BURGHARDT, H.: Über die subjektive Dauer von Schallimpulsen und Schallpausen. Acustica (1973b).

ELLIOTT, L. L.: Backward and forward masking of probe tones of different frequencies. J. acoust. Soc. Amer. **34**, 1116 (1962).

ELLIOTT, L. L.: Changes in the simultaneous masked threshold of brief tones. J. acoust. Soc. Amer. **38**, 738 (1965).

ELLIOTT, L. L.: Development of auditory narrow-band frequency contours. J. acoust. Soc. Amer. **42**, 143 (1967).

FASTL, H.: Über Tonhöhenempfindungen bei Rauschen. Acustica **25**, 350—354 (1971).

FASTL, H.: Transient masking pattern of narrow band maskers. In: ZWICKER, E., TERHARDT, E. (Eds.): Facts and Models in Hearing, Berlin-Heidelberg-New York: Springer 251—257 (1974).

FELDTKELLER, R., OETINGER, R.: Die Hörbarkeitsgrenzen von Impulsen verschiedener Dauer. Akust. Beih. (Acustica) **2**, 489—493 (1956).
FLANAGAN, J. L., GUTTMAN, N.: On the pitch of periodic pulses. J. acoust. Soc. Amer. **32**, 1308 (1960a).
FLANAGAN, J. L., GUTTMAN, N.: Pitch of periodic pulses without fundamental component. J. acoust. Soc. Amer. **32**, 1319 (1960b).
GÄSSLER, G.: Über die Hörschwelle für Schallereignisse mit verschieden breitem Frequenzspektrum. Akust. Beih. (Acustica) **1**, 408—414 (1954).
GOLDSTEIN, J. L.: Auditory nonlinearity. J. acoust. Soc. Amer. **41**, 676 (1967).
GOLDSTEIN, J. L., KIANG, N. Y. S.: Neural correlates of the aural combination tone $2f_1—f_2$. Proc. IEEE **56**, 981 (1968).
GREEN, D. M.: Masking with continuous and pulsed sinusoids. J. acoust. Soc. Amer. **46**, 939 (1969).
GREENWOOD, D. D.: Masking by combination bands. J. acoust. Soc. Amer. **52**, 1144 (1972).
HELLE, R.: Amplitude und Phase des im Gehör gebildeten Differenztones 3. Ordnung. Acustica **22**, 74—87 (1969/70).
HELMHOLTZ, H. F. L., VON: Die Lehre von den Tonempfindungen als physiologische Grundlage für die Theorie der Akustik. Braunschweig: F. Vieweg & Sohn 1863.
HENRY, F. M.: Discrimination of the duration of a sound. J. exp. Psychol. **75**, No. 3, 401 (1962).
ISO-Rec. 226: Equal loudness contours for pure tones (free field), threshold of hearing (binaural free field), age corrections.
ISO-Rec. 131: Relation between phon and sone scales.
ISO-Rec. 532: Method for calculating loudness level.
LEWIS, D., COWAN, M.: Influence of intensity on the pitch of violin and cello tones. J. acoust. Soc. Amer. **8**, 20 (1936).
LICHTE, W. H.: Attributes of complex tones. J. exp. Psychol. **28**, 455 (1941).
LICKLIDER, J. C. R.: Basic correlates of the auditory stimulus. In: STEVENS, S. S. (Ed.): Handbook of experimental psychology. New York: Wiley and Sons 1951.
MAIWALD, D.: Beziehungen zwischen Schallspektrum, Mithörschwelle und der Erregung des Gehörs. Acustica **18**, 69—80 (1967a).
MAIWALD, D.: Ein Funktionsschema des Gehörs zur Beschreibung der Erkennbarkeit kleiner Frequenz- und Amplitudenänderungen. Acustica **18**, 81—92 (1967b).
MAIWALD, D.: Die Berechnung von Modulationsschwellen mit Hilfe eines Funktionsschemas. Acustica **18**, 193—207 (1967c).
MATHES, R. C., MILLER, R. L.: Phase effects in monaural perception. J. acoust. Soc. Amer. **19**, 780 (1947).
NIESE, H.: Vorschläge für die Definition und Messung der Deutlichkeit nach subjektiven Grundlagen. Hochfrequenztechn. u. Elektroakust. **65**, 4 (1956).
NIESE, H.: Die Lautstärkeempfindung von rhythmischen Geräuschen sehr geringer Periodizität. Hochfrequenztechn. u. Elektroakust. **67**, 26 (1958).
OHM, G. S.: Über die Definition des Tones, nebst daran geknüpfter Theorie der Sirene und ähnlicher tonbildender Vorrichtungen. Ann. Phys. Chem. **59**, 513 (1843).
PLOMP, R.: The ear as a frequency analyser. J. acoust. Soc. Amer. **36**, 1628 (1964).
PLOMP, R.: Pitch of complex tones. J. acoust. Soc. Amer. **41**, 1526 (1967).
PLOMP, R.: Timbre as a multidimensional attribute of complex tones. In: PLOMP, R., SMOORENBURG, G. F. (Eds.): Frequency analysis and periodicity detection in hearing. Leiden: A. W. Sijthoff 1970.
PLOMP, R., STEENEKEN, H. J. M.: Effect of phase on the timbre of complex tones. J. acoust. Soc. Amer. **46**, 409 (1969).
PLOMP, R.: Pitch versus timbre. Proc. 7th Int. Congr. on Acoustics, Budapest 20-H-7 (1971).
PORT, E.: Die Lautstärke von Tonimpulsen verschiedener Dauer. Frequenz **13**, 242—245 (1959).
PORT, E.: Über die Lautstärke einzelner kurzer Schallimpulse. Acustica **13**, 212—223 (1963).
PORT, E.: Zur Lautstärkeempfindung und Lautstärkemessung von pulsierenden Geräuschen. Akust. Beih. (Acustica) 224—233 (1963).
RAAB, D. H.: Backward masking. Psych. Bull. **60**, 118 (1963).

Rahlfs, V.: Psychometrische Untersuchungen zur Wahrnehmung musikalischer Klänge. Dissertation Universität Hamburg (1966).

Reichardt, W.: Zur Trägheit der Lautstärkebildung. Acustica **15**, 345 (1965).

Riesz, R. R.: Differential intensity sensivity of the ear for pure tones. Phys. Rev. **31**, 867 (1928).

Ritsma, R. J.: Existence region of the tonal residue. I. J. acoust. Soc. Amer. **34**, 1224 (1962).

Ritsma, R. J.: Existence region of the tonal residue. II. J. acoust. Soc. Amer. **35**, 1241 (1963).

Ritsma, R. J.: Frequencies dominant in the perception of the pitch of complex sounds. J. acoust. Soc. Amer. **42**, 191 (1967).

Ritsma, R., Engel, F. L.: Pitch of frequency modulated signals. J. acoust. Soc. Amer. **36**, 1637 (1964).

Robinson, D. W.: The subjective loudness scale. Acustica **7**, 217 (1957).

Robinson, D. W., Dadson, R. S.: Threshold of hearing and equal loudness relations of pure tones, and the loudness function. J. acoust. Soc. Amer. **29**, 1284 (1957).

Scharf, B.: Partial masking. Acustica **14**, 16 (1964).

Scholl, H.: Das dynamische Verhalten des Gehörs bei der Unterteilung des Schallspektrums in Frequenzgruppen. Acustica **12**, 101—107 (1962).

Schouten, J. F.: The perception of subjective tones. Proc. Konikl. Ned. Akad. Wetenschap. **41**, 1086 (1938).

Schouten, J. F.: The residue, a new component in subjective sound analysis. Proc. Konikl. Ned. Akad. Wetenschap. **43**, 356 (1940).

Schouten, J. F. *et al.*: Pitch of the residue. J. acoust. Soc. Amer. **34**, 1428 (1962).

Schubert, E. D.: The effect of a thermal masking noise on the pitch of a pure tone. J. acoust. Soc. Amer. **22**, 497 (1950).

Stevens, S. S.: Tonal density. J. exp. Psychol. **17**, 585 (1934).

Stevens, S. S.: The relation of pitch to intensity. J. acoust. Soc. Amer. **6**, 150 (1935).

Stevens, S. S.: Handbook of experimental psychology. New York: Wiley and Sons 1951.

Stevens, S. S.: The measurement of loudness. J. acoust. Soc. Amer. **27**, 815 (1955).

Stevens, S. S.: Concerning the form of the loudness function. J. acoust. Soc. Amer. **29**, 603 (1957).

Stevens, S. S. *et al.*: A scale for the measurement of the psychological magnitude pitch. J. acoust. Soc. Amer. 8, 185 (1937).

Stevens, S. S. *et al.*: Loudness a product of volume times density. J. exp. Psychol. **69**, 503 (1965).

Stevens, J. C., Hall, J. W.: Brightness and loudness as functions of stimulus duration. Percept. Psychophys. **1**, 319 (1966).

Study-Group B, ISO/TC43/SC1 (1971).

Stumpf, C.: Tonpsychologie. Leipzig: S. Hirzel Verlag 1890.

Terhardt, E.: Beitrag zur Ermittlung der informationstragenden Merkmale von Schallen mit Hilfe der Hörempfindung. Dissertation TH Stuttgart (1968a).

Terhardt, E.: Über die durch amplitudenmodulierte Sinustöne hervorgerufene Hörempfindung. Acustica **20**, 210—214 (1968b).

Terhardt, E.: Über akustische Rauhigkeit und Schwankungsstärke. Acustica **20**, 215—224 (1968c).

Terhardt, E.: Frequency analysis and periodicity detection in the sensations of roughness and periodicity pitch. In: Plomp, R., Smoorenburg, G. F. (Eds.): Frequency analysis and periodicity detection in hearing, pp. 278—287. Leiden, Niederlande: Sijthoff 1970.

Terhardt, E.: Die Tonhöhe harmonischer Klänge und das Oktavintervall. Acustica **24**, 126—136 (1971).

Terhardt, E.: Zur Tonhöhenwahrnehmung von Klängen. I. Psychoakustische Grundlagen. Acustica **26**, 173—186 (1972).

Terhardt, E.: Zur Tonhöhenwahrnehmung von Klängen. II. Ein Funktionsschema. Acustica **26**, 187—199 (1972).

Terhardt, E.: On the perception of periodic sound fluctuations (roughness). Acustica **30**, 201—213 (1974a).

Terhardt, E.: Pitch, consonance, and harmony. J. acoust. Soc. Amer. **55**, 1061—1069 (1974b).

TERHARDT, E.: Pitch of pure tones: its relation to intensity. In: ZWICKER, E., TERHARDT, E. (Eds.): Facts and Models in Hearing, Springer, Berlin/Heidelberg/New York, 353—360 (1974c).

TERHARDT, E., FASTL, H.: Zum Einfluß von Störtönen und Störgeräuschen auf die Tonhöhe von Sinustönen. Acustica **25**, 53—61 (1971).

VOGEL, A.: Roughness and its relation to the time-pattern of psychoacoustical excitation. In: ZWICKER, E., TERHARDT, E. (Eds.): Facts and Models in Hearing, Springer, Berlin/Heidelberg/New York, 241—250 (1974).

WALLISER, K.: Zusammenhänge zwischen dem Schallreiz und der Periodentonhöhe. Acustica **21**, 319—329 (1969).

WALLISER, K.: Über die Abhängigkeit der Tonhöhenempfindung von Sinustönen vom Schallpegel, von überlagertem drosselndem Störschall und von der Darbietungsdauer. Acustica **21**, 211—221 (1969).

WALLISER, K.: Zur Unterschiedsschwelle der Periodentonhöhe. Acustica **21**, 329—336 (1969).

WALLISER, K.: Über ein Funktionsschema für die Bildung der Periodentonhöhe aus dem Schallreiz. Kybernetik **6**, 65—72 (1969).

WARD, W.D.: Subjective musical pitch. J. acoust. Soc. Amer. **26**, 369 (1954).

WEBSTER, J.C. *et al.*: The masking and pitch shifts of pure tones near abrupt changes in a thermal noise spectrum. J. acoust. Soc. Amer. **24**, 147 (1952).

WEGEL, R.L., LANE, C.E.: The auditory masking of one pure tone by another and its probable relation to the dynamics of the inner ear. Phys. Rev. **23**, 266 (1924).

WELLEK, A.: Musikpsychologie und Musikästhetik. Frankfurt a. Main: Akademische Verlagsgesellschaft 1963.

WEVER, E.G.: Beats and related phenomena resulting from the simultaneous sounding of two tones. Psychol. Rev. **36**, 402 (1929).

WIGHTMAN, F.L.: Pitch and stimulus fine structure. J. acoust. Soc. Amer. **54**, 397—406 (1973).

WRIGHT, H.N.: Loudness as a function of duration. J. acoust. Soc. Amer. **37**, 1174 (1965).

ZWICKER, E.: Die Grenzen der Hörbarkeit der Amplitudenmodulation und der Frequenzmodulation eines Tones. Akust. Beih. (Acustica) H. 3, 125—133 (1952).

ZWICKER, E.: Die Verdeckung von Schmalbandgeräuschen durch Sinustöne. Akust. Beih. (Acustica) **1**, 415—420 (1954).

ZWICKER, E.: Der ungewöhnliche Amplitudengang der nichtlinearen Verzerrungen des Ohres. Akust. Beih. (Acustica) **1**, 67—74 (1955).

ZWICKER, E.: Die elementaren Grundlagen zur Bestimmung der Informationskapazität des Gehörs. Acustica **6**, 365—381 (1956).

ZWICKER, E.: Über psychologische und methodische Grundlagen der Lautheit. Akust. Beih. (Acustica) **1**, 237—258 (1958).

ZWICKER, E.: Ein graphisches Verfahren zur Bestimmung der Lautstärke und der Lautheit aus dem Terzpegeldiagramm. Frequenz **13**, 234—238 (1959).

ZWICKER, E.: Ein Verfahren zur Berechnung der Lautstärke. Akust. Beih. (Acustica) **1**, 304—308 (1960).

ZWICKER, E.: Subdivision of the audible frequency range into critical bands (Frequenzgruppen). J. acoust. Soc. Amer. **33**, 248 (1961).

ZWICKER, E.: Direct comparisons between the sensations produced by frequency modulation and amplitude modulation. J. acoust. Soc. Amer. **34**, 1426—1430 (1962).

ZWICKER, E.: Über die Lautheit von ungedrosselten und gedrosselten Schallen. Akust. Beih. (Acustica) 194—211 (1963).

ZWICKER, E.: Temporal effects in simultaneous masking by white-noise bursts. J. acoust. Soc. Amer. **37**, 653—663 (1965a).

ZWICKER, E.: Temporal effects in simultaneous masking and loudness. J. acoust. Soc. Amer. **38**, 132—141 (1965b).

ZWICKER, E.: Ein Beitrag zur Lautstärkenmessung impulshaltiger Schalle. Acustica **17**, 11—22 (1966).

ZWICKER, E.: Der Kubische Differenzton und die Erregung des Gehörs. Acustica **20**, 206—209 (1968a).

ZWICKER, E.: A model describing temporal effects in loudness and threshold. Reports of the VI. Int. Congr. on Acoustics, Tokyo S. A-21—A-24 (1968b).

ZWICKER, E.: Der Einfluß der zeitlichen Struktur von Tönen auf die Addition von Teillautheiten. Acustica **21**, 16—25 (1969a).

ZWICKER, E.: Subjektive und objektive Dauer von Schallimpulsen und Schallpausen. Acustica **22**, 214—218 (1969b).

ZWICKER, E.: Masking and psychological excitation as consequences of the ear's frequency analysis. In: PLOMP, R., SMOORENBURG, G. F. (Eds.): Frequency analysis and periodicity detection in hearing, pp. 376—394. Leiden, Niederlande: Sijthoff 1970.

ZWICKER, E.: Temporal effects in psychoacoustical excitation. In: "Basic Mechanisms in Hearing". Academic Press, New York/London, S. 809—824 (1973).

ZWICKER, E.: Mithörschwellen und Erregungsmuster stark frequenzmodulierter Töne. Acustica **31**, 243—256 (1974a).

ZWICKER, E.: On a psychoacoustical equivalent of tuning curves. In: ZWICKER, E., TERHARDT, E. (Eds.): Facts and Models in Hearing, Springer, Berlin-Heidelberg-New York, 132—141 (1974b).

ZWICKER, E. *et al.*: Critical band width in loudness summation. J. acoust. Soc. Amer. **29**, 548—557 (1957).

ZWICKER, E., FASTL, H.: On the development of the critical band. J. acoust. Soc. Amer. **52**, 699—702 (1972).

ZWICKER, E. und FASTL, H.: Cubic difference sounds measured by threshold — and compensation — method. Acustica **29**, 336—343 (1973).

ZWICKER, E., FELDTKELLER, R.: Die Lautstärke von gleichförmigen Geräuschen. Acustica **5**, 304 (1955).

ZWICKER, E., FELDTKELLER, R.: Das Ohr als Nachrichtenempfänger, 2. Aufl. Stuttgart: Hirzel-Verlag 1967.

ZWICKER, E., KAISER, W.: Der Verlauf der Modulationsschwellen in der Hörfläche. Akust. Beih. (Acustica) **4**, 239—246 (1952).

ZWICKER, E., SCHARF, B.: A model of loudness summation. Psych. Rev. **72**, 3—26 (1965).

ZWICKER, E. und SCHÜTTE, H.: On the time pattern of the threshold of tone impulse. masked by narrow band noise. Acustica **29**, 343—347 (1973).

ZWISLOCKI, J.: Theory of temporal auditory summation. J. acoust. Soc. Amer. **32**, 1046 (1960).

ZWISLOCKI, J.: Temporal summation in loudness: an analysis. J. acoust. Soc. Amer. **46**, 431 (1969).

Chapter 10

Localization of Sound

LLOYD A. JEFFRESS, Austin, Texas (USA)

Contents

I. Free-Field Studies 449
II. The Stimuli 451
III. Earphone Measurements of Time and Level Differences 452
A. Interaural Time Differences 452
B. Off-Center Source 453
C. Interaural Differences of Level 454
IV. Role of Head Movements 454
V. Role of the Pinna 454
VI. Distance Cues 455
VII. Precedence Effect/Haas Effect 455
VIII. Time-Intensity "Trading" 456
IX. Neurophysiological Implications 457
References 458

From the standpoint of survival, the two most important things to be known about a source of sound are what it is and where it is. Knowledge about the former depends upon the remarkable ability of the hearing mechanism to discriminate subtle differences in the composition of the sound; this is discussed in other sections. The present section is concerned with the problems arising from the location of the source. Histories of early work on the localization of sound are to be found in BORING (1942), TRIMBLE (1928), and MOLINO (1970). The present section is concerned only with recent experiments.

I. Free-Field Studies

The first study to be conducted in a reasonably anechoic environment (free field) was that of STEVENS and NEWMAN (1936). They perched their subject on a chair mounted on the ventilator shaft of one of the Harvard buildings, 9 ft above the roof, and with the nearest vertical surface hundreds of feet away. The source of sound was a small loudspeaker mounted on a boom attached to the subject's chair. The speaker was 6 ft distant from the subject. Tones of various frequencies, clicks, and a hiss caused by escaping air were employed as stimuli. The tones were turned on and off by means of a rheostat to avoid onset transients. The sensation level of the tones were 50 – 60 dB except for the highest tones which were about 30 dB. Sounds were generated at each 15-degree position from 0 – 180° always to the right of the subject. He indicated verbally the 15-degree position from which he thought the sound originated.

The results of the experiment were expressed as average deviations in degrees from the positions of the source. Back-to-front reversals were so frequent that the data for the two quadrants were combined; *i.e.*, 60° was combined with 120°, etc. When all of the errors for a particular frequency were combined, and average error was determined as a function of frequency, the results showed that errors grow with frequency from 1000 Hz up to about 3000 Hz and then diminish. At 10 000 Hz the average error is about that for 500 Hz. The average error as a function of the location of the source, averaged across frequencies, are shown in Table 1.

Table 1. Localization errors as a function of azimuth position

Location	000°	015°	030°	045°	060°	075°	090°
Average error in degrees	4.6	13.0	13.6	16.3	16.2	15.6	16.0

The proportion of front-back reversals proved to be a function of frequency, ranging from about 40% for frequencies up to about 2000 Hz to about 15% for frequencies above 4000 Hz. There was an abrupt drop between 2000 Hz and 4000 Hz.

The complex sounds, the click, and the hiss were localized more accurately than the tones. The average error was 8.0° for clicks, and 5.6° for the hiss. Front-back reversals were also less frequent.

STEVENS and NEWMAN accounted for their data in terms of two binaural cues: difference of level at the two ears, and difference of time. Since the tones were turned on gradually, the only time cue available to the subject was the difference in time between the sinusoidal waveforms at the two ears. To utilize such differences it is necessary that the auditory tracts preserve the temporal relations in the stimuli, that there be a cycle-by-cycle following of the waveform by the nerve impulses. STEVENS and NEWMAN assumed that such following occurred up to 1000 Hz or so and failed to occur at frequencies above 2000 Hz. They assumed that this cue provided the primary basis for localization at the low frequencies. The second cue, interaural difference of level, is present, they believed, at all frequencies but was most pronounced at frequencies above 4000 Hz. STEINBERG and SNOW (1934) had shown that the head casts much denser sound-shadows at the high frequencies than at the low, thus providing much larger differences of level at the high frequencies. Also at the high frequencies the pinna casts stronger shadows, thus making sounds coming from the back weaker than those from the front and so improving front-back discrimination for high-frequency tones and for clicks and hisses.

A second, and very thorough, free-field study of localization was conducted by MILLS (1958). He performed his experiments in a large anechoic chamber. A single source of sound was used; positioned at 0°, 30°, 45°, 60°, or 75° to one side of the subject, and moved from trial to trial from one side to the other of its original position. The subject responded "right" or "left" accordingly as the sound appeared to move to the right or left. Data were recorded as percentage of "right" judgements against degrees of movement of the speaker and the distance in degrees between the value for 25% "right" to 75% "right" divided by two was taken as the measure of the minimal audible angle (MAA). A large number of frequencies from 250 – 10000 Hz was used with three subjects.

The results, which were plotted as a familiy of curves showing MAA vs frequency with the angular position of the fixed speaker as the parameter, supported the findings of STEVENS and NEWMAN. For all positions except 75°, the MAA decreased from 250 Hz to about 1000 Hz, then rose sharply to a maximum near 2000 Hz, fell to a minimum at about 6000 Hz and showed another small maximum at 8000 Hz, falling again at 10000 Hz. The 75° line rose slowly from 250 – 1000 Hz, then rose very sharply to more than 30° at 1250 Hz. It dropped to a minimum at 3000 Hz.

The smallest MAA was slightly less than 1° at 750 Hz for 0° azimuth. The minima for the other azimuth values were about 1.5° for 30°, 2° for 45°, 3° for 60° and 7° for 75°. The left-right procedure could, of course, not be used with an azimuth of 90°.

II. The Stimuli

MILLS next undertook to determine the value of the stimuli responsible for the MAAs that he had obtained. He measured the interaural time differences for various frequencies and azimuth positions by the use of two condenser microphones mounted in a dummy head. The interaural time differences represented by the MAAs at 0° azimuth ranged from about 16 μsec at 250 Hz down to about 9 μsec at 750 Hz and 1000 Hz, and then rose to about 14 μsec at 1250 Hz.

Interaural level differences required for the obtained MAAs, MILLS computed from measurements of levels around the head made by SYVIAN and WHITE (1933). At 0° azimuth, at low frequencies, there is so little change of level with azimuth as to rule this out as a cue for localization. By 1250 Hz the difference for a displacement equal to the MAA is about 0.6 dB. The difference does not get much larger until a frequency of 5000 Hz is reached.

The conclusions to be drawn from MILLS' experiments are similar to those from STEVENS and NEWMAN'S, that frequencies below about 1500 Hz are localized largely on the basis of interaural time differences. Above that frequency the auditory system is insensitive to differences of phase (time) and must depend upon the interaural differences of level. These are too small at the frequencies near 1500 Hz to provide much of a cue. At the higher frequencies, however, the head becomes an appreciable barrier to the transmission of sound to the distant ear, and interaural differences of level become useable cues – localization improves.

The work of STEVENS and NEWMAN left us in doubt about whether the poorer accuracy of localization at large azimuth angles is due simply to the fact that the change in time difference and in level difference with azimuth at these angles is small, or is due in part to a lesser sensitivity of the hearing mechanism at these angles. MILLS' work answers the question for time differences. He found that there is a steady increase in the time difference required for localization as the azimuth angle is increased. The larger MAA at 75° (say) is due in part to the smaller change of time difference with angle and in part to a decreasing sensitivity of the hearing mechanism to time difference as reference time difference is increased. Mills did not present comparable data about the sensitivity to difference of level.

III. Earphone Measurements of Time and Level Differences

A. Interaural Time Differences

Klumpp and Eady (1956) used an electrical delay line to introduce time delays into the stimulus to one of the earphones. Using various delays, they determined the percentage of correct lateralizations for each stimulus used and from the psychometric function thus obtained read off the value of the time delay required for 75% correct judgments. Their results for a variety of stimuli are presented in Table 2.

Table 2. Interaural time difference thresholds in microseconds for various signals

Signal		Threshold in μsec	Number of listeners	Judgements per point per listener
Noise	Broadband[a]	10	10	160
	150—1700 Hz[b]	9	10	80
	425— 600 Hz[b]	14	9	160
	410— 440 Hz[b]	19	8	80
	2400—3400 Hz[b]	44	10	80
	3056—3344 Hz[b]	62	10	80
Clicks (1 msec duration)	Single	28	10	120
	repeated, 30 clicks in a 2 sec burst	11	13	160
Tones	90 Hz	75	10	10
	125 Hz	56	9	80
	250 Hz	27	9	80
	500 Hz	17	9	80
	1000 Hz	11	9	80
	1300 Hz	24	10	10
	1500 Hz	—	10	10
	1800 Hz	—	10	10
	3200 Hz	—	10	80

[a] Random noise limited by the frequency response of the PDR-8 earphones used.
[b] Half-power points.

From the table, it will be seen that the best performance involved bands of noise – the smallest threshold (9 μsec) occurred with the band of noise from 150 – 1700 Hz. Lower frequencies gave poorer performance. For tones, the thresholds are somewhat larger, but diminish with increasing frequency up to 1000 Hz (11 μsec). At 1300 Hz the threshold was considerably larger. These results agree very well with those of Mills. They can be explained if one assumes that there is a threshold for firing in the auditory nerve fibers involved. If the threshold is thought of as exhibiting some variability, one would expect twice as much variation in firing at 500 Hz (say) as at 1000 Hz. The time required to reach maximal displacement for 500 Hz is twice that to reach a maximum at 1000 Hz. The poorer performance at 1300 Hz is owing to the fact that the limit of the system's ability to follow the cycle-by-cycle movement of the stimulus is near 1400 Hz.[1] Presumable as we near the limiting frequency we find fewer and fewer fibers capable of "following".

The fact that lateralization is possible for bands of noise containing no energy at frequencies below 1400 Hz presents us with a new fact. The auditory system is capable of extracting useable temporal information from the *envelope* of the waveform, even though the contents of the envelope are high frequencies. This fact, as well as the 1400 Hz limit for temporal "following," can be readily demonstrated. If a sinusoidal source is connected to earphones by way of a delay line in one channel, shifting the delay will produce a readily perceived movement of the sound in the head so long as the frequency of the sinusoid is below about 1400 Hz (the upper limit depends upon the observer). At higher frequencies there is no sensation of movement. Now if we replace the oscillator by a noise generator with a high-pass filter, we find that there is no setting of the filter pass-band that will prevent the perception of motion. Even when the pass-band is set so high that there is only a barely perceptible hiss in the earphones, changing the delay will shift the position of the sound in the head. This continues to be true even if the larger amplitude peaks are eliminated by inserting a pair of diodes back-to-back in the earphone channels.

Further evidence that temporal information an be extracted from the envelope of the stimulus was presented by LICKLIDER and WEBSTER (1950). They connected the outputs of two oscillators to a mixing network such that they could reverse the phase of either component to one of the ears. If either frequency was below about 1400 Hz, reversing its phase produced a perceptible shift in the sound. If both tones had frequencies in excess of 1400 Hz and they were more than 400 Hz apart in frequency, reversing the phase of either made no noticeable change in the sound. If, however, the frequencies were within 400 Hz of each other, reversing either in phase again produced a perceptible shift in the sound. Thus, even though the tones taken separately are above the limit of following of the system, when the envelope of the combination is of a low frequency, a temporal change in it is readily perceived. Thus, here, as well as in the existence of the "residue" or periodicity pitch (see Vol. V/3, Chapter DE BOER), we have evidence that high-frequency portions of the hearing mechanism are capable of furnishing useable low-frequency information.

B. Off-Center Source

If we take MILLS' data for the minimal phase angle as a function of the time difference associated with a displaced source, and convert his interaural phases to time difference we can obtain the following table (Table 3):

[1] This "following" refers to that for binaural interaction. There is neurophysiological evidence that the auditory nerve fibers can follow the frequency of tones up to much higher frequencies.

Table 3. Minimal audible time difference in microseconds

Interaural displacement in μsec frequency in Hz	0	600
250	16	23
500	11	23
750	9	24
1000	9	33
1250	13	62

The values for an initial time difference of zero are somewhat smaller than those of KLUMP and EADY, but show the same trend. The values for an interaural delay of 600 μsec are substantially larger than those for 0 μsec, indicating that the system is less sensitive to differences of time in the presence of a large time difference than near the median plane.

C. Interaural Differences of Level

In an experiment employing earphones, MILLS (1960) measured the jnds for interaural differences of level for tones in the median plane. He found values of about 0.7 dB at 250 Hz and 500 Hz, a maximum of about 1 dB at 1000 Hz, and a minimum of about 0.5 dB at frequencies in the range from 3000–10000 Hz. The writer has been unable to find similar data for high frequencies for locations away from the median plane.

IV. Role of Head Movements

WALLACH (1940), THURLOW, and RUNGE (1967), and THURLOW, MANGELS, and RUNGE (1967) have studied the influence of head movements upon the accuracy of localization of sources in a free field. The kinesthetic cues from the neck muscles and the changes in the binaural cues of level and time with movement combine to improve the accuracy of localization, especially in regions where the binaural cues are ambiguous. Front-back reversals are virtually eliminated if the subject is permitted to move his head, or is instructed to do so. Accuracy of localization in the vertical plane is greatly enhanced if a tilting movement of the head occurs. One has only to watch a dog attempting to locate the source of a faint sound to realize how important a role head movement and movements of the pinnae play in the localization of sound under normal conditions.

V. Role of the Pinna

Even the modest pinnae with which man is equipped play a substantial role in the localization of noise with high-frequency components. ROFFLER and BUTLER (1968a) found that flattening the pinnae down by means of a plastic band reduced to chance the ability of human subjects to localize high-frequency bands of noise in the vertical plane. With the pinnae in normal position the subjects had been able to indicate with considerable accuracy where, in the vertical, the sounds were coming from. In another study, using tones, ROFFLER and BUTLER (1968b) found that subjects localized tones according to frequency, the highest frequency at the top and the lowest at the bottom with the middle frequencies assigned to locations between in an orderly progression, and irrespective of the actual location.

Probably the most important role of the human pinnae is the attenuation of the high frequencies from sound coming from the back. The consequent change in the spectrum of familiar sounds is an excellent cue to determining whether the source is in front or in back. The large, moveable pinnae of many animals are no doubt of even greater value in localizing a source of sound.

VI. Distance Cues

Apparently the major cue for distance is the alteration of the spectrum of complex sounds with distance. The high frequencies are attenuated more rapidly than the low. Of course, the loudness of sounds of familiar level also provides a cue to distance. COLEMAN (1962) found that even when loudness was ruled out by appropriately adjusting the level of a source at different distances, subjects could *learn* to judge distance with fair accuracy. He used 1 sec bursts of wide-band noise from sources located at distances from 9–27 ft across the ice, and found that subjects in their initial attempts showed very poor accuracy. However, with as many as 100 trials they improved to the point where their errors were only 2–4 ft.

BÉKÉSY (1930) has argued that in judging distance the ear can utilize the varying difference in attenuation with distance between particle velocity and pressure in a spherical wave, differences which vary as a function of frequency. This cue would be operative over distances of a few feet. BÉKÉSY describes a demonstration which illustrates the operation of this cue.

VII. Precedence Effect/Haas Effect

For many years it has been recognized that the echo from a wall at some distance from a source of sound can fuse with the initial sound to form a single image. The echo may suffer a delay of many milliseconds and still blend with the direct sound in creating the impression that all of the sound is coming from the initial source. The phenomenon was given formal recognition by HAAS in 1949 and by WALLACH, NEWMAN, and ROSENZWEIG in the same year. It has come to be known as the "Haas Effect" by sound engineers, and as the "Precedence Effect" by psychologists. WALLACH *et al.* studied the phenomenon first by using phonograph records with two pickups, one directly behind the other in the same groove. When the earlier sound was reproduced in one direction from the listener and the later from another, the listener heard the sounds as fused and as coming from the direction of the earlier sound, even when it was weaker than the "echo". Later WALLACH *et al.* studied the localization of pairs of binaural clicks delivered by way of earphones. The earlier click might lead at one ear and the later at the other and be heard in the direction of the first click, even when there was a delay of several milliseconds in the spacing of the two sets of clicks. They found that the precedence effect was most strikingly demonstrable with impulsive sounds such as occur in speech and in piano music; when long sustained tones are used the sound appears to be coming from all around rather than from the initial source. GARDNER (1968) has written an interesting history of the phenomenon.

Closely allied to the precedence effect is a finding by BLODGETT, WILBANKS, and JEFFRESS (1956) that a wide-band noise delayed at one ear by more than 20 msec can fuse into a single sound that is heard as coming from the undelayed side. Even when the undelayed noise is 9 dB below the delayed, the subject hears the sound as coming from the undelayed side if the time difference is not in excess of 4 msec. This phenomenon, like the precedence effect, indicates the length of time a pattern of excitation can be held in storage by the auditory nervous system.

VIII. Time-Intensity "Trading"

A method of studying the interaction between interaural time differences and interaural differences of level involves introducing a time difference in one direction and a level difference in the other, and asking the subject to adjust one or the other to produce a centered image. The ratio of the time difference in microseconds to the level difference in decibels has come to be known as the "trading ratio". One of the earliest studies to yield a trading ratio was that of SHAXBY and GAGE (1932), who found a value for tones of 1.7 μsec/dB. Later, CHRISTMAN and VICTOR (1955), using click trains, found trading ratios in excess of 120 μsec/dB. SHAXBY and GAGE found a linear relation between the time differences and the level differences required to yield a centered image, whereas other workers such as DEATHERAGE and HIRSH (1959) found that the ratio depended upon the amount of time difference being used.

MOUSHEGIAN and JEFFRESS (1959) employed a new technique in the study of time and level. They asked their subjects to match the lateral position of a tone having a selected interaural difference of time and/or of level with a "pointer" consisting of a broadband of noise reaching the earphones through a calibrated electrical delay line under the subject's control. The subjects adjusted the interaural time difference for the noise until it appeared to be in the same place in his head as the tone. The stimulus tone and pointer alternated in time. They found that a tone having a difference of level but no time difference, the value of the time delay in the "pointer" was a linear function of the difference of level in the tone. The trading ratios thus determined for two of their subjects were in the neighborhood of 20–25 μsec/dB at 500 Hz. Their third subject yielded a ratio of only about 2.5 μsec/dB. Furthermore the trading ratios for the two subjects increased when a time difference as well as a difference of level was introduced into the tonal stimulus. The third subject showed the same small trading ratio for all values of interaural time difference in the tone.

Later, WITHWORTH and JEFFRESS (1961) repeated the earlier work using the same technique, except that the noise pointer was replaced by a tonal pointer of the same frequency as the tone under study. They "rediscovered" a fact mentioned much earlier by BANISTER (1926) that when a tone is presented with conflicting time and intensity differences the subject may hear two images in different locations in his head. They found that one image, which they called the "time image" showed very little movement as a result of a difference of level. The other image, the "intensity image" was much more affected by interaural differences of level. The time images gave trading ratios ranging from 0.29 – 2.44 μsec/dB for the three subjects. The intensity images showed trading ratios from 20 to 26 μsec/dB. In all cases the ratios were constant for a particular subject across time differences. Later, HAFTER and JEFFRESS (1968) found that subjects could distinguish the two images for clicks as well as tones. The trading ratios for clicks ranged from 2 – 35 μsec/dB for the time image, and from 85 – 150 μsec/dB for the intensity. They attribute much of the disparity in the literature to the variety in the stimuli employed and to uncertainty about what image was being centered.

The results of the WHITWORTH and JEFFRESS study help us to understand the results of the earlier experiment. MOUSHEGIAN and JEFFRESS found that when time and intensity were in conflict and where two of the subjects were responding to a composite of the time and intensity images, the two images tended to offset each other and yield a location close to the median plane. When both time and intensity favored the same side, the composite image lay farther out than either the time or intensity image alone. Thus, the greater the amount of time difference, the farther from the median plane the composite lay. This fact resulted in the increase of trading ratio with interaural time difference for the composite image. The third subject in the study by MOUSHEGIAN and JEFFRESS was responding to the time image alone, and his trading ratios remained constant.

In the WHITWORTH and JEFFRESS experiment, a subject who had a trading ratio of 20 μsec/dB for the tone with no time difference also had a ratio of 20 μsec/dB for the tone with a time difference of 270 μsec. When time and intensity were in opposition, and for a 9 dB difference of level, the image was located at a position of $270 - 9 \times 20 = +90$. For a 9 dB difference of level in the same direction as the difference of time, the image was located at $270 + 9 \times 20 = +450$. The corresponding locations for the MOUSHEGIAN and JEFFRESS experiment were +60 and +480. The finding that the composite images is not just the algebraic sum of time and intensity is supported by the work of JEFFRESS and MCFADDEN (1971) who found that for a narrow band of noise centered at 500 Hz, when interaural time and intensity were so adjusted that the composite image fell in the median plane, the stimulus was still readily discriminated from a noise with no difference of time or intensity. The two images did not offset each other in detectability. Subjects differed substantially in their relative response to the time cue and the intensity cue.

IX. Neurophysiological Implications

From the time when it was first realized that the auditory system is sensitive to interaural phase differences in tones below 1400 Hz in frequency, and only to differences of levels at higher frequencies, it should have been apparent that there must be at least two neural mechanisms for localization. One is phase-locked to the stimulus, with fibers that fire at a particular time in the sound cycle. The other has fibers whose firing rate is determined by the level of the stimulus and is independent of the stimulus frequency. The existence of both such fibers has been demonstrated by MOUSHEGIAN (1971), who found in the superior olivary complex fibers which are phase-locked and whose firing rate is wholly determined by the frequency of the stimulus, and other fibers which fire at rates determined by the difference in level of the stimulus and independent of its frequency. Psychophysics tells us that the outputs of the two systems combine in a complex fashion, neither simply adding their outputs when the stimuli are in accord nor completely cancelling when they are in opposition.

A further implication for neurophysiology to be drawn from psychophysical research is to be found in a paper by YOST, WIGHTMAN, and GREEN (1971). They studied the lateralization of high-pass (2000 – 10000 Hz) clicks, and low-pass (4 – 500 Hz) clicks. For single clicks the jnd for the high-pass clicks was about 200 μsec while that for the low-pass clicks was about 22 μsec. When the number of clicks was increased from 1 – 64 at a rate of 50 clicks/sec, the lateralization of the high-pass clicks improved with the number of repetition until at 64, it was equal to that the low-pass at about 20 μsec. The low-pass clicks did not improve in lateralization appreciably through repetition. Apparently adding low-frequency energy to the high-pass clicks through repetition at 50 clicks/sec was responsible for the improvement. That the superior performance was due to the apical portion of the cochlear partition was shown by the fact that high-pass noise did not cause deterioration in the localization of the low-pass click, while low-pass noise seriously affected the lateralization of high-pass clicks. Possibly, the fact that the traveling wave is much flatter for low-pass clicks than it is for high means that more neural fibers would be involved in responding to the low-pass clicks than would be involved with the high. This suggests that the difference in accuracy may be a statistical one based on the number of neural units activated in the two cases.

References

BANISTER, H.: Three experiments on the localization of tones. Brit. J. Psychol. **16**, 226—229 (1926).

VON BÉKÉSY, G.: The spatial attributes of sound. Translated by E. G. WEVER in VON BÉKÉSY. G.: Experiments in hearing, pp. 309—313. New York: McGraw-Hill Book Comp. Inc. 1960, (1930).

BLODGETT, H. C., WILBANKS, W. A., JEFFRESS, L. A.: Effect of large interaural time differences upon the judgment of sidedness. J. acoust. Soc. Amer. **26**, 639—643 (1956).

BORING, E. G.: Sensation and perception in the history of experimental psychology, pp. 381—392. New York: Appleton-Century-Crofts 1942.

CHRISTMAN, R. J., VICTOR, G.: The perception of direction as a function of binaural temporal and amplitude disparity. Rome Air Development Center ARDC, USAF, September 1955, Rome, New York Technical Note RADC-TH-55-302 (1955).

COLEMAN, P. D.: Failure to localize the source distance of an unfamiliar sound. J. acoust. Soc. Amer. **34**, 345—346 (1962).

DEATHERAGE, B. H., HIRSH, I. J.: Auditory localization of clicks. J. acoust. Soc. Amer. **31**, 486—492 (1959).

GARDNER, M. B.: Historical background of the Haas and/or precedence effect. J. acoust. Soc. Amer, **43**, 1243—1248 (1968).

HAAS, H.: The difference of a single echo on the audibility of speech. Literary Commun, No. 363 Dept. of Sci. and Indust. Res. Bld. Res. Sta. Garston, Watford, Herts., England (1949).

HAFTER, E. R., JEFFRESS, L. A.: Two-image lateralization of tones and clicks. J. acoust. Soc. Amer. **44**, 563—569 (1968).

JEFFRESS, L. A., MCFADDEN, D.: Differences of interaural phase and level in detection and lateralization. J. acoust. Soc. Amer. **49**, 1169—1179 (1971).

KLUMPP, R. G., EADY, H. R.: Some measurements of interaural time-difference thresholds. J. acoust. Soc Amer. **28**, 859—860 (1956).

LICKLIDER, J. C. R., WEBSTER, J. C.: Discriminability of interaural phase relations in two-component tones. J. acoust. Soc. Amer. **22**, 191—195 (1950).

MILLS, A. W.: On the minimum audible angle. J. acoust. Soc. Amer. **30**, 237—246 (1958).

MILLS, A. W.: Lateralization of high-frequency tones. J. acoust. Soc. Amer. **32**, 132—134 (1960).

MOLINO, J.: A brief history of research on the physical factors involved in auditory localization. Report PLR-15, Colombia University, 1 July 1970.

MOUSHEGIAN, G.: Personal communication. (1971).

MOUSHEGIAN, G., JEFFRESS, L. A.: Role of interaural time and intensity differences in the lateralization of low-frequency tones. J. acoust. Soc. Amer. **31**, 1441—1445 (1959).

ROFFLER, S. K., BUTLER, R. A.: Factors that influence the localization of sound in the vertical plane. J. acoust. Soc. Amer. **43**, 1255—1259 (1968a).

ROFFLER, S. K., BUTLER, R. A.: Localization of tonal stimuli in the vertical plane. J. acoust. Soc. Amer. **43**, 1255—1257 (1968b).

SHAXBY, J. H., GAGE, F. H.: Studies in the localization of sound. Med. Res. Council (Brit.) Spec. Rept. Ser. No. 166, 1—32 (1932).

STEINBERG, J. C., SNOW, W. B.: Physical factors in auditory perspective. Bell Syst. Tech. J. **13**, 247—260 (1934).

STEVENS, S. S., NEWMAN, E. B.: The localization of actual sources of sound. Amer. J. Psychol. **48**, 297—306 (1936).

SYVIAN, L. J., WHITE, S. D.: Minimal audible sound fields. J. acoust. Soc. Amer. **4**, 288—321 (1933).

THURLOW, W. R., RUNGE, P. S.: Effect of induced head movements on localization of direction of sounds. J. acoust. Soc. Amer. **42**, 480—488 (1967).

THURLOW, W. R., MANGELS, J. W., RUNGE, P. S.: Head movements during sound localization. J. acoust. Soc. Amer. **42**, 489—493 (1967).

TRIMBLE, O. C.: The theory of sound localization: A restatement. Psychol. Rev. **35**, 515—523 (1928).

WALLACH, H.: The role of head movements, and vestibular and visual cues in sound localization. J. exp. Psychol. **14**, 95—124 (1940).

WALLACH, H., NEWMAN, E. B., ROSENZWEIG, M. R.: The precedence effect in sound localization. Amer. J. Psychol. **52**, 315—336 (1949).

WHITWORTH, R. H., JEFFRESS, L. A.: Time vs intensity on the localization of tones. J. acoust. Soc. Amer. **33**, 925—929 (1961).

YOST, W. A., WIGHTMAN, F. L., GREEN, D. M.: Lateralization of filtered clicks. J. acoust. Soc. Amer. (in press) (1971).

Chapter 11

Binaural Analysis

David M. Green, Cambridge, Massachusetts (USA)
and
William A. Yost, Gainesville, Florida (USA)

With 4 Figures

Contents

I. Introduction 461
II. Signal Parameters 465
A. Signal Frequency 465
B. Signal Phase 466
C. Signal Duration 466
D. Interaural Signal Level 466
E. Signal Bandwidth 467
F. Noise Signals 467
G. Speech Signals 468
III. Masker Parameters 468
A. Overall Level 468
B. Interaural Intensity 469
C. Interaural Correlation 469
D. Interaural Time 470
E. Interaural Phase 471
F. Bandwidth 471
G. Gated Masker 472
H. Tonal Masker 472
IV. Assessment of Binaural Improvement Using Methods other than Detection 472
A. Speech Intelligibility 472
B. Loudness 473
C. Frequency and Intensity Discrimination 473
V. Theories 474
A. Webster-Jeffress Theory 474
B. Durlach's Model 476
C. Lateralization Model 477
VI. Summary 477
References 478

I. Introduction

This chapter reviews the phenomena of binaural analysis — the improvement in hearing that results when there are interaural discrepancies in the signal and masker at the two ears. Binaural analysis is an anomalous topic. Although the topic enjoys great popularity in psychoacoustic investigations, it is practically untouched in physiological studies (GREEN and HENNING, 1969). Spatial localization is still the major focus of attention in physiological studies of binaural phenomena (ELDREDGE and MILLER, 1971). While the mechanisms of binaural analysis and localization are probably closely related, it would, nevertheless, be interesting to have physiological measurements taken under the stimulus conditions one encounters in typical binaural analysis experiments. We hope this chapter will make the physiological community aware of this very popular and interesting area of binaural research and promote closer collaboration between these two levels of investigation.

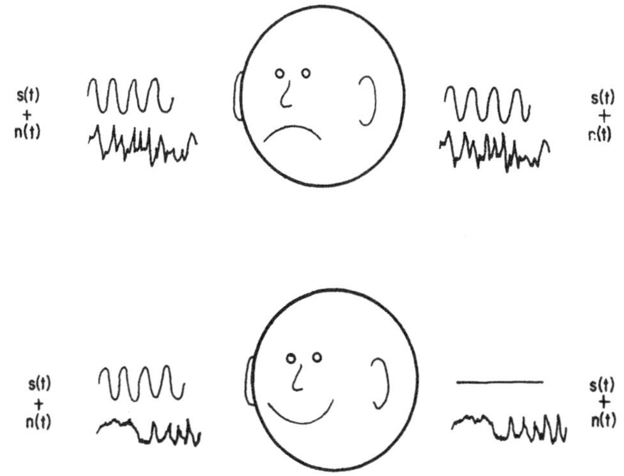

Fig. 1. This figure represents improvement in detection (MLD) obtained when a signal is presented to one instead of both ears, given that the noise masker is present at both ears and in phase. In the top drawing the masked signal is presented at both ears and is difficult to detect as the expression on the figure face indicates. In the bottom the same signal is presented to only one ear and it is easily detected

The basic phenomenon of a binaural analysis experiment is depicted in Fig. 1. A signal and a masker (usually noise) are presented to both ears of a listener (top of Fig. 1) who adjusts the signal level until it is just-detectable. The signal is then attenuated 5 dB. The frown on the figure represents the fact that the signal is undetected. The signal to one ear is then turned off (bottom of Fig. 1), and the smile on the listener's face indicates that the signal is clearly detectable. In certain situations, if the signal is presented to only one ear, it can be detected with as much as 10 dB *less* signal level than if it is presented to both ears. This rather surprising result is typical of those obtained in the area of binaural analysis.

The binaural analysis experiments demonstrate that vastly superior detection performance is possible in many conditions in which some interaural discrepancy exists between the signals or maskers at the two ears. Research in the area of binaural analysis has largely been devoted to investigating those differences in interaural stimulus parameters that lead to improved detection performance.

Many interaural differences in the signal (S) and in the masker (M) have been investigated, and a notation has developed. For instance, the situation shown in the top of Fig. 1 is labeled "M_oS_o"; the subscript "o" indicates that there are no interaural differences between the masker or signal arriving at the ears. The bottom condition of Fig. 1 is labeled "M_oS_m" since the masker is again the same at both ears (subscript "o"), but the signal is presented monaurally (subscript "m"). These and other symbols are described below:

S_o: Signal presented binaurally with no interaural differences.
M_o: Masker presented binaurally with no interaural differences.
S_m: Signal presented to only one ear.
M_m: Masker presented to only one ear.
S_π: Signal presented to one ear 180° out-of-phase relative to the signal presented to the other ear.
M_π: Masker presented to one ear 180° out-of-phase relative to the masker presented to the other ear.
M_μ: Masker presented to one ear uncorrelated (originated from a different noise supply) to the masker presented to the other ear.

For all the binaural conditions described above, except M_μ, the signal or masker is identical in all dimensions except that denoted by a subscript. The M_oS_o and $M_\pi S_\pi$ conditions are sometimes labeled the "homophasic conditions," whereas the M_oS_π and $M_\pi S_o$ conditions are called the "antiphasic conditions." Another classification of the interaural conditions was provided by STUMPF (1916):

1. Monotic: Stimulus presented to only one ear; M_mS_m.
2. Diotic: Stimulus presented the same to both ears; M_oS_o.
3. Dichotic: Stimulus presented differently to the two ears; M_oS_π, M_oS_m, $M_\pi S_\pi$, $M_\pi S_o$, $M_\pi S_m$, $M_\mu S_o$, $M_\mu S_\pi$, $M_\mu S_m$.

In order to compare detection in one binaural condition to that in another, the data are usually presented as the difference between the signal level required for detection in a monotic condition and that required in a diotic or dichotic condition. That is, the signal level required for detection in the pertinent diotic or dichotic condition is subtracted from the signal level required for detection in the M_mS_m condition. Such a difference when expressed in decibels is called a "masking-level difference" (MLD) or a "binaural masking-level difference" (BMLD). Hence, the binaural analysis literature is often referred to as the "MLD Literature" or the "Release-from-Masking Literature".

In most binaural masking experiments, a psychometric function is obtained for each monotic, diotic, or dichotic condition tested. The psychometric function relates some measure of the subject's performance to the signal-to-masker ratio. The MLD is simply the separation in decibels between the psychometric functions associated with diotic or monotic and dichotic conditions. GREEN (1966), EGAN *et al.* (1969), and MCFADDEN and PULLIAM (1971) have shown that the psychometric functions are parallel across the various binaural conditions. Thus, the MLD does not depend on the level of a subject's performance. In addition, intersubject variability appears to be small in a majority of the MLD studies. Many different psychophysical procedures have been used to measure the size of the

MLD, and these estimates are generally in good agreement with one another. Consequently, in many conditions, the MLD appears to be relatively invariant across subjects, level of subjects' performance, and psychophysical method.

The exact value of the MLD depends on a number of factors which will be reviewed systematically in this chapter. To provide some typical results and to review how the size of the MLD depends on the various interaural conditions, a situation is considered in which the masker is a loud, continuous, wideband masker such as white noise (spectrum level 60 dB or more), and the signal a low-frequency sinusoid (say 500 Hz and presented for a brief duration, 10—100 msec). In this situation the following hierarchy of MLDs will probably result:

Interaural condition compared to $M_m S_m$	MLD
$M_m S_m$, $M_0 S_0$, $M_\mu S_m$	0 dB
$M_\mu S_\pi$	3 dB
$M_\mu S_0$	4 dB
$M_\pi S_m$	6 dB
$M_0 S_m$	9 dB
$M_\pi S_0$	13 dB
$M_0 S_\pi$	15 dB

The MLD hierarchy demonstrates, for example, that when a signal and masker are presented to both ears identically ($M_o S_o$), the signal is as easily detected as when the signal and masker are presented to only one ear ($M_m S_m$). However, if the interaural phase of the signal is changed from 0° to 180° ($M_o S_\pi$), then the signal is easier to detect by 15 dB.

Two points regarding this hierarchy should be noted: 1) the MLD is never negative; and 2) the MLD for some binaural conditions is zero dB. These binaural conditions, which yield no masking level difference, are sometimes used as the referent condition for defining the MLD rather than the $M_m S_m$ condition. In particular, many investigators use the diotic condition ($M_o S_o$) as the referent condition. Although there has been a long history of study in the area of binaural interactions of various kinds, MLD's for sinusoidal signals were first observed by Hirsh in 1948, and those for speech signals by Licklider in the same year. Since then, many investigators have systematically studied the signal and masker parameters involved in the MLD (see bibliography at the end of this chapter). Although this research has indicated many of the conditions required to produce an advantage for binaural listening, no general theory has emerged to describe all of the data. Two current theories, however, have generated considerable interest and research. WEBSTER (1951) and JEFFRESS (1948, JEFFRESS *et al.*, 1956) proposed a model, now called the Webster-Jeffress Hypothesis; DURLACH (1963) proposed the Equalization-Cancellation Model for binaural masking. Considerable research in this area has been performed by JEFFRESS and his co-workers at the University of Texas and by DURLACH and his co-researchers at MIT.

The emphasis in this chapter will be on the different stimulus parameters and how they affect the size of the MLD. Starting with sinusoidal signals, the signal parameters and their effects on the binaural analysis are reviewed. Next, the masker parameters are discussed and their effects described. A brief section "Assessment of Binaural Improvement using Methods other than Detection," is included, and, finally, a brief review of the various theories concludes the chapter. The overall emphasis, however, is empirical rather than theoretical.

Throughout this review, the major dependent variable is the MLD, that is, the difference in the signal level required for detection between a reference condition and some other binaural masking condition. A change in the MLD does not indicate in which condition detection has varied: the reference condition, the binaural condition, or both conditions. A change in the MLD shows only that there was a relative change in detection. Notice also that the MLD is defined in terms of detection, not in some other psychological dimension, such as loudness. Thus, in general, the binaural masking literature is an investigation of those parameters and conditions which lead to an improvement in detection due to dichotic listening.

II. Signal Parameters

In the data described below, the MLD's were measured for pulsed signals in the presence of a continuous broadband Gaussian noise. In all experiments, the reference condition for the MLD was either the $M_m S_m$ or the $M_0 S_0$ condition.

A. Signal Frequency

The amount of binaural improvement measured in MLD experiments greatly depends on the frequency of the signal. Hirsh (1948a) showed in his original experiment that the size of the MLD for the $M_0 S_m$ and $M_0 S_\pi$ conditions diminished at frequencies above about 1000 Hz and reached a maximum at about 200 Hz. He showed a relatively small MLD at 100 Hz. A number of later experiments have confirmed the original observations, see Hirsh and Burgeat (1958), Durlach (1964), and Rabiner *et al.* (1966).

Above about 500 Hz, there is good agreement among all of the data. In the $M_0 S_\pi$ and $M_\pi S_0$ conditions, the masking-level difference is large, and diminishes to about 3 dB in the region of 1500 Hz. Above 1500 Hz, there is ample evidence that the MLD does not go to zero, but rather remains at a value near 3 dB. For the other dichotic conditions, the MLD reaches an asymptote at or near 0 dB as the signal frequency increases beyond approximately 1500 Hz.

Data on the size of the MLD at low frequencies are more diverse. It appears from several articles that the discrepancies among the published values for the MLD at low frequencies, below 300 Hz, are almost entirely attributable to differences in the amount of low-frequency experimentally-controlled noise. One of the best studies on this topic is that of Dolan (1968), who systematically varied the level of the experimentally-controlled external noise and measured the MLD for the $M_0 S_\pi$ condition at 150 and 300 Hz. As the external masking noise increased in level, the MLD steadily increased. And, finally, at noise spectrum levels of 50 dB and above, the MLD is about 15 dB at both 150 Hz and 300 Hz. Thus, the relatively small MLD's measured at very low signal frequencies and with a low level of experimental masking noise are presumably caused by other noises not under the experimenter's control, such as those produced by breathing, heartbeat, muscle tonus and room noise. At very low signal frequencies, the level of these other noises is sufficient to obscure the low intensity experimental noise introduced via the headphones. The internal noises at one ear are only partially correlated with those at the other ear. Thus, they resemble to some degree the M_μ condition described earlier. Since a condition like $M_\mu S_\pi$ produces a very small MLD, there is a small MLD measured for these low frequency signals when the experimental noise is low in level. SHAW and PIERCY (1962) directly measured the background level of low frequency noise with a probe tube. The spectrum level of this noise was quite high; approximately 8 dB at 250 Hz and 25 dB at 150 Hz.

Therefore, for high levels of experimental noise, the size of the MLD as a function of frequency can be easily summarized: it is large at the low frequencies and diminishes to 0—3 dB at 1500 Hz, which is its constant value as frequency is increased.

B. Signal Phase

The first systematic study of how signal phase affects the MLD was made by HIRSH (1948a) who varied the phase of a 200-Hz signal in 30-degree steps in both M_o and M_π configurations. The effect of interaural signal phase is profound. The largest MLD's appear when the interaural phase of the signal is 180° different from that of the masker, that is, when the configurations are $M_o S_\pi$ or $M_\pi S_o$. JEFFRESS *et al.* (1952) and COLBURN and DURLACH (1965) confirmed these results concerning signal phase using a 500-Hz signal. For an M_o configuration, the effect of signal phase is pronounced; the graph relating detection performance to signal phase angle is a peaked function. JEFFRESS *et al.* (1962) also studied the interaural phase for a sinewave at 167 Hz. The effects of signal phase are similar at both 167 and 500 Hz.

C. Signal Duration

The duration of the signal changes the masking-level difference very little. As the signal duration is shortened, the signal becomes more difficult to detect; this is true for diotic, monotic and dichotic conditions. For very short durations of less than 50 msec, the MLD for some conditions may increase somewhat, being 1—2 dB larger (BLODGETT *et al.*, 1958). This effect has also been confirmed by GREEN (1966). The shape of the psychometric function, the function relating the percentage of correct detections to the signal level, is approximately the same for many durations and many dichotic and diotic conditions (MCFADDEN and PULLIAM, 1971).

D. Interaural Signal Level

In a number of dichotic conditions, such as $M_o S_\pi$, a signal is present at both ears. That the size of the MLD must depend on the relative signal levels at the two ears can be inferred from the fact that $M_o S_\pi$ produces about 6 dB better detection than $M_o S_m$. In the latter condition, the signal at one ear has been turned off. ZERLIN systematically studied this effect in 1966. In an M_0S_π condition, he employed a low-pass transient stimulus and progressively attenuated the level of the transient stimulus in one of the ears. His data show a near linear relation between the size of the MLD and the amount of attenuation of the signal at one ear. The largest MLD occurs when the signals are equal at both ears. The MLD does not change until there is a 3 dB difference in the level at the ears. Other systematic measurements of the effect of interaural signal intensity were made by COLBURN and DURLACH (1965), who varied both interaural intensity and interaural phase for sinusoidal signals of 500 Hz. These same relations have been confirmed in an experiment by EGAN (1965) who also used a 500 Hz signal.

E. Signal Bandwidth

Several experimenters have used clicks (ZERLIN, 1966), short-duration sinusoids (BLODGETT *et al.*, 1958; WIGHTMAN, 1969; MCFADDEN and PULLIAM, 1971) and pulse trains (FLANAGAN and WATSON, 1966) as signals in the MLD paradigm. All of these stimuli have energy spread over a wide range of frequencies. In general, the results from these studies agree with those involving long-duration sinusoids. The largest MLD's occur when the signals contain energy in the frequency region below 1000 Hz and when, in this spectral region, there is an interaural phase reversal of the signal but not of the masker (M_0S_π).

FLANAGAN and WATSON showed that for a pulse train, the largest MLD's (M_0S_π) were related to the fundamental component of the repetition rate. If the fundamental were eliminated by filtering, the MLD was greatly reduced. Pulse rates around 250 pulses per second (pps) produced the largest MLD's. Since a 250-pps train has energy at 250 Hz, this finding agrees well with those involving sinusoidal signals.

ROBINSON (1971) has recently investigated the effect of using different frequencies for the signal in the S_π condition. In general, the maximum improvement occurs when the signals are the same frequency (400 Hz) and there is little difference in the S_π and S_m conditions once the signals are different by 150 Hz, i.e., 400 Hz in one ear and 550 Hz in the other.

F. Noise Signals

When noise is the signal as well as a masker, the MLD's (M_0S_π) tend to vary over a considerable range, 15—30 dB (HIRSH and WEBSTER, 1949; WEBSTER, 1951; RILLING and JEFFRESS, 1965; JEFFRESS and MCFADDEN, 1971). RILLING and JEFFRESS argued that a crucial variable in the noise-signal experiments is the phase relation between the signal and masker. If the noise-signal and noise-masker are from different noise sources, then the phase angle between the masker and signal is random. JEFFRESS and MCFADDEN were able to control this phase relation by using a noise-signal of the same bandwidth (50 Hz) as the masker and by using both the masker and signal from the same noise supply. In this case, the largest MLD's (M_0S_π) were found when the phase angle between the signal and masker was less than 90°; the MLD decreased as the phase angle was increased beyond 90°.

By controlling the phase relation between the signal and masker, JEFFRESS and MCFADDEN also controlled exactly the interaural intensive and temporal differences in the noise waveforms. The values of these interaural differences agree with those obtained in lateralization and localization studies. Hence, their detection data are compared to the data obtained in a lateralization task.

G. Speech Signals

A large body of literature concerns attempts to study binaural performance using speech as the signal. The binaural improvement has been measured in 2 ways: 1) the speech signal-to-masker ratio required to *detect* the *presence* of the signal is measured in the referent and binaural conditions, or 2) the speech signal-to-masker ratio required to *recognize* or *understand* the speech signal is measured in the referent and binaural conditions. This section deals with the detection of speech. A later section, entitled "Assessments of Binaural Improvement using Methods other than Detection", treats the recognition of speech.

In general, the same set of parameters has been studied for speech signals and for sinusoidal signals. Since the interest in speech MLD's is ultimately in the improvement of speech intelligibility, many different background-maskers or distractors have been employed. Large MLD's (6—15 dB) have been obtained when the speech signal is in the S_π configuration and the masker is either a wideband continuous noise (LICKLIDER, 1948; CARHART *et al.*, 1966), an amplitude-modulated noise (CARHART *et al.*, 1967), an interrupted noise (CARHART *et al.*, 1967), or a competing speech signal (POLLACK and PICKETT, 1958).

If the speech signal is filtered, then the largest MLD's (M_oS_π) occur in the frequency region below 500 Hz (LEVITT and RABINER, 1967 a and b). Sizeable MLD's (up to 12 dB) have been obtained by CARHART *et al.* (1966) when an interaural time delay is introduced between the speech signals arriving at the ears and not between the maskers. Thus, to a first approximation, the MLD's for detection of speech and those for detection of sinusoids seem to vary in the same way as a function of changes in the stimulus parameters.

III. Masker Parameters

In considering the effects of changes in the masker parameter, the MLD will be measured for most conditions as the difference between that dichotic condition under study and M_oS_o or M_mS_m.

A. Overall Level

All available research indicates that the size of the MLD increases as a function of the level of the masker. HIRSH (1948) and BLODGETT *et al.* (1962), systematically investigated this parameter. They showed practically no difference in detection of a signal monaurally (M_mS_m) and detection of a signal in binaural noise (M_oS_m) until the spectrum level of the noise exceeded about 20 dB. From that point, there was a reliable increase in the masking-level difference until the spectrum level of the noise was about 40 dB. Above that level, the MLD is essentially constant. McFadden (1968) has also studied the effect of noise level in the M_0S_π condition, and his results agree with the previous data. The spectrum level of the noise at which the maximum MLD occurs probably depends on the signal frequency. Canahl and Small (1965) and Dolan (1968) have investigated this effect for low-frequency signals. Their results agree with the summary already given (see also the section entitled "Signal Frequency").

Diercks and Jeffress (1962) found that a binaural advantage exists even in the absence of an external masking noise. The absolute threshold for an S_0 or S_π signal is lower than that for an S_m signal. The lower absolute thresholds for the S_0 and S_π signals are presumably caused by the presence of the interaurally uncorrelated internal noise discussed previously in the section on signal frequency.

B. Interaural Intensity

Hirsh (1948) reported that maximum advantage of the binaural system could be achieved when the maskers were equal in level at the two ears. Systematic studies of this effect are numerous, all of the results being in very close agreement with one another. Blodgett *et al.* (1962) have shown that the size of the MLD is lowered if the noise level is reduced to the non-signal ear in the M_0S_m condition. The effect upon the size of the MLD is gradual, but a reduction is apparent even for a 10 dB asymmetry in the level of the noise at the two ears. Egan (1965) and Dolan and Robinson (1967) have replicated these results almost exactly. Binaural improvement approaches a small asymptotic value as the noise is reduced to below its absolute threshold, and these results are consistent with the assumption of a small internal noise having a moderate positive correlation (McFadden, 1968). Weston and Miller (1965) have extended this basic finding with data on conditions where the masker in the non-signal ear is greater in level than the masker in the signal ear. Again, the best detection occurs when the maskers are at the same level.

C. Interaural Correlation

As the MLD hierarchy shows, the maximum advantage of binaural listening depends on the degree to which the noise is correlated at the two ears while the signal is inverted (S_π). Generally, in most practical situations, the ambient acoustic noise is correlated, but the electrical noise containing the signal, when inverted, tends to decorrelate the noise at the two ears. Thus, it is an important issue, both practically and theoretically, to determine how the degree of binaural improvement depends upon the correlation of the masker at the two ears.

One of the first to study this phenomenon systematically was Licklider (1948), who used speech as the signal waveform. Licklider varied the correlation of the noise at the two ears by using three noise sources. The waveform coming from one source was split and added to the waveforms from the other two sources, each of which went to only one ear. The interaural correlation is determined by the relative amount of noise from the common source, which has a perfect positive correlation to the amount of the other noises which were uncorrelated. Licklider's results showed that the correlation had little effect until it was greater than about 0.70. ROBINSON and JEFFRESS (1963) used LICKLIDER's technique to study the effect of noise correlation on the detection of sinusoidal signals. Their results showed a systematic change in the size of the MLD (both the $M_\mu S_0$ and $M_\mu S_\pi$ conditions) as the correlation was changed from —1.0 to +1.0. WILBANKS and WHITMORE (1968) studied the effects of interaural noise correlation for a variety of frequencies ranging from 150 to 4000 Hz. In their discussion, they showed how, by means of a single scaler, all of the results obtained at different frequencies could be reduced to a single function. Their data are consistent with previous investigations of the appropriate frequencies. EGAN and BENSON (1966), DURLACH (1964), and DOLAN and ROBINSON (1967) have also varied the interaural correlation of the noise masker, and their results essentially agree with those of ROBINSON and JEFFRESS.

These results, therefore, tend to confirm the analysis of the effect of overall level of the masker and the results obtained at low signal frequencies in terms of internal and external noise. In fact, one can employ these data on interaural correlation to estimate the degree to which the internal noise is correlated in absolute threshold conditions, DIERCKS and JEFFRESS (1962), ROBINSON and JEFFRESS (1963), MCFADDEN (1968).

D. Interaural Time

For a single sinusoid, there is no difference between a phase shift and a time delay. For a wideband noise, however, a time delay is a phase shift in which the amount of the shift is proportional to frequency. JEFFRESS *et al.* (1952) were the first to investigate how the masking-level difference depended on a time delay between the noise masker at the two ears. The signal in their study was a 500-Hz tone; they found that as the noise was delayed, detection-performance improved to about 10 dB until the delay reached the value of about 1 msec, half the period of the 500-Hz tone. From that point, detection performance deteriorated until the delay was 2 msec, at which point the results were essentially the same as those obtained with no delay. Continuing the amount of the delay produced another maximum of about 8 dB at 3 msec and another local minimum at 4 msec. The same authors, in 1962, studied a delay in the noise when the signal was 167 Hz. The results, however, are quite different, for as the delay increased at 167 Hz, the masking-level difference increased slightly to a value of about 4 dB at 1 msec and remained there for all longer delays. In a later study, LANGFORD and JEFFRESS (1964) studied the effects of delay as long as 10 msec. This study, which is interpreted in terms of an autocorrelation model, showed that performance oscillated with a period equal to that of the signal, as might be expected, but that improvement in detection diminished gradually as longer and longer delays occurred.

The general results obtained with delaying the noise can be anticipated because delaying the noise reduces the interaural correlation of the masker. The exact way in which the correlation decreases depends on how the auditory system filters the waveforms at the two ears. Thus, these data can be interpreted to estimate various filter parameters of the auditory system (JEFFRESS *et al.*, 1952).

E. Interaural Phase

Since noise is composed of many frequencies, a phase shift in the noise produces differential time shifts in the components at the different frequencies. The first to systematically study the effects of phase-shifting the noise interaurally were JEFFRESS *et al.* (1952). In their study, the signal was a 500-Hz sinusoid. The noise was shifted in 30, 6 degree steps from 0 to 180 degrees. The results were as one might expect; best performance occurred when the noise and signal were 180 degrees out of phase from one another, and worst performance resulted when both signal and noise were in phase at the two ears. An interesting result obtained in this experiment was the so-called "flattening effect". The flattening effect refers to a graph relating detection performance and the interaural phase of the signal or the noise. Changes in the interaural phase of the signal, holding the phase of the noise constant, produce a much more pronounced peak in performance than comparable conditions in which the interaural phase of the noise is varied and the phase of the signal is held constant. This same effect was observed by METZ *et al.* (1967), who studied a 250-Hz sinusoid using both a very narrow band (4.2 Hz), and a wider band of noise (250 Hz). Their data show very clearly the flattening effect for both bandwidths of the noise.

F. Bandwidth

The size of the MLD depends upon coherence in the masker at the two ears. For a wideband noise, one might suspect that only a narrow region of the spectrum located near the signal frequency is actually responsible for the binaural improvement in detection. This supposition has been confirmed by MULLIGAN *et al.* (1967). There is one methodological problem, however. As the noise is filtered, the overall level of the masker decreases, and it is known that the size of the MLD depends on level. An ingenious way to overcome this difficulty was utilized by SONDHI and GUTTMAN (1966). They used digital filtering to remove a nearly rectangular band of the noise spectrum. They could systematically vary the width of this noise band, and by inverting this band before adding it to the two ears, they could manipulate the interaural phase relations. The remainder of the noise spectrum from which the band was taken was presented in-phase at the two ears. Thus, in the $M_\pi S_0$ condition only a narrowband of noise centered about the signal frequency was antiphasic (M_π); the rest of the noise was homophasic (M_0). The signal frequency was 250 Hz. SONDHI and GUTTMAN showed that a 15-dB MLD would result as long as the width of this band was approximately 150 Hz wide. Using the Equalization and Cancellation model (see "Theories" section), they estimated that the width of the critical band for the MLD experiment is approximately 125 Hz, at a 250-Hz center frequency.

Another set of experiments involving masker bandwidth and the MLD involves simply narrowing the band of noise present about the signal and determining if the MLD that results when the interaural phases are reversed, varies with the noise bandwidth. The general result seems to be that the narrower the bandwidth of the noise, the larger the MLD. This effect was first observed by BOURBON and JEFFRESS (1965) and has been confirmed in later experiments by METZ *et al.*

(1967) and WIGHTMAN (1971). For a narrow band of noise, about 4.2 Hz wide, METZ *et al.* reported an MLD (M_0S_π) of almost 25 dB, compared with the MLD for wideband noise of about 15 dB. Wightman made a systematic study of how the MLD changes with bandwidth for a signal frequency of 800 Hz. He found that the resulting MLD (M_0S_π) depends heavily upon how one has filtered the noise and signal. He advanced the hypothesis that energy splattered by gating the signal changes considerably the size of the MLD. He also obtained a sizeable MLD (M_0S_π), larger than 20 dB, for a very narrow (about 3 cycle) band of noise.

G. Gated Masker

In practically all detection experiments, the masker is continuous and the signal, a gated sinusoid, is added to the noise. MCFADDEN (1966) was the first to use a procedure in which the masker as well as the signal was turned on only during the observation interval (125 msec). The use of the gating procedure produces essentially no change in detection for the M_0S_0 condition, but a worsening of detection, about 4—6 dB, in the M_0S_π condition. MCFADDEN also investigated how long the noise had to be turned on prior to the signal before detection performance returned to that obtained with a continuous masker. His results showed that presenting the noise about 600 msec before the onset of the signal was equivalent to a continuous masker.

H. Tonal Masker

A number of investigators (JEFFRESS *et al.*, 1956; WIGHTMAN, 1969; HAFTER and CARRIER, 1970; CANAHL, 1970) have used a tone as the masker in studies of the MLD. As was the case with noise signals, the phase relation between the signal masker at each ear can be controlled with the use of a tonal masker of the same frequency and duration as the signal. When the interaural stimulus configuration is M_0S_π, the MLD increases from 0—3 dB at a signal-to-masker phase angle of 0° to 25—30 dB at a phase angle of 90°. When the phase angle between the signal and masker is greater than 90°, the MLD decreases.

In the M_0S_π condition, variations in the phase angle of addition between the signal and the masker will result in changes in the relative magnitudes of the interaural amplitude and temporal differences in the waveforms at the two ears. For this reason, the tonal masker studies of the MLD have been closely tied to studies of lateralization and localization.

IV. Assessments of Binaural Improvement Using Methods other than Detection

A. Speech Intelligibility

The most popular signal, excluding sinusoids, in MLD studies has been speech. That is, investigations have sought ways of presenting speech to two ears such that it would be more easily detected or understood than when presented to only one ear. Two procedures have been used to measure the improvement: 1) the speech signal-to-masker ratio required to detect the presence of the signal is measured in the reference and binaural conditions (these results have been discussed in Section II), or 2) the speech signal-to-masker ratio required to recognize or understand the speech signal is measured in the reference and binaural conditions. The usual measure of speech recognition is intelligibility: the proportion of words correctly repeated or identified from a fixed list of speech signals.

LICKLIDER (1948) was the first to measure the improvement in intelligibility caused by binaural, antiphasic listening. His results showed binaural improvements of about 6 dB when the observer is obtaining scores of 20—30% intelligibility and practically no improvement at higher signal-to-noise ratios. SCHUBERT and SCHULTZ (1962) confirmed this basic finding and also found that the most improvement in intelligibility resulted when the masker was the subject's own voice, a reasonable but interesting result. This change in the amount of binaural improvement as a function of signal level is in marked contrast to the results obtained when detecting sinusoidal signals. A number of investigators have shown that the size of the MLD is largely independent of the level of detection performance (GREEN, 1966; EGAN *et al.*, 1969; MCFADDEN and PULLIAM, 1971).

A reasonable interpretation of this difference is provided by the excellent analysis of LEVITT and RABINER (1967 a and b). In their first paper, they explored how the frequency content of the masker and the signal changes the amount of improvement and how time delay of the signal influences intelligibility in the antiphasic condition. The second paper is theoretical and attempts to derive the improvement in intelligibility score by treating the antiphasic condition as if it simply reduced the noise level over that used in the monaural condition. They demonstrated the effectiveness of this scheme by predicting a wide variety of data. The diminishing increase in improvement in the binaural intelligibility score, as signal level increases, is predicted on the basis of smaller gains in the articulation index as signal-to-noise level increases.

B. Loudness

TOWNSEND and GOLDSTEIN (1972) have also studied the difference between homophasic and antiphasic conditions (M_0S_0 and M_0S_π) on judgments of apparent loudness of the signal. The signal is judged somewhat louder in the antiphasic conditions, but like the results obtained with recognition of speech, the difference is apparent only when the tones are barely detectable in the masker. Once the signals become clearly detectable in the homophasic conditions, they are apparently no louder in the antiphasic conditions.

C. Frequency and Intensity Discrimination

HENNING (1973) has measured the improvement caused by binaural listening when the subject is trying to discriminate a change in frequency of a sinusoid partially masked by noise, and, in a separate experiment, when the subject is attempting to discriminate a change in the amplitude of a sinusoid partially masked by noise. He has studied the difference in performance between M_0S_0 and M_0S_π as the signal-to-noise ratio is systematically varied. For either of these discrimination tasks, the summary is quite simple. As long as the signal energy-to-noise power density (E/N_0) is less than about 5 dB in the S_0 condition, then one can expect improvement in the S_π condition. The amount of this improvement is nearly that obtained using detection measures of performance. For large signal-to-noise ratios, the improvements become smaller and finally vanish. Since large signal-to-noise ratios are needed for difficult discrimination tasks (a small $\triangle f$ or a small $\triangle I$), no improvement is measured in these conditions. Conversely, for large $\triangle f$ or large $\triangle I$, since small signal-to-noise ratios are needed, the measured amount of improvement is large.

It appears that binaural listening is most facilitated when one is trying to detect the presence of a signal. In other tasks, where the signal-to-masker ratio must be large, there appears to be little advantage in using the binaural system.

V. Theories

The following three sections discuss three current theories of why binaural listening improves the detection of signals in certain dichotic conditions. The oldest of these is the Webster-Jeffress hypothesis which dates from Webster's initial observations (1951) of the MLD phenomena. The theory was adapted and elaborated by Jeffress (1956) in some detail and is sometimes known as "the vector model" or the "θ model". About 1960 a relatively formal and mathematical theory was advanced by Durlach (1963) based upon ideas and concepts from radar analysis. Of late, Hafter (Hafter *et al.*, 1969; Hafter and Carrier, 1970) has tried to create a more intimate bridge between the results of the masking level differences and the phenomena of binaural localization and lateralization. These theories will be reviewed very briefly; the pertinent references should be consulted for further details.

A. Webster-Jeffress Hypothesis

Basically, the Webster-Jeffress hypothesis assumes that binaural detection is better than monaural detection because the interaction of the signal and the noise at the two ears produces a time difference in the waveforms arriving at the two ears. To compute the magnitude of this time difference, one needs to make certain assumptions concerning the representation of the signal and the noise. The most popular assumption is that, because only a narrowband of noise is effective in masking the signal, the noise can be treated as a slowly changing sinusoidal process. The signal, when added to such a noise, produces a change in the resulting sinusoid that can be analyzed according to the familiar vector diagrams (see Fig. 2). Computing the magnitude and phase of the vectors at the two ears, one can calculate a phase angle difference between the sinewaves at the two ears and, hence, a temporal delay in the waveforms at that frequency. It is this delay that is presumably responsible for the enhanced detection. Without the signal, there is no delay in the M_0 condition, or a 180^0 phase delay in the M_π

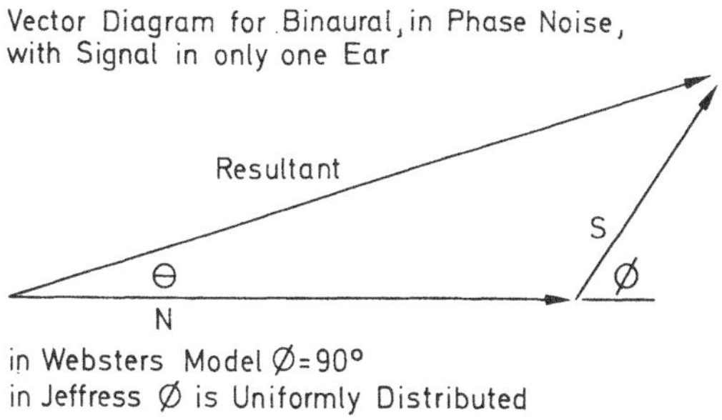

Fig. 2. A vector diagram depicting the addition of a signal of amplitude, S, to a noise masker of instantaneous amplitude, N, at a phase angle of addition, ϕ. The resultant amplitude and resultant phase angle, θ, are a function of the distributions of ϕ and N. In the M_0S_m case the angle θ is the interaural phase difference described in the WEBSTER-JEFFRESS hypothesis

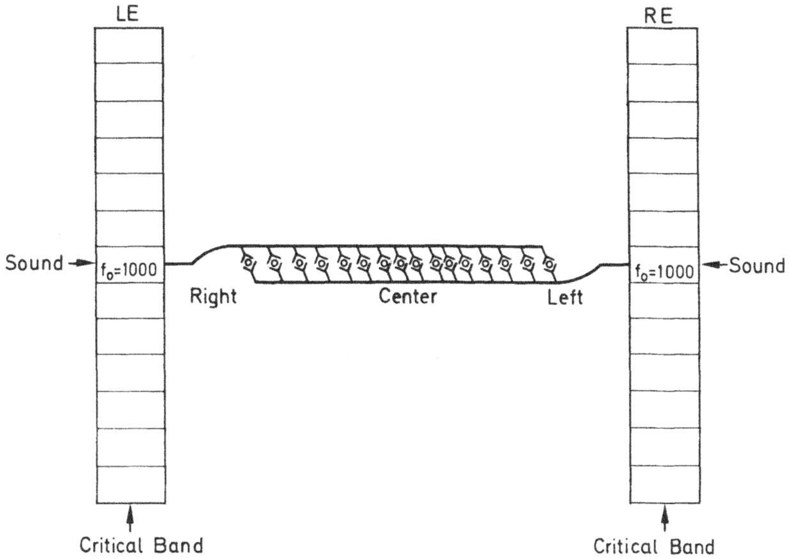

Fig. 3. The neural coincidence network described by JEFFRESS to account for localization and improved detection associated with the introduction of an interaural time difference. Signals entering the Right (RE) and Left (LE) ear are first passed through a critical band filter. At any particular frequency ($f_0 = 1000$ Hz, for instance), there is an array of collaterals which converge on higher neurons (indicated as small circles in the diagram). A signal presented without an interaural time difference will produce maximum activity in the middle of the chain, while signals with an interaural time delay will move the maximum activity toward one or the other side of the chain

condition. When the signal is added to the masker, however, the delay is systematically altered; it is this change in delay that makes the signal easy to detect.

In the 1950's JEFFRESS imbedded this general notion within a theory of localization. He envisions two filter systems (Fig. 3) at the two ears representing the mechanical and neural analysis of the signal, both of which are known to be frequency-independent. After this filtering operation, a hypothetical axon extends toward the other ear and toward the corresponding filter on the opposite side. Thus, as Fig. 3 shows, there is a matrix of hypothetical axons reaching towards each other. Each of these axons carries a signal representing the amount of energy in the acoustic wave at that particular frequency. Collaterals from these axons converge on higher order neurons located at various distances between the two ears. If the waveform were identical at the two ears, the delay in each axon from each side of the head would be roughly the same, and the higher order neurons in the middle of this matrix would be stimulated. As delay is introduced, the neuron impulses meet off center, either to the left or right, depending upon which ear is leading or lagging; consequently, both localization and improved detection can be deduced from this model. A similar model was suggested by Licklider (1959).

B. Durlach's Model

A more abstract and mathematical theory has been advanced by Durlach (see Fig. 4). Durlach's hypothesis is that certain linear operations are performed at the waveforms in the two ears; for example, he assumes either addition or subtraction. By applying these operations, the signal is extracted from the noise. Before the linear operator is applied, however, two aspects of the process must be considered. First, the waveforms at the two ears must be equalized so that they have the same amplitude. This is assumed possible except for a small amplitude error. Secondly, it is assumed that each waveform undergoes some temporal jitter. The failure to achieve perfect equalization, therefore, explains why, for example, in an M_0S_π condition, the signal does not become infinitely more detectable since without the amplitude error the noise would be completely cancelled. The temporal jitter provides an easy way to account for decreases in MLD at higher frequencies. Once the period of the waveform exceeds the range of the jitter, then (on the average) no cancellation is achieved.

A further assumption of the model is that: if the equalization and cancellation processes cannot yield detection better than that achieved by the monaural mechanism, it will not be used. Thus, the model predicts small positive MLD's for some conditions but never negative MLD's.

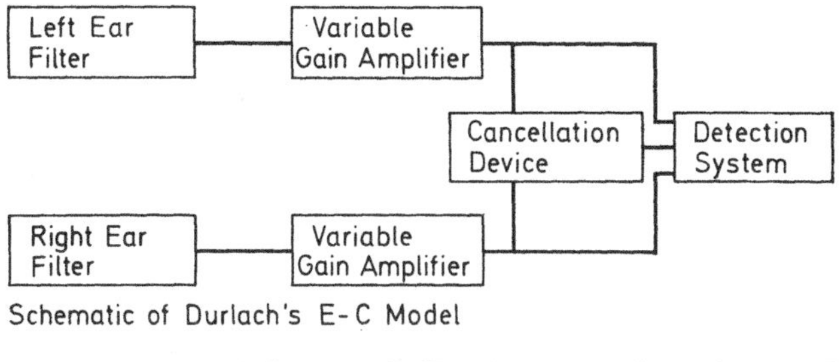

Fig. 4. A schematic diagram of Durlach's Equalization — Cancellation (E-C) model for MLD. Signals entering the right and left ears are filtered and adjusted in gain to equalize the level of the masker at the two ears. The equalized outputs of the two channels are either added or subtracted in order to cancel the masker as much as possible and allow the signal to be detected

The equalization and cancellation steps in the model have lead to its label: "the Equalization-Cancellation" or "E-C Model". Although the model does not describe any physiological mechanism, it has been highly successful in accounting for a large portion of the MLD literature. Other investigators (GREEN, 1966; SCHENKEL, 1967) have used the idea of the E-C Model in proposing other models of the MLD. In general, those models offer predictions for some conditions not accounted for by the DURLACH model although they risk losing the generality of the DURLACH model.

C. Lateralization Model

An MLD occurs only when there are interaural differences in the waveforms arriving at the ears. The interaural differences can be either in amplitude, in time, or in both. Since these are the same interaural differences which enable one to localize or lateralize a sound source, HAFTER (1969) has used a lateralization model to describe the MLD data. He has shown that an observer's ability to lateralize an image in an MLD paradigm is similar to his ability to detect the signal (HAFTER and CARRIER, 1970).

HAFTER's model derives mainly from his work with tonal maskers and the MLD. He has shown that both the interaural amplitude and the interaural temporal differences are important in determining the MLD. His model assumes that interaural amplitude and interaural delay add in a linear fashion to produce a change in the lateral position of an image. This movement results in detection of the signal in a binaural masking experiment. Although the model does not predict the difference in detection between the referent and binaural conditions (MLD), it does predict the possible changes in detection for a dichotic condition. Thus, the model relates the binaural masking data to the data obtained from lateralization and localization studies. The lateralization model has been successful in accounting for a great deal of data, especially that involving tonal maskers.

VI. Summary

This chapter has reviewed those stimulus conditions that lead to an improvement in performance caused by binaural listening. In general, improvement occurs when there is an interaural discrepancy between the masker and signal. The most common means of measuring the improvement is the masking task. Here, the difference in detection performance between the monaural and binaural conditions is known as a masking level difference (MLD). How changes in both signal and masker parameters affect the size of the MLD is reviewed in two major sections of the chapter. Next, the improvement in binaural listening is reviewed in a variety of tasks that do not involve detection. Finally, three theories of the mechanisms responsible for the improvement caused by binaural listening are briefly presented.

The emphasis has been empirical, reviewing those stimulus conditions that cause binaural improvement. It is hoped that this review will facilitate the direct investigation of some of these conditions in physiological studies of the binaural system.

Acknowledgement

This research was supported by the National Institutes of Health, Public Health Service, U.S. Department of Health, Education and Welfare and by a National Science Foundation Postdoctoral Fellowship. The authors would also like to thank Dr. G. B. Henning and Dr. D. McFadden whose comments on an earlier draft of this review were most helpful.

References

Blodgett, H. C., Jeffress, L. A., Taylor, R. W.: Relation of masked threshold to signal-duration for various interaural phase-combinations. Amer. J. Psychol. **71**, 283–290 (1958).

Blodgett, H. C., Jeffress, L. A., Whitworth, R. H.: Effect of noise at one ear on the masked threshold for tone at the other. J. acoust. Soc. Amer. **34**, 979–981 (1962).

Bourbon, W. T., Jeffress, L. A.: Effect of bandwidth of masking noise on detection of homophasic and antiphasic tonal signals. J. acoust. Soc. Amer. **37**, 1180(A) (1965).

Canahl, J. A.: Binaural masking of a tone by a tone plus noise. J. acoust. Soc. Amer. **47**, 476 (1970).

Canahl, J. A., Small, A. M., Jr.: Masking-level differences as a function of masker level. J. acoust. Soc. Amer. **38**, 928 (A) (1965).

Carhart, R., Tillman, T. W., Johnson, K. R.: Binaural masking of speech by periodically modulated noise. J. acoust. Soc. Amer. **39**, 1037–1050 (1966).

Carhart, R., Tillman, T. W., Johnson, K. R.: Release of masking for speech through interaural time delay. J. acoust. Soc. Amer. **42**, 124–138 (1967).

Colburn, H. S., Durlach, N. I.: Time-intensity relations in binaural unmasking. J. acoust. Soc. Amer. **38**, 93—103 (1965).

Diercks, K. J., Jeffress, L. A.: Interaural phase and the absolute threshold for tone. J. acoust. Soc. Amer. **34**, 981—984 (1962).

Dolan, T. R.: Effects of masker spectrum level on masking-level differences at low signal frequencies. J. acoust. Soc. Amer. **44**, 1507 (1968).

Dolan, T. R., Robinson, D. E.: An explanation of masking-level differences that result from interaural intensive disparities of noise. J. acoust. Soc. Amer. **42**, 977—981 (1967).

Durlach, N. I.: Note on the equalization and cancellation theory of binaural masking-level-differences. J. acoust. Soc. Amer. **35**, 1206—1218 (1963).

Durlach, N. I.: Note on binaural masking-level differences at high frequencies. J. acoust. Soc. Amer. **36**, 576—581 (1964).

Durlach, N. I.: Note on binaural masking-level differences as a function of the interaural correlation of the masking noise. J. acoust. Soc. Amer. **36**, 1613—1617 (1964).

Egan, J. P.: Masking-level differences as a function of interaural disparities in intensity of signal and of noise. J. acoust. Soc. Amer. **38**, 1043—1049 (1965).

Egan, J. P., Lindner, W. A., McFadden, D.: Masking-level differences and the form of the psychometric function. Perception and Psychophysics **6** (4), 209—215 (1969).

Egan, J. P., Benson, W.: Lateralization of a weak signal presented with correlated and with uncorrelated noise. J. acoust. Soc. Amer. **40**, 20—26 (1966).

Eldredge, D. H., Miller, J. D.: Physiology of hearing. Ann. Rev. Physiol. **33**, 281—310 (1971).

Flanagan, J. L., Watson, B. J.: Binaural unmasking of complex signals. J. acoust. Soc. Amer. **40**, 456—468 (1966).

Green, D. M.: Interaural phase effects in the masking of signals of different duration. J. acoust. Soc. Amer. **39**, 720—724 (1966a).

Green, D. M.: Signal-detection analysis of equalization and cancellation model. J. acoust. Soc. Amer. **40**, 833—838 (1966b).

Green, D. M., Henning, G. B.: Audition. Ann. Rev. Psychol. **20**, 105—128 (1969).

Hafter, E. R., Bourbon, W. T., Blocker, A. S., Tucker, A.: Direct comparison between lateralization and detection under conditions of antiphasic masking. J. acoust. Soc. Amer. **46**, 1452 (1969).

HAFTER, E. R., CARRIER, S. C.: Masking-level differences obtained with a pulsed tonal masker. J. acoust. Soc. Amer. **47**, 1041 (1970).

HENNING, G. B.: Effect of interaural phase on frequency and amplitude discrimination. J. acoust. Soc. Amer. **54**, 1160—1178 (1973).

HIRSH, I. J.: The influence of interaural phase on interaural summation and inhibition. J. acoust. Soc. Amer. **20**, 536—544 (1948a).

HIRSH, I. J.: Binaural summation and interaural inhibition as a function of the level of the masking noise. Amer. J. Psychol. **56**, 205—213 (1948b).

HIRSH, I. J., BURGEAT, M.: Binaural effects in remote masking. J. acoust. Soc. Amer. **30**, 827—832 (1958).

HIRSH, I. J., WEBSTER, F. A.: Some determinants of interaural phase effects. J. acoust. Soc. Amer. **21**, 496—501 (1949).

JEFFRESS, L. A.: A place theory of sound localization. J. comp. Physiol. Psychol. **41**, 35—39 (1948).

JEFFRESS, L. A., BLODGETT, H. C., DEATHERAGE, B. H.: The masking of tones by white noise as a function of the interaural phases of both components. J. acoust. Soc. Amer. **24**, 523—527 (1952).

JEFFRESS, L. A., BLODGETT, H. C., DEATHERAGE, B. H.: Masking and interaural phase. II. 167 cycles. J. acoust. Soc. Amer. **34**, 1124—1126 (1962).

JEFFRESS, L. A., BLODGETT, H. C., SANDEL, T. T., WOOD, C. L. III: Masking of tonal signals. J. acoust. Soc. Amer. **28**, 416—426 (1956).

JEFFRESS, L. A., MCFADDEN, D.: Differences of interaural phase and level in detection and lateralization. J. acoust. Soc. Amer. **49**, 1169 (1971).

JEFFRESS, L. A., Robinson, D. E.: Formulas for the coefficient of interaural correlation for noise. J. acoust. Soc. Amer. **34**, 1658—1659 (1962).

LANGFORD, T. L., JEFFRESS, L. A.: Effect of noise crosscorrelation on binaural signal detection. J. acoust. Soc. Amer. **36**, 1455—1458 (1964).

LEVITT, H., RABINER, L. R.: Binaural release from masking for speech and gain in intelligibility. J. acoust. Soc. Amer. **42**, 601—608 (1967a).

LEVITT, H., RABINER, L. R.: Predicting binaural gain in intelligibility and release from masking for speech. J. acoust. Soc. Amer. **42**, 820 (1967b).

LICKLIDER, J. C. R.: The influence of interaural phase relations upon the masking of speech by white noise. J. acoust. Soc. Amer. **20**, 150—159 (1948).

MCFADDEN, D.: Masking-level differences with continuous and with burst masking noise. J. acoust. Soc. Amer. **40**, 1414—1419 (1966).

MCFADDEN, D.: Masking-level differences determined with and without interaural disparities in masker intensity. J. acoust. Soc. Amer. **44**, 212—223 (1968).

MCFADDEN, D., PULLIAM, K. A.: Lateralization and detection of noise-masked tones of different durations. J. acoust. Soc. Amer. **49**, 1191—1194 (1971).

METZ, P. J., VON BISMARCK, G., DURLACH, N.: I. Further results on binaural unmasking and the EC model. II. Noise bandwidth and interaural phase. J. acoust. Soc. Amer. **43**, 1085—1091 (1967).

MULLIGAN, B. E., MULLIGAN, M. J., STONECYPHER, J. F.: Critical band in binaural detection. J. acoust. Soc. Amer. **41**, 7—12 (1967).

POLLACK, I., PICKETT, J. M.: Interaural effects upon speech intelligibility at high noise levels. J. acoust. Soc. Amer. **30**, 293—296 (1958).

RABINER, L. R., LAURENCE, C. L., DURLACH, N. I.: Further results on binaural unmasking and the EC model. J. acoust. Soc. Amer. **40**, 62—70 (1966).

RILLING, M. E., JEFFRESS, L. A.: Narrow-band noise and tones as signals in binaural detection. J. acoust. Soc. Amer. **38**, 202—206 (1965).

ROBINSON, D. E.: The effect of interaural signal-frequency disparity on signal detectability. J. acoust. Soc. Amer. **50**, 568—571 (1971).

ROBINSON, D. E., JEFFRESS, L. A.: Effect of varying the interaural noise correlation on the detectability of tonal signals. J. acoust. Soc. Amer. **35**, 1947—1952 (1963).

SCHENKEL, K. D.: Accumulation theory of binaural-masked thresholds. J. acoust. Soc. Amer. **41**, 20—31 (1967).

Schubert, E. D., Schultz, M. C.: Some aspects of binaural signal selection. J. acoust. Soc. Amer. **34**, 844—849 (1962).

Shaw, E. A. G., Piercy, J. E.: Physiological noise in relation to audiometry. J. acoust. Soc. Amer. **34**, 745(a) (1962).

Sondhi, M. M., Guttman, N.: Width of the spectrum effective in the binaural release of masking. J. acoust. Soc. Amer. **40**, 600—606 (1966).

Stumpf, C.: Binaurale Tonmischung, Mehrheitsschwelle und Mitteltonbildung. Z. Psychol. **75**, 340 (1916).

Townsend, T. H., Goldstein, D. P.: Suprathreshold binaural unmasking. J. acoust. Soc. Amer. **51**, 621—624 (1972).

Webster, F. A.: The influence of interaural phase on masked thresholds: I. The role of interaural time-deviation. J. acoust. Soc. Amer. **23**, 452—461 (1951).

Weston, P. B., Miller, J. D.: Use of noise to eliminate one ear from masking experiments. J. acoust. Soc. Amer. **37**, 638—646 (1965).

Wightman, F. L.: Binaural masking with sinewave maskers. J. acoust. Soc. Amer. **45**, 72—78 (1969).

Wightman, F. L.: Detection of binaural tones as a function of masker bandwidth. J. acoust. Soc. Amer. **50**, 623—636 (1971).

Wilbanks, W. A., Whitmore, J. K.: Detection of monaural signals as a function of interaural noise correlation and signal frequency. J. acoust. Soc. Amer. **43**, 785—797 (1968).

Zerlin, S.: Interaural time and intensity difference and the MLD. J. acoust. Soc. Amer. **39**, 134—137 (1966).

Author Index

Page numbers in *italics* refer to the bibliography

Abeles, M. 207
Abeles, M., Goldstein, M. H., Jr. 179, *191*, 200, 201, 203, 205, 213—215, *217*
Abeles, M., see Goldstein, M. H., Jr. 213, *218*
Abraham, F. D., see Worden, F. G. 222, *246*
Ades, H. W. 199, *217*, 320, *390*
Ades, H. W., Brookhart, J. M. 111, 112, *141*, 172, *191*
Ades, H. W., Mettler, F. A., Culler, E. A. 178, *191*
Ades, H. W., see Bredberg, G. 283, 293, 297, *302*
Ades, H. W., see Capps, M. J. 75, *96*, 238, *242*, 381, 382, *390*
Ades, H. W., see Dolan, T. R. 266, 269, 278, 281, 288, *303*
Ades, H. W., see Engström, H. 283, 297, *303*
Ades, H. W., see Kryter, K. D. 324, 325, 364, 365, *395*
Ades, H. W., see Mickle, W. A. 160, *195*
Ades, H. W., see Raab, D. H. 333, 368, 369, *397*
Ades, H. W., see Stewart, J. W. 330, *399*
Adrian, H. O., Goldberg, J. M., Brugge, J. F. 146, 186, *191*
Adrian, H. O., Lifschitz, W. M., Tavitas, R. J., Galli, F. P. 133, *141*, 186, *191*
Adrian, H. O., see Goldberg, J. M. 116—120, 127, 129, *142*
Adrian, H. O., see Woolsey, C. N. 154, 159, 160, *198*
Aitkin, L. M., Anderson, D. J., Brugge, J. F. 120, 121, 124, 128, 129, 133, *141*
Aitkin, L. M., Dunlop, C. W. 181, 184, 186, 187, *191*
Aitkin, L. M., Dunlop, C. W., Webster, W. R. 180, 181—183, *191*
Aitkin, L. M., Webster, W. R. 157, 178—180, 182, 186, 188, *191*, *192*
Aitkin, L. M., see Brugge, J. F. 121, 128, 131, 133—136, 138, 139, *141*, 188, *192*, 205, 207, *217*
Aitkin, L. M., see Dunlop, C. W. 182, 184, 185, *193*
Aitkin, L. M., see Webster, W. R. 190, *198*
Akert, K., Woolsey, C. N., Diamond, I. T., Neff, W. D. 339, *390*
Albe-Fessard, D., Fessard, A. 188, *192*
Albe-Fessard, D., Gillette, E. 188, 189, *192*
Alford, B. R., see Igarashi, M. 75, *101*, 238, *244*, 381, 383, *394*
Allanson, J. T., Whitfield, I. C. 93, 94, *96*
Allen, W. F. 336, 339, 340, *390*
Altmann, F. 277, *302*
Altmann, Ya. A. 177, *192*
Altman, Ya. A., Syka, J., Schmigidina, G. N. 188, *192*
Amaro, J., see Guth, P. S. 235, *244*
Amato, G., La Grutta, V., Enia, F. 190, *192*
Amato, G., see La Grutta, V. 190, *195*
Amon, A. H., see Rosenzweig, M. R. 113, *143*
Andersen, P., Eccles, J. C., Sears, T. A. 181, *192*
Anderson, D. J., Rose, J. E., Hind, J. E., Brugge, J. F. 24, 27—29, 82, *96*, 138, *141*
Anderson, D. J., see Aitkin, L. M. 120, 121, 124, 128 129, 133, *141*
Anderson, D. J., see Brugge, J. F. 7, 30, 35—37, 94, *96*, 121, 128, 131, 133, 134, 135, 136, 138, 139, *141*, 188, *192*, 205, 207, *217*
Anderson, D. J., see Hind, J. E 3, 18, 30, 35, 37, 38, *101*
Anderson, D. J., see Pugh, J. E., Jr. 70, *104*
Anderson, D. J., see Rose, J. E. 3—5, 10, 15, 18—21, 23, 24—27, 35, 37, 38, 40, 93, 94, *105*, 204, *218*
Anderson, H., Wedenberg, E. 268, *302*
Andersson, A., see Engström, H. 283, 297, *303*
Andersson, S., Gernandt, B. 160, *192*
Andersson, S. A., see Carreras, M. 159, 160, *192*
Angell, J. R., Fite, W. 250, *259*
Aran, J. M. 69, 70, *96*
Aran, J. M., Delaunay, J. 71, *96*
Aran, J.-M., see Portmann, M. 69, 91, *104*
Arden, G. B., Sodeberg, V. 189, *192*
Arnott, G. P. 352, *390*
Arnott, G. P., see Neff, W. D. 350—352, *397*
Arthur, R. M., Pfeiffer, R. R., Suga, N. 24, 25, 30—33, 86, *96*
Ascher, P., see Buser, P. 189, *192*
Auth, T. L., see Sanchez-Longo, L. P. 355, *398*
Avanzini, G., Mancia, D., Pelliccioli, G. 241, *242*
Axelrod, S., Diamond, I. T. 358, *390*
Axelrod, S., see Kaas, J. 358—360, *394*

Bach-y-Rita, G., Brust-Carmona, H., Penalozo-Rojas, J., Hernandez-Péon, R. 222, *242*
Backus, A. S. N., see Kiang, N. Y-s. 18, 43, 46, 51, 53, 54, 56, *102*, 121, 122, *142*
Backus, A. S. N., see Wiener, F.M. 13, *107*
Baer, T., see Goldstein, J. L. 28, 29, 81, 83, 84, *100*
Baer, T., see Kiang, N. Y.-s. 23, 30, *102*
Balogh, K., Jr., see Ishii, D. 73, 86, *101*
Banister, H. 456, *458*
Bard, P., Rioch, D. McK. 331, 356, *390*
Barlow, H. B., Levick, W. R. 63, *96*
Barnes, W. T., Magoun, H. W., Ranson, S. W. 112, 120, *141*, 147, *192*
Barnes, H. W., see Massopust, L. C., Jr. 339, *395*
Baru, A. V. 326, *390*
Baru, A. V., Karaseva, T. A. 326, 340, 366, 367, *390*
Baru, A. V., see Gersuni, G. V. 325, 328, 340, *393*
Bassett, I. G., Eastmond, E. J. 41, *96*
Batteau, D. W. 255, *259*
Bauch, H. 439, *444*
Beal, D. D., see Matz, G. J. 273, *304*
Beatty, D. L., see Simmons, F.B. 222, 223, *245*
Bechterev 310
Bechthold, J. K., see Masterton, R. B. 375, *395*
Beekhuis, J., see Schuknecht, H.F. 272, 279, *305*
Békésy, G. von 6, 15, 16, 78, 79, 82, 83, 88, *96*, 223, 231, *242*, 278, 296, 297, *302*, 402, 409, 430, *444*, 455, *458*
Belmont, I., see Karp, E. 327, *394*
Bell, C., see Robinson, P. M. 234, *245*
Benevento, L. A., Coleman, P.D. 139, *141*, 176, 177, *192*
Benevento, L. A., see Loc, P.R. 159, *195*
Benevento, L. A., see Loe, P.R. 241, *244*
Benitez, J. T., Corvera, J., Novoa, V. 366, 367, *390*
Benitez, J.T., Schuknecht, H.F. 277, *302*
Benitez, J. T., Schuknecht, H. F., Brandenburg, J. H. 272, *302*
Benitez, J. T., see Schuknecht, H. F. 272, 279, *305*
Benjamin, R. M., see Hind, J.E. 213, *218*
Benson, W., see Egan, J. P. 470, *478*
Berlin, C. I., Lowe, S. S. 361, *390*
Berlin, C. I., McNeil, M. R. 361, *390*
Berlin, C. I., see Cullen, J. K., Jr. 69, *97*
Berman, A. L. 148, 151, *192*
Bernstein, J. M., Weiss, A. D. 273, *302*
Bernstein, J. M., see Myers, E.N. 266, 269, 273, 283, *305*
Biedenbach, M. A., Freeman, W. J. 113, *141*
Billone, M. C., see Dallos, P. 84, 87, *97*
Birch, H., see Karp, E. 327, *394*
Birks, R. H. E., MacIntosh, F.C. 235, *242*
Bismarck, G. von 433—435, *444*
Bismarck, G. von, see Metz, P.J. 471, 472, *479*
Blakemore, C., Cooper, G. F. 217, *217*
Blau, W. 372, *390*
Blauert, J. 253, 256, *259*
Bloch, E. 248, *259*
Blocker, A. S., see Hafter, E. R. 474, 477, *478*
Blodgett, H. C., Jeffress, L. A., Taylor, R. W. 466, 467, *478*
Blodgett, H. C., Jeffress, L. A. Whitworth, R. H. 468, 469, *478*
Blodgett, H. C., Wilbanks, W. A., Jeffress, L. A. 456, *458*
Blodgett, H. C., see Jeffress, L. A. 464, 466, 470—472, *479*
Bobbin, R. P., Konishi, T. 235, *242*
Boer, E. De 39, 41, 77, 83, 84, 92, 94, *97*, 430, *444*
Boer, E. De, Jongh, H. R. De 39, *97*
Boer, E. De, Jongkees, L. B. W. 39, *97*
Boer, E. De, Kuyper, P. 39, *97*
Boer, E. De, Smoorenburg, G. F., Kuyper, P. 35, *97*
Boerger, G., Gruber, J. 35, 67, *96*
Boerger, G., Gruber, J., Klinke, R. 77, *96*
Boerger, G., see Gruber, J. 94
Boerger, G., see Klinke, R. 76, 77, *102*
Bogdanski, D. F., Galambos, R. 63, *96*
Bohne, B. A. 267, 297, *302*
Bohne, B. A., see Eldredge, D.H. 267, 269, 283, *303*
Bone, R. C., Crowley, D., Rauchbach, E. 90, *96*
Boniny, G. von, Mehler, W. R. 216, *217*
Boord, R. L. 227, *242*
Boring, E. G. 248, *259*, 449, *458*
Bosatra, A. B., Veronese, A. 355, *390*
Boudreau, J. C., Tsuchitani, C. 116, 117, 120, 122, 124, 125, 127, 128, 131, 133, 134, *141*
Boudreau, J. C., see Tsuchitani, C. 113, 116, 122, 123, 127, 131, *144*
Bourbon, W. T., Jeffress, L. A. 471, *478*
Bourbon, W. T., see Hafter, E.R. 474, 477, *478*
Boyle, A. J., see Johnstone, B.M. 15, 16, 78, 79, 88, *101*
Brandenburg, J. H., see Benitez, J. T. 272, *302*
Brandt, J. F., see McClelland, K.D. 87, 93, *103*
Brawer, J. R., see Cohen, E. S. 43, 45, 51, *96*
Bredberg, G. 261, 271, 274, 275—282, 287, 293, *302*
Bredberg, G., Ades, H. W. 283, *302*

Bredberg, G., Engström, H., Ades, H.W. 293, *302*
Bredberg, G., Lindemann, H.H., Ades, H.W., West, R., Engström, H. 297, *302*
Bredberg, G., see Dolan, T.R. 266, 269, 278, 281, 288, *303*
Bredberg, G., see Hunter-Duvar, I.M. 287, *303*
Bredberg, G., see Lindeman, H.H. 297, *304*
Bremer, F., Stoupel, N. 225, *242*
Britt, R., see Starr, A. 57, *106*
Brittain, F.H. 422, *444*
Broadbent, D.E. 221, *242*, 361, *390*
Bromiley, R.B. 331, *390*
Bromiley, R.B., see Galambos, R. 180, 186, *193*
Brookhart, J.M., see Ades, H.W. 111, 112, *141*, 172, *191*
Brown, P.B., see Goldberg, J.M. 113, 114, 115, 117, 119, 120, 122, 124, 125, 127, 128, 130, 131, 133, 134—138, *142*
Brown, R.D., see Daigneault, E.A. 70, *97*
Brown, S.M., Schäfer, E.A. 310, *390*
Brown, T.S., Gedvilas, G.R., Marko, L.A. 337, 338, *390*
Brugge, J.F., Anderson, D.J., Aitkin, L.M. 121, 128, 131, 133, 135, 138, *141*, 188, *192*
Brugge, J.F., Anderson, D.J., Hind, J.E., Rose, J.E. 7, 30, 35—37, 94, *96*
Brugge, J.F., Dubrovsky, N.A., Aitkin, L.M., Anderson, D.J. 134, 135, 136, 139, *141*, 205, 207, *217*
Brugge, J.F., Merzenich, M.M. 216, *217*
Brugge, J.F., see Adrian, H.O. 146, 186, *191*
Brugge, J.F., see Aitkin, L.M. 120, 121, 124, 128, 129, 133, *141*
Brugge, J.F., see Anderson, D.J. 24, 27—29, 82, *96*, 138, *141*
Brugge, J.R., see Hind, J.E. 3, 18, 30, 35, 37, 38, *101*
Brugge, J.R., see Merzenich, M.M. 321, *396*
Brugge, J.F., see Rose, J.E. 3—5, 10, 15, 18—21, 23, 24—27, 35, 37, 38, 40, 93, 94, *105*, 204, *218*
Bruggencate, G. Ten, Sonnhof, U. 220, 235, *242*
Bruner, J.S. 221, *242*
Bruner, J., see Buser, P. 189, *192*
Brunzlow, D. 251, *259*
Brust-Carmona, H., see Bach-y-Rita, G. 222, *242*
Bryan, J.S., see Erulkar, S.D. 163, 164, 168—174, *193*, 342, *392*
Bryan, J.S., see Nelson, P.G. 167, 169, 170, 191, *196*, 342, *397*
Buchwald, J.S., see Holstein, S.B. 189, *194*
Buchwald, J.S., see Kitzes, M. 189, *195*
Bunch, C.C., Wolff, D. 282, *302*
Bunch, C.C., see Guild, S.R. 276, *303*
Bundzen, P.V., see David, E. 191, *193*
Burgeat, M., see Hirsh, I.J. 465, *479*
Burger, J.F. 252, *259*
Burghardt, H. 437, 438, *444*
Burgio, P.A., see Pugh, J.E., Jr. 70, *104*
Burke, W., see Sefton, A.J. 181, *197*
Buser, P., Ascher, P., Bruner, J., Jassik-Gerschenfeld, D., Sindberg, R. 189, *192*
Butler, R.A. 249, 255, *259*
Butler, R.A., Diamond, I.T., Neff, W.D. 159, *192*, 318, 332, 335—337, 343, *390*
Butler, R.A., Honrubia, V., Johnstone, B.M., Fernandez, C. 80, *96*
Butler, R.A., Neff, W.D. 335, *390*
Butler, R.A., see Erulkar, S.D. 63, *98*, 121, 122, *141*, 342, *392*
Butler, R.A., see Gerstein, G.L. 54, *100*, 121, 122, *142*, 167, *193*
Butler, R.A., see Roffler, S.K. 249, 251, *260*, 454, 455, *459*
Butterfield, B.O., see Goldstein, M.H., Jr. 201, 208, 210, *218*
Buzzard, F., see Collier, J. 219, *242*

Cairns, H., see Hallpike, C.S. 272, 279, *303*
Cajal, S.Ry 43, 51, *96*, 110, 118, 119, 124, *141*, 147, 150, 151, 154, 155, 160, *192*
Campbell, F.W., Cooper, G.F., Enroth-Cugell, C. 42, *96*
Canahl, J.A. 472, *478*
Canahl, J.A., Small, A.M., Jr. 469, *478*
Capps, M.J., Ades, H.W. 75, *96*, 238, *242*, 381, 382, *390*
Capranica, R.R., see Frishkopf, L.S. 3, *99*
Carhart, R. 366, 367, *390*
Carhart, R., Tillman, T.W., Johnson, K.R. 468, *478*
Carhart, R., see Noffsinger, D. 366, 369, 377, *397*
Carmel, P.W., Starr, A. 223, *242*
Carreras, M., Andersson, S.A. 159, 160, *192*
Carrier, S.C., see Hafter, E.R. 472, 474, 477, *479*
Casseday, J.H. 366, *391*
Casseday, J.H., Diamond, I.T., Harting, J.K. 322, *391*
Casseday, J.H., Neff, W.D. 365, 371—373, 375, 376, 378, 381, *391*
Casseday, J.H., see Moore, C.N. 372, *396*
Casseday, J.H., see Neff, W.D. 343, 350, 352, *397*
Celesia, G.G., Puletti, F. 326, *391*
Chan, J. 357, *391*

Chain, F., see Lhermitte, F. 330, 332, 334, 340, 345, *395*
Chedru, G., see Lhermitte, F. 330, 332, 334, 340, 345 *395*
Cherry, E.C. 361, *391*
Chorazyna, H. 347, *391*
Chorazyna, H., Stepien, L. 241, *242*, 347, 348, *391*
Chow, K.L. 216, *217*
Chow, K.L., see Diamond, I.T. 159, *193*, 318, 337, 343, *391*, *392*
Christman, R.J., Victor, G. 456, *458*
Churchill, J.A., see Schuknecht, H.F. 234, *245*
Ciocco, A. 271, 278, *302*
Citron, L., Dix, M.R., Hallpike, C.S., Hood, J.D. 277, 281, *302*
Clark, G.M., Dunlop, C.W. 239, *242*
Clark, L.F., see Kiang, N.Y.-s. 3—11, 16, 18, 21—24, 30 32, 43, 56, 69, 70, 78, 84, 91, *102*, 122—124, 132, *142*, 230, *244*
Clark, W.W., see Stebbins, W.C. 266, 267, 269, 279, 306
Coakley, J.D., Culler, E. 178, *192*
Coats, A.C., Dickey, J.R. 69, *96*
Cohen, E.S., Brawer, J.R., Morest, D.K. 43, 45, 51, *96*
Colavita, F.B. 343, 344, *391*
Colavita, F.B., Szeligo, F.V., Zimmer, S.D. 344, *391*
Colburn, H.S., Durlach, N.I. 466, 467, *478*
Cole, G.G., see McGee, T.M. 265, 269, 278, 281, *305*
Coleman, P.D. 455, *458*
Coleman, P.D., see Benevento, L.A. 139, *141*, 176, 177, *192*
Collier, J., Buzzard, F. 219, *242*
Comis, S.D. 76, *97*, 241, *242*
Comis, S.D., Davies, W.E. 76, *97*, 241, *242*
Comis, S.D., Pickles, V.O. 75, *97*
Comis, S.D., Whitfield, I.C. 76, *97*, 236, 241, *242*, 383, *391*
Comis, S.D., see Pickles, J.O. 383, 384, *397*
Cooper, G.F., see Blakemoore C. 217, *217*
Cooper, G.F., see Campbell, F.W. 42, *96*
Coppée, G.E., see Kemp, E.H. 111, *142*
Cordeau, J.P., see Stepien, L.S. 241, *246*, 348, *398*
Cordier, R. 238, *242*
Cornwell, P. 343, *391*
Cortesina, G., see Rossi, G. 120, *143*
Corvera, J., see Benitez, J.T. 366, 367, *390*
Covell, W.P., see Miller, J.P. 15, *103*, 265, 269, *305*, 331, *396*
Cowan, W.M. *391*
Cowan, M., see Lewis, D. 430, *445*
Cowan, W.M., see Powell, T.P.S. 43, 51, *104*
Cowey, A., see Dewson, J.H. 241, *243*
Cranford, J. 360, *391*
Cranford, J., Ravizza, R., Diamond, I.T., Whitfield, I.C. 360, *391*
Cranford, J., see Whitfield, I.C. 360, *400*
Crowe, S.J., Guild, S.R., Polvogt, L.M. 270, 271, 278, *302*
Crowe, S.J., see Guild, S.R. 276, *303*
Crowley, D., see Bone, R.C. 90, *96*
Cullen, J.K., Jr., Ellis, M.S., Berlin, C.I., Lousteau, R.J. 69, *97*
Culler, E.A., Mettler, F.A. 331, *391*
Culler, E.A., see Ades, H.W. 178, *191*
Culler, E.A., see Coakley, J.D. 178, *192*
Culler, E.A., see Mettler, F.A. 332, *396*
Cummings, C.W. 266, 269, 279, *302*
Curtis, D.R. 235, *242*
Curtis, D.R., Duggan, A.W., Johnston, G.A.R. 220, 235, *242*

Dadson, R.S., see Robinson, D.W. 403, *446*
Daigneault, E.A., Brown, R.D., Pruett, J. 70, *97*
Dallos, P. 78, 79, 81, 87, *97*
Dallos, P., Billone, M.C., Durrant, J.D., Wang, C-y., Raynor, S. 84, 87, *97*
Daly, D.D., see Roeser, R.J. 345, *398*
Daly, R.L., see Goldstein, M.H., Jr. 213, *218*
Damasio, A.R., Lima, A., Damasio, H. 327, *391*
Damasio, H., see Damasio, A.R. 327, *391*
Dandy, W.E. 368, *391*
Darwin, C. 248, *259*, 321
David, E., Bundzen, P.V. 191, *193*
David, E., Finkenzeller, P., Kallert, S., Keidel, W.D. 190, *192*, *193*
David, E., see Kallert, S. 191, *194*
David, E.E., Jr., Guttman, N., Van, Bergeijk, W.A. 140, *141*
Davis, W.E., see Comis, S.D. 241, *242*
Davis, P.W., see Erulkar, S.D. 179, *193*, 200, 208, *217*
Davis, P.W., see Hind, J.E. 213, *218*
Davis, W.E., see Comis, S.D. 76, *97*
Davis, H. 85, *97*
Davis, H., Deatherage, B.H., Rosenblut, B., Fernandez, C., Kimura, R., Smith, C.A. 70, *97*
Davis, H., Derbyshire, A.J., Kemp, E.H., Lurie, M.H., Upton, M. 262, 269, *302*
Davis, H., Dworin, S., Lurie, M.H., Katzman, J. 262, 269, *302*
Davis, H., Silverman, S.R. 279, *302*
Davis, H., see Deatherage, B.H. 69, *97*

Davis, H., see Derbyshire, A.J. 69, 70, 89, 90, *97*
Davis, H., see Galambos, R. 3, 43, 53, 54, 57, *99*
Davis, H., see Saul, L.J. 111, *144*
Davis, H., see Teas, D.C. 70, *107*
Deatherage, B.H., Elredge, D.H., Davis, H. 69, *97*
Deatherage, B.H., Hirsh, I.J. 456, *458*
Deatherage, B.H., see Davis, H. 70, *97*
Deatherage, B.H., see Jeffress, L.A. 466, 470, 471, *479*
Decker, R.L., see Parker, W. 367, *397*
Delafresnaye, J.F. 225, *242*
Delaunay, J., see Aran, J.M. 71, *96*
Dell, P., see Dumont, S. 225, *243*
Delwaide, P.J., see Desmedt, J.E. 237, 238, *243*
Demont, D., see Kiang, N. Y.-s. 23, 30, *102*
Dempsey, E.W., see Smith, C. A. 86, *106*
Derbyshire, A.J., Davis, H. 69, 70, 89, 90, *97*
Derbyshire, A.J., see Davis, H. 262, 269, *302*
Desmedt, J.E. 76, *97*, *141*, 220—224, 227, 228, 230—234, 238—241, *242*
Desmedt, J.E., Delwaide, P. J. 237, 238, *243*
Desmedt, J.E., La Grutta, V. 226, 228, 235, *243*
Desmedt, J.E., La Grutta, V., La Grutta, G. 229, 230, *243*
Desmedt, J.E., Mechelse, K. 159, *193*, 220, 221, 224, 238, 241, *243*
Desmedt, J.E., Monaco, P. 72, *98*, 220, 230, 231, 233, 235—237, *243*
Desmedt, J.E., Robertson, D. 231, 234, 236, 237, *243*
Desmedt, J.E., see Rasmussen, G.L. 220, 238
Dewson, J.H. *243*, 362, 363, 382, 383, *391*
Dewson, J.H., Cowey, A., Weiskrantz, L. 241, *243*
Dewson, J.H. III 72, 75, *98*
Dewson, J.H. III, Pribram, K.H., Lynch, J.C. 362, *391*
Dey, F.L., see McCabe, P.A. 273, *304*
Diamond, I.T. 340, *391*
Diamond, I.T., Chow, K.L. 318, *391*
Diamond, I.T., Chow, K.L., Neff, W.D. 159, *193*, 318, 337, 343, *392*
Diamond, I.T., Goldberg, J. M., Neff, W.D. 332, 336, 337, 348, 349, *392*
Diamond, I.T., Jones, E.G., Powell, T.P.S. 160, 161, *193*, 314, 315, 320, *392*
Diamond, I.T., Neff, W.D. 147, *193*, 318, 332, 341—344, 349, *392*
Diamond, I.T., see Akert, K. 339, *390*
Diamond, I.T., see Axelrod, S. 358, *390*
Diamond, I.T., see Butler, R. A. 159, *192*, 318, 332, 335—337, 343, *390*
Diamond, I.T., see Casseday, J.H. 322, *391*
Diamond, I.T., see Cranford, J. 360, *391*
Diamond, I.T., see Goldberg, J.M. 241, *244*, 342, 387, *393*
Diamond, I.T., see Guttman, N. 340, 341, *393*
Diamond, I.T., see Jane, J.A. 366, 371, *394*
Diamond, I.T., see Kaas, J. 358, 359, 360, *394*
Diamond, I.T., see Masterton, R.B. 321, 352, 353, 373, 374—376, *395*
Diamond, I.T., see Neff, W.D. 159, *196*, 317, 318, 337, 343, 352, 356, 372, *397*
Diamond, I.T., see Ravizza, R. 354, 387, *398*
Diamond, I.T., see Whitfield, I.C. 360, 387, *400*
Dickey, J.R., see Coats, A.C. 69, *96*
Diercks, K.J., Jeffress, L.A. 469, 470, *478*
Dix, M.R., Hallpike, C.S., Hood, J.D. 91, *98*, 272, 276, 279, *303*
Dix, M.R., Hood, J.D. 366, 369, *392*
Dix, M.R., see Citron, L. 277, 281, *302*
Dixon, N.T. 163, *193*, 365, 366, *392*
Djupesland, G. 223, *243*
Dolan, T.R. 465, 469, *478*
Dolan, T.R., Bredberg, G., Ades, H.W., Neff, W.D. 266, 269, 278, 281, 288, *303*
Dolan, T.R., Robinson, D.E. 469, 470, *478*
Doran, R., see Schuknecht, H. F. 234, *245*
Dubrovsky, N.A., see Brugge, J.F. 134—136, 139, *141*, 205, 207, *217*
Ducarne, B., see Lhermitte, F. 330, 332, 334, 340, 345, *395*
Duggan, A.W., see Curtis, D. R. 220, 235, *242*
Duifhuis, H. 7, 29, 77, 83, 84, 89, *98*
Dumont, S., Dell, P. 225, *243*
Dumont, S., see Hugelin, A. 225, *244*
Dunker, E.G., Grubel, G., Pfalz, R. 76, *98*
Dunker, D., see Grubel, G. 76, *100*
Dunlop, C.W., Itzkowic, D. J., Aitkin, L.M. 182, 184, 185, *193*
Dunlop, C.W., see Aitkin, L. M. 180—184, 186, 187, *191*
Dunlop, C.W., see Clark, G.M. 239, *242*
Durlach, N.I. 464, 465, 470, 474, 476, 477, *478*
Durlach, N.I., see Colburn, H. S. 466, 467, *478*
Durlach, N.I., see Metz, P.J. 471, 472, *479*
Durlach, N.I., see Rabiner, L. R. 465, *479*
Durrant, J.D., see Dallos, P. 84, 87, *97*
Duvall, A.J., see Ward, W.D. 261, 268, 269, 281, 283, 287, *306*

Dworin, S., see Davis, H. 262, 269, *302*

Eady, H. R., see Klumpp, R. G. 452, *459*
Eastmond, E. J., see Bassett, I. G. 41, *96*
Eccles, J. C. 113, *141*, 220, 235, 237, *243*
Eccles, J. C., see Andersen, P. 181, *192*
Echandia, E., Rodriguez, L. 236, *243*
Edwards, S. B., see Rosenquist, A. C. 386, *398*
Efron, R. 346, *392*
Egan, J. P. 467, 469, *478*
Egan, J. P., Benson, W. 470, *478*
Egan, J. P., Lindner, W. A., McFadden, D. 463, 473, *478*
Eggermont, J. J., see Spoor, A. 70, *106*
Ehmer, R. H. 337, *392*
Elberling, C., Salomon, G. 70, *98*
Eldredge, D. H., Miller, J. D. 267, 269, *303*, 462, *478*
Eldredge, D. H., Mills, J. H., Bohne, B. A. 267, 269, 283, *303*
Eldredge, D. H., see Deatherage, B. H. 69, *97*
Eldredge, D. H., see Engebretson, A. M. 33, *98*
Eldredge, D. H., see Iurato, S. 227, *244*
Eldredge, D. H., see Teas, D. C. 70, *107*
Elliott, D. N. 264, 269, 278, 281, *303*, 325, 334, *392*
Elliott, D. N., McGee, T. M. 266, 269, 278, 279, 281, *303*
Elliott, D. N., Stein, L., Harrison, M. J. 331, *392*
Elliott, D. N., see Frazier, L. 288, 293, *303*
Elliott, D. N., see Hunter-Duvar, I. M. 261, 267—269, 287, 289, 299, 301, *304*
Elliott, D. N., see Trahiotis, C. 75, *107*, 238, *246*, 381, 383, *399*
Elliott, L. L. 408, 414, *444*
Elliot Smith, G. 313, *392*
Ellis, M. S., see Cullen, J. K., Jr. 69, *97*
Engebretson, A. M., Eldredge, D. H. 33, *98*
Engel, F. L., see Ritsma, R. 440, *446*
Enger, P. S. 11, *98*
English, G. M., see Sando, I. 277, *305*
Engström, H., Ades, H. W., Andersson, A. 283, 297, *303*
Engström, H., see Bredberg, G. 293, 297, *302*
Enia, F., see Amato, G. 190, *192*
Enroth-Cugell, C., see Campbell, F. W. 42, 96
Ernstson, S. 268, 269, 283, *303*
Erulkar, S. D. 150, 164, 166, 172—174, *193*
Erulkar, S. D., Butler, R. A., Gerstein, G. L. 63, *98*, 121, 122, *141*, 342, *392*
Erulkar, S. D., Nelson, P. G., Bryan, J. S. 163, 164, 166, 168—174, *193*, 342, *392*
Erulkar, S. D., Rose, J. E., Davies, P. W. 179, *193*, 200, 208, *217*
Erulkar, S. D., see Gerstein, G. L. 54, *100*, 121, 122, *142*, 167, *193*
Erulkar, S. D., see Nelson, P. G. 121, *143*, 150, 166—170, 172, 173, 183, 186, 191, *196*, 342, *397*
Erulkar, S. D., see Powell, T. P. S. 43, *104*
Escourolle, R., see Lhermitte, F. 330, 332, 334, 340, 345, *395*
Evans, E. F. 3—5, 9, 10, 12, 13, 15—17, 21, 22, 26, 43, 57, 71, 78—80, 86, 87, 91, 92, 94, 95, *98*
Evans, E. F., Klinke, R. 18, 80, 87, *98*
Evans, E. F., Nelson, P. G. 3, 44—46, 48—54, 56, 57, 63, 95, *98*, *99*
Evans, E. F., Nelson, R. A. 342, *392*
Evans, E. F., Rosenberg, J., Wilson, J. P. 11, 13, 15, 16, 21, 22, 40, 41, *99*
Evans, E. F., Ross, H. F., Whitfield, I. C. 179, *193*, 213, *217*
Evans, E. F., Whitfield, I. C. 63, *99*, 201, *217*
Evans, E. F., Wilson, J. P. 15, 16, 40—43, 78, 80, 81, 87, 88, 92, *99*
Evans, E. F., see Nelson, P. G. 63, *103*
Evans, J., see Penfield, W. 354, 355, *397*
Evans, E. F., see Whitfield, I. C. 200, 210, *218*, 342, *400*
Evans, E. F., see Wilson, J. P. 41, 43, 88, *108*
Evarts, E. V. 216, *218*, 339, *392*

Faglioni, P., Spinnler, H., Vignolo, L. 362, *392*
Fastl, H. 409, 428, *444*
Fastl, H., see Terhardt, E. 430, *446*
Fastl, H., see Zwicker, E. 89, *108*, 408, 414, 444, *448*
Feferman, M. E., see Mehler, W. R. 160, *195*
Feinmesser, M., see Sohmer, H. 69, *106*
Feldman, M. M., see Harrison, J. M. 375, *393*
Feldtkeller, R., Oettinger, R. 406, *445*
Feldtkeller, R., see Zwicker, E. 403, 404, 408, 410, 411, 413, 415, 421, 425, 428, 429, *448*
Fernald, R. 63, *99*
Fernandez, C. 273, 282, *303*
Fernandez, C., Goldberg, J. M. 117, 119
Fernandez, C., Karapas, F. 112, 120, 122, *141*
Fernandez, C., see Butler, R. A. 80, *96*
Fernandez, C., see Davis, H. 70, *97*
Ferrier, D. 310, *392*
Fessard, A., see Albe-Fessard, D. 188, *192*
Fex, J. 72, 75, 76, 85, 86, *99*, 119, *142*, 220, 228, 230, 231, 239, *243*

Fieandt, H. von, Saxén, A. 282, *303*
Finch, G., see Mettler, F. A. 332, *396*
Finck, A. 90, *99*
Finkenzeller, P., see David, E. 190, 191, *192—194*
Fisher, H. G., see Freedman, S. J. 254, *260*
Fisher, J. D., see Neff, W. D. 159, *196*, 337, 343, 350—352, 356, *397*
Fite, W., see Angell, J. R. 250, *259*
Flanagan, J. L. 6, *99*
Flanagan, J. L., Guttman, N. 430, *445*
Flanagan, J. L., Watson, B. J. 467, *478*
Fleischer, K. 271, *303*
Fletcher, H. *392*
Fletcher, J. L., see Loeb, M. 257, *260*
Flock, Å. 3, 84, *99*
Flock, Å., see Wersäll, J. 236, *246*
Flottorp, G., see Zwicker, E. 87—89, *108*
Flower, R. M., Viehweg, R. 369, *392*
Forster, F. M., see Sanchez-Longo, L. P. 355, *398*
Frazier, L., Elliott, D. N., Hunter-Duvar, I. M. 288, 293, *303*
Freedman, S. J., Fisher, H. G. 254, *260*
Freeman, W. J., see Biedenbach, M. A. 113, *141*
Frishkopf, L. S. 30, *99*
Frishkopf, L. S., Capranica, R. R., Goldstein, M. H., Jr. 3, *99*
Frishkopf, L. S., Goldstein, M. H. 11, 15, 24, 30, 32, *99*
Frishkopf, L. S., see Furman, G. G. 32, 33, 82, *99*
Frost, V., see Massopust, L. C., Jr. 339, *395*
Funkenstein, H. H., Nelson, P. G., Winter, P., Wollberg, Z., Newman, J. D. 200, 212, *218*
Funkenstein, H. H., Winter, P. 200, 212, *218*
Furlong, L. D., see Powell, E. W. 190, *196*
Furman, G. G., Frishkopf, L. S. 32, 33, 82, *99*
Furukawa, T., Ishii, Y. 3, 11, 15, 24, *99*
Furukawa, T., Ishii, Y., Matsuura, S. 85, *99*
Furukawa, T., see Ishii, Y. 85, *101*

Gacek, R., see Spoendlin, H. 225, 236, *246*
Gacek, R. R., see Walsh, B. T. 5, 6, *107*
Gässler, G. 87, *100*, 410, *445*
Gage, F. H., see Shaxby, J. H. 456, *459*
Galambos, R. 30, 54, 58, 67, 72, 75, 90, *99*, 180, 189, *193*, 220, *243*, 381, *392*
Galambos, R., Davis, H. 3, 43, 53, 54, 57, *99*
Galambos, R., Myers, R. E., Sheatz, G. C. 146, *193*
Galambos, R., Rose, J. E., Bromiley, R. B., Hughes, J. R. 180, 186, *193*
Galambos, R., Schwartzkopff, J., Rupert, A. 112, 116, 117, 122, 131—133, 139, 140, *142*, 230, 239, *243*, 373, *392*
Galambos, R., Sheatz, G., Vernier, V. G. 189, *193*, 221, *243*
Galambos, R., see Bogdanski, D. F. 63, *96*
Galambos, R., see Hubel, D. H. 201, *218*
Galambos, R., see Marsh, J. T. 222, *244*
Galambos, R., see Moushegian, G. 43, 54, 56, 58, *103*
Galambos, R., see Rose, J. E. 3, 43, 45, 51, 52, 54—57, *105*, 155, *197*
Galambos, R., see Rupert, A. 3, 10, 24, 93, *105*
Galambos, R., see Sheatz, G. C. 189, *197*
Galley, N., Klinke, R., Pause, M., Storch, W.-H. 72, *99*, 235, *243*
Galley, N., see Klinke, R. 74, *102*
Galli, F. P., see Adrian, H. O. 133, 186, *191*
Gardner, M. B. 456, *458*
Gardner, M. B., Gardner, R. S. 251, *260*
Gardner, R. S., see Gardner, M. B. 251, *260*
Gardner, W. H., see Parker, W. 367, *397*
Gedvilas, G. R., see Brown, T. S. 337, 338, *390*
Geisler, C. D. 77, 83, *100*
Geisler, C. D. Rhode W. S., Hazelton, D. W. 134—136, 138, *142*, 177, *193*
Geisler, C. D., see Robles, L. 6, 83, *105*
Geisler, C. D., see Rose, J. E. 134—136, 138, *143*, 174—177, 188, *197*, 205, *218*
Geniec, P., Morest, D. K. 148—153, 163, *193*
Gernandt, B., see Andersson, S. 160, *192*
Gerstein, G. L., Butler, R. A., Erulkar, S. D. 54, *100*, 121, 122, *142*, 167, *193*
Gerstein, G. L., Kiang, N. Y. S. 3, *100*, 208, *218*
Gerstein, G. L., see Erulkar, S. D. 63, *98*, 121, 122, *141*, 342, *392*
Gerstein, G. L., see Rodieck, R. W. 52, *105*
Gersuni, G. V. 325, 326, 328 *392*
Gersuni G. V. Baru A. V. Karaseva T. A. 325, 328, 340, *393*
Gersuni, G. V., Baru, A. V., Karaseva, T. A., Tonkonogii, I. M. 325, *393*
Giammanco, S., see La Grutta, V. 190, *195*
Gillette, E., see Albe-Fessard, D. 188, 189, *192*
Gilse, V., Roelofs, O. 249—251, 254, *260*
Girden, E. 325, 332, *393*
Girden, E., see Mettler, F. A. 332, *396*
Gisselsson, L. 235, *244*
Glattke, T. J. 65, 67, *100*
Glorig, A., Nixon, J. 271, *303*
Glorig, A., Wheeler, D., Quiggle, R., Grings, W., Summerfield, A. 271, *303*

Glorig, A., see Ward, W.D. 257, *260*
Goblick, T.J., Jr., Pfeiffer, R.R. 7, 9, 36, 37, 81, *100*
Godfrey, D.A., see Kiang, N.Y.-s. 43, 45, 50, *102*
Gold, A., Wilpizeski, C.R. 273, *303*
Gold, T. 82, *100*
Goldberg, J.M., Adrian, H.O., Smith, F.D. 118, 119, 127, 129, *142*
Goldberg, J.M., Brown, P.B. 113—115, 117, 119, 120, 122, 124, 125, 127, 128, 130, 131, 133—138, *142*
Goldberg, J.M., Diamond, I.T., Neff, W.D. 241, *244*, 342, 387, *393*
Goldberg, J.M., Greenwood, D.D. 55, 59, 67, 90, *100*
Goldberg, J.M., Moore, R.Y. 120, *142*, 147, 148, 150, 152, 153, *193*, *393*
Goldberg, J.M., Neff, W.D. 146, 147, 159, *193*, 241, *244*, 336—338, 349, 370, 385, *393*
Goldberg, J.M., Smith, F.D., Adrian, H.O. 116, 117, 120, *142*
Goldberg, J.M., see Adrian, H.O. 146, 186, *191*
Goldberg, J.M., see Diamond, I.T. 332, 336, 337, 348, 349, *392*
Goldberg, J.M., see Fernandez, C. 117, 119
Goldberg, J.M., see Greenwood, D.D. 58, 60, 67, 68, *100*
Goldberg, J.M., see Hind, J.E. 121, 133, 135, *142*, 163, 174, 184, *194*
Goldberg, J.M., see Moore, R.Y. 147, 153, 154, 160, 161, *195*, 313, *396*
Goldberg, J.M., see Rose, J.E. 121, *143*, 150, 162—166, 174, *197*
Goldstein, D.P., see Townsend, T.H. 473, *480*
Goldstein, J.L. 33, 35, 80, 83, 94, *100*, 444, *445*
Goldstein, J.L., Baer, T., Kiang, N.Y.S. 28, 29, 81, 83, 84, *100*
Goldstein, J.L., Kiang, N.Y.S. 33—35, 80, 94, *100*, 444, *445*
Goldstein, J.L., see Houtsma, A.J.M. 94, *101*
Goldstein, M.H., Jr., Abeles, M., Daly, R.L., McIntosh, J. 213, *218*
Goldstein, M.H., Jr., Hall, J.L. II, Butterfield, B.O. 201, 208, 210, *218*
Goldstein, M.H., Jr., Ribaupierre, F. De, Yeni-Komshian, G. 208, 209, *218*
Goldstein, M.H., Jr., see Abeles, M. 179, *191*, 200, 201, 203, 205, 213—215, *217*
Goldstein, M.H., see Frishkopf, L.S. 3, 11, 15, 24, 30, 32, *99*
Goldstein, M.H., Jr., see Hall, J.L. II 133, *142*, 207, *218*
Goldstein, M.H., Jr., see Kiang, N.Y.-s. 53, 70, *102*
Goldstein, M.H., Jr., see Liff, H.J. 11, 30, 32, *102*
Goldstein, M.H., Jr., see Ribaupierre, F. De 201, 208, 210, 216, 217, *218*
Goldstein, R. 327, *393*
Goldstein, R., Goodman, A., King, R.B. 326, *393*
Goldstein, R., see Landau, W.M. 327, *395*
Goodglass, H., see Schulhoff, C. 345, *398*
Goodman, A.C. 369, 371, *393*
Goodman, A., see Goldstein, R. 326, *393*
Gordon, W.P., see Igarashi, M. 75, *101*, 238, *244*, 381, 383, *394*
Goreva, O.E., Kalinina, T.E. 370, 376, *393*
Graf, K. 272, *303*
Gray, P.R. 5, *100*
Graybiel, A.M. 386, *393*
Green, D.M. 414, *445*, 463, 466, 473, 477, *478*
Green, D.M., Henning, G.B. 462, *478*
Green, D.M., see Yost, W.A. 458, *459*
Greene, T.C. 365, *393*
Greenwood, D.D. 87, 88, 90, *100*, 278, 296, 297, *303*, 404, 405, 444, *445*
Greenwood, D.D., Goldberg, J.M. 58, 60, 67, 68, *100*
Greenwood, D.D., Maruyama, N. 30, 54—58, 68, 90, *100*, 121, 122, *142*
Greenwood, D.D., see Goldberg, J.M. 55, 59, 67, 90, *100*
Greenwood, D.D., see Hind, J.E. 121, 133, 135, *142*, 163, 174, 184, *194*
Greenwood, D.D., see Rose, J.E. 121, *143*, 150, 162—166, 174, *197*
Grings, W., see Glorig, A. 271, *303*
Groen, J.J., see Jongkees, L.B.W. 251, *260*
Gross, C.W., see Schuknecht, H.F. 277, *305*
Gross, N.B., Thurlow, W.R. 180, *194*
Gross, N.B., see Rose, J.E. 134—136, 138, *143*, 154, 159, 160, 174—177, 188, *197*, *198*, 205, *218*
Grossman, R.G., see Viernstein, L.J. 52, *107*
Grubel, G., Dunker, D., Rehren, D.V. 76, *100*
Grubel, G., see Dunker, E.G. 76, *98*
Gruber, J., Boerger, G. 94
Gruber, J., see Boerger, G. 35, 67, 77, *96*
Gruber, J., see Klinke, R. 76, 77, *102*
Gudden, B. 154, *194*
Guild, S.R. 270, *303*
Guild, S.R., Crowe, S.J., Bunch, C.C., Polvogt, L.M. 276, *303*
Guild, S.R., see Crowe, S.J. 270, 271, 278, *302*
Guinan, J.J., Jr. 113, 114, 116—120, 122—124, 131—133, *142*
Guinan, J.J., Jr., Peake, W.T. 13, *100*
Guinan, J.J., Jr., see Kiang, N.Y.-s. 43, 45, 50, *102*
Guinan, J.J., Jr., see Li, R.Y.-s. 118, *143*

Guth, P.S., Amaro, J. 235, *244*
Guth, P.S., see Jasser, A. 234, *244*
Guttman, N. 340, *393*
Guttman, N., Diamond, I.T. 340, 341, *393*
Guttman, N., see David, E.E., Jr. 140, *141*
Guttman, N., see Flanagan, J. L. 430, *445*
Guttman, N., see Sondhi, M. M. 471, *480*

Haas, H. 455, *458*
Habermann, J. 273, *303*
Hafter, E.R., Bourbon, W.T., Blocker, A.S., Tucker, A. 474, 477, *478*
Hafter, E.R., Carrier, S.C. 472, 474, 477, *479*
Hafter, E.R., Jeffress, L.A. 141, *142*, 457, *459*
Hall, J.L. 35, *100*, 230, *244*
Hall, J.L. II 116, 131, 133, 139, 140, *142*, 177, *194*
Hall, J.L. II., Goldstein, M. H., Jr. 133, *142*, 207, *218*
Hall, J.L. II, see Goldstein, M.H., Jr. 201, 208, 210, *218*
Hall, J.W., see Stevens, J.C. 426, *446*
Hall, R.D. 189, *194*
Hall, W.C., see Oliver, D.L. 322, *397*
Hallpike, C.S., Cairns, H. 272, 279, *303*
Hallpike, C.S., see Citron, L. 277, 281, *302*
Hallpike, C.S., see Dix, M.R. 91, *98*, 272, 276, 279, *303*
Hansen, C.C., Reske-Nielsen, E. 271, 272, 279, *303*
Harford, E., see Jerger, J. 369, *394*
Harris, G.G. 82, *100*
Harrison, J.M., Feldman, M. M. 375, *393*
Harrison, J.M., Howe, M.E. 375, *393*
Harrison, J.M., Irving, R. 43, *100*, 117, *142*, 321, 375, *393*
Harrison, J.M., Warr, B. 43, 45, 51, *100*
Harrison, J.M., see Elliott, D. N. 331, *392*
Harrison, J.M., see Irving, R. 111, *142*, 375, *394*
Harpman, J.A., see Woollard, H.H. 120, *144*, 147, 160, *198*, 318—320, *400*
Hart, C.W., see Noffsinger, D. 366, 369, 377, *397*
Harting, J.K., see Casseday, J.H. 322, *391*
Hatton, J.B., see Powell, E. W. 160, 190, *196*
Hawkins, J.E., Jr., Stevens, S.S. 88, *100*
Hawkins, J.E., Jr., see Johansson, L.-G. 293, *304*
Hawkins, J.E., Jr., see Stebbins, W.C. 266, 269, 279, 281, 297, *306*
Hazelton, D.W., see Geisler, C.D. 134—136, 138, *142*, 177, *193*
Heath, C.J., Jones, E.G. 159—161, *194*, 319, 343, *393*
Heck, K.H., see Jatho, K. 271, *304*
Heffner, H.E., Masterton, R. B. 353, *393*
Heffner, R., Heffner, H., Masterton, B. 15, *100*
Heffner, H., see Heffner, R. 15, *100*
Heffner, H., see Masterton, B. 374, *395*
Held, H. 219, *244*, 366
Held, R. 46, 95
Held, R., Ingle, D., Schneider, G.E., Trevarthen, C.B. 95, *100*
Helle, R. 442, 444, *445*
Hellerstein, D., see Starr, A. 57, *106*
Hellman, R.P., see Scharf, B. 92, *105*
Helmholtz, H.F.L. von 409, 430, 433, *445*
Hemenway, W.G., see Sando, I. 277, *305*
Henderson, D., see Iurato, S. 227, *244*
Henning, G.B. 473, 478, *479*
Henning, G.B., see Green, D. M. 462, *478*
Henry, F.M. 437, *445*
Hensen, E. 247, *260*
Henson, C.O., see Hubel, D. H. 201, *218*
Hernandez-Péon, R. 189, *194*, 221, 222, 224, *244*
Hernandez-Péon, R., Jouvet, M., Scherrer, H. 224, *244*
Hernandez-Péon, R., Scherrer, H. 189, *194*
Hernandez-Péon, R., Scherrer, H., Jouvet, M. 221, 224, *244*
Hernandez-Péon, R., see Bach-y-Rita, G. 222, *242*
Hicks, L., see Marsh, J.T. 222, *244*
Hiesey, R.W., Schubert, E.D. 82, *101*
Hiltz, F.F. *194*
Hilyard, V.H., see Sando, I. 277, *305*
Hinchliffe, R. 271, *303*
Hind, J.E. 18, 21, 24, *101*, 179, *194*, 335, *393*
Hind, J.E., Anderson, D.J., Brugge, J.R., Rose, J.E. 3, 18, 30, 35, 37, 38, *101*
Hind, J.E., Goldberg, J.M., Greenwood, D.D., Rose, J.E. 121, 133, 135, *142*, 163, 174, 184, *194*
Hind, J.E., Rose, J.E., Brugge, J.F., Anderson, D.J. 35, 37, *101*
Hind, J.E., Rose, J.E., Davies, P.W., Woolsey, C.N., Benjamin, R.M., Welker, W.I., Thompson, R.F. 213, *218*
Hind, J.E., see Anderson, D. J. 24, 27—29, 82, 96, 138, *141*
Hind, J.E., see Brugge, J.F. 7, 30, 35—37, 94, *96*
Hind, J.E., see Neff, W.D. 331, *397*
Hind, J.E., see Rose, J.E. 3—5, 10, 15, 18—21, 23—27, 35, 37, 38, 40, 93, 94, *105*, 121, 134—136, 138, *143*, 150, 162—166, 174—177, 188, *197*, 204, 205, *218*
Hirsch, J.E. 113, *142*
Hirsch, J.E., see Woolsey, C. N. 154, 159, 160, *198*

Hirsh, I. J. 89, *101*, 346, *393*, 464—466, 468, 469, *479*
Hirsh, I. J., Burgeat, M. 465, *479*
Hirsh, I. J., Sherrick, C. E. 346, *393*
Hirsh, I. J., Webster, F. A. 467, *479*
Hirsh, I. J., see Deatherage, B. H. 456, *458*
Hirsh, I. J., see Swisher, L. 347, *399*
Hocart, A. M., McDougall, W. 250, *260*
Hodgson, W. R. 327, 335, *394*
Holstein, S. B., Buchwald, J. S., Schwafel, J. A. 189, *194*
Hood, J. D., see Citron, L. 277, 281, *302*
Hood, J. D., see Dix, M. R. 91, *98*, 272, 276, 279, *303*, 366, 369, *392*
Hood, J. D., see Morales-Garcia, C. 368, *396*
Honrubia, V., see Butler, R. A. 80, *96*
Horn, G. 222, *244*
Hornbostel, E. M. von, Wertheimer, M. 249, *260*
Horwitz, M. R., see Pugh, J. E., Jr. 70, *104*
Hotta, T., Kameda, K. 188, *194*
Hotta, T., Terashima, S. 189, *194*
Houtgast, T. 91, *101*
Houtsma, A. J. M., Goldstein, J. L. 94, *101*
Howe, M. E., see Harrison, J. M. 375, *393*
Hubel, D. H., Henson, C. O., Rupert, A., Galambos, R. 201, *218*
Hubel, D. H., Wiesel, T. N. 200, 212, 214, *218*
Hubel, D. H., see Wiesel, T. N. 217, *218*
Hugelin, A., Dumont, S., Paillas, N. 225, *244*
Huggins, W. H., Licklider, J. C. R. 82, *101*
Hughes, J. R., see Galambos, R. 180, 186, *193*
Hughes, J. R., see Rose, J. E. 3, 43, 45, 51, 52, 54—57, *105*
Hunter-Duvar, I. M. 267, 287, 293, *303*
Hunter-Duvar, I. M., Bredberg, G. 287, *303*
Hunter-Duvar, I. M., Elliott, D. N. 261, 267—269, 287, 289, 299, 301, *304*
Hunter-Duvar, I. M., see Frazier, L. 288, 293, *303*
Huxley, A. F. 79, *101*

Igarashi, M., Alford, B. R., Nakai, Y., Gordon, W. P. 75, *101*, 238, *244*, 381, 383, *394*
Igarashi, M., Schuknecht, H. F., Myers, E. N. 273, *304*
Igarashi, M., see Schuknecht, H. F. 272, *305*
Ingle, D., see Held, R. 95, *100*
Irvine, D. R. F., Webster, W. R. 72, 75, *101*
Irving, R., Harrison, J. M. 111, *142*, 375, *394*
Irving, R., see Harrison, J. M., 43, *100*, 117, *142*, 321, 375, *393*
Ishii, D., Balogh, K., Jr. 73, 86, *101*
Ishii, Y., Matsuura, S., Furukawa, T. 85, *101*
Ishii, Y., see Furukawa, T. 3, 11, 15, 24, 85, *99*
Itzkowic, D. J., see Dunlop, C. W. 182, 184, 185, *193*
Iurato, S. *101*, 227, *244*
Iurato, S., Luciano, L., Pannese, E., Reale, E. 234, *244*
Iurato, S., Smith, C. A., Eldredge, D., Henderson, D. 227, *244*

Jane, J. A., Masterton, R. B., Diamond, I. T. 366, 371, *394*
Jane, J. A., see Masterton, R. B. 353, 373—376, *395*
Jasser, A., Guth, P. S. 234, *244*
Jassik-Gerschenfeld, D., see Buser, P. 189, *192*
Jatho, K., Heck, K. H. 271, *304*
Jeffress, L. A. 133, 140, *142*, 464, 474, 475, *479*
Jeffress, L. A., Blodgett, H. C., Deatherage, B. H. 466, 470, 471, *479*
Jeffress, L. A., Blodgett, H. C., Sandel, T. T., Wood, C. L. III 464, 472, *479*
Jeffress, L. A., McFadden, D. 457, *459*, 467, 468, *479*
Jeffress, L. A., Robinson, D. E. *479*
Jeffress, L. A., see Blodgett, H. C. 456, *458*, 466—469, *478*
Jeffress, L. A., see Bourbon, W. T. 471, *478*
Jeffress, L. A., see Diercks, K. J. 469, 470, *478*
Jeffress, L. A., see Hafter, E. R. 141, *142*, 457, *459*
Jeffress, L. A., see Langford, T. L. 470, *479*
Jeffress, L. A., see Moushegian, G. 456, 457, *459*
Jeffress, L. A., see Rilling, M. E. 467, *479*
Jeffress, L. A., see Robinson, D. E. 470, *479*
Jeffress, L. A., see Whitworth, R. H. 456, 457, *459*
Jerger, J., Jerger, S. 368, *394*
Jerger, J., Lovering, L., Wertz, M. 328, 329, 334, 356, *394*
Jerger, J., Shedd, J. L., Harford, E. 369, *394*
Jerger, J., Weikers, N. J., Sharbrough, F. W. III, Jerger, S. 328, 329, 334, 346, 356, *394*
Jerger, J. E. 330, 346, 356, 366, 367, *394*
Jerger, J. F. 279, 281, *304*
Jerger, J. F., Mier, M. 361, *394*
Jerger, S., see Jerger, J. 328, 329, 334, 346, 356, 368, *394*
Jerison, H. J. 339, 344, *394*
Jerison, H. J., Neff, W. D. 339, 344, *394*
Jones, E. G., Powell, T. P. S. 160, 161, 182, *194*

Jones, E. G., Rockel, A. J. 156, 157, 182, *194*
Jongh, H. R. De 39, *97*
Jongh, H. R. De, see Boer, E. De 39, *97*
Jongkees, L. B. W., Groen, J. J. 251, *260*
Jongkees, L. B. W., see Boer, E. De 39, *97*
Johnson, K. R., see Carhart, R. 468, *478*
Johnsson, L.-G., Hawkins, J. E., Jr. 293, *304*
Johnsson, L.-G., see Stebbins, W. C. 266, 269, 279, 281, 297, *306*
Johnstone, B. M., Sellick, P. M. 78, *101*
Johnstone, B. M., Taylor, K. J., Boyle, A. J. 15, 16, 78, 79, 88, *101*
Johnstone, B. M., see Butler, R. A. 80, *96*
Johnstone, B. M., see Johnstone, J. R. 11, 15, *101*
Johnstone, G. A. R., see Curtis, D. R. 220, 235, *242*
Johnstone, J. R., Johnstone, B. M. 11, 15, *101*
Johnstone, J. R., see Wilson, J. P. 6, 16, 78, 79, 81, 83, *108*
Jones, E. G., see Diamond, I. T. 160, 161, *193*, 314, 315, 320, *392*
Jones, E. G., see Heath, C. J. 159, 160, 161, *194*, 319, 343, *393*
Jones, E. G., see Rockel, A. J. 148, 150—153, 163, *197*
Jørgensen, M. B. 277, *304*
Jørgensen, M. B., Schmidt, R. 273, *304*
Jouvet, M., see Hernandez-Péon, R. 221, 224, *244*
Jungert, S. 95, *101*, 111, *142*

Kaas, J., Axelrod, S., Diamond, I. T. 358—360, *394*
Kaiser, W., see Zwicker, E. 417, *448*
Kallert, S. 190, *194*
Kallert, S., David, E., Finkenzeller, P., Keidel, W. D. 191, *194*
Kallert, S., Keidel, W. D. 191, *194*
Kallert, S., see David, E. 190, *192*, *193*
Kalinina, T. E., see Goreva, O. E. 370, 376, *393*
Kameda, K., see Hotta, T. 188, *194*
Kane, E. C., see Kiang, N. Y.-s. 43, 45, 50, *102*
Kanno, Y., see Katsuki, Y. 3, 4, 10—12, 18, 21, 24, *101*
Kanno, Y., see Maruyama, N. 325, *395*
Kanshepolsky, J., Kelley, J. J., Waggener, J. D. 329, *394*
Kappauf, W. E. 332, *394*
Karapas, F., see Fernandez, C. 112, 120, 122, *141*
Karaseva, T. A. 328, 345, *394*
Karaseva, T. A., see Baru, A. V. 326, 340, 366, 367, *390*
Karaseva, T. A., see Gersuni, G. V. 325, 328, 340, *393*
Karlan, M. S., Tonndorf, J., Khanna, S. M. 87, *101*
Karp, E., Belmont, I., Birch, H. 327, *394*
Katsuki, Y. 164, 179, 180, 190, *194*
Katsuki, Y., Kanno, Y., Suga, N. 3, *101*
Katsuki, Y., Suga, N., Kanno, Y. 3, 4, 10—12, 18, 21, 24, *101*
Katsuki, Y., Sumi, T., Uchiyama, H., Watanabe, T. 162, 164, 179, *195*
Katsuki, Y., Sumi, T., Uchiyama, H., Watanabe, T. 3, 54, *101*
Katsuki, Y., Watanabe, T., Maruyama, N. 179, *195*
Katsuki, Y., Watanabe, T., Suga, N. 16, 58, *101*
Katsuki, Y., see Nomoto, M. 4, 10, 21, 23, 30—33, *104*
Katsuki, Y., see Oonishi, S. 205, 213, *218*
Katsuki, Y., see Tanaka, Y. 235, *246*
Katsuki, Y., see Watanabe, T. 125, 133, *144*
Katz, B. *244*
Katzman, J., see Davis, H. 262, 269, *302*
Kawamura, S., see Niimi, K. 160, *196*
Keidel, W. D. 70, *101*, 190, *195*
Keidel, W. D., see David, E. 190, 191, *192—194*
Keidel, W. D., see Spreng, M. 69, *106*
Kelly, J. B. 342, 343, *394*
Kelly, J. B., Whitfield, I. C. 342, 343, *394*
Kelley, J. J., see Kanshepolsky, J. 329, *394*
Kemp, E. H. 262, *304*
Kemp, E. H., Coppée, G. E., Robinson, E. H. 111, *142*
Kemp, E. H., Robinson, E. H. 111, 113, *142*
Kemp, E. H., see Davis, H. 262, 269, *302*
Kennedy, T. T. 335, *394*
Kennedy, T., Kiang, N., Sutton, S. 335, *394*
Kessel, J. 248, *260*
Khanna, S. M., Sears, R. E. Tonndorf, J. 82, *101*
Khanna, S. M., see Karlan, M. S. 87, *101*
Kiang, N. Y.-s. 3, 5, 10—13, 15, 21, 23, 30, 32, 43, 45, 54, *101*, *102*, 124, *142*, 281
Kiang, N. Y.-s., Baer, T., Marr, E. M., Demont, D. 23, 30, *102*
Kiang, N. Y.-s., Goldstein, M. H., Jr., Peake, W. T. 53, 70, *102*
Kiang, N. Y.-s., Morest, D. K., Godfrey, D. A., Guinan, J. J., Jr., Kane, E. C. 43, 45, 50, *102*
Kiang, N. Y.-s., Moxon, E. C. 23, 81, 84, 85, 93, *102*
Kiang, N. Y.-s., Moxon, E. C., Levine, R. A. 3—5, 11, 13, 18, 71, 74, 86, 91, 92, *102*
Kiang, N. Y.-s., Peake, W. T. 87, *102*
Kiang, N. Y.-s., Pfeiffer, R. R., Warr, W. B., Backus, A. S. N. 18, 43, 46, 51, 53, 54, 56, *102*, 121, 122, *142*

Kiang, N. Y.-s., Sachs, M. B., Peake, W. T. 3, 12, 13, 15, 16, 78, *102*
Kiang, N. Y.-s., Watanabe, T., Thomas, E. C., Clark, L. F. 3—11, 16, 18, 21—24, 30, 32, 43, 56, 69, 70, 78, 84, 91, *102*, 122—124, 132, *142*, 230, *244*
Kiang, N. Y.-s., see Gerstein, G. L. 3, *100*, 208, *218*
Kiang, N. Y.-s., see Goldstein, J. L. 28, 29, 33—35, 80, 81, 83, 84, 94, *100*, 326, *393*, 444, *445*
Kiang, N., see Kennedy, T. 335, *394*
Kiang, N. Y.-s., see Koerber, K. L. 51, 52, *102*
Kiang, N. Y.-s., see Peake, W. T. 69—71, *104*
Kiang, N. Y.-s., see Pfeiffer, R. R. 43, 46, 51, 52, 82, 85, *104*, 122, *143*
Kiang, N. Y.-s., see Rodieck, R. W. 52, *105*
Kiang, N. Y.-s., see Sachs, M. B. 21, 30, 32, *105*
Kiang, N. Y.-s., see Walsh, B. T. 5, 6, *107*
Kiang, N. Y.-s., see Wiederhold, M. L. 6, 15, 70, 72, 73, 86, *107*, 220, 228, 232, 240, *246*
Kimura, D. 361, *394*, *395*
Kimura, R. S., Schuknecht, H. F., Sando, I. 282, 297, *304*
Kimura, R., see Davis, H. 70, *97*
Kimura, D., see Milner, B. 345, *396*
Kietz, H. von 253, *260*
Kitzes, M., Buchwald, J. 189, *195*
Klemm, O. 249, *260*
Klinke, R. 96
Klinke, R., Boerger, G., Gruber, J. 76, 77, *102*
Klinke, R., Galley, N. 74 *102*
Klinke, R., see Boerger, G. 77, *96*
Klinke, R., see Evans, E. F. 18, 80, 87, *98*
Klinke, R., see Galley, N. 72, *99*, 235, *243*
Kleffner, F. R., see Landau, W. M. 327, *395*
Klumpp, R. G., Eady, H. R. 452, *459*
Knight, P. L., see Merzenich, M. M. 321, *396*
Kobrak, H., see Lewy, F. H. 335, *395*
Koerber, K. L., Pfeiffer, R. R., Warr, W. B., Kiang, N. Y.-s. 51, 52, *102*
Kohut, R. L., see Lindsay, J. R. 272, 279, *304*
Konishi, M. 46, 53, *102*
Konishi, T. 235, *244*
Konishi, T., Nielsen, D. W. 84, *102*
Konishi, T., Slepian, J. Z. 228, 231, 232, 240, *244*
Konishi, T., Teas, D. C., Wernick, J. S. 85, *102*
Konishi, T., see Bobbin, R. P. 235, *242*
Konishi, T., see Teas, D. C. 15, 72, 73, 86, *107*, 228, *246*
Konorski, J. 343, 347—349, *395*
Korsan-Bengtsen, M., see Liden, G. 376, *395*
Kos, C. M. *395*
Krames, L., see Matz, G. J. 273, *304*
Kristensen, H. K. 272, 279, *304*
Kryter, K. D., Ades, H. W. 324, 325, 364, 365, *395*
Kryter, K. D., see Licklider, J. C. R. 335, *395*
Kuyper, P., see Boer, E. De 35, 39, *97*

Lagourgue, P., see Portmann, M. 91, *104*
La Grutta, V., Giammanco, S., Amato, G. 190, *195*
La Grutta, V., see Amato, G. 190, *192*
La Grutta, G., see Desmedt, J. E. 226, 228—230, 235, *243*
Landau, W. M., Goldstein, R., Kleffner, F. R. 327, *395*
Landgren, S., Silfvenius, H., Wolsk, D. 160, *195*
Lane, C. E., see Wegel, R. L. 404, *447*
Lane, J. F., see Woolsey, C. N. 154, 159, 160, *198*
Lane, W. C., see Lim, D. J. 297, *304*
Langford, T. L., Jeffress, L. A. 470, *479*
Langford, T. L., see Moushegian, G. 131—133, 135—137, 139, 140, *143*
Larinow, W. 310, *395*
Lashley, K. S. 317, 384, *395*
Latta, J., see Winter, P. 212, *218*
Laurence, C. L., see Rabiner, L. R. 465, *479*
Lawrence, M. 259, *260*
Le Bert, G., see Portmann, M. 69, *104*
Lenn, N. J., Reese, T. S. 118, *143*
Lev, A., Sohmer, H. 71, *102*
Levick, W. R., see Barlow, H. B. 63, *96*
Levine, R. A. 57, *102*, 138, *143*
Levine, R. A., see Kiang, N. Y.-s. 3—5, 11, 13, 18, 71, 74, 86, 91, 92, *102*
Levitt, H., Rabiner, L. R. 468, 473, *479*
Lewis, D., Cowan, M. 430, *445*
Lewy, F. H., Kobrak, H. 335, *395*
Lewis, P. R., see Shute, C. C. D. 234, *245*
Lhermitte, F., Chain, F., Escourolle, R., Ducarne, B., Pillon, B., Chedru, G. 330, 332, 334, 340, 345, *395*
Li, R. Y.-s., Guinan, J. J., Jr. 118, *143*
Liao, T.-T., see Watanabe, T. 125, 133, *144*
Lichte, W. H. 433, *445*
Licklider, J. C. R. 35, *102*, 433, *445*, 468—470, 473, 476, *479*
Licklider, J. C. R., Kryter, K. D. 395, *395*
Licklider, J. C. R., Webster, J. C. 453, *459*

Licklider, J. C. R., see Huggins, W. H. 82, *101*
Lidén, G. 281, *304*
Lidén, G., Korsan-Bengtsen, M. 376, *395*
Liff, H. 15, *102*
Liff, H. J., Goldstein, M. H. Jr. 11, 30, 32, *102*
Lifschitz, W. S., see Woolsey, C. N. 133, *141*, 154, 159, 160, 186, *191*
Lim, D. J. 85, *102*
Lim, D. J., Lane, W. C. 297, *304*
Lindeman, H. H., Bredberg, G. 297, *304*
Lindeman, H. H., see Bredberg, G. 297, *302*
Lindner, W. A., see Egan, J. P. 463, 473, *478*
Lindquist, S. E. 262, *304*
Lindquist, S. E., Neff, W. D., Schuknecht, H. F. 269, *304*
Lindsay, J. R. 272, 279, *304*
Lindsay, J. R., Kohut, R. I., Sciarra, P. A. 272, 279, *304*
Lindsay, J. R., Proctor, L. R., Work, W. P. 272, *304*
Lindsay, J. R., Schultes, G. von 272, 279, *304*
Linehan, J. A., see Simmons, F. B. 3, 93, *106*
Lima, A., see Damasio, A. R. 327, *391*
Loc, P. R., Benevento, L. A. 159, *195*
Locke, S. 147, 159, *195*
Loe, P. R., Benevento, L. A. 241, *244*
Loeb, M., Fletcher, J. L. 257, *260*
Loeb, M., Riopelle, A. J. 258, *260*
Loeffel, R. G., see Molnar, C. E. 6, *103*
Lorente De Nó, R. 4, 43—46, 52, 95, *102*, *103*, 214, 215, *218*, 219, *244*, 335, *395*
Lousteau, R. J., see Cullen, J. K., Jr. 69, *97*
Lovering, L., see Jerger, J. 328, 329, 334, 356, *394*
Lowe, S. S., see Berlin, C. I. 361, *390*
Luciano, L., see Iurato, S. 234, *244*
Lundqvist, P. G., see Wersäll, J. 236, *246*
Lurie, M. H., see Davis, H. 262, 269, *302*
Lynch, J. C., see Dewson, J. H. III 362, *391*
Lynn, P. A., Sayers, B. McA. 84, 86, *103*

Mach, E. 248, 255, *260*
MacIntosh, F. C., see Birks, R. H. E. 235, *242*
Magoun, H. W. 221, 224, *244*
Magoun, H. W., see Barnes, W. T. 112, 120, *141*, 147, *192*
Maiwald, D. 417—419, *445*
Majkowski, J. 146, *195*
Majkowski, J., Morgades, P. P. 146, *195*
Makishima, K. 271, *304*
Mancia, D., see Avanzini, G. 241, *242*
Mangles, J. W., see Thurlow, W. R. 454, *459*
Manley, G. A. 46, 53, 54, 78, *103*
Marko, L. A., see Brown, T. S. 337, 338, *390*
Marr, E. M., see Kiang, N. Y.-s. 23, 30, *102*
Marsh, J. T., McCarthy, D. A., Sheatz, G., Galambos, R. 222, *244*
Marsh, J. T., Smith, J. C., Worden, F. G. 67, 90, *103*
Marsh, J. T., Worden, F. G. 57, *103*
Marsh, J. T., Worden, F. G., Hicks, L. 222, *244*
Marsh, J. T., Worden, F. G., Smith, J. C. 57, *103*
Marsh, J. T., see Worden, F. G. 221—223, *246*
Maruyama, N., Kanno, Y. 325, *395*
Maruyama, N., see Greenwood, D. D. 30, 54—58, 68, 90, *100*, 121, 122, *142*
Maruyama, N., see Katsuki, Y. 179, *195*
Massopust, L. C., Jr. Barnes, H. W. Verdura, J. 339, *395*
Massopust, L. C., Jr., Barnes, H. W., Meder, J., Meder, R. 339, *395*
Massopust, L. C., Jr., Wolin, L. R., Frost, V. 339, *395*
Mast, T. E. 46, 53, 56, 76 77, *103*
Masterton, R. B., Heffner, H., Ravizza, R. 374, *395*
Masterton, R. B., Diamond, I. T. 321, 352, 375, *395*
Masterton, R. B., Jane, J. A., Diamond, I. T. 353, 373—376, *395*
Masterton, R. B., Thompson, G. C., Bechthold, J. K., Robards, J. 375, *395*
Masterton, R. B., see Heffner, R. 15, *100*, 353, *393*
Masterton, R. B., see Jane, J. A. 366, 371, *394*
Masterton, R. B., see Ravizza, R. J. 325, 353, 354, *398*
Mathes, R. C., Miller, R. L. 430, *445*
Matsuura, S., see Furukawa, T. 85, *99*
Matsuura, S., see Ishii, Y. 85, *101*
Matz, G. J., Beal, D. D., Krames, L. 273, *304*
Matz, G. J., Wallace, T. H., Ward, P. H. 273, *304*
Matzker, J. 355, 377, *396*
McCabe, P. A., Dey, F. L. 273, *304*
McCarthy, D. A., see Marsh, J. T. 222, *244*
McClelland, K. D., Brandt, J. F. 87, 93, *103*
McDonald, D. M., Rasmussen, G. L. 240, *244*
McDougall, W., see Hocart, A. M. 250, *260*
McFadden, D. 469, 470, 472, 478, *479*
McFadden, D., Pulliam, K. A. 463, 466, 467, 473, *479*
McFadden, D., see Egan, J. P. 463, 473, *478*
McFadden, D., see Jeffress, L. A. 457, *459*, 467, 468, *479*
McGee, T. M., Olszewski, J. 265, 269, 283, *305*

McGee, T. M., Cole, G. G., Van Den Ende, H. 265, 269, 278, 281, *305*
McGee, T. M., see Elliott, D. N. 266, 269, 278, 279, 281, *303*
McIntosh, J., see Goldstein, M. H., Jr. 213, *218*
McNeil, M. R., see Berlin, C. I. 361, *390*
Mechelse, K., see Desmedt, J. E. 159, *193*, 220, 221, 224, 238, 241, *243*
Meder, J., see Massopust, L. C., Jr. 339, *395*
Mehler, W. R. 160, 161, *195*
Mehler, W. R., Feferman, M. E., Nauta, W. J. H. 160, *195*
Mehler, W. R., see Bonin, G. von 216, *217*
Melnick, W. 258, *260*
Merzenich, M. M. 321, *396*
Merzenich, M. M., Brugge, J. R. 321, *396*
Merzenich, M. M., Knight, P. L., Roth, G. L. 321, *396*
Merzenich, M. M., see Brugge, J. F. 216, *217*
Mettler, F. A. 219, *245*
Mettler, F. A., Finch, G., Girden, E., Culler, E. A. 332, *396*
Mettler, F. A., see Ades, H. W. 178, *191*
Mettler, F. A., see Culler, E. A. 331, 391
Metz, P. J., Bismarck, G. von, Durlach, N. I. 471, 472, *479*
Meyer, D. R., Woolsey, C. N. 335, 336, 350, *396*
Mickle, W. A., Ades, H. W. 160, *195*
Mier, M., see Jerger, J. F. 361, *394*
Miki, M., see Niimi, K. 160, *196*
Miller, J. D., Watson, C. S., Covell, W. P. 265, 269, *305*, 331, *396*
Miller, Josef M. 201, 216 *218*
Miller, J. P., Watson, C. S., Covell, W. P. 15, *103*
Miller, J. B., see Walsh, B. T. 5, 6, *107*
Miller, J. D., see Eldredge, D. H. 267, 269, *303*, 462, *478*
Miller, J. D., see Weston, P. B. 469, *480*
Miller, J. M., see Stebbins, W. C. 266, 269, 279, 281, 297, *306*
Miller, R. L., see Mathes, R. C. 430, *445*
Mills, A. W. 450—454, *459*
Mills, J. H., see Eldredge, D. H. 267, 269, 283, *303*
Milner, B. 334, 340, 344, 345, *396*
Milner, B., Kimura, D., Taylor, L. B. 345, *396*
Møller, A. R. 19, 53—55, 58, 59, 61—63, 65—67, *103*
Molino, J. 449, *459*
Molnar, C. E., Loeffel, R. G., Pfeiffer, R. R. 6, *103*
Molnar, C. E., Pfeiffer, R. R. 46, *103*
Molnar, C. E., see Pfeiffer, R. R. 27—30, 51, 52, *104*, 138, *143*
Monaco, P., see Desmedt, J. E. 72, *98*, 220, 230 231, 233, 235—237, *243*
Monakow, C. von 366
Moore, C. N., Casseday, J. H., Neff, W. D. 372, *396*
Moore, J. K., Moore, R. Y. 111, *143*
Moore, R. Y., Goldberg, J. M. 147, 153, 154, 160, 161, *195*, 313, *396*
Moore, R. Y., see Goldberg, J. M. 120, *142*, 147, 148, 150, 152, 153, *193*, *393*
Moore, R. Y., see Moore, J. K. 111, *143*
Moore, R. Y., see Tarlov, E. C. 147, 153, *197*
Morales-Garcia, C., Hood, J. D. 368, *396*
Morest, D. K. 118, 119, 124, *143*, 148, 150, 151, 153, 154—157, 161, 163, 178, 181, *195*, *196*, 313, *396*
Morest, D. K., see Cohen, E. S. 43, 45, 51, *96*
Morest, D. K., see Geniec, P. 148—153, 163, *193*
Morest, D. K., see Kiang, N. Y.-s. 43, 45, 50, *102*
Morgades, P. P., see Majkowski, J. 146, *195*
Morgan, C. T., see Stevens, S. S. 334, *399*
Mott, F. W. 310, *396*
Mountcastle, V. B. 214, *218*
Mountcastle, V. B., see Poggio, G. F. 159, 188, *196*, 313, *397*
Mountcastle, V. B., see Walzl, E. M. 160, *197*
Moushegian, G. 458, *459*
Moushegian, G., Jeffress, L. A. 456, 457, *459*
Moushegian, G., Rupert, A., Galambos, R. 43, 54, 56, 58, *103*
Moushegian, G., Rupert, A. L., Langford, T. L. 131—133, 135—137, 139, 140, *143*
Moushegian, G., Rupert, R. L., Stillman, R. D., Weiss, I. P. 32, *103*
Moushegian, G., Rupert, A., Whitcomb, M. A. 116, 122, 124, 125, 127, 131, 133, 135, 137, 139, 140, *143*, 177, *196*
Moushegian, G., see Rupert, A. 3, 10, 24, 57, 93, *105*, 116, 131—133, 139, 140, *143*, 230, 239, *245*
Moxon, E. C. 10, *103*
Moxon, E. C., see Kiang, N. Y.-s. 3—5, 11, 13, 18, 23, 71, 74, 81, 84—86, 91—93, *102*
Müller, J. 247, *260*
Münzer, E., Wiener, H. 154, *196*
Mulligan, B. E., Mulligan, M. J., Stonecypher, J. F. 471, *479*
Mulligan, M. J., see Mulligan, B. E. 471, *479*
Munchel, M., see Neff, W. D. 380, *397*
Munk, H. 310, *396*
Myers, E. N., Bernstein, J. M. 266, 269, 273, 283, *305*
Myers, E. N., see Igarashi, M. 273, *304*
Myers, R. E., see Galambos, R. 146, *193*

Nager, G. T. 277, *305*
Naito, F., see Niimi, K. *397*
Nakai, Y., see Igarashi, M. 75, *101*, 238, *244*, 381, 383, *394*
Nauta, W. J. H., see Mehler, W. R. 160, *195*
Neff, W. D. 241, *245*, 262, 269, *305*, 309, 336, 339, 344, 348, 349, 352, 355, 362, 368, 372, *396*, *397*
Neff, W. D., Arnott, G. P., Fisher, J. D. 350—352, *397*
Neff, W. D., Casseday, J. H. 350, 352, *397*
Neff, W. D., Casseday, J. H., Neff, P. L. 343, *397*
Neff, W. D., Diamond, I. T. 317, 318, 337, 343, 372, *397*
Neff, W. D., Fisher, J. F., Diamond, I. T., Yela, M. 159, *196*, 337, 343, 352, 356, *397*
Neff, W. D., Hind, J. E. 13, 331, *397*
Neff, W. D., Munchel, M. 380, *397*
Neff, W. D., Neff, P. A. 364, *397*
Neff, P. A., see Neff, W. D. 364, *397*
Neff, P. L., see Neff, W. D. 343, *397*
Neff, W. D., see Akert, K. 339, *390*
Neff, W. D., see Butler, R. A. 159, *192*, 318, 332, 335—337, 343, *390*
Neff, W. D., see Casseday, J. H. 365, 371—373, 375, 376, 378, 381, *391*
Neff, W. D., see Diamond, I. T. 147, 159, *193*, 318, 332, 336, 337, 341—344, 348, 349, *392*
Neff, W. D., see Dolan, T. R. 266, 269, 278, 281, 288, *303*
Neff, W. D., see Goldberg, J. M. 146, 147, 159, *193*, 241, *244*, 336—338, 342, 349, 370, 385, 387, *393*
Neff, W. D., see Jerison, H. J. 339, 344, *394*
Neff, W. D., see Lindquist, S. E. 269, *304*
Neff, W. D., see Moore, C. N. 372, *396*
Neff, W. D., see Oesterreich, R. E. 333, 334, 369, *397*
Neff, W. D., see Scharlock, D. P. 349, 350, *398*
Neff, W. D., see Schuknecht, H. F. 263, 269, *305*
Neff, W. D., see Stoughton, G. S. 330, *399*
Neff, W. D., see Strominger, N. L. 355, *399*
Nelson, P. G., Erulkar, S. D. 121, *143*, 150, 166—168, 172, 173, 183, 186, *196*
Nelson, P. G., Erulkar, S. D., Bryan, J. S. 167, 169, 170, 191, *196*, 342, *397*
Nelson, P. G., Evans, E. F. 63, *103*
Nelson, P. G., see Funkenstein, H. H. 200, 212, *218*
Nelson, P. G., see Erulkar, S. D. 163, 164, 166, 168—174, *193*, 342, *392*
Nelson, P. G., see Evans, E. F. 3, 44—46, 48—54, 56, 57, 63, 95, *98*, *99*, 342, *392*
Newman, E. B., see Stevens, S. S. 449—451, *459*
Newman, E. B., see Wallach, H. 455, *459*
Newman, J. D., see Funkenstein, H. H. 200, 212, *218*
Newman, J. D., see Wollberg, Z. 200, 210, 212, 216, *218*
Nieder, P. 82, *103*
Nieder, P., Nieder, I. 70, 72, 75, 86, 87, *103*
Nieder, P. C., Strominger, N. L. 146, *196*
Nieder, I., see Nieder, P. 70, 72, 75, 86, 87, *103*
Nielsen, D. W., see Konishi, T. 84, *102*
Nielsen, D. W., see Teas, D. C. 15, 72, 73, 86, *107*, 228, *246*
Niese, H. 421, 426, 427, *445*
Niimi, K., Naito, F. *397*
Niimi, K., Miki, M., Kawamura, S. 160, *196*
Nixon, J., see Glorig, A. 271, *303*
Noffsinger, D., Olsen, W. O., Carhart, R., Hart, C. W., Sahgal, V. 366, 369, 377, *397*
Nomoto, M., Suga, N., Katsuki, Y. 4, 10, 21, 23, 30—33, *104*
Novoa, V., see Benitez, J. T. 366, 367, *390*

Oda, D. 271, 278, *305*
Oesterreich, R. E., Strominger, N. L., Neff, W. D. 333, 334, 369, *397*
Oesterreich, R. E., see Strominger, N. L. 375, 376, *399*
Oettinger, R., see Feldtkeller, R. 406, *445*
Ohashi, T., see Yoshie, N. 69, 70, *108*
Ohgushi, K., see Watanabe, T. 63, *107*
Ohm, G. S. 439, *445*
Oliver, D. L., Hall, W. C. 322, *397*
Olsen, W. O., see Noffsinger, D. 366, 369, 377, *397*
Olszewski, J., see McGee, T. M. 265, 269, 283, *305*
Oonishi, S., Katsuki, Y. 205, 213, *218*
Osen, K. K. 43, 45, 46, 51, *104*, 113, 116, 122, *143*, 150, *196*
Osen, K. K., Roth, K. 240, *245*

Paillas, N., see Hugelin, A. 225, *244*
Palmer, L. A., Rosenquist, A. C., Sprague, J. M. 386, *397*
Palmer, L. A., see Rosenquist, A. C. 386, *398*
Pandya, D. N., Sanides, F. *397*
Pannese, E., see Iurato, S. 234, *244*
Papez, J. W. 95, *104*, 147, *196*
Parker, W., Decker, R. L., Gardner, W. H. 367, *397*
Parker, W., Decker, R. L., Richards, N. G. 367, *397*
Pause, M., see Galley, N. 72, *99*, 235, *243*

Pavlov, I.P. 262, 269, *305*, 310, 385, *397*
Peake, W.T., Kiang, N.Y.-s. 69—71, *104*
Peake, W.T., see Guinan, J. J., Jr. 13, *100*
Peake, W.T., see Wiederhold, M.L. 72, 73, *107*, 228, 231, *246*
Peake, W.T., see Kiang, N. Y.-s. 3, 12, 13, 15, 16, 53, 70, 78, 87, *102*
Pearson, R.D., see Stebbins, W.C. 266, 267, 269, 279, *306*
Pelliccioli, G., see Avanzini, G. 241, *242*
Penalozo-Rojas, J., see Bach-y-Rita, G. 222, *242*
Penfield, W., Evans, J. 354, 355, *397*
Penfield, W., Perot, P. 327, *397*
Perekalin, W.E. 251, 254, *260*
Perl, E.R., Whitlock, D.G. 188, *196*
Perl, E.R., see Whitlock, D.G. 160, *198*
Perlman, H.B. 258, *260*
Perlman, H.B., see Schuknecht, H.F. 263, 269, *305*
Perot, P., see Penfiield, W. 327, *397*
Persky, A., see Sloane, P. 366, *398*
Petri, J. 255, *260*
Pfalz, R.K.J. 76, *104*, 240, *245*
Pfalz, R., see Dunker, E.G. 76, *98*
Pfeffer, A.Z., see Wortis, S.B. 355, *400*
Pfeiffer, R.R. 33, 43, 46, 51, 53, 56, 57, 81, *104*, 118, 121, 122, *143*
Pfeiffer, R.R., Kiang, N.Y.-s. 43, 46, 51, 52, *104*, 122, *143*
Pfeiffer, R.R., Kim, D.O. 82, 85, *104*
Pfeiffer, R.R., Molnar, C.E. 27—30, 51, 52, *104*, 138, *143*
Pfeiffer, R.R., see Arthur, R. M. 24, 25, 30—33, 86, *96*
Pfeiffer, R.R., see Goblick, T. J., Jr. 7, 9, 36, 37, 81, *100*
Pfeiffer, R.R., see Kiang, N. Y.-s. 18, 43, 46, 51, 53, 54, 56, *102*, 121, 122, *142*
Pfeiffer, R.R., see Koerber, K. L. 51, 52, *102*
Pfeiffer, R.R., see Molnar, C. E. 6, 46, *103*
Pfeiffer, R.R., see Wiener, F. M. 13, *107*
Pickett, J.M., see Pollack, I. 468, *479*
Pickles, J.O., Comis, S.D. 383, 384, *397*
Pickles, V.O., see Comis, S.D. 75, *97*
Piddington, R.W. 77, *104*
Piercy, J.E., see Shaw, E.A. G. 466, *480*
Pillon, B., see Lhermitte, F. 330, 332, 334, 340, 345, *395*
Plomp, R. 68, 87, 93, *104*, 430, 433, 440, *445*
Plomp, R., Steeneken, H.J.M. 433, 440, *445*
Ploog, D., see Winter, P. 212, *218*
Poggio, G.F., Mountcastle, V. B. 159, 188, *196*, 313, *397*
Pokov, A.V., see Radionova, E.A. 46, 50, *104*
Politzer, A. 249, *260*
Poljak, S. 43, 95, *104*
Pollack, I., Pickett, J.M. 468, *479*
Polvogt, L.M., see Crowe, S.J. 270, 271, 278, *302*
Polvogt, L.M., see Guild, S.R. 276, *303*
Port, E. 426, 427, *445*
Portmann, M., Aran, J.-M., Lagourgue, P. 91, *104*
Portmann, M., Le Bert, G., Aran, J.-M. 69, *104*
Powell, E.W., Furlong, L.D., Hatton, J.B. 190, *196*
Powell, E.W., Hatton, J.B. 160, *196*
Powell, T.P.S., Cowan, W.M. 43, 51, *104*
Powell, T.P.S., Erulkar, S.D. 43, *104*
Powell, T.P.S., see Diamond, I.T. 160, 161, *193*, 314, 315, 320, *392*
Powell, T.P.S., see Jones, E. G. 160, 161, 182, *194*
Pribram, K.H., see Dewson, J.H. III 362, *391*
Price, G.R. 78, *104*
Proctor, L.R., see Lindsay, J. R. 272, *304*
Pruett, J., see Daigneault, E. A. 70, *97*
Pugh, J.E., Jr., Anderson, D. J., Burgio, P.A., Horwitz, M.R. 70, *104*
Puletti, F., see Celesia, G.G. 326, *391*
Pulliam, K.A., see McFadden, D. 463, 466, 467, 473, *479*

Quiggle, R., see Glorig, A. 271, *303*

Raab, D.H. 408, *445*
Raab, D.H., Ades, H.W. 333, 368, 369, *397*
Raab, D.H., see Taub, H.B. 70, *107*
Rabiner, L.R., Laurence, C. L., Durlach, N.I. 465, *479*
Rabiner, L.R., see Levitt, H. 468, 473, *479*
Radionova, E.A. 53, 70, *104*
Radionova, E.A., Pokov, A. V. 46, 50, *104*
Raffaele, R., Santangelo, F., Sapienza, S., Urbano, A. 188, *196*
Rahlfs, V. 433, *446*
Rainbolt, H., Small, A.M. 90, *104*
Ramon-Moliner, E. 157, *196*
Randall, W.L. *398*
Ranson, S.W., see Barnes, W. T. 112, 120, *141*, 147, *192*
Rasmussen, G.L. 43, 45, 51, 72, 76, *104*, *105*, 116, 120, 122, *143*, 147, 153, 160, *196*, 219, 220, 224, 225, 238, 240, *245*
Rasmussen, G.L., Desmedt, J. E. 220, 238,
Rasmussen, G.L., see McDonald, D.M. 240, *244*

Rasmussen, G.L., see Smith, C.A. 236, *245*
Rasmussen, T., see Stepien, L. S. 241, *246*, 348, *398*
Rasmussen, G.L., see Walther, J.B. 220, *246*
Ratliff, F. 42, *105*
Rauchbach, E., see Bone, R.C. 90, *96*
Ravizza, R., Diamond, I.T. 354, 387, *398*
Ravizza, R.J., Masterton, B. 353, 354, *398*
Ravizza, R.J., Masterton, R. B. 325, *398*
Ravizza, R., see Cranford, J. 360, *391*
Ravizza, R., see Masterton, B. 374, *395*
Ravizza, R., see Whitfield, I. C. 360, *400*
Rayleigh, O.M. 248, *260*
Raynor, S., see Dallos, P. 84, 87, *97*
Reale, E., see Iurato, S. 234, *244*
Reese, T.S., see Lenn, N.J. 118, *143*
Reger, S.N. 258, *260*
Rehren, D.V., see Grubel, G. 76, *100*
Reichardt, W. 426, 427, *446*
Reske-Nielsen, E., see Hansen, C.C. 271, 272, 279, *303*
Rhode, W. 79, 80, *105*
Rhode, W.S. 16, 79, 82, *105*
Rhode, W.S., see Geisler, C.D. 134—136, 138, *142*, 177, *193*
Rhode, W.S., see Robles, L. 6, 83, *105*
Ribaupierre, F. De, Goldstein, M.H., Jr., Yeni-Komshian, G. 201, 208, 210, 216, 217, *218*
Ribaupierre, F. De, see Goldstein, M.H., Jr. 208, 209, *218*
Richards, N.G., see Parker, W. 367, *397*
Riesz, P.R. *446*
Rilling, M.E., Jeffress, L.A. 467, *479*
Rinvik, E. 160, 161, *196*
Rioch, D.McK. 154, *196*
Rioch, D.McK., see Bard, P. 331, 356, *390*
Riopelle, A.J., see Loeb, M. 258, *260*
Riss, W. 356, *398*
Ritsma, R.J. 428, 430, *446*
Ritsma, R., Engel, F.L. 440, *446*
Robards, J., see Masterton, R. B. 375, *395*
Robertson, D., see Desmedt, J.E. 231, 234, 236, 237, *243*
Robinson, D.E. 467, *479*
Robinson, D.E., Jeffress, L.A. 470, *479*
Robinson, D.E., see Dolan, T.R. 469, 470, *478*
Robinson, D.E., see Jeffress, L.A. *479*
Robinson, D.W. 419, *446*
Robinson, D.W., Dadson, R.S. 403, *446*
Robinson, E.H., see Kemp, E. H. 111, 113, *142*
Robinson, P.M., Bell, C. 234, *245*
Robles, L., Rhode, W.S., Geisler, C.D. 6, 83, *105*
Rockel, A.J. 148, 150, 163, *196*
Rockel, A.J., Jones, E.G. 148, 150—153, 163, *197*
Rockel, A.J., see Jones, E.G. 156, 157, 182, *194*
Rodieck, R.W., Kiang, N.Y.-s., Gerstein, G.L. 52, *105*
Rodriguez, L., see Echandia, E. 236, *243*
Roelofs, O., see Gilse, V. 249—251, 254, *260*
Roeser, R.J., Daly, D.D. 345, *398*
Roffler, S.K., Butler, R.A. 249, 251, *260*, 454, 455, *459*
Rose, J.E. 3, 35, *105*, 154, *197*, 316, *398*
Rose, J.E., Brugge, J.F., Anderson, D.J., Hind, J. E. 3, 10, 18, 24—27, 35, 38, 93, 94, *105*, 204, *218*
Rose, J.E., Galambos, R. 155, *197*
Rose, J.E., Galambos, R., Hughes, J.R. 3, 43, 45, 51, 52, 54—57, *105*
Rose, J.E., Greenwood, D.D. Goldberg, J.M., Hind, J.E. 121, *143*, 150, 162—166, 174, *197*
Rose, J.E., Gross, N.B., Geisler, C.D., Hind, J.E. 134—136, 138, *143*, 174—177, 188, *197*, 205, *218*
Rose, J.E., Hind, J.E., Anderson, D.J., Brugge, J.F. 4, 5, 15, 18—21, 23, 24, 26, 37, 38, 40, *105*
Rose, J.E., Woolsey, C.N. 147, 154, 155, 158, 159, 178, 179, *197*, 313, 316—318, 337, 343, *398*
Rose, J.E., see Anderson, D. J. 24, 27—29, 82, *96*, 138, *141*
Rose, J.E., see Brugge, J.F. 7, 30, 35—37, 94, *96*
Rose, J.E., see Erulkar, S.D. 179, *193*, 200, 208, *217*
Rose, J.E., see Galambos, R. 180, 186, *193*
Rose, J.E., see Hind, J.E. 3, 18, 30, 35, 37, 38, *101*, 121, 133, 135, *142*, 163, 174, 184, *194*, 213, *218*
Rosenberg, J., see Evans, E.F. 11, 13, 15, 16, 21, 22, 40, 41, *99*
Rosenblith, W.A. 70, *105*
Rosenblut, B., see Davis, H. 70, *97*
Rosenquist, A.C., Edwards, S. B., Palmer, L.A. 386, *398*
Rosenquist, A.C., see Palmer, L.A. 386, *397*
Rosenzweig, M.R. 113, *143*, 333, *398*
Rosenzweig, M.R., Amon, A. H. 113, *143*
Rosenzweig, M.R., Sutton, D. 112, 113, *143*
Rosenzweig, M.R., Wyers, E. J. 112, 113, *143*, 172, *197*
Rosenzweig, M.R., see Wallach, H. 455, *459*
Ross, H.F., see Evans, E.F. 179, *193*, 213, *217*
Rossi, G. 234, *245*
Rossi, G., Cortesina, G. 120, *143*

Roth, G. L., see Merzenich, M. M. 321, *396*
Roth, K., see Osen, K. K. 240, *245*
Rüedi, L., Spoendlin, H. 277, *305*
Runge, P. S., see Thurlow, W. R. 454, *459*
Rupert, A. L., Moushegian, G. 57, *105*, *245*
Rupert, A., Moushegian, G., Galambos, R. 3, 10, 24, 93, *105*
Rupert, A., Moushegian, G., Whitcomb, M. A. 116, 131—133, 139, 140, *143*, 230, 239, *245*
Rupert, A., see Galambos, R. 112, 116, 117, 122, 131—133, 139, 140, *142*, 230, 239, *243*, 373, *392*
Rupert, A., see Hubel, D. H. 201, *218*
Rupert, A., see Moushegian, G. 43, 54, 56, 58, *103*, 116, 122, 124, 125, 127, 131, 133, 135, 137, 139, 140, *143*
Rupert, A. L., see Moushegian, G. 131—133, 135—137, 139, 140, *143*, 177, *196*
Rupert, R. L., see Moushegian, G. 32, *103*

Sachs, M. B. 30, 32, 33, *105*
Sachs, M. B., Kiang, N. Y.-s. 21, 30, 32, *105*
Sachs, M. B., see Kiang, N. Y.-s. 3 12, 13, 15, 16, 78, *102*
Sahgal, W., see Noffsinger, D. 366, 369, 377, *397*
Salomon, G., Starr, A. 223, *245*
Salomon, G., see Elberling, C. 70, *98*
Saltzman, M. 371, *398*
Saltzman, M., see Sloane, P. 366, *398*
Sanchez-Longo, L.P., Forster, F.M. 355, *398*
Sanchez-Longo, L.P., Forster, F.M., Auth, T.L. 355, *398*
Sandel, T. T., see Jeffress, L. A. 464, 472, *479*
Sando, I. 45, 46, *105*
Sando, I., Hemenway, W. G., Hilyard, V. H., English, G.M. 277, *305*
Sando, I., see Kimura, R. S. 282, 297, *304*
Sanides, F. 321, 324, *398*
Sanides, F., see Pandya, D. N. *397*
Santangelo, F., see Raffaele, R. 188, *196*
Sapienza, S., see Raffaele, R. 188, *196*
Saul, L. J., Davis, H. 111, *144*
Sayers, B. McA., see Lynn, P. A. 84, 86, *103*
Saxén, A., see Fieandt, H. von 282, *303*
Schäfer, E. A., see Brown, S. M. 310, *390*
Scharf, B. 89, *105*, 425, *446*
Scharf, B., Hellman, R. P. 92, *105*
Scharf, B., see Zwicker, E. 425, *448*
Scharlock, D. P., Neff, W. D., Strominger, N. L. 349, 350, *398*
Scharlock, D. P., Tucker, T. J., Strominger, N. L. 342, 350, *398*
Schenkel, K. D. 477, *479*
Scherrer, H., see Hernandez-Péon, R. 189, *194*, 221, 224, *244*
Schlegel, P., see Suga, N. 241, *246*
Schmigidina, G. N., see Altman, Ya. A. 188, *192*
Schmidt, R., see Jørgensen, M. B. 273, *304*
Schneider, G. E., see Held, R. 95, *100*
Scholl, H. 414, *446*
Schouten, J. F. 428, 430, *446*
Schouten, J. F. et al. (1962) 430, *446*
Schubert, E. D. 429, *446*
Schubert, E. D., Schultz, M. C. 473, *480*
Schubert, E. D., see Hiesey, R. W. 82, *101*
Schultes, G. von, see Lindsay, J. R. 272, 279, *304*
Schultz, M. C., see Schubert, E. F. 473, *480*
Schütte, H., see Zwicker, E. 409, *448*
Schuknecht, H.F. 263—265, 271—273, 277—279, 281, 282, *305*
Schuknecht, H. F., Benitez, J. T., Beekhuis, J. 272, 279, *305*
Schuknecht, H. F., Churchill, J. A., Doran, R. 234, *245*
Schuknecht, H. F., Gross, C.W. 277, *305*
Schuknecht, H. F., Igarashi, M. 272, *305*
Schuknecht, H. F., Neff, W. D. 263, 269, *305*
Schuknecht, H. F., Neff, W. D., Perlman, H. B. 263, 269, *305*
Schuknecht, H. F., Sutton, S. 263, 269, *305*
Schuknecht, H. F., Woellner, R. C. 263, 264, 269, 281, *305*
Schuknecht, H. F., see Benitez, J. T. 272, 277, *302*
Schuknecht. H. F., see Igarashi, M. 273, *304*
Schuknecht, H. F., see Kimura, R. S. 282, 297, *304*
Schuknecht, H. F., see Lindquist, S. E. 269, *304*
Schulhoff, C., Goodglass, H. 345, *398*
Schwafel, J. A., see Holstein, S. B. 189, *194*
Schwartzkopff, J., see Galambos, R. 112, 116, 117, 122, 131—133, 139, 140, *142*, 230, 239, *243*, 373, *392*
Sciarra, P. A., see Lindsay, J. R. 272, 279, *304*
Sears, R. E., see Khanna, S. M. 82, *101*
Sears, T. A., see Andersen, P. 181, *192*
Seaton, W. H., Trahiotis, C. 89, *105*
Sefton, A. J., Burke, W. 181, *197*
Sellick, P. M., see Johnstone, B. M. 78, *101*
Selters, W., see Ward, W. D. 257, *260*
Shankweiler, D. 345, 361, *398*

Shankweiler, D., see Studdert-Kennedy, M. 361, *399*
Sharbrough, F. W. III, see Jerger, J. 328, 329, 334, 346, 356, *394*
Shaw, E. A. G., Piercy, J. E. 466, *480*
Shaxby, J. H., Gage, F. H. 456, *459*
Sheatz, G. C., Vernier, V. G., Galambos, R. 189, *197*
Sheatz, G., see Galambos, R. 146, 189, *193*, 221, *243*
Sheatz, G., see Marsh, J. T. 222, *244*
Shedd, J. L., see Jerger, J. 369, *394*
Sherrick, C. E., see Hirsh, I. J. 346, *393*
Sherrington, C. S. 384, 385, *398*
Shute, C. C. D., Lewis, P. R. 234, *245*
Siebert, W. M. 77, 84, 92, 93, *105*, *106*
Silfvenius, H., see Landgren, S. 160, *195*
Silverman, S. R., see Davis, H. 279, *302*
Simada, Z., see Watanabe, T. 68, 69, 91, *107*, 166, *197*
Simmons, F. B. 259, *260*
Simmons, F. B., Beatty, D. L. 222, 223, *245*
Simmons, F. B., Linehan, J. A. 3, 93, *106*
Sindberg, R. M., Thompson, R. F. 241, *245*
Sindberg, R., see Buser, P. 189, *192*
Sjöstrand, F. S., see Smith, C. A. 236, *245*
Slepian, J. Z., see Konishi, T. 228, 231, 232, 240, *244*
Sloane, P., Persky, A., Saltzman, M. 366, *398*
Small, A. M., Jr. 89, 90, *106*
Small, A. M., Jr., see Canahl, J. A. 469, *478*
Small, A. M., see Rainbolt, H. 90, *104*
Smith, C. A. 236, 238, *245*
Smith, C. A., Dempsey, E. W. 86, *106*
Smith, C. A., Rasmussen, G. L. 236, *245*
Smith, C. A., Sjöstrand, F. S. 236, *245*
Smith, C. A., Takasaka, T. 227, 236, *245*
Smith, C. A., see Davis, H. 70, *97*
Smith, C. A., see Iurato, S. 227, *244*
Smith, F. D., see Goldberg, J. M. 116—120, 127, 129, *142*
Smith, F. D., see Woolsey, C. N. 154, 159, 160, *198*
Smith, H. E., see Thompson, R. F. 338, *399*
Smith, J. C., see Marsh, J. T. 57, 67, 90, *103*
Smith, R. L., Zwislocki, J. J. 46, 55, *106*
Smoorenburg, G. F. 80, 83, 94, *106*
Smoorenburg, G. F., see Boer, E. De 35, *97*
Snow, W. B., see Steinberg, J. C. 450, *459*
Sodeberg, V., see Arden, G. B. 189, *192*
Sohmer, H. 228, *245*
Sohmer, H., Feinmesser, M. 69, *106*
Sohmer, H., see Lev, A. 71, *102*
Sondhi, M. M., Guttman, N. 471, *480*
Sonnhof, U., see Bruggencate, G. Ten 220, 235, *242*
Sousa-Pinto, A. 320, 343, *398*
Spencer, W. A., see Thompson, R. F. 189, *197*
Spinnler, H., Vignolo, L. 362, *398*
Spinnler, H., see Faglioni, P. 362, *392*
Spoendlin, H. 4, 32, 33, 70, 85—87, *106*, 225, 227, 236, *246*, 281, 305, *306*
Spoendlin, H., Gacek, R. 225, 236, *246*
Spoendlin, H., see Rüedi, L. 277, *305*
Spoor, A. 271, *306*
Spoor, A., Eggermont, J. J. 70, *106*
Sprague, J. M., see Palmer, L. A. 386, *397*
Spreng, M., Keidel, W. D. 69, *106*
Starr, A. 57, *106*
Starr, A., Britt, R. 57, *106*
Starr, A., Hellerstein, D. 57, *106*
Starr, A., Wernick, J. S. 76, *106*, 241, *246*
Starr, A., see Carmel, P. W. 223, *242*
Starr, A., see Salomon, G. 223, *245*
Starr, A., see Wernick, J. S. 113, *144*
Stebbins, W. C., Clark, W. W., Pearson, R. D., Weiland, M. G. 266, 267, 269, 279, *306*
Stebbins, W. C., Miller, J. M., Johnsson, L.-G., Hawkins, J. E., Jr. 266, 269, 279, 281, 297, *306*
Steele, C. R. 82, *106*
Steeneken, H. J. M., see Plomp, R. 433, 440, *445*
Stein, L., see Elliott, D. N. 331, *392*
Steinberg, J. C., Snow, W. B. 450, *459*
Stepien, L. S., Cordeau, J. P., Rasmussen, T. 241, *246*, 348, *398*
Stepien, L., see Chorazyna, H. 241, *242*, 347, 348, *391*
Stevens, J. C., Hall, J. W. 426, *446*
Stevens, S. S. 258, 260, 402, 419, 422, 423, 429, 430, 437, *446*
Stevens, S. S., Newman, E. B. 449—451, *459*
Stevens, S. S., Morgan, C. T., Volkmann, J. 334, *399*
Stevens, S. S. et al. (1937, 1965) 428, 433, 436, *446*
Stevens, S. S., see Hawkins, J. E., Jr. 88, *100*
Stevens, S. S., see Zwicker, E. 87—89, *108*
Stewart, J. W., Ades, H. W. 330, *399*
Stillman, R. D., see Moushegian, G. 32, *103*
Stonecypher, J. F., see Mulligan, B. E. 471, *479*
Stopp, P. E., Whitfield, I. C. 54, 58, *106*

Storch, W.-H., see Galley, N. 72, *99*, 235, *243*
Stotler, W.A. 43, 51, *106*, 110, 112, 117, 118, 120, *144*
Stoughton, G.S. 325, 330, 350, *399*
Stoughton, G.S., Neff, W.D. 330, *399*
Stoupel, N., see Bremer, F. 225, *242*
Strominger, N.L. 352, *399*
Strominger, N.L., Neff, W.D. 355, *399*
Strominger, N.L., Oesterreich, R.E. 375, 376, *399*
Strominger, N.L., see Oesterreich, R.E. 333, 334, 369, *397*
Strominger, N.L., see Nieder, P.C. 146, *196*
Strominger, N.L., see Scharlock, D.P. 342, 349, 350, *398*
Studdert-Kennedy, M., Shankweiler, D. 361, *399*
Stumpf, C. 433, *446*, 463, *480*
Suga, N. 63, *106*, 212, *218*, 378, *399*
Suga, N., Schlegel, P. 241, *246*
Suga, N., see Arthur, R.M. 24, 25, 30—33, 86, *96*
Suga, N., see Katsuki, Y. 3, 4, 10—12, 16, 18, 21, 24, 58, *101*
Suga, N., see Nomoto, M. 4, 10, 21, 23, 30—33, *104*
Sumi, T., see Katsuki, Y. 3, 54, *101*, 162, 164, 179, *195*
Summerfield, A., see Glorig, A. 271, *303*
Sutton, D., see Rosenzweig, M.R. 112, 113, *143*
Sutton, S., see Kennedy, T. 335, *394*
Sutton, S., see Schuknecht, H. F. 263, 269, *305*
Suzuki, T., see Yoshie, N. 69, 70, *108*
Swisher, L.P. *399*
Swisher, L., Hirsh, I.J. 347, *399*
Syka, J., see Altman, Ya.A. 188, *192*
Symmes, D. 362, *399*
Syvian, L.J., White, S.D. 451, *459*
Szeligo, F.V., see Colavita, F. B. 344, *391*
Szentágothai, J. 182, *197*

Takasaka, T., see Smith, C.A. 227, 236, *245*
Tanaka, Y., Katsuki, Y. 235, *246*
Tarlov, E.C., Moore, R.Y. 147, 153, *197*
Tasaki, I. 3, 10, 24, 54, 69, 70, 78, *106*, 231, *246*
Tashiro, S. 219, *246*
Taub, H.B., Raab, D.H. 70, *107*
Tavitas, R.J., see Adrian, H. O. 133, *141*, 186, *191*
Taylor, K.J., see Johnstone, B.M. 15, 16, 78, 79, 88, *101*
Taylor, L.B., see Milner, B. 345, *396*
Taylor, R.W., see Blodgett, H.C. 466, 467, *478*
Teas, D.C., Eldredge, D.H., Davis, H. 70, *107*
Teas, D.C., Konishi, T., Nielsen, D.W. 15, 72, 73, 86, *107*, 228, *246*
Teas, D.C., see Konishi, T. 85, *102*
Tebecis, A.K. 185, *197*
Ten Cate, J. 356, *399*
Terashima, S., see Hotta, T. 189, *194*
Terhardt, E. 87, *107*, 415, 428—431, 441, *446*, *447*
Terhardt, E., Fastl, H. 430, *446*
Thomas, E.C., see Kiang, N. Y.-s. 3—11, 16, 18, 21—24, 30, 32, 43, 56, 69, 70, 78, 84, 91, *102*, 122—124, 132, *142*, 230, *244*
Thompson, R.F. 336—338, 343, *399*
Thompson, R.F., Smith, H.E. 338, *399*
Thompson, R.F., Spencer, W. A. 189, *197*
Thompson, R.F., Welker, W. I. 357, *399*
Thompson, S.P. 255, *260*
Thompson, G.C., see Masterton, R.B. 375, *395*
Thompson, R.F., see Hind, J. E. 213, *218*
Thompson, R.F., see Sindberg, R.M. 241, *245*
Thorndike 310
Thurlow, W.R., Mangles, J. W., Runge, P.S. 454, *459*
Thurlow, W.R., Runge, P.S. 454, *459*
Thurlow, W.R., see Gross, N. B. 180, *194*
Tillman, T.W., see Carhart, R. 468, *478*
Tonndorf, J. 82, *107*
Tonndorf, J., see Karlan, M.S. 87, *101*
Tonndorf, J., see Khanna, S. M. 82, *101*
Tonkonogii, I.M., see Gersuni, G.V. 325, *393*
Townsend, T.H., Goldstein, D.P. 473, *480*
Toynbee, J. 248, *260*
Trahiotis, C., Elliott, D.N. 75, *107*, 238, *246*, 381, 383, *399*
Trahiotis, C., see Seaton, W. H. 89, *105*
Treisman, A.M. 221, *246*
Trevarthen, C.B., see Held, R. 95, *100*
Trimble, O.C. 449, *459*
Tsuchitani, C., Boudreau, J.C. 113, 116, 122, 123, 127, 131, *144*
Tsuchitani, C., see Boudreau, J.C. 116, 117, 120, 122, 124, 125, 127, 128, 131, 133, 134, *141*
Tucker, A., see Hafter, E.R. 474, 477, *478*
Tucker, T.J., see Scharlock, D.P. 342, 350, *398*
Tunturi, A.R. 335, 340, *399*

Uchiyama, H., see Katsuki, Y. 3, 54, *101*, 162, 164, 179, *195*
Upton, M., see Davis, H. 262, 269, *302*
Urbano, A., see Raffaele, R. 188, *196*

Van Bergeijk, W.A., see David, E.E., Jr. 140, *141*

Van Den Ende, H., see Mc Gee, T. M. 265, 269, 278, 281, *305*
Van Noort, J. 112, 113, 116—118, 120, 122, *144*
Verdura, J., see Massopust, L. C., Jr. 339, *395*
Vernier, V. G., see Galambos, R. 189, *193*, 221, *243*
Vernier, V. G., see Sheatz, G. C. 189, *197*
Veronese, A., see Bosatra, A. B. 355, *390*
Victor, G., see Christman, R. J. 456, *458*
Viehweg, R., see Flower, R. M. 369, *392*
Viernstein, L. J., Grossman, R. G. 52, *107*
Vignolo, L. A. 362, *399*
Vignolo, L., see Faglioni, P. 362, *392*
Vignolo, L., see Spinnler, H. 362, *398*
Vogel, A. 433, *447*
Vogt, O. 310, *399*
Volkmann, J., see Stevens, S. S. 334, *399*

Waggener, J. D., see Kanshepolsky, J. 329, *394*
Walker, A. E. 317, *400*
Wallace, T. H., see Matz, G. J. 273, *304*
Wallach, H. 454, *459*
Wallach, H., Newman, E. B., Rosenzweig, M. R. 455, *459*
Waller, W. H. 159, *197*
Walliser, K. 428—430, *447*
Walsh, B. T., Miller, J. B., Gacek, R. R., Kiang, N. Y.-s. 5, 6, *107*
Walsh, E. G. 376, *400*
Walther, J. B., Rasmussen, G. L. 220, *246*
Walzl, E. M., Mountcastle, V. B. 160, *197*
Walzl, E. M., see Woolsey, C. N. 335, *400*
Wang, C-y., see Dallos, P. 84, 87, *97*
Ward, W. D. 256, 257, *260*, 268, *306*, 430, *447*
Ward, W. D., Duvall, A. J. 261, 268, 269, 281, 283, 287, *306*
Ward, W. D., Selters, W., Glorig, A. 257, *260*
Ward, P. H., see Matz, G. J. 273, *304*
Warr, W. B. 95, *107*, 112, 113, 116, 120, 122, *144*, 372, *400*
Warr, B., see Harrison, J. M. 43, 45, 51, *100*
Warr, W. B., see Kiang, N. Y.-s. 18, 43, 46, 51, 53, 54, 56, *102*, 121, 122, *142*
Warr, W. B., see Koerber, K. L. 51, 52, *102*
Watanabe, T. 94, *107*
Watanabe, T., Liao, T.-T., Katsuki, Y. 125, 133, *144*
Watanabe, T., Ohgushi, K. 63, *107*
Watanabe, T., Simada, Z. 68, 69, 91, *107*, 166, *197*
Watanabe, T., see Katsuki, Y. 3, 16, 54, 58, *101*, 162, 164, 179, *195*
Watanabe, T., see Kiang, N. Y.-s. 3—11, 16, 18, 21—24, 30, 32, 43, 56, 69, 70, 78, 84, 91, *102*, 122—124, 132, *142*, 230, *244*
Watson, B. J., see Flanagan, J. L. 467, *478*
Watson, C. S. 88, 89, *107*, 310
Watson, C. S., see Miller, J. D. 15, *103*, 265, 269, *305*, 331, *396*
Webster, F. A. 464, 467, 474, 475, *480*
Webster, F. A., see Hirsh, I. J. 467, *479*
Webster, J. C. et al. (1952) 429, *447*
Webster, J. C., see Licklider, J. C. R. 453, *459*
Webster, W. R. 189, *197*
Webster, W. R., Aitkin, L. M. 190, *198*
Webster, W. R., see Aitkin, L. M. 157, 178—183, 186, 188, *191*, *192*
Webster, W. R., see Irvine, D. R. F. 72, 75, *101*
Wedenberg, E., see Anderson, H. 268, *302*
Wegel, R. L., Lane, C. E. 404, *447*
Wegener, J. G. 325, 353, *400*
Weikers, N. J., see Jerger, J. 328, 329, 334, 346, 356, *394*
Weiskrantz, L., see Dewson, J. H. 241, *243*
Weiss, A. D., see Bernstein, J. M. 273, *302*
Weiss, I. P., see Moushegian, G. 32, *103*
Weiss, T. F. 7, 77, 78, *107*
Weiland, M. G., see Stebbins, W. C. 266, 267, 269, 279, *306*
Welker, W. I., see Hind, J. E. 213, *218*
Welker, W. I., see Thompson, R. F. 357, *399*
Wellek, A. 433, *447*
Wernick, J. S., Starr, A. 113, *144*
Wernick, J. S., see Konishi, T. 85, *102*
Wernick, J. S., see Starr, A. 76, *106*, 241, *246*
Wersäll, J. 86
Wersäll, J., Flock, A., Lundqvist, P. G. 236, *246*
Wertheimer, M., see Hornbostel, E. M. von 249, *260*
Wertz, M., see Jerger, J. 328, 329, 334, 356, *394*
West, R., see Bredberg, G. 297, *302*
Weston, P. B., Miller, J. D. 469, *480*
Wever, E. G. 276, 282, *306*, *447*
Wheeler, D., see Glorig, A. 271, *303*
Whitcomb, M. A., see Moushegian, G. 116, 122, 124, 125, 127, 131, 133, 135, 137, 139, 140, *143*, 177, *196*
Whitcomb, M. A., see Rupert, A. 116, 131—133, 139, 140, *143*, 230, 239, *245*
White, S. D., see Syvian, L. J. 451, *459*
Whitfield, I. C. 74, 77, 78, 92, 94, *107*, 342, 381, *400*
Whitfield, I. C., Cranford, J., Ravizza, R., Diamond, I. T. 360, *400*

Whitfield, I. C., Diamond, I.T. 387, *400*
Whitfield, I. C., Evans, E. F. 200, 210, *218*, 342, *400*
Whitfield, I. C., see Allanson, J.T. 93, 94, *96*
Whitfield, I. C., see Comis, S. D. 76, *97*, 236, 241, *242*, 383, *391*
Whitfield, I. C., see Cranford, J. 360, *391*
Whitfield, I. C., see Evans, E. F. 63, *99*, 179, *193*, 201, 213, *217*
Whitfield, I. C., see Kelly, J.B. 342, 343, *394*
Whitfield, I. C., see Stopp, P. E. 54, 58, *106*
Whitlock, D. G., Perl, E. R. 160, *198*
Whitlock, D. G., see Perl, E.R. 188, *196*
Whitmore, J. K., see Wilbanks, W. A. 470, *480*
Whittlesey, J. R. B., see Worden, F. G. 222, *246*
Whitworth, R. H., Jeffress, L. A. 456, 457, *459*
Whitworth, R. H., see Blodgett, H. C. 468, 469, *478*
Wickelgren, W. O. 189, *198*
Wiederhold, M. L. 15, 23, 72—74, 86, *107*, 228, 230, *246*
Wiederhold, M. L., Kiang, N. Y.-s. 6, 15, 70, 72, 73, 86, *107*, 220, 228, 232, 240, *246*
Wiederhold, M. L., Peake, W. T. 72, 73, *107*, 228, 231, *246*
Wiener, F. M., Pfeiffer, R. R., Backus, A. S. N. 13, *107*
Wiener, H., see Münzer, E. 154, *196*
Wiesel, T. N., Hubel, D. H. 217, *218*
Wiesel, T. N., see Hubel, D. H. 200, 212, 214, *218*
Wightman, F. L. 94, *107*, 440, *447*, 472, *480*
Wightman, F. L., see Yost, W. A. 458, *459*
Wilbanks, W. A., Whitmore, J.K. 470, *480*
Wilbanks. W. A., see Blodgett, H. C. 456, *458*
Wilpizeski, C. R., see Gold, A. 273, *303*
Wilson J. P. 41, 94, 96, *107*
Wilson, J. P., Evans, E. F. 41, 43, 88, *108*
Wilson, J. P., Johnstone, J. R. 6, 16, 78, 79, 81, 83, *108*
Wilson' J. P., see Evans, E. F. 11, 13, 15, 16, 21, 22, 40—43, 78, 80, 81, 87, 88, 92, *99*
Winter, P., Ploog, D., Latta, J. 212, *218*
Winter, P., see Funkenstein, H.H. 200, 212, *218*
Wittmaack, K. 262, 282, *306*
Woellner, R. C., see Schuknecht, H. F. 263, 264, 269, 281, *305*
Wolff. D., see Bunch, C. C. 282, *302*
Wolin, L. R., see Massopust, L. C., Jr. 339, *395*
Woollard, H. H., Harpman, J. A. 120, *144*, 147, 160, *198*, 318—320, *400*
Wollberg, Z., Newman, J. D. 200, 210, 212, 216, *218*
Wollberg, Z., see Funkenstein, H. H. 200, 212, *218*
Wolsk, D., see Landgren, S. 160, *195*
Wood, C. L., III, see Jeffress, L.A. 464, 472, *479*
Woolsey, C.N. 154, 159, 179, *198*
Woolsey, C. N., Adrian, H. O., Gross, N. B., Hirsch, J. E., Lane, J. F., Lifschitz, W. S., Smith, F. D. 154, 159, 160, *198*
Woolsey, C. N., Walzl, E. M. 335, *400*
Woolsey, C. N., see Akert, K. 339, *390*
Woolsey, C. N., see Hind, J. E. 213, *218*
Woolsey, C. N., see Meyer, D. R. 335, 336, 350, *396*
Woolsey, C. N., see Rose, J. E. 147, 154, 155, 158, 159, 178, 179, *197*, 313, 316—318, 337, 343, *398*
Worden, F. G. 222, 223, *246*
Worden, F. G., Marsh, J. T. 221, 223, *246*
Worden, F. G., Marsh, J. T., Abraham, F. D., Whittlesey, J. R. B. 222, *246*
Worden, F. G., see Marsh, J. T. 57, 67, 90, *103*, 222, *244*
Work, W. P., see Lindsay, J. R. 272, *304*
Wortis, S. B., Pfeffer, A. Z. 355, *400*
Wright, H.N. 426, *447*
Wurtz, R.H. 216, *218*
Wyers, E. J., see Rosenzweig, M.R. 112,113,*143*,172,*179*

Yamakawa, K. 272, *306*
Yela, M., see Neff, W. D. 159, *196*, 337, 343, 352, 356, *397*
Yeni-Komshian, G., see Goldstein, M. H., Jr. 208, 209, *218*
Yeni-Komshian, G., see Ribaupierre, F. De 201, 208, 210, 216, 217, *218*
Yerkes 310
Yoshie, N. 69—71, *108*
Yoshie, N., Ohashi, T., Suzuki, T. 69, 70, *108*
Yost, W. A., Wightman, F. L., Green, D. M. 458, *459*

Zerlin, S. 467, *480*
Zimmer, S. D., see Colavita, F. B. 344, *391*
Zwicker, E. 33, 83, 87, 88, 93, *108*, 406—410, 414, 415, 417—419, 422, 423, 425—428, 438, 440—442, 444, *447*, *448*
Zwicker, E., Fastl, H. 89, *108*, 408, 414, 444, *448*
Zwicker, E., Feldtkeller, R. 403, 404, 408, 410, 411, 413, 415, 421, 425, 428, 429, *448*
Zwicker, E., Flottorp, G., Stevens, S. S. 87—89, *108*, 410, 421, *448*
Zwicker, E., Kaiser, W. 417, *448*
Zwicker, E., Scharf, B. 425, *448*
Zwicker, E., Schütte, H. 409, *448*
Zwislocki, J. 91, *108*, 406, 426, *448*
Zwislocki, J. J., see Smith, R. L. 46, 55, *106*

Subject Index

AI, see auditory area I
A II, see auditory area II
ablation of auditory centers 309
—, cortical 317, 323ff., 369, 382
—, one-stage 330
— of subcortical centers 323, 364ff.
—, two-stage 330
— -behavior method 311, 315
— techniques 310
abnormal adaptation 367
— critical bands 92
absolute frequency shifts 417
— thresholds 324, 365, 376, 381, 402
— time difference, detection by units of inferior colliculus 176
absolutely unresponsive period 173
acetylcholine, collicular neurons 166
—, geniculate neurons 185
acetylcholinesterase, cochlear nucleus 240
—, OCB axons 234
acoustic, see also auditory
— gratings 88
— habituation 222
— input, reticular and extrareticular influences on 224f.
— neurinoma 277
— noise, background 6
— pathways, dorsal ascending 95
— —, effects of COCB stimulations on 233f.
— —, ventral ascending 95
— patterns 95
— shadow 374
— stimuli, short term memory of 216
— —, intense, hearing loss by 273, 274, 276
— stimulus parameters, coding of 92
— striae 112, 270, 272
— trauma 256, 257, 273, 281
acoustical law, Ohm's 439
action potentials, cochlear nerve — 69, 70, 90
acuity, grating 42, 43
adaptation, abnormal 367
—, cochlear nerve — 32, 38, 84
—, cochlear nucleus 46, 57
—, middle ear reflex 257
—, SOC neurons 127
adjustment, method of 402
adrenergic systems 315
AES, see ectosylvian gyrus, anterior
airborne sound 269
alcaloids, effect on OCB axons 235
alteration of complex sounds 455
AM, see amplitude modulation
amnesia 330
amplitude, just noticeable differences in 415ff.
— fluctuation 431, 433
— modulated tones 63, 167, 430, 431, 439
— modulation 415, 416, 419, 433, 440, 441
anaesthetic agents, effect on neurones in dorsal cochlear nucleus 45, 50
anatomical frequency scale 265, 278f.
anechoic environment 449
angiosclerotic degeneration 282
animals, behavioral test of hearing 262f., 283, 310
—, conscious, efferent recurrent auditory system 221f.
anion conductance, inner ear 236
anoxia 18
anterior "ascending" branch, cochlear nerve 45
anterodorsal region of anteroventral cochlear nucleus 46
antibiotics, effect on inner ear 265, 272
anticholinergic drugs, effect on OCB axons 235
antiphasic condition, interaural 463
— listening 473
AP, see action potentials
aphasia, congenital 327
aphasic patients 346, 347
area, auditory, see auditory area
—, Clare-Bishop 387
—, cuneiform, posterior colliculus 148
areas, association —, auditory cortex 216
—, primary sensory, of the cortex 385
—, projection —, auditory cortex 216
arousal system, nonspecific reticular 225
ascending auditory pathways 95, 110, 337
— auditory system 311, 312, 366
— pathway from ear to cortex 311
— pathway from the inferior colliculus 376
aseptic techniques 310
association areas 216
associative cortex, auditory 241

atrophy of cochlear nerve 270
— of the spiral ganglion 272
— of stria vascularis 270, 272
atropine, effect on cochlear nucleus neurons 241
—, effect on OCB axons 235, 383, 384
attention to auditory stimuli 221, 222, 224, 371
— to one ear 359
—, neurons in auditory cortex 201, 216
audiogram and hemispherectomy 326
—, behavioral 15, 265, 268, 287f.
—, "dip" in 282
audiometers, clinical 324
auditory, see also acoustic
— afferents, ascending, inferior colliculus 163
— area I of the cortex 179, 200, 316, 317, 318, 319, 321, 325, 331, 338, 342, 352, 357, 376
— area II of the cortex 316, 317, 318, 325, 331, 338, 342, 352, 357
— area IV 241
— —, primary 158
— axons, activation by crossed olivo-cochlear bundles 228
— —, high characteristic frequency — 130
— centers, ablation of 309
— control system, centrifugal extra-reticular 220
— cortex 146, 153, 234, 311, 313, 314, 315, 316, 317, 320, 321, 386
— —, ablation of 323ff., 369
— —, anterograde denegeration studies 318
— —, binaural neurons 133
— — in carnivores 387
— —, chronic lesions in 241
— —, columnar organization 214f.
— —, comparative anatomy 321
— —, definition of 316ff.
— —, descending fibers from 160
— —, evoked potentials 112, 113
— —, functional architecture 212f.
— —, limits of 313, 316
— —, phase-sensitive neurons 135, 139
— — in primates 387
— —, retrograde degeneration studies 317, 318
— —, single unit activity in 199ff.
— —, tonotopic organization 213f., 335, 340
— cues, anticipated 241
— —, selective perception of 221
— discrimination, behavioral studies 307ff.
— —, historical review 309ff.
— —, influence of external and middle ear 247ff.
— —, learned 385
— fields of the temporal cortex 147
— input to the brain, control of 221
— —, interactions with visual inputs 189
— koniocortex 314, 316, 321
— masking 259
— midbrain 313
— nerve, CERACS 239
— —, phase constancy 138
— nuclei of the pons, evoked potentials 111ff.
— — —, morphology 110f.
— — —, physiology 109ff.
— — —, single-unit studies 113ff.
— pathways, ascending 110, 337
— —, descending 110, 120, 381
— —, thalamic region of 158f.
— —, primary, to thalamic levels 147
— —, —, tonotopic projection 179
— patterns, discrimination of 241
— projection area, primary 326, 328
— radiation, evoked potentials 112
— space perception 249
— system, ascending 311, 312, 366
— —, descending 314
— —, efferent 3, 10, 72, 219
— —, recurrent 72, 219
— —, tonotopic organization of 315
— thalamus, see thalamus, auditory
— threshold, pure tone 13, 238
aural reflex 256, 257, 259
autolysis, postmortem 282, 296, 297
AVN, see cochlear nucleus, anteroventral
avoidance conditioning 333
— response 358
axis, interaural 254
axons, see also fibres
—, passing to the MSO 118
—, auditory, see auditory axons
—, centrifugal 219
—, corticofugal 220

background-maskers 468
— noise 6, 18, 423, 425
back-to-front reversals, sound localization 450
backward masking 68, 91, 408, 414, 428
band, critical, see critical band
— -pass filtering characteristic 78
— of noise 452, 453
— -width, critical 419
— —, effective 40, 54, 81, 87, 88, 89
— —, noise — 60, 61
— — as masker parameter 471

Bark 410, 419, 423
basal turn of the cochlea 69
basilar membrane 315, 316
— —, characteristics of 78
— — filtering properties 87
— — frequency response 79, 82
— —, movements of 7
— —, static "downwards" displacement 84
— — vibration 79, 80, 81, 82, 86
bats, auditory cortex 212
—, auditory system of 320
—, cochlear nerve responses 30
—, echo-location 212, 241, 378
—, evoked potentials 378
—, pinna 248
BC, see brachium conjunctivum
"beam" hypothesis 82
beats 404, 405
behavior, conditioned 262
—, learned 310, 385
—, localizing 351
behavioural audiogram 15
— phenomena of frequency coding 87 ff.
— studies of auditory discrimination 307 ff.
— —, combination with unit recordings 216
— tests of hearing and inner ear damage 261 ff.
— thresholds 13
Békésy tracing 367
belt cortex 331
—, cortical, see cortical belt
—, thalamic, see thalamic belt
best frequencies 116, 163, 179, 213, 215, 315
BFs, see best frequencies
BIC, see brachium of the inferior colliculus
binaural analysis 461 ff.
— clicks 140, 455
— convergence 111, 113
— discriminations after cortical ablations 358
— facilitation 186
— hearing 110
— intelligibility 473
— intensity functions 130
— interaction 362, 373
— —, evoked potentials 113
— — in lower centers 386
— latent period 130
— localization, theories of 474
— masking 463
— —, equalization-cancellation model 464, 471, 476, 477
— — -level difference 463
— neurons 114, 125, 133
— occlusion 186, 187
— phenomena, geniculate body 186
— —, inferior colliculus 172
— representation, auditory cortex 215
— response areas 125
— sound localization in the horizontal plane 254
— stimulation, response of cortical units 206
— time cues 375
— — differences 374, 376, 377
— — interactions 230
— tonal stimuli, response in geniculate body 182, 186
— unmasking 89
bird, cochlear nucleus 46, 53, 58
"blackbox" models of cochlear nerve activity 6
— — of the inner ear mechanisms 77
BMLD, see binaural masking-level difference
BP, see brachium pontis
brachium conjunctivum 110
—, inferior, evoked potentials 112
— of the inferior colliculus (BIC) 146, 181, 238, 312, 313, 365, 369, 370, 371, 375, 376, 385
— pontis 110
brain, auditory input to the 221
brainstem, auditory centers of 364
—, lesions of 369
—, sclerosis in 376
—, lateral, CERACS 239
—, lower, in the cat 373
—, upper, tumor in 376
— auditory pathways 111
— level 342
— reflex centers 385
— thrombosis 376
breathing noise 466
brightness sensation 433
broadband noise 38, 40, 407, 412, 420, 465
— —, sharpness of 435
bromide anion, inner ear 234
brucine, effect on cochlear microphonics 226
—, effect on OCB axons 235
—, effect on olivo-cochlear inhibition 220
"build up" pattern, DCN units 51
— time course 48
— units 56
bulbs of Held 46
bullfrog, cochlear nerve 11, 24
bushbaby, sound localization 354, 357

calculation of loudness 419 ff., 425 f.
calyceal synapse 118
calyces of Held 46, 51, 118

carnivores, auditory cortex in 387
—, evolution of sensory cortex 387
carrier frequency 431, 432
cat, ablation of auditory cortex 324, 330, 335, 341, 349, 350, 351
—, absolute thresholds 324, 365
—, audiogram to pure tones 266, 288, 291, 293
—, auditory areas in the cortex 317, 318, 319, 320, 321
—, auditory cortex 316, 321, 386
—, behavioural audiogram 15, 238, 266
—, binaural discriminations 358f.
—, cochlear nerve 4, 10, 13, 15, 16, 30
—, cochlear nucleus 46
—, conditioned reflex 262
—, decortication 331
—, discrimination of changes in duration 349f.
—, — of frequency changes 335, 370
—, — of intensity changes 333, 368
—, — of onset of sound 324, 365
—, — of sounds with complex spectra 362
—, frequency discrimination 349, 358, 370, 371
—, — scale of the cochlea 264, 278
—, hair cell losses 284, 288, 291, 293
—, inner ear damage 269
—, medial geniculate body 312
—, pattern discrimination 341
—, reflex responses 356, 357
—, relation between — and other species 387
—, retention of learned discriminations 330
—, sound localization 351f., 357, 360, 371, 372, 375
—, ultrasonic stimulation 265, 266, 269
CC, see corpus callosum
cell, see also neuron, unit
— region, granular, cochlear nucleus 46
cells, elongate 119
—, ganglion 264, 265, 281
—, glial 216
—, Golgi type II 181, 185
—, idle 216
— of the inhibitory response areas, VCN 54
—, large spherical of Osen 46
—, pillar 263
—, "primary-like" 53
—, principal 118, 121
—, stellate 119, 150
center frequencies 431
central gray, posterior colliculus 148
— masking 91
— mechanisms of phase-sensitivity 139
— nervous system, behavioral studies of auditory discrimination 307ff.
— neurons, receptive fields of 124
— —, response areas of 124
— nucleus, see nucleus, central
centrifugal axons 219
— blocking 222
— extra-reticular auditory control system (CERACS) 220, 234, 238f., 241f.
— — — — —, driving of OCB neurons 239f.
— gating 222, 232
centrum medianum 188
CERACS, see centrifugal extra-reticular auditory control system
cerebellopontine angle, lesions of 371
cerebellum 226
cerebral cortex, auditory projection areas 323
— —, somatic area II of 317, 318, 325, 331
— peduncle 155
— vascular accident 328, 329
CFs, see characteristic frequencies
CG, see central gray
characteristic frequencies (CFs) 4, 8, 11
chemical ototoxicity 273
— transmission process 85
— transmitter, COCB terminals 234, 236
chinchilla, cochlear nucleus 46, 54
—, hair cell loss 287, 288
—, histology of cochlea 297
—, inner ear damage 267, 268, 269, 293
—, olivo-cochlear bundle neurones 227
choline-acetyltransferase, OCB axons 234
cholinergic endings, olivocochlear bundle 74, 234
— neurons, geniculate 185
— systems 315, 383
cholinomimetic compounds, effect on OCB axons 235
"chopper-type" neurones 49, 51, 53, 55, 57
CIC, see commissure of the inferior colliculus
cilia, giant 297, 298, 299
circuits, inhibitory, inferior colliculus 166
Cl, see claustrum
Clare-Bishop area 387
click-evoked responses 366
clicks 345, 353, 358, 415
—, binaural 140, 455
—, contralateral 132
—, high-pass 458
—, low-pass 458
—, rarefaction — 9
—, coding by cortical neurons 208f.
click stimuli 81, 131, 140, 172
— —, effect of noise on responses to 43, 67
— —, responses of cochlear nerve to 6, 43 70

— —, responses of cochlear nucleus to 52, 61, 62
CN, see cochlear nucleus
COCB, see olivo-cochlear bundles, crossed
cochlea, action potentials 70
—, basal turn of 69
—, frequency localization in 271, 279
—, frequency scale of 264, 278
—, response to click stimuli 70
—, scanning electron microscopy of 296
cochlear complex, single unit studies 121
— damage 263, 264
— deafness 92
— dendrites, radial 236
— division of VIIIth nerve, fibres in 5
— dysfunction 369
— efferent system, background noise 75
— — —, FCTs 72
— excitation function 93
— fibres, synchronization of responses 25, 26
— filter function 21, 38, 39, 41, 77
— innervation 73, 85
— linearities 81
— locations for threshold responses 267
— mechanisms 77
— microphonics 28, 79, 80, 81, 84, 223, 226, 228, 230
— —, increase by COCB activation 230, 236
— — generator potential 236
— nerve 1ff.
— —, action potentials 69, 70, 90
— —, branches in cochlear nuclei 45
— —, efferent pathways 72
— —, homogeneity of 11, 85
— —, interaction of responses to two tones 35
— —, interruption of 51
— —, response to click stimuli 6, 43, 70
— —, response as a function of intensity 21
— —, response as a function of time 23
— —, response to complex sounds 30
— —, response to low frequency tones 24
— —, response to single tonal stimuli 10
— —, section of 263, 264, 265, 269
— —, spontaneous activity 4, 6, 10
— —, tonotopic organization 4
— — discharges, generation of 83
— — frequency threshold curves 10, 12, 72, 78, 81
— — response, latency of 9
— non-linearities 80, 81, 83
— nuclei, projection upon central nucleus of inferior colliculus 150
— nucleus 1ff., 43ff., 226, 234, 311, 312, 313, 314, 315, 384
— —, CERACS 239
— —, efferent pathways to 76
— —, functional organization 46, 95
— —, masking 90
— —, pathway from — to lateral lemniscus 112
— —, phase-sensitive neurons 138
— —, response to click stimuli 52
— —, response to complex sounds 58
— —, response to single tonal stimuli 53
— —, sensory input to neurones 51
— —, spontaneous activity 46, 48, 51, 52
— —, tonotopic organization 44, 45, 114 116, 117, 120
— —, anteroventral (AVN) 43, 46, 48, 51, 52, 76, 113, 116, 167, 239, 241
— —, dorsal (DCN) 43, 45, 48, 49, 50, 51, 52, 54, 55
— —, —, evoked potentials 112
— —, posteroventral (PVN) 43, 48, 50, 51
— —, ventral (VCN) 43, 46, 52, 54
— — action potentials 70
— — neurons, efferent effects on 240f.
— pathology 69, 71, 266, 270
cochleographs 284, 286, 287, 288, 291, 293
coding of acoustic stimulus parameters 92ff.
— of clicks, cortical units 208f.
— of tonal stimuli, cortical units 201
collaterals of COCB in cochlear nucleus 240
collicular neurons, see neurons, collicular
colliculus, anterior 149
—, inferior 91, 109, 110, 160, 234, 311, 312, 313, 314, 315, 320
—, —, ablation of 365, 368, 369, 371, 376, 378
—, —, axons of 376
—, —, binaural neurons 133
—, —, brachium of, see brachium of the inferior colliculus
—, —, CERACS 238
—, —, corticofugal axons to 220
—, —, degeneration of 367
—, —, electrophysiology 162f.
—, —, evoked potentials 112, 113
—, —, lesions in 370, 378
—, —, model of synaptic organization 170ff.
—, —, morphology 147f.
—, —, phase-sensitive neurons 135, 138
—, —, phoneme-related responses 190
—, —, physiology 145ff.
—, —, primary path from 386
—, —, response to transients 139
—, —, single unit studies 121
—, —, tonotopic organization 162
—, posterior 148, 149
—, superior 152

columnar organization, auditory cortex 214f.
"comb-filtered" noise 41, 42, 61, 81, 88
combination tone 34, 38, 83, 94, 441
— —, psychophysical 35, 91
— of tone and noise bands 67
commissure of the inferior colliculus 365, 372, 378
commissures, intercollicular 122, 148
—, lemniscal 122
— of Held 366, 372
— of von Monakov 366, 372
— of Probst 120, 366
comparison, paired, method of 402
competing stimuli 371
compression phenomenon 282
conditioned reflex 262
condensation transients 7
congenital aphasia 327
conscious sensation 385
constant detectors 191
continuous noise stimulation 18
contractions of the middle ear muscles 223, 225, 227
contralateral response 112
— threshold shift (CTS) 257
— tuning curves, LSO neuron 125
control of the auditory input to the brain 220, 221
convergence, binaural 111, 113
core, cortical, see cortical core
— area of the auditory system 321, 324, 328, 339
—, thalamic, see thalamic core
corpus callosum, transection of 372
— restiforme 226
cortex, functional relation between core- and belt- 387
—, layer IV of 153
—, relation to subcortex 384ff.
—, S II in 159, 317, 318
—, association, evolution of 386
—, auditory, see auditory cortex
—, belt — 331, 339
—, cerebral, see cerebral cortex
—, dorsal, inferior colliculus 148
—, inferotemporal 344, 382
—, insular, see insular cortex
—, insular-temporal 159, 220, 240, 241, 331, 333, 339, 342
—, internuclear, inferior colliculus 147, 150, 151
—, lateral, inferior colliculus 147, 150
—, midtemporal 362
—, occipital 189
—, parietal 339
—, pericruciate 343
—, primary auditory 330, 386
—, sensory 384
—, —, evolution of 386f., 387
—, sensory-motor 385
—, temporal, see temporal cortex
—, temporo-insular 159, 220, 240, 241
Corti, organ of 263, 265, 270
cortical ablation 317, 323ff., 335, 382
— —, recovery from 332
— belt 319, 320, 321, 324
— core 320, 321, 324
— damage 355
— lesion, contralateral 326
— subdivisions, cytoarchitectonic of 313, 316, 317
cortico-cortical projections 386
corticofugal axons, from temporal cortex 220
corticotectal efferents 315
cortico-thalamic projection 320
critical band 81, 87, 88, 89, 92, 93, 384, 409ff., 421, 432, 433
— — levels 425
— — noise 405, 423
— — rate 409ff., 414, 424, 435
— — scale 410, 425, 429
— band-width 419
— ratio 88, 89
crocodile, cochlear nucleus 46
"crossed-ramp" type of rate-intensity function 23
CTS, see contralateral threshold shift
cubic difference tones 441, 442, 443
— distortion 442
cues, auditory, perception of 221, 241
—, binaural time — 375
—, distance — 455
—, non-auditory 366
—, tactual 366
cuneiform area, colliculus 148
curarizing drugs 227, 235
currents, local, interaction of 87
cyanide, effect on cochlea 80
cycle-by-cycle movement, following the 453

dB method, equivalent 228
DCN, see cochlear nucleus, dorsal
deaf, nerve action potentials in — 69
deafness, cochlear 92
—, sensorineural 91
decortication, effect on sound localization 331
definition of sensory discrimination 309
degeneration, angiosclerotic 282
—, anterograde, auditory cortex 318
—, hypotonic 282
— of pillar cells 263

—, retrograde, auditory cortex 317, 318, 385
—, —, in the medial geniculate bodies 333, 339, 343
—, thalamic 317, 318, 330, 338, 342, 343
— of unmyelinated fibres 319
delay, characteristic 138, 176, 177
—, interaural 138, 177, 188, 207
delayed noise 470
delays, transmission — 138
de-modulation 65
dendrites, cochlear 236
—, geniculate body 157
dendritic laminae on central nucleus 149
density sensation 433, 436
depression, maintained 32
— of spontaneous activity 85
depth, organization of auditory cortex according to 213
descending auditory pathways 110, 120
— — —, lesions in 381ff.
— — system 314
— fibers, see fibers, descending
desynchronization by noise stimuli 67
detection of speech 468
detectors, constant — 191
—, formant — 191
—, phoneme — 191
—, transient — 191
—, vowel — 190
dichotic condition, interaural 453, 474
— listening 361, 465
— tests of speech 361
difference, just-noticeable 402, 410, 415, 416, 417, 429, 437, 439, 440
— of level at the two ears, see interaural difference of level
— of time, see interaural time difference
— in the time of arrival of sounds 316
— limens 332
— threshold 332, 334
— -tones 404
— —, cubic 441, 442, 443
— —, quadratic 441
differential threshold 402
digits 345
dihydro-β-erythroidine, effect on cochlear microphonics 226
—, effect on cochlear nucleus neurons 241
—, effect on collicular neurons 166
—, effect on OCB axons 235
dihydrostreptomycin, effect on inner ear 265
diotic condition, interaural 463
"dip" in the audiogram 282
diplacusis 94, 371
direction of the sound source 331
directional hearing 248
— selectivity, collicular neurons 168
— —, geniculate neurons 188
discharge, "jitter" of 85
—, maintained, cochlear nerve 10
— rate, evoked by combined stimuli 37
— —, synchronized 20
discharges, "phase-locked" 24, 26, 33
—, spontaneous, cochlear nucleus 46, 48
discrimination of changes in duration 349ff.
— of changes in frequency 335, 370, 381
— of changes in intensity 332ff., 368ff., 381
— of onset of sound 324, 365
— of temporal order, after cortical ablations 346ff.
—, auditory, see auditory discrimination
—, binaural, after cortical ablations 358
—, learned, loss of 342
—, —, retention of 330, 342
disease, Ménière's 272, 279, 281
displacement, static "downwards", of basilar membrane 84
distance between particle velocity and pressure 455
— cues, sound localization 455
distortion, cubic 442
—, nonlinear 441ff.
—, quadratic 442
DL's, see difference limens
dog, absolute thresholds 325, 326
—, auditory cortex 323
—, behavioral tests of hearing 262
—, decortication 331
—, discrimination of changes in frequency 339
—, — of onset of sound 325
—, memory after cortical ablations 347
double-grill apparatus 333, 360
drugs, anticholinergic 235
—, curarizing 227, 235
—, ototoxic 265, 266, 269, 272, 273, 279
D-tubocurarine, effect on OCB axons 235
duration of an auditory signal 325, 326, 328, 332, 345, 349
—, just-noticeable differences in 437
—, subjective 436ff.
Durchlach's model for binaural analysis 476
dynamic range of auditory system 21
dysfunction, cochlear 369

ear, external, see external ear
—, inner, see inner ear
—, middle, see middle ear
—, outer, see external ear, pinna
eardrum, tactual cues from 366

earphone measurements of time and level differences 452ff.
E-C Model, see Equalization-Cancellation Model
echo 455
— -location 212, 241, 378
ectosylvian gyrus 159, 189, 348
— —, anterior 323
— —, middle 316, 318, 323
— —, posterior 316, 317, 318, 323, 325, 331, 338, 340, 342, 343, 352, 357
EE-neurons, see neurons, excitatory-excitatory
effective bandwidth 40, 81, 87, 88, 89
efferent auditory system 3, 10, 72, 219
— cholinergic mechanism, OCB axons 234
— effects on cochlear nucleus neurons 240f.
— innervation of OHC 86
— motor commands to middle ear muscles 224
— pathway, descending (inhibitory) 32
— pathways of the cochlear nerve 72
— — to cochlear nucleus 76
efferents, corticotectal 315
EI-neurons, see neurons, excitatory-inhibitory
electrophysiology of the inferior colliculus 162f.
— of the medial geniculate body 178f.
electrotonic conduction 86
— interaction 32, 33
endings, cholinergic 74
endocochlear potential 231, 234, 236
endolymphatic flow 85
envelope of the stimulus 453
Ep, see ectosylvian gyrus, posterior
equal loudness contours 420, 421
— — ratios 423
Equalization-Cancellation Model for binaural masking 464, 471, 476, 477
equivalent dB method 228
"essential" non-linearity 83
ethacrynic acid, effect on inner ear 273
evoked potential studies 316
— potentials, auditory nuclei of the pons 111ff.
— — in bat cortex 378
— response 326, 366
— —, cortical neurons 201
evolution of association cortex 386
— of auditory thalamus 388
— of sensory cortex 386f., 387
excitation, accessory 413, 418, 419
—, cortical neurons 201
—, main 413, 418
—, psychoacoustical 409ff.
— factor 411, 412
— function, cochlear 93
— level 411, 412, 414, 417, 423, 424
— model 423
— pattern 413, 414, 425, 428, 439, 444
— — for broad band noise 412
— — of different sounds 422
— — and just noticeable differences 417
— — for narrow band noise 412
— — model 440
excitatory-excitatory neurons 114ff., 133
— frequency threshold curves 54
— -inhibitory influences, units in inferior colliculus 164f., 169
— — interactions, geniculate body 180f., 183
— — neurons 114ff., 133, 140
— response areas 125
— transmitter, outer hair cell 236
— tuning curve, LSO neuron 125
experimental animals, training of 310
— surgery-behavior method 311
external ear, see also pinna
— —, influence on auditory discrimination 247ff.
— —, mechanical properties of 77
— noise 465
extra-reticular influences on acoustic input 224

facial paralysis 258
facilitation, in-phase 136
factory noise 425, 426
fasciculus, medial longitudinal 148
feedback, positive 82
fibre branch of cochlear nerve, anterior "ascending" 45
— — of cochlear nerve, posterior "descending" 45
— pathway, intranuclear, AVN to DCN 52
fibres, see also axons
—, cochlear, synchronization in 25, 26
—, descending, from the auditory cortex 160
—, high-threshold — 11, 12, 70
—, inner radial 86
—, low-threshold — 11, 70
—, phase-locked 458
—, olivocochlear 119
—, spiral 85, 86
—, unmyelinated, degeneration of 319
— in the cochlear division of VIIIth nerve 5
filter function, cochlear 21, 38, 39, 41, 77
— system, linear 81
filtering characteristic, band-pass 78

— properties of basilar membrane 87
fish, cochlear nucleus 77
—, frequency threshold curve 15
—, VIIIth nerve 11
—, response of cochlear fibres 24
flattening effect 471
flaxedil, effect on OCB axons 235
fluctuations of the excitation-critical band rate pattern 433
fluent aphasics 347
fluid-hair cell coupling 82
"following", temporal 453
— the cycle-by-cycle movement 453
formant detectors 191
— frequencies 92, 93
forward masking 68, 69, 91, 408, 409, 414
FRCs, see frequency response curves
free field 426
— — studies in localization of sound 449
frequencies, best (BFs) 315
—, —, LSO neuron 116
—, center — 431
—, characteristic 4, 8, 11, 17, 230
—, formant 92, 93
—, modulation — 431, 432
frequency of an auditory signal 325, 332
—, carrier — 431, 432
—, threshold as a function of 53
— bands, inhibitory 58
— —, suppressive 30
— changes, discrimination of 335ff., 370ff.
— coding 92
— cut offs, high 16
— — —, low 16
— differences, just noticeable 416ff.
— discrimination 87, 92, 238, 314, 358, 370, 371, 386
— —, binaural analysis 473
— — after cortical ablations 342, 344, 349
— —, geniculate body 178, 180
— —, inferior colliculus 164
— following response 57, 90
— localization in the cochlea 271, 279
— modulated tones 63, 65, 167, 191, 200, 210, 212, 342, 343, 440
— modulation 419, 440
— response, basilar membrane 79, 82
— — area 12, 16, 36, 48, 50
— — curves 18, 19, 20
— — —, inferior colliculus 164
— — function 79
— responses of units in DCN 48, 50, 53
— scale, anatomical 265, 278f.
— — of the cochlea 264
— selectivity 19, 41, 87, 88, 92, 410
— sharpening process 82
— shifts, absolute 417
— —, relative 417
— threshold curve 10, 12, 46, 54, 72, 78, 81
— — —, sharpness of 15, 16, 72
Frequenzgruppen 426
fringe, suprasylvian 159
—, sylvian 316, 317, 319, 387
frog, cochlear nerve 11, 15, 30
front-back discrimination 450
— — reversals 454
frontal lobe, ablation of 361
front-rear discrimination of sound sources 253
Frusemide, effect on cochlea 18, 80
FTC, see frequency threshold curve
fullness sensation 433
functional architecture, auditory cortex 212f.
funnelling mechanism 164, 180

GABA, collicular neurons 166
Galago senegalensis, sound localization 354, 357
gallamine, effect on cochlear nucleus neurons 241
—, effect on OCB axons 235
ganglion cell population and speech discrimination 281f.
— — losses 264, 265, 281
gated masker 472
gating, centrifugal 222, 232
—, OCB-induced 230
Gaussian noise 465
generalization, gradients of 340
— curves 341
generation of cochlear nerve discharges 83
generator, probabilistic 85
— potential 236
genetics of inner ear damage 269
geniculate body, medial 311, 312, 313, 314, 317, 318, 319, 327, 337
— —, —, binaural neurons 133
— —, —, caudal division of 318, 342, 343
— —, —, degeneration in 333, 339, 342
— —, —, dorsal division of 317
— —, —, electrophysiology 178
— —, —, evoked potentials 112
— —, —, magnocellular division of 317, 318, 319
— —, —, medial division of 317
— —, —, neuronal structure 153f.
— —, —, phoneme-related responses 190
— —, —, principal division of 318, 333, 343
— —, —, retrograde degeneration 333, 339
— —, —, tonotopic organization 157, 178
— —, —, ventral division of 317, 318, 321, 333, 376
— complex, medial, physiology 145ff.
— nucleus, medial, CERACS 238

geniculo-striate system 387
gerbil, cochlear nucleus 46
giant cilia, cochlea 297, 298, 299
glial cells, auditory cortex 216
glioblastoma 367, 377
glycine, effect on OCB axons 235
glycinergic post-synaptic inhibition 220
GM, see geniculate body, medial
GMc, see geniculate body, medial, caudal division of
GMd, see geniculate body, medial, dorsal division of
GMm, see geniculate body, medial, medial division of
GMmc, see geniculate body, medial, magnocellular division of
GMp, see geniculate body, medial, principal division of
GMv, see geniculate body, medial, ventral division of
Golgi type II cells, geniculate body 181, 185
go situation 340
gradients of generalization 340
granular cell region, cochlear nucleus 46
— layers of the DCN 45
grating acuity 42, 43
gratings, acoustic 88
gray, central, posterior colliculus 148
—, peri-aqueductal 366
groupers, cortical units 209, 210
guinea pig, audiograms 15, 268, 270
— —, cochlear nerve 4, 5, 10, 13, 15, 16, 70, 79
— —, conditioned reflexes 262
— —, inner ear damage 269
gyri, transverse, of Heschl 334, 361
gyrus, anterior sylvian 241
—, ectosylvian, see ectosylvian gyrus
—, lateral 338, 343
—, middle temporal 330
—, post central 327
—, sigmoid 325
—, superior temporal 330, 344
—, — —, corticofugal axons from 220
—, suprasylvian 160, 189, 331, 338, 343
—, sylvian, see sylvian gyrus
—, ventral sylvian 241

Haas effect 455
habituation 358, 360
—, acoustic 222
—, auditory cortex 201, 216
hair cell polarities 84
— — populations and recruitment 279f.
— — -synapse area 85
— cells 220, 301
— —, damage of 91, 281
— —, loss of 263, 267, 279, 281, 287, 288
— —, "resonance" of electrochemical nature 82
— —, inner 11, 16, 70, 84, 85, 86, 87, 91, 227, 236
— —, outer 11, 13, 16, 70, 84, 85, 86, 91, 231, 236
harmonic complex tone sounds, sharpness of 435
— spectra, pitch of 430
harmonics 35
head blows, inner ear damage by 269
— movements, effect on sound localization 252, 454ff.
— shadow, effect on location judgements 250
— —, effect on stimulus size 223
hearing, binaural 110
—, directional 248
— impairment, behavioral studies 261ff.
— loss and conditioned reflexes 262
— — and hair cell loss 279, 287
— —, high frequency 271, 366
— — by intense acoustic stimuli 273, 274 276
— —, low frequency 271
heartbeat noise 466
hedgehog, sound localization 354, 357, 375
Held, calyces (bulbs) of 46, 51, 118
—, commissures of 366, 372
hemicholinium HC-3, effect on OCB axons 235
hemiplegia, infantile 326
hemispherectomy 326, 335, 355
hemisphere lesions 363
Heschl, transverse gyri of 334, 361
hierarchy of masking-level differences 464
— of responses 357
high frequency hearing loss 271, 366
— — tones, response to 133
— -pass clicks 458
— — filtered noise 418
— — noise 417
— threshold fibres 11, 12, 70
histogram, interspike interval — 5
—, post-stimulus time — 6, 7, 8
homogeneity of cochlear nerve 11, 85
homologies 388
homophasic condition, interaural 463
Hörmesser, Politzer's 249
horse, pinna 248
human, see man
hydrops 272
hyoscine, effect on cochlear microphonic 226
—, effect on OCB axons 235

hyperirritability to sounds 258
hyperpolarization, collicular neurons 167
hypersensitivity 334, 335, 369
hypothesis, "beam" — 82
—, Webster-Jeffress — 464, 474
hypotonic degeneration of the inner ear 282
hypoxia, effect on cochlea 18, 80, 87

I, see insular cortex
IC, see colliculus, inferior
identification-situation 358
idle cells, auditory cortex 216
IHC, see inner hair cells
impedance, middle ear 256
incitation, psychoacoustical 410, 411, 417
— factor 412, 423, 424
— level 412
infection of the middle ear 284, 285
inferior colliculus, see colliculus, inferior
inferotemporal cortex 344, 382
influences, excitatory-inhibitory 164, 169
information, temporal 95, 132
inhibition, asymmetrical 63
—, cochlear nucleus 58
—, contralateral 116
—, glycinergic post-synaptic 220
—, lateral 41, 42, 63, 82, 90, 91
—, off- 63
—, olivo-cochlear 86, 220, 228, 230
—, out-of-phase 136
—, postsynaptic, collicular neurons 166
—, —, inner ear 237
—, presynaptic, collicular neurons 166
—, —, inner ear 237
—, recurrent 181
— of spontaneous activity of unit in DCN 48
— of the spontaneous activity, cochlear nerve 10
—, by stimulation of OCB 72
—, two-tone — 30
inhibitory circuits, inferior colliculus 166
— DCN unit 49, 50
— efferent pathway 32
— frequency bands, two-tone 58
— influences, lateral 30, 32
— interaction, interaural 207
— interneuron 116
— processes, inferior colliculus 166
— response areas, cells of 54, 124, 125
— sensory input to the DCN 52
— side bands 90, 124, 125
— transmitter, COCB terminals 232
— tuning curve, LSO neuron 125
inner ear, action of OCB axons in 236f.
— —, hypotonic degeneration 282
— —, ionic mechanism of efferent COCB 234f.
— —, mechanical properties of 77
— —, phase microscopy of 297
— —, scanning electron microscopy of 293, 296
— —, transmitter in 235
— — anatomy, ototoxic stimuli on 269, 277
— — damage and behavioral tests of hearing 261ff.
— — histology 269, 282
— — lesions and intensity discrimination thresholds 266
— — mechanisms, "blackbox" models 77
— — pathology 264, 266, 270
— hair cells 11, 16, 70, 84, 85, 86, 87, 91, 227, 236
— — —, loss of 263, 281
— radial fibres 86
innervation of crossed OCB 86
— pattern models 82
—, cochlear 73, 85
—, efferent, of OHC 86
in-phase facilitation 136
input, excitatory, from one ear 133, 141
—, inhibitory, from one ear 133, 141
—, sensory, to cochlear nucleus 51
— -output models, cochlear nucleus 51
insular cortex 159, 160, 317, 318
— cortical areas 325, 331
— -temporal areas of the cortex 317, 318, 325, 331, 333, 338, 342, 343, 352, 357, 358, 359, 360
— — cortex 159, 220, 331, 333, 339, 342
— — zone 319, 337, 344
integration of different sensory modalities 188
intelligibility of speech 472
intensity, response of cochlear nerve to 21
—, response as a function of 54
— of an auditory signal 325, 332
— change, interaural 177
— —, discrimination of 332ff., 368ff.
— coding 94
— differences, interaural 133, 134, 139, 141, 176, 186, 207, 249, 259
— discrimination, binaural analysis 473
— — thresholds after inner ear lesions 266
— functions, binaural 130
— —, monaural 127
— image 457
— information, geniculate body 184

interaction, of auditory and visual inputs 189, 375
—, binaural 362, 373
—, electrotonic 32, 33
—, excitatory-inhibitory 180, 183
—, interaural 207
— of responses to two tones, cochlear nerve 35
—, somatic-auditory 188
interaural axis 254
— conditions 463ff.
— delay 135, 177, 188, 207
— discrepancy 462
— inhibitory interaction 207
— intensity change 177, 467
— — differences 133, 134, 139, 141, 176, 186, 207, 259
— — of maskers 469
— — sensitive unit 176, 177, 188
— interactions, auditory cortex 207
— level differences 450, 451, 453ff., 454, 456, 457, 458
— noise correlation 469, 470
— phase 120, 453, 464, 467
— — differences 133, 134, 137, 207, 259, 457
— — as masker parameter 471
— — reversal 467
— signal level, binaural analysis 466
— — phase 466
— time delay 177, 188, 475
— — differences 133, 141, 207, 450, 451, 452ff., 456, 457, 475
— — as masker parameter 470
— — relationships 175, 176
— — sensitive unit 176, 177, 188
interaurally uncorrelated internal noise 469
intercollicular commissures 122, 148
interconnections, nonreciprocal 314
—, tactile and auditory 159
—, reciprocal 314
internal noise 83, 469, 470
interneuron, inhibitory 116
interspike interval 52
— — histograms 5
interval distribution, spontaneous activity of cochlear nerve 6, 52
— —, spontaneous activity of cochlear nucleus 52
— scale 402
intracellular potentials, auditory cortex 208
intracochlear neurinoma 277
intranuclear fibre pathway between AVN and DCN 52
"inverted" spectrum, noise stimuli 42
ionic mechanism of efferent COCB in the inner ear 234f.
ipsilateral response 112
— tuning curves, LSO neuron 125
iso-intensity frequency response curves 19, 20
— -rate response contours 19, 21
I-T, see insular and temporal areas in the cortex

"jitter" of discharge 85
—, temporal 476
JND, see just-noticeable difference
just-noticeable differences 402, 410, 415ff., 416ff., 417ff., 429, 437, 439, 440
— — increment in level for critical band noises 423

kanamycin, effect on cochlea 266, 272, 273, 285
kangaroo rat, cochlear nucleus 57
KCN 18
kernicterus 366, 369
Klang theory of sound localization 255
koniocortex, auditory 314, 316, 321

labile units, auditory cortex 216
laminae, dendritic, central nucleus 149
laminar arrangement, geniculate body 179
— — of neurons in the inferior colliculus 163
— organization in the central nucleus 150
latency of cochlear nerve response 9
— during COCB inhibition 230
latent period, binaural 130
— —, monaural 130
lateral inhibition 30, 32, 41, 42, 63, 82, 90, 91
— lemniscus, see lemniscus, lateral
lateralization 453
— of sound after cortical ablations 350ff.
— — — after subcortical ablations 372ff.
—, theories of 474, 477
— model for binaural analysis 477
— test 355
learned auditory discrimination 385
— behavior 310, 385
— discrimination, loss of 342
— —, retention of 330, 342
learning 338
lemniscal commissures 122
lemniscus, lateral 76, 109, 147, 315, 365, 366
— —, CERACS 238
— —, degeneration of 367
— —, dorsal nucleus of, corticofugal axons to 220

— —, evoked potentials 112, 113
— —, fibers to inferior colliculus 152, 153
— —, nuclei of 109, 148
— —, stimulation of 76
— —, transection of 365, 372, 375
—, medial 155
leopard frog, cochlear nerve 11, 15
lesions, cortical 326
—, chronic, in auditory cortex 241
—, retrocochlear 369
— of the brainstem 369
— of the cerebellopontine angle 371
— in the descending auditory pathways 381ff.
— in the medial geniculate body 319
— of the nucleus of the trapezoid body 117
— in the pons 367
— in the posterior nuclei of thalamus 319
— of the spiral ganglion 264
— of the temporal lobe 346
— of the ventral temporo-insular cortex 241
level differences, interaural 450, 451, 453ff., 454, 456, 457, 458
linear aspects of responses to two tones 35
— filter system 81
linearities, cochlear 81
line-busy effect 90
linguistic, see also speech
— communication 190, 191
listening, antiphasic 473
—, dichotic 361, 465
lizard, cochlear nerve 11, 15
LL, see lemniscus, lateral
LLN, see nuclei of the lateral lemniscus
local currents, interaction at the IHC level 87
localization, theory of 475
— of low frequency sound 375
— of noise 454
— of sound, see sound localization
— of transients 133, 139
localizing in the ipsilateral field 375
— in the contralateral field 375
lockers, cortical units 208ff., 210
longitudinal shear model 82
loss of ganglion cells 264, 265, 281
— of hair cells 13, 86, 263, 279, 281, 287, 288
— of learned discrimination 342
loudness 420, 421, 439, 440
—, binaural analysis of 473
— of complex sounds 425
—, equal 420, 421, 423
—, just-noticeable differences in 439
— of narrow band noises 420
—, sensation of 92
—, specific 422, 423, 424, 425, 435
— of tones 420
—, total 425, 426
— calculation 419ff., 425f.
— contours, equal 420, 421
— —, specific 423
— formation 425
— —, model of 422ff.
— function 419
— level 419, 420, 421, 422, 440
— —, duration of a tone 426
— —, temporal effects on 426
— recruitment 91, 369
— summation 422
low frequency hearing loss 271
— — sound, localization of 375
— — tones 24, 57, 133
— -pass clicks 458
— — noise 417
— threshold fibres 11, 70
LPO, see nucleus, lateral preolivary
LRO, see retroolivary region, lateral
LSO, see olive, lateral superior

MAA, see minimal audible angle
Macaca mulatta 339, 344, 362, 382
— *speciosa* 362
macaque, cochlear nerve 4
—, cochlear pathology 266, 269
maintained depression 32
— discharge, cochlear nerve 10
mammalian auditory cortex, subdivision of 313, 321
mammals, auditory projection areas of the cerebral cortex 323
—, auditory system of 320, 321
—, olivo-cochlear system 225ff.
man, absolute thresholds 326, 366
—, auditory cortex 326, 327
—, binaural discriminations after cortical ablations 361f.
—, cochlear pathology 270
—, discrimination of changes in frequency 340, 371
—, — of changes in intensity 334, 369
—, — of onset of sound 326, 366
—, — of sounds with complex spectra 362
—, — of temporal order 346
—, evoked response 326
—, gross nerve action potentials 71
—, hair cell loss 287
—, hearing loss 270, 271
—, inner ear damage 270f.
—, middle ear muscle contractions 223
—, nerve atrophy 270

man, pattern discrimination 344
—, posterior colliculus 148
—, scanning electron microscopy of inner ear 293, 296
—, sound localization 354, 357, 360, 376
—, surface specimen technique 293
—, temporal bone pathology 270, 278
Marchi method 311
masked signal 462
— threshold 89, 402ff., 444
— — for a narrow-band noise 384
— — for a wide-band noise 384
masker, gated 472
—, small band 408
—, tonal 472
—, binaural analysis of 463
—, overall level of 468, 470
— parameters, binaural analysis 468ff.
masking 89f., 93, 223, 259
—, backward — 68, 91, 408, 409, 414, 428
—, binaural 463
—, central 91
—, contralateral 257
—, forward — 68, 69, 91, 408, 409, 414
—, non-simultaneous 91
—, — psychophysical 68
—, simultaneous 402, 408, 409
— of click and tone responses 43
— and descending auditory pathways 382ff.
— by extraneous noise, effect on stimulus size 223
— by self-generated noise, effect on stimulus size 223
— of tones by noise 68, 87
—, two-tone — 410
— by white noise 403, 404, 408
— effect, contralateral 360
— —, "overshoot" 408, 409, 414
— -level difference 463, 464, 465, 466, 467, 468, 470
— — —, hierarchy of 464
— — —, models of 477
— noise 68, 87, 223, 465, 466, 467
— pattern 413
— sound 429
mechanical insult, inner ear damage by 269
— properties of outer, middle and inner ear 77
— tuning 78
medulla 381
—, auditory commissures of 366
—, efferent pathways in 382
medullary pyramid 110
mel 428
— scale of pitch 410
membrane, basilar 7, 78, 79, 80, 84, 86, 87, 315, 316
—, Reissner's 272, 283
—, tectorial 85, 283
— potential, noisy 83
— resistance, inner ear 236
memory 339, 343, 344
—, dynamic 347
—, recent, after cortical ablations 347
—, short term, of acoustic stimuli 216, 345, 347
Ménière's disease 272, 279, 281, 369
meningioma 277, 355
MES, see ectosylvian gyrus, middle
method, ablation-behavior 311
— of adjustment 402
—, behavioral testing — 262, 277, 283
— of constant stimuli 402
—, experimental surgery-behavior — 311
—, Marchi — 311
— of paired comparison 402
— of tracking 402
metrazol, effect on OCB axons 235
microphonics, cochlear 28, 79, 80, 81, 84, 223, 226, 228, 230, 236, 237
midbrain, auditory 313
— tegmentum 152
middle ear, influence on auditory discrimination 247ff., 256f.
— —, mechanical properties of 77
— — damage 287
— — impedance 256
— — infection 284, 285
— — muscle activities, non-acoustic 225
— — — contractions 223, 225, 227
— — muscles 3
— — —, efferent motor commands to 224
— — reflex 256, 257, 259
minimal audible angle 451, 452
— — time difference 454
— phase angle 453
"missing fundamental" signal 35
MLD, see masking-level difference
model, "blackbox", of inner ear mechanisms 77
—, Durlach's, for binaural analysis 476
—, equalization-cancellation-, for binaural masking 464, 471, 476, 477
—, excitation — 423
—, excitation-pattern — 440
—, lateralization — 477
— of loudness formation 422ff.
— of the masking-level difference 477
— of the pinna 255
— for the sharpening of the basilar membrane frequency response 82
—, stochastic 84

— of synaptic organization in inferior colliculus 170ff.
—, Theta —, binaural analysis 474
—, vector —, binaural analysis 474
modulation, just-noticeable 440, 441
—, threshold of 410, 441
— depth 168f.
— frequency 417, 431, 432
— rate 168f.
molecular layers of the DCN 45
von Monakov, commissures of 366, 372
monaural intensity functions 127
— latent period 130
— response area, neurons of pontine nuclei 122ff.
— sound localization 249, 250
— tests 331
monkey, see also macaque, squirrel monkey
—, absolute thresholds 325
—, auditory cortex 216, 317, 320, 321
—, behavior studies of OCB system 238
—, cochlear nerve response 4
—, — pathology 266, 269
—, discrimination of changes in frequency 339
—, — of intensity changes 334
—, — of onset of sound 325
—, — of sound with complex spectra 362
—, frequency discrimination 344
—, — location in the cochlea 279
—, memory after cortical ablations 348
—, pattern discrimination 344, 362
—, retention of learned discriminations 330
—, sound localization 353, 357
—, speech discrimination 382
monotonic condition, interaural 463
Morest's subdivisions 320
morphology of the pontine auditory nuclei 110f.
motor commands, efferent, to middle ear muscles 224
movement, perception of 453
—, sensation of 453
movements of the basilar membrane 7
MPO, see nucleus, medial preolivary
MRO, see retroolivary region, medial
MSO, see olivary nucleus, superior medial
multicomponent signals 35
multiple-component spectra, stimuli with 41, 61
— sclerosis 366, 369
muscles, middle ear — 3, 224
muscle tonus noise 466
music 414, 456
Myotis, see bat
narrow-band noise 384, 405, 407, 412, 420
— —, sharpness of 434
narrowing of the tuning curve 180
neocortex 312, 340, 386
neocortical sensory systems 386
neomycin, effect on cochlea 266, 272
nerve, cochlear, see cochlear nerve
—, eighth, fibre groups in 11
—, —, fibres in the cochlear division of 5
—, —, lesions of 367, 371
—, —, neurofibroma of 276
—, —, tumor of 281
neural centrifugal gating process 222
— coding, auditory cortex 200f.
— frequency threshold curves 78
— inhibition, cochlear nucleus 58
— pathways, transection of 309
— tuning 78
neurinoma, acoustic 277
neurochemical mechanisms of OCB axons 234f.
neurofibroma of the eighth nerve 276
neuronal structure of the medial geniculate body 153f.
neuron, see also cell, unit
neurons, binaural, MSO 114, 133, 140
—, central 124
—, cholinergic 185
—, "chopper-type" 49, 51, 53, 55, 57
—, collicular, acetylcholine content 166
—, —, GABA content 166
—, —, laminar arrangement 163
—, —, nicotinic cholinoceptive receptors in 166
—, —, pre- and postsynaptic inhibition 166
—, —, sound localization 176
—, crossed OCB 225
—, disc-shaped, posterior colliculus 149, 150
— in dorsal cochlear nucleus, effect of anaesthetic agents on 45
—, excitatory-excitatory (EE) 114ff., 133
—, excitatory-inhibitory (EI) 114ff., 133, 140
—, nicotinic-muscarinic 185
—, olivo-cochlear 220
—, —, and CERACS 239f.
—, phase-sensitive 135, 136, 138
—, somato-sensory 161
—, spherical cell — 122
—, thalamo-cortical 157
—, time-sensitive 141
neuron types, geniculate body 155, 156
neurophysiology of sound localization 457ff.
nicotinic cholinoceptive receptors in collicular neurons 166
— -muscarinic neurons, geniculate 185

nitrogen mustard, ototoxicity by 266, 273, 279
no-go situation 340
noise, background — 6, 18, 423, 425
—, breathing — 466
—, broad band 38, 40, 407, 420
—, — —, excitation pattern for 412
—, — —, sharpness of 435
—, "comb-filtered" 41, 42, 61, 81, 88
—, critical band — 405, 423
—, delayed 470
—, external 465
—, factory- 425
—, Gaussian 465
—, heartbeat — 466
—, high intensity-, and inner ear damage 273, 274, 276
—, high-pass 417
—, high-pass filtered 418
—, internal 83, 469, 470
—, low-pass 417
—, masking 68, 87, 223, 465, 466, 467
—, muscle tonus — 466
—, narrow-band 384, 405, 407
—, — —, excitation pattern for 412
—, — —, loudness of 420
—, — —, sharpness of 434
—, room — 466
—, suprathreshold 58
—, uniform exciting 422
—, white 416
—, —, level for 415
—, —, masking by 403, 404, 408
—, wideband 81, 384, 456
—, bands of 452, 453
—, effect on behavioral audiograms after damage of the organ of Corti 265
—, effect on responses to click and tone stimuli 43, 67
—, just noticeable differences in 417
—, localization of 454
—, roughness in 431
— bandwidth 60, 61
— bursts, wide-band 210
— correlation, interaural 469, 470
— masker 462
— signal, see also signal
— —, binaural analysis 467
— —, inhibition by 67
— —, response of cortical units 210
— stimulation, continuous 18
— stimuli 23, 38, 58, 67
— —, broadband 38
— thresholds 40
— -tone comparisons 59
noisy membrane potential 83
non-auditory cues 366
— -linear aspects of responses to two tones 35
— — distortion 441 ff.
— — selectivity 413
— — threshold detector mechanism 35
— -linearities, cochlear 80, 81, 83
— — in the ear 404, 405
— -linearity, "essential" 83
— —, transducer — 82
— —, cochlear nerve response 35, 38
— -monotonic intensity functions, DCN 55
— reciprocal interconnections 314
novel stimuli, response to 201
novelty, cortical units 216
NTB, see nucleus of the trapezoid body
nuclei, auditory 109 ff.
—, dorsal column, projections to pericollicular tegmentum 153
— of the lateral lemniscus 109, 110
— — —, binaural neurons 133
— — —, binaural intensity functions 131
— — —, monaural intensity functions 128, 129
— — —, monaural response area 122, 124
— — —, phase sensitive neurons 135, 137
— — —, single unit studies 120
—, medial geniculate 188
—, olivary 110
—, posterior, see posterior nuclei
—, preolivary, binaural neurons 133
—, —, single unit studies 120
—, retroolivary, binaural neurons 133
— of the superior olivary complex, single unit studies 120
—, thalamic 337
nucleus, central, of inferior colliculus 147, 148, 149, 150, 151, 314
—, cochlear, see cochlear nucleus
—, dorsal, of lateral lemniscus, corticofugal axons to
—, dorsomedial, inferior colliculus 148
—, external, inferior colliculus 150, 152
— interstitialis 45, 48
—, lateral geniculate 189
— lateralis 45
—, lateral posterior, of thalamus 155, 188
—, lateral preolivary 110, 111
— magnocellularis 54
—, medial preolivary 110, 111, 118
—, — —, phase-sensitive neurons 137
—, pericentral, inferior colliculus 151, 152
— reticularis 189
—, S-shaped 116
—, spinal trigeminal 110
—, superior olivary (SON) 72, 76
—, suprageniculate 155, 188

— of the trapezoid body 110, 111, 116, 117
— — —, elongate cells 119
— — —, monaural intensity functions 127, 129
— — —, monaural response area 122, 123
— — —, principal cells 118, 121
— — —, response to clicks 131, 132
— — —, stellate cells 119
— — —, single unit studies 117
—, ventral, of the medial geniculate body 314
—, ventrolateral, inferior colliculus 148
—, —, of medial geniculate 155
—, vestibular 226

OCB, see olivocochlear bundles
occlusion, binaural 186, 187
octave, third, band levels 430
— enlargement 430
"octopus cell" region of the PVN 50
off-center source 453
— -inhibition 63
— suppression 56
OHC, see outer hair cells
Ohm's acoustical law 439
olivary complex, superior, see superior olivary complex
— nuclei 110
— nucleus, superior lateral 320, 375
— —, superior medial 320, 373, 375
— —, — —, sound localization 375
olive, lateral superior 110, 111, 112, 375
—, — —, binaural neurons 133
—, — —, binaural intensity functions 131
—, — —, binaural response areas 125
—, — —, monaural intensity functions 127, 128
—, — —, monaural response area 122
—, — —, single-unit studies 116, 121
—, medial superior 110, 111, 225, 372, 373
—, — —, binaural intensity functions 130, 131
—, — —, binaural response-areas 125
—, — —, evoked potentials 112
—, — —, monaural intensity functions 127
—, — —, monaural response area 122
—, — —, phase-sensitive neurons 135, 136, 137
—, — —, single-unit studies 114, 121
—, — —, response to clicks 131, 132, 139
—, — —, response to transients 139
—, superior 226, 234, 241, 373
—, —, evoked potentials 112, 113
olivocochlear bundle (OCB) 18, 72, 225, 315
— — axons, action in the inner ear 236f.
— — —, neurochemical mechanisms 234f.
— — effects, parameters of 231f.
— — —, titration of 228
— — neurones 225, 226
— — system, psycho-physiological significance 238f.
— — terminals, transmitter released from 232, 234, 236
— bundles as an inhibitory mechanism 240
— —, stimulation of 72, 73, 227f., 229, 230
— —, transection of 381, 382, 383
— —, crossed 72, 73, 75, 86, 225, 228, 230
— —, —, effect of stimulation on acoustic pathway 233f.
— —, —, ionic mechanism in the inner ear 234
— —, uncrossed 225, 226
— fibres 119
— inhibition 220
— neurons 220
— system 220, 226
— tracts 32
on suppression 203
— unit 50, 51, 53, 55
opossum, auditory cortex 321
—, sound localization 325, 354, 357
optic tract 155
organ of Corti damage 263, 265, 270
oscillatory activity, cochlear nerve 7
Osen, large spherical cells of 46
ossicular chain, interruption of 267, 293
otosclerosis 277
ototoxic agents, effect on cochlea 18, 87
— drugs, effect on inner ear 265, 266, 269, 272, 273, 279
— stimuli on inner ear anatomy 277
ototoxicity, chemical 273
— by nitrogen mustard 266, 273
out-of-phase inhibition 136
outer ear, see external ear, pinna
— hair cell density and pure tone sensitivity 278
— — cells 11, 13, 16, 70, 84, 85, 86, 91, 231, 236
— — —, loss of 263, 279, 281
— spiral fibres 85, 86
overall level of the masker 468, 470
"overshoot" masking effect 408, 409, 414
— phenomenon 414

paradoxial trading 140
Paraechinus hypomelas 354
"parallel-ramp" type of rate-intensity function 23
paralysis, facial 258
parietal cortex 339

pars lateralis, geniculate body 156
— magnocellularis, geniculate body 154, 188
— ovoidea, geniculate body 156
— parvocellularis, medial geniculate 147
— principalis, geniculate body 154, 160
path, primary, from inferior colliculus 386
—, visual tecto-pulvinar 386
pathology, cochlear 69, 71
pathway, ascending, from ear to cortex 311
—, descending efferent 32
—, spinothalamic 147, 155
—, subcortical, transection of 364ff.
— from cochlea to cortex 386
— from the cochlear nucleus to the lateral lemniscus 112
— outside of the lateral lemniscus 366
pattern, excitation — 412, 413, 414, 417, 422, 425, 428
—, masking — 413
—, time variable excitation-critical band rate — 414
— discrimination after cortical ablations 341f., 349, 362, 386
— — after subcortical lesions 378 ff
pause, subjective duration 438
"pauser" time course 48
— type unit, DCN 51, 53, 55
peduncle, cerebral 155
pentobarbital anesthetic 365
perception 386
— of auditory cues 221
— of movement 453
— of pitch 94
— of short sound signals 345
— of space 249
— of speech 92, 93, 94
perceptual signal-to-noise ratio 238
peri-aqueductal gray 366
perilymph chloride 234
period, unresponsive 173
periodicity coding 94
— mechanisms 92
— theory 93
PES, see ectosylvian gyrus, posterior
phase, interaural 120, 453, 464
—, relative, of two stimulus components 36, 37
— of stimulus cycle 27
—, reversing of 453
— angle, minimal 453
— changes 439ff.
— constancy 138
— differences, interaural 133, 134, 137, 207, 259, 457
— lag, cochlear fibres 27, 28, 29
— locked discharges, cochlear fibres 24, 26, 33, 39
— — —, phase-sensitive neurons 138
— — fibres, superior olivary complex 458
— — responses, cochlear nucleus 57
— locking, cortical units 208, 209
— —, non-linearity of 38
— — to tonal stimuli 93, 204, 205, 207
— relation 467, 468
— -sensitive neurons 135, 136, 138
— sensitivity 135, 139
— — at threshold of modulation 410
— shift 470, 471
— synchronization, cochlear fibres 26
phon 419
phoneme detectors 191
— -related responses 190
physiology of the auditory nuclei of the pons 109ff.
— of the efferent recurrent auditory system 219ff.
— of the inferior colliculus 145ff.
— of the medial geniculate complex 145ff.
picrotoxin 166, 235
pigeon, cochlear nucleus 58
—, efferent cochlear bundle 237
pillar cells, degeneration of 263
pinealoma 368
pinna, action potentials from the region of 71
—, influence on auditory discrimination 247ff.
—, models of 255
—, role in sound localization 255f., 450, 454ff.
— as sound reflector 255
— orientation, effect on stimulus size 223
pinnae, artificial 253, 254
pitch 428ff.
—, spectral 430
—, virtual 430
—, just-noticeable differences in 439
—, mel scale of 410
— discrimination after cochlear insult 264
— measurements, and phase changes 440
— perception 94
— sensation 41
— shifts 430
place code 139
— mechanisms 92
— theory 93
plane, horizontal, binaural sound localization in 254
—, median sagittal, sound localization in 251
Po, see posterior nuclei
polarities of hair cells 84

Politzer's Hörmesser 249
pons, auditory nuclei of 109ff., 110
—, lesions in 367
pontine tumors 376
positive feedback models 82
posterior "descending" branch, cochlear nerve 45
— nuclei of thalamus 313, 317, 319
postmortem autolysis 282, 296, 297
post-stimulus time (PST) histogram 6, 7, 8
postsynaptic inhibition, OCB axons 237
— potentials, cortical units 208
potential, action —, see action potential
—, endocochlear 231, 234, 236
—, generator 236
—, membrane — 83
—, microphonic, see microphonic
potentials, evoked, see evoked potentials
—, intracellular 208
—, slow-wave 113
—, post-synaptic 208
power law, Steven's 422, 423
precedence effect 360, 455ff.
prepotency of auditory over visual stimuli 371
prepotent stimulus 371
prepotentials 46
presbyacusis 271, 272, 282
presynaptic inhibitory effect, COCB axons 237
"primary-like" cells 53, 55
— — time course 48
primates, auditory cortex in 387
probabilistic generator 85
Probst, commissures of 120, 366
projection, cortico-cortical 386
—, cortico-thalamic 320
—, sustaining 318, 319
—, thalamo-cortical 317, 320
— of auditory cortex to medial geniculate body 314
— of auditory cortex to inferior colliculus 314, 315
— of the basilar membrane 315
— of caudal medial geniculate body on insular-temporal areas 342
— from the thalamic core GMv to the cortical core A I 320
— areas 216
— —, auditory, of the temporal lobe 345
pro-koniocortex 321
pseudorandom noise waveform 39
PSPs, see post-synaptic potentials
PST, see post-stimulus time
psychoacoustical excitation 409ff., 417
— incitation 410, 411, 417
psychophysical combination tone 33, 35, 91
— masking 68
— phenomena of frequency coding 87f.
— significance of the OCB system 238f.
pure tones, exposure to 266, 288, 291, 293
— —, sensitivity to 278
— — thresholds 13, 238, 281
PVN, see cochlear nucleus, posteroventral
pyramid, medullary 110

Q_{10}, see sharpness of tuning
quadratic difference tones 441
— distortion 442

radial fibres, inner 86
rarefaction clicks 9
— transients 7
rat, auditory cortex 317
—, olivo-cochlear bundle neurons 227
—, sound localization 375
rate-intensity function, types of 23
ratio, critical 88, 89
— scale 402
receptive fields of central neurons 124
reciprocal interconnections 314
recognition of speech 468, 473
recovery from cortical ablation 332
recruitment 369
—, and hair cell populations 279f.
"rectangular grid", pericentral nucleus 152
recurrent auditory system 72, 219
— collaterals from thalamocortical axons 180
— inhibition 181
reflex, conditioned 262, 385
—, unconditioned 385
— centers in the brainstem 385
— orientation 357
— responses after cortical ablations 356
refractory mechanisms, cochlear nerve 7
Reissner's membrane 272, 283
relation between cat and other species, auditory cortex and sound localization 387
— of cortex to subcortex 384ff.
— of structure and function in auditory discrimination 311
—, functional, between core and belt cortex 387
relative frequency shifts 417
release-from-masking 463
removal of area I of the cortex 317
representation of the basilar membrane 315, 316
reptile, cochlear nucleus 53
Residuum 430, 453

"resonance" of electrochemical nature within hair cells 82
— theory 262
response as a function of intensity 21, 54
— as a function of time 23, 56
— area, asymmetrical 16
— —, binaural 125
— —, inhibitory 54
— —, monaural, neurons of pontine nuclei 122
— —, symmetrical 16
— contours, iso-rate 19, 21
— patterns, cortical neurons 201f.
— —, medial geniculate body 180f.
— — of neurons in inferior colliculus 164
— time of the cochlear filter 29
restiform body, see corpus restiforme
retention of learned discriminations 330, 342
reticular formation 221, 385
— —, CERACS 238
— influences on acoustic input 224
— system, non-specific 220, 221, 224, 225
retrocochlear lesions 369
retrograde degeneration 317, 318, 333, 339, 343, 385
retroolivary region, lateral 110, 111
— —, medial 110, 111
— —, single unit studies 120
reverberatory responses, geniculate body 180, 181
reversing the phase 453
rhesus monkey, auditory cortex 324
roof cortex, inferior colliculus 150
room acoustics, effect on stimulus size 223
— noise 466
roughness 430ff.
Round Robin Test 427

S I, see somatic area I of cerebral cortex
S II, see somatic area II of cerebral cortex
saccule, ruptures of Reissner's membrane 272
Saimiri, see squirrel monkey
salicylate intoxication 266, 273
saturation, cochlear nerve 84, 85
SC, see colliculus, superior
scale, critical band — 429
—, interval — 402
—, nominal 402
—, ordinal 402
—, ratio 402
—, just-noticeable-difference — 429
scaling 401ff.
sclerosis, multiple 366, 369, 377
— in the brainstem 376
Seashore Measures of Musical Talents 334
— Pitch test 340
— Test of Tonal Memory 344
section of the cochlear nerve 263, 264, 265, 269
— of crossed OCB 75
selectivity, nonlinear 413
semantic contents of words 191
sensation 386
—, conscious 385
— of loudness 92
— of movement 453
— of pitch 41
sensorineural deafness 91
sensory control of acoustic pathway 224
— cortex 384, 386, 387
— discrimination, definition of 309
— input to neurones of cochlear nucleus 51
— -motor cortex 385
— systems, neocortical 386
SF, see sylvian fringe
shadow, acoustic 251, 374, 450
—, head — 223, 250
sharpening of the basilar membrane frequency response 82
sharpness 433ff., 434f.
— of the FTCs 15, 16, 72
— of tuning 123
shear model, longitudinal 82
sidebands, inhibitory 90, 124, 125
side tones 433
sigmoid gyrus 325
signal, see also noise signal, sound signal
—, masked 462
—, "missing fundamental" 35
— bandwidth, binaural analysis 467
— duration 325, 326, 328, 332, 345, 349
— —, binaural analysis 466
— frequency 325, 332
— —, binaural analysis 465
— intensity 325, 332
— parameters, binaural analysis 465ff.
— phase, binaural analysis 466
— -to-masker ratio of speech 473
— -to-noise ratio, perceptual 238
— transfer after cortical ablations 359
signals, multicomponent 35
simultaneous masking 402, 408, 409
single unit activity, auditory cortex 199ff.
— — studies, auditory nuclei of the pons 113ff.
slow-wave potentials 113
small band masker 408
— — test sound 408
SOC, see superior-olivary complex

somatic area I of cerebral cortex 331
— area II of cerebral cortex 317, 318, 325, 331, 338, 350, 352
— -auditory interaction 188
somato-sensory system 161
SON, see nucleus, superior olivary
sone 419, 420, 423
sound, harmonic complex tone — 435
—, masking 429
—, onset of 324, 365
—, airborne, inner ear damage by 269
— complexes, coding of 94
— localization 95, 133, 314, 325, 331, 345, 449ff.
— —, auditory cortex 207, 386
— — after cortical ablations 350ff.
— —, free-field studies 449
— — in the horizontal plane 254f.
— —, inferior colliculus 172, 174
— — after lesions in the descending auditory pathways 382
— — in the median sagittal plane 251f.
— —, role of pinna 249f., 450, 454
— — after subcortical ablations 372ff.
— —, theory of 255, 475
— shadow 251, 450
— signal, see also signal
— —, short, perception of 345
— source, direction of 331
— —, front-rear discrimination of 253
sounds, complex, alteration of 455
—, —, loudness of 425
—, —, pitch of 428, 430ff.
—, —, response to 30, 58
—, —, threshold of 410
—, different, excitation patterns of 422
—, fluctuating 417
—, partially masked 425
—, steady state — 435
— with complex spectra, discrimination after cortical ablations 362ff.
—, hyperirritability to 258
—, time relationships between 345
space perception, auditory 249
spatial excitation pattern 444
— localization 462
special responders, cortical units 209
spectra, harmonic 430
—, multicomponent, stimuli with 41, 61
—, unharmonic 430
spectral pitch 430
— resolution of CN units 59, 61
speech 430, 456, 464
—, see also linguistic
—, detection of 468
—, recognition of 468, 473
— discrimination 272, 281f., 376, 382
— — after cortical lesions 361
— intelligibility 472
— masking-level difference 468
— perception 92, 93, 94
— signal-to-masker ratio 473
— signals, binaural analysis 468
— sounds 362, 414
— tests 361
spherical-cell neurons 122
spike generator, stochastic 84
spinal cord, projections to pericollicular tegmentum 153
— trigeminal nucleus 110
spinothalamic pathways 147, 155
spiral fibers, outer 85, 86
— ganglion, atrophy of 272
— —, lesions of 264
— —, tumor of 277
spontaneous activity, cochlear nerve 4, 6, 10
— —, cortical neurons 201, 208
— —, depression of 85
— discharges, cochlear nucleus 46, 48, 51, 52
squirrel, auditory cortex 321
— monkey, audiograms 268, 273
— —, cochlear nerve response 4, 5, 10, 13, 15, 16, 18, 19, 20, 24, 35, 36, 75, 79
— —, cochlear pathology 266, 267, 268, 269, 293
— —, frequency discrimination 381
— —, hair cell loss 287, 289
— —, histology of cochlea 297, 298, 299, 301
— —, inferior colliculus 153
— —, vocalization 200, 212
SS, see suprasylvian gyrus
S-segment, SON 76
ST, see sulcus, superior temporal
stapedectomized patients 258
stapedius muscle 256, 258
— reflex 258
stapes displacement 3
steady state, critical band 409
— —, loudness calculation 419
— — condition, masking 402
— — sounds, sharpness of 435
— — stimulation, cortical units 201
stellate cells, central nucleus 150
stereocilia 296
Steven's power law 422, 423
stimulation, double click 61, 62
— of COCB, effect on acoustic pathway 233f.
stimuli, competing 371
—, constant, method of 402
— responsible for minimal audible angle 451

stimulus, prepotent 371
—, envelope of 453
— dominance 90
stochastic model 84
— spike generator 84
streptomycin, effect on inner ear 265, 272
stria vascularis, atrophy of 270, 272
striae, acoustic 112
strychnine, effect on cochlear microphonics 32, 226
—, effect on cochlear nucleus neurons 241
—, effect on OCB axons 235, 237
—, effect on olivo-cochlear inhibition 220
subcortex, relation to cortex 384ff.
subcortical centers 385
— —, ablation of 323, 361, 364ff.
— pathways, transection of 364ff.
sulcus, superior temporal 324, 327
superior-olivary complex 109, 110, 111, 120, 122, 311, 313, 314, 316, 320, 365, 372, 374, 375, 458
— — —, binaural neurons 133
— — —, monaural intensity functions 127
— — —, monaural response area 122
— — —, phase sensitive neurons 135, 138
— — —, response to clicks 131, 132
— — —, response to transients 139
suppression, cortical neurons 201
—, due to one tone 58
—, off — 56, 203
—, on — 203
—, through — 203
—, two-tone — 30, 37, 58, 81, 83, 90
suppressive frequency bands 30
suprageniculate-magnocellular complex 160
suprasylvian fringe 159
— gyrus 160, 189, 331, 352
suprathreshold noise 58
surface specimen technique 283, 293
sustaining projection 318, 319
swept tones 200, 212
SYL, see sylvian gyrus
sylvian fringe 316, 317, 319, 387
— gyrus 159, 160, 189, 241, 323, 324
synapse, calyceal 118
synapses, tonotopic organization of 63
synaptic aggregates, geniculate body 182
— contact between outer spiral and inner radial fibres 86
— nests, geniculate body 182
— organization, inferior colliculus 166, 170ff.
— termination of efferent OCB axons 237
synchronization of responses of cochlear fibres 25
synchronized discharge rate 20
— isorate contour 19
T, see temporal cortex
tactile and auditory interconnections 159
tactual cues from the eardrum 366
TB, see trapezoid body
tecto-pulvinar path, visual 386
tectorial membrane 85, 283
tectum 152, 364
tegmental area, lateral 313, 376
tegmentum, midbrain — 152
—, pericollicular 152, 153
temporal bone pathology 270, 278
— cortex 317, 318, 382
— —, auditory fields 147, 159
— —, cortifugal axons from 220
— cortical areas 325, 331
— effects on critical bands 414
— — on loudness 426
— — on masking 406
— "following" 453
— information 95, 132, 453
— jitter 476
— lobe 339
— —, auditory projection areas 345
— —, lesions of 346
— —, superior, belt cortex of 339
— — damage, bilateral 328, 356
— — surgery 326, 328, 330, 361
— lobectomy 334, 340, 345, 354
— meningioma 355
— order tests 346, 347
— patterns, discrimination of 341, 349, 358
temporary threshold shift (TTS) 256, 257
temporo-insular cortex 159, 240, 241
— — —, corticofugal axons from 220
test sounds, white-noise 408
— —, small band 408
thalamic belt, auditory 314, 319, 320
— core 320, 321
— degeneration 317, 318, 330, 338, 342, 343
— nuclei 337
— regions of the auditory pathway 158f.
— subdivisions, cytoarchitectonic 312, 313
— tumor 345, 376
thalamocortical axons, recurrent collaterals from 180
— neurons 157
— projections 317, 320
— relations 178
thalamus, antero-lateral region of 189
—, lateral posterior nucleus 155, 188
—, relay nuclei of 386
—, auditory 160, 313, 314, 317, 318, 321
—, —, evolution of 388
theories of binaural analysis 474
theory of localization 475
Theta-model, binaural analysis 474

thin spike units, auditory cortex 208
thiosemicarbazide, effect on OCB axons 235
third octave band levels 425
threshold, broad band noise — 40, 407
—, cochlear nerve fibres 11
— of complex sounds 410
—, differential 402
— as a function of frequency, cochlear nucleus 53
— of modulation 410, 441
— of narrow band noise 407
— for noise stimuli 58, 59
— in the quiet 402, 404, 406
— detector mechanism, non-linear 35
— response curves, inferior colliculus 164
— shift, contralateral 257
— —, temporary 256, 257
— trigger mechanism 83
thresholds, absolute 324, 365, 376, 381, 402
—, behavioural 13, 238
—, intensity discrimination — 266
—, masked 89, 384, 402ff.
—, pathological 18, 69
—, pure tone — 13, 238, 281
—, subjective 69
through response 203
— suppression 203
timbre 433, 440
—, just-noticeable differences in 439
— theory of sound localization 255
time, response of cochlear nerve to 23
—, response as function of — 56
— code 139
— course, "build-up" 48
— —, "pauser" 48
— —, "primary like" 48
— — of units in DCN 48
— delay 470
— difference detecting units, inferior colliculus 176
— differences, binaural 374, 376, 377
— —, interaural 133, 141, 207, 249, 450, 451, 452ff., 456, 457, 475
— image 457
— -intensity trading 140, 177, 456
— interactions, binaural 230
— relationships, interaural 175, 176
— — between sounds 345
— -sensitive neurons 141
— variable excitation-critical band rate pattern 414
tinnitus 371
titration of OCB effects 228
tonal dip 276, 282
— masker 472
— Memory, Seashore Test of 344
— stimuli, coding by cortical units 201
— —, parameters of 204
— —, single, response of cochlear nerve to 10
— —, —, response of cochlear nucleus to 53
tone, falling, discrimination of 342
—, periodically changed in level 418
—, rising, discrimination of 342
— -bursts 406
— —, response of cortical units 202
— combinations, excitation of cochlear nerve by 33, 37, 83
— decay test 367
— -impulses 406
— and noise band, combination of 67
— pairs 343
— sequences 345
— stimuli, effect of noise on responses to 43, 67
— thresholds 40
tones, amplitude modulated 63, 167, 430, 431, 439
—, combination — 441
—, difference — 404, 441, 442
—, frequency modulated 63, 65, 167, 191, 200, 210, 212, 342, 343, 440
—, high-frequency, response to 133
—, low frequency, response to 24, 57, 133
—, pure 13, 238, 266, 278, 281, 288, 291, 293
—, —, pitch of 428
—, side — 433
—, swept 200, 212
—, loudness of 420
—, masking by noise 68
tonotopic localization within the central nucleus 150
— organization, auditory cortex 213f., 335, 340
— —, cochlear nerve 4
— —, cochlear nucleus 44, 45, 63, 114, 116, 117, 120
— — of the inferior colliculus 162
— — of medial geniculate body 157
topographic projection of the cochlear nuclei upon the central nucleus 150
total loudness 425, 426
tracing, type III 367
tracking, method of 402
tract, optic 155
tracts, olivocochlear 32
trading, paradoxial 140
—, time-intensity — 140, 177, 456ff.
— ratio 456, 457
training of experimental animals 310
transducer nonlinearity 82
— process 85

transection of neural pathways 309, 323
— of subcortical pathways 364ff.
transfer of signal 359
transient detectors 191
— stimuli, response of cortical units 201
transients, condensation 7
—, rarefaction 7
—, localization of 133, 139
—, response to 139
—, fast, responses of cortical units 207
transmission, chemical 85
— delays 138
transmitter in the inner ear 235
—, chemical, COCB terminals 234, 236
—, excitatory, outer hair cell 236
—, inhibitory, COCB terminals 232
trapezoid body 76, 110, 111, 226
— —, evoked potentials 112
— —, nucleus of 110, 111, 116, 117
— —, transection of 365, 366, 371, 372, 374, 386
trauma, acoustic 256, 257, 273, 281
travel time 28
tree shrew, auditory cortex 321, 322
— —, auditory thalamus 321
— —, sound localization 375
trigger mechanism 83
TTS, see temporary threshold shift
tumor of the cerebellopontine angle 276
— of the eighth nerve 281
— on the floor of the fourth ventricle 367, 376
—, pontine 376
— of the spiral ganglion 277
—, thalamic 345, 376
— in the upper brainstem 376
tuning, sharpness of 123
—, mechanical 78
—, neural 78
— characteristics, NTB units 123
— curves, broad 205
— —, contralateral (inhibitory) 125
— —, cortical units 205f., 213
— —, geniculate body 178, 180
— —, inferior colliculus 162
— —, ipsilateral (excitatory) 125
— —, multi-peaked 205
— —, narrow 205, 215
two-tone combination, absolute level 38
— — inhibition 30, 58
— — masking 410
— — stimuli, response to 30, 35, 36, 58
— — suppression 30, 37, 58, 81, 83, 90

ultrasonic stimulation 265, 266, 269
ultrasound, inner ear damage by 269
unharmonic spectra, pitch of 430
unit, see also cell, neuron
—, inhibitory 49, 50
—, interaural intensity sensitive 176, 177, 188
—, interaural time sensitive 176, 177, 188
—, "on" — 50, 53, 55
—, "pauser type" 51, 53, 55
— recordings, combination with behavioral studies 216
units of the cochlear nucleus, responses of 49, 50, 51, 52, 53, 54, 58, 59, 65
—, labile, auditory cortex 216
—, thin spike, in auditory cortex 208
— unresponsive to sounds, cortex 200, 209, 210
unmasking, binaural 89
unresponsive period, absolutely 173
— units, auditory cortex 200, 209, 210
UOCB, see olivo-cochlear bundles, uncrossed
utricule, ruptures of Reissner's membrane 272

VCN, see cochlear nucleus, ventral
vector model, binaural analysis 474
vestibular nucleus 226
vibration, basilar membrane — 79, 80, 81, 82, 86
vibrato 417
virtual pitch 430
visual tecto-pulvinar path 386
vocalization, squirrel monkey 200, 212
— and middle ear muscle contractions 223
volume sensation 433
vowel detectors 190

Webster-Jeffress Hypothesis 464, 474
white noise 416
— —, level for 415
— —, masking by 403, 404, 408
wideband noise 81, 210
— —, localization 456
words, semantic contents of 191

Handbook of Sensory Physiology
Outline

Distribution rights for India:
UBS, New Delhi

Vol. I: Principles of Receptor Physiology
Edited by W. R. Loewenstein
262 figures. 612 pages. 1971
ISBN 3-540-05144-9

Vol. II: Somatosensory System
Edited by A. Iggo
240 figures. 862 pages. 1973
ISBN 3-540-05941-5

Vol. III/1: Enteroceptors
Edited by E. Neil
91 figures. 241 pages. 1972
ISBN 3-540-05523-1

Vol. III/2: Muscle Receptors
Edited by C. C. Hunt
128 figures. 320 pages. 1974
ISBN 3-540-06891-0

Vol. III/3: Electroreceptors and Other Specialized Receptors in Lower Vertebrates
Edited by A. Fessard
118 figures. 341 pages. 1974
ISBN 3-540-06872-4

Vol. IV: Chemical Senses
Edited by L. M. Beidler
Part 1: Olfaction
212 figures. 526 pages. 1971
ISBN 3-540-05291-7
Part 2: Taste
176 figures. 418 pages. 1971
ISBN 3-540-05501-0

Vol. V: Auditory System
Part 1: Anatomy and Physiology (Ear)
Edited by W. D. Keidel, W. D. Neff
305 figures. 744 pages. 1974
ISBN 3-540-06676-4
Part 2: Physiology (CNS), Behavioral Studies Psychoacoustics
Edited by W. D. Keidel, W. D. Neff
Part 3: Clinical and Special Topics
Edited by W. D. Keidel, W. D. Neff
ISBN 3-540-07129-6
In preparation

Vol. VI: Vestibular System
Part 1: Basic Mechanisms
Edited by H. H. Kornhuber
251 figures. 684 pages. 1974
ISBN 3-540-06889-9
Part 2: Psychophysics, Applied Aspects and General Interpretations
Edited by H. H. Kornhuber
198 figures. 688 pages. 1974
ISBN 3-540-06864-3

Vol. VII/1: Photochemistry of Vision
Edited by H. J. A. Dartnall
296 figures. 822 pages. 1972
ISBN 3-540-05145-7

Vol. VII/2: Physiology of Photoreceptor Organs
Edited by M. G. F. Fuortes
342 figures. 775 pages. 1972
ISBN 3-540-05743-9

Vol. VII/3: Central Processing of Visual Information
Edited by R. Jung
A: Integrative Functions and Comparative Data
208 figures. 786 pages. 1973
ISBN 3-540-05769-2
B: Visual Centers in the Brain
216 figures. 746 pages. 1973
ISBN 3-540-06056-1

Vol. VII/4: Visual Psychophysics
Edited by D. Jameson, L. M. Hurvich
297 figures. 822 pages. 1972
ISBN 3-540-05146-5

Vol. VII/5: The Visual System in Evolution
A: Vertebrates
Edited by F. Crescitelli
In preparation
B: Invertebrates
Edited by H. Autrum
In preparation

Vol. VIII: Perception
Edited by R. Held, H. W. Leibovitz, H. L. Teuber
In preparation

Vol. IX: Development of Sensory Systems
Edited by M. Jacobson
In preparation

GPSR Compliance

The European Union's (EU) General Product Safety Regulation (GPSR) is a set of rules that requires consumer products to be safe and our obligations to ensure this.

If you have any concerns about our products, you can contact us on ProductSafety@springernature.com

In case Publisher is established outside the EU, the EU authorized representative is:

Springer Nature Customer Service Center GmbH
Europaplatz 3
69115 Heidelberg, Germany

Batch number: 10659826

Printed by Printforce, the Netherlands